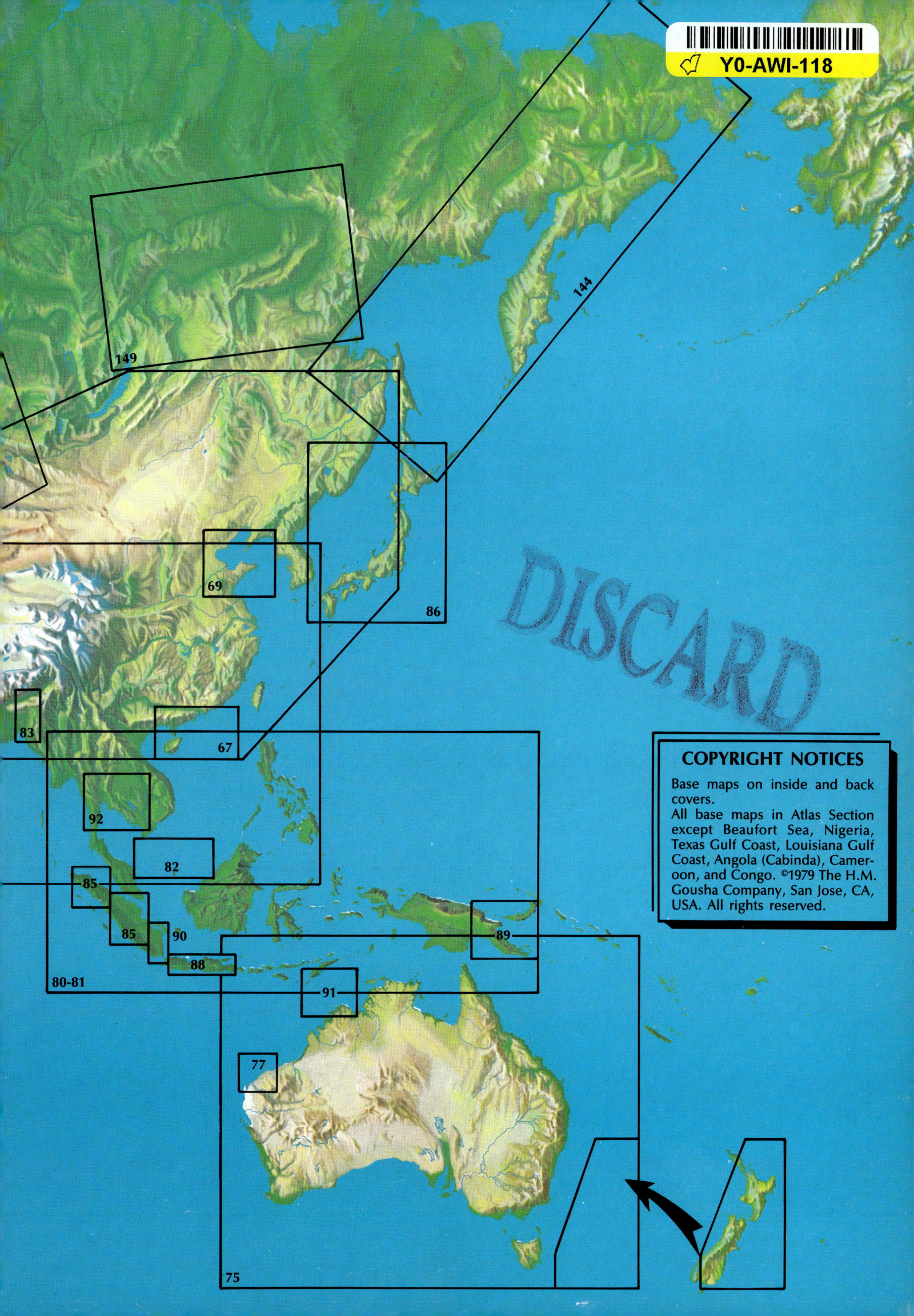

Property of Anadarko

International Petroleum Encyclopedia

Price $115

Copyright, 1991 by
PennWell Publishing Co.
Box 1260, Tulsa, OK 74101

Library of Congress
Catalog Card Number 77-76966
International Book
Numbering System 0-87814-369-6
Printed in U.S.A.

ALL RIGHTS RESERVED

*This book or any part thereof
may not be reproduced
in any form without written
permission of the publisher.*

INTERNATIONAL PETROLEUM ENCYCLOPEDIA

TABLE OF CONTENTS

Use of horizontal drilling is increasing rapidly.............5

NORTH AMERICA
United States 20

Canada 55

CHINA 62

ASIA/PACIFIC
Australia............................... 74
Bangladesh.......................... 75
India..................................... 76
Indonesia............................. 79
Japan................................... 82
Korea, South 83
Laos..................................... 83
Malaysia 84
Mongolia.............................. 84
Myanmar.............................. 87
New Zealand 87
Pakistan............................... 88
Papua New Guinea 88
Philippines 89
Taiwan 90
Thailand............................... 91
Viet Nam 93

MIDDLE EAST
Bahrain 94
Iran 94
Iraq 97
Jordan.................................. 97
Kuwait.................................. 99
Oman................................... 99
Qatar.................................. 101
Saudi Arabia 102
Syria 104
United Arab Emirates 105
Yemen 105

LATIN AMERICA
Argentina	106
Belize	107
Bolivia	107
Brazil	107
Chile	111
Colombia	112
Costa Rica	112
Cuba	113
Dominican Republic	114
Ecuador	114
Guyana	115
Mexico	115
Paraguay	118
Peru	118
Suriname	121
Trinidad and Tobago	122
Uruguay	122
Venezuela	123

AFRICA
Algeria	124
Angola	125
Cameroon	126
Congo	126
Egypt	126
Ethiopia	126
Equatorial Guinea	126
Gabon	128
Kenya	130
Libya	130
Madagascar	132
Morocco	132
Nigeria	132
Somalia	138
Tanzania	138
Tunisia	138
Zimbabwe	139

U.S.S.R. **140**

EASTERN EUROPE **154**

EUROPE
Austria	158
Belgium	158
Denmark	158
Finland	160
France	160
Greece	163
Ireland	163
Italy	165
Malta	165
Netherlands	166
Norway	166
Portugal	168
Spain	170
Sweden	174
Switzerland	174
United Kingdom	174
Germany	177
North Sea	177

SPECIAL FEATURES

Chronology of events	190	Production survey	282
Liquefied natural gas	192	Industry prices	314
Kuwait roundup	201	Data tables—production	316
Tankers and mooring	210	Data tables—consumption	320
Gulf of Mexico	218	Data tables—refining	321
Enhanced oil recovery	227	World oil flow	326
Environment	250	World refining update	328
Gas	256	Refining survey	336
Coal	264	Refining throughput	349
World well completions	269	Historical reserves	349
Gas production	269	Pipelines	350
International active rigs	270	Worldwide gas processing	357
Spending	272	Index of contents	358
Stratigraphic charts	274	Ad index	368
Oil and gas at a glance	280		

International Petroleum Encyclopedia

Editor
Jim West

Managing Editor
Robert G. Lair

Presentation Editor
Max L. Batchelder

Art Director
Kay Wayne

Atlas Editors
Patrick Crow
Jim Stilwell

Staff Editors
Robert J. Beck
A.D. Koen
Sandra Meyer
Guntis Moritis
G. Alan Petzet
Lou Ann Thrash
Bob Tippee
Warren True
Roger Vielvoye
Bob Williams

International Petroleum Encyclopedia is published annually by the Energy Group of PennWell Publishing Co., 1421 S. Sheridan Rd., Box 1260, Tulsa, Okla. 74101. Phone: (918) 835-3161. Telex: 20-3604.
  Fax: (918) 831-9555.

Chairman and Chief Executive
  Philip C. Lauinger Jr.

President
  Joseph A. Wolking

Senior Vice President, Energy Group
  Carl J. Lawrence

Vice President/Finance
  John Maney

Publisher
  A. Donald Karecki

UNION Pacific Resources Co. operated this horizontal hole among cotton fields along the Austin chalk trend in Burleson County, Tex.

Use of horizontal drilling is increasing rapidly

APPLICATIONS OF HORIZONTAL DRILLING WERE RAPidly being implemented in the world's oil and gas basins in 1990.

A round of horizontal drilling that got under way about 4 years earlier had begun to spread to numerous oil fields and geologic formations and, despite sluggish gas prices, was beginning to be tried in several gas producing areas.

International operators have tried the technique in or off at least 20 countries: Denmark, France, Italy, Netherlands, U.K., West Germany, Oman, Neutral Zone, Angola, Gabon, Australia, China, New Zealand, Indonesia, Malaysia, Trinidad, Brazil, Venezuela, Canada, and the U.S. (Fig. 1).

Horizontal drilling was being contemplated for many other countries, including an extensive program in several western Siberian fields in the Soviet Union.

In the U.S., the Austin chalk play in Texas still stood far above any other in terms of numbers of horizontal wells drilled and completed and cumulative production.

Two other major U.S. horizontal drilling plays were taking shape, and various horizontal drilling techniques were being tried in nearly every oil producing state.

The Texas Railroad Commission issued permits to drill horizontal wells in more than 50 counties in Texas in 1990. In the year's first 11 months, TRC issued 985 horizontal well drilling permits, compared with only 200 during the previous 6 years.

Of the nearly 1,200 horizontal permits issued in Texas, 973 were in Pearsall and Giddings fields.

What follows looks at reports of horizontal wells from around the world, technology and regulatory considerations, and other developments in horizontal drilling during 1990.

Significant chalk oil recovery expected

Union Pacific Resources Co. (UPRC), Fort Worth, one of the most active Austin chalk horizontal well operators, said horizontal drilling will result in production of at least 1-1.5 billion bbl of additional oil from the Austin chalk trend.

An Austin chalk well with 2,500 ft of horizontal displacement will drain about 75 million cu ft of formation, while a vertical well has an effective drainage radius of about 500 ft and drains about 23 million cu ft of formation.

A well with 4,000 ft of horizontal displacement can drain four to five times the rock volume as a vertical well can, the company said.

In May 1990, TRC applied to the entire state of Texas similar special rules on spacing and allowables it had ordered for Pearsall field in November 1989.

The rules provide for 320 acre spacing for horizontal wells, compared with 80 acres for vertical wells, an allowable of 1,312 b/d/well or four times that for vertical wells, and spacing and allowable based on variance from a horizontal displacement of 2,461 ft.

Texas' Pearsall field: horizontal headquarters

Oil operators have drilled about 1,500 wells in Pearsall field since its discovery in 1936.

The field covers parts of Frio, Zavala, Dimmit, LaSalle, Atascosa, and Maverick counties.

It had a production peak in the late 1970s, but by the mid-1980s production had fallen to an average of about 5 b/d/well.

Permits have been issued to drill 740 horizontal wells in the field since 1984.

Winn Exploration Co., Eagle Pass, Tex., tested its 25 Leta Glasscock well in the Zavala County portion of Pearsall field flowing at the rate of 19,576 b/d of oil, 11.7 MMcfd of gas, and 2,176 b/d of water from Austin chalk.

The well, 19 miles southeast of Batesville, produced 2,447 bbl of oil in 3 hr. Horizontal displacement was about 3,272 ft.

Oryx chalk operations hike production, reserves

Oryx Energy Co., Dallas, the most active Austin chalk operator, in July 1990 briefed TRC on sustained production performance of seven of its most successful Pearsall field horizontal wells (Table 1). The company estimated reserves at 450,000 bbl of oil equivalent (BOE)/well based on hyperbolic decline curves.

At yearend 1990 Oryx had identified about 220 future Austin chalk horizontal drillsites and expected to complete its drilling program in 2-3 years.

The company expected production to reach as much as a gross 38,000 b/d of oil, of which 30,000 b/d would be net to Oryx.

In late 1989 Oryx disclosed plans to drill 85 horizontal Austin chalk wells by yearend 1990, but it cut that to 75 in July 1990 because increased lateral displacement was permitting greater recovery of oil with fewer wells. In late

HORIZONTAL DRILLING

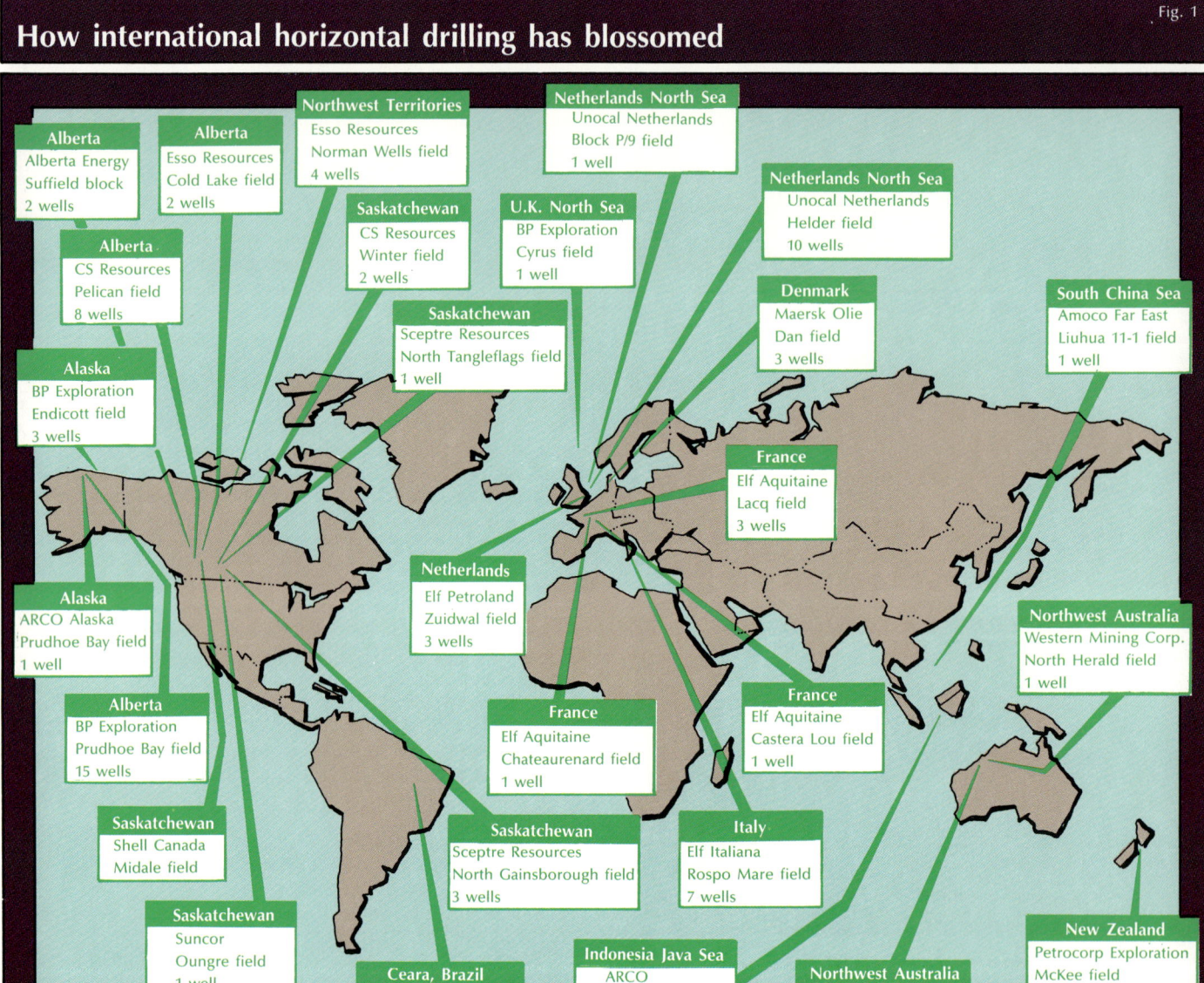

Fig. 1 How international horizontal drilling has blossomed. Horizontal drilling's international appeal.

December 1990 it was within five wells of reaching the revised goal.

Drilling time was 40 days/well with an average lateral length of more than 3,000 ft.

Drilling costs were about $1.2 million/well, but the cost per foot of the lateral portion had dropped 25% since the start of the program.

Oryx placed its average lifting cost at $1.56/BOE, but by late 1990 lifting costs had fallen to $1.38/BOE since January 1990. Oryx placed combined finding, development, and acquisition costs at about $3/BOE.

The company's cumulative production from all its Pearsall field horizontal wells was nearly 3 million bbl, including nearly 400,000 bbl from 1 Heitz, the well with the greatest cumulative production.

Oryx's gross production from all its Pearsall field horizontal wells was more than 22,000 b/d of oil and 12 MMcfd of gas. Net production was about 80% of gross production.

The gross production figure was 2.3 times what it was in January 1990.

The company operated more than 170 Austin chalk wells, including 12 horizontal wildcats and 43 horizontal development wells. The other 115 were vertical wells.

Oryx's success ratio was better than 70% for exploratory Austin chalk wells and more than 90% for development wells.

Gross reserves averaged 450,000 bbl of oil and 225 MMcf of gas, or a combined 488,000 BOE/well.

Through September 1990 Oryx had booked 68 million net equivalent bbl of proved reserve additions.

All but five of Oryx's horizontal Austin chalk wells were economic successes, and none was a dry hole.

Oryx simulations show that its Austin chalk horizontal wells may produce two thirds of estimated ultimate recovery in the first 2 years of production and have ultimate economic lives of 15-20 years.

UPRC building up Austin chalk program

UPRC operated 57 horizontal wells on production and participated in 13 nonoperated horizontal producing wells by yearend 1990.

Its Austin chalk oil equivalent production was a gross 24,400 b/d and net 14,700 b/d.

Through mid-October 1990 UPRC estimated average gross reserves for its Austin chalk horizontal wells at

HORIZONTAL DRILLING

Horizontal completion technology
Fig. 2

Past, present technology

1985 technology
- Open hole
- Slotted liner

1990 technology
- Noncemented selective completions with external casing packers, sliding sleeves
- Multiple zone hydraulically fractured wells with selective production control
- Gravel packed horizontal wells

Completion challenges

U.S.
- Technology available not being widely used
- Driver is accelerated production at minimum short term cost

Europe
- Operators pushing development of new completion technology
- Driver is long term reservoir management and lowest overall cost

Future technology

Drivers
- Simpler, cost effective zonal isolation
- Multiple lateral completion systems
- Workover systems requiring no logs

What will be required
- Reservoir management, drainage
- Incremental recovery of oil
- Minimize overall project cost

Source: Eastman Christensen

Completions today
- Understanding the reservoir
- Conveying tools and mechanical forces to extend horizontal lengths
- Long term sealing technology
- Perforating
- Non-rig workovers

236,000 BOE/well, including a combined 4.5 million BOE net to the company. The figures were based on data from 26 completed horizontal wells.

UPRC held about 220,000 acres in the trend at yearend 1990 in Burleson, Brazos, Robertson, Lee, Fayette, Washington, Gonzales, Wilson, Frio, and LaSalle counties.

At yearend it was running 15 rigs in the chalk play and planned to spend $150 million on Austin chalk development during 1991.

UPRC also planned to exploit the slightly deeper Cretaceous Buda and Georgetown zones, where initial horizontal tests yielded very encouraging results.

The operator's first Georgetown horizontal well was 1 Elaine Hester, 7 miles northwest of Bryan in Brazos County.

It flowed 374 b/d of oil and 258 Mcfd of gas through a $^{28}/_{64}$ in. choke with 225 psi flowing tubing pressure. True vertical depth (TVD) was 8,130 ft, and the well had a 2,061 ft horizontal section in Georgetown.

Williams completes 50th horizontal well

Clayton W. Williams Jr. Inc., Midland, Tex., in late 1990 completed its 50th operated horizontal well in the Austin chalk trend.

The 1-E Burns, in the LaSalle County portion of Pearsall field, flowed 2,039 b/d of oil, 773 Mcfd of gas, and 1,072 b/d of load water through two $^{24}/_{64}$ in. chokes with 565 psi flowing tubing pressure. Horizontal displacement was 3,994 ft.

Williams had operated or participated in 701 chalk wells since 1976, including 60 horizontal efforts, and at yearend owned more than 200,000 net acres in the trend.

The operator planned to continue Austin chalk development drilling through 1991 using eight to 10 rigs.

Other operators ply chalk with successes

Many other companies and combines were successfully playing the Austin chalk trend.

HDP Inc., Palo Alto, Calif., completed a horizontal well in

Typical horizontal well applications
Fig. 3

Application	Short radius	Medium radius	Long radius
Water coning	X	X	X
Naturally fractured	X	X	X
Low permeability	X	X	X
Gas coning	X	X	X
Thin reservoirs	X	X	X
Low energy	X	X	
Irregular formations	X	X	
Extended reach			X
Gas development		X	X
Solution mining	X	X	
Coal degasification	X	X	
Increased md-ft	X	X	X

Source: Eastman Christensen

August 1990 producing Austin chalk oil from beneath a basaltic cone in the Uvalde volcanic area of Dimmit County, Tex. (Fig. 4)

The 1 Autumn Unit, 9 miles northeast of Carrizo Springs in Elaine field in the greater Pearsall area, potentialed flowing 1,632 b/d of oil and 288 b/d of water with 50 Mcfd of gas without stimulation from about 1,500 ft of horizontal and deviated hole in Austin chalk B-1.

Correlation of electric logs from two vertical Austin chalk wells in Elaine field showed a spontaneous potential anomaly at the contact between the basaltic cinder cone and the underlying Austin chalk formation.

The assumption was that the SP anomaly might indicate a more permeable, fractured zone caused by the explosive intrusion and deposition of the basaltic volcano.

Elaine field was the most prominent among a cluster of oil fields in the greater Pearsall area associated with buried

7

HORIZONTAL DRILLING

Pearsall field drainage efficiency

Table 1

Well	Lateral chalk exposure ft	Assigned proration unit size acres	Estimated OOIP st-tk bbl/acre ft	Estimated net pay thickness ft	Estimated OOIP 1,000 st-tk bbl	Estimated ultimate oil recovery 1,000 st-tk bbl	% OOIP
9 John B. Baggett	950	160	76.6	176	2,157.8	430.7	20.0
13 John B. Baggett	2,255	280	76.6	173	3,719.0	359.9	9.7
3 E.B. Jones B	2,327	280	76.6	179	3,849.9	534.0	13.9
2 Stroman Harris	4,043	384.57	77.9	144	4,325.5	753.1	17.4
11 Panther Hollow	1,709	240	85.5	141	2,897.2	198.3	6.8
1 J.R. Avant	2,623	320	84.9	129	3,506.7	231.1	6.6
9 Frank Bracewell	2,323	280	76.7	97	2,089.6	144.6	6.9
					22,545.7	2651.7	11.8

Source: Oryx Energy Co.

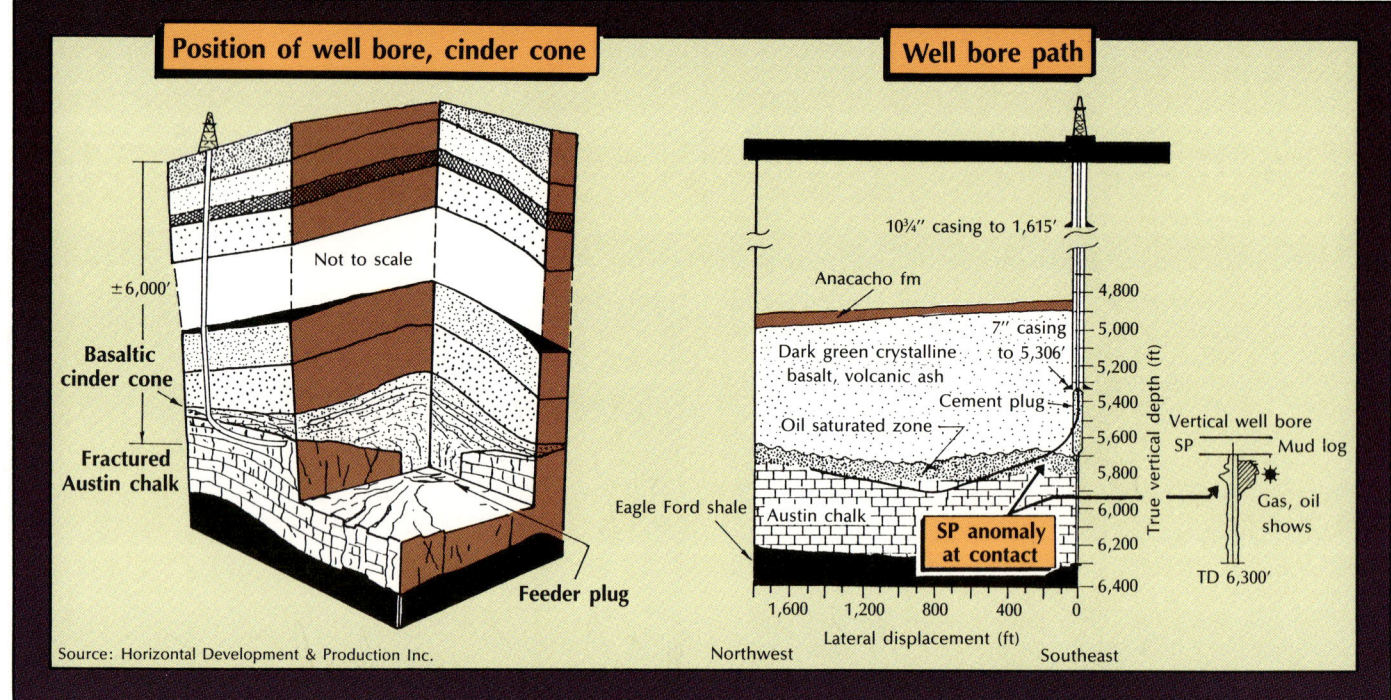

Two views of the HDP Inc. 1 Autumn Unit — Fig. 4
Source: Horizontal Development & Production Inc.

volcanic mounds. The well bore of the 1 Autumn Unit entered the chalk about 500 ft from the south end of the volcanic plug.

The well was drilled on a farmout from American Exploration Co., Houston.

Meanwhile, ARCO Oil & Gas Co. signed an agreement in early 1990 with Petro-Hunt Corp., Dallas, to pursue Austin chalk development on about 50,000 acres in Wilson and Gonzales counties, Tex.

The companies hoped to drill eight to 12 wells during the year.

And Paramount Petroleum Co. Inc., Houston, had acquired more than 30,000 acres by mid-1990 in Gonzales, Wilson, and Karnes counties and was seeking partners for an Austin chalk exploratory program.

Attention to detail important in the chalk

The performance of Austin chalk horizontal wells is not evenly distributed.

The data that lead to that conclusion are skewed by the concentration of horizontal wells in the more heavily developed portions of Pearsall field, particularly in Frio County, and operator competence, said Petrie Parkman & Co., Denver investment consulting firm.

The degree of Austin chalk horizontal drilling success appeared to be operator specific and correlated with the degree of technical sophistication employed in all phases of the project: wellsite selection, drilling, completion, and reservoir management/producing operations.

Individual horizontal well performance in Pearsall and Giddings fields (Fig. 5) was heavily influenced by:

- Reservoir depletion by vertical wells and/or offsetting horizontal wells.
- Placement of the horizontal well bore relative to fracture orientation.
- The number and geometry of macrofractures intersected by the horizontal well bore.
- Length of the completed interval within specific zones containing a high proportion of clean or brittle chalk.

Any Time...
Any Place...
Anywhere.
...Well, Almost Anywhere

During the past 82 years Willbros has laid over 200,000 kilometers of pipeline in over 50 countries around the world under some of the most difficult conditions imaginable. With only a few exceptions Willbros will undertake & complete any project...Any Time...Any Place... Anywhere.

WILLBROS GROUP INC.
Contact: Lonnie Hamilton, Willbros USA, Inc., (918) 748-7000, 2431 E. 61 St. Ste 700, Tulsa, OK 74136

WILLBROS International, Inc. • WILLBROS USA, Inc. • WILLBROS Energy Services Co. • WILLBROS Colombia, S.A. • WILLBROS (Overseas) Ltd. • WILLBROS Drilling, Inc. WILLBROS West Africa, Inc. • WILLBROS Middle East, Inc. • WILLBROS (Nigeria) Ltd. • WILLBROS Butler Engineers, Inc. • The Oman Construction Co.

HORIZONTAL DRILLING

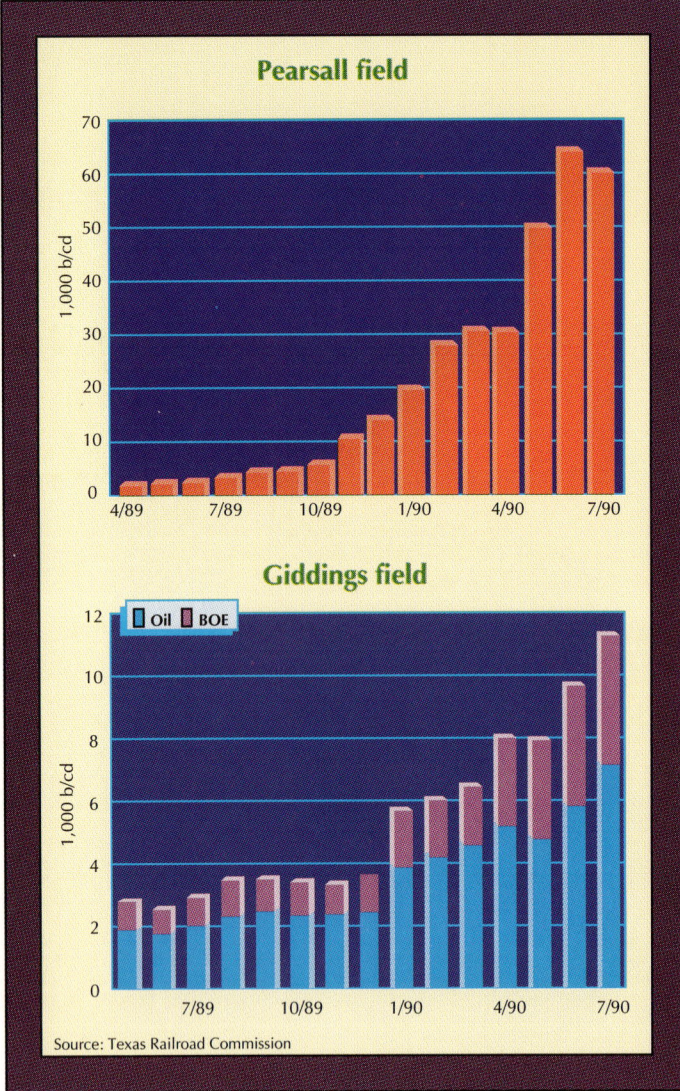

Fig. 5 Austin chalk horizontal well production

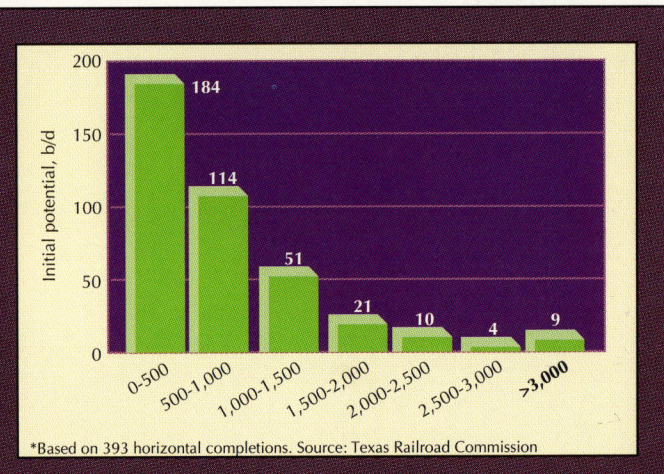

Fig. 6 Pearsall field initial potentials*
*Based on 393 horizontal completions. Source: Texas Railroad Commission

East Texas, Louisiana chalk action picks up

East Texas and North Louisiana in 1990 began to see horizontal exploration in Austin chalk, Cretaceous Saratoga chalk, and Eocene Wilcox.

Maersk Energy Co., Houston, completed 1 ARCO in Brookeland field of southern Sabine County, Tex., flowing 711 b/d of 43.6° gravity oil, 2.061 MMcfd of gas, and 556 b/d of water from Austin chalk open hole at 8,904-84 ft.

The well, 13 miles south-southeast of Hemphill, was 150 miles northeast of the nearest horizontal Austin chalk well in Brazos County.

ARCO, which owned a royalty interest in the well, held a large amount of fee acreage in the area and was said to be planning horizontal drilling there.

Maxwell Oil & Gas Co., Kilgore, completed 1 Hawkeye in Center field of Shelby County in late 1990.

It pumped 80 b/d of 43° gravity oil with a trace of water and 16 Mcfd of gas from Cretaceous Saratoga chalk at 2,210-60 ft measured depth (MD) in the curved portion of the well bore. TVD was 2,190 ft.

Saratoga Resources, Houston, in early 1991 planned to drill 110 Temple-Eastex, a horizontal Saratoga chalk test in Hemphill field of Sabine County. It was to be a short radius test with a proposed horizontal displacement of 200-300 ft at TVD 6,000 ft.

OXY U.S.A. Inc., Houston, which operated nearly one third of the 200 vertical producing wells in Olla field of LaSalle Parish, La., was planning more horizontal development in the region in 1991.

The company in late 1990 completed a horizontal well producing oil from the Cruse sand member of Eocene Wilcox. The test was the first horizontal well drilled in the North Louisiana basin.

OXY's 13-1 Tremont H produced 605 b/d of oil and no formation water on a submersible pump from an approximately 300 ft section in the Cruse F-1/Reservoir B sand at 2,310 ft TVD, 3,243-3,542 ft MD. The field is about 40 miles south of Monroe, La.

A typical vertical Cruse sand well in the field produced 50-100 b/d of oil. OXY said the horizontal well was intended to improve recovery and reduce water encroachment.

Wolfe & Magee Inc. and JWR Exploration Inc. completed 1 Sonat Minerals B-5 in Pendleton-Many field of Sabine Parish, La. Saratoga chalk at about 2,320 ft TVD flowed 104 b/d of 41.4° gravity oil and 15 Mcfd of gas but later began producing mostly water.

Late in the year, Oryx staked 1 Sonat Minerals, a Saratoga chalk test in Fort Jessup field about 5 miles southeast of the Pendleton-Many field well.

Smackover target in Alabama, Arkansas

Alabama and Arkansas operators tried horizontal drilling to Jurassic Smackover limestone, with the Alabama well setting a depth record for horizontal wells.

And Mississippi got its first two horizontal completions producing oil from Eocene Wilcox.

Smackco Ltd., Brewton, Ala., set the depth record in deepening a former vertical well in Huxford field, Escambia County (Fig. 7).

Its 1 Cruit 26-15 flowed 300 b/d of oil and 475 Mcfd of gas through a $^{13}/_{64}$ in. choke with 1,350 psi flowing tubing pressure.

The well reached horizontal at 14,936 ft TVD, surpassing the previous world record of 11,484 ft set at a North Dakota well.

After reaching horizontal, the well continued to build and hold angle of about 100° for a length of 405 ft in Smackover.

The well is a recompletion of Smackco's 2 ATIC 35-2, which produced more than 250,000 bbl of oil before water encroached. Using medium radius lateral drilling, Smackco displaced the bottom of the hole about 800 ft from the old vertical well bore.

The horizontal leg cost about half the price of a new

HORIZONTAL DRILLING

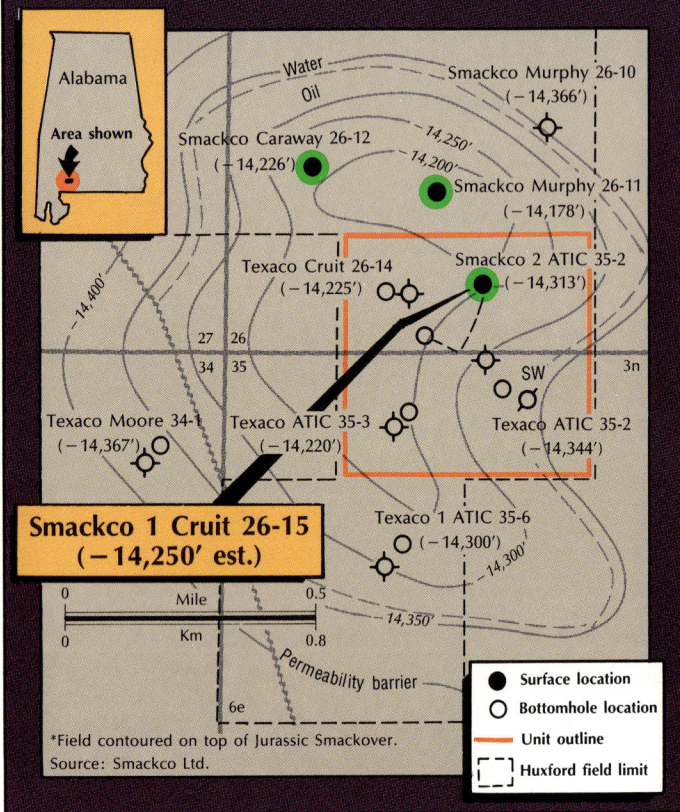

Fig. 7 — Huxford, Ala., horizontal well*
*Field contoured on top of Jurassic Smackover.
Source: Smackco Ltd.

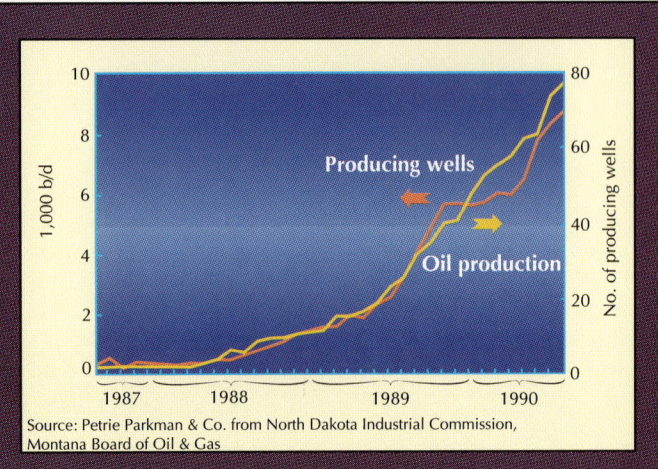

Fig. 8 — How Bakken drilling, production has grown
Source: Petrie Parkman & Co. from North Dakota Industrial Commission, Montana Board of Oil & Gas

vertical hole from surface and was an economic success, the operator said.

Production averaged 215 b/d of oil for 114 days. The operator had to shut the well in after lightning struck a gas processing plant serving the field Aug. 30, 1990. The field's vertical wells were returned to production within 30-45 days, but nitrogen jetting with coiled tubing was required at the horizontal hole.

Arkansas, which hosted its first two horizontal drilling attempts in 1989, had its third well drilling as 1991 began.

It was the American Exploration 2-15 Midway Field Unit in Northeast Lafayette County. The operator was seeking 1,400 ft of displacement with about 1,000 ft of horizontal section in Jurassic Smackover limestone at about 6,400 ft TVD.

Well bore would be in the top 20-40 ft of Smackover, which was 150-200 ft thick in the area and typically contained an oil column as thick as 40 ft.

The only previous horizontal drilling in Arkansas occurred in Columbia County in 1989. It was an unsuccessful Upper Cretaceous Buckrange (Ozan) oil attempt in Stephens field at about 2,200 ft TVD.

Amoco Production Co. late in 1990 was starting the first of three more planned horizontal wells in Clear Springs field, Franklin County, Miss., after completing two wells there earlier in the year.

The goal was to reduce water coning in Wilcox at about 4,700 ft.

Amoco potentialed 17 USA in the field for 320 b/d of oil, 103 Mcfd of gas, and 21 b/d of water through an $^{18}/_{64}$ in. choke. Potential test at the 18 USA was 472 b/d of oil and 43.5 Mcfd of gas through a $^{15}/_{64}$ in. choke.

After 6 weeks of production, each well was being produced at a rate of about 300 b/d of fluid. Oil cuts were 70-75% at the 17 USA well and 45% at the 18 USA.

Coho Resources Inc., Houston, staked the first horizontal well in Soso field, Jasper County, Miss. Its 1 Soso H was to evaluate Lower Cretaceous Washita-Fredericksburg at TVD 9,110 ft. The field extended into Smith and Jones counties.

Coho planned a 1,300 ft lateral in an attempt to reduce the water:oil ratio. Vertical wells produced 50-60 b/d of oil and 400-500 b/d of water.

During July 1990, Soso field produced 12,120 bbl of oil, 20.14 MMcf of gas, and 67,250 bbl of water from 15 conventional wells.

Bakken horizontal drilling forges ahead

Operators drilled 117 horizontal holes to Mississippian Bakken shale in North Dakota and Montana portions of the Williston basin during 1988-90 (Fig. 9).

All but five were in North Dakota.

Two other North Dakota horizontal wells were drilled in Mississippian Mission Canyon. One was dry, and the other extended Hay Draw field.

Montana Oil Journal (MOJ), Billings, tallied 81 reported horizontal Bakken producing wells in North Dakota and one intermittent Bakken horizontal producing well in Montana.

Ten horizontal wells in North Dakota and one in Montana were drilling at yearend and therefore excluded from the totals.

More horizontal wells were drilled in the Williston basin in 1990 than in the previous 2 years combined. In 1990's first 50 weeks, 79 were completed for the record or undergoing completion.

Horizontal wildcats in 1990 were drilled as far as 70 miles from the Bakken fairway, but in general the expansion effort was disappointing.

The play was carried out by 18 operators.

Meridian Oil Inc., which completed the basin's first Bakken horizontal hole in 1987, had completed 66 wells in North Dakota and two in Montana.

Other operators included American Hunter Exploration Ltd., Columbia Gas Development Corp., Conoco Inc., Samuel Gary & Associates, Leede Oil & Gas, Maxus Energy Corp., Oryx, Pacific Enterprises Oil Co., Shell Western E&P Inc., Slawson Exploration Inc., and Texaco Inc.

Horizontal drilling accounted for 27.6% of total completions of record in North Dakota in 1990.

As the Bakken play expanded beyond the Bakken fairway in Billings, Golden Valley, and McKenzie counties, there was a sobering reality that some refinement were necessary before the benefits of the technology were applicable in other sectors of the Williston basin, MOJ observed.

Wells drilled in regions of thicker Bakken sections found fewer and wider spaced regional fracture systems.

A lot depends on the control system you choose. In fact, it controls the profitability of your entire operation. To keep profits moving upward, you need a tool that allows you to continuously increase your product quality. And nothing helps you do it better than a Rosemount distributed control system. Like you, we are committed to continuous quality improvement. And as a result, our control system has earned a reputation as the most reliable, effective

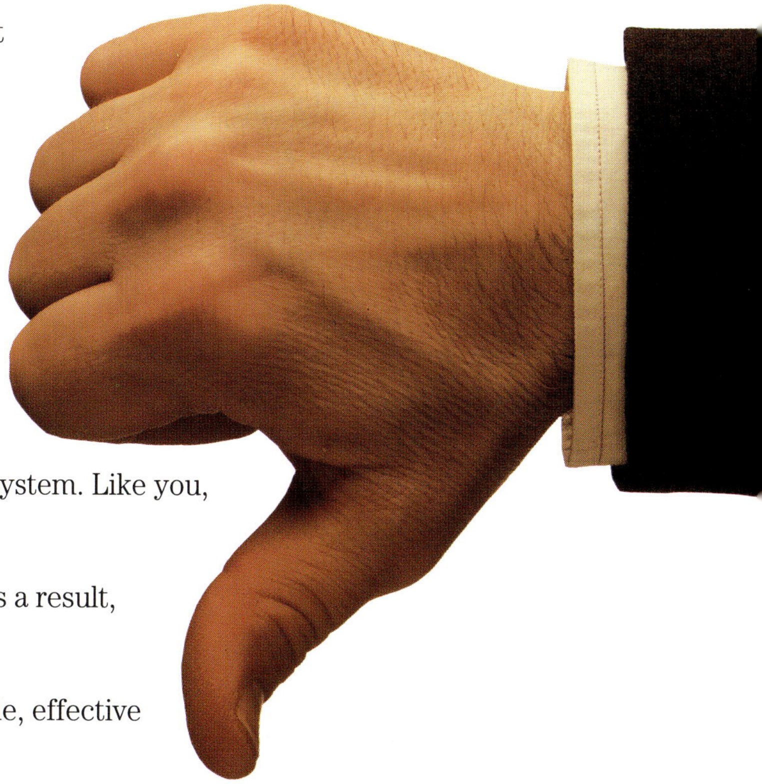

THE CONTROL SYSTEM YOU CHOOSE DEFINES THE DIRECTION OF YOUR PROFITS.

system on the market. A performance record that could only come from one of the world's largest instrumentation and process control companies. To find out why more and more of your colleagues are moving up to our control systems, contact the Rosemount Control Systems Division, 12000 Portland Avenue South, Burnsville, MN 55337. Call 612-828-3568 or Fax 612-828-3236.

©Rosemount Inc. 1991

ROSEMOUNT® Measurement Control Analytical Valves

HORIZONTAL DRILLING

One operator's Bakken learning curve*

Table 2

	Drilling/ completion costs $/BOE	Reserves BOE	Initial potential B/DOE	Finding and development cost $/BOE	After tax rate of return† %
First 10 wells	1.52	252,000	252	6.83	30.6
Second 10 wells	1.26	234,000	227	5.84	44.2
Third 10 wells	1.09	269,000	240	4.36	66.6
After 30 wells	1.29	241,000	240	5.77	43.3
After 50 wells	1.23	200,000	199	6.71	31.0

*Adapted from Meridian Oil Inc. †Based on $18/bbl wellhead netback.

Operators completed 67 horizontal wells for the record during 1990. Of those, 48 were oil and gas wells, nine were shut in awaiting further evaluation, and 10 were either abandoned or temporarily abandoned. Most of the unsuccessful or noncommercial wells were in Dunn and Mountrail counties.

Ash Coulee field in Billings County was one of the sweet spots being developed with horizontal wells, with at least nine Bakken wells completed there during 1990, MOJ reported.

Initial potential at one well was reported as 1,900 b/d of oil, and another made more than 1,350 b/d. Early in the fourth quarter the average production for each Ash Coulee horizontal well was 107 b/d of oil and a trace of water. Two wells were flowing about 430 b/d of oil with no water.

Most active Ash Coulee field horizontal well operators were Slawson, Columbia, and Leede.

Meanwhile, Pennzoil Exploration & Production Co. in early 1991 was starting 26-13H Candee, in Richland County, Mont., the state's first horizontal well to target Mississippian Madison.

The company planned to evaluate the fractured, porous Ratcliff reservoir and a fractured limestone in Mission Canyon on the north side of Ridgelawn field. Projected depths were 12,350 ft MD, 9,350 ft TVD.

On the Bearpaw arch in Blaine County, North Central Montana, Texaco completed two horizontal oil wells in Bowes-Sawtooth Unit from the Bowes member of the Piper formation (Jurassic Sawtooth).

The A502 Bowes-Sawtooth Unit pumped 127 b/d of 18° gravity oil and 54 b/d of water from Sawtooth perforations at 3,617-5,620 ft MD. The 1,500 ft horizontal segment of the hole entered Sawtooth at 3,485 ft TVD.

The D806 Bowes-Sawtooth Unit, 2 miles east of the first well, pumped 83 b/d of 18.9° gravity oil and 71 b/d of water from perforations at 3,460-5,199 ft MD. Horizontal displacement was 1,557 ft.

Texaco in 1983 drilled six vertical infill wells in the unit that pumped 12-90 b/d of oil and 2-52 b/d of water.

Niobrara plays expand in Colorado, Wyoming

Exploitation of oil in fractured Cretaceous Niobrara limestone got a jump start in mid-1990 with completion of a well in Silo field in the Denver basin in Laramie County, Wyo. (Fig. 10).

Cowan Oil Co., Greeley, Colo., 1 Warren was completed in June flowing at the rate of 2,000 b/d of oil and 500 Mcfd of gas from Niobrara. The well had produced about 21,000 bbl of oil in its first 4 months on production, and pump rate had declined to about 50 b/d.

UPRC plugged 1 Lazy D 21-3, in Weld County, Colo., its first Niobrara horizontal test. The well was delayed by many mechanical problems involving drilling and testing. It apparently found very few open, natural fractures in Niobrara and recovered oil at noncommercial rates from a 2,000 ft open hole interval, Petroleum Information reported.

Later in the year UPRC's 1H Antelope 9-11 in Silo field flowed at a rate of 1,462 b/d of oil on a 21 hr test through a $^{44}/_{64}$ in. choke with 225 psi flowing tubing pressure.

Horizontal displacement was 2,529 ft, of which 2,037 ft were within Niobrara.

Snyder Oil Corp., Fort Worth, completed its first Silo field well, 2H Willett, pumping at a rate of 216 b/d of oil, but later production and pressure tests were disappointing. At yearend the company was planning to stimulate the well.

It was also spudding a second well, 10-21H Hutton.

About 30 miles southeast of Silo field, Snyder staked 20-3H Burbach, a horizontal Niobrara wildcat 1 mile southeast of Hereford field in Weld County, Colo. Location was 24 miles east of a Weld County wildcat UPRC plugged in 1990 after recovering noncommercial volumes of oil from a horizontal Niobrara section.

Westoil Production Co., Denver, staked two horizontal Niobrara wells in Wellington field, Larimer County, Colo.

The 1 Thompson 4-7H was to have a 1,440 ft horizontal section at 4,100 ft TVD in a N. 15° W. direction from the vertical portion of the hole.

Site was about 1½ miles south of a vertical well that produced about 42,000 bbl of oil from Niobrara in the late 1920s.

Westoil also planned 1 Box I4-30H, in 30-10n-68w, about 3 miles north of the first well.

The two wells were part of an eight well horizontal drilling program Westoil planned on its 35,000 acre Front Range leasehold in Northeast Colorado and Southeast Wyoming.

Other areas in which the company planned horizontal Niobrara tests were in Boulder field, Boulder County; Berthoud field, Larimer County; South Borie-Rawhide prospect, Larimer County; and Wyoming's Silo field.

Gerrity Oil & Gas Corp., Denver, also staked two horizontal Niobrara tests in Silo field.

Economics of the Niobrara appeared to be attractive at about $20/bbl if operators were able to establish reserves of about 125,000 bbl/horizontal well and completed well cost could be reduced to $1.2 million/well.

HORIZONTAL DRILLING

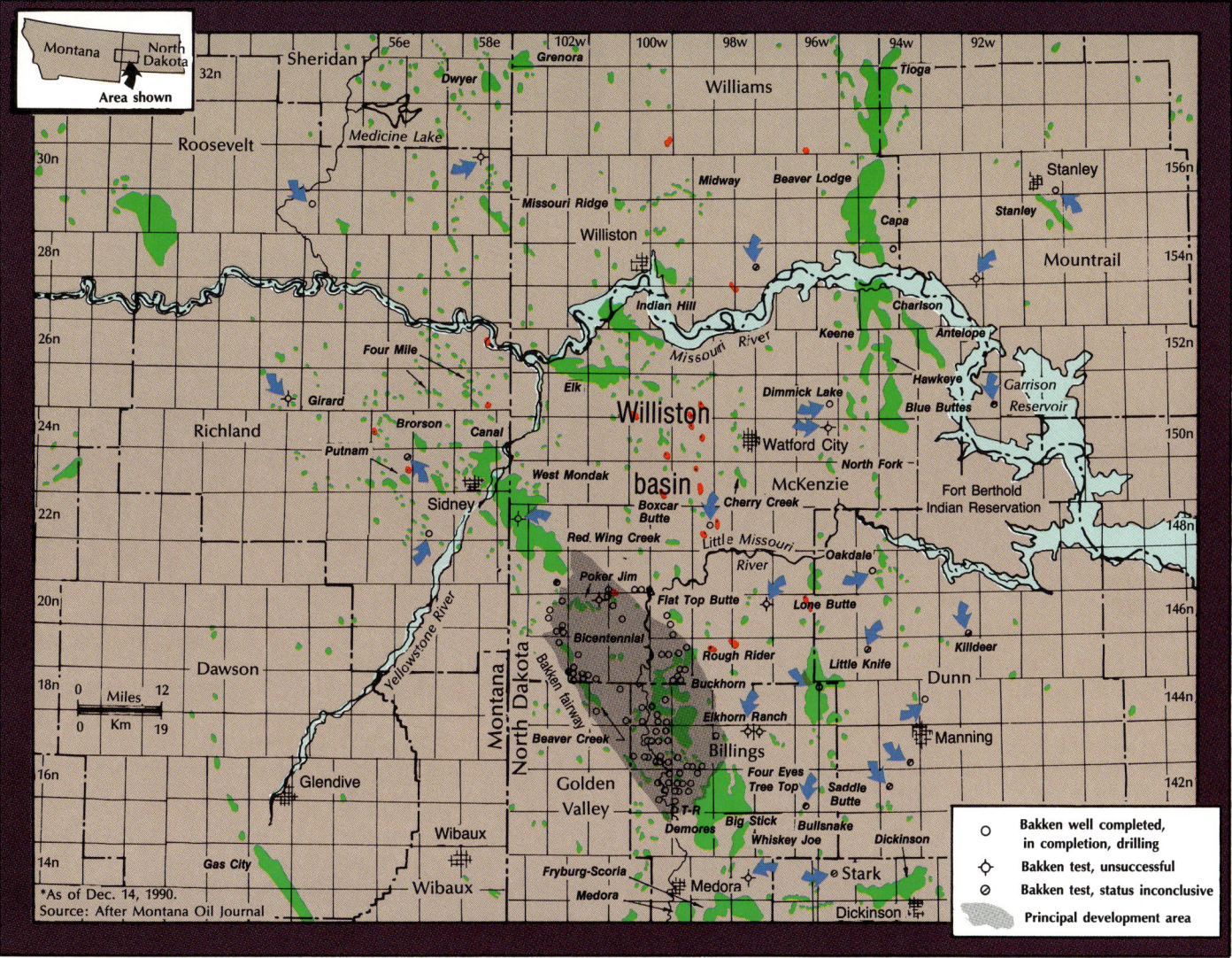

Horizontal Bakken drilling in the Williston basin*

That reserve figure was three to four times vertical well reserves.

Other horizontal well operators working in the Sand Wash, Denver, and Hanna basins included Snyder, Wolverine Exploration Co., Meridian Oil Inc., Oryx, Pacific Enterprises, Hondo Oil & Gas Corp., UPRC, Gerrity, and Wainoco Oil Corp. (Fig. 11).

New Mexico horizontal successes mounting

The San Juan basin yielded successful horizontal oil completions in Jurassic Entrada sandstone and Cretaceous Mancos shale.

Merrion Oil & Gas Corp., Farmington, N.M., planned to drill several more horizontal holes after what it termed "a highly commercial success" at a horizontal well in Lower Jurassic Entrada sandstone in Papers Wash field, McKinley County, N.M.

The San Juan basin reentry, 15-2H Federal, cut 434 ft of horizontal hole in Entrada.

The well averaged 309 b/d of oil and 5,046 b/d of water during a 40 day test on a submersible pump with essentially no decline rate. Pumping fluid level was 3,000 ft.

Before the test, the well was beam pumped at low volumes, averaging 60 b/d of oil and 800 b/d of water for 5 months.

Merrion, which operated six Entrada fields, believed the horizontal well greatly increased reserves in Papers Wash field, which it considered to be 95% depleted by vertical wells.

The company had been conducting a seismic exploration program for 5 years, geared mainly toward Entrada prospects. It said horizontal drilling will not only help squeeze more oil out of depleted reservoirs but also will play a role in optimum development of future discoveries.

Elsewhere in New Mexico, BASF Corp. started a horizontal Cretaceous Gallup siltstone well in Verde field, San Juan County, and Samuel Gary & Associates was preparing to horizontally drill four Mancos wells in Rio Puerco field, Sandoval County.

Richmond Petroleum Co., Farmington, planned to drill 1 Apache, a horizontal test in Gavilan-Mancos field, Rio Arriba County. Proposed depths were 9,715 ft MD, 7,500 ft TVD.

Oklahomans searching wide for success

Economically successful horizontal completions had all but eluded Oklahoma operators, which had drilled or staked attempts in many areas of the state.

At yearend 1990, horizontal wells had been drilled in Carter, Garfield, Love, Murray, and Osage counties and were pending in Creek, Grant, Nowata, Osage, Pontotoc, Rogers, Tulsa, and Woods counties.

Conoco completed a horizontal development well, 8

HORIZONTAL DRILLING

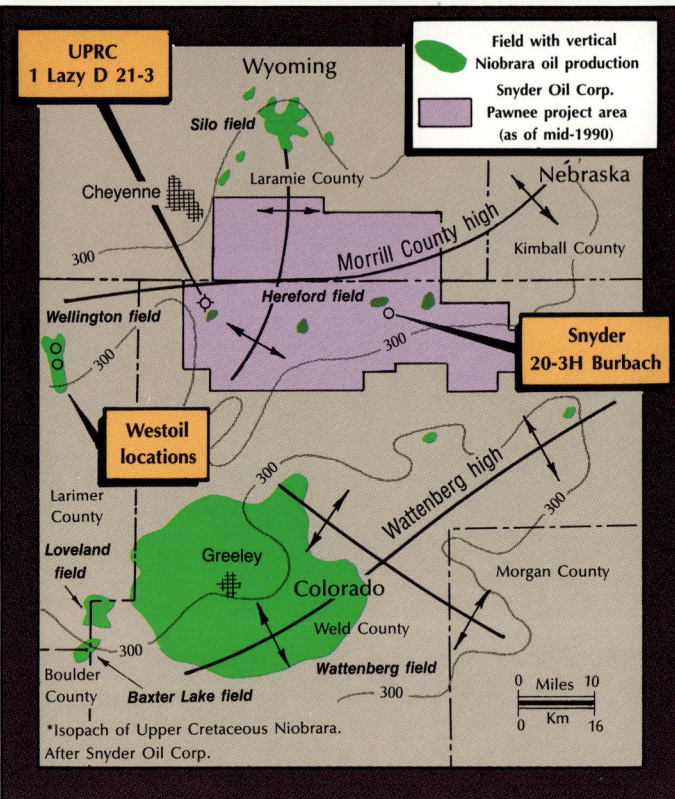

Fig. 10 Denver basin Niobrara horizontal layout*

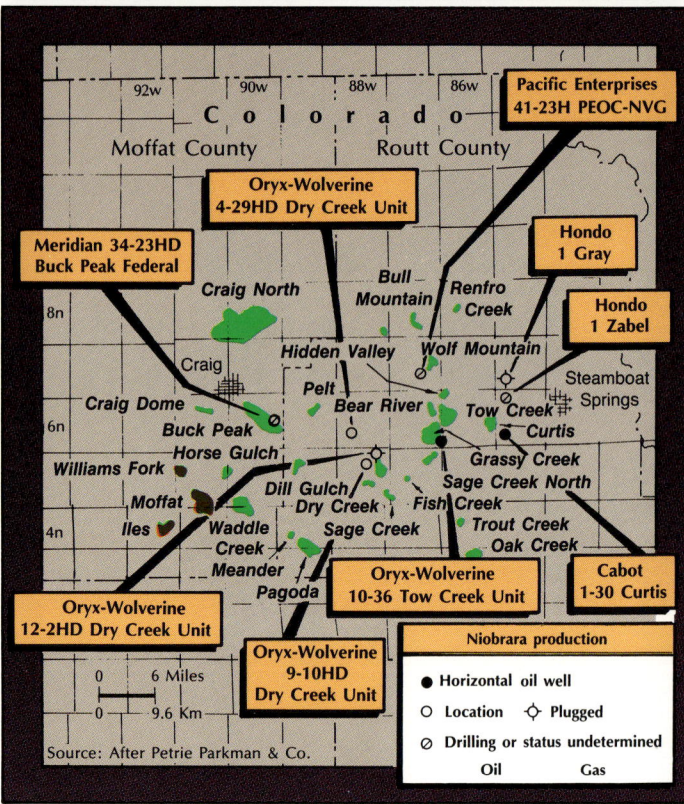

Fig. 11 Sand Wash basin Niobrara action

Greer, in Davis field of Murray County. It flowed 284 b/d of oil through a 6/64 in. choke with 350 psi flowing tubing pressure from Ordovician Third Bromide sand open hole at 4,120-4,484 ft MD, 3,867-3,879 ft TVD.

Samson Gas Producing Co., Tulsa, staked 2-4 Dietz in Oakdale field, Woods County.

It was to be the state's longest horizontal well if drilled as planned.

Proposed length was 2,500 ft directed S. 14° E. in Mississippi limestone at about 6,800 ft TVD.

Brown & Borelli Inc., Kingfisher, Okla., drilled a 500 ft lateral in Siluro-Devonian Hunton at 1 Big Four in the Sooner Trend, Kingfisher County, early in 1990. It pumped 46 b/d of oil and 150 Mcfd of gas on initial completion.

Ladd Petroleum Corp., Tulsa, drilled 2-27 Leon in Lorena field, Beaver County, in the Oklahoma Panhandle. It cut a 1,000 ft horizontal hole in Mississippian Chester at 6,900 ft TVD.

The well produced 106 MMcf of gas from January 1989 through February 1990.

At yearend 1990, Esco Exploration Inc., Tulsa, was seeking creation of horizontal well spacing units for several sections to produce oil from Ordovician Viola limestone in Southeast Cornish field, Jefferson County.

Canadian application frees heavy crudes

A horizontal drilling success story in Canada resulted from a technology license agreement between Institute Francaise du Petrole (IFP), Paris, and CS Resources Ltd., Calgary.

CS Resources had completed 18 drainholes in thin oil pays in Pelican Lake field of Alberta and Winter field of Saskatchewan.

The 1990-91 program called for 12 more long horizontal wells to be drilled in Winter field, operated by provincially owned Saskoil in association with CS Resources and others.

Pelican Lake, a former farmout of Gulf Canada Resources Ltd. now operated by CS Resources, covers about 50,000 acres.

IFP said hundreds of horizontal wells could be drilled after a complete assessment of the results of 13 such wells during the previous 2 years.

Both heavy oil fields were marginally profitable to unprofitable under conventional production systems.

IFP assisted CS Resources in covering all aspects of field development using its new technologies such as reservoir evaluation and simulation, production monitoring, adjustment of French made cavity pumps placed at high inclination, and water control treatment of some sections of drainholes.

Further, recent research and development programs such as sand cleaning devices and three phase production logging systems were about to be tested.

Reaching profitability in difficult conditions

One of the significant achievements of the IFP-CS Resources joint efforts was the ability to reach profitability in those two difficult fields, which produced 14° gravity oil from shallow, unconsolidated sands.

The Cretaceous Wabiskaw reservoir at Pelican Lake field was about 1,300 ft deep and 9-13 ft thick with permeability of about 3,000 md.

Oil viscosity was about 1,000 cp.

Long to medium radius curvature wells were drilled in an average of less than 10 days with 8½ in. hole in the reservoir varying from 1,476 ft long for the first well to more than 3,280 ft for the most recent.

The Cretaceous Cummings fluvial sands of Winter field are complex deposits of limited extent, 33-49 ft thick, above an active water table with very high permeabilities. Oil viscosity is 3,000 cp.

The five horizontal wells were 1,800-2,460 ft long inside the upper section of the oil pay.

For both fields, initial production rates were about 200 b/d,

HORIZONTAL DRILLING

Fig. 12 Well bore schematic*

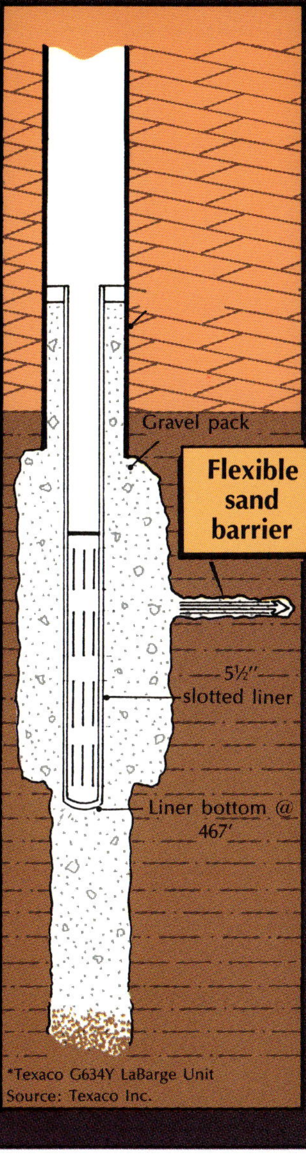

- Gravel pack
- Flexible sand barrier
- 5½" slotted liner
- Liner bottom @ 467'

*Texaco G634Y LaBarge Unit
Source: Texaco Inc.

Fig. 13 Horizontal drilling planning information

Proposed horizontal well data
- Overall project objectives
- Location, lease size
- Detailed target description
- Formations
- Proposed bit program
- Unusual well conditions
- Horizontal length, direction, hole size
- New well or recompletion
- Survey requirements
- Logging program
- Rig specifications
- Drill string sizes, tool joints
- Casing program
- Mud program
- Planned completion technique and pump placement

Drilling rig considerations
- Top drive or power swivel recommended
- Provide for adequate pumping capacity
- Short radius wells can be drilled with portable workover rigs

Offset well data
- Penetration rates
- Bit program
- Mud program
- Problem formations
- Bottomhole temperature
- Well surveys
- Bottomhole assemblies used

Drilling fluids, hole cleaning considerations
- Use full range of drilling fluids
- Plan for short trips, wiper runs
- Monitor mud system carefully
- Use higher flow rates and annular velocities
- Flexible mud program—adaptable to problems
- Avoid filter cake build-up to minimize torque and drag

Source: Eastman Christensen

---

a four fold increase, compared with vertical well initial rates.

U.S.S.R. joint venture eyes horizontal drilling

Elsewhere, a Houston company formed a joint venture with the production association of a Soviet republic to study the feasibility of drilling horizontal wells in several western Siberian oil fields. Anglo-Suisse Inc., Houston, and Varyegan Oil & Gas Association of the U.S.S.R. Ministry of Oil and Gas formed the White Night joint venture to study the feasibility of drilling a series of horizontal wells in West Varyegan and Tagrin oil fields. An addendum signed at midyear added Roslaval oil field to the project.

Addition of Roslaval's estimated 200 million bbl of reserves brought the project scope to 1.1 billion bbl. Ultimate plan was to drill as many as 400 horizontal wells in West Varyegan and Tagrin fields and about 50 in Roslaval.

Anglo-Suisse said the first step might be to redrill some West Varyegan and Tagrin wells horizontally in early 1991.

The agreement committed the association to add more fields to the venture to assure that Anglo-Suisse's share amounted to 550 million bbl of recoverable oil.

Varyegan arch fields, which yield oil from Cretaceous-Jurassic reservoirs at 8,500-9,000 ft, were among the most disappointing in western Siberia in terms of productivity.

U.K., Dutch North Sea wells successful

Conoco units achieved good flow rates during 1990 at medium radius horizontal wells in the U.K. and Dutch North Sea.

Conoco (U.K.) Ltd. drilled the U.K. southern gas basin's first horizontal well in Block 49/16 in North Valiant field. The well flowed 47 MMcfd of gas through a choke, more than three times the rates reported for other wells in the field.

Conoco dubbed the well Dunehunter because the horizontal section was targeted to penetrate several large dunes in the Permian Rotliegendes sandstone at 8,100 ft TVD.

The angle of deviation, built at an average rate 12°/100 ft, was more than five times that achieved by standard directional drilling. Conoco used specially designed steerable motors supplied by Sperry Sun Drilling Services, Great Yarmouth.

The well cost 35% more than previous conventional wells in the field.

Conoco was considering more horizontal wells, including one in Vulcan gas field and another from the Murchison oil field platform in the East Shetlands basin.

Conoco Netherlands' Dutch North Sea horizontal well doubled production to 25,000 b/d from Kotter oil field in Block K18.

The company sidetracked its K8/1A horizontal well from the abandoned K8/1 producer using a jack up rig. Stabilized flow was 12,000 b/d of oil following 1 month of gradually increasing rate.

Conoco built angle at 8-11.5°/100 ft, and the hole reached horizontal at 6,280 ft MD, 5,760 ft TVD. It continued horizontally for 620 ft. Sidetracking from an existing well cut drilling/completion cost about 50%, compared with a new well from surface.

Conoco said horizontal drilling offers the best opportunity to boost oil production without increasing water production or impairing ultimate recovery from the relatively thin, heavily faulted, Lower Cretaceous Vlieland sandstone reservoir.

The company was monitoring performance and investigating whether Kotter or nearby Logger field could benefit from further horizontal drilling.

**New Zealand field gets
two horizontal wells**

Petrocorp Exploration Ltd. successfully completed two horizontal wells in McKee field on the north island of New Zealand, said a report in Petroleum Exploration in New Zealand News.

McKee field, discovered in 1979, produces about 10,500 b/d of oil and 11.4 MM cfd of gas from the McKee formation, a late Eocene to early Oligocene age sandstone, within an east dipping overthrust structure.

The structure is a structural trap similar to that found in the Ahuroa/Tariki gas-condensate fields and Waihapa (Tikorangi formation) oil and gas field.

Petrocorp's first horizontal well, 12 McKee, was drilled in the central part of the field in late 1989. Horizontal displacement in McKee was nearly 1,200 ft.

The second well, 4 Tuhua, was drilled in late 1989 and early 1990 in the north part of the field, where a relatively thin oil rim underlies a large gas cap.

The well was intended to increase oil recovery by producing at low drawdown, thereby reducing the tendency for gas coning or cusping.

Of the 1,073 ft of horizontal displacement, about 715 ft was in McKee.

Petrocorp was assessing long term production capabilities and economic viability of the two wells.

**Technology evolution,
improvements assisting**

Medium radius was the dominant type of horizontal drilling technology being implemented, but operators said an ultrashort radial system was achieving a good degree of success in wide ranging applications.

The ultrashort radius radial system (URRS) was being implemented by Petrolphysics Inc., San Francisco.

Through 1990 the company had placed more than 700 horizontal radials in commercial and test wells. Petrolphysics, alone or in partnership with Bechtel Corp., had installed more than 30,000 ft of multiple 1¼ in. radials.

Radials had been placed in California, Canada, Louisiana, and Wyoming.

Louisiana work focused on light oil reservoirs characterized by water drives.

Here are benefits of radial completions based on more than 1 year of production data reported by two operators, W.C. Allen and Flex-Tech, Shreveport:

• Wells with four horizontal radials incorporating Petrolphysics' flexible sand barrier completions produced four to 10 times as much oil per day as nearby control vertical wells.

• Wells with four horizontal radials reduced the water:oil ratio by more than four times, compared with nearby vertical control wells.

• Wells with radials showed a much reduced decline or no decline in oil production during this period, compared with vertical wells.

• Radials appeared to reduce near well bore damage.

Petrolphysics planned to begin radial placement operations in Texas in 1991.

The first Wyoming well with multiple radials was completed in a light oil reservoir with stabilized production four times that of an average vertical well in the area. The second Wyoming well's production was twice that of a vertical well.

Petrolphysics completed a deep well cleanout program for British Petroleum in Prudhoe Bay field, Alaska, and the contractor was asked to consider expanded work there.

Large projects were under negotiation or discussion for Petroleos de Venezuela's Corpoven, Lagoven, and Maraven

Horizontal well trends — Fig. 14

- Potential use of horizontal drilling will be considered for every oilfield
- Improved reservoir knowledge/description will enhance applications
- Increased use of complex completion and zonal isolation techniques
- Increased use of ofe SRM techniques in reentry wells
- Enhanced steerable system drilling capabilities; enhanced measurement while drilling
- Increased use of retrievable and casing whipstocks—multilaterals
- Create synergy between horizontal drilling, completion technology

Source: Eastman Christensen

units. Work was to start early in 1991. Petrolphysics was discussing a similar project with Indonesia's national oil concern, Pertamina.

**Ultrashort radials
aid in steamflooding**

More than 500 ultrashort radials were placed in California wells alone.

The greatest concentration of the radials was in unconsolidated heavy oil fields in the San Joaquin Valley and in light oil fields of Louisiana and Wyoming.

Of 181 radials placed in California wells greater than 300 ft in depth, 60% exceeded 100 ft in length.

Petrolphysics' work in California demonstrated three important benefits associated with radials in heavy oil steamflood operations:

• For injection purposes, the radials serve as an effective conduit to disperse steam a considerable distance from the injection well, as monitored by remote observation wells, thereby increasing the overall thermal efficiency of the injected steam.

• During the production cycle, the efficacy of the steam dispersion into the reservoir resulted in a gradual temperature rise in the injection/production well followed by an increased level of oil production from the well.

• The well with radials produced as much from a 30 ft interval as did a nearby well from an identical 300 ft interval.

**Ultrashort radials
tried in Wyoming**

Texaco completed a shallow well producing oil by gravity drainage through three horizontal legs about 120° apart on the La Barge platform in Wyoming.

The well, G634Y La Barge Unit in Sublette County, produced from the low pressure Almy member of Eocene Wasatch with three laterals kicked from the same vertical well bore from the same kickoff depth.

It was on conventional vertical rod pump making about 30 b/d of oil and 3 b/d of water with no decline from a 24 in. underreamed, gravel packed, unstimulated zone in Almy at 415-520 ft. Most vertical Almy wells produced less than 10 b/d of oil.

The laterals with approximate direction and length were north-northeast 133 ft, east-southeast 155 ft, and south-southwest 70 ft.

The 4 in. diameter laterals were jetted into Almy using a nonrotating nozzle through which water was pumped from surface through coiled tubing at 175 gal/min at about 10,000 psi.

The laterals, made with an ultrashort (1-15 ft) radius

HORIZONTAL DRILLING

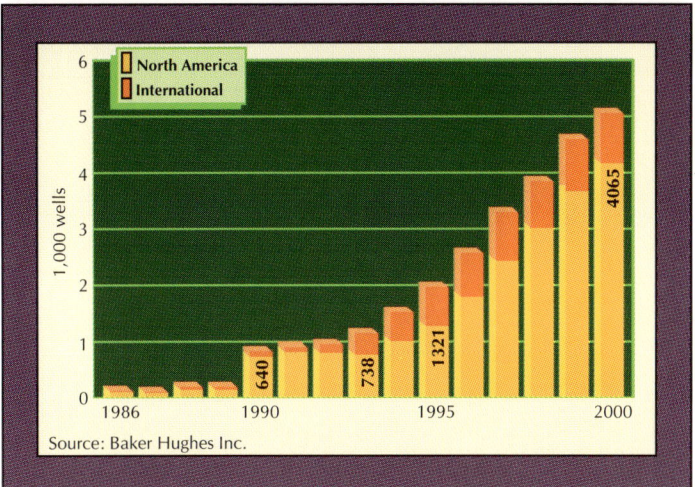

Fig. 15 — How horizontal drilling will grow. Source: Baker Hughes Inc.

system provided by Petrolphysics, were at angles of 90° or more to vertical.

After drilling, the laterals were lined with flexible sand barrier, a tube that served as a production conduit (Fig. 12).

Making each radial required 12-14 hr. Horizontal drilling contractor's cost was $30,000-40,000/radial.

Laterals averaged 40 ft in length at a reentry in the same section that was not as successful, Texaco said.

Texaco was evaluating results, believed the wells showed horizontal drilling is applicable in the area, believed costs could be reduced, and was considering further horizontal drilling in early 1991.

Drilling, completion concerns voiced

Here is a selection of technology concerns voiced by a horizontal drilling specialist:

Most of the early wells were drilled diagonally across target formations. Reservoir modeling clearly shows that there is an optimum elevation for most horizontal well applications.

With a slant hole, the portions that are too high and too low will have limited productivity. Eventually, all horizontal wells will be drilled at the optimum elevation. A noticeable shift in this direction is probably already started in the Austin chalk.

The critical keys required for a successful horizontal hole are the vertical continuity of the reservoir and the control of drilling and completion fluid damage. The completion technique can have a significant impact on these two parameters. The Austin chalk boom did not start until someone discovered that the incredibly damaging effect of dry drilling could be essentially eliminated by underbalanced drilling. The result, in August 1989, was the first Austin chalk well that potentialed for more than 1,000 b/d.

In the rest of the industry's horizontal testing efforts, the norm has been overbalanced drilling and slotted liner completions. This technique offers the worst possible way for obtaining vertical continuity and controlling damage.

Completions that utilize cemented strings and selective perforation and stimulation techniques offer the best methods for improving vertical continuity and eliminating damage effects. The number of aggressive completions is increasing but is still a small fraction of the total number drilled.

Gas applications slow but increasing

Horizontal drilling for gas was much slower in developing, and few resounding successes were reported outside Conoco's North Valliant test in the U.K. North Sea.

The U.S. Department of Energy was conducting an extensive research and field testing program on horizontally drilled wells in the Appalachian basin and publishing large volumes of data on planning, drilling, and testing of the resulting wells.

Early field results showed that horizontal wells can yield flow rates greater than vertical wells, drilling and completion costs are extremely high compared with vertical wells, and multiple frac jobs may be required to maximize production.

BDM Corp., McLean, Va., CNG Development Co., Pittsburgh, and DOE spudded West Virginia's third horizontal well and the first in Calhoun County late in 1990.

It was 3997 Hunter Bennett, 44 miles northwest of Charleston in Millstone quadrangle, Lee district. It was on a CNG lease in Russet field, which produces oil and gas from Mississippian Big Injun. The plan was to drill vertically to 2,810 ft, then kick off at about 2,000 ft and drill a 2,000 ft horizontal leg through Devonian Fifth sand.

The well was within 2 miles of Devonian gas production in three fields: Dekalb, Grantsville, and Yellow Creek.

Two horizontal wells were drilled in Wayne County, W.Va., in the mid-1980s.

Many companies see potential in horizontal gas drilling if gas prices improve.

For instance, National Fuel Gas Co. drilled and completed only one vertical development gas well in Northeast Clay field during fiscal 1990 because of low gas prices.

However, it planned to pursue further development of the acreage as gas prices rose and as it continued to evaluate the potential for horizontal drilling in the region.

Columbia Gas Development pointed out that one third of its extensive Appalachian acreage holdings have Devonian shale potential, and successful development of horizontal drilling techniques in the basin could add considerable value.

Oryx late in the year staked a horizontal test of Cretaceous Cozzette, a gas bearing horizon, in the Piceance basin in northeastern Mesa County, Colo.

Its 1 HD Acapulco-Federal was to be drilled to 11,700 ft MD and about 9,150 ft TVD. Location was about 6 miles west of Divide Creek field.

More spectacular growth: supportable?

Horizontal drilling's growth in the first few years of widespread commercial applications has been at the rate of several hundred percent per year.

Many forecasts are available. One, by Baker Hughes Inc., indicates that U.S. horizontal drilling will more than double to 1,321 wells in 1995, compared with 640 in 1990. In 2000 Baker Hughes expects horizontal drilling to reach 5,000 wells internationally, including 4,065 in the U.S.

Such growth rates provided tremendous advantages, such as the application of a vast array of brainpower, and created enormous problems in hiring and training people and providing/maintaining equipment. One concern in medium radius drilling in the Austin chalk trend has been the supply of downhole motors. One large horizontal drilling service company increased its inventory of 4¾ in. motors to 110 in fourth quarter 1990 from 90 in the third quarter and said economics provided for no further increases.

Maurer Engineering Inc., Houston, had been conducting a joint industry Drilling Engineering Association project on horizontal drilling since 1987. About 104 companies from 19 countries joined the $3.5 million, 36 month first phase of the project.

The project was successful, resulting in distribution of 12 manuals and 10 computer programs on horizontal well technology.

Participants asked the engineering firm to follow up with a second phase project that started Dec. 1, 1990. ⬜ IPE

IPE ATLAS

NORTH AMERICA

UNITED STATES

CAPITAL: Washington
MONETARY UNIT: Dollar
REFINING CAPACITY: 15,558,923 b/d
PRODUCTION: 7.225 million b/d
RESERVES: 26.177 billion bbl

THE U.S. INDUSTRY CLOSED OUT 1990 AMID PREDICtions that 1991 will be the sixth consecutive year of declining oil production.

Among those sounding that warning was the Independent Petroleum Association of America.

IPAA's supply/demand committee disclosed late in 1990 disclosed its forecast of U.S. production of 7.043 million b/d for 1991, down about 2.5% from the expected 1990 average (Table 1).

Natural gas liquids production will rise slightly more than 1% to stand at 1.526 million b/d in 1991, yielding a decline of a little less than 2% to 8.569 million b/d in total liquids production, IPAA said.

Domestic demand meantime will inch up to 16.968 million b/d, leaving an 8.474 million b/d gap in consumption and stocks to be filled by imports of crude oil and petroleum products. At that level, imports will account for slightly less than 50% of U.S. oil supply for the year, not counting volumes imported for the Strategic Petroleum Reserve (SPR).

The Department of Energy (DOE) completed a test sale of almost 4 million bbl of oil from the SPR near yearend 1990. With an eye on the Middle East crisis, President Bush ordered the test in part to show that the SPR is ready for use in the event of an oil supply disruption.

The IPAA committee pointed out that since 1985 U.S. crude oil production has been declining at an average 4%/year, or about 340,000 b/d/year.

IPAA's annual meeting in October also heard reports that U.S.:

• Natural gas consumption will total 18.7 tcf in 1991, an increase of 2.5% from 1990.

• Average weekly tallies of active rotary rigs will show a gain of 15% in 1991. That will further bolster demand for oil country tubular goods, which showed a sharp jump in first half 1990 from the same period of 1989.

• Horizontal drilling will increase dramatically, more than doubling in the next 5 years. Among other things, the technique will create a host of redevelopment opportunities in older fields.

Oil and gas supply/demand

IPAA's supply/demand committee predicted a decline in 1991 crude oil production from the 1990 level of 7.225 million b/d, off 5.1% from 1989.

The relatively sharp slide for 1990 stemmed in part from maintenance work in Alaska that resulted in a reduction in North Slope production.

In 1991, improved economic returns from wells in the Lower 48 will encourage workovers and servicing. This will reduce abandonments and slow the production decline rate.

The committee also said it expected the crude oil production decline rate in Alaska to slow due to increased natural gas handling capacity and a hydraulic fracturing program.

Domestic consumption will reverse the trend of the past 5 years and decline marginally in 1991. IPAA estimated demand was down 1.3% in 1990 at 17.093 million b/d and will be down another 0.7% in 1991.

Imports as a share of domestic demand were an estimated average 49.1% in 1990 and will inch up to 49.9% in 1991. During second half 1991 imports will average 50.9% of domestic demand, IPAA said.

Demand for gasoline will continue to fall slowly in 1991. Demand rose during 1982-88 but then slipped slightly—by 8,000 b/d—in 1989 to 7.328 million b/d. The supply/demand committee estimated 1990 gasoline demand of 7.259 million b/d, down 0.9%, and 1991 demand of 7.15 million b/d, down 1.5%.

Continued improvement in fuel efficiency lies ahead, IPAA predicted. And personal travel by cars and light trucks will decrease in 1990-91 due to sluggish economic growth. Sharp jumps in retail gasoline prices also will rein demand.

An expected return to normal weather in 1991 will result in a slight increase in distillate demand—up 0.2% to 3.105 million b/d.

Residual fuel oil demand will slide 3.2% in 1991 to 1.251 million b/d. That's because of an increase in the contribution of natural gas, hydropower, and nuclear power to electricity consumption.

IPAA estimated U.S. natural gas production of 17.3 tcf for

Where the U.S. SPR and related facilities are

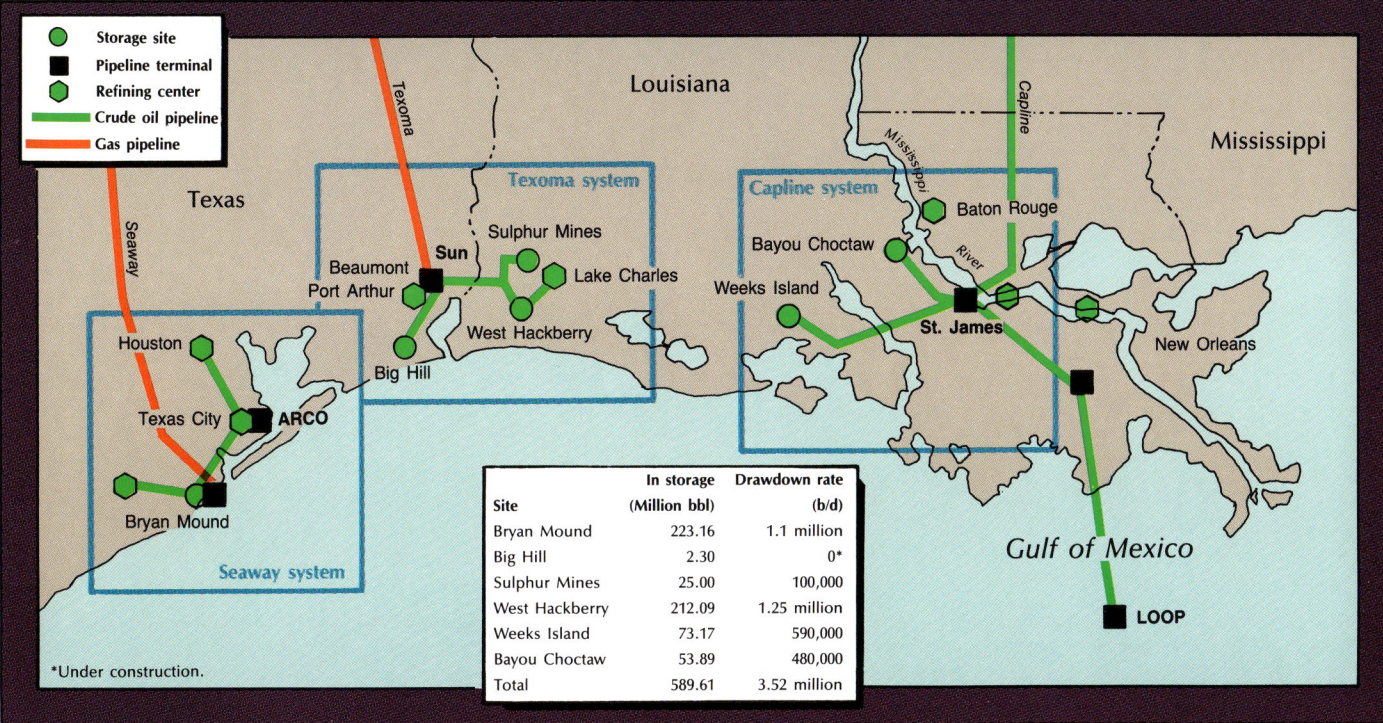

Storage site
Pipeline terminal
Refining center
Crude oil pipeline
Gas pipeline
Texas
Louisiana
Mississippi
Texoma
Seaway
Capline
Mississippi River
Texoma system
Capline system
Seaway system
Sulphur Mines
Lake Charles
Beaumont
Port Arthur
Sun
West Hackberry
Big Hill
Houston
Texas City
ARCO
Bryan Mound
Baton Rouge
Bayou Choctaw
Weeks Island
St. James
New Orleans
Gulf of Mexico
LOOP

Site	In storage (Million bbl)	Drawdown rate (b/d)
Bryan Mound	223.16	1.1 million
Big Hill	2.30	0*
Sulphur Mines	25.00	100,000
West Hackberry	212.09	1.25 million
Weeks Island	73.17	590,000
Bayou Choctaw	53.89	480,000
Total	589.61	3.52 million

*Under construction.

Schematic of dual drainhole well

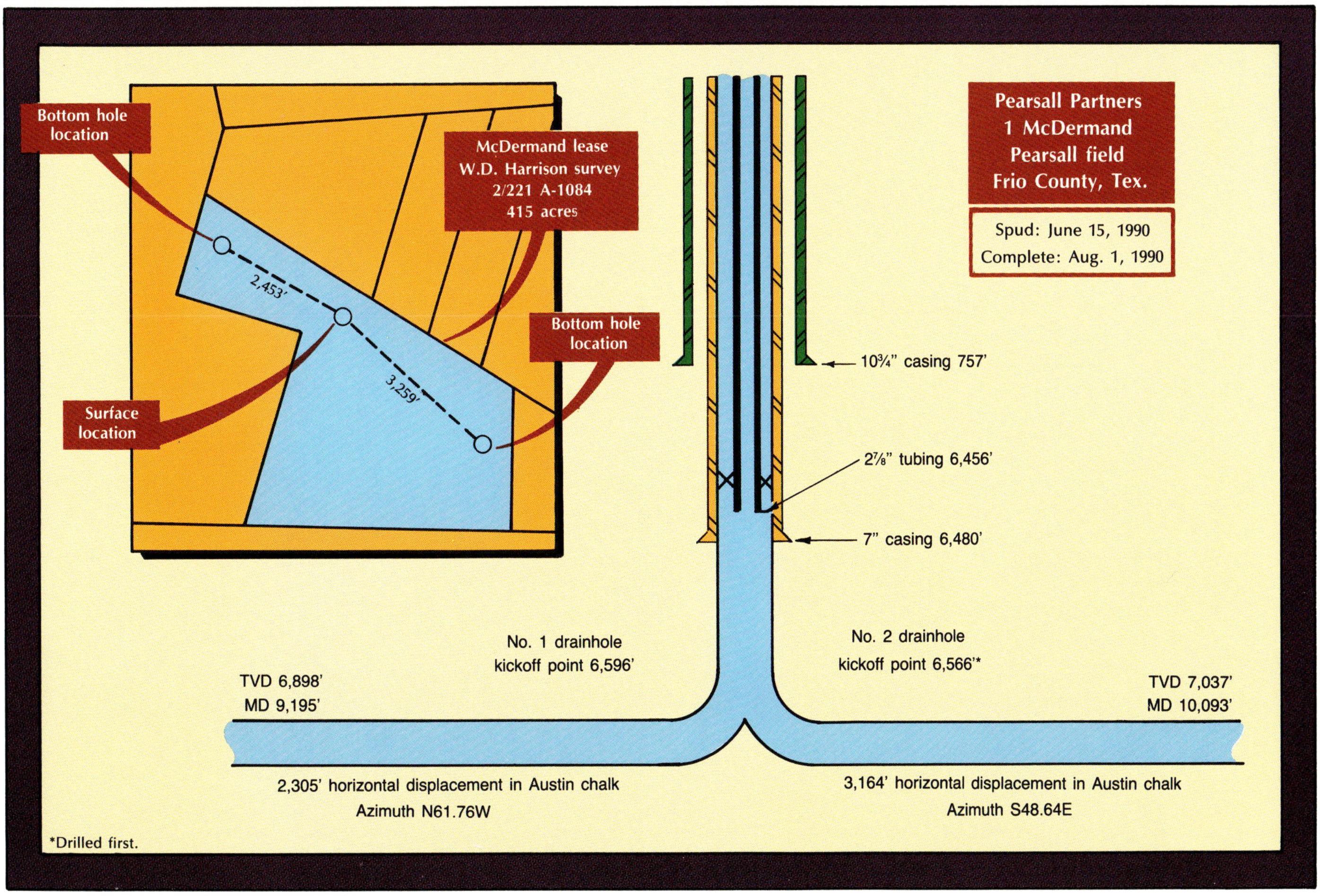

Bottom hole location
Surface location
McDermand lease
W.D. Harrison survey
2/221 A-1084
415 acres
Bottom hole location
2,453'
3,259'
Pearsall Partners
1 McDermand
Pearsall field
Frio County, Tex.
Spud: June 15, 1990
Complete: Aug. 1, 1990
10¾" casing 757'
2⅞" tubing 6,456'
7" casing 6,480'
No. 1 drainhole
kickoff point 6,596'
No. 2 drainhole
kickoff point 6,566'*
TVD 6,898'
MD 9,195'
TVD 7,037'
MD 10,093'
2,305' horizontal displacement in Austin chalk
Azimuth N61.76W
3,164' horizontal displacement in Austin chalk
Azimuth S48.64E

*Drilled first.

1990, up a bare 0.3% from 1989. It projected a climb to 17.5 tcf in 1991.

The production increase in 1990 reflected storage buildup after the sharp stock drawdown during the extremely cold weather in December 1989.

Imports of Canadian gas and Algerian liquefied natural gas will continue to grow at a relatively rapid pace, IPAA predicted. Total imports were up 7.6% in 1990 at 1.486 tcf. In 1991, imports will increase another 5.3% to 1.565 tcf.

The highly seasonal nature of U.S. natural gas demand plays a major role in supply requirements. In winter, demand peaks at more than 2 tcf/month. In spring and summer, usage bottoms out at about 1.2 tcf/month.

Storage withdrawals supply more than one third of consumption during winter months. This is in contrast to summer months, when storage additions account for as much as one fourth of natural gas production.

The economy, total energy

Underlying IPAA's outlook was the supply/demand committee's projection of weak economic growth and sluggish demand for energy in 1991.

The committee found that events in the Middle East placed further pressure on a weakened U.S. economy. Real economic growth slowed to 1.1% in 1990 and will slip to only 0.5% in 1991. The latter will repre-

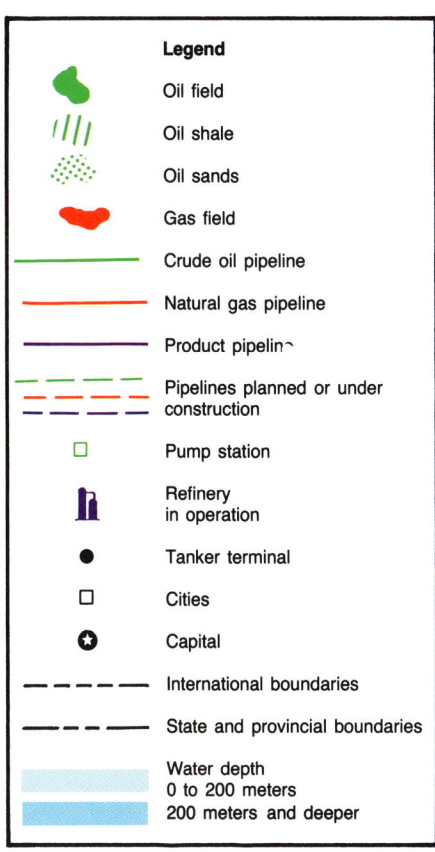

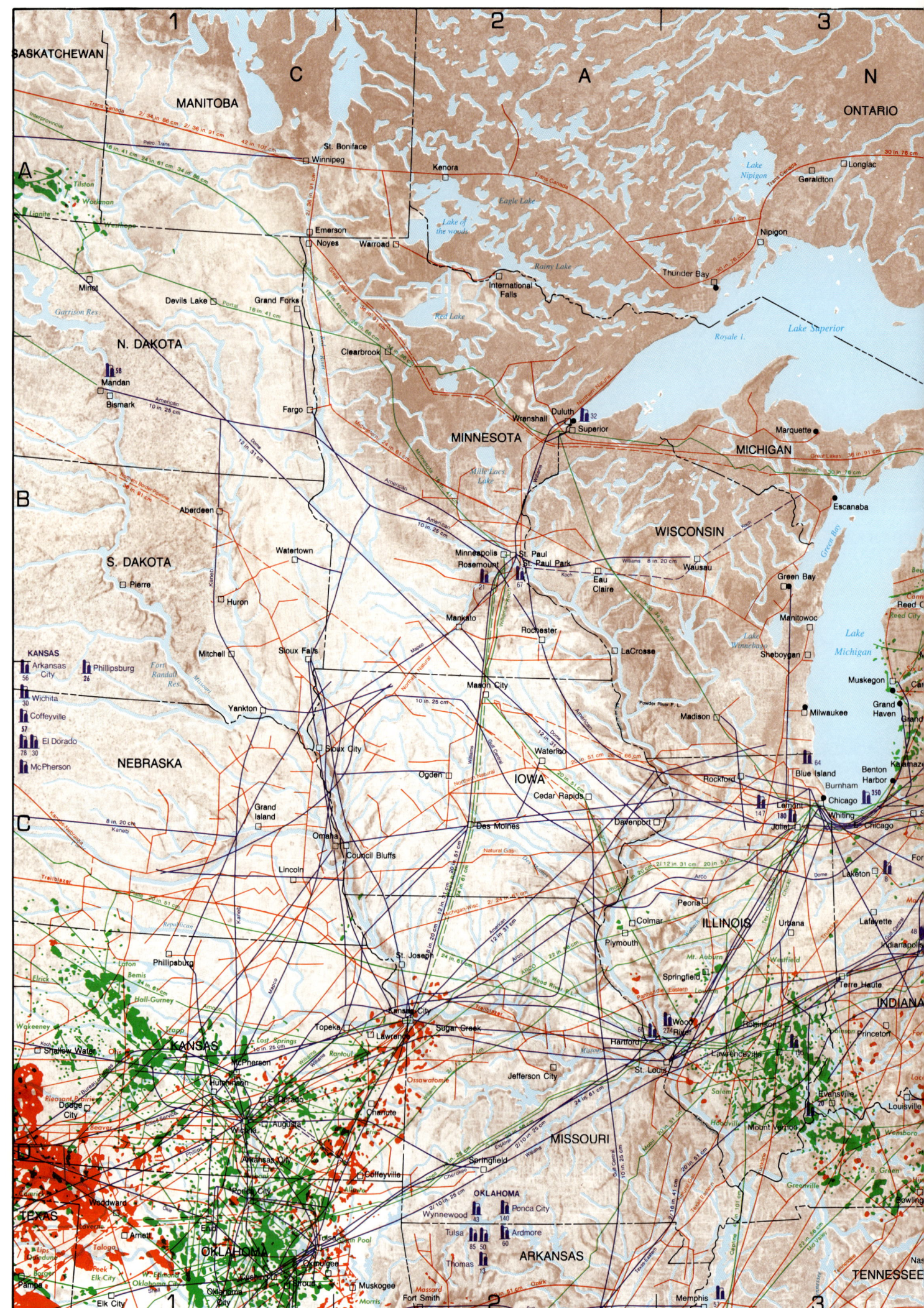

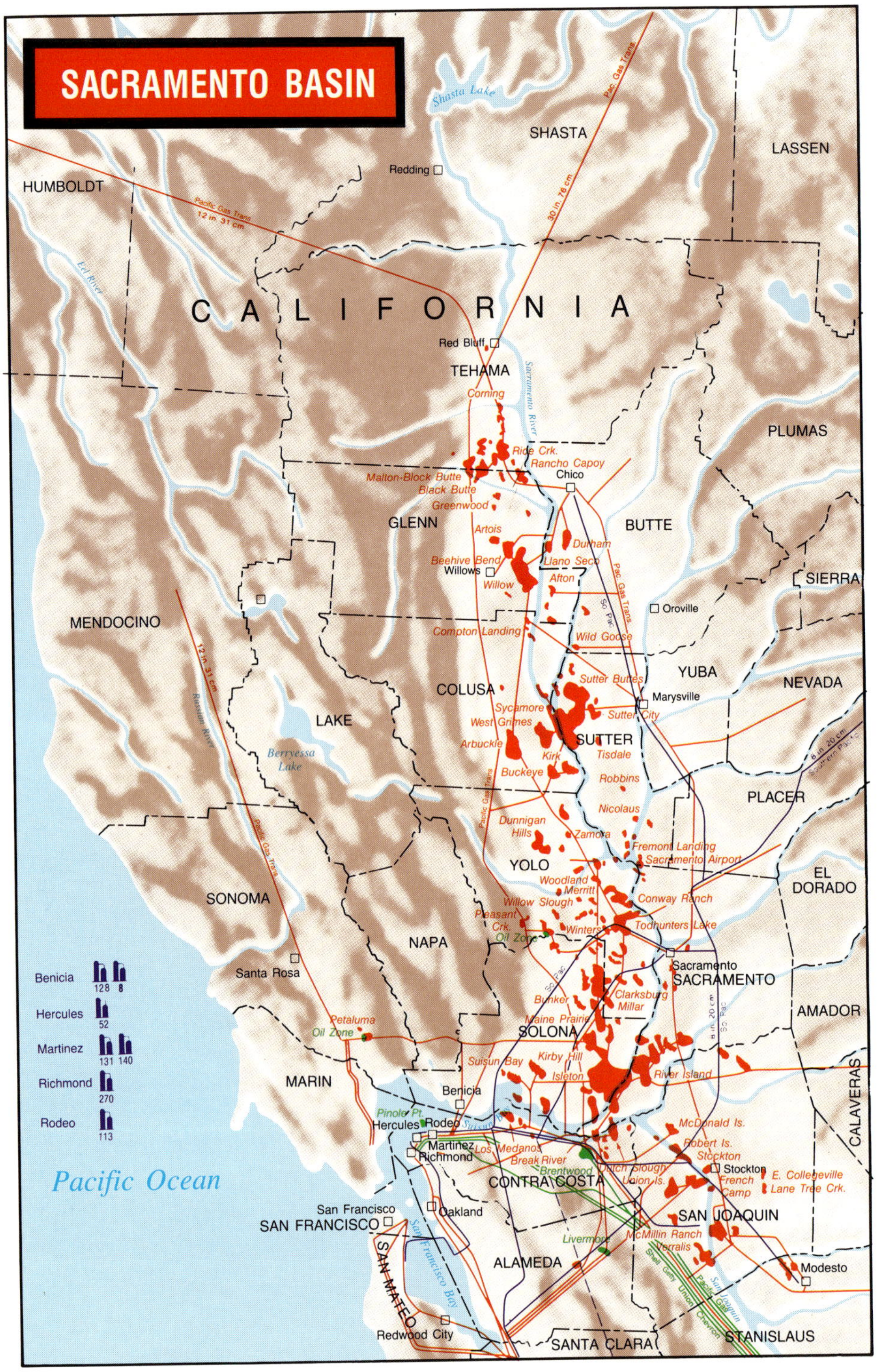

SACRAMENTO BASIN

sent the lowest economic growth since 1982.

As a result, total U.S. energy consumption was off 0.4% in 1990 at 80.91 quadrillion BTUs. It will move up only 0.5% in 1991 to 81.28 quads (Table 2).

The level and pattern of U.S. energy consumption during 1991 will depend on the duration and intensity of the Persian Gulf crisis. Demand for energy in general and oil in particular was reduced in 1990 by the combined effects of higher prices and slower economic growth. Energy consumption in 1990 also was reduced by exceptionally warm weather in the first quarter.

Energy consumption from petroleum is projected to fall to 33.46 quads in 1991. Natural gas energy consumption will increase 2.5% in 1991 to 19.26 quads.

Energy consumpion from coal will increase slowly, advancing 0.3% in 1991.

IPAA's supply/demand committee said the pattern of increasing natural gas use at the expense of oil and coal could intensify during the next several years as environmental concerns and legislative mandates begin to swing the fuel mix toward natural gas.

Use of nonfossil energy sources will grow as hydroelectric power gen-

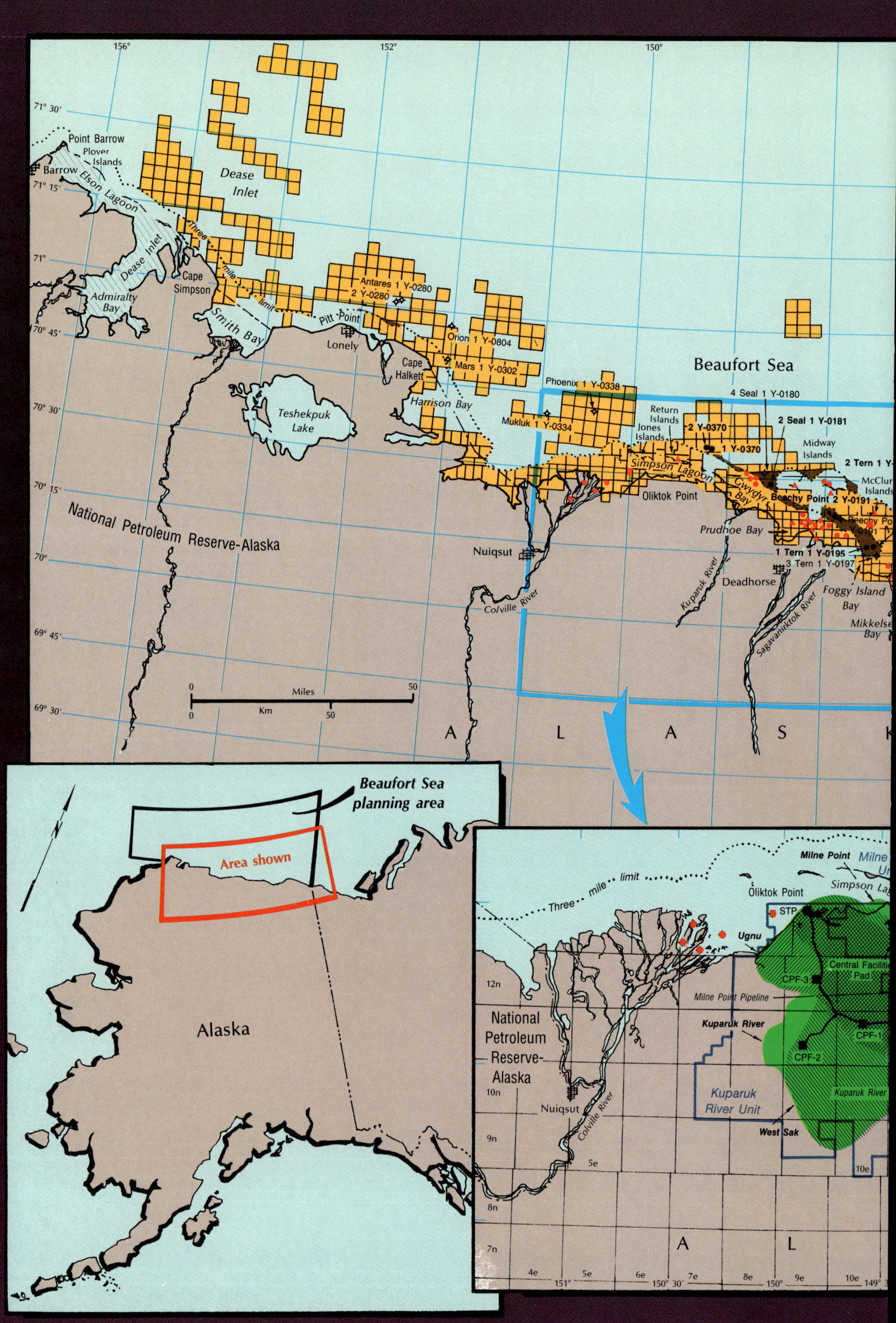

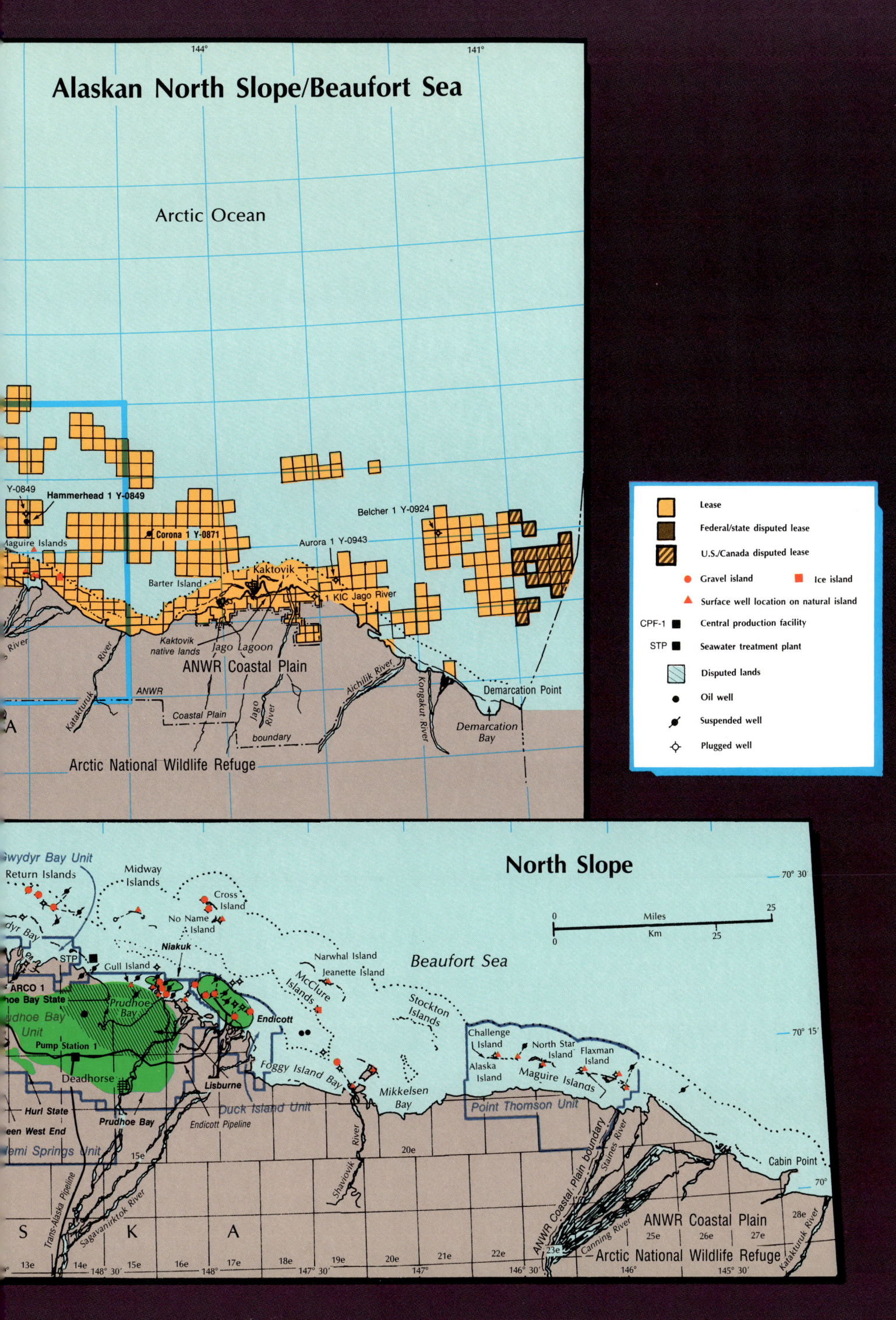

NORTH AMERICA

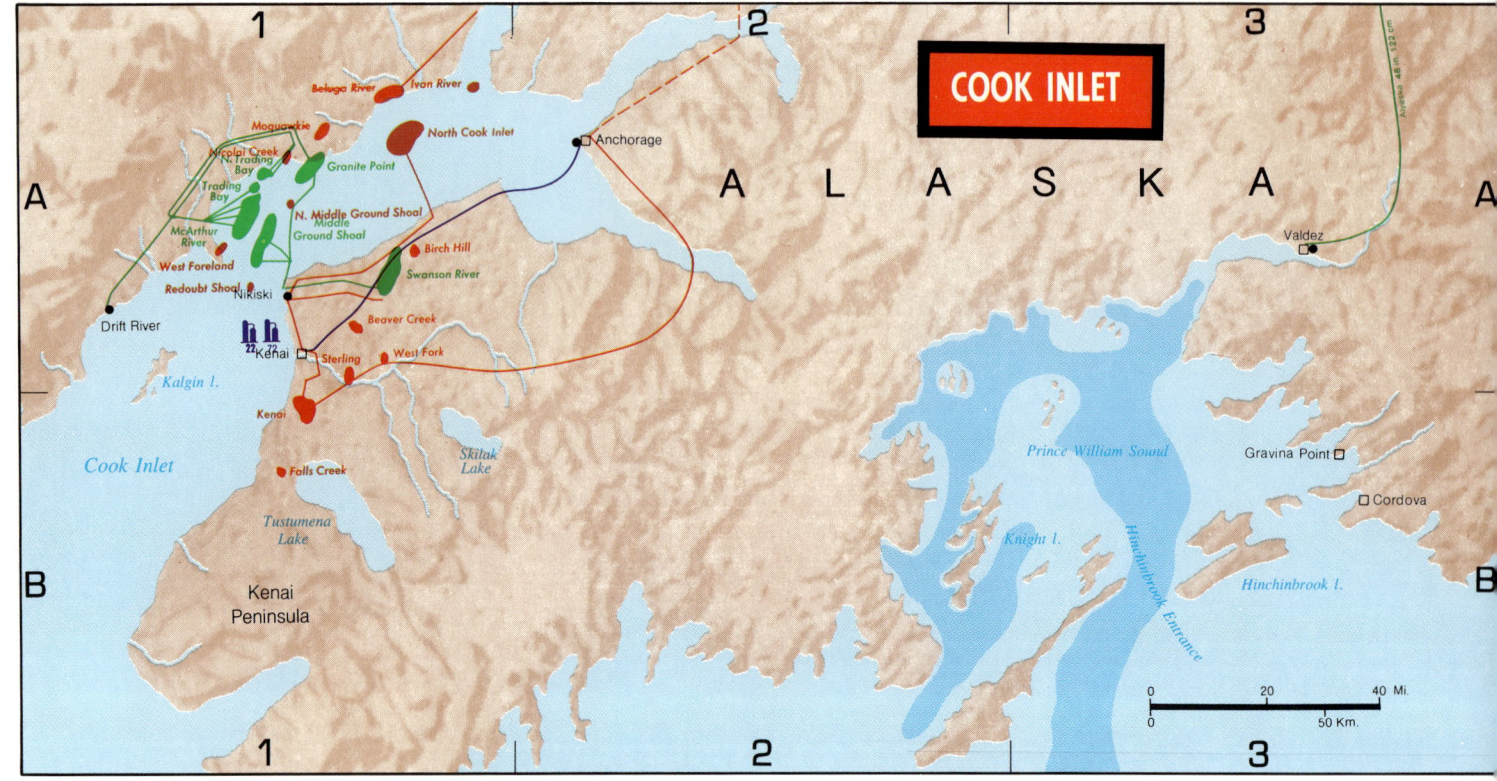

eration recovers from drouths of the late 1980s. Energy from hydro and geothermal power will move up 2.9% in 1991.

The growth in nuclear power will slow as the last few planned nuclear plants go on line. It was up 6.4% in 1990 after little growth in 1989 but will increase only 1.9% in 1991.

The committee projected a continued decline in the ratio of energy use per unit of real gross national product. This will fall to 19,400 BTU/GNP dollar in 1990 from 19,700 BTU in 1989 as energy efficiency continues to improve. The ratio is expected to remain at about 19,400 BTU/GNP dollar in 1991 if there is a return to normal weather.

Outlook for drilling activity

Baker Hughes Inc. told IPAA's cost study committee the U.S. tally of active rotary rigs will show a gain to 1,160 in 1991 from 1,011 in 1990.

The company predicted 35,800 well completions in 1991, an increase of 17.5% from an estimated 30,470 wells in 1990. Footage drilled in 1991 will be up 12% at 165.9 million ft from 148.125 million ft in 1990.

Average well depth will fall to 4,635 ft in 1991 from 4,862 ft in 1990.

Higher prices for oil will trigger a shift toward oil well drilling. However, for the longer term, gas drilling will dominate, Baker Hughes said.

IPAA's cost study committee said increased drilling activity resulted in rate increases by drilling contractors. Contractors' operating costs rose, but contractors were experiencing positive cash flows and minor net profits.

A lack of trained personnel continued to be a problem. Wages were being increased as a means to attract quality drilling crews. However, workmen's compensation costs were a major problem for contractors in most states.

The committee reported across the board price increases by virtually all suppliers of goods and services. Most of the price rise was due to increased fuel costs.

There was a dramatic turnaround in the oil country tubular goods market.

In first half 1990 U.S. mill shipments of tubulars increased by 49%, compared with first half 1989 (Table 3).

In 1989 the reverse was true. There was a decline of 49% in the first half.

Exports of U.S. tubular goods for first half 1990 were down 57% from the same period of 1989. Imports of tubular goods had not begun to increase with the pickup in drilling activity. They were down 1.1% for first half 1990.

Total supply of tubular goods in the U.S. for first half 1990 was up 74% from first half 1989.

Paul Leibman, a principal in Petrie Parkman & Co., Denver, cited a Baker Hughes forecast that horizontal drilling will more than double from the current level to 1,321 wells during 1995. What's more, such drilling will gain even more strength during the ensuing 5 years, accounting for 4,065 wells in 2000.

The surge flows from technological advances in things such as steerable drilling systems, reservoir modeling, 3D seismic and borehole geophysics, coalbed methane development, and advanced computerized fracturing.

Leibman said if horizontal drilling yields good results in a given reservoir, the economics are attractive at oil prices of $15-20/bbl.

Bush orders SPR drawdown

Bush's order for a small drawdown of the SPR was designed largely to calm crude oil futures markets, where prices were spiking in response to Middle East war jitters.

The law permits the president to conduct a 5 million bbl drawdown test without declaring a supply disruption is near, the normal requirement for a drawdown.

Bush ordered the test sale after crude futures prices approached $40/bbl and a group of congressmen urged him to sell oil to calm the market.

Under the Sept. 26, 1990, order DOE was authorized to sell 5 million bbl, with deliveries beginning in 15 days and lasting 30-60 days. At the time of the order the SPR held 590 million bbl of oil at six sites—caverns leached in salt domes—on the Texas-Louisiana Gulf Coast (Fig. 1).

Bush said he took the action to thwart speculators who might try to drive up the price of oil.

NORTH AMERICA

Table 1
1991 U.S. oil and gas supply/demand

	(1,000 b/d)	% change 1990-91		(bcf)	% change 1990-91
PETROLEUM			**NATURAL GAS**		
DEMAND			**DISPOSITION**		
Motor gasoline	7,150	−1.5	Consumption		
Aviation fuels	1,543	1.8	Residential	4,625	4.2
Distillate fuel	3,105	0.2	Commercial	2,675	3.8
Residual fuel oil	1,251	−3.2	Industrial	6,910	1.9
All other	3,919	−0.2	Electric utilities	2,730	0.6
Domestic demand	16,968	−0.7	Total to consumers	16,940	2.6
Exports	782	−0.4			
Total demand	17,750	−0.7	Lease and plant fuel	1,190	0.2
			Pipeline fuel	590	4.6
SUPPLY			Total consumption	18,720	2.5
Production					
Crude oil	7,043	−2.5	Exports	120	13.2
Natural gas liquids	1,526	1.3	Additions to storage	2,375	1.5
Total production	8,569	−1.9	Unaccounted for	330	40.4
			Total disposition	21,545	2.9
IMPORTS					
Motor gasoline	413	−5.7	**SUPPLY**		
Aviation fuels	99	−6.6	Production (dry)	17,500	1.1
Distillate fuel	301	−0.7	Imports	1,565	5.3
Residual fuel oil	531	−3.3			
All other	793	−0.8	Withdrawals from storage	2,325	16.2
Total products	2,137	−2.6	Supplemental gaseous fuels	155	0.6
Crude oil imports*	6,337	2.1	Total supply	21,545	2.8
Total imports	8,474	0.9			
Other supply†	702	−17.6	**INVENTORY**		
Total supply	17,745	−1.3	Base natural gas▪	3,850	0.0
			Working natural gas	2,900	1.8
Change in stocks	−5	—	Total inventory	6,750	0.7
Refinery crude runs§	13,728	−1.1			
ENDING STOCKS		(Million bbl)			
Motor gasoline	222.5	3.1			
Aviation fuels	44.1	−0.9			
Distillate fuel	118.6	−4.0			
Residual fuel oil	46.2	0.9			
All other	268.8	1.4			
Total products	700.2	0.8			
Crude oil	339.9	−2.0			
Total stocks	1,040.1	−0.2			
Imports as % of domestic demand	49.9	—			

*Excludes Strategic Petroleum Reserve. †Refinery processing gain, unaccounted crude oil, and other hydrocarbons. §Crude oil, natural gas liquids, unfinished oils, and other hydrocarbons. ▪The volume required for permanent inventory to maintain underground reservoir pressure and deliverability. √The volume above design level of base natural gas.

Source: Independent Petroleum Association of America supply/demand committee

---

"While the oil market is very tight with little spare capacity," he said, "there is sufficient oil to meet current needs. The oil market has simply not taken into account the additional production coming on stream from a variety of sources or available commercial stocks. There is no justification for the intensive, unwarranted speculation in oil futures."

DOE sent sale notices Sept. 28 to more than 300 U.S. refiners, traders, cooperatives, and state energy offices.

Although 5 million bbl was authorized for sale, 11 companies were high bidders for only 3.925 million bbl.

Average high bid prices ranged from $34.14/bbl for West Hackberry sour to $38.68 for Bayou Choctaw sweet. Including rejected offers, bids ranged from a low of $27.85 to a little more than $39.06.

Amoco Oil Co. was high bidder on the most oil, 1.12 million bbl.

The sale attracted 33 bidders with 40 offers totaling 10.4 million bbl.

DOE said no bids for one of the six types of crude offered resulted in a little less than 4 million bbl being sold.

It said the lack of bids for sour crude from the SPR's Weeks Island site southwest of New Orleans was most likely the result of an increase in the flow of Alaskan North Slope crude oil to the Gulf Coast expected in November. DOE had offered 800,000 bbl of sour crude from Weeks Island.

The last 250,000 bbl of the test sale was delivered Dec. 2 to Phibro Energy Inc., Greenwich, Conn.

Energy Sec. James Watkins said, "The test sale should remove any doubts the SPR bidding and distribution system can function effectively and expeditiously. The system and the people who ran it performed virtually flawlessly and on schedule throughout every step of the sales process.

"Equally important, the private sector now has valuable experience in working with the SPR should we ever be forced to use it during an energy emergency."

During the test, 26 shipments of crude were made from three of the reserve's six storage sites. The first shipment took place Oct. 19, 2 days after the first contracts were awarded, to meet an expedited request from one of the purchasers. Most deliveries were in November under normal scheduling practices.

About 3.7 million bbl of the 3.9 million bbl were shipped by pipeline and the remaining 255,000 bbl by barge.

Although all payments had not been computed by mid-December, DOE said the purchase price will average about

PREVENTABLE!

World opinion is now forcing a stop to inefficient processing and industrial pollution.

Western Research's team of experienced analysts is ready to respond quickly to help prevent sulphur recovery problems anywhere in Europe.

Using our unique approach to problem solving, through process stream analysis and our world-recognized process simulator, **Sulsim**®, we perform a comprehensive engineering evaluation. This evaluation will result in improved sulphur recovery through the implementation of correct operating procedures, accurate process control, and state-of-the-art emission monitoring.

Prevent this threat to the environment. Call on Western Research's 25 years of world-wide problem-solving experience in sulphur recovery.

Western Research
A member of the BOVAR group

Head Office
Western Research
1313 - 44 Ave. N.E.
Calgary, Alberta
Canada T2E 6L5
Tel: 1 (403) 291-1313
FAX: 1 (403) 250-2610

European Office
Western Research GmbH
Voltastrasse 7
Postfach 1440
D-6234 Hattersheim 1
Germany
Tel: 49-6190-8591
Fax: 49-6190-73560

U.S. Office
Western Research & Development
#122 - 1300 S. Potomac
Aurora, CO.
USA 80012
Tel: 1 (303) 751-8990
FAX: 1 (303) 751-8994

25 Years and Beyond...

NORTH AMERICA

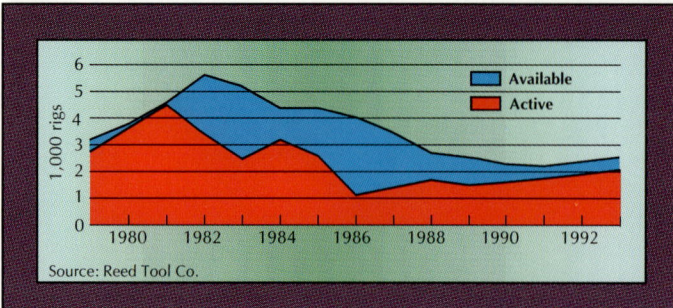

Fig. 3
The rise and fall of U.S. drilling activity

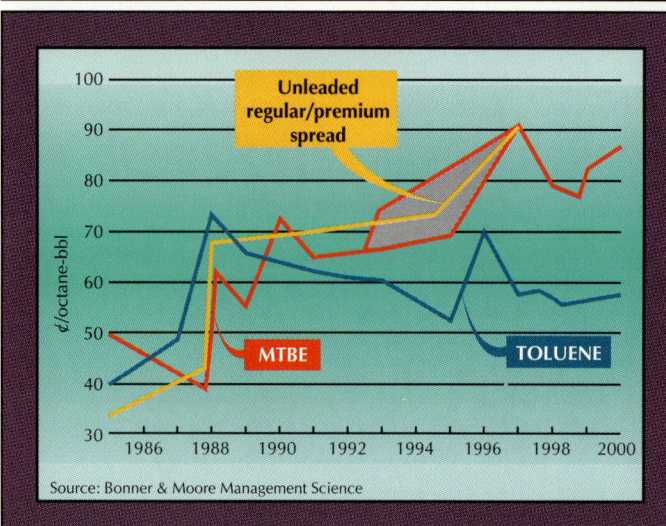

Fig. 4
U.S. Gulf Coast octane costs, values

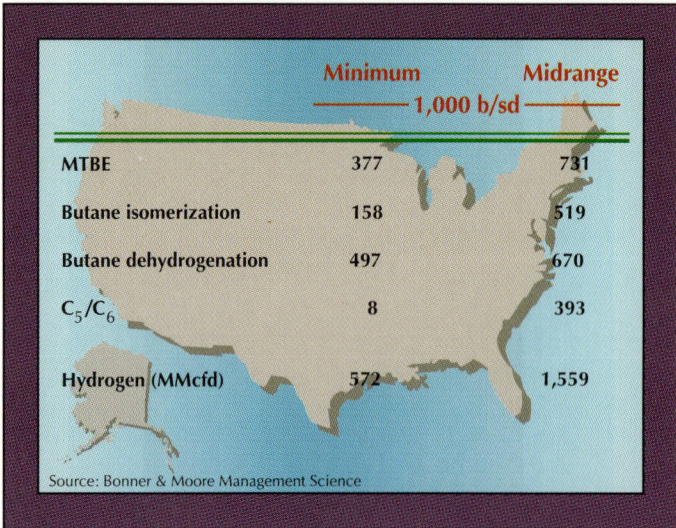

Fig. 5
Required U. S. capacity additions

SANTA MARIA BASIN

$30.10/bbl for sour crude to $33.30 for sweet. Those prices were about 13% below average winning bid prices because of adjustments under a price index system DOE used to track changes in the market price of various crudes.

The drawdown was the second for the SPR. In 1985 Congress ordered a 1 million bbl test sale. It went smoothly.

DOE has invested $3 billion in facilities and $16 billion in crude, paying an average $27.26/bbl.

SPR sites are served by barges, tankers, and pipelines. For example, the Bryan Mound site, 3 miles west of Freeport, Tex., is linked to Houston and Texas City refineries through a 42 in. DOE pipeline that connects to the area pipeline grid. It is in DOE's distribution system designated "Seaway" because it formerly was served by Seaway Pipeline, now converted to gas from crude oil service.

DOE's "Texoma" system once was served by a pipeline of that name, also now in gas service. Pipeline interconnections enable SPR crude from the Texoma system sites to move to Upper Midwest refineries.

The third cluster of SPR sites is served by Capline Pipeline.

U.S. not likely to hike oil flow

The SPR is a valuable asset in U.S. energy supply because only Alaska and California can make a sizable near term contribution to boosting U.S. oil production, Oil & Gas Journal reported in October 1990.

In terms of current reserves, all other U.S. oil producing states combined could not ramp up production enough in 1 year to make much of a dent in the natural decline in U.S. oil fields, OGJ said.

Even with a strong effort from Alaska and California

GAS CORROSION?
SERVO HAS THE INHIBITORS AND THE EXPERIENCE

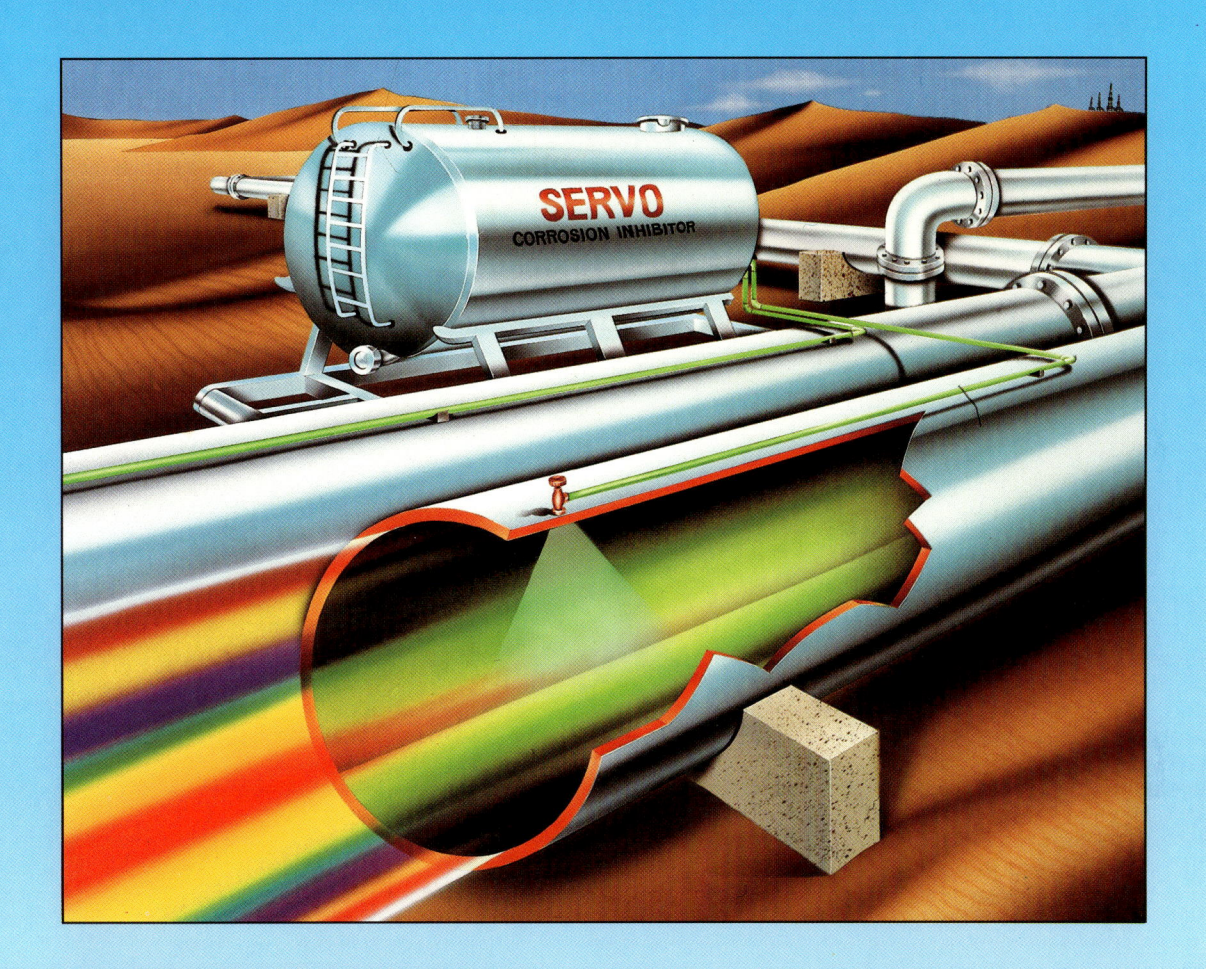

We have been involved for many years in a world-wide campaign against corrosion. Our experience is at your disposal. Please contact:

OIL
GAS
WATER
servo division

servo delden bv

Head office:
P.O.Box 1,
7490 AA Delden,
The Netherlands

Phone: 05407-63535*
Telex: 44347
Telefax: 05407-64125

NORTH AMERICA

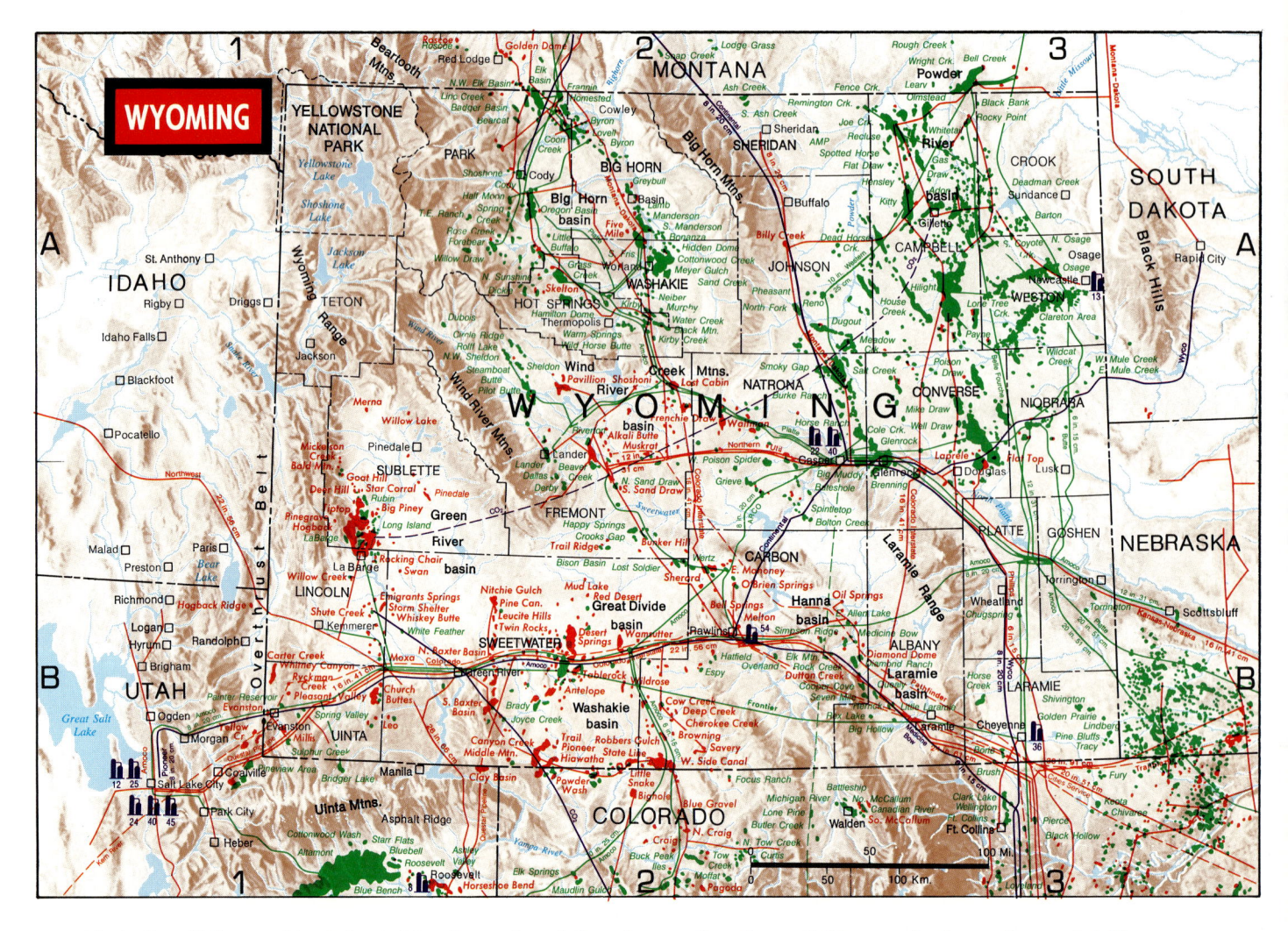

added, the U.S. would at best trim the rate of its oil production decline in 1 year.

With sustained higher oil prices and fast track permitting, however, the two states could make a significant difference in U.S. production in a 5-10 year span, based on undeveloped oil reserves.

Judging from responses to a survey of producing states by the Interstate Oil Compact Commission (IOCC), even a sustained oil price increase through 1995 would have little effect on U.S. ability to increase oil production from the 1990 level.

Survey responses suggested the U.S. could hike oil flow by about 360,000 b/d in 1 year. However, respondents did not mention taking into consideration field decline rates, which probably would peg the net gain in 1 year at less than 20,000 b/d, OGJ said.

Further, some assumed sustained higher oil prices and restoration of all shutin production. The exceptions for the 5 year outlook, again, were Alaska and California, where more than several billion barrels of reserves remained blocked from development by environmental or economic reasons or a combination of both.

That did not include the postulated huge potential of the Arctic National Wildlife Refuge Coastal Plain on Alaska's North Slope or the region off northern California, both off limits to drilling. Even if drilling were allowed in those areas, production probably could not go on stream before the end of the 1990s.

The rest of the U.S. Outer Continental Shelf might also contribute to U.S. oil production some day, but leasing bans and technological constraints will push that contribution into the next century. An exception to that rule was the Gulf of Mexico deepwater play, where the potential for finding giant oil and gas fields was fueling a flurry of drilling.

However, it was California's offshore and Alaska's North Slope and nearshore Beaufort Sea that offered the biggest identified reserves capable of being placed on production in the 1990s.

The IOCC survey and industry responses to the question of where the U.S. could ramp up oil flow quickly in a crisis conflicted with estimates by DOE that if U.S. producers were given certain fiscal incentives, U.S. oil production could jump by almost 200,000 b/d by 1995.

Assuming suspension of production allowables, a sustained higher oil price of about $25/bbl in 1991, and restart of half or more of shut in stripper production, the U.S. could boost oil flow by 424,280 b/d within 1 year, IOCC survey responses showed. However, only Alabama and Nevada specified their estimates as net gains from current production.

Assuming a natural field decline of 5% from OGJ's projected U.S. production of 7.2 million b/d in 1990 otherwise yields a drop of 360,000 b/d. Using the IOCC survey reponses, that leaves the net gain in U.S. oil production in 1 year at about 60,000 b/d.

OGJ estimated the gain from a best efforts response would more likely be only about one fifth that level.

But a best efforts response was not likely, given the decimation of the U.S. oil service and supply sector in the wake of the 1986 oil price plunge, uncertainty over oil prices, and limited capital. So OGJ instead projected a fall in U.S. oil production in 1991 of about 145,000 b/d from projected 1990 levels. Using a hypothetical projected decline rate of 5%/year, the U.S. could otherwise be expected to lose about 1.6 million b/d of production during 1990-95. That compares with a loss of almost 1.9 million b/d of production in the

NORTH AMERICA

LOS ANGELES BASIN

OVERTHRUST BELT

Lower 48 in 1973-85, a span that covers two periods of big oil price jumps.

U.S. gains in oil production during the 1970s and 1980s came essentially from Alaska. In 1973-85, even with expectations of indefinitely higher oil prices and a record drilling surge, Lower 48 oil production could muster gains of only 0.3% in 1983 and 2.6% in 1984 before slipping into decline again. Further, Alaskan North Slope oil flow, which had softened the Lower 48 slide, is in decline.

So for a net gain of 200,000 b/d by 1995 from 1990 levels, the U.S.—without a 2 million b/d surge in Alaskan flow—would somehow have to hike gross production by as much as the Lower 48 lost in 1973-85. And that was when the oil industry infrastructure had built to unprecedented levels, not lying in tatters after 4 years of industry depression.

Production potential in Alaskan fields

Alaska, with three of the six biggest U.S. fields in terms of production, can account for more than half of the 1991 U.S. production increase covered by IOCC survey responses, OGJ reported.

By early 1991, the North Slope could hike production about a net 125,000-170,000 b/d—after accounting for Prudhoe Bay's natural decline—from levels in early 1990, North Slope operators said. The increases would come mainly in Prudhoe Bay field, where gross production would jump about 225,000 b/d from a 1990 low of about 1.1 million b/d in early September.

In addition to Prudhoe Bay's natural decline, production in the giant field fell about another 100,000 b/d in summer 1990 because of operating inefficiencies in warmer weather and to accommodate tie-in of facilities for Prudhoe Bay's gas handling expansion (GHX) project.

Partly because of the downtime, affecting five of six flow stations and temporary shutdowns of the central gas facility, Prudhoe Bay averaged only about 1.3 million b/d in the first 7-8 months of 1990 vs. 1.45 million b/d in late 1989.

BP Exploration (Alaska) Inc. said North Slope production could jump by about 70,000 b/d of oil before yearend 1990.

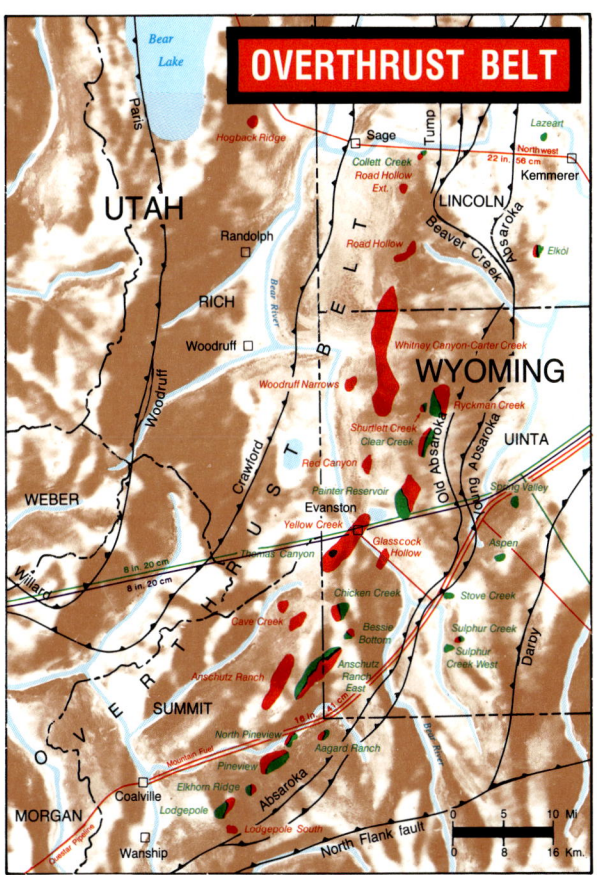

Of that total, more than 50,000 b/d would come from an accelerated hydraulic fracturing program in the east and west sectors of Prudhoe Bay field. North Slope operators earlier had planned 80-100 fracture treatments in Prudhoe Bay for 1990. After DOE officials asked them to push up

WESTERN U.S.

SAN JOAQUIN VALLEY

Legend

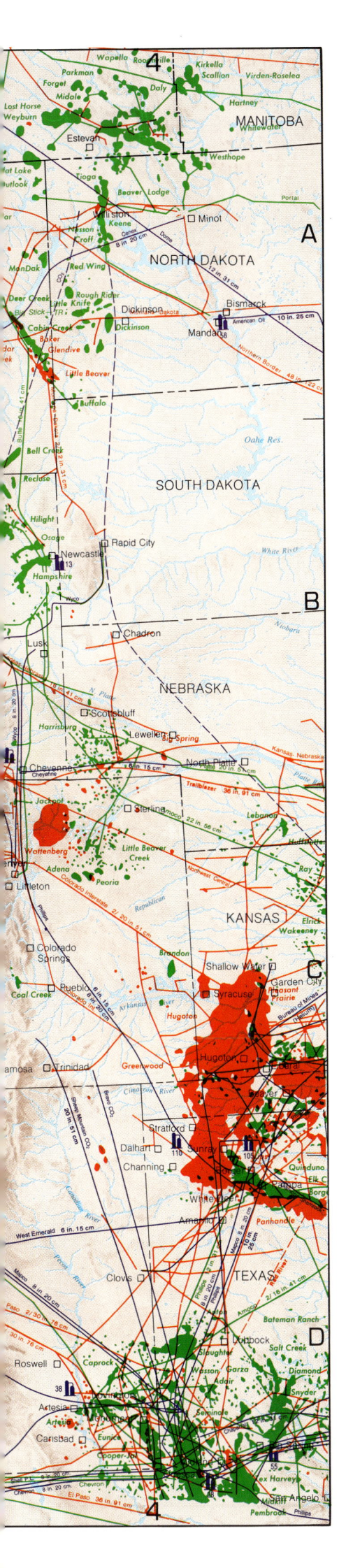

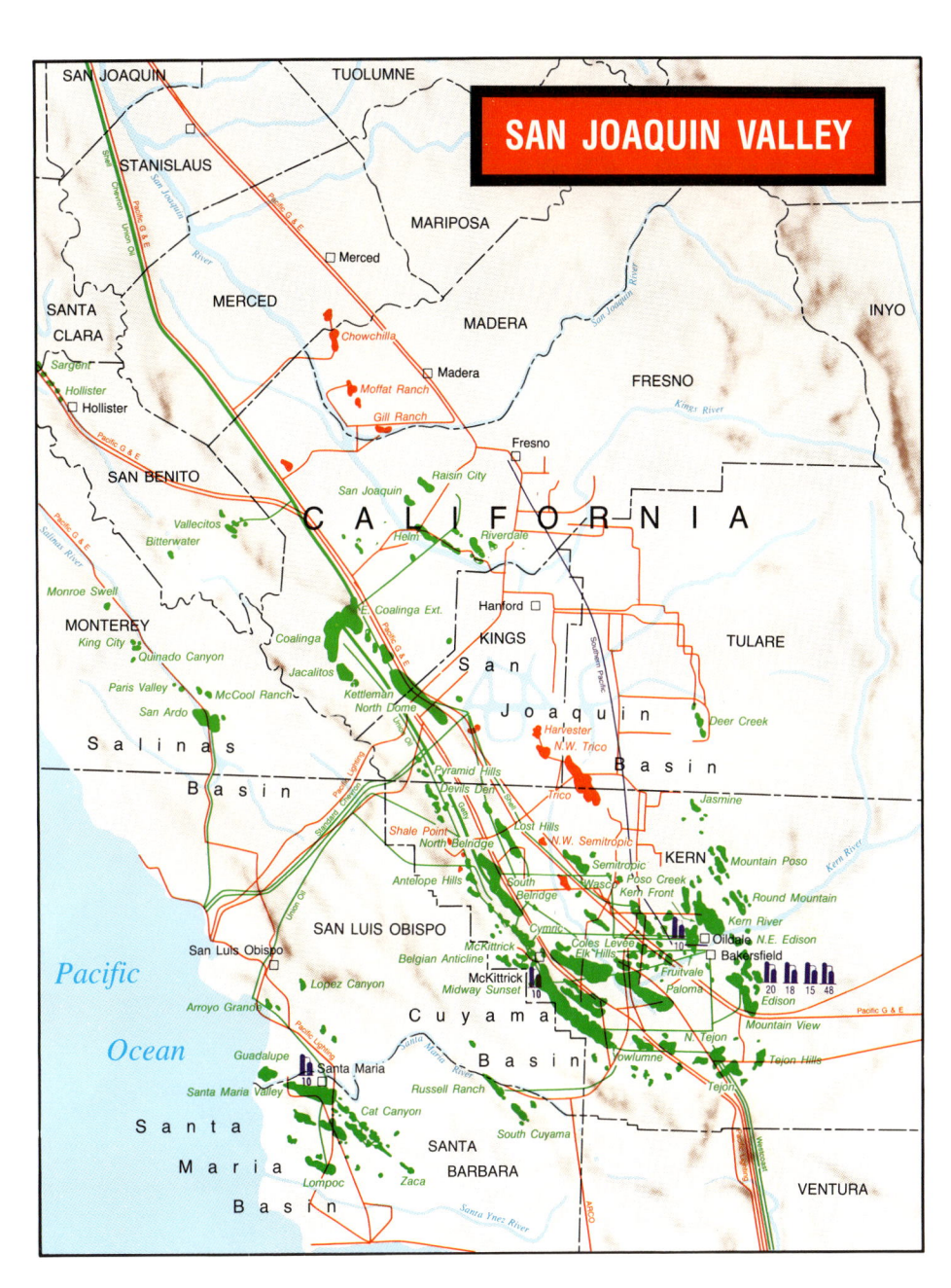

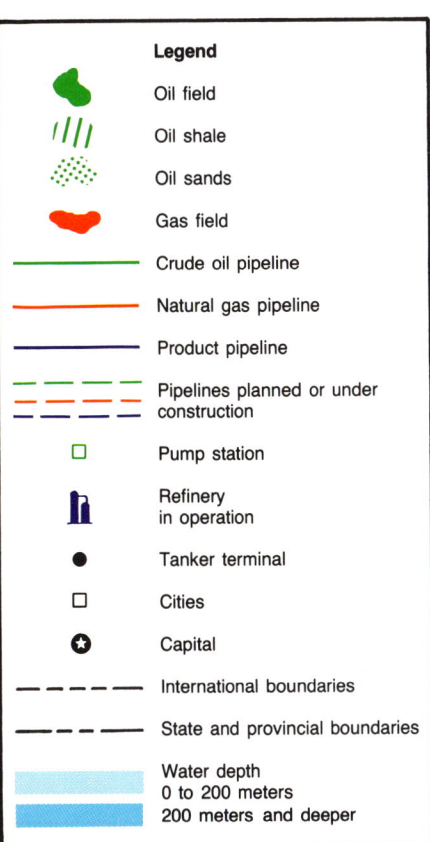

- Oil field
- Oil shale
- Oil sands
- Gas field
- Crude oil pipeline
- Natural gas pipeline
- Product pipeline
- Pipelines planned or under construction
- Pump station
- Refinery in operation
- Tanker terminal
- Cities
- Capital
- International boundaries
- State and provincial boundaries
- Water depth 0 to 200 meters
- 200 meters and deeper

production, operators jumped that program to 130-140 frac jobs for 1990.

BP said fracking a Prudhoe Bay well can quickly boost its flow by 100-500%.

Of the remaining expected increase, Endicott field would provide about 10,000-12,000 b/d in addition to current flow of about 100,000 b/d. Sag Delta, a satellite to Endicott, was expected to increase to 10,000 b/d from 4,000 b/d, BP said.

Increasing gas handling capacity with GHX-1 would add a further 90,000-100,000 b/d to North Slope production. BP said the project could be on stream by late this year or early 1991.

ARCO Alaska Inc. also estimated accelerated fracking would boost Prudhoe flow by 50,000 b/d and expected GHX-1 to add 75,000 b/d of production by early 1991.

"At Kuparuk River and Lisburne fields, we are essentially at capacity," an ARCO official said. "We were doing everything possible to lift as

Localized service through an international network of convenient locations

You're on the go throughout the world. And we're staying right up with you. That's why wherever you operate you'll find trained Halliburton men nearby with service, tools and materials selected and proven for each locale.

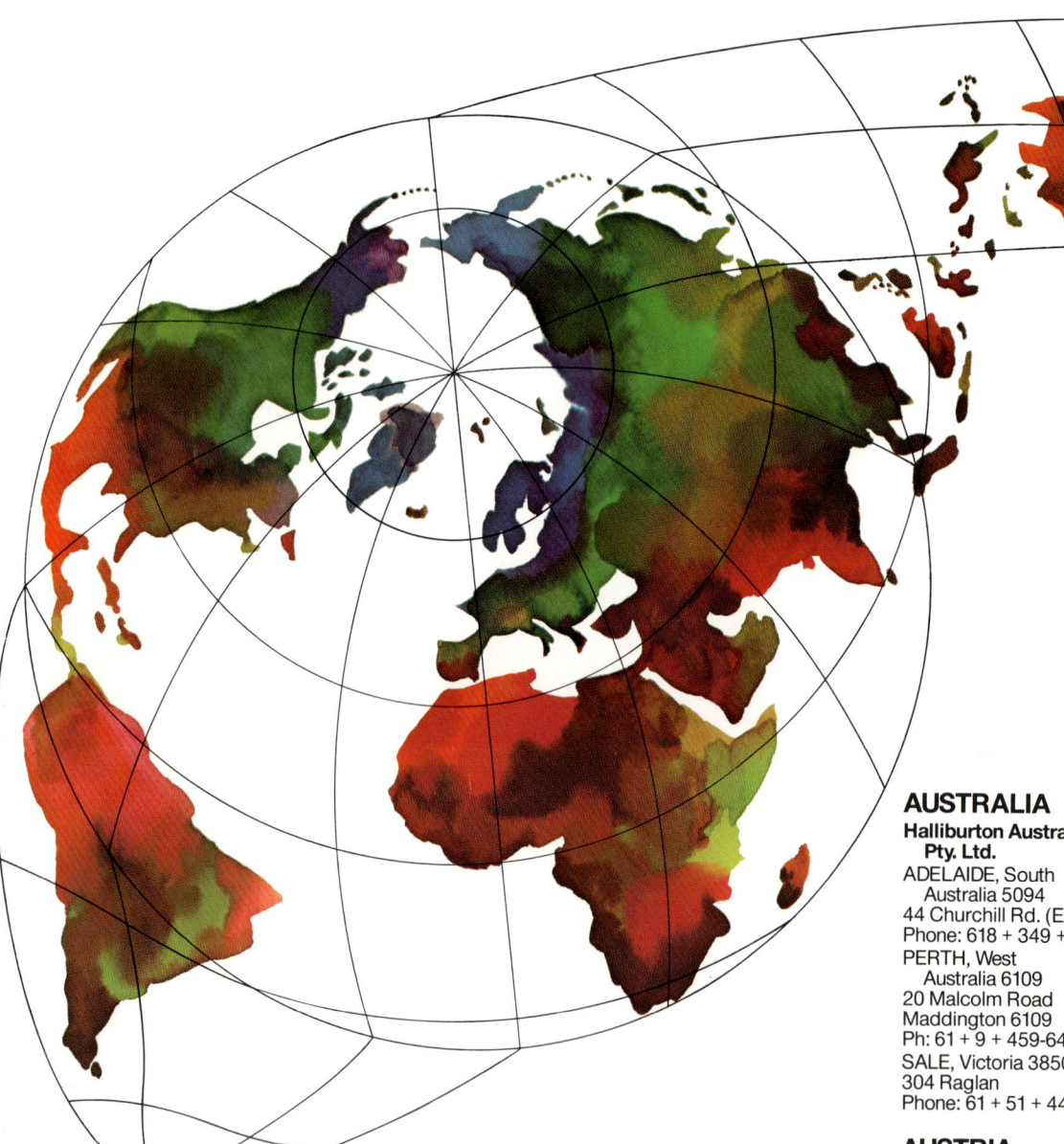

ALGERIA
Halliburton Limited
ALGIERS
33 Blvd. Mohammoud
2nd Floor
Ph: 213 2-648428

ANGOLA
Halliburton Overseas Ltd.
LUANDA
Av. 4 de Fevereiro No. 42
Edificio Secil, Andar 11
Ph: 2442-373758/373459

ARGENTINA
Halliburton Argentina S.A.
(Division Office)
1340 BUENOS AIRES
Maipu 942, Piso 14
Phone: 541 + 312-8413
8300 NEUQUEN
Casilla De Correo 166
Phone: 54 + 943-23470
9000 COMODORO
   RIVADAVIA, Chubut
Casilla 177,
Barrio Industrial
Ph: 54 + 967-24919
5507 LUJAN de CUYO, Prov.
   de Mendoza
Casilla de Correo 32
Ph: 54 + 61 + 981869

AUSTRALIA
Halliburton Australia
   Pty. Ltd.
ADELAIDE, South
   Australia 5094
44 Churchill Rd. (Extension)
Phone: 618 + 349 + 4588
PERTH, West
   Australia 6109
20 Malcolm Road
Maddington 6109
Ph: 61 + 9 + 459-6444
SALE, Victoria 3850
304 Raglan
Phone: 61 + 51 + 443484

AUSTRIA
Halliburton Company
   Austria GmbH
A-1213 VIENNA
Postfach 107
Ph: 43 + 2246-4333

BAHRAIN
Halliburton Limited
MANAMA (Division Office)
Box 515
Phone: 973 + 258866

BOLIVIA
Halliburton Company
SANTA CRUZ
Casilla 482
Phone: 591 + 33 + 47117

BRAZIL
Halliburton — do
   Brasil-Servicos
   Comercio e Industria,
   Ltda.
20.000 RIO DE JANEIRO, RJ
   (Division Office)
Caixa Postal 2038
Ph: 55 + 21 + 220-6125
40000 SALVADOR BAHIA
Caixa Postal 939
Ph: 55 + 71 + 246-2433

BRUNEI
Halliburton Limited 6003
KUALA BELAIT
P.O. Box 393
Phone: 6733 + 22156

CANADA
Halliburton Services
   Limited
CALGARY, Alberta T2P OS2
   (Division Office)
Ste. 1100
333-5th Ave. S.W.
Phone: (403) 231-9300
Technical Centre
CALGARY, Alberta T2V 6V1
5140 Skyline Way N.E.
Phone: (403) 231-9300

CHILE
Halliburton-Servicios
   Petroleros
   (Chile) Ltda.
PUNTA ARENAS,
   Magallanes
Calle Roca Nbr 817
Phone: 5661 + 213598

CHINA, PEOPLE'S REPUBLIC OF
Halliburton Overseas
   Limited
SHEKOU, Shenzhen
Room 316
China Merchants Bldg.
Shekou Industrial Zone
Shenzhen, P.R.C.
Ph: 86 + 755-92470
BEIJING, P.R.C.
No. 508, 5th Floor
Block A-2
Lido Commercial Bldg.
Lido Center
Jinchang Road
Ph: 861-500-6662

COLOMBIA
Halliburton Company
BOGOTA
Apartado Aereo 4549
Phone: 571 + 292-0211

DENMARK
Halliburton Company Germany GmbH
DK-6705 ESBJERG
P.O. Box 2066
Phone: 45 + 7514-5444

ECUADOR
Halliburton Company
QUITO
83 A Casilla
Phone: 593 + 2 + 520079

EGYPT, A.R.E.
Halliburton Limited
CAIRO
P.O. Box 1227
Ph: 20 + 2 + 352-4776

FRANCE
Halliburton Company
ST. QUENTIN EN YVELINES
  (Division Office)
Le Florestan
2 Blvd. Vauban
78180 Montigny-Le-Bretonneux
Ph: 331-3043-5000
64142 BILLERE CEDEX
B.P. 209, Lons
Phone: 33 + 59 + 321446
91801 BRUNOY
B.P. 18, Rue de La Foret
Zone Industrielle
Epinay Sous Senart
Epinay, France
Phone: 331 + 60465550

GABON
Halliburton — IMCO Gabon S.A.R.L.
PORT GENTIL
B.P. 507
Ph: 241 + 752196

WEST GERMANY
Halliburton Company Germany GmbH
D-3100 CELLE
Bruchkampweg 7
Postfach 380
Phone: 49 + 5141 + 88560

GREECE
Halliburton Ltd.
65110 KAVALA
P.O. Box 1525
Phone: 30 + 51 + 831842

GUATEMALA
Halliburton Company
GUATEMALA CITY
2DA. Calle 15-96 Zona 13
Phone: 502 + 2 + 318185

HOLLAND
Halliburton Services BV
Leiderdorp
Weversban 1-3
2352 BZ
Phone: 31-71-899-271
1976 BE IJMUIDEN
Amperestraat 1/G
Ph: 31 + 2550 + 34924

INDIA
Halliburton Offshore Services, Inc.
BOMBAY 400 093
Sanghi Oxygen Compound
Mahakali Caves Road
Andheri (East)
Phone: 91 + 22-6351921

INDONESIA
P.T. Halliburton Indonesia
JAKARTA (Division Office)
Cilandak Commercial Estate
  Facility #110 NGE
J.L. Raya Cilandak KKO
Phone: 62-21-780-2218
JAKARTA (District Office)
Wisma Pe De, 4th Floor
Jln. M.T. Haryono Kav. 17
Phone: 6221 + 829-7108
BALIKPAPAN
Kampung Damai
RT. XIII-A/58
Sepinggan Bypass
Phone: 62 + 542 + 23278

ITALY
Halliburton Italiana S.P.A.
MILANO
Via L. Tolstoi 86
20098 Zivido
(San Giuliano Milanese)
Phone: 392 + 98491451

JAPAN
Halliburton Overseas Ltd.
TOKYO 106
5th Floor, Maruyama Bldg.
3-8, 2-Chome Azabudai
Minato-Ku
Phone: 81 + 3 + 586-9271

KENYA
Halliburton Ltd.
NAIROBI
P.O. Box 62216
Phone: 254 + 2 + 762-191

KUWAIT
Halliburton Limited
AHMADI
P.O. Box 9022
Phone: 965 + 3984801

EAST MALAYSIA
Far East Oilwell Services Sdn. Bhd.
MIRI, SARAWAK 98007
P.O. Box 383
Phone: 60 + 85 + 651043

WEST MALAYSIA
Far East Oilwell Services Sdn. Bhd.
KUALA LUMPUR 50250
Suite 708, 7th Floor
Pernas Intl.
Letter Box #35
Jalan Sultan Ismail
Phone: 60 + 3 + 2619244

MEXICO
Halliburton de Mexico S.A. de C.V.
11300 MEXICO D.F.
  (Division Office)
Bahia de San Hipolito
51, I-ER, PISO
Colonia Veronica Anzure
Ph: 525 - 531 - 2867
CUIDAD, del Carmen
Calle 31 #254
Avenida Aviacion
Phone: 52938 + 213-71
REFORMA, Chis.
Km. 10.4 Carretera a
  Reforma
Phone: 52932 + 80111
TAMPICO, Tamps
Km. 146.5 Carretera
Tampico-Mante
Phone: 52 + 121 + 381-20
VILLAHERMOSA, Tabasco
Cesar Sandina, No. 603
Esq. A. Quevedo
Phone: 52 + 931 + 30347
REYNOSA, Tamps
Km. 103.5 Carretera
Monterrey Reynosa
Phone: 52 + 892 + 300-86

NEW ZEALAND
Halliburton Overseas Ltd.
NEW PLYMOUTH
P.O. Box 7160
Phone: 64 + 67 + 72405

NIGERIA W.A.
Halliburton Nigeria Limited
LAGOS
P.O. Box 3694
Phone: 234 + 1 + 615444
WARRI
P.O. Box 359
Phone: 234 + 53 + 231300
PORT HARCOURT
P.O. Box 462
Phone: 234 + 84 + 335619

NORWAY
Halliburton Overseas Ltd.
4056 TANANGER
P.O. Box 67
Phone: 47 + 4 + 696733

SULTANATE OF OMAN
Halliburton Limited
MUSCAT
Box 9081
Phone: 968 + 603246

PAKISTAN
Halliburton Limited
ISLAMABAD
P.O. Box 1136
Phone: 92 + 51 + 410172

PERU
Empresa de Servicios Tecnicos Petroleros S.A. (ESTEPSA)
TALARA
Apartado 7A
Phone: 51 + 74 + 334-668
LIMA
Apartado 4988
Phone: 51 + 14 + 420258
IQUITOS
Apartado 563
Phone: 51 + 94 + 236426

PHILIPPINES
Halliburton Services
MAKATI, METRO MANILA
Ground Floor Makati
  Tuscany
6751 Ayala Ave.
Phone: 63 + 2 + 871 + 892

QATAR
Halliburton Limited
DOHA
P.O. Box 3036
Phone: 974 + 671111

SAUDI ARABIA
Halliburton Company
DHAHRAN
Box 657
Phone: 966 + 3 + 856-1616

SINGAPORE
Halliburton Limited
SINGAPORE 9123
  (Division Office)
P.O. Box 156
Orchard Point P.O.
Phone: 65 + 533-7741

THAILAND
Halliburton Company
BANGKOK, 10900
67144 Soi Suparpong
Lar Prao Soi 35 Road
Phone: 66-2-513-69858

TRINIDAD, W.I.
Halliburton Trinidad Limited
SAN FERNANDO
P.O. Box 57
Phone: 809 + 6579181

TUNISIA
Halliburton Limited
SFAX
Zone Industrielle
Plage Poudriere
Route Sidi Mansour Km 3
Phone: 216 + 4 + 27487
TUNIS
10, Rue 7000 Centre 7000
IER ETAGE Appt 13
1002
Phone: 2161 - 281 - 635

TURKEY
Halliburton Company
ANKARA
Tunus Caddesi 50-6
Kavaklidere
Phone: 90 + 41 + 282921
DIYARBAKIR
P.K. 187
Phone: 90 + 831 + 26020

UNITED ARAB EMIRATES
Halliburton Limited
ABU DHABI
P.O. Box 57
Phone: 971 + 2 + 553000
DUBAI
P.O. Box 3111
Phone: 971 + 4 + 361588

UNITED KINGDOM
Halliburton Manufacturing & Services Ltd.
U.K. LONDON
Halliburton House
3 Putney Bridge Approach
Phone: 44-71-371-5500
GREAT YARMOUTH,
  Norfolk NR3O 1QF
South Denes Road
Ph: 44 + 493 + 330300
ABERDEEN ABI IDL,
  Scotland
(Division Office)
Belmont House
No. 1 Berry St.
Phone: 44 + 224 + 626-902
ARBROATH, DD11,
  2NF, Scotland
P.O. Box 9
Phone: 44 + 241 + 77333

VENEZUELA
Cia. Halliburton de Cementacion y Fomento
CARACAS 1060
  (Region Office)
Apartado 61229 Chacao
Phone: 58 + 2 + 752-6098
EL TIGRE, Edo. Anzoategui
Apartado 221
Phone: 58 + 83 + 351114
LAS MOROCHAS
Apartado 698
Phone: 5865 + 27477

YEMEN
Halliburton Ltd.
P.O. Box 481
Phone: 9672-215 731

HALLIBURTON
A Halliburton Company

NORTH AMERICA

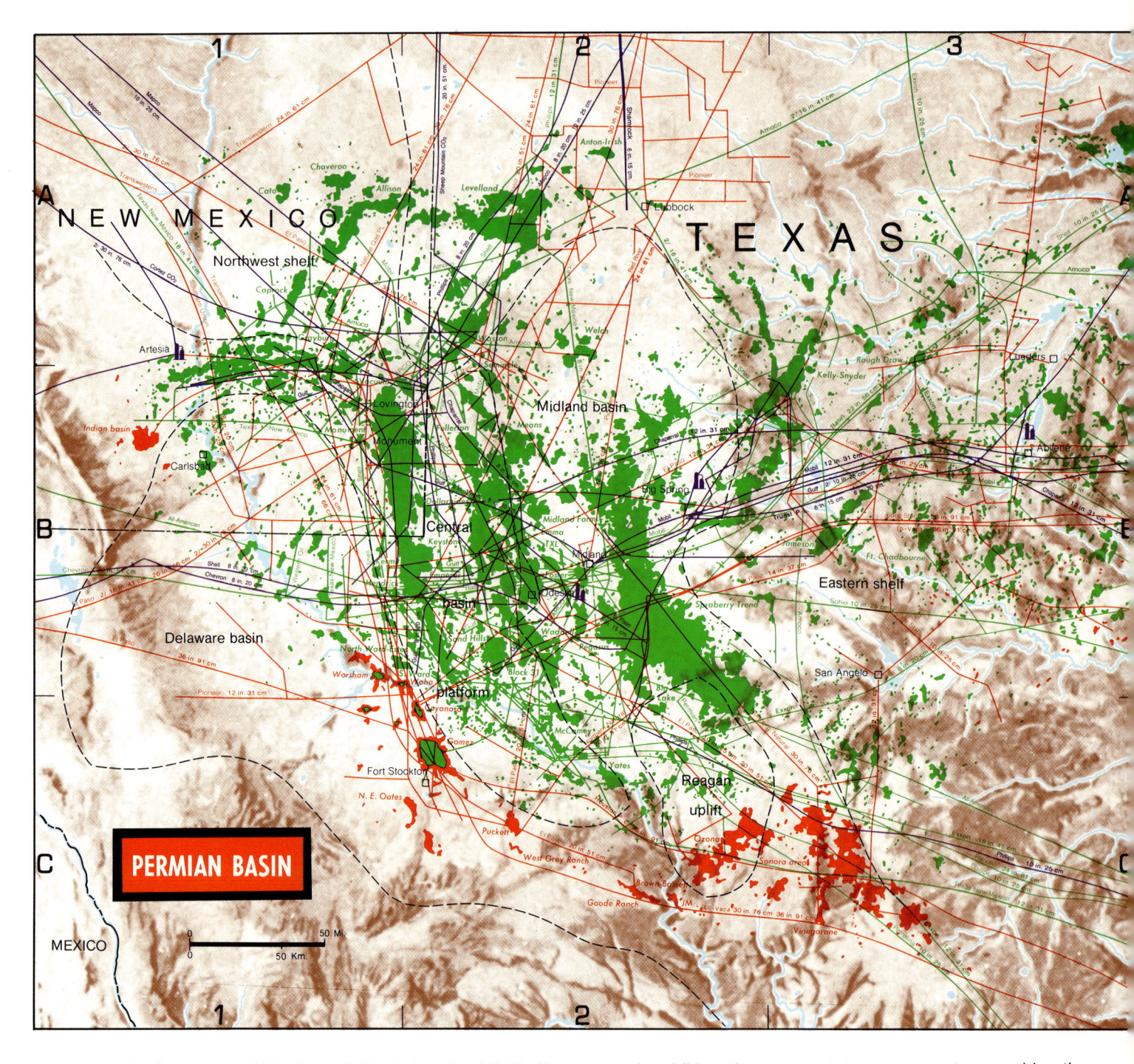

PERMIAN BASIN

much oil as we could in those fields before the Middle East crisis."

Capacity to boost North Slope production further in 1991 or beyond in the near term is limited.

A principal concern is pipeline capacity.

Tests of drag reducing agents showed the Trans-Alaska Pipeline System throughput could be pushed to more than 2 million b/d, but it is not certain how long that could be sustained.

BP rebuffed suggestions that flaring gas might provide a quick jump in North Slope oil production.

The concern expressed was the threat of losing reserves in the long term by not maintaining reservoir pressure with gas injection. In addition, there would be loss of the gas itself, which could become a valuable commodity if a feasible market emerged for North Slope gas.

The best midterm prospects for hiking North Slope oil flow as Prudhoe Bay declines remained hamstrung by economic or environmental concerns.

In addition, there were state severance tax considerations that raised other questions about the economic feasibility of any new development in Alaska. BP earlier shelved an $80 million development project, Hurl State in the west end of Prudhoe Bay field, because of changes in Alaska's severance tax calculations approved after the Exxon Valdez tanker spill of March 1989.

Those fiscal questions and environmental issues don't bode well for developing other North Slope reservoirs, such as West Sak, Seal Island/Northstar, Colville Delta, Gwydyr Bay, and Point Thomson-Flaxman Island, OGJ said.

Niakuk and Point McIntyre, the two "developed" fields DOE pointed to as contributing 100,000 b/d in 12-18 months, in fact remain undeveloped because of economic and environmental concerns.

What's possible in California

IOCC survey responses showed California could hike oil

NORTH AMERICA

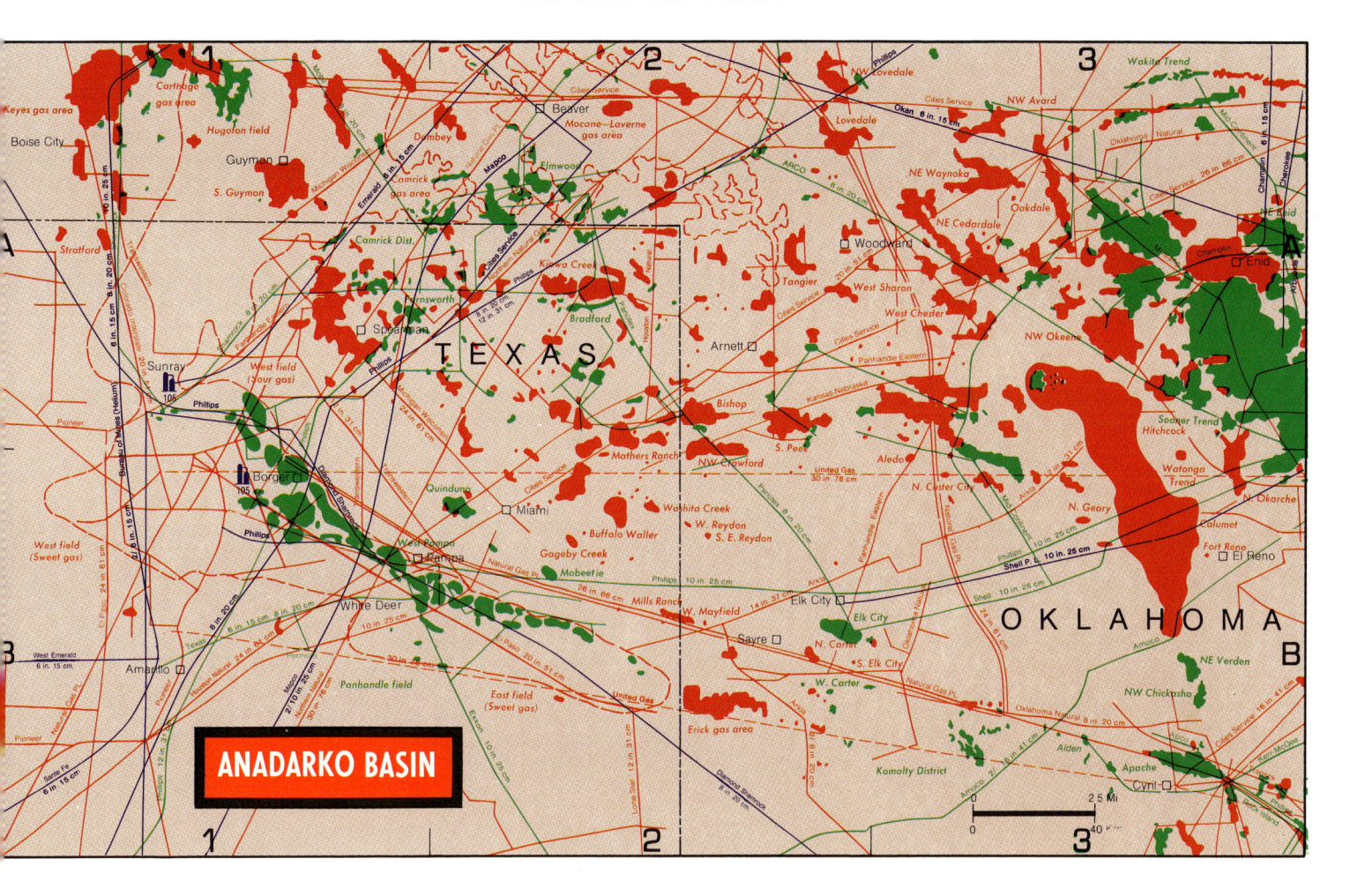

ANADARKO BASIN

production by 225,000 b/d by 1995. That assumes, however, that giant Point Arguello field off Santa Barbara is producing at peak by then.

Chevron Corp. and partners in November 1990 gave a green light to a midyear 1991 start-up of Point Arguello oil field despite their inability to win a tanker permit to move the oil to Los Angeles.

Instead, 20,000 b/d of Point Arguello crude will move via offshore and onshore pipelines to California refineries outside the Los Angeles area.

The project is permitted to produce as much as 100,000 b/d of crude, a volume Chevron pointed out is equal to about half the U.S. imports of Kuwaiti crude before Iraq invaded that country Aug. 2, 1990. A 1981 discovery with a likely minimum reserve of 300 million bbl, it is the biggest oil field found in U.S. federal waters.

Chevron hoped to accelerate its timetable to reactivate the three platforms, onshore and offshore pipelines, onshore oil and gas processing plants, storage tanks, and marine terminal built for the project. Chevron earlier said demothballing could take as much as 6 months, with another 6 months to build production to minimum expected peak of 75,000-80,000 b/d.

The $2.5 billion development project had been mothballed for 3 years because of environmental opposition.

Although California has a substantial amount of shutin productive capacity, the state's decimated service sector prevents it from stepping up production significantly in the near term.

Low oil prices left more than 14,000 wells idle in California for 5 years or more, said Marty Mefferd, the state's oil and gas supervisor.

"A lot of the wells are sitting there with production equipment intact," he said. "A number of those are still hooked to pumping units, and we're trying to get them started. For a lot of them, that just means pulling rods and

U.S. energy consumption by source

Table 2

	1986	1987	1988	1989	1990	1991	% change 1989-90	% change 1990-91
			Quadrillion BTU					
Petroleum*	32.20	32.87	34.23	34.21	33.83	33.46	−1.1	−1.1
Natural gas	16.71	17.75	18.55	19.33	18.80	19.26	−2.8	2.5
Coal	17.26	18.01	18.85	18.92	18.94	19.00	0.1	0.3
Hydro and geothermal	3.60	3.32	2.91	3.11	3.31	3.41	6.4	2.9
Nuclear	4.47	4.91	5.66	5.69	6.04	6.15	6.2	1.9
Total	74.24	76.85	80.20	81.25	80.91	81.28	−0.4	0.5

*Includes natural gas liquids, still gas, and liquefied refinery gas.

Source: Independent Petroleum Association of America supply/demand committee

NORTH AMERICA

tubing. And there are a lot of steam generators shut down. For a lot of the wells and steam generators, the permits are already in hand."

Most of California's crude oil production is low gravity and thus especially sensitive to oil price declines. As a result, with expectations of higher oil prices in the 1990s gaining strength, operators rushed into infill and workover programs in California.

The state underwent a sizzling pace of new wells, workovers, and abandonments. The workovers and abandonments were spurred in part by the state's tightening rules on idle wells. Even before the Middle East crisis took shape in August 1990, many California operators were running economic litmus tests of their shutin wells forced by the tighter idle well rules and finding more of them passing the test than they had expected.

"We were starting to see some stability in prices," Mefferd told OGJ. "Major operators decided they weren't going to fool around with offshore California any more and started to put their money in proven areas. With $20/bbl for Kern River (13°) crude, we're going to see some of that production we've lost since 1985."

Mefferd placed that production decline at about 150,000 b/d. It entailed mostly shutin stripper wells in areas of very low gravity crudes such as the onshore Santa Maria basin and shutdown thermal enhanced oil recovery projects in the southern San Joaquin Valley.

"We're reasonably sure 75% of that can come back on stream in 1-1½ years—maybe 75,000-100,000 b/d within a year," he said.

The state could not speed that process because of heavy losses among drilling and service rig contractors.

Beyond the infrastructure constraints and uneasiness about future oil prices, California operators are hamstrung by environmental concerns as well.

In addition to government bans on further federal leasing off California, operators continued to be plagued by environmental opposition to development of 1980s discoveries—some of them giant fields—in the offshore Santa Maria and Santa Barbara basins.

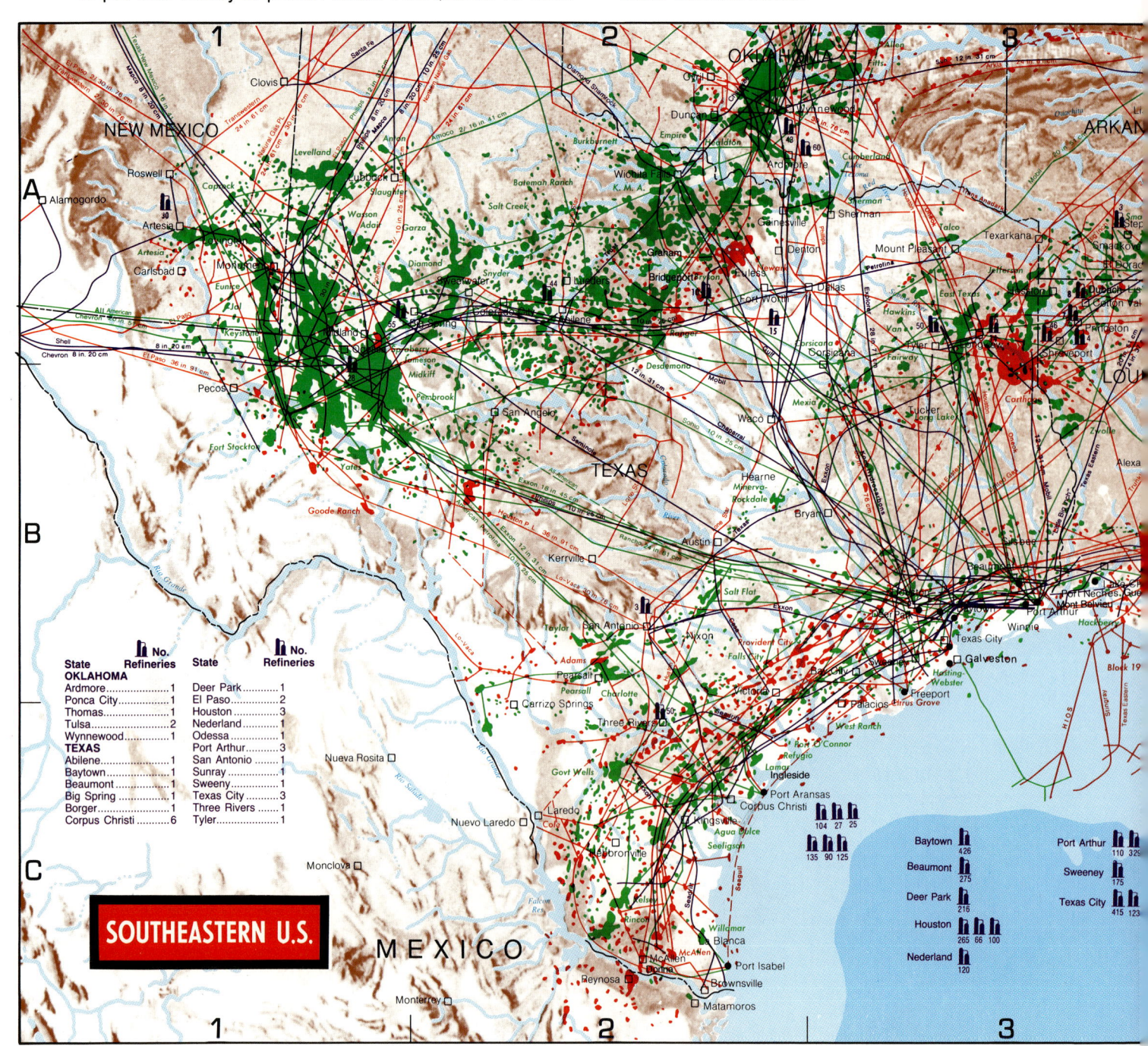

SOUTHEASTERN U.S.

State	No. Refineries	State	No. Refineries
OKLAHOMA		Deer Park	1
Ardmore	1	El Paso	2
Ponca City	1	Houston	3
Thomas	1	Nederland	1
Tulsa	2	Odessa	1
Wynnewood	1	Port Arthur	3
TEXAS		San Antonio	1
Abilene	1	Sunray	1
Baytown	1	Sweeny	1
Beaumont	1	Texas City	3
Big Spring	1	Three Rivers	1
Borger	1	Tyler	1
Corpus Christi	6		

NORTH AMERICA

Of those projects not on stream, only Exxon Corp.'s further development of Santa Ynez Unit was moving ahead.

Earlier estimates of California offshore production in the 1990s climbed to as high as 500,000 b/d, although that assumed the unlikely prospect of concurrent development. More recent estimates placed that level at about 200,000-300,000 b/d. Outside the San Joaquin Valley, there was little prospect for increased production in California.

"The industry has given up on the Los Angeles basin," Mefferd said. "There are so many constraints there it is too much overcome. With the current political climate, nothing is going to change in state waters, either."

On the other hand, said Mefferd, the Middle East crisis sharply focused Californians' attention on energy security issues, a development that might provide some long term support for drilling off California.

Horizontal drilling fanning out

An October 1990 report by OGJ showed that horizontal drilling programs were mushrooming in many U.S. fields and reservoirs to tap multiple vertical fractures.

Most oil and gas producing states had horizontal wells, although the South Texas Cretaceous Austin chalk play was not likely to be matched soon by any other horizontal play in number of wells and production, OGJ said.

Operators reported completion or drilling of horizontal oil wells in Alabama, Louisiana, Mississippi, Michigan, North Dakota, Montana, Colorado, Wyoming, New Mexico, Oklahoms, and several regions of Texas.

Figures showed 847 horizontal drilling permits were issued in the U.S. during January-September 1990, up from 257 in all of 1989.

Horizontal well permits numbered 97 in September, down from 119 in August and 123, the highest monthly total, in May 1990.

The Texas Railroad Commission (TRC) had granted 929 horizontal well drilling permits through September 1990 since it issued the first to Exxon Co. U.S.A. for a horizontal well in Giddings field in Burleson County Aug. 1, 1984.

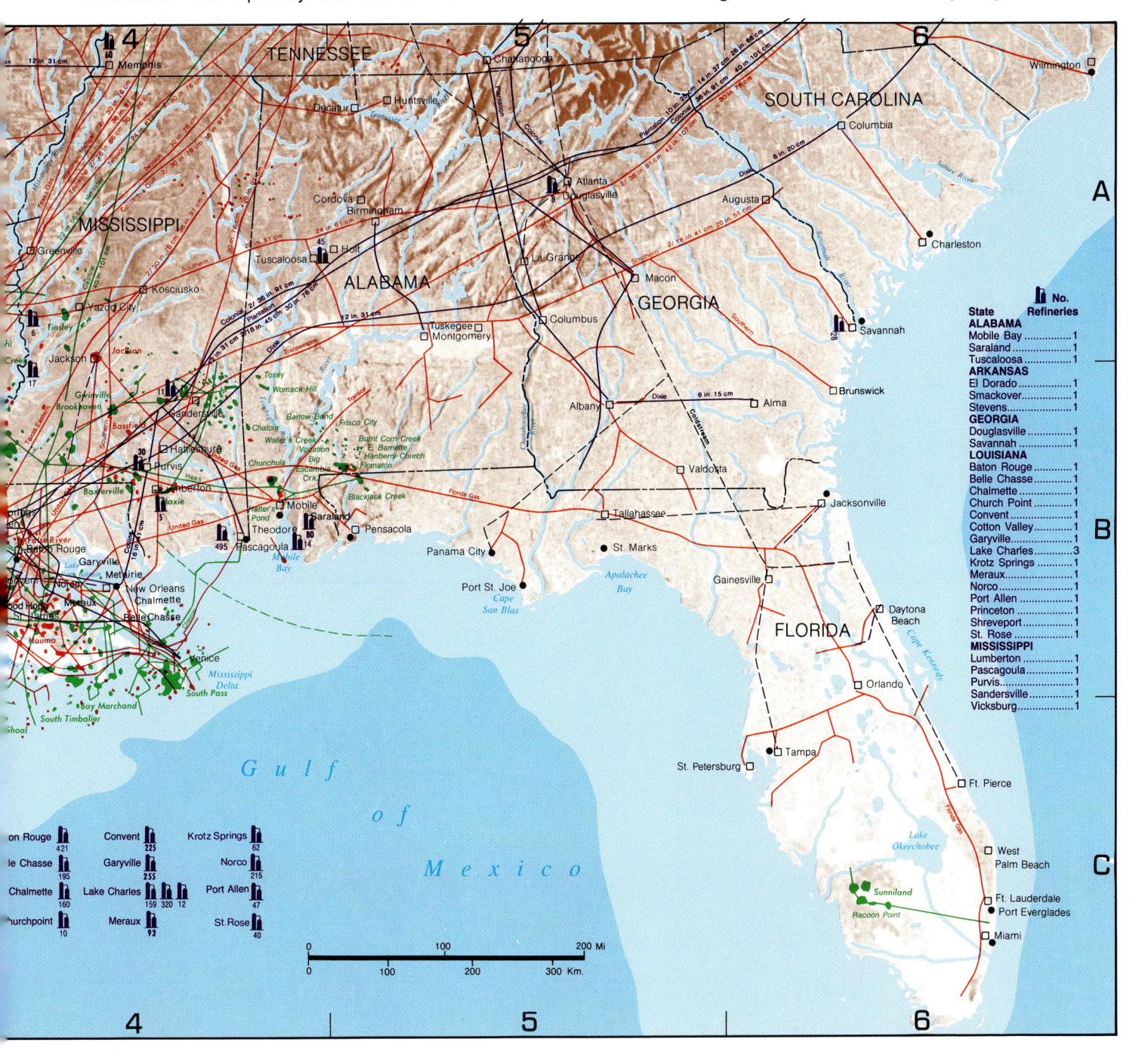

NORTH AMERICA

Northwest Arkansas

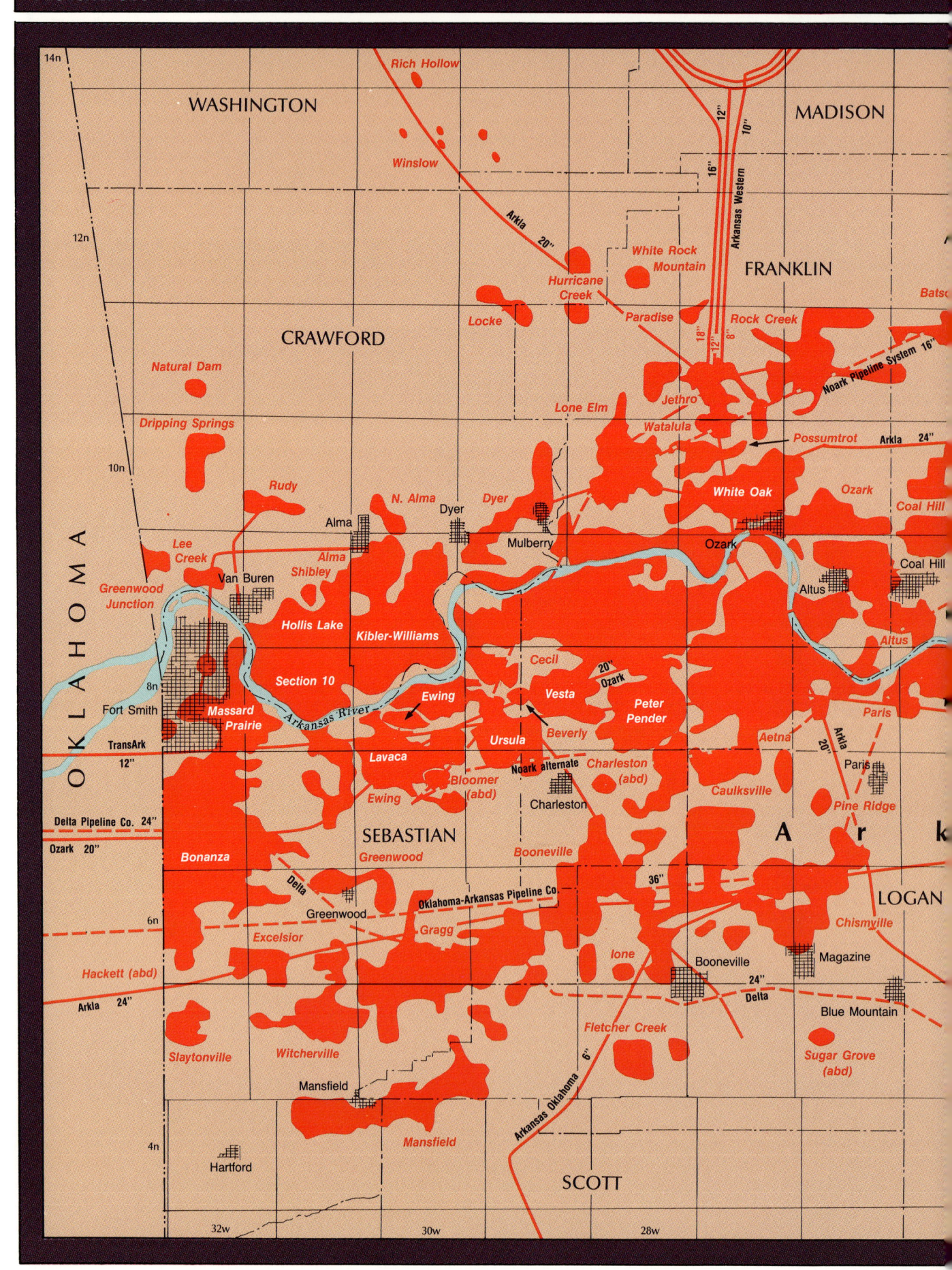

NORTH AMERICA

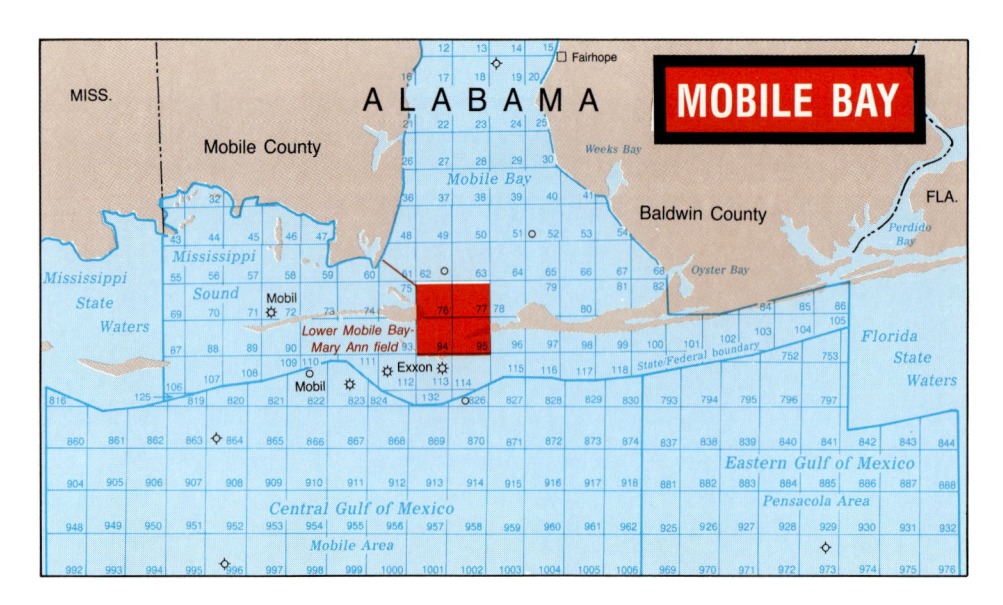

Giddings field and Pearsall field of Frio County were the horizontal drilling hotspots of 1990.

Baker Hughes listed 88 rigs active Dec. 3, 1990, in TRC Dist. 1, center of chalk activity, compared with only 17 Dec. 4, 1989.

Chalk activity also was growing to the east in TRC Dist. 3, where the latest count was 44 active rigs, compared with 29 at the same time in late 1989. Dist. 3 includes Lee, Fayette, Burleson, Brazos, and Washington counties.

The use of horizontal drilling techniques was limited only by industry's ability to innovate.

An example of that ability showed up in October 1990 when Pearsall Partners, of which Petro-Hunt Corp., Dallas, was managing partner, detailed completion of a well in Pearsall field with two horizontal legs producing Austin chalk oil (Fig. 2).

The partnership's 1 McDermand, about 4 miles west of Dilley, had an initial potential flow of 2,291 b/d of oil and 1.65 MMcfd of gas.

The well produces from horizontal drainholes with a combined length of about 5,700 ft drilled in opposite directions from a vertical hole. The legs were drilled without a whipstock or kickoff plug.

Jim McGowen, Petro-Hunt vice-president, said, "We were constrained by lease lines, and the dual drainhole approach used by Charles Rigdon, our chief engineer, was the only practical way to get the length of drainhole we wanted and control necessary to stay within field rules."

Elsewhere, Smackco Ltd., Brewton, Ala., reported good production at its 1 Cruit 26-15, a horizontal extension of a former vertical Jurassic Smackover oil producer in Huxford field, Escambia County, Ala.

The well reached horizontal at 14,936 ft, surpassing the previous world record of 11,484 ft, then continued to build and hold angle of about 100° for a length of 405 ft in Smackover. The horizontal leg cost about half the price of a new vertical hole from surface and has been an economic success, the operator said. Production averaged 215 b/d of oil

Service Fracturing Co., a Pampa, Tex., pumping services company, had limited success at a project in West Panhandle-Carson field to connect a new high angle well with an existing vertical oil well. The company planned to try again.

The Carson County project was successful mechanically, but a frac job penetrated into bottom water in Permian Brown dolomite, said Jerry Guinn, Service Fracturing president.

The idea was to use the existing No. 14 well's vertical rod and beam pump system to drain oil from 871 ft of open hole inclined at about 80° in Brown dolomite in the 14S high angle well.

Brown in an adjacent vertical well has permeability of 200 md to more than 2 darcies, average porosity of more than 20%, and reservoir pressure is less than 5 psi. Cores in the 14S confirmed those parameters and showed the natural fractures to be 5° off vertical and oriented E. 10° N.

Guinn said the 14S was the first horizontal well permitted in Texas to deliberately intersect a vertical well, first drainhole permitted in Texas to produce from a second well at the bottom of the drainhole, and first drainhole permitted for completion in Panhandle field Brown dolomite.

It also was believed to be the first drainhole fracture treated from each end and, to minimize screenouts, both ends at the same time.

The frac job took place July 27, 1990, and propagated many vertical fractures along 600 ft of the inclined wellbore.

Pump tests continued at the rate of 300 b/d of fluid. Oil rate declined from 30 b/d just after the frac.

Texaco Inc. completed a shallow

TEXAS GULF COAST

TEXAS
BEAUREGARD
ALLEN
ST. LANDRY
LOUISIANA
CALCASIEU
JEFFERSON DAVIS
ACADIA
LAFAYETTE
ST. MARTIN
IBERVILLE
CAMERON
VERMILION
IBERIA
ST. MARTIN
ST. MARY
ASSUMP
Lake Charles
Jennings
Lafayette
New Iberia
Abbeville
Baton Rouge
Port Allen
Cameron
Calcasieu Lake
Sabine Lake
Grand Lake
White Lake
Lac Blanc
Vermilion Bay
West Cote Blanche Bay
Hackberry
Big Lake
Mud Lake
Holly Beach
Pecan Island
Avery Island
Weeks Island
Cote Blanche Island
Garden City
Bayou Sale
Lake Sand
Rabbit Island
Tiger Shoal
Mosquito Bayou
WEST CAMERON
SABINE PASS
HIGH ISLAND
EAST CAMERON
SOUTH MARSH ISLAND NORTH ADDITION
WEST CAMERON WEST ADDITION
VERMILION
SOUTH MARSH ISLAND
EUGENE ISLAND
HIGH ISLAND EAST ADDITION
WEST CAMERON SOUTH ADDITION
VERMILION SOUTH ADDITION
EAST CAMERON SOUTH ADDITION
SOUTH MARSH ISLAND SOUTH ADDITION
EUGENE ISLAND SOUTH ADDITION
HIGH ISLAND EAST ADDITION SOUTH EXTENSION
GARDEN BANKS NG 15-2
THREE LINE
A
B
C
D
1
2
3

LOUISIANA GULF COAST

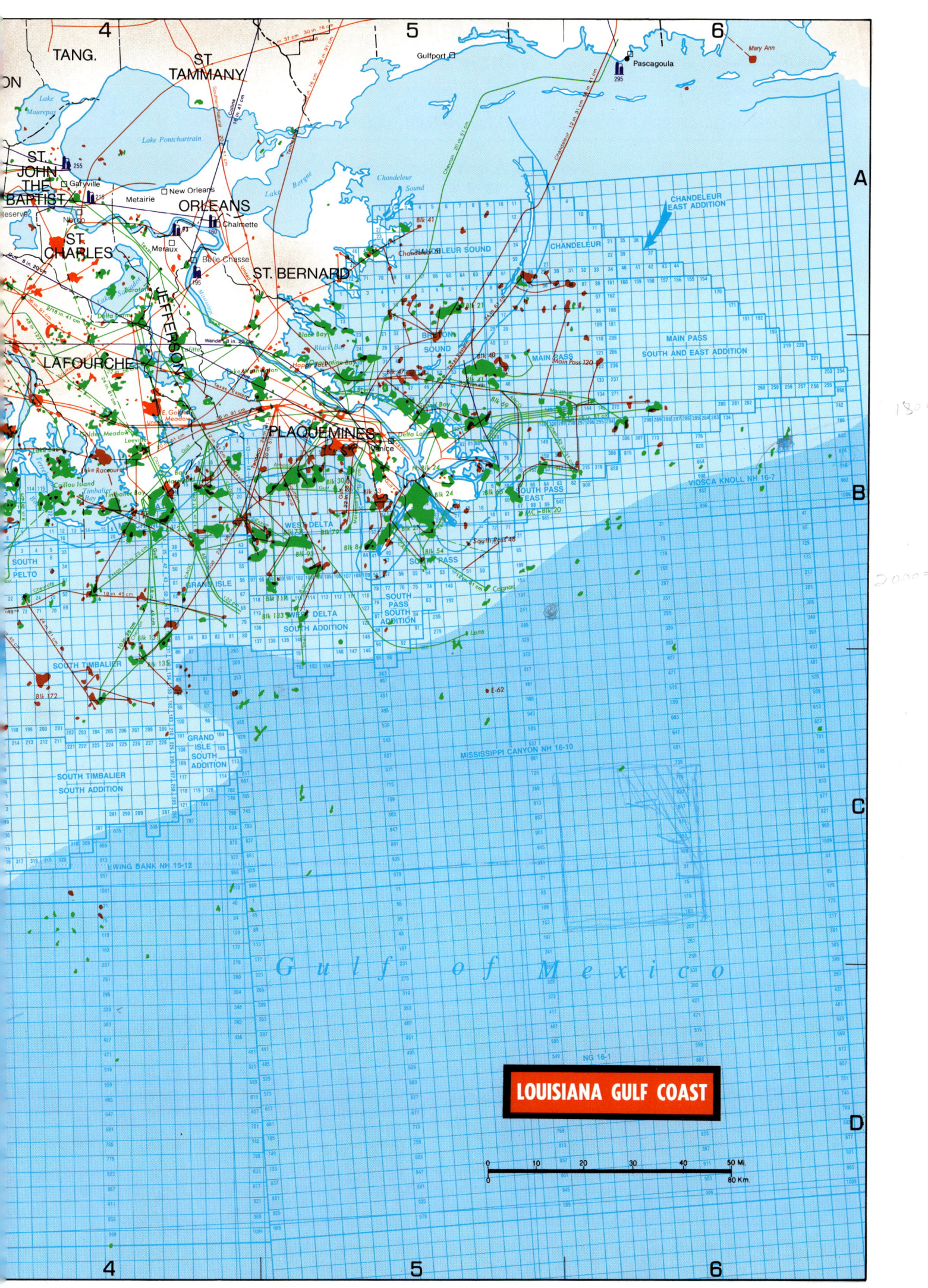

NORTH AMERICA

Table 3
U.S. tubular goods supply, drilling activity, and footage drilled

Tubular goods	1980	1981	1982	1983	1984	1985	1986	1987	1988	1989	January 1989	June 1990	% change 1989-90
							1,000 tons						
Mill shipments	3,612	4,241	1,759	677	1,406	1,299	483	919	1,130	909	392	583	48.7
Less exports	−133	−128	−67	−8	−12	−14	−36	−30	(74)	(320)	−176	−76	−56.8
Plus imports	1,036	2,389	2,180	565	2,214	1,506	617	570	988	429	176	174	−1.1
U.S. supply	**4,515**	**6,502**	**3,872**	**1,234**	**3,608**	**2,791**	**1,064**	**1,459**	**2,044**	**1,018**	**392**	**681**	**73.7**
Rotary rigs running*	2,912	3,970	3,106	2,232	2,428	1,980	964	936	936	869	841	953	13.3
Wells drilled†	69,486	89,234	84,054	76,280	84,983	70,806	39,015	36,253	34,015	30,401	13,689	14,973	9.4
							1,000 ft						
Footage drilled†	311,445	406,520	375,706	317,996	368,796	316,778	177,641	163,848	158,883	137,492	62,839	72,421	15.2
							Tons/1,000 ft						
Tubular goods used	14.5	16.0	10.3	3.9	9.8	8.8	6.0	8.9	13	13	6.2	9.4	50.7
							ft						
Average depth of wells	4,482	4,556	4,470	4,169	4,340	4,474	4,553	4,520	4,671	4,523	4,590	4,837	5.4
Rig census§	3,672	4,803	5,644	5,273	4,580	4,409	3,993	3,331	2,752	2,542	2,542	2,320	−8.7
Stacked	130	100	2,419	2,734	1,490	1,784	2,941	1,943	1,220	1,098	1,098	643	−41.4
Active	**3,542**	**4,703**	**3,225**	**2,539**	**3,090**	**2,625**	**1,052**	**1,388**	**1,532**	**1,444**	**1,444**	**1,677**	**16.1**

* Baker Hughes Inc. count of rigs drilling. † Excludes stratigraphic tests and service wells. § Reed Tool Co. count, excludes rigs not capable of drilling to more than 3,000 ft.

Source: Independent Petroleum Association of America cost study committee based on data from several sources

---

well producing oil by gravity drainage through three horizontal legs about 120° apart on the La Barge platform in Wyoming. Completion was in the low pressure Almy member of Eocene Wasatch with the laterals kicked from the same vertical well bore from the same kickoff depth.

Texaco's G634Y La Barge Unit, 34-27n-113w, Sublette County, was producing with a conventional vertical rod pump making about 30 b/d of oil and 3 b/d of water with no decline from a 24 in. underreamed, gravel packed, unstimulated zone in Almy at 415-520 ft. Most vertical Almy wells produced less than 10 b/d of oil.

The laterals with approximate direction and length were: north-northeast 133 ft, east-southeast 155 ft, and south-southwest 70 ft.

The 4 in. diameter laterals were jetted into Almy using a nonrotating nozzle through which water was pumped from surface through coiled tubing at 175 gal/min at about 10,000 psi.

Cretaceous Niobrara, a formation of interbedded chalks and calcareous shales, was the top horizontal target in Colorado and Wyoming. Niobrara horizontal drilling was planned or being attempted in the Denver, Sand Wash, and Hanna basins in the two states.

Rig utilization rate increases

The number of active rotary rigs in the U.S. continued a 4 year upward trend during 1990, while the number of available rigs continued to slide.

A midyear survey by Reed Tool Co., Houston, found 1,677 rigs working out of 2,320 available, a utilization rate 15 percentage points higher than at the same time in 1989 (Fig. 3).

In 1989, Reed counted 1,444 rigs working out of 2,542 available, a 57% utilization rate.

U.S. rig utilization increased for the fourth straight year since the rate plummeted to 26% with the crude oil price collapse of 1986, when only 1,052 of 3,993 rigs were active. The number of available rigs counted yearly by Reed has decreased since 1982, when the U.S. fleet climbed to a record 5,644.

Reed considered a rig available if it worked during the preceding 3 years and can be returned to service with an investment of $50,000 or less. It considered a rig active if it worked during Reed's 15 day census or during the 30 days preceding the summer survey period.

Reed Pres. Roy Caldwell said increasing drilling will trim attrition of the rig fleet during 1991 to half that of the past 2 years.

Reed's census found that the number of drilling contractors in the U.S. declined to 500 in 1990 from 558 in 1989. Of the 191 contractors who left the drilling business since 1987, 181 owned fewer than 10 rigs.

Of the 1990 count, 221 contractors owned two to five rigs and 170 owned only one. The average contractor owned a little less than five rigs.

Ninety percent of 273 companies surveyed by Reed cited low contract rates as a major impediment facing drilling contractors, 70% cited crew shortages, 55% drill pipe shortages, and 32% lack of financing.

Yet companies reported an average of 18,000 ft of drill pipe/rig, and 2.7 crews/rig, a 10% personnel shortfall. Thus, Reed concluded that low contract rates and cash flows were the greatest threats to maintaining an adequate U.S. drilling industry.

Reed found that 382 rigs had left the U.S. fleet since last year and 160 had been added for a net decline of 222 rigs.

Among the losses were 82 rigs that had been stacked for 3 years or longer, 105 rigs costing $50,000 or more to return to work, 166 cannibalized, 23 moved out of the U.S., and six destroyed or vandalized.

On the plus side, 116 rigs that had fallen from earlier Reed counts were put back in service, 40 rigs were built from available components, and four were moved into the U.S.

"There were no newly manufactured rigs reported in this year's fleet," Caldwell said. "Some contractors apparently found it cheaper to refurbish older rigs than pay transportation costs to move in newer rigs from other areas."

Availability of U.S. offshore rigs reached its lowest level since 1976, when 240 rigs were available.

Reed's 1990 census showed the U.S. offshore rig fleet decreased since 1989 by 34 to 259 rigs, while activity climbed by three rigs to 205. Rig use increased to 79%, a 10 percentage point rise since the 1989 census.

NORTH AMERICA

But Reed said low natural gas prices had caused Gulf of Mexico drilling to slow, causing many contractors to consider moving rigs to other areas. The moratorium on offshore drilling except for Alaska and the Gulf of Mexico likely will cause 15 rigs available off California to be moved, sold, or scrapped.

In terms of regional highlights, Reed described the surge of horizontal drilling in the Austin chalk trend of South Texas as reminiscent of the nationwide drilling frenzy of 1981. In that year, Reed found 4,703 of 4,803 U.S. rigs working, for a 98% utilization rate.

Reed said coalbed methane drilling had caused the rig fleet to increase in northern Alabama to 47 from 26. Utilization rate there is 91%.

Results of Reed's 1990 census, its 38th, cannot be compared directly with other counts of rig activity.

For its weekly count, Baker Hughes considers a rig active only if it is drilling when surveyed. Smith Tool Co. counts as active any rig engaged in any operation between start of rigging up and end of rigging down.

Gas pipelines dominate construction

Gas pipeline projects represented by far the largest segment of the U.S. pipeline industry's 1990 construction schedule.

Laying of more than 13,000 miles of gas transmission lines was planned in 1990 and beyond, compared with 9,284 miles planned in 1989.

The large increase in U.S. gas line construction more than made up for a decrease in planned crude oil pipeline construction. One major reason for the large drop was completion of All American Pipeline Co.'s California-West Texas system.

Big transmission projects were taking shape to move more gas to California, aiming for thermal enhanced recovery, industrial, and residential markets.

Among the biggest—one that will extend into 1991—Kern River Gas Transmission Co. was to begin construction in November 1990 on its 904 mile Wyoming-California pipeline project.

Associated Pipeline Contractors (APC), Houston, and H.B. Zachry Co., San Antonio, were scheduled to start work in southern Utah and Nevada on spreads five and six of the eight spread project.

Kern River also let contracts to APC and four other companies to lay pipe in the six remaining spreads in 1991.

As of late November, shippers had committed firm transmission volumes of more than 600 MMcfd on the 700 MMcfd capacity system. Remaining firm capacity was expected to be filled soon.

Kern River, a joint venture between units of Tenneco Inc., Houston, and Williams Cos. Inc., Tulsa, expected to be moving gas from Southwest Wyoming to near Bakersfield, Calif., by Jan. 1, 1992.

At Daggett, Calif., near Barstow, Kern River will merge with Mojave Pipeline Co. for the final distance to its terminus in Kern County, Calif. Mojave, a partnership of Enron Corp. and El Paso Natural Gas Co., received an optional expedited certificate from FERC for its California pipeline project in summer 1990.

Kern River's progress on its project killed Coastal Corp.'s Wyoming-California Pipeline Co. (WyCal) proposal. A competitor to Kern River, it was to have been a $576 million, 670 mile, 600 MMcfd Wyoming-California transmission line. WyCal held a Federal Energy Regulatory Commission construction permit and firm transportation contracts covering 1.2 bcfd of gas.

Elsewhere, Natural Gas Pipeline Co. of America's Arkoma basin pipeline construction project in Southeast Oklahoma was about 50% complete in November 1990.

The $51 million project, which got under way the previous September, included construction of a 4,000 hp compressor station in northern Atoka County and 105 miles of 24 in. line. The route extended from Amoco Production Co.'s Red Oak gas gathering system in Latimer County to NGPL's A/G line in Bryan County, near Bennington.

Several competing projects were proposed to transport gas from a series of discoveries in the Arkoma basin.

Here are examples of other gas transmission projects under way or planned:

• Transcontinental Gas Pipe Line Corp. (TGPL) received FERC approval to build its $60 million southern expansion project. It will provide 167 MMcfd in added mainline capacity for customers in Georgia, South Carolina, North Carolina, and Virginia during peak winter months.

• Northern Natural Gas Co. received FERC approval to expand the east leg of its system in Iowa, Illinois, and Wisconsin by 199 MMcfd and signed seven firm transportation contracts.

• Northern Border Pipeline Co. asked FERC for a permit to add compression to its 822 mile system and extend it by 368 miles in Iowa and Illinois.

• Williams Field Services, Salt Lake City, completed the first segment of its $95 million Manzanares pipeline project being built exclusively to transport San Juan basin coalbed methane.

Initial capacity is 390 MMcfd. A planned expansion in 1991 will boost capacity to 490 MMcfd. The first segment connects the Sims Mesa central delivery point into the system of Northwest Pipeline Corp. for delivery at the outlet of Northwest's Ignacio, Colo., gas processing plant.

• FERC approved a project by Consolidated Natural Gas Co.'s CNG Transmission Corp. to build underground gas storage for utilities in the Northeast U.S.

• Williams Natural Gas Co. was studying construction of a 24-26 in., 450 MMcfd line from Blackwell to Hartshorne, Okla., to link with the planned Oklahoma-Arkansas Pipeline.

• Six interstate pipeline companies disclosed a plan to consolidate competing projects to help speed development of gas reserves in Mobile Bay off Alabama. A letter of intent signed by the companies called for as much as 1.2 bcfd of gas to move through jointly owned lines. Participants in the settlement were ANR Pipeline Co., Florida Gas Transmission Co., Southern Natural Gas Co., Tennessee Gas Pipeline Co., Texas Eastern Transmission Corp., and TGPL.

Refining industry under pressure

U.S. refiners found themselves under increasing pressure to produce cleaner burning motor fuels at a profit as 1990 progressed.

The pressure came from, among other things:

• An August order issued by the Environmental Protection Agency for an 80% reduction in the sulfur content of U.S. diesel fuel beginning Oct. 1, 1993.

• Passage of Clean Air Act (CAA) amendments requiring oxygenated gasoline. President Bush signed the amendments into law Nov. 15. The petroleum industry blasted the measure, warning of substantial increases in the cost of gasoline with little environmental benefit even if the required volumes of motor fuel can be made.

• A 5¢/gal increase to 14¢/gal in the federal tax on gasoline, refiners' big ticket item. The increase, effective Dec. 1, was part of a law enacted Nov. 5 aimed at cutting the burgeoning federal budget deficit. Refiners feared it will discourage gasoline consumption.

EPA Administrator William K. Reilly called the diesel fuel order "an important step" for cleaner air. He said the reduction will have the largest effect in urban areas where particulate levels are the highest and the number of people exposed the greatest.

"This action, along with the stringent heavy duty particulate reduction standards for 1991 and later vehicles, will

NORTH AMERICA

virtually eliminate the black smoke from diesel tailpipes, cutting particulates by 90%," Reilly said.

EPA estimated the rule will add 1.8-2.3¢/gal to diesel's cost, depending on volume controlled. It requires refiners to reduce the sulfur content of diesel used on the road to 0.05 wt % from 0.25 wt %.

EPA said the rulemaking will cost refiners $380-910 million in 1995 and from $540 million to $1.3 billion by 2010.

Partially offsetting this cost will be the diesel manufacturers' ability to use less costly emission control devices, resulting in consumer savings of $60-325/vehicle, the agency said. It expected total savings, including fuel savings from the use of catalytic devices, to be $200 million in 1995 and $465 million in 2010.

If the diesel fuel sulfur cuts result in significantly reduced engine wear, as some information suggested, consumer savings could outweigh the increased costs per gallon for the low sulfur fuel, EPA said.

Diesel fuel aromatics will be controlled at near current levels by requiring a minimum cetane index specification or an optional maximum aromatics content standard.

The rule gives small refiners until 1995 to phase in the new standards, but they will be required to meet interim standards.

Many refiners were turning to reformulated gasoline as a means to reduce emissions. But refiners aren't likely to revamp processing units quickly enough to meet projected demand for reformulated gasoline, said a November 1990 report by Bonner & Moore Associates, Houston.

An inadequate supply of oxygenates, mainly methyl tertiary butyl ether (MTBE), will cause EPA to delay or relax oxygenate levels spelled out in CAA amendments.

The amendments require an oxygen content of at least 2.7 wt % during 4 winter months beginning in 1992 in gasoline sold in 44 cities where EPA has measured excessive CO levels. Also, by 1995 gasoline sold in nine severe ozone contamination areas must have at least 2 wt % oxygen content year round.

CO nonattainment areas account for about 27% and ozone nonattainment areas about 22% of U.S. gasoline consumption.

Bonner & Moore estimated about one third of U.S. gasoline markets will be required by CAA amendments to sell reformulated gasoline during at least a portion of the year.

Because of tight supplies, MTBE cost will become an important variable in gasoline pricing, often exceeding the spread between prices of regular and premium unleaded gasolines (Fig. 4).

Historically, toluene and MTBE competed as octane enhancers. But as limits are placed on total aromatics, toluene prices will fall significantly below the octane equivalent level for MTBE and regular/premium unleaded spread, Bonner & Moore predicted.

CAA amendments will cause extreme seasonal fluctuations in demand for oxygenates by winter 1992. By then, Bonner & Moore said, U.S. MTBE capacity will be about 160,000 b/d, leaving U.S. suppliers about 150,000 b/d short of peak demand.

At a minimum, even if only nonattainment areas were supplied with reformulated gasoline, an added 377,000 b/d of MTBE capacity would be needed (Fig. 5).

Bonner & Moore estimated outlays of $13-23 billion for new instrumentation and fractionation units will be needed to achieve reformulation as required by CAA amendments. Substantial investment will be needed to increase storage, install sophisticated control equipment, and expand U.S. butane isomerization and dehydrogenation units in new MTBE plants.

Gasoline reformulation will increase refinery operating costs by 2.4-3.4¢/gal, with much MTBE stock bought from third party producers. The cost of MTBE feedstock alone could add as much as 10¢/gal to the cost of blending reformulated gasoline if long term supply contracts recover essential capital expenses.

Reducing benzene content to 1 vol % in gasoline sold in nonattainment areas will require fractionation of reformer feed and/or benezene extraction.

Cat gasoline and reformate will be fractionated as a secondary control for aromatics, limited by CAA amendments to 25 vol %.

Reformer throughput rates and operating severities will be reduced to control aromatics contents, at the same time limiting an important hydrogen source for desulfurization. So refiners will be forced to invest in costly hydrogen plants to make up the loss.

OSHA responds to accidents

An outbreak of refinery and petrochemical plant accidents spurred the Occupational Safety and Health Administration (OSHA) to begin surprise safety inspections at 26 big U.S. processing plants in summer 1990.

OSHA inspectors began verifying adequacy of safety precautions in plants in 10 states with high concentrations of hazardous work involving refining, petrochemical, and gas processing operations.

Targeted were Texas, Louisiana, Oklahoma, New Jersey, Illinois, Pennsylvania, New York, Ohio, West Virginia, and Delaware.

Plants owned by companies with more than 2,500 employees were to be subject to unannounced visits. OSHA Deputy Assistant Sec. Alan McMillan said surprise inspections were to be scheduled the next several months, as manpower allowed.

OSHA officials said each inspection in the special emphasis program (SEP) will cover an entire plant and include "an in depth analysis of one process unit."

To select a process unit for inspection, compliance officers will review past leaks, explosions, and related incidents, as well as the types and volumes of chemicals involved. McMillan said particular attention was to be paid to repair and maintenance activities, turnarounds, start-ups, and shutdowns.

Supplemental contract workers are often hired to assist in that kind of work, and organized labor officials have sought to link the increased use of nonunion contract workers with the string of accidents.

Here are some of them:

• An explosion and fire Oct. 23, 1989, killed 22 workers, injured more than 130 others, and heavily damaged Phillips 66 Co.'s Houston Chemical Complex in Pasadena, Tex. One year later, Phillips had finished building Plant 6, the first of three polyethylene plants it planned to complete at its Houston complex by mid-1992 to replace 1.8 billion lb/year capacity lost to the accident.

• A July 5, 1990, explosion and fire at ARCO Chemical Co.'s Channelview, Tex., petrochemical complex killed 17 workers and leveled the plant's utilities area. The complex, temporarily shut down after the accident, produced almost 15% of U.S. stryrene monomer supplies and was the world's largest producer of MTBE at 20,000 b/d. The complex's first MTBE unit resumed operations the following September, and other process units were scheduled to resume runs by yearend.

• An explosion and fire Aug. 23, 1990, at Shell Oil Co.'s Deer Park refining/petrochemical complex near Houston injured two workers and knocked out a 165,000 b/sd crude distillation unit. The unit was the only process equipment damaged by the accident. Another 62,000 b/sd distillation unit continued to operate at the plant.

The Channelview blast followed a string of lesser accidents in previous weeks, flaring anew a controversy over worker safety at refineries and petrochemical plants, expe-

NORTH AMERICA

cially in the heavily industrialized Houston area.

A fire erupted in the fin fan cooler of the crude unit at Crown Central Petroleum Co.'s Pasadena refinery, injuring one worker July 3. That occurred when a charge cup blew a seal.

There were fires at two plants June 8. At the Pearland, Tex., chemical plant of Solvents & Chemicals Inc., a forklift driver punctured a barrel, touching off a blaze that burned for 20 hr, causing multiple explosions and injuring two workers.

At the same time, a fire broke out at Phillips' Pasadena plant, sending eight persons to the hospital with minor injuries.

Plant operators faced a new layer of federal regulatory oversight under safety rules proposed in July 1990 by OSHA.

Although many practices included in the proposed standard were in effect at many processing plants as a course of generally accepted engineering practices or existing safety and health codes, the new standard will give those practices the force of federal law.

The likely upshot will be an increase in recordkeeping by process plant operators and a recognition that operators not following generally accepted safety practices will face fines and other penalties from the federal government.

In April 1990, OSHA revealed what those fines can amount to. It proposed fines totaling nearly $6.4 million for alleged safety violations that led to the fatal accident at Phillips' Houston complex.

OSHA issued citations for alleged willful and serious violations of federal safety law and regulations against Phillips and Fish Engineering & Construction Inc., a service contractor long associated with Phillips.

The agency charged that Phillips committed 566 willful violations—one for each employee at the complex—carrying proposed penalties of $10,000 each and nine serious violations totaling $6,200, for a combined total of $5,666,200.

Phillips Petroleum Co. Pres. Glenn A. Cox said, based on preliminary information, Phillips 66 planned to contest a number of OSHA's alleged violations.

The proposed penalties against Phillips were the second largest OSHA has proposed against a company for a single inspection. In November 1989 it issued penalties totaling $7.3 million against USX Corp. for alleged safety violations at two steel plants.

OSHA charged that Fish Engineering committed 181 willful violations at $4,000 each, 12 serious ones at $5,500, and one other than serious at $100 for a total of $729,600.

CANADA

CAPITAL: Ottawa
MONETARY UNIT: Dollar
REFINING CAPACITY: 1,882,060 b/d
PRODUCTION: 1.508 million b/d
RESERVES: 5.8 billion bbl

THE CANADIAN INDUSTRY ENTERED 1991 WITH strong expectations of increased income from natural gas—especially from increased gas exports to the U.S.

Those expectations flowed in part from the U.S. Federal Energy Regulatory Commission's unanimous approval in November 1990 of construction of the Iroquois Gas Transmission System to serve Connecticut and New York markets.

Canadian factions and U.S. independent producers had opposed the $582.6 million (U.S.) project, designed to import Canadian gas at a 575.9 MMcfd clip. The pipeline is to connect with TransCanada PipeLines Ltd.'s system near Waddington, N.Y., and extend through Connecticut to South Commack, N.Y., on Long Island.

About 80% of the gas is to go to local distribution companies in Connecticut, Massachusetts, New York, New Hampshire, New Jersey, and Rhode Island, and the rest to electric power generating firms.

FERC gave the project its general approval in July 1990 but because of objections raised by U.S. independent producers ordered an administrative law judge to conduct a hearing on the line's rates and demand for its gas.

FERC Judge Walter Alprin reported back that Iroquois was needed, had reasonable growth rate projections and capital and rate structures, and would not displace U.S. gas supplies.

New York and Connecticut issued permits for the line.

Iriquois had negotiated easements for about 40% of its route on a voluntary basis and planned to continue to seek voluntary easements with the remaining landowners, although the FERC certificate gave it the right of eminent domain.

Elsewhere, TransCanada cleared a hurdle on its rate structure, the U.S. Department of Energy approved still more imports of Canadian gas, and the Canadian industry was moving toward first development of offshore oil.

All that was taking place in 1990 after the federal Petroleum Monitoring Agency reported that 1989 was the Canadian industry's second year of decline in internal cash flow and net income (Table 1).

NEB settles
TransCan toll issue

Canada's National Energy Board (NEB) in November 1990 upheld TransCanada's current method of assessing pipeline tolls in a major regulatory battle involving a $2.6 billion natural gas pipeline expansion.

The expansion by TransCanada was linked with the Iroquois pipeline project designed to export gas from western Canada.

The U.S. is the only customer for Canadian gas exports. In 1989, the last full year for which data are available, shipments to the U.S. amounted to 37.9 billion cu m, or almost 40% of total Canadian deliveries of 96.5 billion cu m (Fig. 1). An NEB ruling said rolled in tolling under which all present and future users would share the expansion cost will be maintained.

Major gas buyers in Central Canada argued for incremental tolling under which only shippers benefiting from deals using the new line should pay for it. They said 75% of volumes to be shipped in the 989 mile loop of the TransCanada system will go to U.S. markets but gas buyers in Ontario will pay toll increases of as much as $128 million/year if costs are rolled in.

NEB said it is the demand of all shippers that creates the need for added pipeline capacity. The board said its decision may not necessarily apply to future expansions, but there would have to be a compelling reason for it to reconsider the toll issue.

As 1990 drew to an end, NEB still had to deal with the economic viability of the project and conduct hearings on several gas export applications tied to it.

The full expansion will allow shipment of an added 831 MMcfd of gas from western Canada to markets in Central Canada and the U.S. TransCanada hopes to have the full line in operation in late 1992.

Alberta Energy Minister Rick Orman said the pipeline expansion, if approved, would trigger a major increase in exploration activity in 1991.

Consumers Gas Co. Ltd., Toronto, the largest gas distributor in Central Canada, said the NEB ruling means Ontario gas users will be forced unfairly to subsidize a pipeline expansion to serve the U.S. The company said gas markets in the U.S. Northeast are not as stable for Canadian producers as markets in Central Canada.

The NEB ruling followed an October 1990 decision by the

NORTH AMERICA

Where Canada's oil and gas shipments go
Fig. 1

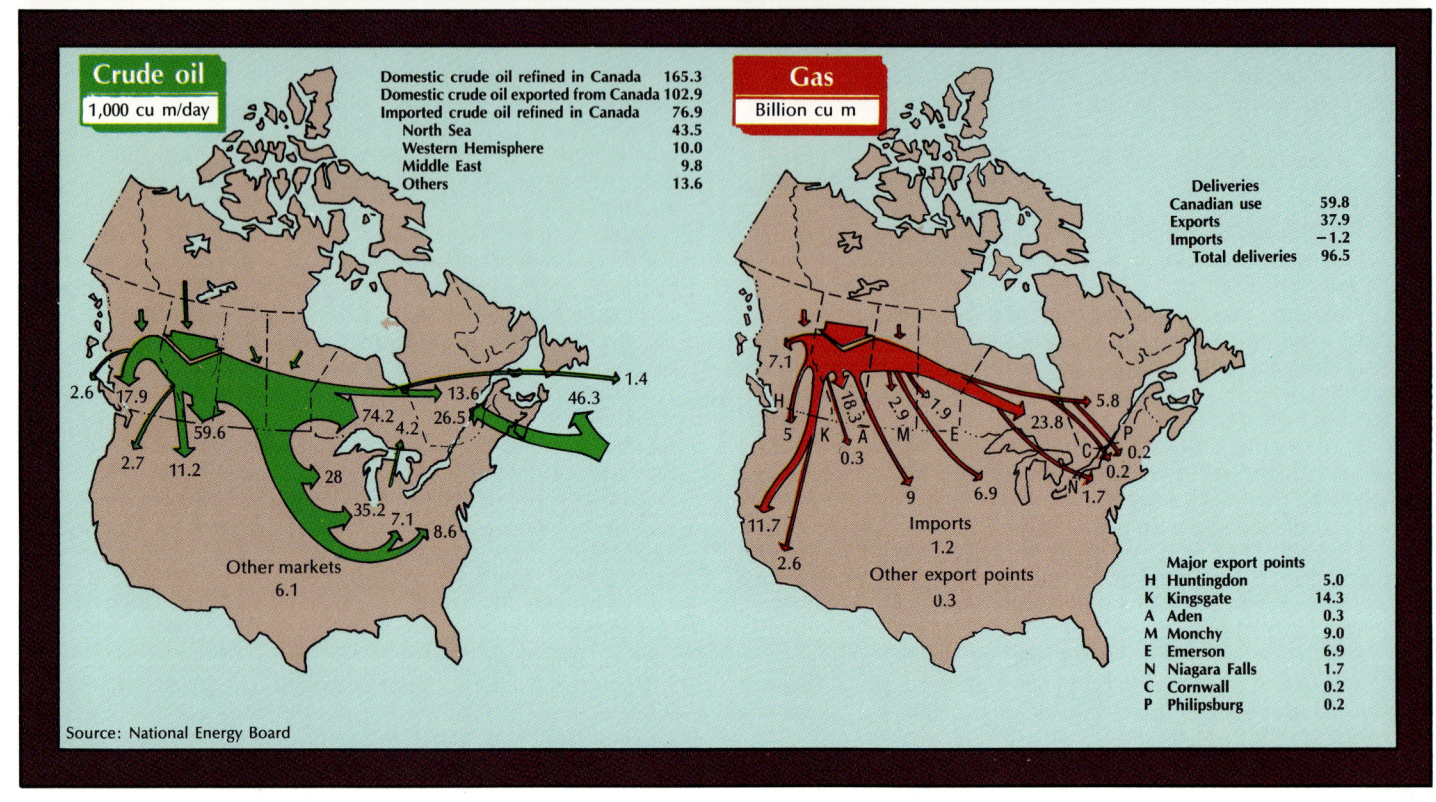

Crude oil — 1,000 cu m/day

Domestic crude oil refined in Canada	165.3
Domestic crude oil exported from Canada	102.9
Imported crude oil refined in Canada	76.9
North Sea	43.5
Western Hemisphere	10.0
Middle East	9.8
Others	13.6

Other markets 6.1

Gas — Billion cu m

Deliveries	
Canadian use	59.8
Exports	37.9
Imports	−1.2
Total deliveries	96.5

Imports 1.2
Other export points 0.3

Major export points		
H	Huntingdon	5.0
K	Kingsgate	14.3
A	Aden	0.3
M	Monchy	9.0
E	Emerson	6.9
N	Niagara Falls	1.7
C	Cornwall	0.2
P	Philipsburg	0.2

Source: National Energy Board

U.S. DOE giving U.S. companies permission to import Canadian gas through the Niagara Import Point Projects (Nipps) to the U.S. Northeast.

Nipps involves construction of 545 miles of line by seven U.S. interstate pipeline companies, increasing the capacity of a pipeline system that begins at the Minnesota-Manitoba border, crosses Wisconsin and Michigan back into Canada, then returns to the U.S. at Niagara, N.Y.

About 545 miles of loops by seven U.S. interstate pipeline companies, along with 61,300 hp of added compression, will open the way for an additional 346.4 MMcfd of gas to be shipped to U.S. customers as far south as Virginia.

Canada's 1989 gas exports were up 4.5% from 1988's 5.9 billion cu m. California accounted for 37% of Canada's export sales, the central region of the U.S. 40%, Pacific Northwest 13%, Northeast 8%, and the mountain region 2%.

Export prices averaged $1.90 (U.S.)/MMBTU, flat with 1988's $1.88.

Access to U.S. pipeline systems continued to improve, and by yearend 1989 almost all major U.S. gas pipelines were open access carriers.

This aided rapid growth in Canadian short term sales, which grew to 13.8 billion c m in 1989 from 11.3 billion cu m in 1988. Short term sales accounted for about 37% of total gas exports in 1989, compared with 30% in 1988 and only 22% in 1987.

British Gas purchase of Consumers

Among other gas developments, British Gas plc, London, was closing in on its planned $1.1 billion (Canadian) takeover of Canada's largest gas utility.

The purchase of Consumers Gas Co. Ltd., Toronto, won approval in November 1990 from Investment Canada, a federal foreign investment screening agency. That action followed approval by the Ontario Energy Board, an agency of the province's New Democratic government.

Consumers serves more than 1 million natural gas customers in Ontario, Quebec, and New York state. The company was being sold by GW Utilities Ltd., controlled by the Reichmann family.

Among other things, British Gas agreed to make a public offering of 15% of Consumers shares by 1992, spend $30 million by Dec. 31, 2000, in Ontario for research and development focusing on efficient, "environmentally friendly" use of natural gas, make available at least $50 million (Canadian) for investment during the next 10 years in Ontario cogeneration projects to be chosen by British Gas, and create a $5 million venture capital fund to invest in Ontario energy and environmental technology during the next 5 years.

If the fund is fully spent at the end of that time, British Gas will invest another $5 million during the next 5 years.

The provincial approval was seen as a test of the New Democratic party government's relationship with the business community.

The party's previous policy called for Canadian ownership

Hibernia production system
Fig. 2

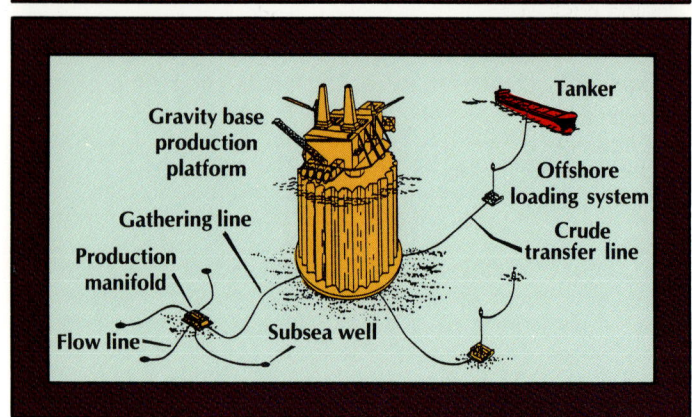

Gravity base production platform, Tanker, Offshore loading system, Crude transfer line, Gathering line, Production manifold, Flow line, Subsea well

NORTH AMERICA

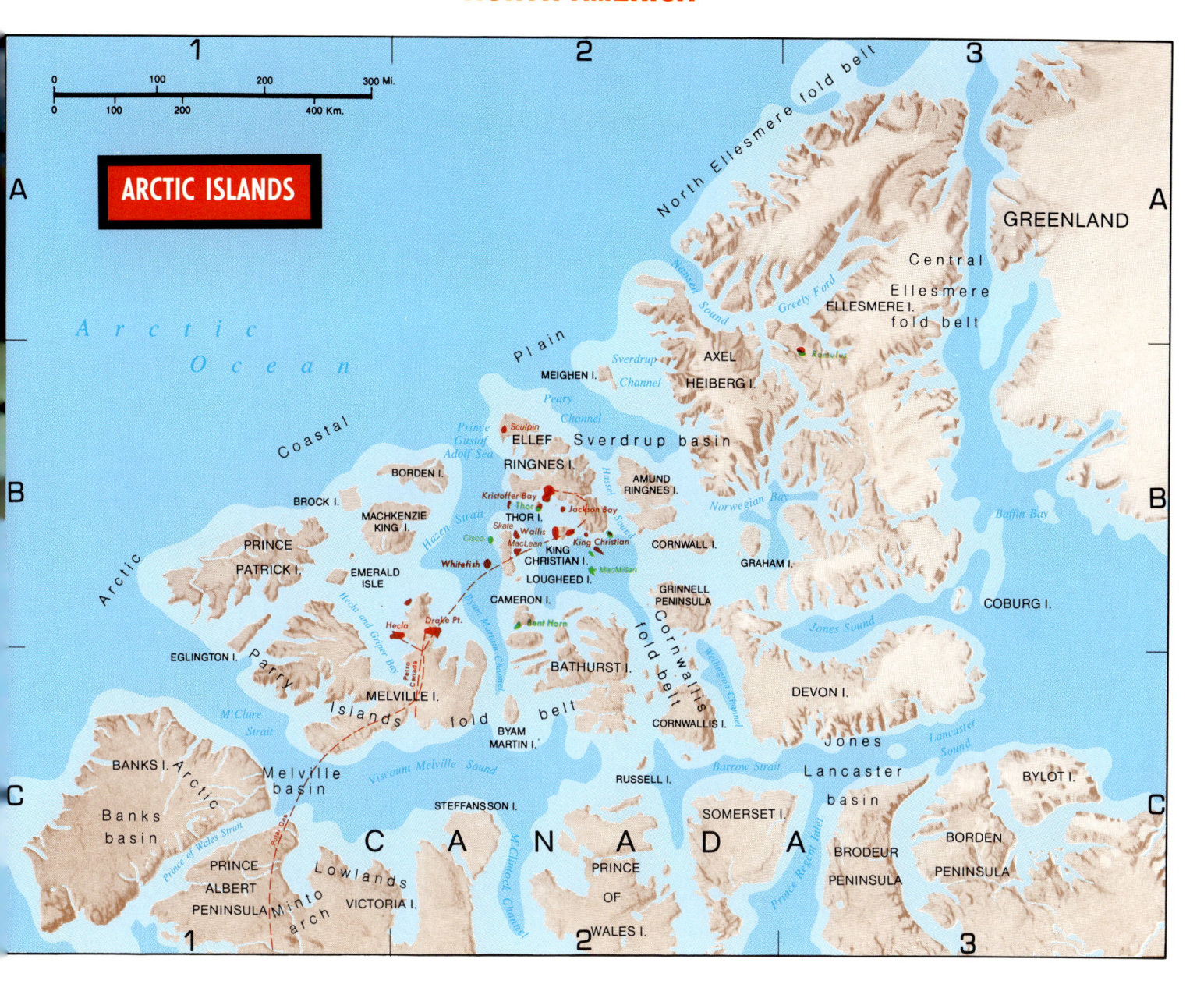

ARCTIC ISLANDS

of public utilities and resource companies. Ontario Premier Bob Rae said the cost of ownership of Consumers would be too high for his government, which had a deficit of $2.5 billion.

British Gas, privatized by the British government in December 1988, also purchased an equity interest in Bow Valley Industries, Calgary, late in 1988. British Gas, with 17 million customers, is the western world's largest integrated gas business.

Canada's first offshore development

Canadian operators were moving toward their first offshore field development programs, both off eastern Canada, as 1990 drew to a close.

In quick succession in mid-September a group led by Mobil Oil Canada Ltd. decided to go ahead with the $5.2 billion (Canadian) Hibernia oil field project on the Grand Banks off Newfoundland after Ottawa agreed to provide $95 million in interim financing.

The federal government also agreed to provide $2.7 billion in grants and loan guarantees.

Production start-up is scheduled by 1996, aiming for peak flow of 110,000 b/d by 1998 from reserves of 525-650 million bbl in the geologically complex Lower Cretaceous Ben Nevis-Avalon and Hibernia sandstones. Canada's Department of Energy, Mines, and Resources estimated Hibernia will account for about 12% of total Canadian light oil production in 2000.

The government financing arrangement received final legislative approval in November 1990 in a vote by the Canadian Senate.

In addition, the Lasmo Nova Scotia Ltd.-Nova Scotia Resources Ltd. combine received regulatory approval to develop Cohasset and Panuke oil fields on the Scotian Shelf.

The Canada-Nova Scotia Offshore Petroleum Board and federal and Nova Scotia energy ministers approved development of Cohasset and Panuke fields. Further approvals were required.

The Lasmo-Nova Scotia Resources combine expects seasonal production of 30,000 b/d for 6 years, beginning in 1992, via a converted jack up rig and storage tanker at the field site, 41 km southwest of tiny Sable Island off Nova Scotia. Start-up on or before the target date will make the project Canada's first offshore hydrocarbons production.

NORTH AMERICA

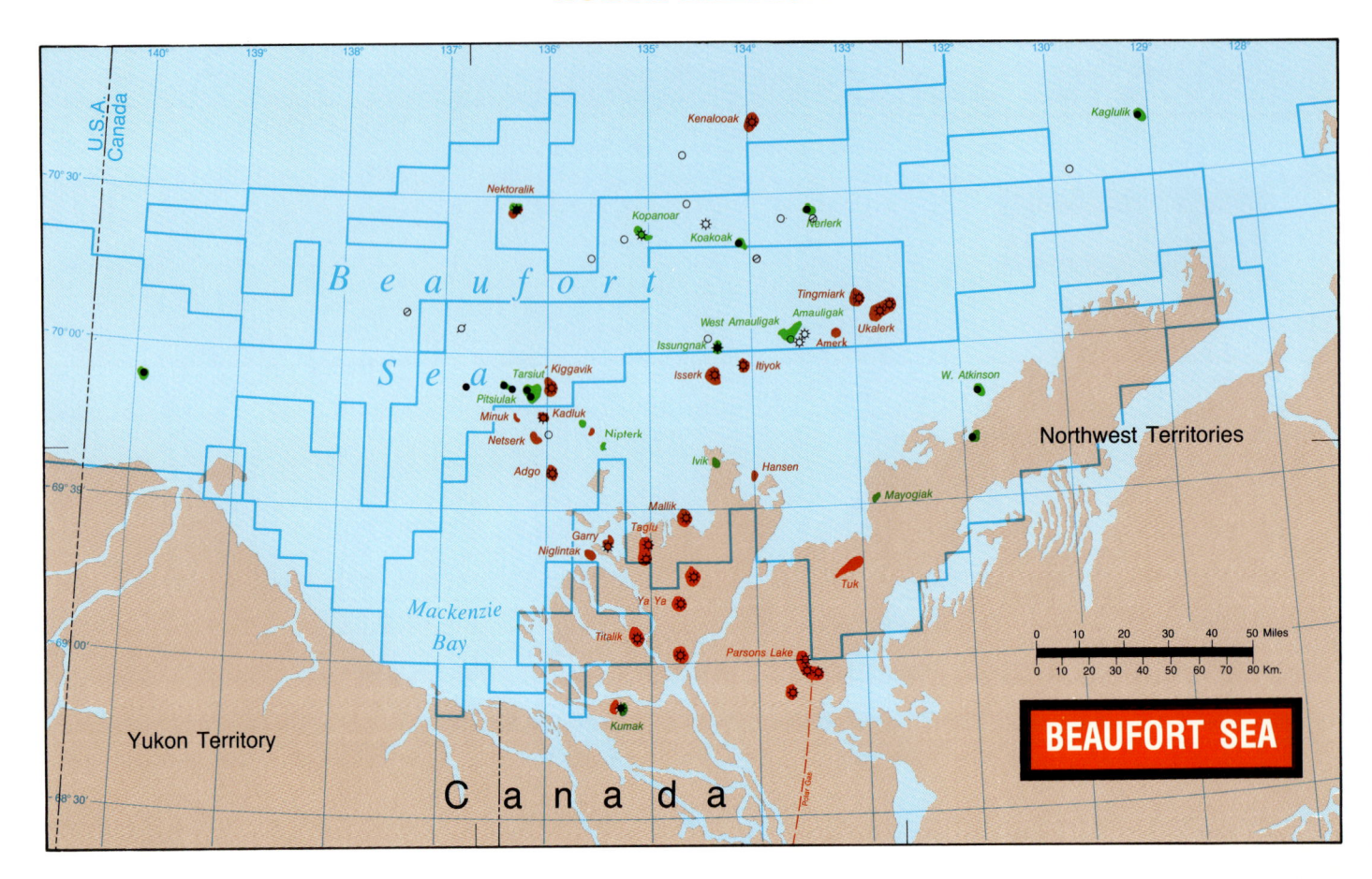

Combined reserves of Cohasset and Panuke fields are 35-40 million bbl of 50° gravity oil.

Process facilities will be on board the Gorilla III jackup, held under contract with Rowan Cos. Inc. Before conversion to production in late summer 1991, the rig will drill two wildcats for Lasmo-Nova Scotia Resources on prospects northeast of Cohasset.

Many other candidates for development lie off eastern Canada and in the Beaufort Sea.

Gulf Canada Ltd., operator, expects Amauligak field, with reserves of 500 million bbl, to be the lead project for any Beaufort Sea oil development.

Gulf holds interests in seven oil and three gas discoveries in the Beaufort Sea, as well as interests in four gas strikes on the Mackenzie Delta and Tuktoyaktuk Peninsula.

Mobil group's Hibernia project

Construction began quickly for the Hibernia development project, in which the $5.2 billion sum amounts to preproduction capital outlays.

Another $3.3 billion (Canadian) in capital outlays will be incurred after production begins. Operating phase spending is estimated to total $10 billion during the projected 18 year life of the field.

Construction was under way on an access road to a site at Bull Arm, Newf., where a concrete gravity base will be built for the production platform (Fig. 2), to be installed 193 miles southeast of St. John's, Newf., in about 80 m of water.

The installation is scheduled for late summer 1995, among the key dates in Hibernia's development schedule (Table 2).

Mobil, with a 28.125% interest in the project, planned shortly to call tenders for about $500 million in supplies and equipment.

Its partners were Gulf and Petro-Canada 25% each and Chevron Canada Resources 21.875%.

Hibernia's gravity base structure will consist of a base slab of reinforced concrete supporting a 105 m diameter caisson that will reach from the seabed to 5 m above sea level. Reaching above the caisson will be four shafts that will support topside facilities for drilling, oil and gas processing, and accomodation.

The base, to be built using 100,000 metric tons of cement and 55,000 metric tons of reinforcing bars and prestressed cables and rods, will contain water and solid ballast. It will be able to store 1.3 million bbl of oil.

The topside will consist of five large modules weighing 3,900-6,700 metric tons and eight smaller units weighing 100-1,300 metric tons. Total topside weight will be about 33,000 metric tons.

The platform's two rigs will drill 48 wells into Hibernia field's pay zones. In addition, as many as 35 subsea wells—25 in the BenNevis-Avalon reservoir and 10 in the Hibernia pay zone—may be drilled by floating rigs at sites beyond the reach of the platform rigs.

Oil will move from the platform's storage cells to shuttle tankers on station at two loading systems about 2 km from the platform. Each loading system will consist of a crude oil transport line, seabed terminal, vertical flexible pipe linked to a subsurface mooring buoy, and a flexible hose to the tanker.

Three 120,000 dwt tankers, each able to carry 850,000 bbl of oil, will be built to deliver crude oil to markets. The vessels will be double hulled and equipped with iceberg detection and surveillance equipment.

Three multipurpose vessels will assist tanker operations and provide standby and supply services for the platform.

Chevron's Hibernia P-15 wildcat found Hibernia field in 1979 on a farmout of the Mobil block. Eight of nine delineation wells drilled in 1980-84 encountered volumes of oil.

Development was delayed in part by a jurisdictional dispute between the Newfoundland and federal governments.

The dispute was settled in February 1985 by the Atlantic

NORTH AMERICA

Table 1
Breakout of the Canadian industry's financial performance

	1980	1981	1982	1983	1984	1985	1986	1987	1988	1989
	\- Billion current $ (Canadian) \-									
Revenues										
Total operations*	44.2	53.0	57.2	59.3	62.7	66.3	47.6	49.2	47.0	50.2
Upstream	16.1	17.3	21.0	24.0	26.7	28.0	17.0	17.9	15.8	17.2
Downstream	24.8	30.5	31.3	30.7	31.2	33.4	26.2	27.0	27.0	28.8
Internal cash flow										
Total operations*	9.6	8.4	7.4	8.7	10.8	11.6	8.0	9.4	9.0	8.9
Upstream	6.7	5.3	5.8	7.4	8.5	9.5	5.5	6.8	5.8	6.4
Downstream	2.2	2.1	1.3	0.8	1.1	1.1	1.7	1.7	2.3	1.9
Net income										
Total operations*	4.7	3.2	1.8	1.6	3.5	3.3	−2.8	3.2	2.2	1.3
Upstream	2.9	1.7	2.1	2.5	2.9	3.3	−1.9	2.0	0.8	0.6
Downstream	1.5	1.4	0.4	−0.1	0.4	0.2	0.8	0.9	1.5	0.9
Capital expenditures										
Upstream	6.9	6.9	6.5	6.8	8.0	9.3	6.3	5.5	6.8	5.4
Downstream	0.8	1.6	2.2	1.6	0.9	0.8	0.7	1.1	1.2	1.2
Other	0.6	1.7	1.9	1.2	0.9	0.9	0.4	0.4	0.6	0.6
Total in Canada	8.3	10.2	10.6	9.6	9.8	10.9	7.4	7.1	8.5	7.2
Expenditures abroad	1.1	1.8	1.2	0.8	0.7	1.2	0.6	0.5	0.6	0.5
Total capital outlays	9.4	12.0	11.8	10.4	10.5	12.1	7.9	7.6	9.1	7.7
Dividend payments	1.0	1.2	1.1	1.3	1.4	1.8	2.6	1.5	2.5	1.4
Total capital employed	43.9	59.9	66.1	69.0	81.2	82.9	73.4	73.7	76.0	78.0
	\- Percent \-									
Total reinvestment rate	100	136	138	101	81	87	88	75	96	85
Upstream reinvestment rate	103	114	86	70	72	77	98	74	110	82
GNP deflator	100.0	111.8	124.2	132.0	135.8	139.5	142.6	147.6	151.7	156.6

*Includes other Canadian and foreign operations.

Source: Petroleum Monitoring Agency

Accord between the two governments.

E&P efficiency slide continues

Canada's oil and gas industry's efficiency in terms of finding costs and production replacement continues to falter, said Canadian Oil Patch Analyst Inc. (COPA) in a late 1990 report.

A COPA study based on analysis of 46 companies, including most major Canadian producers, showed the 5 year rolling average finding and development cost for 1985-89 was more than $9/bbl of oil equivalent (BOE), 4% higher than the previous 5 year average (Fig. 3).

Further, the 1989 replacement cost of $8/BOE was higher than in the previous 2 years.

COPA found that highgrading of plays common after the 1986 oil price crash seems to have reached its limits, as seen in declining production replacement rates. For the first time in 5 years, revisions of reserves in 1989 exceeded discoveries and extensions.

The 46 companies added 1.755 billion bbl of liquids reserves during 1985-89, replacing 85% of production but just 46% through discoveries and extensions.

Companies are finding it more difficult to replace oil production with discoveries and extensions, COPA said, in part because of low prices, diversion of funds to debt payments or acquisitions, and lack of elephant scale discoveries.

"As a consequence, the industry—especially the senior sector—has begun to rely more on engineering generated revisions to replace production, similar to the U.S," COPA said.

The companies' gas reserve additions during 1985-89 totaled 7.1 tcf, replacing 63% of production. The low gas replacement figure, however, is due to large reserve writedowns and has little effect on the deliverability surplus because of the relatively long reserve life of Canadian natural gas, COPA said.

Table 2
Key dates in Hibernia schedule

1990
    October—Work begins at Bull Arm, Newf., construction site

1991
    August—Bull Arm site completed

1992
    April—Gravity base structure drydock completed
    June—Module fabrication begins

1993
    April—Engineering and construction of offshore loading system begins
    April—Order shuttle tankers

1994
    April—Module assembly and hookup begins

1995
    March—Mating of base structure and topside begins
    May—Inshore hookup begins
    August—Tow-out to production site
    September—Installation of offshore loading system begins

1996
    April—Development drilling begins
    July—Delivery of first two shuttle tankers
    October—Production begins

1997
    September—Delivery of third shuttle tanker

NORTH AMERICA

Canadian oil, gas performance

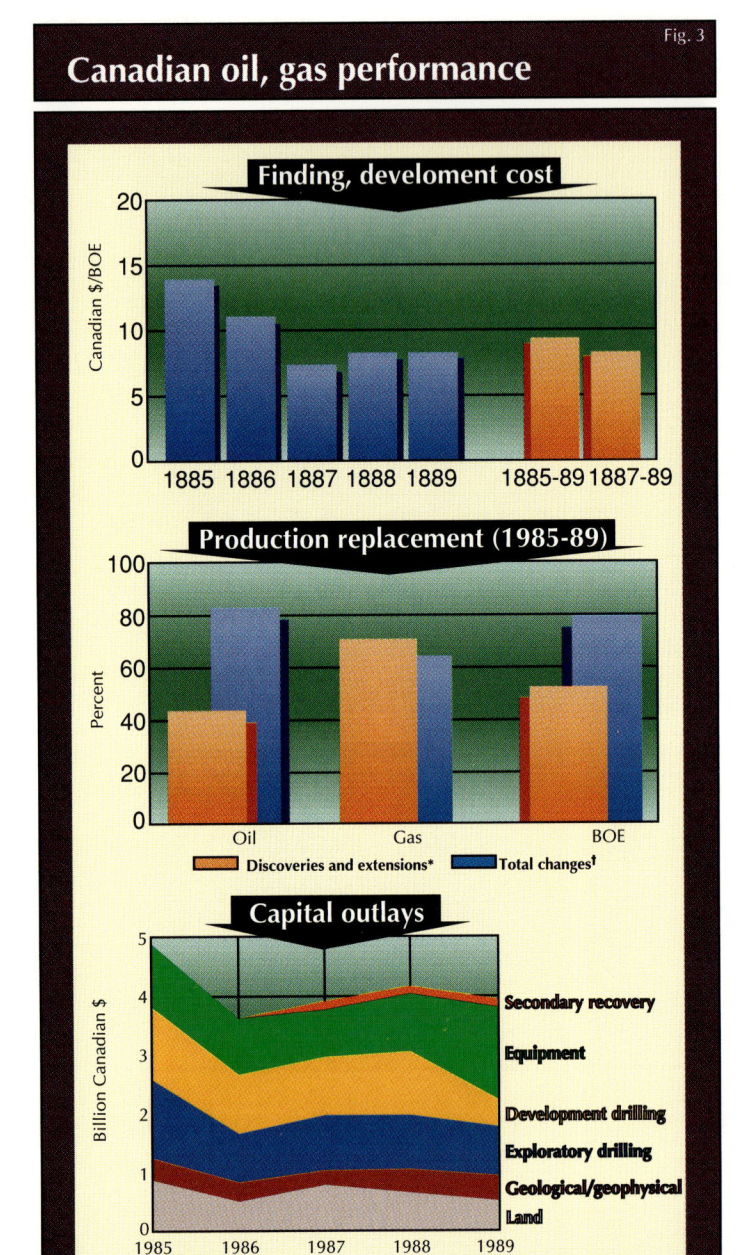

Fig. 3

Finding, development cost (Canadian $/BOE): 1885, 1886, 1887, 1888, 1889, 1885-89, 1887-89

Production replacement (1985-89) (Percent): Oil, Gas, BOE
- Discoveries and extensions*
- Total changes†

Capital outlays (Billion Canadian $): 1985, 1986, 1987, 1988, 1989
- Secondary recovery
- Equipment
- Development drilling
- Exploratory drilling
- Geological/geophysical
- Land

*Based on 41 of 46 companies surveyed. †Based on all 46 companies surveyed.
Source: Canadian Oil Patch Analyst Inc.

In 1989, combined hydrocarbon production replacement on an oil equivalent basis was 79%, but discoveries and extensions accounted for only 36%.

COPA said major producers benefited from revisions because that enabled them to greatly reduce the replacement cost gap between them and medium sized producers to less than $1/BOE.

Among individual companies, replacement costs were $5-20/BOE.

The study showed that less than half of the companies surveyed were able to replace their production below the "going concern value" of proved reserves. "This may explain the significant effort and money spent recently by the industry on acquiring reserves vs. finding them," COPA said. The exploratory finding cost also continued to climb, averaging $6/BOE during the 5 year period.

CO₂ prospects grow brighter

Prospects were brightening for Canadian carbon dioxide flood projects as 1990 drew to a close.

For example, Shell Canada Ltd. disclosed plans for a $40 million CO_2 flood in a Southeast Saskatchewan oil field that would be the first phase of a project designed to recover as much as an incremental 80 million bbl of crude oil.

In addition, Vikor Resources Ltd., Calgary, was considering expansion of its CO_2 flood near Red Deer, Alta.

Shell planned to begin work on its Midale field flood project in early 1991, pending government approval, with start-up scheduled for third quarter 1992.

Plans called for Shell at first to flood about 10% of the field to stimulate oil remaining after waterflooding. If successful, it may expand the CO_2 flood to a much larger portion of the field by the late 1990s.

Shell began a pilot CO_2 project in Midale field in 1984 and said results were so promising they led to the new proposal.

Midale has produced about 100 million bbl of oil since Shell discovered the field in 1953. The company estimated an incremental 20 million bbl recoverable from waterflood operations.

It believed CO_2 flooding of the entire field could yield another 80 million bbl.

Shell said a decision on expanding the CO_2 flood to the rest of Midale field and into nearby fields will be made after several years of operation on the initial phase.

Shell will operate the project for Midale Unit working interest owners.

Vikor said its CO_2 flood in Joffre Viking pool, Alberta, has accounted for more than 20% of the original oil in place, boosting overall recovery to more than 62%. It estimated commercial scale CO_2 flooding of the field could produce another 12 million bbl.

The project produced its 1 millionth bbl in early November 1990.

Discovered in 1953, Joffre Viking pool was put under waterflood in 1957 but abandoned in the mid-1960s after recovery of about 42% of OOIP.

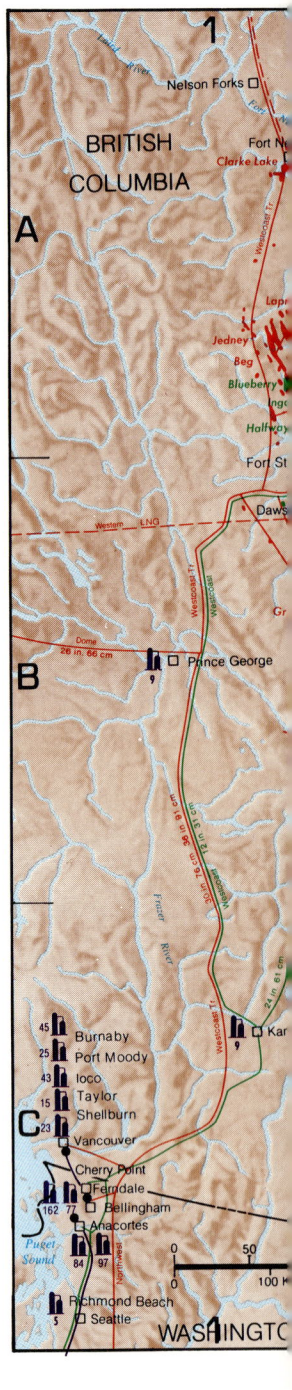

After Alberta Oil Sands Technology & Research Authority promised financial support in 1982, Vikor started the Joffre Viking Tertiary Oil Unit as a field experiment of CO_2 flooding in the pool.

In the pilot project, water and CO_2 were alternately injected into center of each pattern, producing from six wells. The unit completed an eight well expansion in mid-1985 about 3 miles from the pilot. Earlier, Shell received regulatory approval for an $825 million sour gas development project in the Caroline area of Alberta, 75 miles northwest of Calgary.

The provincial Energy Resources Conservation Board (ERCB) chose a Shell proposal for the 2 tcf field over a competing plan submitted by Husky Oil Ltd.

Shell held a 61.2% interest in the field, which it discovered in 1986. Its main partners were Husky 11.6%, Gulf 8.5%, and Union Pacific Resources Ltd. 6.7%. The remaining 12% is shared by 10 other companies.

The project was to include development drilling, an under-

NORTH AMERICA

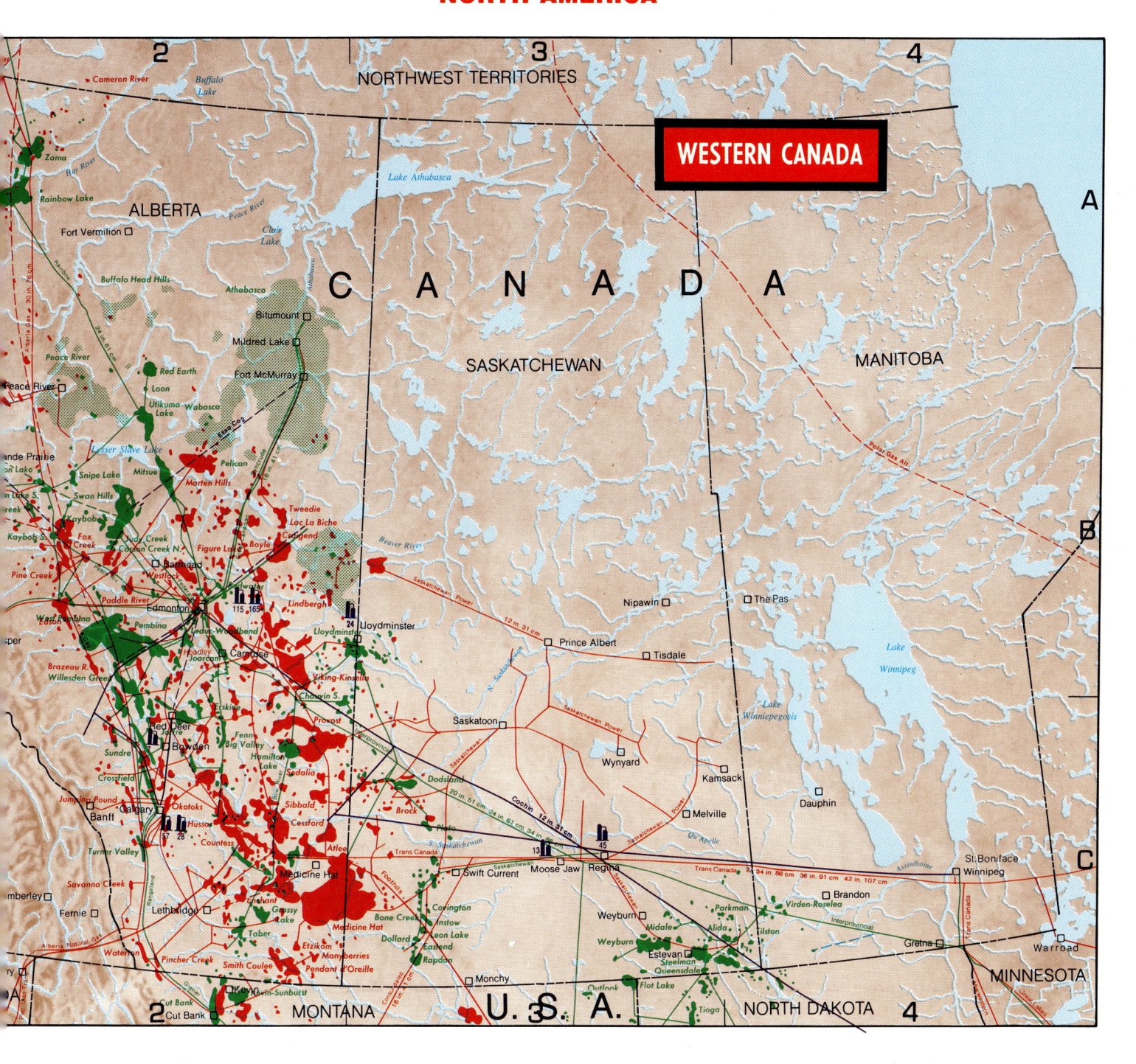

WESTERN CANADA

ground molten sulfur pipeline, gathering facilities, and a gas processing plant. The field was expected to produce 100 MMcfd of gas, 17,500 b/d of condensate, 28,000 b/d of natural gas liquids, and 4,000 metric tons/day of sulfur at full development early in 1993.

The Husky proposal involved upgrading its 630 MMcfd gas processing plant at Ram River, Alta., and laying a 35 mile pipeline from the field to the plant.

The ERCB report said the Shell proposal was the best in terms of the overall public interest.

It was supported by more people in the area and provided greater economic benefits.

Husky expressed disappointment at ERCB's decision but said it was prepared to "help expedite speedy development of this huge gas resource." The Shell plan drew opposition from environmental groups and owners of recreational properties in the area.

Meantime, oil sands development waxed and waned in 1990. For example, withdrawal of federal funding set back a $4 billion oil sands project in northern Alberta by at least 8 years, NEB said in December.

Virtually at the same time, Chevron unveiled plans for a new oil sands recovery pilot project on a 49,000 acre lease in the Fort McMurray region of northern Alberta.

An NEB draft study on oil supply said construction of the OSLO oil sands plant is unlikely until 1999. The 77,000 b/d project by a group led by Esso Canada Resources Ltd. had a targeted construction start in 1991 with completion in 1996. Ottawa withdrew financial support of as much as $1 billion in February 1990.

Industry estimates placed Canada's crude production decline at about 50,000 b/d/year as a result of depleting conventional oil reserves and delays in exploration and development projects due to unstable prices.

Chevron's 5 year project is designed to test a new steam injection process that could lead to a 10,000 b/d bitumen recovery operation. It had conducted preliminary tests on a heated annulus steam drive at an Alberta test facility.

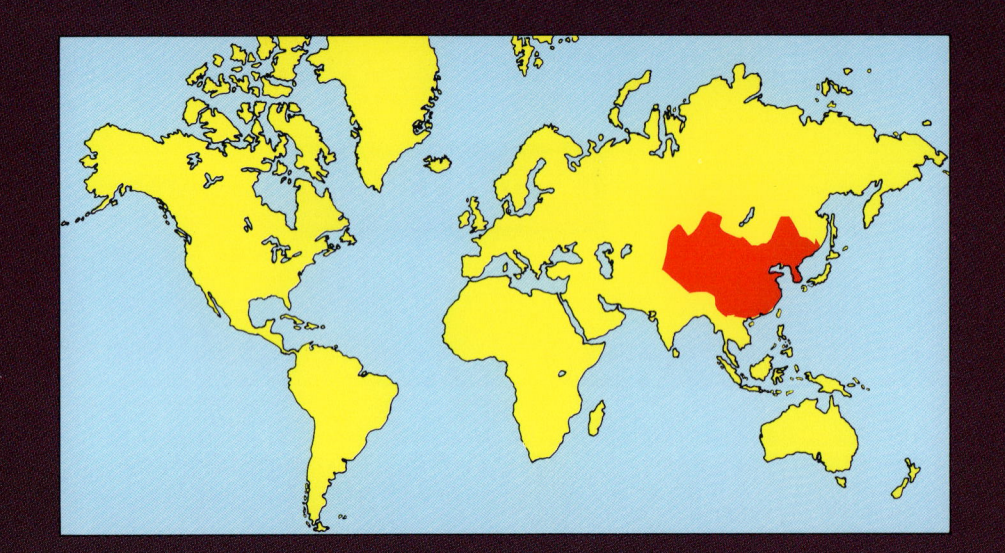

IPE ATLAS

CHINA

CHINA

CAPITAL: Beijing
MONETARY UNIT: Yuan Renminbi
REFINING CAPACITY: 2,200,000 b/cd
PRODUCTION: 2.755 million b/d
RESERVES: 24,000,000 billion bbl

CHINA BEGAN 1991 BY STEPPING UP EFFORTS TO boost exploration and development.

Foreign companies are likely to play an increasingly important role in that effort.

Faced with rising domestic demand for oil and a declining rate of oil production replacement, the country is refocusing its E&D efforts to rebuild its reserves base. In addition, China wants to sustain a strong level of oil exports to ensure a steady flow of hard currency. That calls for an increase in crude production of more than one third this decade from current levels.

Beijing newspaper China Daily reported the government will spend more than 100 billion yuan (more than $20 billion U.S.) for oil and gas E&D during its eighth 5 year plan, 1991-95.

China Oil & Natural Gas Exploration & Development Corp. (Congedc) estimated China would have to prove another 5.6-7 billion bbl in that period to maintain "a rational proportion" between oil reserves and production, China's official Xinhua News Agency reported.

Until 1985, exploration accounted for 45% of total Chinese investment in its petroleum sector, Xinhua said. However, exploration spending has fallen drastically since then.

History

In the 1980s, China's oil policy was keyed to rapid increases in offshore production and maintaining output in its major onshore producing areas. That approach was intended to underpin China's industrial base and provide a major source of hard currency from crude oil exports.

The need for advanced technology to work offshore spurred China's efforts to attract foreign companies as operators in a region many had thought rivaled the North Sea in hydrocarbon potential. However, offshore results have been disappointing, and Chinese petroleum officials are targeting onshore potential, notably the massive Tarim basin in Northwest China. Executives with multinational companies have long viewed the area as one of the world's most prospective untapped basins, often comparing it with Alaska's North Slope.

Tarim basin E&D at first was to be reserved solely for Chinese oil agencies, with only technology and consultation coming from foreign companies. But recent indications from the government are of some headway toward allowing foreign companies to take a more direct role in Tarim basin E&D.

Those indications came about the time the government stepped back from earlier claims it would achieve record production levels in 1990 and boost oil exports sharply in the wake of the Middle East crisis.

Meantime, China is still concentrating on stemming declines in its major producing areas and stepping up E&D offshore. Further, the government late in 1990 signed two agreements covering joint venture E&D with foreign companies, representing the second such deal on Hainan Island and the first on the mainland.

China's dilemma

China's oil production will fall 200,000 b/d short of demand by 2000—assuming production at 4 million b/d by then and no change in export levels—if China takes no steps to reverse the decline, says Ensearch Corp. Vice Pres. Robert E. Ebel.

He estimates China's 1990 crude production at 2.72-2.74 million b/d.

Ebel, writing in Geopolitics of Energy, said the potential for an expanded oil production base in China exists but that potential probably won't be realized without greater investment by non-Chinese companies.

Geopolitics of Energy is a publication of Conant & Associates Ltd., Washington.

Ebel cited domestically subsidized oil prices, increasing production costs, excessive oil exports, and inadequate exploration spending as factors in the worsening outlook for Chinese oil self-sufficiency.

In particular, Ebel noted China's declining reserves base, blaming it on high rates of production during the late 1970s and the absence of recent large oil discoveries.

CHINA

FIRST OIL produced commercially in the South China Sea came in September 1990 from HZ 21-1 platform operated by ACT Operators Group, a venture of units of Agip SpA, Chevron Corp., and Texaco Inc. Photo by Dennis Harding, courtesy of Chevron.

In 1988, China's ratio of reserves-to-production was 17:1, which Ebel deems inadequate with current levels of demand growth and exploration.

China's major fields are mature and their rate of decline is increasing, Ebel wrote. In addition, Chinese production costs increased at the rate of 10%/year during 1980-87 and will continue to rise, he contends.

In recent years, oil production hikes have been based on ever increasing investment levels, and China now finds it difficult to reduce such outlays, Ebel wrote. But exploration has taken a back seat to other priorities, which Ebel contends involve first sustaining welfare benefits of Chinese state agency employees and, second, field development.

In addition, China's controlled domestic crude price, at 21% of the international price, is too low and must be raised if it hopes to recoup costs and generate cash flow for E&D, Ebel wrote.

China also should rethink its export policies to trim crude exports and increase refinery runs for domestic markets, Ebel contends, adding that only crude deemed excess to local needs should be exported.

China needs to reject central planning and adopt a free market economy, along with some political reform, to attract such investment, he wrote.

Crisis effects, production

The Persian Gulf crisis spurred China to boost its crude exports late in 1990 and led some officials to predict China will achieve record production levels.

At the same time, however, implementation of United Nations sanctions against Iraq and Kuwait cost China $2 billion in trade, transport, and civil aviation losses, the official China Features news service reported.

In October 1990, Xinhua quoted Chinese oil industry officials as predicting oil production would top 2.76 million b/d in 1990, surpassing the previous record of 2.75 million b/d in 1989. Production in 1989 was up 13,000 b/d from 1988.

However, China Daily in November 1990 reported, "Experts said the oil industry now stands a very good chance of meeting the year's target of 137 million tons (2.74 million b/d)."

During the first 10 months of 1990, China's crude production totaled 83.7 million bbl, up 0.6% from the same period in 1989, China Daily said.

Chinese crude production slipped 1.3% in October 1990, due mainly to heavy flooding in the Daqing area, the newspaper reported.

Exports surge

The Persian Gulf crisis created an opportunity for China to boost its oil exports, wrote Xu Yihe for China Features.

Xu quoted Jiang Yunlong, vice-president of China National Chemical Import & Export Corp. (Sinochem) as predicting Chinese crude exports would jump to as much as 556,000 b/d in fourth quarter 1990, compared with Sinochem's target of 460,000 b/d for the quarter. Sinochem also predicted refined products exports would average 14,000 b/d in fourth quarter 1990.

Xu reported Sinochem had planned to export an average 480,000 b/d of crude and 100,600 b/d of products in 1990.

In first quarter 1990, Sinochem exported 466,000 b/d of crude, compared with its target of 425,000 b/d. Chinese crude production usually slides in the first quarter, when domestic demand for crude jumps, Xu wrote. International oil prices at the time were a little less than $20/bbl, and exports of Daqing crude to Japan netted about $19.21/bbl.

Second quarter 1990 price declines sliced Daqing crude to $16.09/bbl, spurring Sinochem to cut exports by 40,500 b/d from the 483,000 b/d planned for the period.

After Iraq's Aug. 2 blitz of Kuwait and subsequent loss of both countries' oil exports and spike in prices, the export price for Chinese oil jumped 52% to about $35/bbl and for products 90% to $380/metric ton by November from first quarter 1990 levels.

In third quarter 1990, Sinochem exported 528,000 b/d of crude and 127,000 b/d of products, up 21,000 b/d and 2,600 b/d, respectively, from the quarter's plan, Xu wrote.

Japan buys most of China's oil exports, about 240,000 b/d. Other buyers of Chinese crude and/or products include the U.S., Singapore, Philippines, Thailand, and South Korea.

Sinochem's Jiang told China Features the company plans to increase its oil storage capacity to take greater advantage of future price fluctuations and give it more flexibility.

CHINA

"Expansion of domestic refineries in recent years has been taking an increasing amount of crude, leaving smaller amounts for export," he said.

Refining constraints

China's refineries are pushing capacity to the limit, Ebel wrote.

"There is very little flexibility left. Any substantial increases in future runs will be possible only if capacity is expanded. Western assistance is desirable because of capital shortages."

Refining capacity reached 2.2 million b/d in 1988, having jumped by an average 62,000 b/d/year since 1978, Ebel noted.

As crude exports dropped in 1985-88, however, refinery throughput increased by an average 100,000 b/d/year.

If refinery construction lags, Ebel said, crude exports may be maintained to a degree but would be offset by increased product imports.

The government allocates about 1.726 million b/d of crude to the 37 refineries administered by China Petrochemical Corp. (Sinopec), Xu wrote for China Features. China's total refining capacity is 2.48 million b/d.

Sinochem usually imports about 160,000 b/d of crude to accommodate refinery slates to meet domestic demand for products. But crude imports plunged to 70,000 b/d after the second half runup in oil prices, Xu wrote.

The government encourages China's refineries to process imported crude for reexport as products to the crude suppliers. China processed 90,000 b/d of such crude imports in 1990, up 4,000 b/d from 1989's level, Xu wrote.

Export capability

China is expanding its export capability with new storage and transportation capacity.

An oil transshipment terminal is under construction at Aoshan Island off Zhejiang province, Southeast China. It will be the country's first commercial marine oil terminal, Xinhua reported.

In May 1990, Zhoushan Xingzhong Oil Transfer Corp. Ltd., a joint venture of Zheijian province and Hong Kong interests, agreed to spend $30 million for the terminal.

It will be able to handle oil tankers of 200,000 dwt capacity and have two 630,000 bbl and two 315,000 bbl capacity oil tanks.

When completed, the terminal will be able to store and move oil at the rate of 52,000 b/d. It is expected to be on stream early in 1993.

Plans call for the terminal to receive crude from large foreign tankers. Small Chinese tankers will take crude from the terminal for transport to China's eastern coastal cities or reexport.

Of the 19 major harbors along China's coast, none has the capability for oil transshipment.

Meanwhile, China launched its first domestically built tanker under a foreign contract, Xinhua reported.

The 95,000 dwt Wilomi Eira petroleum products tanker has a shallow draft and a length of 790 ft. Norway's Anders Wilhelmsen ordered the tanker under a contract with Dalian shipyard that calls for two more similar tankers to be built at the shipyard.

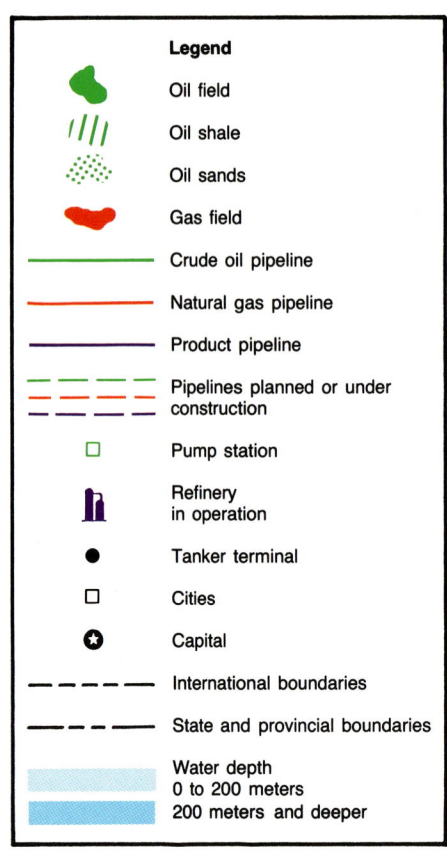

Tarim basin

China's ambitious target of producing 4 million b/d by 2000 hinges on bringing major offshore discoveries or the huge postulated oil resources of the Tarim basin on stream by then.

Generally disappointing results offshore put increasing pressure on the need for stepped up Tarim E&D. The government in 1990 earmarked 1.5 billion yuan ($319 million) for Tarim basin E&D during 1991-92.

Ebel compares the Tarim basin with the Soviet Union's Tengiz field: The potential is there, but developing it will require technology, equipment, and expertise beyond China's current capability.

There are many hurdles to overcome in the area: deep pay, high drilling costs, complex geology, high subsurface pressures and temperatures, a harsh climate, and lack of infrastructure. In addition, moving crude to market from the remote basin in the Xinjiang Uygur autonomous region would require a costly 1,500 mile pipeline.

Earlier indications were that non-Chinese companies would not be permitted to take equity interests in Tarim

CHINA

basin projects. However, Chinese Premier Li Peng indicated otherwise in a meeting with Mitsubishi Pres. Shinroku Morohashi, Xinhua reported.

Morohashi told Li of Mitsubishi's willingness to participate in oil development in the Tarim basin.

"Li responded by saying these items hold great potential for cooperation," Xinhua reported.

Li gave "a very good impression" arrangements could be worked out for joint projects in the Tarim basin, but officials did not discuss details or a timetable.

China spent 3 billion yuan ($636 million) in 1989-90 for exploration in the Tarim basin, where it expects production to jump sevenfold to 20,000 b/d in 1991, China Features said.

Onshore potential

Onshore fields in mature areas still are the mainstay of China's oil production, and China is pressing gains in technology to boost output and discover more reserves in proven basins.

About 75% of China's crude production comes from three onshore producing complexes in Northeast China: Daqing, Shengli, and Liaohe.

CHINA

ACT group fields

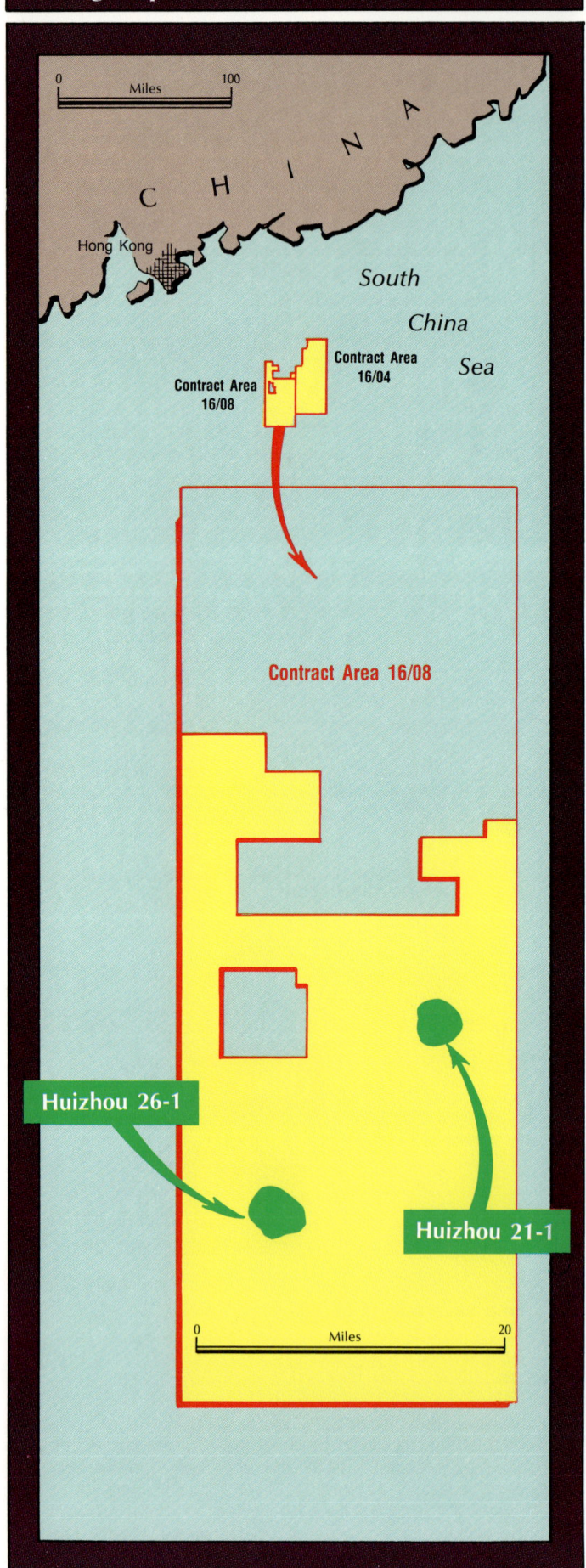

China is mounting a major effort to stave off a decline in Daqing production, which has averaged more than 1 million b/d for 12 straight years.

Chinese oil officials last fall projected Daqing would meet its target of 1.113 million b/d for 1990, Xinhua reported.

They also estimated Liaohe production at 272,000 b/d for 1990, exceeding plan by 2,000 b/d. Flow from Liaohe, a subbasin of the Bohai basin, climbed to as much as 274,480 b/d in October 1990, an increase of 3,650 b/d from the prior month. Liaohe produced 240,000 b/d in 1989.

Elsewhere, Dagang oil field production in October 1990 jumped 2,387 b/d to 77,528 b/d from the previous month.

Northeast China has yielded another major oil and gas discovery, Xu reported.

The first development well was spudded by yearend 1990 in the unnamed field in Kangping county, Liaoning province. The field, about 100 km north of provincial capital Shenyang, has an oil resource estimated at 730 million bbl and gas resource estimated at 706 bcf.

When development is complete in 1993, production is expected to reach 10,000-20,000 b/d of oil and 48-96 MMcfd of gas, Xu reported.

Xinhua also reported full start-up of Eren oil field in the Inner Mongolia autonomous region. Full production of more than 17,000 b/d represented a 257% increase from early production levels in 1989.

In addition, Chinese explorationists have identified a third petroliferous basin in East Xinjiang Uygur, in addition to the Junggar and Tarim basins.

Teams from Congedc's Yumen petroleum administration identified 10 likely hydrocarbon bearing structures in the Turpan-Hami depression. The basin covers 48,000 sq km surrounding Lanzhou-Xinjiang railway.

Shengli campaign

China is targeting an ambitious program to find and develop more oil and gas reserves in the Yellow River delta area of Shandong province, where the Shengli complex is China's second major producing area.

China Daily reported a plan that calls for the Shengli petroleum administration to target reserve additions of at least 730 million bbl during 1991-95, pushing average Shengli oil production to 672,000 b/d during that period. The Shengli administration will spend 4.6 billion yuan ($978 million) to drill 1,200 wells and explore 50 prospects as part of that effort.

Shengli produced an average 668,000 b/d during the previous 5 year plan. The 37,000 sq km area holds 61 oil and gas fields

Of special interest is the coastal plain bordering the Bohai Sea. Estimates of the ultimate potential oil resource is 46.7 billion bbl, China Daily reported.

In addition, a long term plan proposed by officials in nearby Dongying, on the Yellow River delta, calls for construction of "massive infrastructural facilities" there, along with oil treatment facilities and ethylene plants, China Daily reported.

As part of the plan, Dongying Mayor Li Diankui hopes to attain special economic zone status for his city, founded in 1983 to provide for delta area industrial development. "We're lobbying the central government to open the area to foreign investors," China Daily quoted Li as saying.

In another key coastal region, marine carbonate strata recently were found in the Yangtze River Valley, People's Daily said. More than 100 prospective structures have been identified in a 100 sq km area out of a total 1.5 million sq km area from east of Yunnan province to Shanghai at the Yangtze River mouth.

Of special interest are gas prospects in Sichuan province and the area along the lower Yangtze between Nanjing and Nantong.

CHINA

Offshore prospects

China's offshore region is still an area of strong interest.

Beijing Review said China oil agencies plan to accelerate development of nine oil fields discovered in the South China and Bohai seas during 1991-95. Cnooc is aiming for production of 100,000 b/d of crude and 116 MMcfd of gas from offshore fields within 2 years.

That calls for an investment of $1 billion, more than half of which will come from foreign partners, the magazine quoted Cnooc as saying.

Non-Chinese oil companies have spent about $2.8 billion in searching for oil off China to date, but that effort has fallen far short of expectations.

China's offshore oil production in 1989 averaged less than 18,000 b/d. Production from four fields was estimated at 24,000 b/d in 1990—exceeding plan by 20%—and is projected to climb to 60,000 b/d in 1991, 100,000 b/d by 1992 and 150,000 b/d by 1995, based on expected flow from fields under development.

Cnooc also hopes five other offshore oil fields will be producing by 1995. They include Shuizhong 36-1 in Bohai Sea and Liuhua 11-1 in the South China Sea. Combined resource potential of the two heavy oil fields is more than 7 billion bbl.

To utilize heavy crude from the two fields, Cnooc and Royal Dutch/Shell Group plan to build a $2 billion refining/petrochemical complex at Huizhou.

Cnooc subsidiary Nanhai East Oil Corp. since 1983 has been involved in joint exploration ventures with 27 foreign oil companies in the eastern South China Sea, China Features' Xu wrote. They have found 18 hydrocarbon bearing structures in the area.

In the Beibu Gulf, Cnooc and France's Total Cie. Francaise des Petroles have placed a small oil field on stream.

To date, Cnooc has signed 43 agreements with 45 oil companies from 12 countries, resulting in discovery of 36 hydrocarbon bearing structures, Xinhua reported. More than 170 wells have been drilled offshore since 1979.

Cnooc estimates China's offshore potential reserves at 6.2 billion bbl of oil and 4.9 tcf of gas.

Huizhou area

The big jump in production off China by 1992 will stem from the industry's first oil field developments in the South China Sea, in the Huizhou area.

Operators have spent about $1.5 billion to date in exploring for oil and gas in the South China Sea. About 100 wells have been drilled there since 1984.

Cnooc expects about 75,000 b/d in combined peak flow from Huizhou 21-1 and Huizhou 26-1 oil fields in the eastern South China Sea south of Hong Kong. The former started up in 1990, and the latter is to start up in the fall of 1991.

The two fields are being developed by Cnooc in concert with ACT Operators Group, a venture of Agip (Overseas) Ltd., Chevron Overseas Petroleum Ltd., and Texaco Petroleum Mij. Nederland BV.

Chevron said the group will install an eight leg platform in 360 ft of water in May 1991 to develop Huizhou 26-1. It will be linked via subsea pipeline to Huizhou 21's four leg platform in 380 ft of water.

Handling production from the two platforms will be a 250,000 dwt, 1.5 million bbl capacity floating processing/storage/offloading vessel anchored 1 mile from the Huizhou 21-1 platform.

After peaking within 1½ years, production from the two fields will settle at a sustained rate of about 40,000 b/d, Chevron said. The life of both is estimated at 7-10 years.

Huizhou 21-1 has 15 wells, and Huizhou 36-1 is planned for 20 wells, all predrilled through a subsea template. Huizhou 21-1 reserves are estimated at 35 million bbl. Combined cost of developing both fields is about $498 million, Chevron said.

Cnooc has a 51% interest in the Huizhou projects, with ACT splitting the remaining interests equally among partners.

ACT early in 1990 tested oil and was completing operations in the Huizhou 32-2-1 wildcat.

The strike flowed oil from four main zones at a combined rate of 15,000 b/d of 27-36° oil.

Results of tests and other well data were being evaluated to determine the potential of the structure.

Bohai action

In the Bohai Sea, Sino-Japanese operated oil fields produced about 22,000 b/d of crude in 1990, up from about 12,000 b/d in 1989.

Bohai oil production from all fields on stream and under development is expected to reach 50,000 b/d, China Features' Xu reported.

The most recent of those to go on stream is Bozhong 34-2.

Bozhong 34-2, about 180 km west of Tanggu Port near Tianjin, has reserves estimated at 168 million bbl and is expected to produce about 8,000-10,000 b/d, Xu quoted Cnooc Pres. Zhong Yiming as saying.

It is the third field to go on stream in the Bohai Sea area. Chengbei and Bozhong 28-1 started up in 1989.

Foreign oil companies have invested $1.14 billion for exploration in the Bohai Sea since China opened its continental shelf for international bidding in 1979, China Daily reported.

During the past 10 years, Cnooc's Bohai Oil Corp. (BOC) signed six contracts with companies from the U.S., Japan, France, and the U.K.

In all, BOC and partners have identified 77 structures, drilled 283 wells, 47 of which were discoveries.

BOC estimated Bohai reserves at 2.19 billion bbl of oil and 7 tcf of gas, China Daily reported.

BOC and Japan's Japan-China Oil Development Corp. have in 10 years drilled Bohai Sea 38 exploratory and development wells, finding 11 hydrocarbon bearing structures.

Meanwhile, BOC, operating alone, has found an oil field in the Liaodong Bay area of Bohai Sea, Xu reported.

CHINA

New exploration area in China

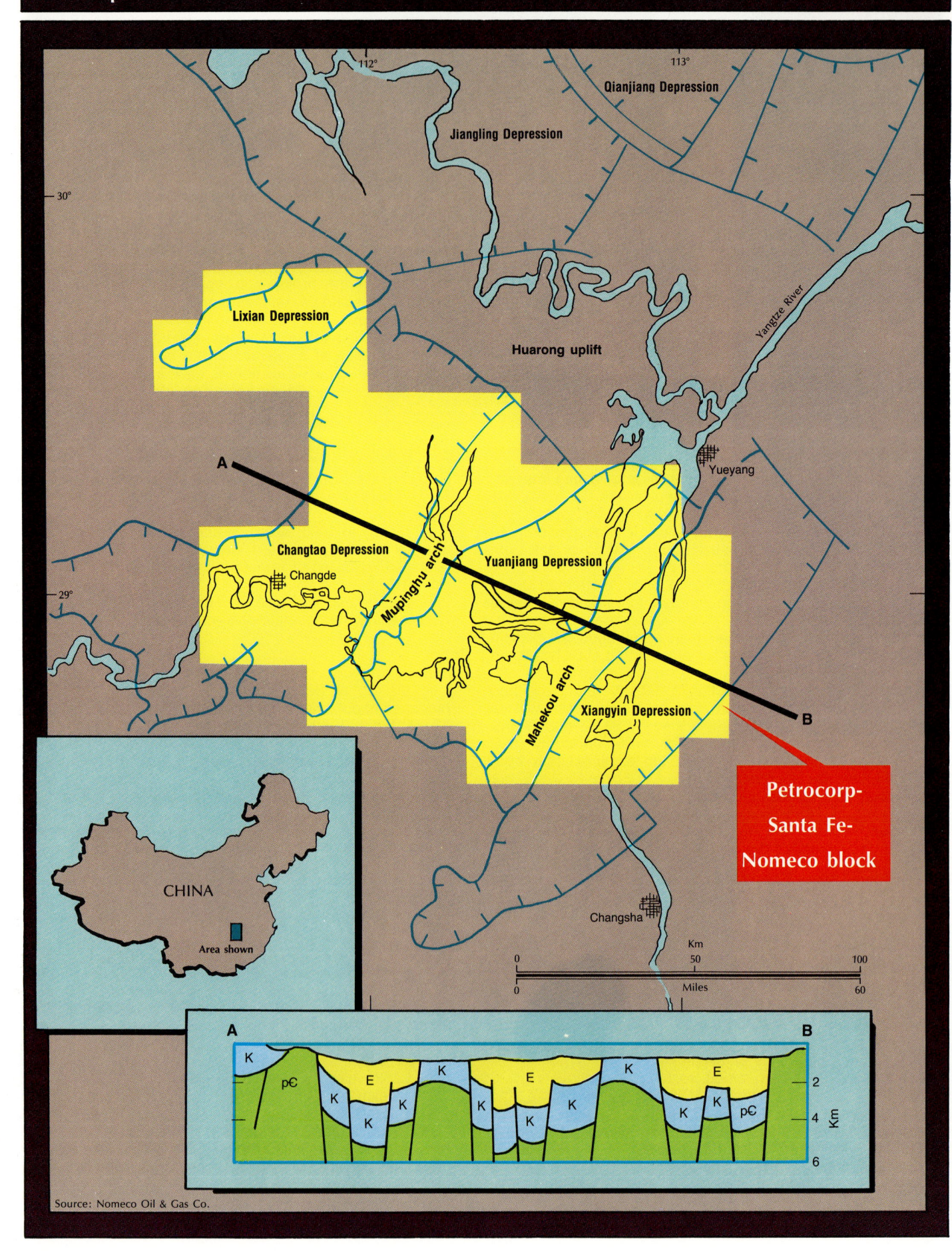

CHINA

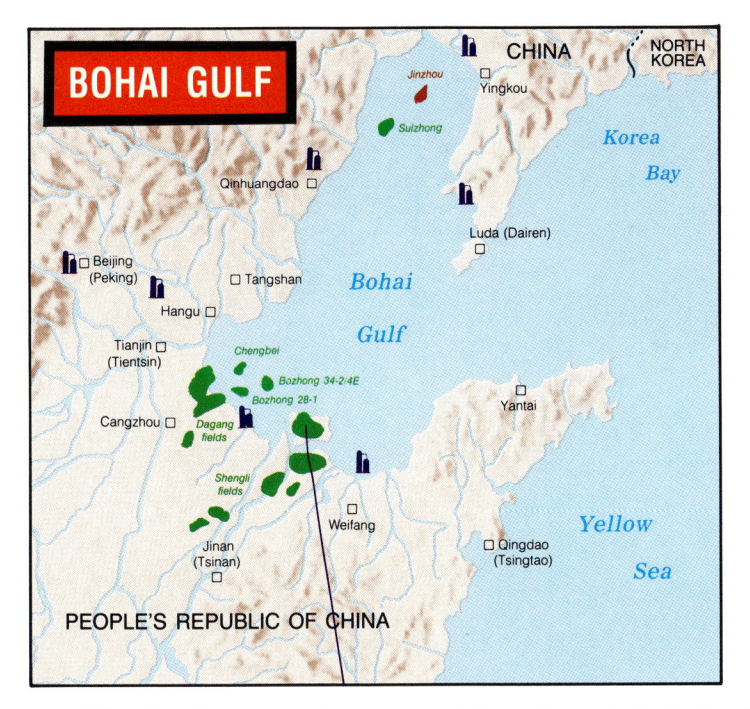

BOHAI GULF

The Jingzhou 21-1 discovery well flowed 734 b/d of oil, 223 b/d of condensate, and 10.5 Mcfd of gas. The structure covers 30 sq km.

BOC is developing Jinzhou 20-2 and Suizhong 36-1, both in Liaodong Bay. Jinzhou 20-2 production is targeted at 20,000 b/d of oil and 48 MMcfd of gas. Suizhong 36-1 reserves are pegged at more than 700 million bbl.

Both are expected to start up in 1992, Xu wrote.

Dongting basin action

A joint venture in which Nomeco China Oil Co. is a member has signed a contract with state owned China National Oil Development Corp. (Cnodc) for exploration, development, and production in China's Dongting basin.

Nomeco China officials believe it to be the first contract issued to non-Chinese for exploration on the Chinese mainland. The area covered by the agreement was the focus of an earlier attempt at mainland China exploration by a U.S. company.

The joint venture is composed of operator Petrocorp Exploration China Ltd. 60%, Santa Fe Energy Resources of China Inc. 25%, and Nomeco China Oil Co. 15%.

Petrocorp is a unit of Fletcher Challenge Ltd., Wellington, N.Z., Santa Fe Energy Resources China is a unit of Santa Fe Energy Resources Inc., and Nomeco China is a unit of Nomeco Oil & Gas Co., in turn a subsidiary of CMS Energy Corp., Jackson, Mich.

Their contract calls for an initial 3 year term, during which the joint venture is required to conduct geophysical surveys and drill two test wells. No further details are disclosed on the combine's obligations.

Cnodc retained the option to participate in development of any oil fields discovered in the contract area for as much as a 51% interest.

The contract area of about 3.9 million acres lies on the southern edge of the East China rift system, about 450 miles west of Shanghai, in Hunan Province.

IEMC Ltd., Golden, Colo., earlier negotiated basic terms of a production sharing contract with Cnodc covering the block. There was no later report of activity on the block.

Hainan Island contract

A small Tulsa independent is the first U.S. company to sign an onshore production sharing contract in China.

The contract Myung & Associates Inc. signed covers joint venture exploration and development on Hainan Island with state owned Hainan Oil Corp.

The contract covers about 20,000 acres encompassing the Jinfeng field area in the Fushan basin of northern Hainan Island, where an Australian group led by CSR Orient Oil Pte. Ltd. found small volumes of oil, gas, and condensate in 1989.

After spending more than $15 million, completing about 1,200 line km of seismic surveys, and drilling six wells, the CSR group decided the finds were not commercial for it at 1989 oil prices and relinquished its contract acreage, essentially all of the 600,000 acre Fushan basin, to HOC.

Myung Pres. John Myung contends, however, that his proposed exploration/development project involving the Jinfeng discoveries could lead to production of about 10,000 b/d of oil, 20-30 MMcfd of gas, and more than 3,500 b/d of condensate. He believes such a program could be commercial if oil prices remain more than $20/bbl.

Further, Myung says, the deal reflects a strong interest Beijing has in attracting more U.S. and other foreign independents to participate in shallow, low risk, modest potential plays in onshore provinces. Previously, the government's emphasis on foreign joint ventures targeted major companies offshore.

Project details

Myung's contract calls for a $2 million minimum commitment to drill two wells offsetting the Jinfeng discoveries to delineate the reservoir and probe for deeper pay and to reprocess the CSR group's seismic data.

Myung planned to spud the first well before the required 6 month time limit.

The wells would target Tertiary pay at 6,500-7,500 ft to assess downdip potential of the Jinfeng strikes, where Myung thinks the oil pay pinches out.

Initial flow from CSR's 1 Jinfeng South discovery well was 1,630 b/d of oil and 1.8 MMcfd from pay at 7,262-70 ft and 1,004 b/d and 1.1 MMcfd 7,286-7,308 ft during an 18 hr drillstem test.

Combined oil flow in the well then dropped to 511 b/d of oil while gas production rose to 3.84 MMcfd after 14 days, Myung said.

The CSR group's 1 Jinfeng flowed 2.5 MMcfd from pay at 7,283-7,333 ft.

Myung estimates the Jinfeng area contains about 8,000 net acres prospective for development, holding potential reserves of more than 30 million bbl of oil and 200 bcf of gas.

Assuming all that acreage covers commercial pay, a 20-50 well development program that has an average production per well of about 300 b/d of liquids and 2 MMcfd of gas, and oil prices remain at more than $20/bbl, payout could occur within a few years, Myung said.

His company has a 51-49 production sharing deal with HOC.

Myung also has an option to acquire more acreage in the Fushan basin.

Improving prospects for marketing Hainan oil production in the near term is a 115,000 b/d export refinery planned at Haikou.

Myung plans to produce Jinfeng liquids and reinject the gas until a gas market develops.

With ARCO's planned development of Yacheng gas field off southern Hainan Island leading to a pipeline across the island to Haikou, Myung has plans to begin gas sales by 1992.

Other prospects for gas sales in the Haikou area are a distribution grid Hainan Province is considering and a joint venture industrial/commercial port development. Jinfeng is 6 km from Haikou.

CHINA

China's oil production, exports

	Crude production Million b/d	Exports Crude 1,000 b/d	Products 1,000 b/d
1975	1.541	198	44
1976	1.743	170	39
1977	1.873	182	39
1978	2.081	226	43
1979	2.123	269	61
1980	2.119	266	84
1981	2.024	277	92
1982	2.042	304	105
1983	2.121	298	108
1984	2.296	446	120
1985	2.498	601	124
1986	2.622	570	109
1987	2.683	545	99
1988	2.740	550	94
1989	2.750	490	90
1990	*2.756	391	—

*From Central Intelligence Agency, International Energy Statistical Review.
Source: Geopolitics of Energy, after other sources

China's 1990 oil and gas production*

Producing area	Oil 1,000 b/d	% of annual plan	Gas MMcfd	% of annual plan
Daqing	1,124	58.4	218	58.9
Shengli	672	57.8	138	56.4
Huabei	109	59.3	24	68.5
Liaohe	273	58.1	171	59.8
Xinjiang	134	56.7	48	—
Dagang	77	54.8	34	56.2
Henan	50	56.7	4	—
Zhongyuan	127	53.0	136	59.7
Jilin	72	62.3	10	—
Changqing	30	61.0	3	—
Jianghan	17	57.7	7	—
Yumen	11	68.2	1	—
Jiangsu	17	64.5	4	—
Qinghai	15	41.8	4	—
Sichuan	—	75.8	636	61.0
Jidong	7	41.6	3	—
Offshore	20	56.7	—	—
Total	**2,755**		**1,441**	

*First 7 months.
Source: China National Petroleum Corp.

Low risk venture

Myung is intent on keeping the Hainan venture as low risk as possible.

China declared Hainan Island, formerly part of mainland Guangdong Province, a special economic zone and granted it provincial status in 1988.

For the commitment wells, Myung can reduce drilling day rates to $6,000-7,000/day with an Oklahoma rig vs. a typical international rate of $10,000-12,000/day. The toolpusher, geologist, and engineer on the job will be Myung staff, the drilling crew Chinese.

Myung planned to ship a truck mounted rig and logging truck from Oklahoma for the two wells and possible future drilling.

Natural gas activity

China is stepping up its efforts to find and utilize more natural gas.

Xinhua reported the Ministry of Geology and Mineral Resources' Institute of Petrogeology increased the estimate of China's potential gas resource to 1.4 quadrillion cu ft.

Previous estimates ranged from 918 tcf to 1.16 quadrillion cu ft.

Those estimates exclude a coalbed methane potential resource of 582 tcf, Xinhua reported.

China produced 1.4 bcfd of gas in 1989.

New potential

Although most Chinese gas production comes from Sichuan province, explorationists are stepping up efforts to find more gas in China's frontier areas.

Garnering increased interest for gas potential is Northwest China's Shaanxi-Ninqxia-Inner Mongolia basin, reported Xinhua.

Gas was first discovered there in 1986, and 43 wells have been drilled in the eastern and central parts of the basin, which covers parts of Inner Mongolia autonomous region and Shaanxi province. Of 24 discovery and extension wells drilled there, half yielded at least 7 MMcfd. The biggest flowed 10.6 MMcfd.

Pay is relatively shallow at 10,500 ft, with the gas of good quality and production relatively uncomplicated, Xinhua reported.

The regional Changqing petroleum administration has formed 10 exploration teams to further delineate the extent of pay there to pave the way for Congedc development.

Sichuan potential

Sichuan province accounts for 43% of China's gas output, producing about 606 MMcfd in 1989, Chen Gengtao reported for China Features.

Runnerup Daqing complex produced about 218 MMcfd that year.

With oil production in China concentrated in the Northeast and North coastal areas, gas provides fuel or feedstock to one third of industrial operations and many residential customers in Sichuan, according to Fan Youzhen, deputy director of Sichuan Oil Administrative Bureau (SOAB). Sichuan is China's most populous province.

SOAB faces the task of ramping up production to meet projected demand in the province during 1991-95, Chen wrote. Early estimates put Sichuan gas demand at 822 MMcfd and provincial production at only 726 MMcfd by 1995.

Sichuan's natural gas resource base is estimated at 282 tcf. However, proven reserves total only 9.9 tcf, reflecting a low level of exploration, Chen wrote.

Modern exploration for gas in Sichuan did not begin until 1949. Since then, SOAB has identified 240 structures, discovered 73 fields, and by yearend 1989 had produced 4.2 tcf of gas.

Sichuan production

SOAB has five production areas that yield a combined 530-600 MMcfd of gas, Chen wrote.

The southern area, centering on Luzhou, was first to go on stream, and is declining at a rate of 7%/year. Current production is 117 MMcfd.

The southwestern area around Zigong also is in decline, producing about 102 MMcfd.

Efforts are under way to boost production in the northwestern area, around Jiangyou. Production there is about 69 MMcfd. SOAB discovered a large gas field in Qionglai county west of provincial capital Chengdu.

The central area, around Nanchong, produces mainly

CHINA

New Hainan Island production sharing contract

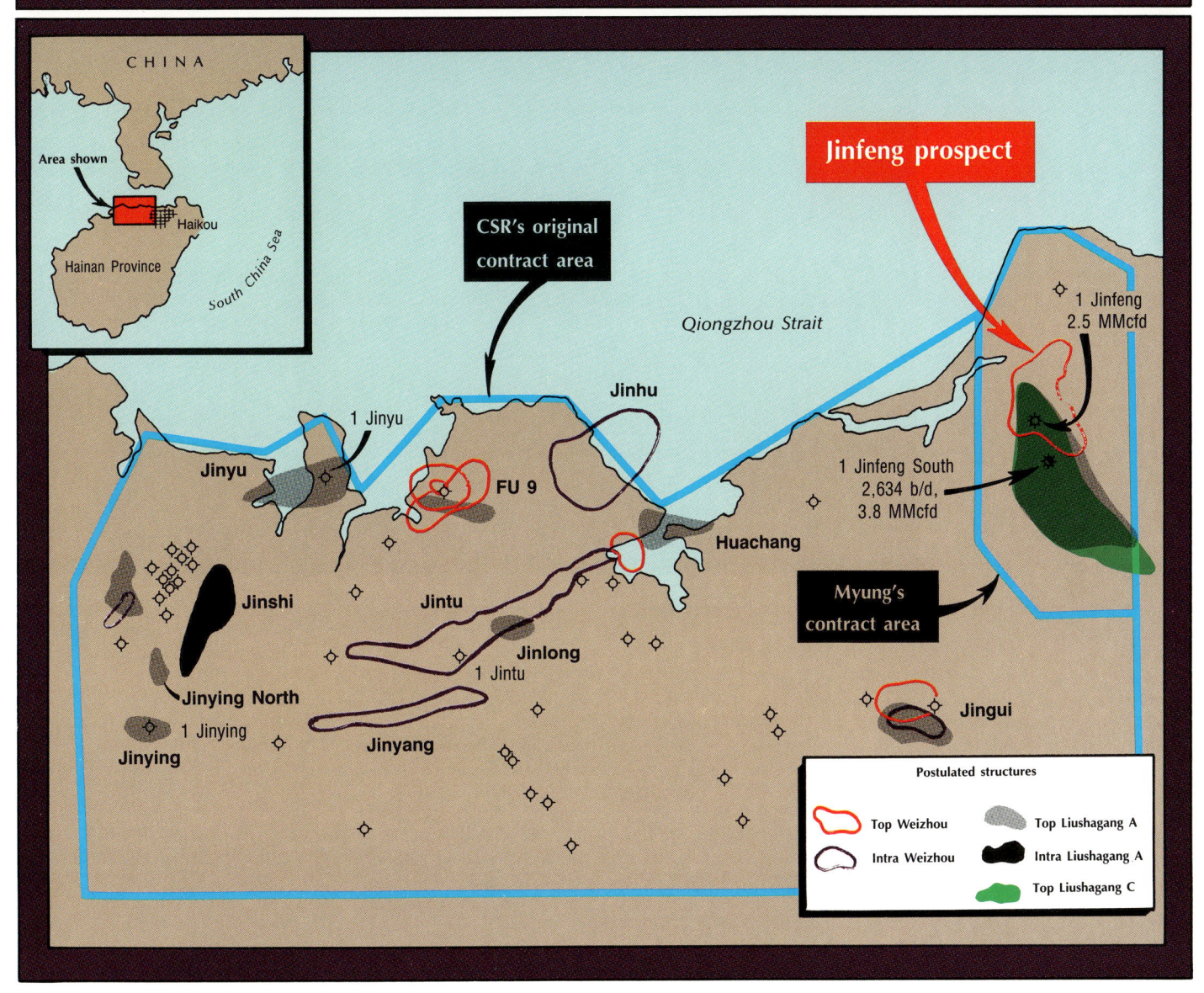

crude—2,000 b/d.

The eastern area accounts for more than half of Sichuan's gas flow. Production is about 318 MMcfd. In 1990, SOAB drilled more than 10 wells there with production of more than 35 MMcfd each. It also discovered two fields in Kaijiang county, with reserves of more than 350 bcf each.

Fan said eastern Sichuan will continue to be SOAB's priority area during 1991-95, potentially offsetting declines in southern and southwestern Sichuan.

As part of that strategy, SOAB built a 328 km gas pipeline from Quxian county in the east to Chengdu in western Sichuan. Completed in July 1990, the line's throughput capacity is 193 MMcfd.

The new line completes a gas grid totaling more than 8,000 km in Sichuan that includes another new line in the north and an existing trunk line that runs from Chengdu to Chongqing, the province's chief industrial center, in the south.

Plans call for increases in gas supplies moved to the south and west from eastern Sichuan.

High costs

SOAB must cope with high costs in producing Sichuan gas. In addition to complex geology, the gas is high in sulfur, Chen wrote.

SOAB estimates about half of all Sichuan gas has sulfur content averaging about 70,000 mg/cu ft.

The bureau has installed six 141 MMcfd desulfurization plants at the main pipeline facilities. They recover about 60,000 metric tons/year of sulfur.

In addition, SOAB imported a small modular desulfurization plant from Canada's Delta Projects Inc. at a cost of $4.17 million, Chen wrote. Installed near Jiangyou, the new unit has a capacity of 14 MMcfd.

Sichuan gas produces mainly from Ordovician carbonates at an average depth of more than 13,000 ft. Drilling costs average 6,560 yuan ($1,263)/ft, and the successs ratio is only about 50%.

SOAB also must grapple with water coning problems. More than 80% of SOAB's producers have problems with water incursion.

Water has blocked full development of giant Weiyuan gas field, where reserves are estimated at 1.4 tcf. More than 50% of the wells there have been shut in, and production has been cut to only 24.7 MMcfd from a targeted 141 MMcfd.

Except for gas reservoirs with serious water coning

CHINA

OIL SEEP, found along a fault near Karamay, in China, suggests the prolific potential of oil in China. (Photo courtesy Esso Exploration.)

problems, average recovery rate in Sichuan fields is 72.5% of proven reserves.

Sichuan gas use

Gas use is growing in Sichuan and neighboring Guizhou and Yunnan provinces.

SOAB estimates 60% of the gas produced in Sichuan is used as feedstock for chemical fertilizers, mainly urea. The province produces about 10.6 million metric tons/year of standard fertilizer, about half of which is exported to other provinces.

Gas feeds four large fertilizer complexes in Southwest China, two in Sichuan and one each in Guizhou and Yunnan. Capacities are 500,000-660,000 tons/year of urea.

Another two big urea plants, each with capacity of 500,000 tons/year, are under construction in eastern Sichuan.

In addition, 76 small and medium sized chemical fertilizers in Sichuan use gas as fuel and feedstock.

Other industrial concerns and residential consumers compete for limited supplies of Sichuan gas, so SOAB has instituted controls on the number of users. About 1 million Sichuan households are hooked up, consuming about 5.6% of the province's gas output.

Ethylene projects

China has approved an ethylene project by Taiwan Plastic Corp. (TPC) in Beijing's biggest deal with a foreign firm, Japan's Kyodo News Agency reported. China has completed building four ethylene plants with combined capacity of 1.2 million metric tons/year, said Xinhua.

Quoting sources in Hong Kong, Kyodo said the deal with TPC is seen as pivotal to expanding ties between China and Taiwan. Chinese authorities and TPC Chairman Wang Yung-Ching signed an agreement covering construction during the next 3 years of a 1.27 million metric ton/year ethylene plant in Fujian province, Kyodo said.

Chinese Premier Deng Xiaoping earlier approved the project in an investment zone in Haicang, Xiamen district, Fujian, aimed exclusively at Taiwanese businessmen.

It is not clear whether the project cost, pegged by Kyodo at $7 billion, includes about $5 billion the Chinese government has spent to improve the industrial infrastructure in and around the investment zone.

Completion by China Petrochemicals Corp. (CPC) of four 300,000 metric ton/year ethylene plants makes China the ninth biggest ethylene producer in the world, Xinhua claimed. When fully on stream, the plants in Shanghai and Heilongjiang, Shandong, and Jiangsu provinces will boost China's ethylene capacity to 1.925 million metric tons/year from 725,000 metric tons/year in 1985. With the added ethylene capacity, CPC will be able to hike its capacity, in metric tons per year, for plastics 1.43 million, rubber 140,000, fibers 680,000, and other derivatives 2.26 million.

Other processing activity

China's biggest foreign funded refinery is under construction in Hainan province.

The $530 million, 115,000 b/d plant at Haikou, funded solely by U.K. company Haikou (Ko Fung) Comprehensive Refinery Co. Ltd., is to be complete in 1992.

Feedstock will be Middle East crudes, with products earmarked for export to Europe. Yangzi Petrochemical Corp. started up an aromatics complex at Nanjing, Jiangsu province, China.

Designed by UOP and using several UOP designed or licensed units, the complex can produce 196,000 metric tons/year of benzene, 450,000 metric tons/year of paraxylene, and about 80,000 metric tons/year of orthoxylene.

It is China's fifth aromatics complex licensed by UOP.

Royal Dutch/Shell Group, in partnership with five Chinese state companies, is seeking permission for a feasibility study for a $2.5-3 billion refining/petrochemical complex at Huizhou, Guangdong Province, China.

The State Planning Authority was expected to take about a year to rule on the application for the 100,000 b/d oil processing unit, specially designed to handle heavy indigenous crude. Shell's partners are China National Offshore Oil Corp., China Merchants, China National Petroleum Corp., China National Petrochemical Corp (Sinopec), and Guangdong Province.

OUR WINDOW ON THE WORLD

From the North Pole to Tierra del Fuego, from America to Asia, from Europe to the Middle East, we are a company constantly looking for new areas to expand in.

Our teams are at work on numerous major prospects worldwide helping to produce, process and market both oil and gas.

Committed to research and innovation, we are creating today the products of tomorrow - in gasolines, fuels and specialty chemicals.

We never cease to advance.

Corporate Communications
Tour Total. CEDEX 47
92069 Paris La Défense. France

TOTAL

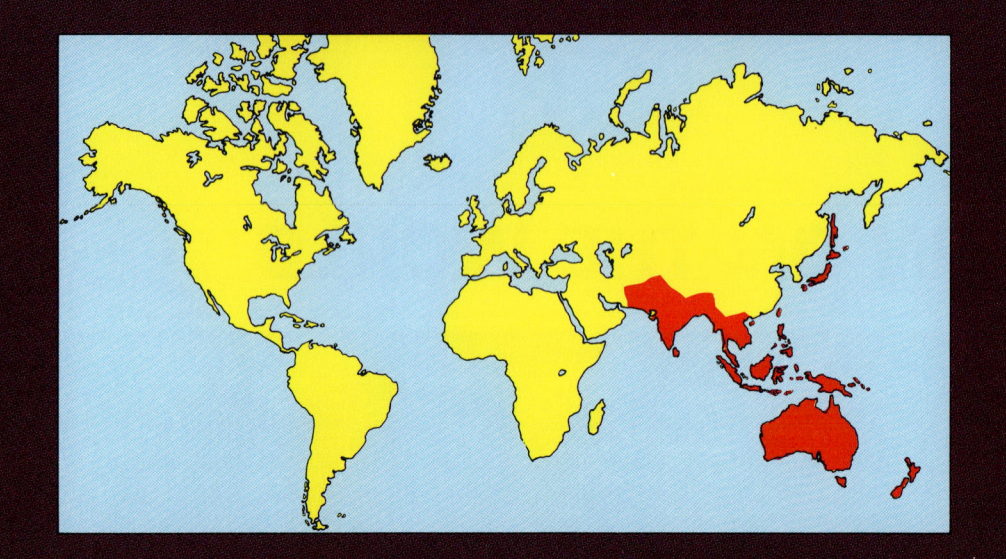

IPE
ATLAS

ASIA—PACIFIC

AUSTRALIA

CAPITAL: Canberra
MONETARY UNIT: Dollars
REFINING CAPACITY: 705,500 b/cd
PRODUCTION: 582 Mb/d
RESERVES: 1,566,163 Mbbl

EXPLORATION IN WESTERN AUSTRALIA IS WELL ON the way to recovery from its mid-1980s slump, says Jeff Carr, the Australian state's Minister for Mines, Fuel, and Energy.

Western Australian exploratory drilling was expected to total 36 wells in 1990 compared with 18 in 1989.

This was to be complemented by an increase in seismic surveys to 50,000 line km in 1990 from 22,000 line km in 1989.

Looking to 1991, Carr said the industry expects to conduct about 80,000 line km of seismic survey in Western Australia, far exceeding the record in 1982.

The likelihood of commerciality for a series of oil discoveries on Australia's Northwest Shelf is growing.

A group led by BHP Petroleum Pty. Ltd. drilled a second apparently successful appraisal to its Griffin discovery 59 miles southwest of Barrow Island oil field.

Reserves are estimated at about 100 million bbl. Combined with preliminary reserve estimates of the nearby Chinook and Scindian discoveries, that would put the total area reserves at 150 million bbl.

Timor Sea
production system

BHP let two contracts valued at a combined £1.5 million to Peter Brotherhood Ltd., Peterborough, U.K., for a steam turbine and two gas compressors for installation on a floating production system (FPS) in the Timor Sea off northern Australia.

BHP uses the converted oil tanker Jabiru Venture as an FPS to produce and store for tanker loading oil produced from Jabiru field.

In the Cooper and Eromanga basins, Vamgas Ltd., subsidiary of Santos Ltd., Adelaide, South Australia, was restoring production from wells which was disrupted by heavy flooding early in 1990.

Western Australia
gas exploration

American Shoreline Inc. and J. Paul Rainey, both of Corpus Christi, planned to begin the first phase of a gas exploration program in Western Australia by yearend 1990.

As equal partners, Amshore and Rainey were awarded exploration permit EP-349 by the state's Department of Mines.

The 400,000 acre permit area is in the Perth basin, about 30 miles northeast of Perth. It contains the Gingin, or Bullsbrook, anticline, which is about 40 miles long and 5-6 miles wide. Rainey said the anticline is "highly gas prone" from 11,000 to below 15,000 ft.

Skua
development plan

A BHP-led approved a $175 million (Australian) development plan for Skua, the third commercial oil field in the Timor Sea off northwestern Australia.

Plans call for the BHP group to install a floating production/storage system, similar to that used at nearby Jabiru field, to develop Skua field on Permit AC/P2.

Expected Skua production rates were not disclosed, but BHP estimates the field's reserves at 30.5 million bbl and the group projects a productive life of 8-9 years. Start-up is expected in December 1991.

Wapet
development plans

West Australian Petroleum Pty. Ltd. (Wapet) plans to develop two small oil fields on Australia's Northwest Shelf simultaneously for $30 million (Australian).

Wapet declared Cowie and Yammaderry oil fields commercial.

The two fields were to go on stream in February 1991 at a combined rate of about 6,500 b/d of 49° gravity oil. Wapet estimates a 4 year production life and pegs each field's reserves at 4 million bbl of oil.

Site is immediately southwest of Wapet's producing Saladin oil field and 20 km north of Onslow, Western Australia.

Asia-Pacific/IPE Atlas

Great Barrier Reef

The final steps leading to putting Australia's Great Barrier Reef off limits to oil exploration began in 1990.

Chevron Corp.'s Chevron Asiatic Ltd. voluntarily relinquished its four permits in the region. The move came in response to a request by Australia's government for companies to follow that course.

The issue was considered nonnegotiable after Minister for Resources Alan Griffiths told oil companies he could see no future for exploration in the marine park surrounding the reef off the central eastern coast of Queensland.

One other company, Petroz Ltd., Perth, is involved in the region.

The company holds a single permit off the tip of Cape York Peninsula.

BANGLADESH

CAPITAL: Dhaka
MONETARY UNIT: Taka
REFINING CAPACITY: 31,200 b/cd
PRODUCTION: 0.9 Mb/d
RESERVES: 500 Mbbl

TITAS GAS LET A $30 MILLION CONTRACT TO SPIE Batignolles unit Spie Capag to lay a 78 mile, 24 in. and 61 mile, 12 in. gas pipeline in Bangladesh from a gas field near Ashunganj to a fertilizer plant under construction near Tarakandi.

ASIA-PACIFIC/IPE ATLAS

Another Spie Batignolles unit, Horizontal Drilling International, will drill five river crossings for the project, which is financed by the Asia Bank for Development.

Work was to have begun by the end of October 1990 and be completed by June 30, 1991.

INDIA

CAPITAL: New Delhi
MONETARY UNIT: Rupees
REFINING CAPACITY: 1,122,360 b/cd
PRODUCTION: 679 Mb/d
RESERVES: 7,997,100 Mbbl

INDIA'S STATE OIL AND GAS COMPANIES ARE stepping up efforts to find and develop more hydrocarbons.

The state companies and petroleum ministry are shifting their strategy on exploration in India for the early 1990s. Impetus for the shift comes mainly from the recognition that India's reserves additions in recent years have lagged production.

India projects its energy demand will soar as a result of population growth and efforts to maintain the pace of industrialization. Thus, the country faces a rising oil import bill while it struggles to keep its economy afloat.

The problems

Trade barriers and political instability have made the country a difficult place for foreign operators.

Efforts to open more exploration acreage to foreign participation previously have stumbled but may move off high center in 1991 as the new government places a higher priority on energy development.

The Indian government postponed to yearend 1990 the fourth bidding round for exploration permits. The new National Front government was reviewing programs of the petroleum ministry and state owned Oil & Natural Gas Commission (ONGC). In addition, Indian oil officials were trying to focus government attention on gas utilization. In the western offshore area, a costly gas development project was partially shut in and gas flared for lack of demand.

This thrust may put efforts at developing a gas pipeline grid in high gear again, thus boosting domestic demand and drilling for gas.

Eighth plan

Indian oil officials continue to formulate a detailed strategy for the country's eighth energy plan (fiscal 1990-95).

The new plan is centered on India's projected patterns of energy demand. India posted a growth rate of 6-7%/year in demand for petroleum products during the seventh plan. Domestic supply met 66% of that demand. That compares with an oil import dependence level of 40% in fiscal 1988-89 and 30% in 1984-85.

Officials predict India's crude oil production will reach 1.02 million b/d by 1995. Indian oil demand, however, is projected to climb to 1.56 million b/d by then.

The government hopes to keep India's tab for crude and products in fiscal 1990-91 at the previous fiscal year's $4.265 billion. The oil import bill for 1990-91 was projected at $5.735 billion, assuming domestic crude production of 700,000-720,000 b/d.

Total demand for the 1990-91 fiscal year was estimated at 1.18 million b/d, a 10% increase from the last fiscal year. To boost Indian oil reserve additions, petroleum officials call for a sweeping reassessment of Indian geological knowledge buttressed by new theoretical approaches, extensive seismic surveys, and further advances in exploration technology such as 3D basin modeling.

To that end, India's oil industry has earmarked about $19.5 billion for exploration in 1990-95, compared with $7.1 billion spent in 1985-90.

ONGC plans, performance

ONGC hopes to boost its crude oil production in the eighth plan to an average 940,000 b/d in 1994-95 from an average 640,000 b/d in 1989-90, despite likely government cuts in its proposed 5 year budget of $18 billion.

Overall, ONGC is targeting a production volume increase of 387 million bbl for the current 5 year planning period vs. the seventh plan.

In exploration, ONGC emphasized improving its wildcat success ratio while trimming costs. ONGC boosted total combined hydrocarbon reserves to 10.15 billion bbl of oil equivalent (BOE) in the 5 year plan against a target of 8.68 billion BOE.

ONGC found 59 oil and/or gas fields during the seventh plan, with the southern basins emerging as a major oil and gas producing province. The state company opened plays in the Meghalaya, Nagaland, and Mizoram areas.

OIL efforts

Oil India Ltd. (OIL) had its best year in fiscal 1989-90, said OIL Chairman Surajit Chaliha.

OIL's profits jumped 17% from the previous fiscal year as revenues shot up to $448 million from $222.5 million. Revenues to the government in taxes and royalties climbed

A sample of E&D action in India

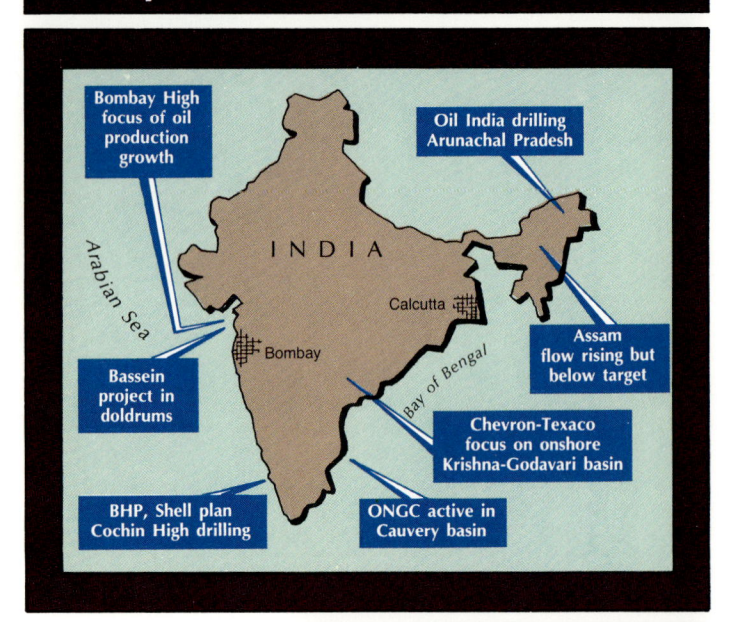

- Bombay High focus of oil production growth
- Oil India drilling Arunachal Pradesh
- Bassein project in doldrums
- Assam flow rising but below target
- Chevron-Texaco focus on onshore Krishna-Godavari basin
- BHP, Shell plan Cochin High drilling
- ONGC active in Cauvery basin

Exploration permit

EP-349 — Perth-Dampier gas pipeline, Indian Ocean, Perth, Australia (Area shown)

ASIA-PACIFIC/IPE ATLAS

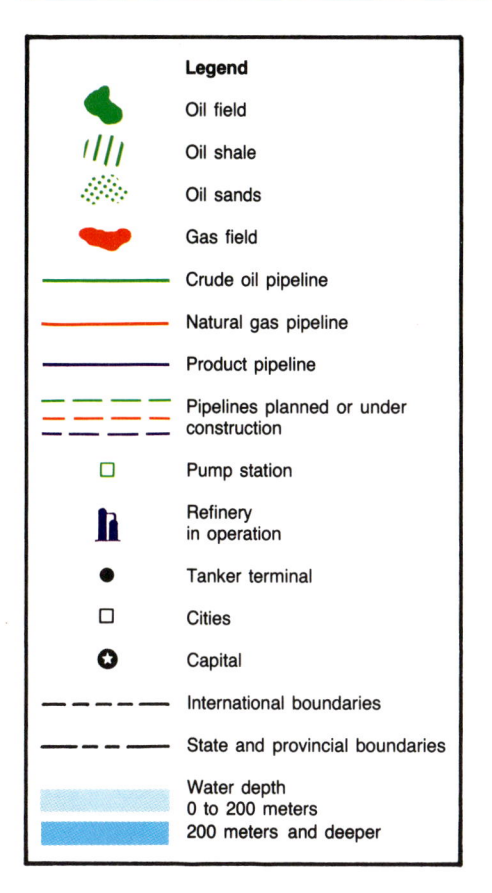

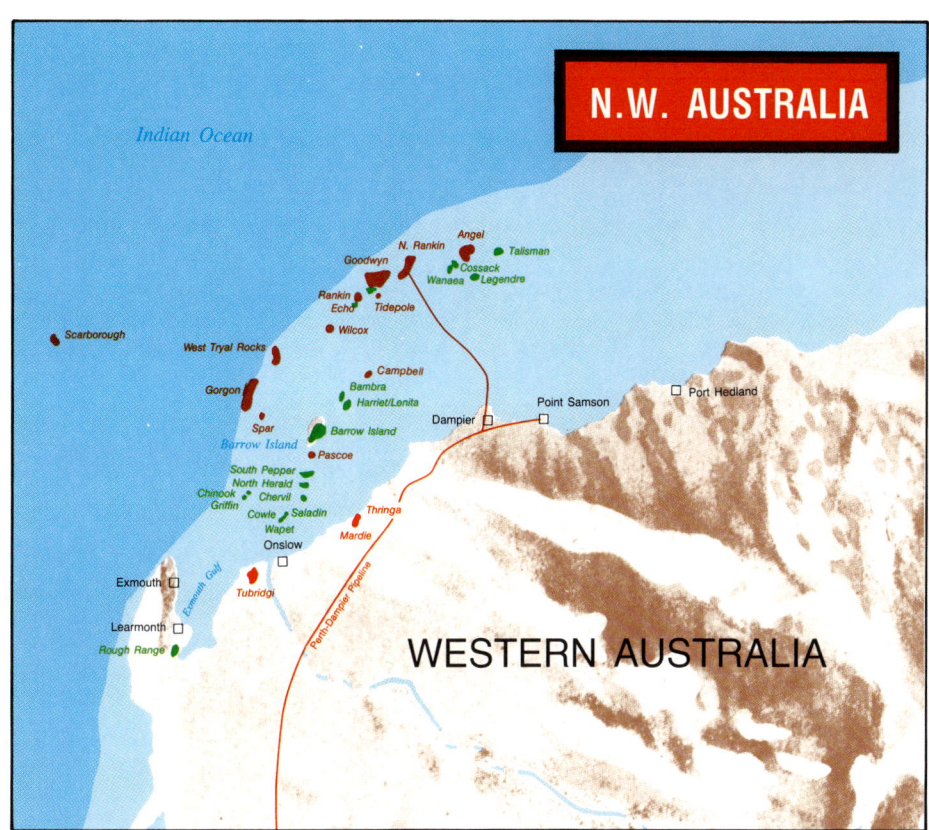

Legend
- Oil field
- Oil shale
- Oil sands
- Gas field
- Crude oil pipeline
- Natural gas pipeline
- Product pipeline
- Pipelines planned or under construction
- Pump station
- Refinery in operation
- Tanker terminal
- Cities
- Capital
- International boundaries
- State and provincial boundaries
- Water depth 0 to 200 meters
- 200 meters and deeper

N.W. AUSTRALIA

to $26 million in 1989-90 from $17 million in the previous period.

OIL's crude production in 1989-90 rose to 54,000 b/d from 48,000 b/d the previous fiscal year, although total volume for the 5 year plan fell short of the targeted 109.5 million bbl at 94.1 million bbl. Declining oil production in mature fields such as Nahorkatiya and Moran and civil strife in Assam contributed to a general decline in production for several years during the seventh plan, Chaliha said.

OIL has ventured outside its base operations in Assam to explore in Rajasthan, Saurashtra, and Arunachal Pradesh.

OIL has pulled out of exploration along the northeast coast of India and off Mahanadi and Anadaman because of a lack of success.

Gas utilization push

India's new minister for petroleum and chemicals, M.S. Gurupadaswamy, asked Gas Authority of India Ltd. (GAIL) to study the feasibility of moving gas from western offshore fields to states in the south for power generation.

Proponents of a national gas grid point out that laying the 1,700 km Hazira-Bijaipur-Jagdishpur (HBJ) pipeline from South Bassein field off the western coast has spawned industrial development such as fertilizer and power production in areas of Gujarat, Madhya Pradesh, and Uttar Pradesh.

India has only 2,571 km of gas pipelines, including the HBJ line.

ONGC proposed a national gas grid in 1984. S.P. Wahi, former chairman of ONGC, in July 1989 put India's gas reserves at 16.95 tcf and then current production at 1.16 bcfd. India consumes only about 70% of its gas productive capacity, however, Wahi said.

Bassein woes

Although Bassein gas has spurred industrial development in western India, slow progress in that regard has forced ONGC to shut in big volumes in the field for lack of demand.

As a result, the ambitious $400 million offshore gas development project is suffering economically. About 10% of the investment in the project is by foreign partners.

ONGC is producing about 424 MMcfd at Bassein, formerly South Bassein, against a capacity of about 706 MMcfd.

ONGC wants to slice Bassein flow by a further 71-106 MMcfd following an increase in associated gas production at Bombay High.

The Bassein-Hezira pipeline extends to Bombay High, allowing the area's increasing associated gas production to supplant the shutin flow at Bassein in the pipeline.

India's government had set a gas production target for ONGC of 636-706 MMcfd by the end of the seventh plan.

Cochin High action

Drilling by ONGC and foreign operators in the Cochin High area is expected to increase in the next 3 years.

ONGC estimates Cochin High's potential oil resource at more than 1 billion bbl in the Cochin-Mangalore region.

Drilling plans in the Cochin High area call for three wells by ONGC and one well each by BHP and Shell International Petroleum Co. Ltd.

Other E&D programs

Among other action, ONGC has drawn up a development plan for Beghraji field in North Gujarat, which it considers the best discovery in the state after the find at Gandhar.

Plans call for drilling an additional 140 wells to push production to 7,500 b/d by 1992-93.

Beghraji produces about 400 b/d of oil from six wells.

In the Cauvery basin's Narimanam field in the Thanjavur district of Tamil Nadu, ONGC has installed a sour crude stripping plant, India's first such facility onshore.

ONGC also has proposed a $150 million project to develop the Tatipaka-Pasarlapudi structure in the onshore Krishna-Godavari basin. Production is estimated to climb to about 88 MMcfd, with much of it destined for feedstock for

77

ASIA-PACIFIC/IPE ATLAS

SOUTHERN ASIA

Nagarjuna Fertilisers & Chemicals Ltd.

Elsewhere, ONGC is stepping up the pace of wildcat drilling in the onshore and offshore Krishna-Godavari basin area.

ONGC plans to increase the number of rigs working in the Krishna-Godavari basin to 11 onshore and four offshore by 1995 from the current eight onshore rigs and two offshore.

ONGC also has identified seven wildcat sites in new areas in the Vindhyan, South Rewa, Satpura, and Pranhita-Godavari basins.

Refining efforts

India is pressing efforts to involve foreign companies in upgrading its refining industry.

A government board chose British Petroleum plc and

ASIA-PACIFIC/IPE ATLAS

Caltex Petroleum Corp. to provide technical assistance to India's refineries for expansion and upgrading plans. More such involvement by foreign companies in India's downstream industries is likely in the 1990s as the country seeks to cut its imports of petroleum products.

India is placing greater emphasis on increased domestic production of certain refined products that are in sharply rising demand.

India's imports of petroleum products for fiscal 1990-91 were forecast to jump to 240,000 b/d from 142,800 b/d in fiscal 1989-90. That compares with crude oil imports projected roughly flat at 390,800 b/d in fiscal 1990-91 vs. 392,000 b/d in fiscal 1989-90. Government estimates place India's total demand for petroleum products at 2 million b/d by the end of the 1990s.

India's first hydrocracker was being installed at the Koyali refinery, which also was being debottlenecked to boost capacity to 190,000 b/d from 146,000 b/d, the report said. In addition, the refinery at Mathura was being debottlenecked to hike capacity to 150,000 b/d from 120,000 b/d.

INDONESIA

CAPITAL: Jakarta
MONETARY UNIT: Rupiahs
REFINING CAPACITY: 813,600 b/cd
PRODUCTION: 1,274 Mb/d
RESERVES: 11,050,000 Mbbl

A SUBSTANTIAL INCREASE IN EXPLORATORY drilling in Indonesia during the next 3 years will result from the signing of new production sharing contracts.

Far East Oil Report, published by County Natwest Woodmac, Edinburgh, predicts the number of wildcats and appraisal wells in Indonesia to increase from 120 in 1990 to a peak of about 190 in 1993.

Exploratory drilling fell from 217 wells in 1985 to 82 in 1987 because of the collapse in crude oil prices and a lack of new production sharing contracts.

In 1979-82 state owned Pertamina awarded an average of 11 contracts/year. In 1983-86 the average was only three per year. The number of new contracts has since increased, with 10 awarded in 1988, 19 in 1989, and 20 were expected in 1990. After the peak of 1993, the report says, the drilling rate will decline quite rapidly unless the current contract impetus can be maintained. And it adds that exploration levels will ultimately depend on the success of wildcat drilling, particularly in Indonesian frontier areas where most of the virgin acreage is still to be licensed.

Indonesian oil production, about 1.25 million b/d, will rise during 1990-91 but decline through the rest of the decade. Production from future discoveries will only help stabilize production levels into the mid-1990s.

Production will fall to less than 1.3 million b/d in 1998, then decline quite sharply. Barring future exploration successes, Indonesia is set to become a net importer of oil in 1998, the report said. Gas production will continue at its current high level of 4.179 bcfd until 1997. But production will then decline unless new LNG contracts are signed or current contracts are rolled over.

Camar field production

Enterprise Oil plc, London, planned to start oil flow from Camar field off Indonesia by midyear 1991.

Camar, in the Java Sea 65 miles northeast of Surabaya, East Java, holds reserves of 12 million bbl and was expected to produce about 20,000 b/d.

Production from the unitized field will flow through a satellite wellhead and a process platform to a central production platform through a 6 mile, 6 in. pipeline. From there, crude will move about 1 mile to a floating storage and accommodation tanker with about 800,000 bbl of storage.

Intan, Widuri production hikes

Maxus Energy Corp., Dallas, was ahead of schedule with plans to increase oil production to 200,000 b/d from Intan and Widuri fields in the Southeast Sumatra production sharing contract area.

Following installation of a permanent processing facility, Maxus began producing a gross 80,000 b/d of oil from Widuri Platform A.

Intan Platforms A and B already were producing a gross 45,000 b/d. Maxus planned to start production from Widuri Platforms B, C, and D, lifting Intan and Widuri production to 200,000 b/d.

MALAYSIA-INDONESIA

4
5
6
SABAH
Champion
Magpie
Fairley
Ampa
Baram
S.W. Ampa
Tali
Seria
10
W. Lutong
BRUNEI
SARAWAK
A
B
C
D
LIPPINES
Philippine
Sea
Mindanao I.
Davao
Republic of Palau
Celebes
Sea
Pacific
Ocean
Manado
Soasiu
Halmahera I.
Molucca
Sea
Sorong
Salawati I.
1-X TBN
Kasim
TBC-1X
TBA-2X
Jaya
Walio
Sele
Klamono
VOGELKOP
Winiagar
Wasian
Mogo
Manokwari
Steenkool
Ceram Sea
Teluk Bintuni
Teluk
Irian
Ceram I.
Bula
Fakfak
Buru I.
Ambon
Banda Sea
Hollandia
IRIAN JAYA
Wewak
PAPUA
NEW GUINEA
Juha
Hides
Agogo
Iagifu
S.E. Hedinia
Jehi
Kuru
Barikewa
Bwata
Omati
Lake Murray
Puri
Uramu
Pasca
Gulf of Papua
Port Moresby
Lae
New Britain I.
Aru I.
ESIA
Sea
Tanimbar I.
Dili
Ossulari
Aliambata
Matai
TIMOR
Kupang
Sunrise and Troubadour
Arafura
Sea
Jabiru
Challis
Cassini
Evan Shoa.
Puffin
Skua
Darwin
AUSTRALIA

ASIA-PACIFIC/IPE ATLAS

JAVA

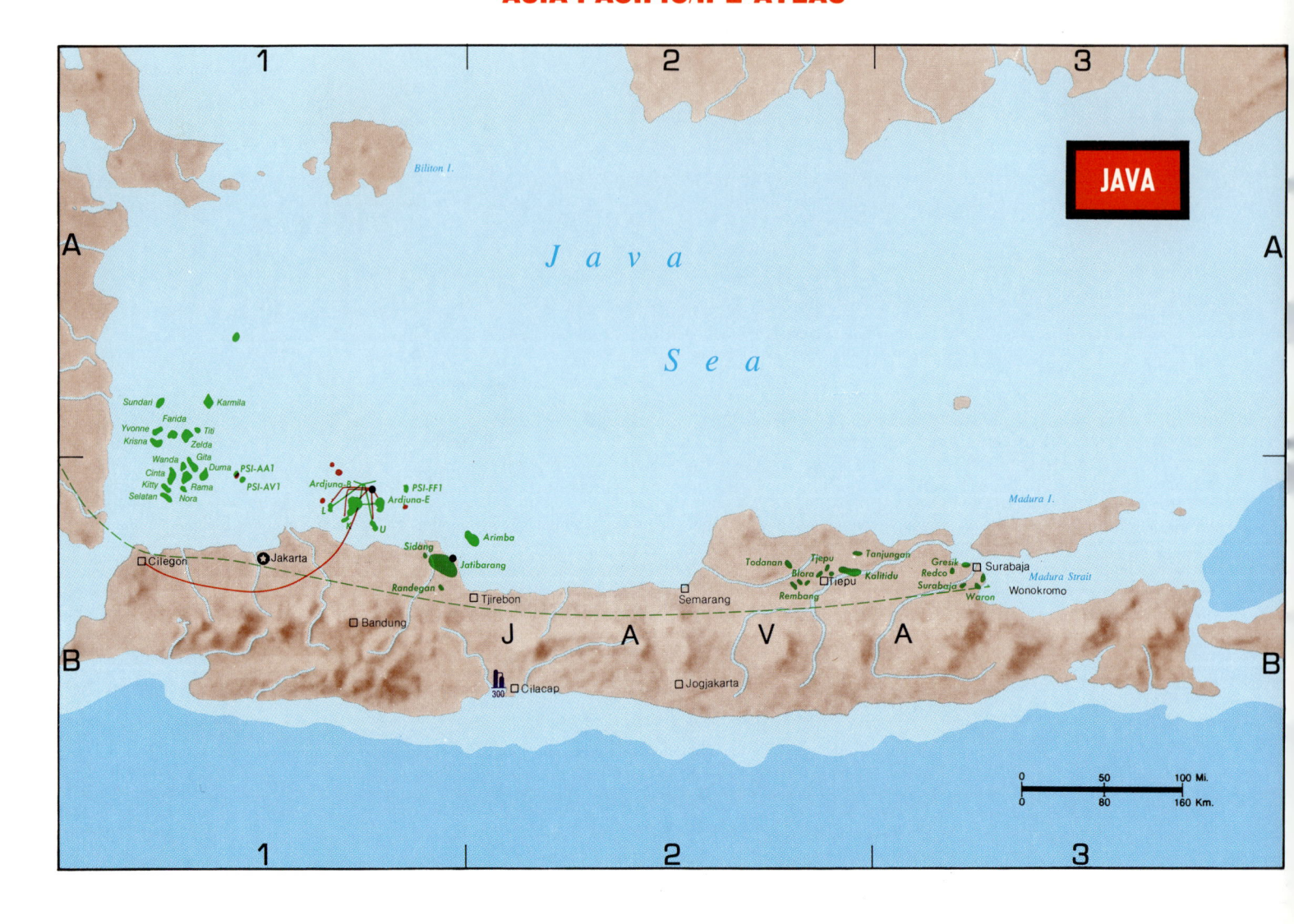

Processing activity

Pertamina signed an advanced payment agreement with Java Petroleum Investment Co. Ltd. (Japic) to finance construction of the $1.8 billion EXOR-1 refinery at Balongan on West Java's northeast coast.

The contract will allow release of funds to build the 125,000 b/d refinery, about 250 km east of Jakarta, which its backers claim will have the world's largest residue catalytic cracking (RCC) complex of its kind.

Scheduled for 1994 start-up, EXOR-1 is being developed by an Anglo-Japanese engineering and construction group made up of Foster Wheeler Corp., Clinton, N.J., the BP Group of the U.K., and JGC Corp. and Mitsui Engineering & Shipbuilding Co. Ltd., both of Japan.

Privately owned Chandra Asri let contract for construction of Indonesia's first olefin plant at Serang, West Java.

JAPAN

CAPITAL: Tokyo
MONETARY UNIT: Yen
REFINING CAPACITY: 4,383,400 b/cd
PRODUCTION: 10.5 Mb/d
RESERVES: 63,019 Mbbl

A HOST OF PROCESSING ACTIVITY TOOK PLACE in Japan in 1990.

Repsol and Japan's Idemitsu Petrochemical Co. and Sumitomo Corp. were negotiating to form a joint venture to produce and sell polystyrene in Spain. The new company, to be majority owned by Repsol, would build a 50,000 ton/year plant using Idemitsu technology.

Start-up is targeted for mid-1992. Chevron Research & Technology Co. licensed its technology for manufacturing styrene acrylonitrile, crystal polystyrene, and high impact polystyrene to Daicel Chemical Industries Ltd. of Japan. Daicel plans to use the technology in its styrenic polymers and fabricated products businesses. Shell Sekiyu KK commissioned a 16,000 b/sd reformate treater at its Kawasaki, Japan, refinery using the Fiber-Film Aquafining corrosion prevention process licensed by Merichem Co., Houston.

Idemitsu Petrochemical Co. Ltd. licensed the ProCam software system from ChemShare, Houston, to model operations at ethylene plants in Chiba and Tokuyama, Japan. ChemShare was expected to install the system at Chiba and Idemitsu at Tokuyama, both by first quarter 1991.

Koa Oil Co. of Japan placed the first order for the SURE sulfur recovery process developed jointly by Ralph M. Parsons Co., Pasadena, Calif., and BOC Group of England.

Koa will install a double combustion version of the process at its 80,000 b/d Osaka refinery and upgrade the Claus sulfur recovery unit, increasing sulfur recovery to 90 metric tons/day from 60 metric tons/day.

Showa Denko KK will build a 60,000 metric ton/year Unipol process polyethylene plant at Oita, Kyushu, Japan.

ASIA-PACIFIC/IPE ATLAS

The grassroots plant, to be completed in third quarter 1991, will produce Showa's first slate of linear low density polyethylene as well as high density polyethylene (HDPE).

KOREA, SOUTH

CAPITAL: Seoul
MONETARY UNIT: Won
REFINING CAPACITY: N.A.
PRODUCTION: N.A
RESERVES: N.A.

STEADY, ROBUST ECONOMIC growth and a rapidly expanding automobile fleet during the next 20 years will boost Korea's oil demand to 1.9 million b/d by 2010 from 787,000 b/d in 1989.

Total primary energy demand in Korea will increase to 188 million tons of oil equivalent by 2010 from 82 million tons in 1989. Oil will have about a 50% energy market share at the end of 20 years, as it does now.

Those estimates were outlined by Bok-Jae Lee, director, Oil and Gas Policy Division, Korea Energy Economics Institute, in a paper delivered at the sixth Asia-Pacific Petroleum Conference in the fall of 1990 in Singapore.

The estimates are from an interim report, Korean Energy Strategies Toward the 21st Century, published in July 1990 by Korea Energy Economics Institute.

The Korean economy is assumed to grow at 6.7%/year in the 1990s and 4.5%/year during the first 10 years of the next century. The manufacturing sector will account for a steadily increasing share of gross national product, rising to 41.4% in 2010 from 34.6% in 1989, Lee predicted.

Another key factor in consumption is an expected 13%/year increase in the car fleet during the 1990s, although it will slow to 3% during the next decade. Transportation's share of energy demand will grow to 27% in 2010 from 19% in 1989.

Residential and commercial demand will account for a declining share of market, to 22% in 2010 from 30% in 1989.

Domestic oil demand gained 14.7% to 787,000 b/cd in 1989 despite slow economic growth.

Gasoline gained more than 34% to 50,000 b/cd, and other products, such as diesel, propane, and jet fuel, had healthy increases.

To help meet increasing product demand, the private sector is scheduled to expand its refining capacity to 1.235 million b/d by 1992 from the current 840,000 b/d.

Cracking capacity will soar to 154,000 by 1993 from the present 34,000 b/d.

Refinery utilization rate during the year of 1989 was more than 95%.

Consumption of unleaded gasoline accounted for 48% of total consumption in 1989.

And low sulfur diesel fuel—0.4% or less—made up 80% of consumption.

The private sector, with financial support from the government, has a plan to build 120,000 b/d of desulfurization capacity by 1992.

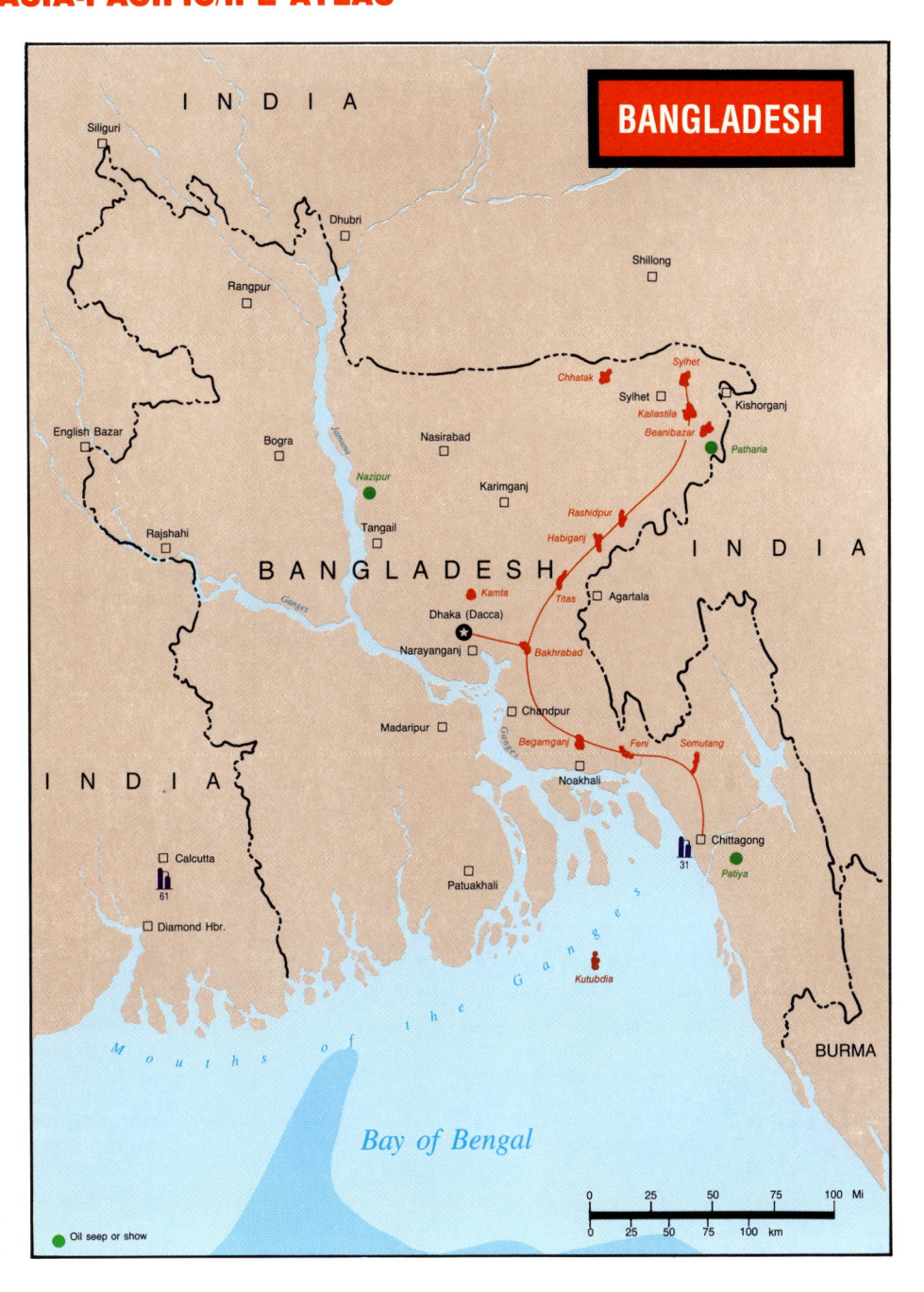

LAOS

CAPITAL: Vientiane
MONETARY UNIT: New kips
REFINING CAPACITY: N.A.
PRODUCTION: N.A.
RESERVES: N.A.

THE LAOTIAN GOVERNMENT AND LAOS HUNT Oil CO. signed a production sharing contract covering 6.4 million acres in the southern part of Laos.

The agreement provides for an initial 2 year exploration term with options for four more 2 year terms.

Laos Hunt, a subsidiary of Hunt Oil Co., Dallas, will conduct geological and geophysical studies in the first 2 years of operations and acquire at least 500 line km of seismic data during the second 2 years.

The company will drill an unspecified number of wildcats to maintain the contract beyond the fourth year.

ASIA-PACIFIC/IPE ATLAS

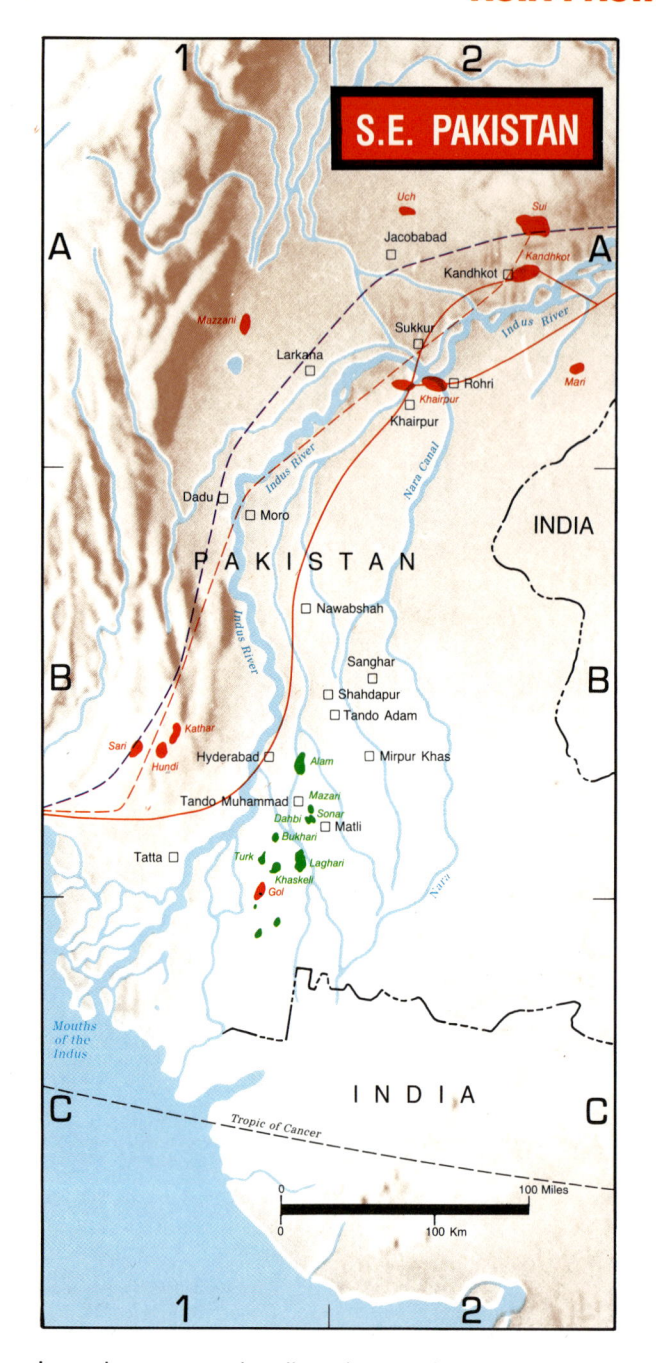

S.E. PAKISTAN

Laos has no crude oil and natural gas legislation but began negotiating contracts broadly based on its foreign investment laws.

In 1989 Laos awarded a 70-30 partnership of Enterprise Oil plc, London, and Cie. Francaise des Petroles, Paris, a production sharing contract on about 5 million acres, a northwest offset to Hunt's contract area.

MALAYSIA

CAPITAL: Kuala Lumpur
MONETARY UNIT: Ringgit
REFINING CAPACITY: 209,500 b/cd
PRODUCTION: 605 Mb/d
RESERVES: 2,900,000 Mbbl

AN EXXON CORP. UNIT AND MALAYSIA'S STATE OIL company agreed to gas supply terms in the latest stage of that country's $4 billion peninsular natural gas utilization project.

Esso Production Malaysia Inc. and Petroliam Nasional Bhd. (Petronas) extended a 1989 agreement in principle covering gas development and sales terms for the second phase of Malaysia's Peninsular Gas Utilization (PGU-II).

Under terms of the deal, Esso will be the major gas supplier to the project. Esso has supplied gas to the Malay Peninsula since 1984 under PGU-I, currently limited to customers on the peninsula's eastern coast.

Under PGU-II, Esso will supply gas from Malaysia's offshore east coast fields to electrical power plants and industries throughout the Malay peninsula and to Singapore.

Petronas is laying a 450 mile onshore gas transmission grid on the peninsula that will handle gas demand expected to reach 1 bcfd by 2000. To meet that demand growth, Esso is developing Jerneh gas field, at 3 tcf the biggest gas field Esso has found in Malaysia. Jerneh, which lies in about 195 ft of water about 60 miles offshore in the South China Sea, is scheduled to start up in April 1992.

Initial productive capacity at Jerneh will be 450 MMcfd, to be hiked to 750 MMcfd as demand grows during the 1990s.

Texaco exploration program

Texaco Exploration Penyu Inc. plans an exploration program that could involve outlays of $25.25 million in a 5 year campaign in the South China Sea off Malaysia.

Site will be on Block PM 14, acquired in a production sharing contract with Malaysia's state owned Petronas and Petronas subsidiary Petronas Carigali Sdn. Bhd.

Block PM 14 covers 6,711 sq miles in 32.8-229.7 ft of water. It is bounded on the north by Block PM 12, on the east by the Malaysia-Indonesia boundary, and on the south by the Singapore platform.

Shell's spending

Shell Malaysia plans to step up its spending program to $5 billion Malaysian ($1.85 billion U.S.) in the upstream sector during the next 5 years. That compares with capital spending of $8 billion ($2.96 billion) during the past 10 years.

About $750 million ($277.8 million) will be spent during the 5 year period for exploration and appraisal on the SB1 contract area in Sabah and the SK5 area in Sarawak.

Shell said if wildcatting is successful, about 40 wildcats and appraisal wells will be drilled in 1990-92 at a cost of more than $500 million ($185 million). About $2 billion ($740.7 million) has been allocated for the Baram Delta gas gathering project and on boosting recovery from older fields.

Production outlook

Malaysia's crude oil and condensate production will peak at 670,000 b/d in 1993 from its 1990 level of 590,000 b/d, predicts County NatWest Woodmac, Edinburgh, Scotland.

Production will then decline to about 600,000 b/d and, depending on how the government's depletion policy is implemented, could remain at that level until 2000.

Exploratory drilling will remain at about 30 wells/year during the next 5 years, says Woodmac's Far East Report.

Malaysian gas production will rise from 1.26 bcfd to a little less than 3 bcfd by the turn of the century.

MONGOLIA

CAPITAL: Ulaanbaatar
MONETARY UNIT: Tugriks
REFINING CAPACITY: N.A.
PRODUCTION: N.A.
RESERVES: N.A.

MONGOLIAN PETROLEUM CO. (MGT), FORMERLY THE Ministry for Energy, Mining, and Geology, disseminated

ASIA-PACIFIC/IPE ATLAS

information in 1990 about Mongolia's oil and gas potential in a bid to attract foreign operators.

Exploration Associates International of Texas Inc., Houston, prepared a report on the country's potential, mainly by translating reports prepared by geoscientists from the Soviet Union, Hungary, Romania, Poland, and the former East Germany. The Mongolian government plans to award exploration areas to western companies in 1991 or 1992.

Geological setting

Soviet and Mongolian geoscientists defined 13 major sedimentary basins in Mongolia. They divided the basinal systems into 59 subbasins.

Outside the basins, the surface is metamorphic, granitic, or volcanic rock.

Mongolia contains Cenozoic and Mesozoic rift basins developed in accretionary terrane on the southern margin of the Siberian craton.

Successive tectonostratigraphic terrane accreted to the Laurasian, Siberian craton during the Proto-Tethys, Paleo-Tethys, and Neo-Tethys.

In northern Mongolia the terrane involves Caledonian accretion of Devonian and younger sediments. In southern Mongolia, the terrane is made up of the Hercynian accretion of Permian and younger sediments.

Extensive, postaccretionary rifting developed during extensional periods of opening and closing of Paleo-Tethys and Neo-Tethys.

Rift and graben deformation was initiated in the Upper Jurassic.

Rift fill included conglomerates and breccias, as well as volcanics.

During the Lower Cretaceous, continental deposits with lacustrine clastics continued as the rift fill sequence. The Upper Cretaceous continental sediments are related to the wrench rifting, compressive phase.

The Upper Jurassic-Lower Cretaceous lacustrine shale-sandstone sequence contains the known hydrocarbon source rocks and reservoir rocks in the southeastern and eastern Mongolian basins.

Oil occurrences

Oil occurrences in Mongolia's basins and subbasins are surface oil seeps, impregnated outcrops, and oil shows in cores and samples.

Two oil fields, Zuunbayan and Tsagaan-Els, have been discovered in the Zuunbayan subbasin of the East Govi basin.

Zuunbayan field, about 286 miles southeast of Ulaanbaatar, was discovered in late 1941 by core hole drilling of surface oil seeps.

Tsagaan-Els field, about 12 miles southwest of Zuunbayan field, was discovered in 1953 by exploratory drilling on an anticline defined by seismic data.

Zuunbayan field was developed, and oil was produced until the on site refinery was destroyed by fire in December 1969.

Tsagaan-Els field was not developed because of lack of facilities to lift and transport the high paraffin content oil.

Zuunbayan field, a faulted anticline with about 3 sq miles of closure, had a total of 200 wells that produced a combined 1,040 b/d during 1955, 840 b/d in 1957, and 710 b/d in 1959.

Cumulative production is reported to be about 3.85 million bbl.

Exploration history

About 200 exploratory wells have been drilled in Mongolia, about 80% of them in the East Govi basin.

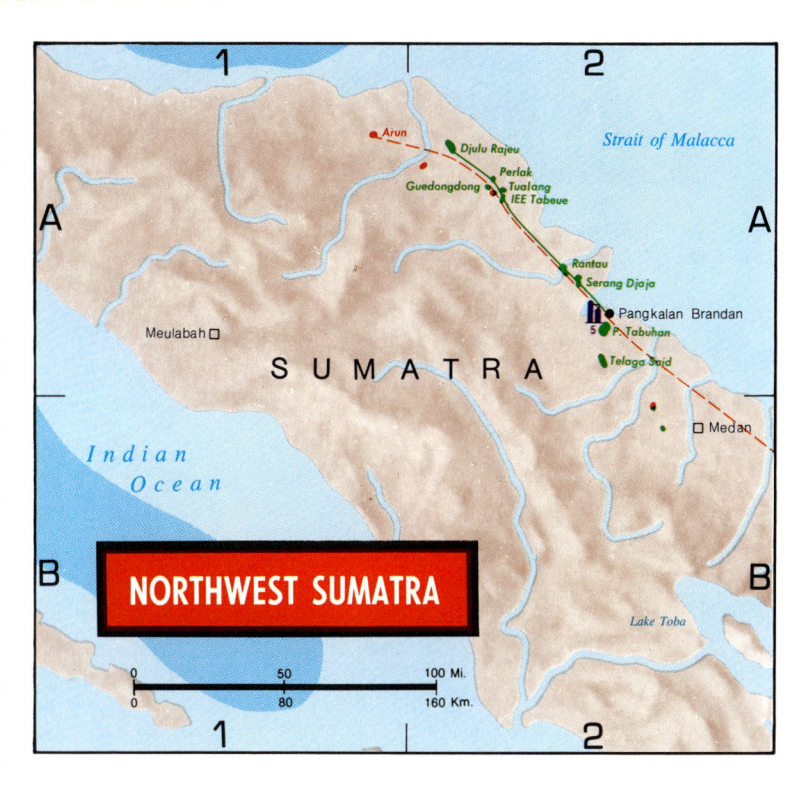

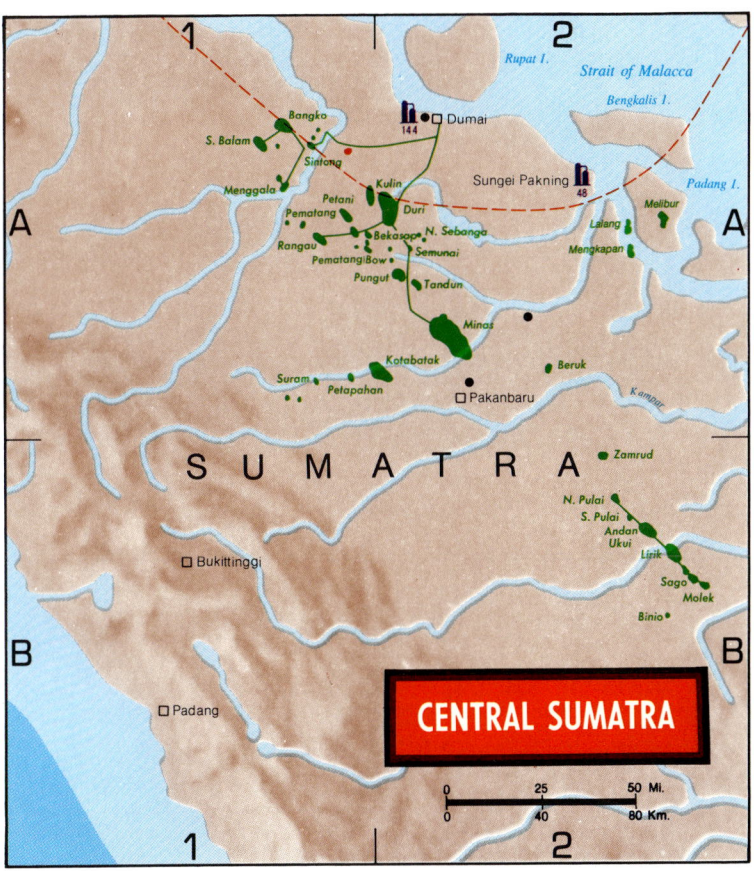

Several thousand shallow core holes related to oil and gas exploration have been been drilled in Mongolia, mostly in the East Govi basin.

More than 60% of Mongolia has been covered by surface geological mapping, at various scales, supplemented by geochemical and geophysical surveys.

There are estimates that more than 1,240 line miles of analog single fold seismic data were acquired by Soviet and Hungarian geophysical expeditions in the 1940s to 1960s, mainly in eastern Mongolian basins.

JAPAN

Location	No. Refineries	Location	No. Refineries
Sakaide	1	Mizushima	2
Yokohoma	1	Okinawa	3
Chiba	4	Toyama	1
Sakai	2	Chita	2
Yokkaichi	2	Funakawa	1
Sodegaura	1	Niigata	1
Hyogo	1	Muroran	1
Kashima	1	Negishi	1
Marifu	1	Yamaguchi	2
Osaka	1	Ehime	1
Oita	1	Wakayama	1
Tomakomai	1	Owase	1
Kawasaki	5	Sendai	1

ASIA-PACIFIC/IPE ATLAS

Recent activity

During late 1989 and early 1990, the Mongolian Geophysical Expedition acquired about 150 line miles of 1,200% digital common depth point reflection seismic data in the Nyalga basin.

In early 1990, the new Ministry of Heavy Industry, formerly the Ministry of Energy, Mining, and Geology, formed Mongolian Petroleum Co. (Mongol Gazryn Tos or MGT).

The company was formed to administer all petroleum related matters for the ministry and the government.

Subsequently, MGT staff members have been conducting geological exploration surveys in the field.

The geophysical expedition staff also has been conducting field exploration.

MGT is continuing a major effort to formulate a petroleum law with supplemental regulations.

It was the intent of the government to have the petroleum law and regulations submitted for debate and adoption by the Great People's Khural (congress) before yearend 1990.

In September 1990, during the reorganization of the government, the Ministry of Heavy Industry was dissolved. MGT, which was responsible to that ministry, now is an independent organization responsible to the government.

MYANMAR

CAPITAL: Rangoon
MONETARY UNIT: Kyats
REFINING CAPACITY: 32,000 b/cd
PRODUCTION: 13 Mb/d
RESERVES: 51,000 Mbbl

Myanmar issued licenses for all tracts in its first onshore licensing round, received bids for second round tracts, and was expected to invite bids for improved production in existing fields.

Foreign oil companies last operated onshore in Myanmar, formerly Burma, in 1962.

The licenses are expected to result in a combined $363 million work commitment during the first 3 years. Work commitments are $12-70 million/block for the 3 year exploration period.

A unit of Idemitsu of Japan signed a contract to study the feasibility of developing Martaban gas fields, with reserves of 4.1 tcf.

Prospects fertile

Myanmar's production has dropped to less than 15,000 b/d of oil from twice that volume in 1982.

The state oil company's production averaged 14,713 b/d of oil and 103.5 MMcfd of gas in 1989.

The industry response to the licensing offers results from good geological potential, relatively favorable fiscal terms, and encouragement toward industry from the Ministry of Energy, energy planning department, and the state company.

The Myanmar tertiary geosyncline covers about 140,000 sq miles, of which 111,000 sq miles could be considered to have hydrocarbon potential.

Many fields were discovered based on surface features, several contain five to 40 pay sands, and average producing depth is about 4,000 ft.

The potential for large discoveries is probably greatest in the Central basin (Salin basin) and southern Chindwin basin, Johnston said.

Almost all production in Myanmar is from Oligocene and lower Miocene sandstones.

Crudes are often high in paraffin content with pour points of 80-100° F.

Myanmar's main fields, production

Field	Year discovered	Cumulative Oil Million bbl	Cumulative Gas bcf	Present Oil b/d	Present Gas MMcfd
Mann	1960	19	—	6,800	6
Htaukshabin	1960	88	—	4,900	7
Chauk/Lanywa	1902	128	10	700	12
Yenangyaung	1800	192	—	2,500	2
Ayadaw	1910	1	10	—	10
Pyalo	1972	—	—	—	2
Prome	1965	7	—	300	5
Shwepyitha	1967	—	—	—	10
Myanaung	1964	21	—	—	—
Total				15,200	54

Source: Daniel Johnston & Co.

NEW ZEALAND

CAPITAL: Wellington
MONETARY UNIT: Dollars
REFINING CAPACITY: 95,100 b/cd
PRODUCTION: 39 Mb/d
RESERVES: 209,000 Mbbl

PETROCORP EXPLORATION (WAIHAPA) LTD. AND partners at yearend 1990 were proceeding with second stage development of Waihapa gas/condensate field in New Zealand despite continuing legal and political problems over oil and gas issues in that country. Delivery of first gas from Waihapa, in the onshore Taranaki basin, to a power plant at Stratford, N.Z., in late September 1990 marked the end of 2½ years of flaring during long term production tests.

First stage development, costing $8.4 million, entailed laying 4 mile, 7⅝ mile parallel oil and gas pipelines to Stratford, installing permanent wellhead facilities and flow lines for five producing wells, and linking the wells to a manifold at the Waihapa B site where the gas is collected for market. Waihapa joint venture partners, led by Petrocorp, are selling Waihapa gas to Electricorp for the Stratford power plant under a short term contract of 10.4 MMcfd.

Second stage development plans, costing $30 million, call for constructing permanent production facilities and extending the oil pipeline to the Omata tank farm at New Plymouth, N.Z. Condensate from long term production testing has been trucked to the tank farm. The new production center is to go on stream late in 1991 with initial capacity of 7,500 b/d, expandable to 15,000 b/d if needed for production from other fields in the Ngaere license contiguous to Waihapa or adjoining petroleum licenses such as Tariki and Ahuroa.

The Waihapa project has been stymied by lengthy licensing procedures as well as a legal dispute over ownership of the contiguous Ngaere license. Gas flaring, which began in February 1988 under an initial term petroleum mining license (PML), persisted while Petrocorp sought a market amid uncertainties over gas deregulation.

Flaring was reduced to minimal levels after the Waihapa joint venture received a specified term PML in July 1990. The joint venture had negotiated the gas sales contract with Electricorp in 1989 but was forced to await the permit before sales could begin. Production testing during the year ended June 30, 1990, averaged 5,871 b/d of liquids with a gas:oil ratio of 1,000:1. Waihapa reserves are estimated at 19 bcf of gas and 12.5 million bbl of liquids.

ASIA-PACIFIC/IPE ATLAS

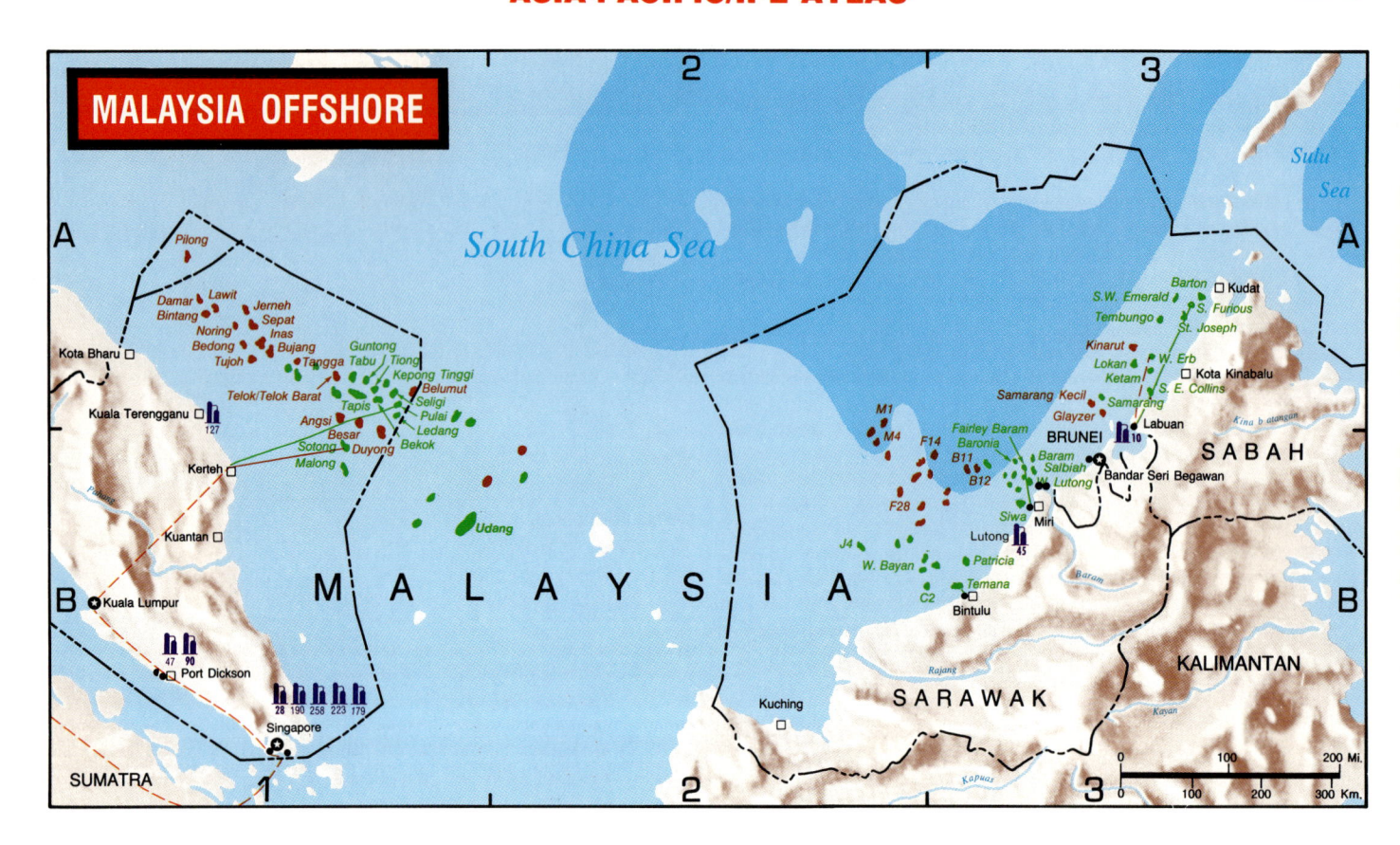

MALAYSIA OFFSHORE

License dispute

There has been no drilling or seismic work by operator Petrocorp on the Ngaere license for 2½ years because of a dispute over the license's status. David Butcher, minister of energy in 1989, granted title to the Ngaere license to himself, representing the New Zealand government.

His actions were upheld by the New Zealand High Court in October 1989. The New Zealand Court of Appeals reversed the high court's judgment, invalidating the minister's action and validating the Waihapa joint venture's original title to the license, thus allowing production to proceed in 1990.

However, Butcher announced he would appeal the appellate court ruling to the Privy Council in London just before New Zealand's general election. The new minister of energy, John Luxton, took office after his National Party scored a landslide victory in the Oct. 27 election.

"The crown is the major beneficiary in the Waihapa joint venture, where it is the largest participant, holding a 38.36% interest, and this interest would carry through if the extension of the Waihapa license over the Ngaere area were granted," said John Holdsworth, chairman of Southern Petroleum NL, a 22.6% partner in the Waihapa license.

Holdsworth urged the government to reconsider its appeal to the Privy Council and allow the joint venture to proceed immediately with appraisal and development of the Ngaere area.

An agreement with Pakistan's state owned Oil & Gas Development Corp. and the Pakistani president covers Block 17, northwest of Indus basin oil and gas fields in southern Pakistan and 75-125 miles northeast of Karachi.

Additionally, a four company group acquired Blocks 22, 25, and 26 covering 14,200 sq km. OMV (Pakistan) Exploration GmbH will operate Blocks 22 and 26 and Pakistan Petroleum Ltd. (PPL) Block 25. The group will conduct 3,500 line km of seismic surveys and drill three wells on the blocks in the Indus basin, west and southwest of Sui gas field.

British Gas plc also signed an exploration agreement covering the 4,500 sq km Block 34 in Central Pakistan. In the first 3 year phase of the license British Gas will conduct field geological studies and seismic surveys and drill one well. Premier Consolidated Oilfields plc reported the 2 Qadirpur well flowed 21 MMcfd of gas, confirming a major gas reservoir immediately south of Kandkhot field in Pakistan's North Indus basin. The well is north of the 1 Qadirpur discovery well, which flowed 19 MMcfd in early 1990. Premier, which has a 9.5% interest in the well in partnership with state owned Oil & Gas Development Corp., Pakistan Petroleum Ltd., and Burmah Oil plc, said early production is feasible through Kandkhot gathering and transmission facilities. Iran and Pakistan plan a joint venture to build an 80,000 b/d, $400 million refinery at Port Qasim near Karachi.

Pakistan plans to add a $14 million oil terminal at Karachi by 1992 to boost its oil handling capacity there.

PAKISTAN

CAPITAL: Karachi
MONETARY UNIT: Rupees
REFINING CAPACITY: 120,975 b/cd
PRODUCTION: 60 Mb/d
RESERVES: 162,087 Mbbl

TEXACO EXPLORATION PAKISTAN INC. ACQUIRED a 926.4 sq mile concession in Sind province, Pakistan.

PAPUA NEW GUINEA

CAPITAL: Port Moresby
MONETARY UNIT: Kina
REFINING CAPACITY: N.A.
PRODUCTION: N.A.
RESERVES: 200,000 Mbbl

THE GOVERNMENT OF PAPUA NEW GUINEA GAVE final approval to BP Petroleum Development Ltd. to begin its

ASIA-PACIFIC/IPE ATLAS

Hides gas development project in the Southern Highlands.

Hides, the first commercial hydrocarbons development in Papua New Guinea, will deliver as much as 15 MMcfd from two wells to a 60,000 kw power plant that will provide electricity during the 19 year life of the new Porgera gold mine.

BP will lay a 5.1 mile, 4 in. pipeline from the two Hides wells down the steep sides of Mount Tumbudu to a gas processing plant in the Hanimu-Tagari Valley.

First gas is scheduled for delivery to the new power plant late in 1991.

Chevron development plans

Chevron Niugini Pty. Ltd. group's Kutubu oil development and export project won approval. It will mean the start-up of the nation's first crude exports in mid-1992.

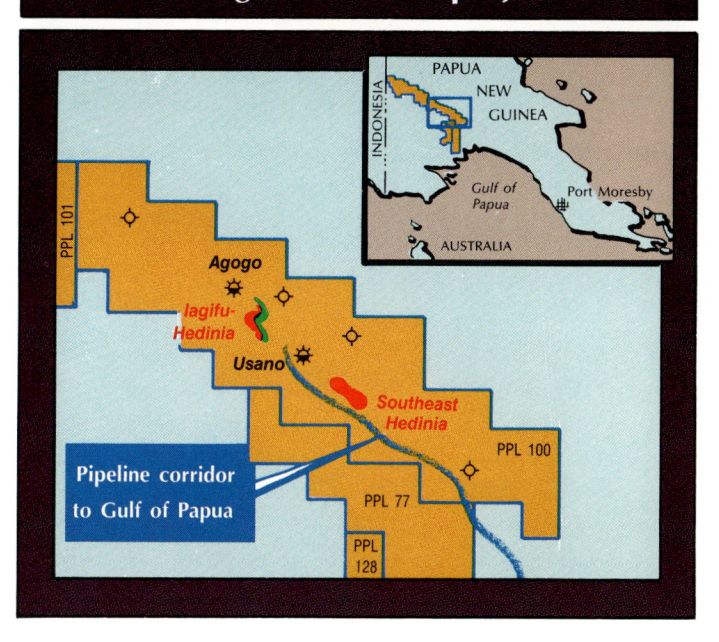

Plans call for production to peak at 128,000 b/d. The project's total cost is pegged at about $950 million, of which $730 million will have been spent before first oil is produced.

The Kutubu area in the south central highlands provided Papua New Guinea its first commercial hydrocarbon discoveries.

Chevron and partners signed licenses PDL2 and PLL2 covering petroleum development and pipeline construction.

Papua New Guinea's government has exercised its right to take a 22.5% carried interest in the project, initially based on production from Iagifu-Hedinia, Usano, and Agogo oil and gas fields in the central highlands.

Kutubu oil will be gathered from the three fields and moved to a processing plant near Lake Kutubu. From there, the crude oil will move via a 270 km pipeline from the highlands to a loading buoy in the Gulf of Papua for delivery to export tankers. Project operator Chevron estimates proven reserves at a minimum of 200 million bbl of light, sweet crude.

BP discovery

BP's 1 Elevala, drilled to 3,216 m on License PPL 81, flowed 11.9 MMcfd of gas and 634 b/d of 54° gravity condensate through a ½ in. choke from perforations at 3,060-72 m.

The strike lies 50 km east of Kiunga and 750 km northwest of Port Moresby. It is the first discovery in the foreland region of the Papuan basin and the first well drilled in the region in 9 years.

PHILIPPINES

CAPITAL: Manila
MONETARY UNIT: Pesos
REFINING CAPACITY: 279,300 b/cd
PRODUCTION: 5 Mb/d
RESERVES: 38,688 Mbbl

ALCORN (PRODUCTION) PHILIPPINES INC. found a major oil field in the South China Sea off Philippines.

Alcorn's 1 West Linapacan wildcat recorded substantial flows of 30.1° oil with no water on three drillstem tests of Miocene-Oligocene Nido-Galoc platform carbonate rock at 5,350-5,896 ft.

Biggest flow was 6,708 b/d through a 2 in. choke on a test that covered a dolomite interval about 40 ft thick. The well also flowed at a stabilized rate of 5,051 b/d a through a $^{48}/_{64}$ in. choke with 552 psi flowing wellhead pressure and 342:1 gas:oil ratio.

Site is in 1,150 ft of water, 250 miles southwest of Manila and northwest of Palawan Island, on Alcorn's Block C held under Service Contract 14.

Preliminary estimates place recoverable reserves in excess of 100 million bbl and oil in place at 1 billion bbl.

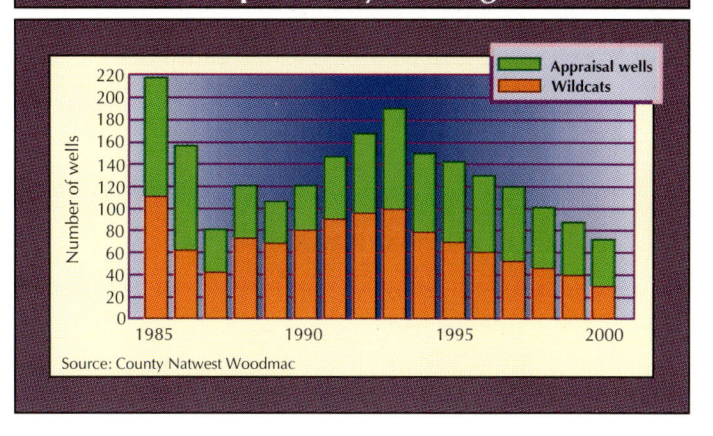

ASIA-PACIFIC/IPE ATLAS

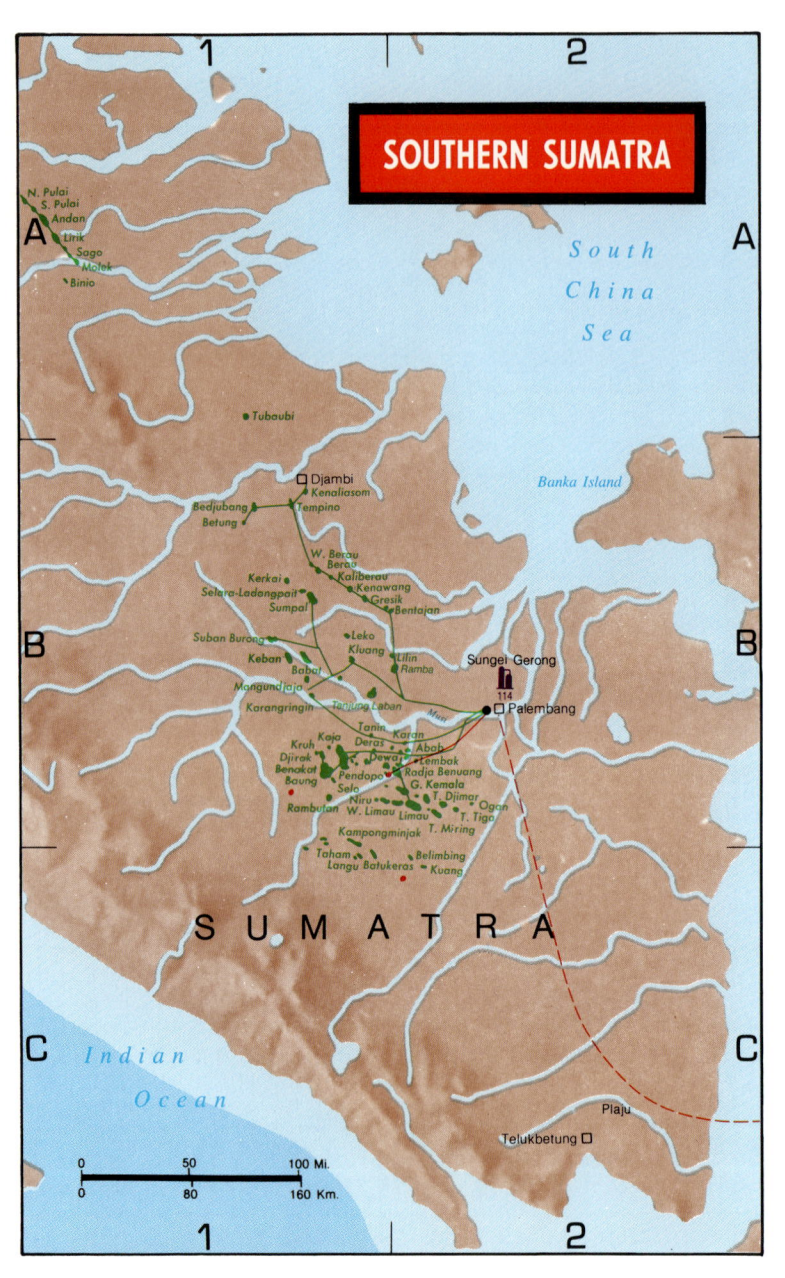

SOUTHERN SUMATRA

BP plans

BP Petroleum Development Ltd. planned a wildcat on a block off Palawan in the Philippines on a farmout from Crestone Energy Corp., Denver.

Under the farmout agreement, BP, at its sole cost and risk, was to acquire at least 200 line km of seismic data and drill a well on the Cliff Head reefal prospect in 2,156 ft of water by February 1991. BP plans to drill the well to about 11,155 ft or about 820 ft below the base of the Nido platform carbonate sequence, which is the commercial pay in Northwest Palawan. Crestone estimates the structure's prospective resource at 450 million bbl of oil equivalent.

TAIWAN

CAPITAL: Taipei
MONETARY UNIT: New Taiwan dollar
REFINING CAPACITY: 542,500 b/cd
PRODUCTION: 2.5 Mb/d
RESERVES: 4,500 Mbbl

OIL AND LNG WILL CLAIM A GREATER SHARE OF energy consumption in Taiwan during the 1990s than was forecast 5 years ago, says Resource Systems Institute's East-West Center, Honolulu.

In addition, Taiwan will have to expand refining capacity to meet a corresponding demand increase for gasoline and diesel.

Its renewed growth in demand for imported low sulfur fuel oil will then subside as domestic production, diesel and gasoline demand, and LNG imports combine to squeeze—possibly end—fuel oil imports, RSI says.

Taiwan began implementing an agreement to import 1.5 million tons/year of LNG from Indonesia in 1990.

Energy forecast

In 1985, Taiwan's Ministry of Economics predicted that by 2000 the island's total energy demand would be 1.29 million b/d of oil equivalent (BOE). Of that amount, it forecast that petroleum would account for 42%, coal 29%, nuclear 20%, hydroelectric power 4%, and gas and LNG 3.4%.

The ministry's revised April 1989 forecast showed a 12% increase in total demand, compared with its 1985 figures. It scrambled the energy mix, with fossil fuels gaining 230,000 BOE/d over nuclear, hydropower, and alternatives. Oil is still expected to maintain its 42% share.

Oil may, however, increase its share if Taiwan's fourth nuclear plant, with 2 million kw, is not built, RSI says. The plant originally was planned to be operating in 1992-93, but because of hardened Taiwanese opposition it has been pushed back and is now projected for start-up in 1999-2000.

Similar environmental concerns have led to problems in building coal fired plants, RSI reports.

Gasoline, diesel surge

Oil rebounded to account for 27% of power generation in 1989, but its share of total energy supply has fallen 6 percentage points from a high of 50% in 1980.

The Ministry of Economics predicts any delays building coal fired power plants, combined with nuclear's problems, will mean oil will continue to provide about half of Taiwan's total energy at the end of the century.

Growth in oil demand, RSI reports, will be centered on gasoline and diesel fuel. It says CPC expects gasoline demand to be 112,000 b/d in 2000, accounting for 27% of total oil demand. Its oil demand share was 19% in 1989 and 10% in 1980.

Diesel's share, now 18.5%, is forecast to hit 25% at 104,000 b/d in 2000.

Fuel oil fall

The changing energy mix in fossil fuels will be at the expense of fuel oil demand, projected to tumble to a pre-1973 low of 42%, or 174,000 b/d.

At first glance, it appears fuel oil is undergoing a rally of sorts. After falling from a peak of 69% in 1977 to 46% in 1987, fuel oil's share of total petroleum fuel demand surged to 53% in 1989.

RSI predicts the inflow of Indonesian LNG will signal the end of the fuel oil rebound as gas displaces it in power generation.

In 1990, fuel oil demand was expected to fall by more than 20,000 b/d. A "slow decline" will ensue as lower power plant use offsets increased industrial demand.

Refining crunch

To keep pace with increasing demand for transport fuels, RSI says, Taiwan will need to expand refining capacity. Since 1986, its gasoline demand has risen an average 10%/year and diesel 8.3%/year.

Present refining capacity is 600,000 b/d, which includes a 100,000 b/d tower kept in backup reserve. Increased processing of lighter Asian crudes has had the effect of derating capacity to about 450,000 b/d.

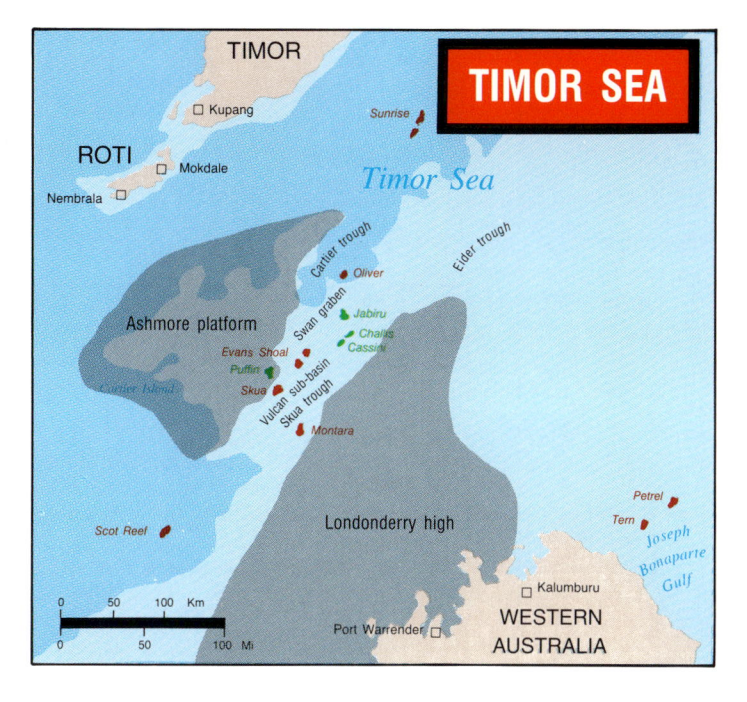

The 100,000 b/d added capacity at Kaohsiung will be partly offset by closure of an old 50,000 b/d tower. CPC plans to expand its Taoyuan refinery by 70,000 b/d and has proposed a site in northern Taiwan for a new refinery. Opposition to the latter project has been strong.

THAILAND

CAPITAL: Bangkok
MONETARY UNIT: Baht
REFINING CAPACITY: 220,550 b/cd
PRODUCTION: 41 Mb/d
RESERVES: 150,000 Mbbl

THAILAND PLANS AN AMBITIOUS ARRAY OF upstream and downstream petroleum projects to help meet its soaring energy demand.

Deregulation

The Thai government plans to deregulate oil prices in phases possibly within the first half of the decade.

The first steps in the process have been taken. As of August 1990, domestic oil prices are being set by senior civil servants using a new formula to partially deregulate oil prices, adjusting prices "in line with world market movement."

Trade barriers also will be relaxed or lifted, allowing international marketers increased access.

Cutting oil imports

Because much of Thailand's hydrocarbon potential is gas prone, natural gas will be a key to reducing Thai dependence on imported oil in the years ahead. Thailand produces about 600 MMcfd of gas but only 25,000 b/d of oil.

Gas flow is expected to reach 1.1 bcfd by the end of the 1990s as new fields are developed.

Tongchat Hongladaromp, former PTT governor and now senior PTT adviser, projects Thai demand for natural gas to climb 9.1%/year, reaching 350,000 b/d of oil equivalent by 2000. That level is about triple current gas production.

With limited indigenous energy resources and rising energy demand, Thailand will continue to rely heavily on imported petroleum the next few decades, Tongchat said.

Gas grid expansions

An ambitious expansion of gas gathering and distribution networks in and outside the country is the key to efforts to increase Thai gas use.

Thailand is studying plans to develop an extensive transnational gas grid linking its expanding national network with three neighboring countries to boost imports of gas.

A study by PTT, which also holds the domestic gas distribution monopoly, envisages a massive investment in laying natural gas pipelines from Malaysia, Indonesia, and Myanmar. Tongchat suggests the entire regional grid could be developed during the next 3 decades in stages, likely to start with importing gas from Malaysia via pipeline.

In addition, Thailand is considering two other possible transnational pipelines in the 1990s.

One line would run 500 km from gas fields in Myanmar's Martaban Bay southeast to the western Thai province of Kanchanaburi and possibly to Bangkok.

The next 2 decades could also see installation of a subsea line from Indonesia's giant Arun gas field in northern Sumatra across the Strait of Malacca to the southwestern Thai province of Krabi. From Krabi it would extend farther across the Thai southern peninsula to Khanom.

PTT also plans to lay another trunkline to ship gas from the south to the Thai central plains in the 1990s.

Joint development

Legislation allowing the start of joint development of hydrocarbons in the Thai-Malay disputed territorial claims is now in effect.

The 7,300 sq km JDA area is thought to contain potential gas resources estimated at at least 7.5 tcf, or about equal to half Thailand's proved reserves.

One of the bills calls for establishment of a two nation joint authority (JA) responsible for administering development of hydrocarbons to be shared equally by the two countries.

Another prospect for JDA development involves an ambitious liquefied natural gas export project.

Thai LNG International (TLI), a Thai-Japanese joint venture, is eyeing the JDA as a potential source of gas supply that will help its $9 billion LNG export project in Thailand get off the ground.

Myanmar venture?

Thailand is seeking a joint venture to develop potential gas resources in the Gulf of Martaban off Myanmar as another source of Thai gas supply.

Pttep estimates the Martaban basin could provide as much as 500 MMcfd of gas to Thailand via a 500 km pipeline to be laid from the Gulf of Martaban to the western Thai province of Kanchanaburi.

Pipeline green light

PTT's board gave a green light to the $360 million, 331 km offshore system to bring more Gulf of Thailand gas to landfall on the southern peninsula.

The project entails first laying a 32 in., 170 km line from the B structure gas field, under development by a Pttep-led group, to Unocal's Erawan gas field.

A second, 27 in. line will extend west from Erawan field to landfall at Khanom, Nakhon Si Thammarat Province, where two 300,000 kw, gas fired, combined cycle plants are planned.

PTT said the 700 MMcfd design capacity B structure-Erawan line is to be ready by yearend 1993 to coincide with

ASIA-PACIFIC/IPE ATLAS

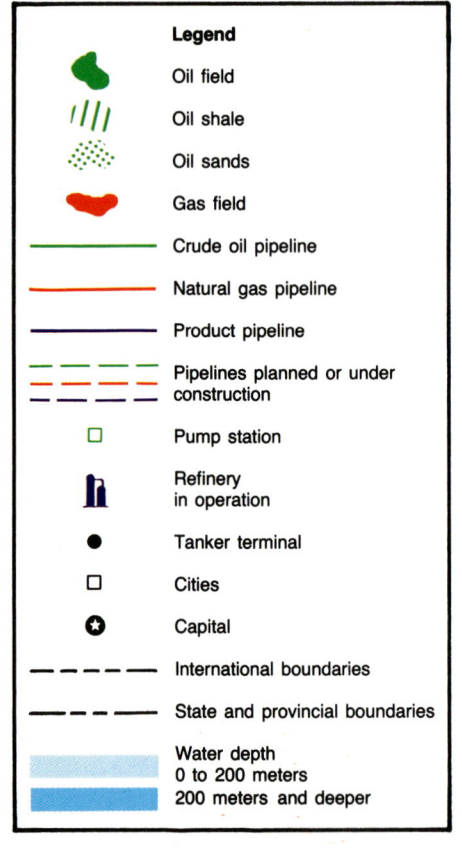

production start-up of B structure field at an initial rate of 150 MMcfd.

Completion of the Erawan-Khanom pipeline is expected about mid-1994, timed to coincide with start-up of the first stage of EGAT's power plants, which are expected to take 100-150 MMcfd.

EGAT seeks Malay gas

EGAT has agreed to purchase gas from Malaysia to fuel a major power plant it plans to build in southern Thailand.

EGAT Gen. Manager Paopat Javanalikikorn said natural gas from peninsular Malaysia offers strong potential as an alternative source of fuel for the planned Sabayoi plant if attempts to secure other fuel sources fail.

The 900,000 kw power plant originally was intended to burn Thai lignite mined in Songkhla Province's Sabayoi district, where preliminary exploration showed substantial lignite reserves.

But opposition from environmentalists prompted EGAT to turn to other sources of fuel, particularly gas, to meet the scheduled 1998 start-up.

Unocal exploration activity

Encouraged by recent promising results, Unocal Thailand and partners planned to enter the next phase of a campaign to assess extent of gas/condensate deposits on the gas prone Block B-12/27 in the Gulf of Thailand.

Initially, that called for a major 3-D seismic survey expected to begin late in 1990. It would be followed by a drilling program tentatively set to begin near yearend 1991.

But Unocal Thailand, which operates the 13,550 sq km concession, had not obtained approval for those plans from partners Britoil plc, Amerada Hess Corp., and Mitsui Oil Exploration.

The Unocal group's drilling program was wrapping up as its 1 Kaimuk, the last wildcat in the four well program, approached planned total depth of 10,820 ft.

The group has found gas in the first three wells: 1 Morakot, 1 Pailin, and 2 Pailin.

Downstream activity

Thai refined products consumption totals about 400,000 b/d now and is expected to rise to 900,000 b/d by the end of the decade, said H.E. Korn Dabbaransi, minister to the office of the Thai prime minister at the Asia-Pacific Petroleum Conference in Singapore in September.

Capacity at three existing Thai refineries will jump to a combined 370,000 b/d by 1993 from 230,000 b/d. A fourth, 134,000 b/d refinery is scheduled to be on stream in 1994 at Map Ta Phut, Rayong Province.

Approval by NEPC, the chief energy policymaking body chaired by Thai Prime Minister Chatichai Choonhavan, allows Caltex to proceed with plans to build a $520 million refinery in the eastern seaboard province of Rayong rather than on the southern peninsula under an earlier government plan.

It also means Shell can go ahead with its proposed $760

ASIA-PACIFIC/IPE ATLAS

Thailand's gas grid options

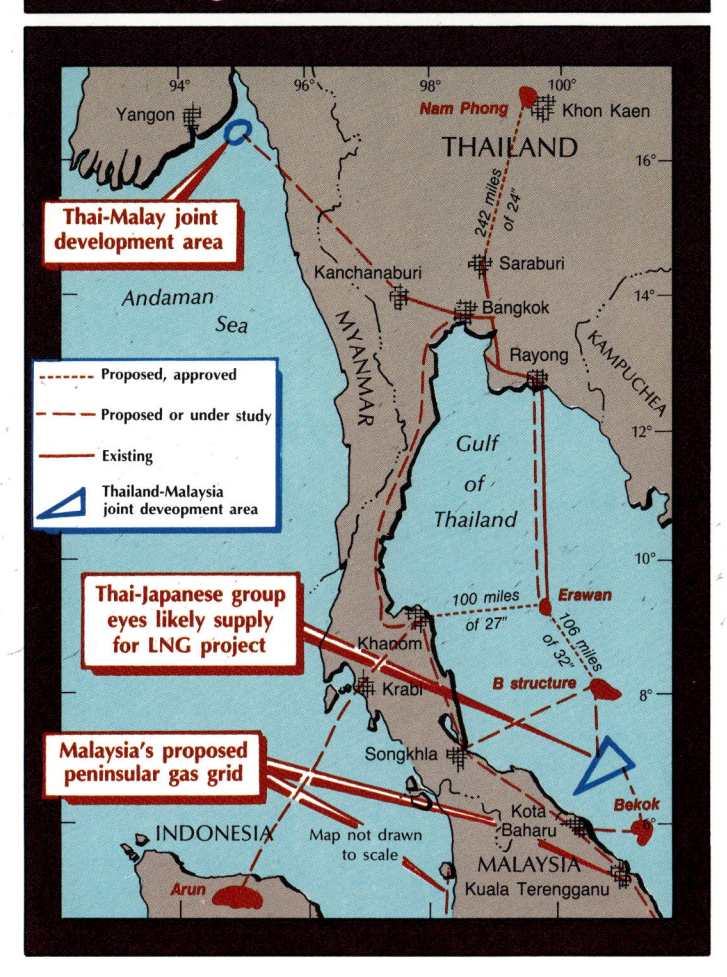

million refinery based on a shareholding structure contained in a draft agreement.

VIET NAM

CAPITAL: Hanoi
MONETARY UNIT: Dong
REFINING CAPACITY: N.A.
PRODUCTION: 40 Mb/d
RESERVES: 500,000 Mbbl

THE INTERNATIONAL PETROLEUM INDUSTRY continued to be busy in Viet Nam in 1990.

At yearend the country reportedly had increased oil flow from its only producing South China Sea field to more than 60,000 b/d, Soviet sources said.

If true, Viet Nam substantially exceeded its 1990 target of 50,000 b/d, and production from Bach Ho (White Tiger) field rose sharply since the first half of 1990.

The Moscow newspaper Trud said the joint Soviet-Vietnamese venture Vietsovpetro produced more than 2.5 million metric tons of crude during the first 10 months of 1990. That's an average 60,033 b/d, compared with about 20,000 b/d during the first 6 months of the year.

A later report in the Moscow newspaper Pravda datelined Hanoi said Viet Nam had started exporting crude to the U.S.S.R. In the past, Viet Nam has imported most of its petroleum products from the Soviet Union, buying more than 37,000 b/d in 1989.

Vietnamese oil exports "to the U.S.S.R." apparently are not reaching the Soviet Union itself but are being delivered to third countries with which Moscow has contracts to provide crude.

Viet Nam has only one very small refinery producing just 800 b/d of products. Work is under way to increase the plant's refining capacity to 5,000 b/d.

The nation has no known onshore oil production, but it reportedly has several undeveloped South China Sea fields. The most recent discovery, near White Tiger field, is said to have tested about 300 b/d.

Sovereignty issue

Protective of its oil resources in the South China Sea, Viet Nam warned China, Taiwan, Philippines, and Malaysia not to challenge its sovereignty over the Spratly Islands. These small islands, believed to be in a petroleum prone area, lie about 400 km southeast of the Vietnamese coast.

Hanoi's warning against intrusion by other nations into the Spratly area followed alleged overflights by Philippine military planes during maneuvers. Vietnamese officials said the Philippine government plans to improve an airstrip on one of the Spratly islands it already controls.

Despite the purported big hike in White Tiger oil production, Viet Nam is experiencing a severe shortage of petroleum products.

Gasoline prices have soared since the Persian Gulf crisis began.

Imports of automobiles and motor scooters were banned during fourth quarter 1990. The ban was slated to continue in 1991, with citizens advised to use "other forms of transportation."

Other activity

Four Japanese companies were studying feasibility of a $600-800 million petrochemical complex, the first major Japanese investment there in 16 years. Nissho Iwai Corp. submitted a draft proposal, with C. Itoh & Co., Meiwa Trading Co., and Toyo Engineering Corp. also under consideration.

Vying to operate the project are British Petroleum Co. plc, Royal Dutch/Shell Group, and Total CFP. The government was to choose a project manager in first half 1991, with construction to begin later in 1991 and completion set for 1995. Enterprise Oil plc's first wildcat off Viet Nam flowed 300 b/d of oil on test. Enterprise suspended 17-C-1X while it decides whether to stimulate the well, drilled with Lauritzen's Dan Duchess drillship on Block 21 in 131 ft of water 50 miles south of Vung Tau.

Vietsovpetro, a Vietnamese-Soviet joint venture, let contract to Simon Geophysical Services, Swanley, England, for reprocessing, special processing, and interpretation of about 800 line km of marine seismic data.

ONGC (Videsh) Ltd., India, spudded its 1 A-1 well in Offshore Viet Nam Block 6, off Vung Tau harbor, the first under its Vietnamese production sharing contract.

Fina Exploration Minh Hai acquired industry's first seismic option off Viet Nam. Covering Blocks 45, 49, and 50 and parts of Blocks 46, 51, 53, 54, and 55, the agreement with Petrovietnam allows Fina to select an area of about 11,000 sq km, subject to a production sharing contract, after completion of a seismic survey in about 18 months.

Two companies took farmouts on South China Sea blocks off Viet Nam operated by International Petroleum Ltd. Overseas Petroleum Investment Corp. acquired a 20% interest in Block 115, Cairn Energy plc a 10% interest in Block 22. IPL completed 2,510 line km of seismic survey on Block 22 and expected to complete 3,500 line km on Block 115. A Petro-Canada group planned to drill three wildcats, conduct seismic surveys, and reprocess existing seismic data under a 3 year exploration contract on a 6,959 sq mile block off Con Son Island, southern Viet Nam.

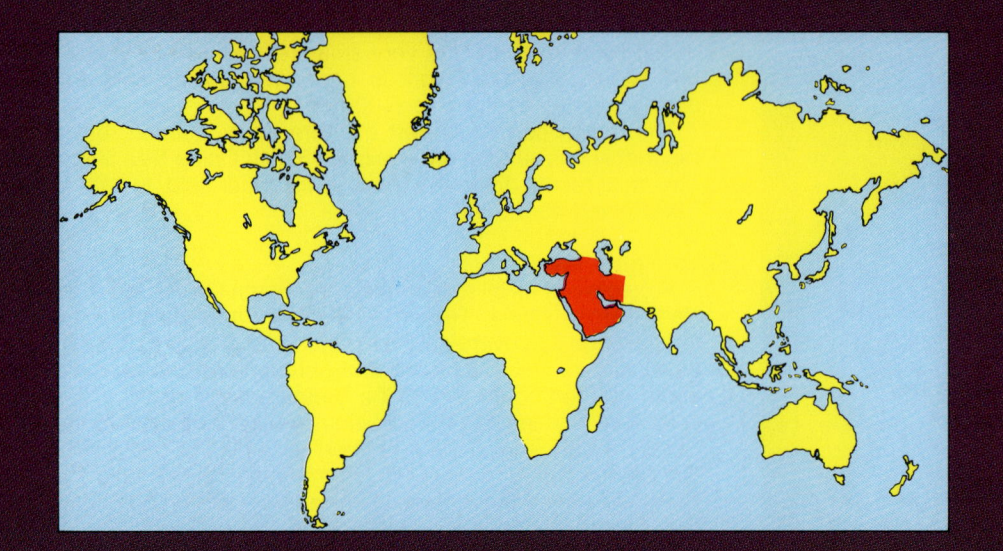

IPE ATLAS

MIDDLE EAST

BAHRAIN

CAPITAL: Manama
MONETARY UNIT: Dinar
REFINING CAPACITY: 243,000 b/cd
PRODUCTION: 42,000 b/d
RESERVES: 97.46 million bbl

OVERNIGHT, THE WORLD OIL MARKET CHANGED dramatically in 1990 with Iraq's Aug. 2 invasion and takeover of Kuwait.

Nevertheless, plans for work in the Persian Gulf region continue. Harken Enery Corp. has its sights set on early spudding of the first wildcat under terms of a new production sharing contract in the gulf off Bahrain. The contract, which covers almost all of the country's offshore acreage, requires the Dallas independent to spud a wildcat to test Permian Khuff carbonate within 2 years.

The first wildcat will be drilled at Jarim reef, a seabed anomaly exposed at low tide. The reef, about 15 miles northwest of Manama, is a manifestation of a seismically controlled, closed anticline. Its surface expression is 6-7 miles wide and 12 miles long. Harken will decide whether to drill from a shallow draft jack up or dredge and build a pad for a land rig. The well, expected to go to 12,500 ft, will cost $6-8 million.

Harken's postulated structure lies on trend between Awali field, the only production in Bahraini territory, and Abu Safah field off Saudi Arabia. Abu Safah, discovered in 1963, contained about 6 billion bbl of oil reserves in Jurassic Arab limestone at 8,000-8,500 ft and is believed to have deep Khuff potential. Awali, with 1.5 billion bbl of original reserves and 10 tcf of gas, produces oil from Cretaceous Bahrain zone at 2,500 ft and Arab at 6,000 ft. The field's Khuff gas at 9,000-11,000 ft is the nearest Khuff production to Harken's structure about 20 miles north.

The Cretaceous zone contains about 70% of the field's oil, but analyses have shown Cretaceous and Jurassic oils to be alike. Intense folding and fracturing of Upper Jurassic Hith anhydrite is thought to have allowed Jurassic oil to escape to Cretaceous in Awali field. Harken said it believes its structure to be more similar to Abu Safah than Awali because no fracturing of Hith anhydrite appears in seismic data.

Harken's production sharing contract covers Muharraq Island, Umm at Nassan Island, and all offshore territory except a small area southeast of Bahrain island. The company may activate a 3 year final exploration term if it makes no commercial discovery during a 2 year initial exploration term. A commercial discovery triggers a 35 year contract term. The contract gives Harken rights to its share of oil and nonassociated gas and Bahrain National Oil Co. rights to all associated gas as well as its share of oil and nonassociated gas.

Large Bahraini consumers of gas include Bahrain Petroleum Co.'s 243,000 b/d refinery on Sitra Island, aluminum and desalination plants, and other businesses. Bahrain also has a large gas processing plant. Harken has the right to export its share of oil and expects to market any nonassociated gas in Bahrain.

IRAN

CAPITAL: Tehran
MONETARY UNIT: Rial
REFINING CAPACITY: 720,000 b/cd
PRODUCTION: 3.12 million b/d
RESERVES: 92.85 billion bbl

IRAN IS ONCE AGAIN INVESTIGATING GAS EXPORTS to Europe through an exchange with the Soviet Union. A National Iranian Oil Co. marketing official told a London conference that countries in eastern and western Europe alike could be supplied by such an exchange.

The joint Iran-Soviet Economic Commission (ISEC) agreed to pursue the project and get it operational during 1991, reports IRNA, the official Iranian news agency.

Before the revolution, Iran was involved in detailed discussions on a deal to deliver gas to southern regions of the Soviet Union. European customers would be supplied with similar volumes from other Soviet sources. Although ISEC wants to activate exports to Europe, Iran will have to upgrade its gas delivery systems.

The IGAT-2 gas pipeline from Kangan to the Soviet border

MIDDLE EAST

MIDDLE EAST

Saudi Aramco's capital and operating budget

	1990	1991	1992	1993	1994	1995	1996	1997	1998	1999	2000	Total
						Million current U.S. $						
Capital costs												
E&P	545	1,069	1,901	2,793	2,920	3,478	4,045	4,331	4,816	4,851	5,379	36,128
MS&T*	332	466	430	337	275	301	365	373	282	142	102	3,405
Support	112	196	298	401	409	484	564	602	653	639	701	5,059
Total capital	989	1,731	2,629	3,531	3,604	4,263	4,974	5,306	5,751	5,632	6,182	44,592
Operating costs												
E&P	499	526	556	620	734	828	897	990	1,121	1,258	1,347	9,376
MS&T*	818	887	969	1,052	1,131	1,208	1,291	1,383	1,479	1,573	1,661	13,452
Support	2,372	2,474	2,593	2,759	2,984	3,257	3,501	3,796	4,160	4,530	4,813	37,239
Total operating	3,689	3,887	4,118	4,431	4,849	5,293	5,689	6,169	6,760	7,361	7,821	60,067
Total	4,678	5,618	6,747	7,962	8,453	9,556	10,663	11,475	12,511	12,993	14,003	104,659

*Manufacturing, supply, and transportation.

was to have been an integral part of an earlier proposed delivery system, but work was suspended during the revolution. Completion of the line would be required for a new proposal, along with a substantial increase in capacity of the Kangan gas processing plant.

Meanwhile, Iran also is exploring the concept of bringing in international companies to assist in development of its oil and gas fields during the 1990s. Iran's oil minister said his country is talking to international companies about development of oil and gas reserves on a contract basis. The minister said the companies will not be offered equity or production sharing agreements. Companies will be paid from production of fields they develop or from other reservoirs.

Iran's main effort will be concentrated on developing new oil and gas reservoirs, although there are negotiations for help with secondary recovery projects. Pars gas field and the Mond very heavy oil reservoir might be developed with international help. The minister didn't rule out the possibility of foreign companies becoming involved in exploration, although he said, "Up to now we have not gone into this matter very seriously."

Oil production capacity is to be raised to 4 million b/d during 1991. And the minister said he expects Iran within 5 years to have ability to sustain production of 5 million b/d.

One of the key elements in Iran's reconstruction program is rebuilding Kharg Island terminal in the Persian Gulf, which was seriously damaged during Iran's 8 year war with Iraq in the 1980s. As 1991 neared, ETPM of France was awarded a $225 million contract to rebuild the terminal. Work is expected to take 2 years.

First contracts also have been placed for replacing storage capacity at the terminal. Sangyong Construction will build five 1 million bbl tanks and one 500,000 bbl tank. Throughout the war, Kharg Island remained Iran's major oil export terminal.

The project to link Iran with the Mediterranean through a 1,150 mile, 1 million b/d pipeline through Turkey has been dropped and there has been little activity on plans for a pipeline to a new terminal on Iran's Indian Ocean coast. ETPM also was awarded a $45 million contract to repair and expand Nasr platform in Sirri field. Prewar output of 50,000 b/d will be increased to 80,000 b/d.

Other fields were also badly damaged in the fighting. Salman field, on the border with Abu Dhabi, was hit by Iraqi air attacks and U.S. naval action. Before the war it was producing about 150,000 b/d.

Under a $300 million contract, two Japanese companies will restart production at 50,000 b/d and then install facilities to boost production to 220,000 b/d. South Korean companies are expected to participate in the offshore reconstruction program as a result of closer ties between Seoul and the Iranian government.

Korean companies also will participate in the second stage of the Vali Asr gas processing plant, where capacity is being raised to 2.8 bcfd from 1.2 bcfd to handle production from Nar and Kangan gas fields and boost gas export capacity.

Iran to build new refinery at Arak

NIOC has anounced that it let a contract in April 1989 for the construction of a grassroots, 150,000 b/d refinery to be sited some 200 miles southwest of Tehran near a pipeline at Arak, according to the Iranian Ministry of Petroleum. Start-up of the refinery is scheduled for early 1993.

The contract for the refinery was awarded to a joint venture comprising JGC Corp., Tokyo, and Tecnologie Progetti Lavori SpA (formerly Technipetrol SpA), Rome. The contract price is $1.3 billion.

The refinery will include distillation and downstream processes, making it suitable to produce a full complement of light products and asphalt. The refinery will be designed to charge Ahwaz Asmari crude oil of 31.7° gravity and 1.66 wt % sulfur. The refinery will include steam and power generation facilities.

In other processing action:

• Iran let a $150 million contract to South Korea's Daelim Industrial Co. for a 300,000 metric ton/year ethylene plant at Bandar Khomeini in southern Iran. Construction is to begin this year. The project was begun in 1976 by a joint venture between an Iranian state company and a Japanese group led by Mitsui. It was interrupted by the Iran-Iraq war, then hobbled by a dispute between Iran and the Mitsui group, with the latter pulling out.

• Iran started up an 84,000 metric ton/year methanol plant at Shiraz petrochemical complex. Methanol will feed a planned $70.6 million, 16,000 metric ton/year melamine plant and possibly later a methyl tertiary butyl ether plant under study.

• Iran let a $28.9 million contract to Air Liquide, Paris, for design and construction of three industrial gas units at two petrochemical complexes, all due on stream by yearend 1991. Shiraz will get a 5,000 ton/year argon unit. Arak's ethylene oxide plant will get 450 ton/day oxygen and nitrate units.

• NIOC let a contract to Chiyoda Corp. of Japan and Snamprogetti SpA of Italy to build a 240,000 b/d refinery at Bandar Abbas on the approaches to the Strait of Hormuz. The $1.2 billion plant is to be on stream in 1993.

MIDDLE EAST

IRAQ

CAPITAL: Baghdad
MONETARY UNIT: Dinar
REFINING CAPACITY: 318,500 b/cd
PRODUCTION: 2.083 million b/d
RESERVES: 100 billion bbl

EVEN BEFORE ITS AUG. 2, 1990, military invasion and takeover of Kuwait, Iraq was emerging as the second ranked oil producer in the Middle East. It has productive capacity of 4.5 million b/d, and its oil minister says another 1.5-2 million b/d is planned through exploitation of proved but undeveloped reserves.

Iraq hasn't revealed the fields that will be developed using foreign contractors. The minister said Iraq is anxious to develop these reserves using oil companies and engineering companies as contractors and paying them in crude oil.

The foreign companies would be responsible for all surface facilities, including development drilling, and will connect the newly developed fields to the infrastructure for transportation and export.

The foreign companies will be expected to provide all the required hard currency for the projects, while Iraq will provide local funds for work undertaken by Iraqi companies and Iraqi labor. In addition to crude to repay initial outlays, foreign companies will be assured of longer term contracts for oil at commercial prices negotiated between the two parties.

The minister said the new development program will include sweet and sour crudes. It is likely to take in the large Yamama formation with its light crudes, which lie at about 9,800 ft. That's much deeper than Iraq's other producing fields.

Iraq has made considerable progress in adding productive capacity. Last year, the country inaugurated Suba field, about 60 miles west of Basrah. Production is 60,000 b/d. Another 80,000 b/d of capacity became available when Khabbaz and Saddam fields in Tamin province, north of Baghdad, went on stream.

Elsewhere, installation of new gas oil separation capacity and revamping of existing capacity in Zubair field, combined with a new drilling program, should increase the field's production to more than 230,000 b/d from 70,000 b/d.

The first stage of West Qurna field development was due to produce oil in 1990, the latest report said. The Mishrif reservoir will add 200,000 b/d to productive capacity. Iraq has longer term plans for West Qurna's Ratawi reservoir, which could add 400,000 b/d when needed to meet demand later in the decade.

Iraq has made considerable progress in repairing its Iran war damaged tanker export facilities through the Persian Gulf. The Mina al-Bakr terminal now has an export capacity of 800,000 b/d, although shipments early in 1990 were only about 350,000 b/d. Outline plans are being made to restore the unit to its pre-Iran war capacity of 1.6 million b/d and to repair the Khor al-Amaya offshore facility that was more seriously damaged in the fighting.

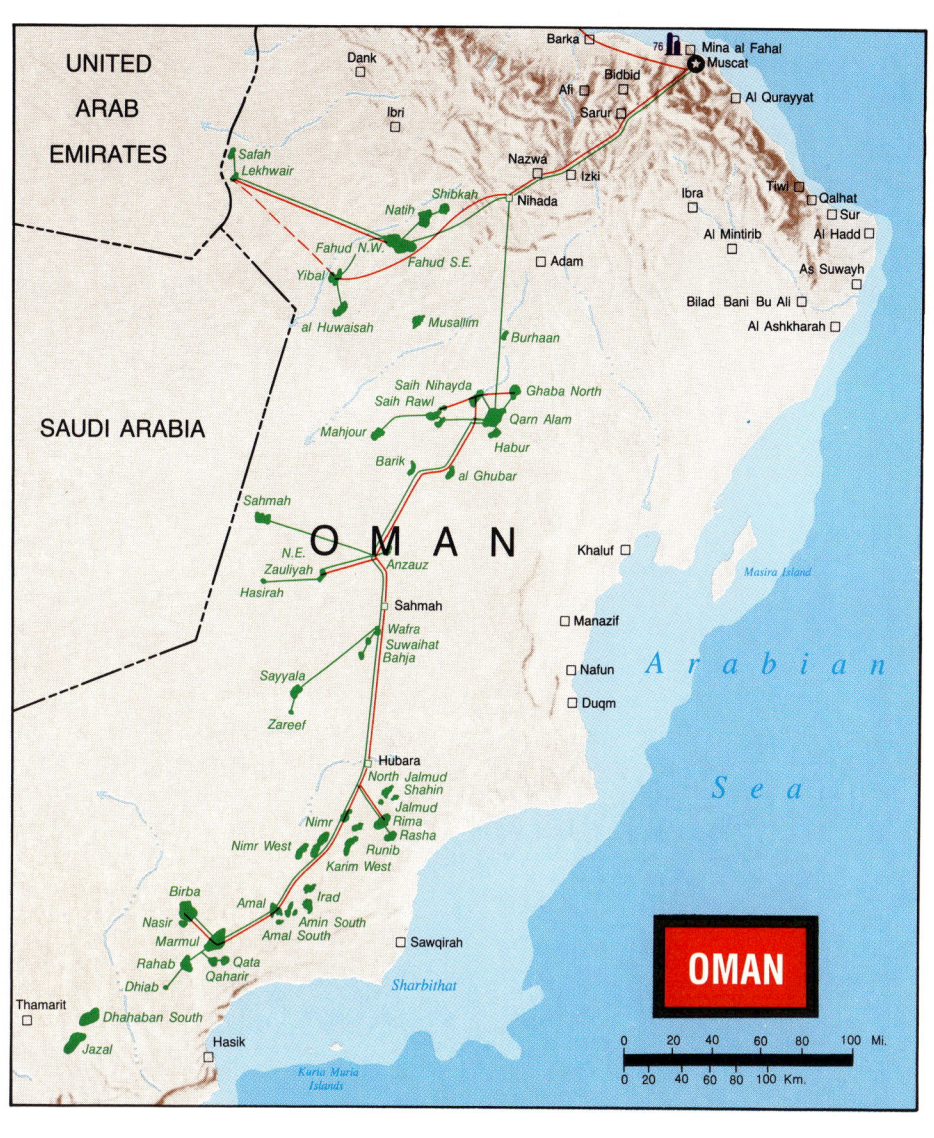

Persian Gulf tanker terminals supplement Iraq's considerable investment in pipeline outlets to the Mediterranean Sea through Turkey and to the Red Sea through Saudi Arabia. The IPSA-2 pipeline project to newly dedicated export facilites on the Red Sea at Yanbu was dedicated in January 1990, hiking Iraq's total export capacity to a little less than 5 million b/d.

IPSA-2 was the second stage of a 978 mile pipeline moving crude from Zubair in southern Iraq to a new terminal at Muajjiz, about 30 miles south of Yanbu on Saudi Arabia's Red Sea coast. The terminal has 10 million bbl of storage and two berths that can handle tankers as large as 300,000 dwt and a third that can accommodate 500,000 dwt vessels. In addition to IPSA-2 and the 800,000 b/d capacity of the Persian Gulf terminal, Iraq's pipeline outlet to the Mediterranean through Turkey has a capacity of 1.5 million b/d.

JORDAN

CAPITAL: Amman
MONETARY UNIT: Dinar
REFINING CAPACITY: 100,000 b/cd
PRODUCTION: 400,000 b/d
RESERVES: 20 million bbl

A TECHNICAL ASSISTANCE PROJECT IN JORDAN HAS been extended for 2 years by Petro-Canada International

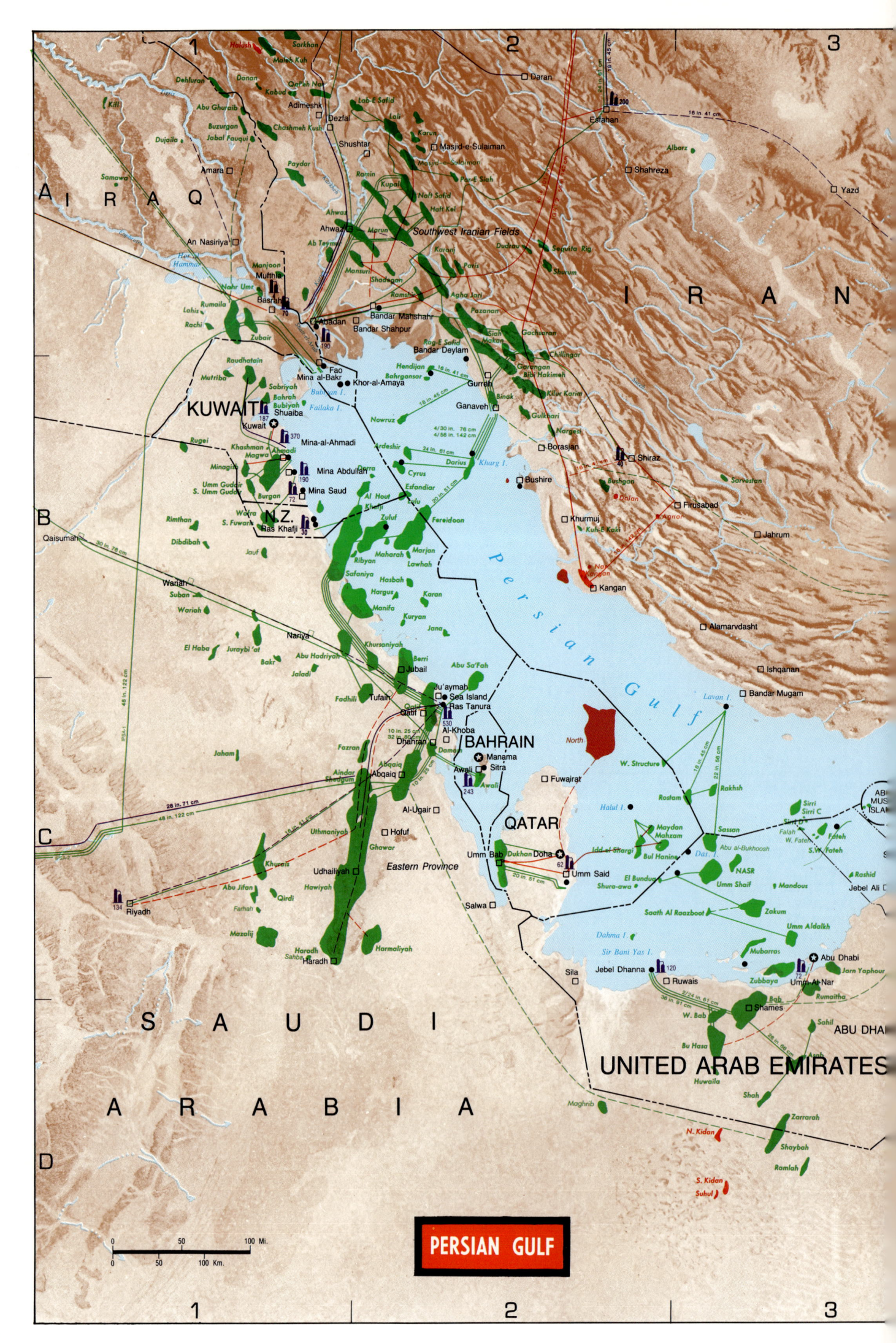

PERSIAN GULF

MIDDLE EAST

Assistance Corp. (Pciac). The $15 million extension boosts Pciac's commitment to Jordan's Natural Resources Authority (NRA) to a total of as much as $47 million in goods and services drawn mainly from the western Canadian petroleum industry.

The added funds will provide equipment, training, advisers, and technical support for NRA's hydrocarbon exploration programs. A Canadian drilling rig and associated equipment, including critical spare parts and a field maintenance and machine shop, were donated to Jordan in 1990. Canadian drilling advisers last year were posted to Jordan for 18 months to work with NRA to assist management and field operations. Advisers will use seismic data acquired by Pciac's Canadian contractors on behalf of NRA covering the Risha and Sirhan areas to help direct development and delineation of hydrocarbon structures.

Pciac last year raised its estimate of proved reserves in Northeast Jordan's Risheh gas field to 500 bcf from its November 1989 estimate of 260 bcf. Evidence of additional pools could increase that estimate still further, Pciac said. Studies will be conducted on how best to enhance its production further.

Pciac's project extension will include assistance in computerization of data storage and retrieval, a Jordan-wide program to interpret all gravity and magnetic data, and assistance to the geochemical and core analysis functions. More assistance will be provided for evaluation of well results from the authority's drilling programs, with emphasis on training and technology transfer.

Pciac and NRA signed an initial agreement worth $19.1 million in March 1987, and Pciac committed a further $13 million late in 1988. A Canadian company under contract to Pciac has acquired 4,600 line km of seismic data in the Risha, Sirhan, and western highlands areas of Jordan. Canadian geophysical companies have processed those and other data in Calgary.

Pciac assigned a team of geologists and geophysicists to Amman to assist in interpretation and training. Several drillsites in the Sirhan area have been identified.

KUWAIT

CAPITAL: Kuwait City
MONETARY UNIT: Kuwaiti Dinar
REFINING CAPACITY: 819,000 b/cd
PRODUCTION: NA
RESERVES: 94.525 billion bbl

FOLLOWING THE AUG. 2, 1990, INVAsion and occupation of Kuwait by Iraq, Kuwait Petroleum Co. rebounded from the crisis by setting up an office in exile in London to keep its extensive European downstream operations running smoothly.

KPC administration was set up in the offices of international affiliate Kuwait Petroleum International

Kuwait Oil Tanker Co. also moved its headquarters to London, handling fleet operations from there. KOTC controls 30 crude and products tankers and gas carriers with a combined capacity of 2.5 million dwt.

Under its Q8 brand, KPI has about 24% of the Danish market, 12% in Sweden, 10% in Italy, 7% in Belgium, and a little more than 2% in the U.K.

KPI also operates a 100,000 b/d refinery in Naples, a 75,000 b/d refinery in Rotterdam, and a 56,000 b/d refinery in Denmark.

KPI said it solved operational problems stemming from the decision of many European countries to freeze Kuwaiti assets to prevent them from falling into the hands of the Iraqis.

OMAN

CAPITAL: Muscat
MONETARY UNIT: Omani Riyal
REFINING CAPACITY: 76,932 b/cd
PRODUCTION: 658,000 b/d
RESERVES: 4.3 billion bbl

PETROLEUM DEVELOPMENT OMAN has approved a $500 million waterflood to turn Lekhwair field into one of Oman's biggest oil and gas producers. The project will boost production to a little more than 100,000 b/d by 1994 from a last reported 24,000 b/d for the field in Northwest Oman.

The field is linked by pipeline to the transportation system for oil from central and southern Oman but will require a new pipeline to handle gas shipments. PDO is 60% owned by the government of Oman.

The Royal Dutch/Shell Group has 34%, Total-CFP 4%, and Partex (Oman) Corp. 2%. PDO, the biggest producer in Oman, will play a major part in plans to boost national oil production from 640,000 b/d to 700,000 b/d and sustain that level.

Since 1984, a pilot waterflood has been under way in Lekhwair field with 17 injectors and 16 producers. PDO said the pilot demonstrated the feasibility of the larger waterflood that will now proceed. Boosting Lekhwair production to 100,000 b/d will require the drilling of 126 more production wells and 47 water injectors. Drilling will continue until 1995.

The project also will require 10 remote manifold stations to collect oil and natural gas. Glass reinforced epoxy flow lines will be laid to move the mixture to a central production station, where water

MIDDLE EAST

treatment units for high pressure injection will be installed. PDO said the program is based on a successful high pressure injection test in one of the injection wells. This test success showed that sufficiently high injection rates can be achieved to allow an inverted nine spot pattern of wells. PDO said the pattern yields major savings, with an average one injection well to every three producers.

The new program will produce 141 MMcfd of gas that will be used for gas lift in the field with four large four stage compressors, each driven by a 9,200 kw motor with a variable speed gearbox. Other gas will move through a new 68 mile, 16 in. pipeline to the Oman government's processing plant at Yibal, where it will be used to back out gas from Yibal and Natih fields.

Lekhwair gas also will move from Yibal through an existing pipeline to Fahud oil field, where it will be used for gas lift. PDO will use a number of new techniques in the expanded waterflood. These include a computerized distributed control system to monitor production and injection systems. Another innovation for PDO will be the use of a Sulferox gas sweetening plant to remove hydrogen sulfide from the natural gas stream and reduce it to pure sulfur.

PDO let the main contract for further development of Lekhwair to Brown & Root Inc. And PDO let another contract to National Drilling Services Co., an Omani company owned 70-30 by Galfar Group of Oman and Forasol-Foramer, Paris. NDSC will mobilize two rigs to begin drilling by August 1991, and provide ancillary services. Separate contracts will be awarded for gas pipeline construction and for a 50,000 kw power station to be combined with a similar project for power generation in Saih Nihayda field in Central Oman.

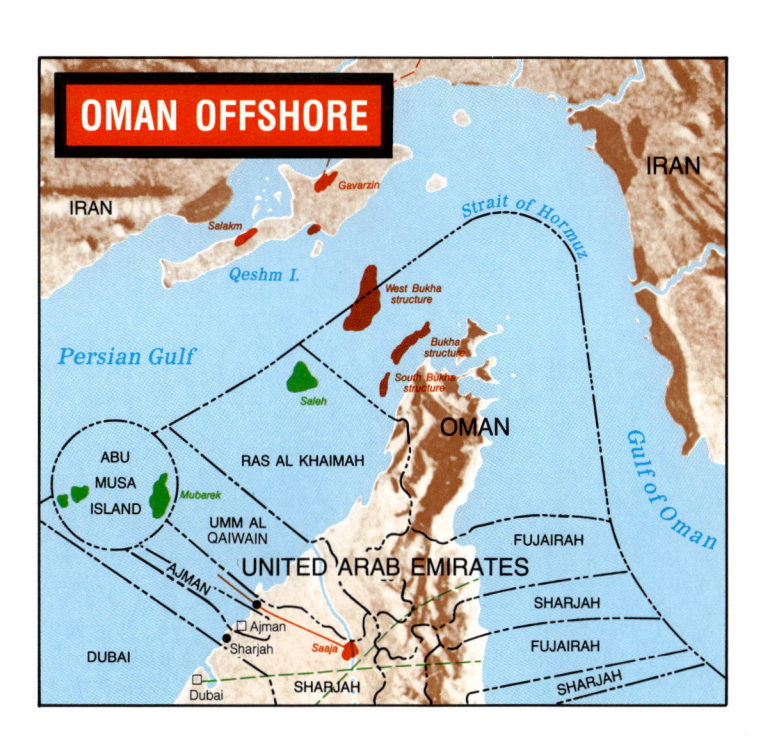

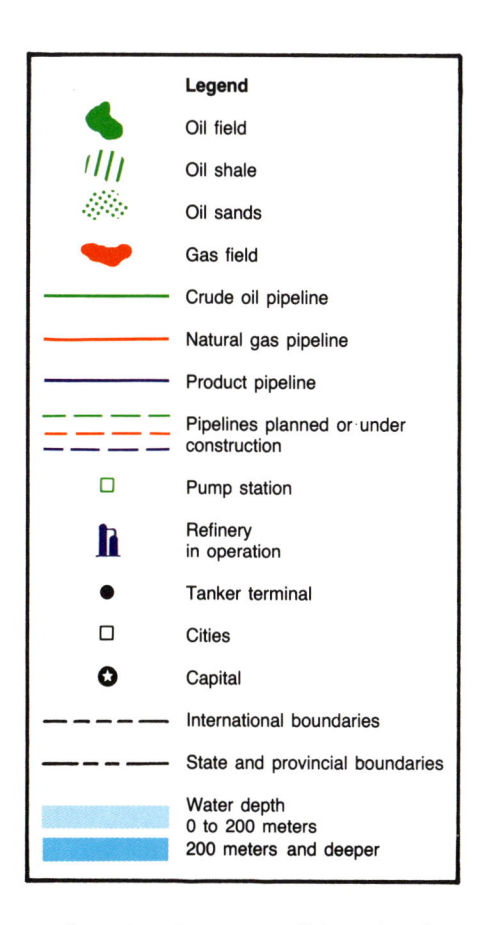

IPC eyes field development project off Oman

Bukha gas/condensate feld off Oman in the Strait in Hormuz could be on stream by the end of 1991, says field operator International Petroleum Corp., Dubai. Gas production might be used as feedstock for the world's first floating methanol plant, which entered final planning stages last year. Or it could be sold into the gas distribution grid of the nearby United Arab Emirates and used in the northern emirates.

IPC said because the methanol plant could not be ready to handle gas until at least 1 year after Bukha start-up, it is likely the gas will be sold in the U.A.E.

IPC proposes a $50-60 million project to connect three existing Bukha wells through a wellhead platform to an onshore gas processing plant in nearby Ras al Khaimah, U.A.E. The plant was built to serve Saleh offshore field, previously operated by IPC. The first phase of the Bukha project is to produce 75 MMcfd of gas and 9,000 b/d of condensate. Highlands Petrochemicals Ltd. and Ocean Phoenix, which have a 10 year agreement with the Omani government covering 224 bcf of gas as feedstock for the methanol plant, said the floating unit could be ready for processing in early 1993. The contract with the Omani government gives Highlands and Ocean Phoenix the option to moor the unit off Sohar in the Indian Ocean and take gas from Oman's main natural gas pipeline network. IPC, which operates the Bukha concession in partnership with Wintershall Oman BV, said the Ras al Khaimah processing facilities are handling about 20 MMcfd from Saleh field. Saleh will be almost depleted before Bukha goes on stream. The plant might be expected to handle 95 MMcfd of gas.

Bukha is to be connected to the plant by an 18.6 mile, 16 in., two phase flow pipeline. Condensate will be exported through a line that runs from the plant to an offshore loading buoy. Bukha, IPC said, is likely to be the first of several development projects on the permit. The acreage holds a number of prospects, including the West Bukha-Henjam structure that extends into Iranian waters.

Onshore, Occidental of Oman Inc. plans further development of Safah oil field in Northeast Oman. The program includes plans to drill another 30 development wells, along with start-up of gas injection. Safah, a 65-35 joint venture with Gulf Oman, a Chevron Corp. unit, produced an average 19,100 b/d in 1989, and about 22,000 b/d in 1990.

As 1991 approached Chevron Corp. put its Omani exploration and production assets on the auction block. Chevron said the move represents the first time producing properties have been offered for sale in Oman. It offers a "unique entry opportunity for a company looking for a stable source of Middle East crude," Chevron said. The offering is part of Chevron's restructuring.

Chevron is offering its Suneinah concession in Northwest Oman about 300 miles west of Muscat. Less than 1% of the 2,856,000 gross acre concession has been developed. The concession contains Chevron's key asset in Oman, Safah oil field, which accounts for all production in the concession.

Chevron also made two gas/condensate discoveries in the concession. Wadi Rafash, about 50 miles southeast of Safah, has gross potential reserves of about 188 bcf of gas and 11 million bbl of condensate. Shams, 22 miles northeast of Safah, has gross potential reserves of about 134 bcf of gas and 5 million bbl of condensate.

QATAR

CAPITAL: Doha
MONETARY UNIT: Riyal
REFINING CAPACITY: 62,000 b/cd
PRODUCTION: 387,000 b/d
RESERVES: 4.5 billion bbl

QATAR ACTIVITY CENTERS ON CONTINUED DEVELOPment of the world's biggest offshore gas reservoir, the 150

MIDDLE EAST

tcf North field. The $1.3 billion first phase of the project is designed to deliver 800 MMcfd, starting this year. However, Qatar General Petroleum Corp. says first stage production will be only 600 MMcfd. Initial flow will be used domestically.

QGPC is pressing attempts to sell liquefied natural gas to customers in Japan. Industry sources say the upturn in prospects for sales to the Far East has improved Qatar's chances of achieving a commercial LNG project. QGPC also is pursuing the idea of North field becoming a long term source of gas to the entire Gulf Cooperation Council zone through construction of a transmission system linking Qatar with Kuwait. Qatar is negotiating to sell gas to Dubai in a separate deal.

North field development is right on schedule, QGPC said. A 16 well drilling program is complete. The wells were drilled from two wellhead platforms, which were bridge linked to four other platforms for living quarters, utilities, riser treatment, and flaring. Production will move ashore through a 34 in. dry gas line and a 12 in. liquids line to a landfall at Ras Laffan. Gas and liquids will then move by pipeline to gas processing and export facilities under construction at Umm Said. About 50,000 b/d of gas liquids will be available for export.

QGPC has broadened the scope of the project to include a gas sweetening plant at Umm Said and a 60,000 ton/year sulfur processing unit at the nearby Qatar Petrochemical Co. complex. Petrochemical units will be built to use gas liquids from North field. QGPC has signed an outline agreement with Montedison of Italy for a 900,000 ton/year methanol plant. It also is considering construction of a major ethylene project with Phillips Petroleum Co. and expansion of output from the Qafco fertilizer joint venture between QGPC and Norsk Hydro.

In addition, the state company is studying the feasibility of building a 500,000 ton/year MTBE plant and has agreements for two metals projects in the Umm Said area.

Although Qatar's long term financial future rests with the huge North field, QGPC is looking for short term increases in oil production to match any rises in its Organization of Petroleum Exporting Countries quota. On the schedule is development of Diyab onshore oil field. It is to produce about 50,000 b/d during 1992 using upgraded facilities in nearby Dukhan field.

Offshore production facilities in Idd el-Shargi, Maydam Mahzam, and Bul Hanine fields are being upgraded.

SAUDI ARABIA

CAPITAL: Riyadh
MONETARY UNIT: Riyal
REFINING CAPACITY: 1,862,500 b/cd
PRODUCTION: 6.215 million b/d
RESERVES: 257.504 billion bbl

CAPITAL SPENDING BY SAUDI ARABIAN OIL CO. could total as much as $44.6 billion during 1990-2000 in Saudi Arabia, with another $60 billion spent in the same period. The figures appear in Saudi Aramco's quarterly capital programs bulletin from November 1989 published by Business International's Saudi Arabian Monitor, London. Since yearend 1989 the $44.6 billion figure may have been cut to $36 billion, the report said.

Dominating the spending will be a program to boost sustainable productive capacity to 10 million b/d by 2000. Saudi Aramco estimates exploration and production spending during the 1990s at $36.1 billion. Installing new production units and reactivating onshore and offshore facilities, mothballed since the early 1980s, to achieve the productive capacity target will tax Saudi Aramco's organizational skills and purse strings, says Saudi Arabian Monitor. The program was put into place very speedily, it added, because only the year before Saudi Aramco was told by the government to reduce its operating budget by $1 billion in a cost cutting move.

Much of Saudi Aramco's expansion program centers on existing fields. However, the program includes $724 million for expansion of the east-west crude oil pipeline to Yanbu and $968 million to increase crude oil export capacity at the Red Sea port. Last year, Anlagenbau, a unit of Mannesmann GmbH, Dusseldorf, let contract to Fluor Daniel Inc. for engineering services for pump stations, support facilities, and loops on the east-west line. Saudi Aramco earlier let contract to Mannesmann for system expansion. The project calls for additions of horsepower and upgrading of control systems at 11 pump stations.

On another pipeline action front, Arabian Petroleum Pipelines will boost capacity of the Sumed pipeline from the Gulf of Suez to the Mediterranean Sea to 2.4 million b/d from 1.6 million b/d during 1990-93.

Saudi Aramco's spending program also provides the first indication that development of the al-Hawtah, Dilam, and al-Raghib discoveries in Central Saudi Arabia south of Riyadh is on the agenda. The budget for 1990-95 includes $500 million for projects in Central Saudi Arabia.

The biggest single new project will be development of Shaybah field, close to the borders with the United Arab Emirates and Oman and far from production facilities. First phase of the project has a 1999-2002 timetable and includes $1 billion for a crude oil pipeline to Ras Tanura and $400 million for bringing infrastructure to that remote area. Also in the first phase are crude handling and gas separation facilities with a capacity of 300,000 b/d and gas gathering and reinjection facilities.

In the second phase, 2000-2003, another 200,000 b/d of capacity will be installed. Total cost of the two phases is estimated at $2.96 billion.

Saudi Arabian Monitor says reports indicate Saudi Aramco is fairly bullish about its capacity to fund this vast program from internal resources, although there have been indications that the Saudi French bank has been considering putting together a financing package for the company with other banks.

At the end of 1989 the Saudi Aramco board approved a $17.5 billion capital budget for 1990-95.

Other projects in the program to return to 10 million b/d level

Current sustainable capacity is thought to be about 7-7.5 million b/d.

Among the first projects in the program to restore productive capacity to 10 million b/d involve spending $550-600 million to expand offshore Safaniya field and Uthmaniyah, which is part of onshore Ghawar field.

In addition, Saudi Aramco has a program of remedial work on wells and gathering systems that were shut in during the early 1980s when Saudi production slumped in response to surplus supply.

Other projects will follow.

In the boom period of the early 1980s Saudi Arabia produced about 10 million b/d but was forced to mothball many producing facilities when demand for Organization of Petroleum Exporting Countries oil collapsed. Saudi Aramco pushed production back to about 10 million b/d for a few days in fourth quarter 1988 during OPEC's production free-for-all, but this volume cannot be sustained for a lengthy period.

Internaitonal companies have been asked to bid on a $400 million project in Safaniya, which will require constructon of four large gas oil separation plants (GOSPs). Saudi Aramco awarded a $300 million contract to a group of three Italian companies for further development of the field. Snamprogetti will design the fourth GOSP to be built and

MIDDLE EAST

A SEISMIC PARTY searches for oil in Oman, which conducts one of the most intensive exploration programs in the Middle East, a region that is assured of maintaining its key role in world oil supplies for a long time to come. Photo courtesy of Petroleum Development Oman.

installed by Belleli Saudi Heavy Industries and Saipem SpA. The unit is scheduled for completion in early 1994.

Saudi Aramco has had plans for the three platform complex on its books for a number of years, but the project has remained on the back burner as OPEC production quotas restrained demand for offshore crude. The unit will have a capacity of 270,000 b/d of oil and 100 MMcfd of gas. Oil will be linked into the existing oil infrastructure, and gas will provide feedstock for the gas processing plant at Juaymah.

Saudi Aramco will spend about $120 million on two GOSPs in Uthmaniyah. After the units have been upgraded and expanded, production should rise to about 300,000 b/d. Saudi Arabian Bechtel Co. and Consolidated Contractors Co., Athens, were awarded a lump sum turnkey contract to complete the work in 2 years. Saudi Aramco also plans a smaller upgrading on another of the field GOSPs at a cost of $25 million.

Saudi Aramco let a 5 year contract to Fluor Daniel to manage the expansion/demothballing program in the country's northern onshore and offshore areas. The contract, value of which isn't disclosed, has options for 4 more years of work. Involved are gas processing plants, pipelines, production facilities, and offshore platforms. Work got under way last year.

Elsewhere, Saudi Aramco has started to step up its exploration program after a lean period when there was little requirement for more drilling or seismic surveys. The company, at last report, had four seismic crews at work and had access to five rigs. Most of the exploration effort is concentrated outside the area originally awarded to the former combine of four foreign companies.

Saudi Aramco has drilled a series of discoveries in a

previously unexplored area southeast of Riyadh. Three successful wells about 70 miles apart raised speculation that Saudi Aramco had found yet another giant oil field. However, there is no firm evidence to show the fields are connected although Saudi sources indicate that the area has considerable potential. Initial estimates place reserves at 2 billion bbl for the area, which could go higher with further drilling.

Reserves are light sweet crude, which is much in demand and could lead to an early development. The new fields are close to the east-west pipeline to Yanbu.

Meanwhile, Eastern Petrochemical Co. (Sharq) delayed until early 1991 scheduled maintenance at its ethylene glycol plant at Jubail because of high EG demand. A catalyst change and related maintenance had been scheduled for October 1990 at the 390,000 metric ton/year EG, 140,000 metric ton/year polyethylene plant. Sharq is a joint venture of Saudi Basic Industries Corp. (Sabic) and a Japanese group led by Mitsubishi.

In other activity involving processing:

- Sabic reports plans for a 500,000 ton/year methyl tertiary butyl ether plant at Jubail. Sabic affiliate National Methanol Co. will build the plant, in which Texas Eastern Corp. and Hoechst-Celanese each will have a 25% interest, and provide feedstock from its 770,000 ton/year methanol plant at Jubail.
- Sabic plans to build a grassroots polypropylene plant at Jubail using Union Carbide Chemicals & Plastics Co. Inc. technology. When complete in early 1993, the single train, two reactor plant will have design capacity of 200,000 metric tons/year. It will be owned and operated by Saudi European Petrochemical Co., a joint venture 70% owned by Sabic.
- Phillips Petroleum Co. completed a Hydrisom process design for an isomerization unit for Petromin-Mobil Yanbu Refinery Co. Ltd.'s refinery at Yanbu.
- Petrokemya, a Sabic affiliate, estimated its ethylene production at 700,000 tons in 1989 (the year for which latest data are available), up from 650,000 tons in 1988, OPEC News Agency reported. Production of polystyrene was 90,000-95,000 tons and butane 30,000 tons vs. 50,000 tons of butane capacity.

SYRIA

CAPITAL: Damascus
MONETARY UNIT: Syrian Pound
REFINING CAPACITY: 237,394 b/cd
PRODUCTION: 385,000 b/d
RESERVES: 1.7 billion bbl

SYRIA IS ENCOURAGING LOCAL AND FOREIGN COMpanies to step up exploration work in anticipation for an expected increase in world demand during the 1990s. In the past 3-4 years, the government has capitalized on light crude discoveries in Northeast Syria by the Royal Dutch-Deminex combine's al-Furat Petroleum Co. to attract additional foreign operators that have licensed most of the prospective acreage in the country.

With exploration activity on the rise, the government in Damascus has turned its attention to the country's underexploited gas resources. Further appraisal of two potentially interesting strikes made in the early 1980s could allow four power generating stations to switch from burning heavy oil to gas.

Cherrife and Ash Shaer fields were discovered by Marathon Petroleum Syria Ltd. about 150 miles northeast of Damascus in 1982 and 1985. But Marathon's production sharing contract expired before appraisal was complete. State owned Syrian Petroleum Co. has gas discoveries in the same area, around Palmyra. Syrian authorities estimate the region holds gas reserves of 2.7 tcf, about half the country's total of 5.54 tcf.

At the end of 1988 Marathon signed a new style production sharing contract designed to encourage gas exploration by foreign companies by allowing them to participate not only in development of reserves but in projects to use the gas. As a result of the deal, Marathon is committed to spend $60 million on exploration on the 9,952 sq km permit and drill 12 wells during an 8 year period. At first, however, the program was aimed at appraising the two earlier gas finds. The program is off to a good start with the first well, 3 Ash Shaer, flowing 40.6 MMcfd.

If the two fields are declared commercial, Marathon will provide as much as $20 million for power stations conversion and pipelines to move 210 MMcfd that will be needed to fuel the staitons. Cost of the pipelines will be recovered as part of the field development project. Longer term, the new gas contract also allows Marathon to get involved in other distribution projects, new power generation, and petrochemicals.

And if there are enough reserves, Marathon could even get into the gas export business.

In other drilling-production events, Ste. Nationale Elf Aquitaine will develop its North Attala discovery in Northeast Syria after receiving government approval for a $100 million project.

The initial 3 year phase will require drilling 21 wells and laying a 12½ mile pipeline into the al-Furat Petroleum pipeline system.

Dair al-Zur Oil Co., owned equally by Elf and Syrian Petroleum, will conduct development operations. Production will be less than 50,000 b/d.

Last year al-Furat Petroleum linked two small fields, Sijan and Ratka, to Omar field production facilities in northern Syria.

The two fields are producing about 11,500 b/d. Two other small fields, Saban and Shdeha, were tied into Omar facilities last year.

On the exploration scene, Occidental Petroleum Corp. unit Damascus Petroleum Ltd. and Syrian Petroleum signed a production sharing agreement covering the 1.7 million acre Al-Nabk block in Southwest Syria. Oxy plans a 1,000 line km seismic survey and four wells the first 3 years on the block, which is immediately north of the 2.8 million acre Bosra block an Oxy group acquired in 1988.

Total Syria and partners found oil on the 9,330 sq km Al Bishri permit in the Euphrates Valley in eastern Syria. The Wadi Aabeid well flowed 6,800 b/d of 15-20° gravity oil from an interval at 6,396-6,757 ft. Total owns a 60% interest in the permit, and units of Petrofina SA and Lasmo plc have 20% each.

In downstream activity, Syrian Petroleum plans modernization on that side of the business, which revolves around two old manufacturing units: the 126,000 b/d Banias refinery on the Mediterranean coast and the 117,000 b/d Homs plant in northern Syria.

The first stage of the modernization process is under way. UOP has won a $1 million contract to conduct a study on how to revamp, expand, and upgrade the two units to ensure they can meet Syria's demand for products throughout the 1990s.

The study, in which Bechtel and Chem Systems, London, will participate, will also take a more wide ranging look at the general hydrocarbon situation in Syria and the domestic supply/demand balance.

Meantime, a new long term customer has been signed for light crude from Syria.

As much as 20,000 b/d will be supplied to Lebanon once the shutdown Tripoli refinery is ready to resume operations.

The crude will be delivered through the final section of the Kirkuk-Tripoli pipeline, which has been closed to Iraqi exports since 1976 by a dispute between Iraq and Syria.

MIDDLE EAST

UNITED ARAB EMIRATES

CAPITAL: Abu Dhabi Town
MONETARY UNIT: Dirham
REFINING CAPACITY: 192,500 b/cd
PRODUCTION: 2.101 million b/d
RESERVES: 98.1 billion bbl

ABU DHABI, BIGGEST OF THE EMIRATES, HAS STARTed a demothballing program that will increase its productive capacity.

Production has been running consistently at about 1.6 million b/d.

The emirate has approved a $500 million program to hike onshore oil production capacity by about 360,000 b/d to 1.125 million b/d by 1995.

A program to hike offshore capacity by about 200,000 b/d to a little more than 1 million b/d is currently under way in Abu Dhabi.

First focus will be Asab field, slated to jump by 60,000 b/d to 280,000 b/d in 1991. A major infill drilling program in Bab field will increase flow to 210,000 b/d in 1995 from the current 40,000 b/d.

Water injection capacity in Bu Hasa field will be upgraded, pushing production capacity by 100,000 b/d to 550,000 b/d by 1994.

Another 15,000 b/d is targeted in Shah and Sahil fields, increasing the combined capacity of the two fields to 45,000 b/d.

Elsewhere offshore, Abu Dhabi National Oil Co. (Adnoc) at last report was working on plans to increase sustainable productive capacity of Upper Zakum oil field to 500,000 b/d from 320,000 b/d by 1992.

Adnoc has an 88% interest in Upper Zakum, where its partner is Jodco of Japan. The project will be one of the biggest offshore operations in the Middle East, requiring installation of five production platforms and drilling of 100 wells.

That will increase the number of Upper Zakum wells to 360.

An integral part of the program will be the start of gas lift using gas from Abu Al-Bukhoosh's nonassociated Khuff reserves, also under development. Khuff gas production from Abu Al-Bukhoosh is to start this summer from a production platform that will be installed under a contract with Technip Geoproduction of France.

Meanwhile, Abu Dhabi has approved plans for a $1.5 billion expansion of its Das Island LNG/LPG chain. A new train will produce 2.3 million metric tons/year of LNG and 250,000 tons/year of LPG beyond existing capacity of 2.3 million tons/year of LNG, 750,000 tons/year of LPG, and 300,000 tons/year of condensate. Expansion is due for completion in 1993-94, and talks are under way with Das Island sole LNG/LPG customer Tokyo Electric to lift extra output.

Abu Dhabi Gas Liquefaction Co. let a $600 million construction contract for expansion of the Das Island LNG plant to Chiyoda Corp. of Japan.

Chiyoda will build a 2.3 million ton third LNG train to double capacity of the plant and add that extra 250,000 tons/year to existing LPG capacity.

In other emirate action, Sharjah awarded a 1,018 sq km onshore exploration concession to a unit of Amoco Corp., OPEC News Agency reported.

Amoco Sharjah is to conduct a seismic survey. The company operates Sharjah's only oil field, Sajaa, which produces about 60,000 b/d.

Dubai, the other substantial producer in the U.A.E., produces a steady 400,000 b/d from its offshore fields.

YEMEN

CAPITAL: Sanaa
MONETARY UNIT: Rial
REFINING CAPACITY: 171,500 b/cd
PRODUCTION: 179,000 b/d
RESERVES: 4 billion bbl

ONE OF THE HOTTEST EXPLORATION PLAYS ON THE Arabian Peninsula is under way in Yemen. The 2,200 sq km area lies between Yemen Hunt Oil Co.'s Marib-Jawf fields and the Shabwa area, where Soviet contractors are placing a series of fields on stream.

In May 1990, North and South Yemen merged into the Republic of Yemen. The countries had been jointly administering a common zone for oil and gas exploration by a combine of Soviet, Kuwaiti, French, and U.S. companies. In 1989 the two former Yemen states ended a long dispute over their border zone.

They created a jointly administered area that attracted many bids from international groups when it was offered for exploration licensing.

A combine led by Hunt-Exxon Corp. and including Total CFP, Kuwait Foreign Petroleum Exploration Co. (Kufpec), and Machinoexport and Zarughgeologia of the Soviet Union acquired the acreage.

The group was obligated to conduct a 1,700 line km seismic survey and spend at least $37 million to drill seven wells during the next 5 years.

Soviet development of Shabwa field moved into a crucial stage last year. A 118 mile, 20 in. pipeline from the fields to Bir Ali on the Gulf of Aden was prepared for start-up. Initial throughput will be 20,000-30,000 b/d, ending the 10,000 b/d tank truck shuttle between Shabwa and the Aden refinery. Technoexport, Soviet operator for field development, has negotiated through the Soviet Ministry of Geology a change of its status from contractor to operator of a production sharing contract on a large tract northwest of the Shabwa producing area, where Yemen will remain as the sole owner.

Last year, a Total CFP, Unocal, and Kufpec combine drilled its second well on the East Shabwa permit after reporting hydrocarbon shows in the first.

Yemen is awarding new acreage. Petro-Canada at last report was at the point of signing an agreement covering the Habrout block near the border with Oman. The tract adjoins the North and South Sanau blocks, where the Ministry of Energy and Minerals has negotiated with a combine of the Bin Ham Group of Abu Dhabi, Tullow Oil plc of Ireland, and Coplex (Yemen) Ltd.

Elsewhere, Yemen Hunt's gross production from the Marib-Jawf fields appears to be steady at about 200,000 b/d following start-up of a fifth field in the area, Asad al-Kamil gas/condensate field.

On the exploration front, Tullow Oil, Dublin, signed a production sharing contract covering North Sanau block. The 14,000 sq km tract is next to a block awarded to Ste. Nationale Elf Aquitaine and near blocks awarded to Total CFP and Occidental Petroleum Corp. Tullow's partners are Bin Ham Group and Coplex (Yemen). Exploration, to get under way this year, is expected to last 3½ years. Work will include field geological studies, seismic surveys, and drilling of two wells.

Meantime, Sun Oil Co. signed a production sharing contract for Block 1 in the East Shabwa area.

On the refining scene, Aden Refinery Co. began a new phase in the revamp of its 161,000 b/d Little Aden refinery. It let contract for design verification and construction supervision of 16 new crude and product storage tanks and refurbishment of seven existing gasoline tanks to Rendel, Palmer & Tritton, England. The 2 year contract also includes pipelines, pumps, and instrumentation.

IPE ATLAS

LATIN AMERICA

ARGENTINA

CAPITAL: Buenos Aires
MONETARY UNIT: Australes
REFINING CAPACITY: 696,285 b/cd
PRODUCTION: 473,000 b/d
RESERVES: 2.28 billion bbl

NOWHERE HAS THE PUSH TOWARD PRIVATIZATION in South America been more dramatically highlighted than in Argentina.

Argentine President Carlos Saul Menem, elected in 1989 as head of the ultranationalist Peronist Party shocked supporters and pleased conservatives by undertaking a drive for free market economic reform and privatization that was a 180° turnabout from the nationalization/labor/populist Peronist policies of the past.

The party's founder, General Juan D. Peron, nationalized Argentina's oil industry—along with most other key industries—after World War II.

Menem claims his policy of allowing private local and foreign ownership of state oil assets is a first for Latin America. From the 1920s until until 1990, about 70% of oil and gas production had been in the hands of the state oil company, Yacimientos Petroliferos Fiscales, and the rest owned by private companies.

Menem's administration has targeted an increase in YPF production to almost 700,000 b/d in 8 years from 448,000 b/d. But YPF is strapped for cash and has limited technological resources, which has opened the door for joint ventures with foreign operators.

New tender areas

YPF was offering four major oil producing areas with identified total combined crude reserves estimated at 256 million bbl.

Vizcachueras produces 13,000 b/d and has 58 million bbl of reserves in Mendoza province 1,000 km west of Buenos Aires. El Huemul produces 13,000 b/d and has 44 million bbl of reserves in the Santa Cruz province of Patagonia. El Tordillo produces 16,000 b/d and has 47.8 million bbl of reserves in Chubut province 1,530 km from Buenos Aires. Puesto Hernandez produces 39,000 b/d and has 106 million bbl of reserves 1,200 km southwest of Buenos Aires.

In all, the country has more than 9,000 producing oil wells. Total proved reserves are about 2.28 billion bbl of oil and about 28 tcf of gas.

Of the 19 sedimentary basins YPF has identified in the country, only five have oil production: Noroeste, Cuyana, Neuquen, San Jorge Gulf and Austral—which laps over into Chile where it is called Magallanes. Not until June 1989 did production start up in the offshore Austral basin, off Tierra del Fuego.

Previous activity

YPF mostly has focused its exploration and production in the Cuyana, Neuquen, and San Jorge Gulf basins.

Up to 1989, YPF had drilled one well per 101 sq km in Cuyana basin, one per 84 sq km in Neuquen basin, and 1 well per 86 sq km in San Jorge Gulf basin.

Exploration activity offshore has been slight. There is one well for every 467 sq km in Austral basin and only one well per 33,800 sq km in the Rawson basin.

Gas action

In the late 1970s, YPF began to step up its search for natural gas reservoirs.

It turned up the 1.32 tcf Loma de la Lata field in Neuquen basin. In the Noroeste basin, YPF tapped Paleozoic pay to open Ramos field, still under development with reserves estimated at more than 600 bcf. The state company also proved another 600 bcf in Aguaragua field, also in Noroeste basin.

In the southernmost tip of Argentina, YPF has developed an important complex of natural gas reservoirs, Candon Alfa and San Sebastian on Tierra del Fuego and various fields tied into the gas processing complex at El Condor south of Santa Cruz province.

LATIN AMERICA

Production data

At yearend 1989, Argentine production was split 69% YPF and 31% private companies.

Production flowed from 9,280 wells with an average daily output of 43 b/d/well in 1989.

About 84% of crude production was primary recovery and 16% secondary recovery.

Argentina became self-sufficient in oil and natural gas in 1982.

In 1989, due to diminishing domestic demand, crude and oil products became the second biggest export after steel products.

Texaco

The biggest player in Argentina's recent push to attract foreign investment is Texaco Petrolera Argentina SA.

Texaco in 1989 signed nine exploration contracts with Argentina's YPF for 11 blocks covering more than 19 million acres.

The contracts were offered by YPF under the fourth round of its Houston Plan, the broad licensing program that began in 1985 but only has been implemented under Menem's administration.

"The Houston Plan is considered the main impact for increased exploration/development activity in Argentina," said S. R. Gregori, Texaco Argentina executive vice-president.

Gregori said, "The drive for privatization is real. Based on government information as of Aug. 31, 1990, about $174 million out of a total commitment of $292 million made through the Houston Plan, has been spent for exploration by private industry."

Texaco Argentina has drilled two discoveries since it acquired its Argentine acreage, on its Petrolera Argentina San Jorge Curamched (CNQ-8) prospect and Pluspetrol Cuchuma (CNQ-11) prospect.

Total combined reserves of oil, gas, and condensate in the former are estimated at 20 million bbl of oil equivalent.

Texaco plans to develop CNQ-8 with 30 development wells to produce about 9,000 b/d of oil. It hasn't set a timetable yet. Reserves are estimated at 15 million bbl of crude.

BELIZE

CAPITAL: Belize City
MONETARY UNIT: Dollars
REFINING CAPACITY: NA
PRODUCTION: NA
RESERVES: NA

PENTAGON PETROLEUM INC., BATON ROUGE, LA., signed an exploration license covering 268,000 acres in Belize.

Pentagon believes its Block V14 in the northwest corner of the country to be within the prolific Peten basin of Mexico and Guatemala.

Pentagon's 4 year exploration license calls for a wildcat to spud by the fourth year.

Consulting geologist Jean Cornec, Benque Viejo, Belize, contends the Pentagon block lies within the most prospective area of the country.

About 100 exploratory wells have been drilled in the 15 million acre Peten basin, and two thirds of the basin has no seismic coverage, Cornec wrote in a report on the concession's hydrocarbon potential.

Cornec cites geologic similarities with Mexico and Guatemala, with prospective targets in Cretaceous carbonate/anhydrite alternations.

EXPERIENCE and advances in technology helped cut Petrobras's drilling costs and time in the Rio Urucu area of the Upper Amazon basin.

BOLIVIA

CAPITAL: La Paz
MONETARY UNIT: Bolivianos
REFINING CAPACITY: 45,250 b/cd
PRODUCTION: 19,200 b/d
RESERVES: 119.182 million bbl

YACIMIENTOS PETROLIFEROS FISCALES BOLIVIANOS (YPFB), Bolivia's state oil company, and private operators completed 35 wells in 1989 (14 exploratory and 21 development) in the country. Bolivia's drilling surge that year posted the most footage in 8 years at 282,191 ft.

Natural gas is the main hydrocarbon resource in the country, providing 78% of the country's hydrocarbon output at 92,367 b/d of oil equivalent.

Of all wells drilled, contractor companies operating in Bolivia accounted for three with total footage of 25,397 ft. Two of those wells, Taiguati-X2 and Escondido-X5, were gas/condensate discoveries in Tarija state.

BRAZIL

CAPITAL: Brasilia
MONETARY UNIT: New Cruzados
REFINING CAPACITY: 1,411,520 b/cd
PRODUCTION: 633,000 b/d
RESERVES: 2.84 billion bbl

BRAZIL'S STATE OIL COMPANY IS EMBROILED IN

LATIN AMERICA

NAMORADO 1 platform off Brazil contributes to Campos basin production of 433,000 b/d, more than two thirds of the country's total output.

a national dispute over its role in the country.

The central issue is whether to continue the monopoly established for Petroleos Brasileiro SA on Oct. 3, 1954, or continue moves toward privatization and free market reform.

The new president of Petrobras, Eduardo de Freitas Teixeira, at 36 the company's youngest head ever, has continued the tone of his embattled predecessor by attacking the state oil monopoly concept and defending the return of risk contracts for private domestic and foreign companies.

Although a wave of deregulation and economic reform swept President Fernando Collor de Mello to power, monopoly is still a sacred cow in Brazil, supported by powerful interests.

Petrobras is Brazil's largest company with gross annual revenues of $16 billion and a net stockholder equity of $8 billion.

Company sources estimate that more than 10% of Brazil's gross domestic product is directly or indirectly related to Petrobras activities.

Petrobras and subsidiaries employ 67,000 people, which translates to about 3 million indirect jobs. The Association of Petrobras Engineers (Aepet) claims that 10 million Brazilians depend directly or indirectly on the petroleum industry.

Financial woes

Brazil's financial woes at the beginning of 1991 had worsened sharply since the beginning of the Persian Gulf crisis.

Teixeira, an economist, estimates Brazil's oil import bill in 1990 at about $5 billion. About $2 billion of that total would have represented the increase in oil prices beginning in summer 1990.

Petrobras estimated that during January-October 1990 Brazil spent $3.6 billion for imported crude and products, up about 26% from the same period in 1989. The country was importing about 180,000 b/d of crude from Iraq and Kuwait before an international embargo removed those countries' exports from the market. Most of those imports had been replaced with supplies from Iran, Africa, and South America—particularly Venezuela, which increased its exports to Brazil to 92,000 b/d from 20,000 b/d.

The problem with Venezuela's crude is that it is ultimately more expensive for Brazil than the crude from the Persian Gulf. That's because the heavier gravity crude is costlier to process and requires increased runs to get the same products yield, freight costs are higher because Venezuela uses smaller capacity tankers than do the gulf exporters, and Brazil is unable to get barter deals for its goods and services—especially the case with Iraq—because Venezuela insists on payment in dollars.

Petrobras cash squeeze

Defenders of increased domestic investments say that if the federal government, which controls the company's budget, had allowed a budget hike of $350 million Petrobras asked for 2 years ago for investment in deepwater Campos basin projects, Brazil would now be producing an extra 200,000 b/d.

Petrobras is still being squeezed by subsidies to support government controlled fuel prices.

Since oil prices jumped in fourth quarter 1990, the average price Petrobras paid for crude shot to as high as $33.56/bbl while the average price it received for domestic products sales was $15.62/bbl.

The company complains that despite price hikes its oil products price levels are still not compatible with adequate return on capital invested. But the government is obliged to keep a lid on prices in line with its anti-inflationary measures.

Petrobras' spending target of $16.9 billion breaks out as $2.5 billion in 1991, $3.1 billion in 1992, $3.4 billion in 1993, $3.7 billion in 1994, and $4.2 billion in 1995. Of that total, Petrobras has earmarked $12.2 billion for E&D.

Spending of $9 billion for 1991-93 correspond to the $3 billion/year Petrobras spent in 1981-83, when Campos basin discoveries helped Brazil pass the 500,000 b/d mark.

Petrobras pointed out that after 1994, cash flow generated by new projects going on stream could then sustain the added E&D investment.

Reducing dependence

Brazil's oil focus now is on reducing dependence on oil imports by increasing domestic production.

After 2 years of near stagnation on E&P work because of deep budget cuts, the government reversed itself on oil policy in the wake of the Persian Gulf crisis by approving revived spending for development of supergiant Marlim field—hitherto the victim of budget cuts—in the deepwater Campos basin.

Brazil at the beginning of 1991 was producing 670,000 b/d of crude, up 8.6% from 1989's average, and imports about 600,000 b/d.

Its proved reserves of crude total about 2.8 billion bbl. Including probable and possible reserves in existing fields roughly doubles that number, yielding an estimate of sustainable production of 900,000 b/d.

Petrobras says that without increased E&D spending, Brazil's crude output will drop to about 400,000 b/d by 1995. However, domestic consumption is expected to rise to 1.4 million b/d by that time.

Critical to Brazil's production outlook is development of

SOUTH AMERICA

ARGENTINA	NO. REFINERIES
Zamora	1
Compana	1
Galvan	1
Bahia Blanca	1
Buenos Aires	1
Duran	1
Dock Sud	1
LaPlata	1
Lujan deCuyo	1
Plaza Huincul	1
San Lorenzo	1

SOUTH AMERICA, SOUTHERN PORTION

BRAZIL	NO. REFINERIES
Parana	1
Betim	1
Rio Grande do Sul	2
São Paulo	4
Rio de Janeiro	2
Fortaleza	1
Manaus	1
Bahia	1

LATIN AMERICA

Marlim and another Campos basin supergiant field, Albacora.

E&P renewal

Despite Brazil's problems, there are signs of renewed oil and gas exploration and production activity in the country.

Brazil has the fifth largest sedimentary basin area in the world and there remains much to be explored.

Petrobras onshore exploration will focus in the Potiguar basin in Rio Grande do Norte State, Solimoes basin in Amazonas state, and Parana basin.

Petrobras is anxious to discover oil and gas in Parana basin, where it will return after an absence of 9 years, because it is in the south central portion of the country, which accounts for most of the nation's oil and gas consumption.

In 1990, Petrobras drilled 11 wells in Sergipe-Alagoas basin, 14 in Espirito Santo basin, and two in the Lower Amazon basin. In all, 122 wells—88 onshore and 34 offshore—were drilled in 1990 vs. 111 in 1989.

At year-end 1990, more than 4,500 b/d of oil were produced in the Amazon. Petrobras hopes to increase Amazon production to 12,000 b/d by yearend 1991. That calls for another 17 wells, bringing the Amazon total to 35.

Santos basin action

Petrobras is highgrading oil prospects in the Santos basin, where it discovered reserves estimated at 400 million bbl.

The 350,00 sq km Santos basin lies off Rio de Janeiro state's southern coast, Sao Paulo and Parana states, and northern Santa Catarina state. This represents a total area of 350,000 square km. The area considered prospective for hydrocarbons totals 130,000 square km.

Oil and gas exploration in the Santos basin got under way in the late 1960s with the first seismic surveys undertaken by Petrobras.

The first Santos basin wildcat, 1-PRS-1, was drilled off Parana state in 1970.

A new exploration phase started in 1976 with the signing of the first risk contract in the Santos basin. Since then, foreign companies have drilled 29 wildcats, resulting in one commercial gas/condensate strike, Merluza, and five noncommercial hydrocarbon shows.

Pecten activity

A Pecten Brazil official said Merluza production is expected to start up in the early 1990s with a $370 million project.

Merluza lies in 430 ft of water about 115 miles from Sao Paulo.

Merluza produces from Upper Cretaceous Itajai sandstone at a depth of about 14,764 ft.

Condensate gravity is 44°, and the gas has 9% methane content.

Pecten didn't disclose reserves or production estimates, but Petrobras estimated Merluza reserves at 303 bcf of gas and 10.6 million bbl of condensate.

Other Santos basin action

When foreign companies relinquished their risk contract areas to Petrobras in 1986, the Brazilian company renewed its exploration campaign in the Santos basin.

That led to the 1988 discovery of Tubarao field off Santa Catarina state on a block previously held by a foreign operator.

Petrobras estimates reserves at 50 million bbl of crude and 212 bcf of gas.

In 1990, Petrobras drilled a discovery 12 km south of Tubarao on a block previously held by BP. Petrobras estimates reserves in the field, named Estrela do Mar, at 75 million bbl of crude and 106 bcf of gas.

The third field in Santos basin, Coral, was tapped by 1-BSS-56 170 km from the coast and 13 km south of Estrela do Mar on a former BP block.

Petrobras estimates Coral reserves at 155 million bbl of crude and 212 bcf of gas.

CHILE

CAPITAL: Santiago
MONETARY UNIT: Chilean Peso
REFINING CAPACITY: 145,800 b/cd
PRODUCTION: 20,300 b/d
RESERVES: 300 million bbl

CHILE'S STATE OWNED EMPRESA NACIONAL de Petroleo (ENAP) historically has concentrated its oil exploration efforts in the Magallanes basin, site of its only commercial hydrocarbon production.

The Magallanes basin produces about 22,000 b/d of oil. Of that total, 65% comes from fields under development by 36 platforms in the Straits of Magellan in Chile's southern tip.

In 1989, ENAP sharply boosted its use of Magallanes natural gas with start-up of a gas processing plant there. Natural gas volumes sold in 1989 increased 107% from the previous year.

Domestic oil production supplies only 15% of Chile's needs.

In 1988, Chile produced an average 24,397 b/d. In 1989, Chile imported about 95,000 b/d of oil at a cost of $642 million. ENAP also has taken on foreign partners in exploration ventures

Hunt Oil Co. is exploring a 5,000 sq km concession at Salar de Atacama.

A second Hunt concession covers exploration in the Altiplano of Arica.

Pecten also has a joint venture with ENAP to explore rank wildcat acreage in the Salar de Pedernales area.

Its concessions cover the 3,000 sq km Chiuchiu block north of Calama and the 5,000 sq km Salar de Pedernales block.

ENAP also plans to explore the 300 sq km Depression Intermedia de Arica.

LATIN AMERICA

COLOMBIA

CAPITAL: Bogota
MONETARY UNIT: Pesos
REFINING CAPACITY: 247,400 b/cd
PRODUCTION: 445,000 b/d
RESERVES: 2 billion bbl

COLOMBIA PLANS TO STAY THE COURSE ON OIL policies that turned it into a significant crude exporter in the 1980s.

A strong exploration and development program pushed by state oil company Empresa Colombiana de Petroleos (Ecopetrol) separately and in concert with foreign operating companies have made energy the mainspring of Colombia's economic growth the past few years, broadening its economic base significantly.

Colombia's gross domestic product grew by about 3% in 1989, among the highest growth rates in Latin America. Crude oil output increased by 7.6% to 404,283 b/d in 1989. Average 1990 oil production to December was 435,000 b/d of oil and 390 MMcfd of gas, Vergara Munarriz said.

Colombia is lagging its targets for replacing that increased production, however. Ecopetrol and associate contract partners have added 140 million bbl of reserves since 1987, compared with an official target of 400 million bbl for the period. Increased production sliced Colombian reserves during 1987-89 to 1.75 billion bbl from 2.2 billion bbl.

As a result, Ecopetrol is stepping up E&D, targeting another 350 million bbl with a 5 year, $540 million program. Estimates of the foreign contribution to that effort run as high as $5 billion by 2000.

Ecopetrol faces the same problem plaguing other state oil companies in Latin America and elsewhere: a revenue drain caused by domestic fuels and other subsidies.

The problem is certainly not one of petroleum potential. Cano Limon, the Llanos basin field in eastern Colombia responsible for the country's newly found crude exporter status, is one of only a tiny handful of supergiant oil fields discovered in the 1980s. Colombia's proven oil reserves are estimated at about 2 billion bbl.

Exploration potential

Colombia holds large, highly prospective areas that have not been exploited for hydrocarbon potential.

Revised data published in February 1990 by Ecopetrol indicate Colombia's sedimentary basins cover 695,000 sq km onshore and 50,800 sq km offshore. That means more than half of Colombia's territory has hydrocarbon potential. Current plans call for about 70 exploratory wells/year in Colombia.

Llanos update

Colombia's Llanos basin continues to be its most important oil producing province.

The Llanos sedimentary basin covers an area of more than 200,000 sq km. Heavy rains greatly affect activity in the region.

Cano Limon, discovered in 1983 by a unit of Occidental Petroleum Corp., has produced more than 277 million bbl of crude through June 1990. Its original estimated reserves were slightly more than 1 billion bbl.

Production at Cano Limon has ramped up from 35,000-40,000 b/d from six wells in 1985. First exports from Cano Limon began in April 1986.

By June 1987, Cano Limon was producing more than 180,000 b/d. For the whole of 1987 the field produced 187,000 b/d. Cano Limon production slipped in 1988 to about 163,000 b/d, then rebounded to 172,000 b/d in 1989.

Other areas

The eastern Llanos continues to be the most promising oil region, but there are said to be good expectations for areas such as the Upper Magdalena Valley, Middle Magdalena Valley, Putumayo, and on the Atlantic Coast area mainly for natural gas.

First oil exploration in Colombia took place in the Middle and Upper Magdalena River valleys, with several large heavy oil discoveries and some smaller light oil fields.

The Middle Magdalena is Colombia's second most important producing basin. Infantas field, discovered in 1918, is the oldest in the country and has produced 226 million bbl of oil. With more than 300 wells drilled, Infantas still produces from about 200 wells.

La Cira, discovered in 1926 to the northeast of Infantas, has produced more than 450 million bbl and still has 600 active wells.

Also in the Middle Magdalena basin, Casabe field is undergoing a major waterflood. Casabe has produced more than 240 million bbl and still produces from about 330 wells.

Velasques field is producing about 2,900 b/d. The 1946 discovery has produced about 170 million bbl. In the same Middle Magdalena area, Provincia field is currently the top producer, yielding more than 14,500 b/d. Discovered in 1962, it has yielded a cumulative 165 million bbl.

Gas fields dominate in the Lower Magdalena basin. A key exception was the 1985 discovery of San Francisco field. It produces more than 22,000 b/d from some 60 wells in Cretaceous pay, with total production to date 42 million bbl.

Ecopetrol plans

In the near term, Ecopetrol plans increases in petroleum sector spending in addition to increased E&D spending. Specific sums weren't disclosed.

At the top of the list is added outlays for construction of a new refinery, with an initial capacity of 75,000 b/d, to start up in 1996. Ecopetrol spending increases also will go towards expanding the country's crude and oil products pipeline network.

The state company also will hike investment in the government's natural gas grid expansion, directly or through joint ventures and/or through loans to pipelines and natural gas distributors. Another priority for Ecopetrol is renewed growth in the petrochemical sector.

COSTA RICA

CAPITAL: San Jose
MONETARY UNIT: Colones
REFINING CAPACITY: 15,000 b/cd
PRODUCTION: NA
RESERVES: NA

INSTITUTO COSTARRICENSE DE ELECTRICIDAD (ICE) of Costa Rica received a $182.8 million loan from Inter-American Development Bank for a $264.1 million electrical power project.

ICE will build the 55,000 kw Miravalles II geothermal plant, the 24,000 kw Toro I and 66,000 kw Toro II hydroelectric plants, and a substation and expand six other substations.

First bidding round

Costa Rica received bids from seven independent oil companies for offshore and onshore acreage under its first formal round of bidding for exploration acreage.

The sparsely drilled nation received bids for exploration rights covering 11 blocks, mostly in its northeastern interior

LATIN AMERICA

and coastal regions.

The bidding was the first since the government established Refinadora Costarricense de Petroleo SA (Recope) to coordinate hydrocarbon exploration, development, and refining in Costa Rica. The country has no oil or gas production.

Typical work program under a concession would involve running regional seismic surveys and drilling a wildcat in the next 3 years.

Blocks 1 and 3 in the Limon Sur basin drew bids from a group made up of Albion International Resources Inc., Laguna Beach, Calif.; Aberdeen Petroleum plc, London; Blackland Exploration Ltd., London; Overseas Exploration Corp., Yorba Linda, Calif.; and Davis & Namson Consulting Geologists, Glendale, Calif.

Blocks 1 and 3 are in an overthrust play area on the western edge of the Panama deformed belt.

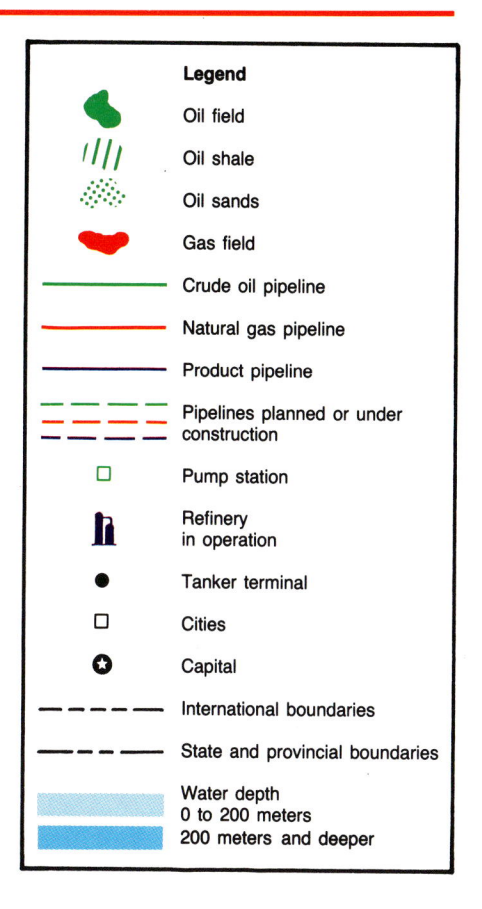

Although most of Costa Rica's wells have been drilled in the Limon Sur basin, studies suggest several new targets for exploration, said Western Atlas International (WAI), Houston.

EXOK Inc., Oklahoma City, bid on Blocks 7, 11, and 12 covering most of the onshore and shallow water areas of the San Juan River Delta, an undrilled area.

Mallon Oil Co., Denver, bid on two groups of blocks: 4, 5, and 6 in the Limon Norte basin, which is an area of broad basement anticlines, and 8, 9, and 10 in the San Carlos basin near the Pataste oil seeps.

Recope offered 27 blocks covering more than 54,000 sq km of acreage onshore and in generally shallow waters offshore in the first round.

Future bidding will involve, in addition to reoffered blocks, another 18 offshore blocks in waters generally deeper than 200 m.

CUBA

CAPITAL: Havana
MONETARY UNIT: Pesos
REFINING CAPACITY: 280,000 b/cd
PRODUCTION: 15,000 b/d
RESERVES: 100 million bbl

A TOTAL UNIT, ALONG WITH ANOTHER FRENCH concern signed the first production sharing agreement between Communist Cuba and a foreign company.

Total Exploration Cuba in tandem with Cie. Europeene des Petroles (CEP), a subsidiary of the trading group

LATIN AMERICA

Interagra, signed their production sharing agreement with Cuba's state owned Union del Petroleo de Cuba.

Under the contract, Total and CEP are committed to conduct 1,800 line km of seismic surveys and drill four wells during a 6 year period.

Their 2,000 sq km block is in the Santa Clara area off Cuba's northern coast, about 160 km east of Havana. It is adjacent to a block in Cardenas Bay where Soviet agencies reportedly discovered oil and gas.

Most of Cuba's 16,000 b/d of oil production comes from small fields along the north coast. Cuba has been heavily dependent on imports of oil from the Soviet Union, which were curtailed sharply in 1990.

New products pipeline

Cuba completed a 29 mile, 12 in., 50,000 b/d products line from the port of Matanzas on its north coast to a power plant in northern Havana.

Spain's Empresa Nacional de Ingenieria y Tecnologia SA handled engineering services.

DOMINICAN REPUBLIC
CAPITAL: Santo Domingo
MONETARY UNIT: Peso
REFINING CAPACITY: 48,000 b/cd
PRODUCTION: NA
RESERVES: NA

MOBIL CORP. OBTAINED AN OFFSHORE/ONSHORE concession in the Dominican Republic, spurring the first exploration in that country in almost a decade.

Mobil Exploration Dominican Inc.'s agreement covers 5.6 million acres in the southern Dominican Republic, covering about one third of the country.

The company will conduct geological and geophysical field work that includes gathering, processing, and interpreting 1,500 line km of seismic data in the first 2 years of the concession agreement.

Focus of the exploration work will be with oil and gas prospects onshore and offshore in the Azua and San Pedro basins. In the first 2 years of the contract, Mobil will spend about $3 million for concession rights and geological and geophysical data.

Mobil must await ratification of the contract by the Dominican congress. Mobil expected to begin the work within 3 months of ratification.

Superior Oil Co., which Mobil acquired in 1984, drilled 7-8 noncommercial wells in the Dominican Republic's Enriquillo basin. The last well, thought to be the island nation's deepest, was drilled in 1981.

ECUADOR
CAPITAL: Quito
MONETARY UNIT: Sucres
REFINING CAPACITY: 141,800 b/cd
PRODUCTION: 287,000 b/d
RESERVES: 1.42 billion bbl

A LITTLE MORE THAN A YEAR AFTER ITS RESTRUCturing state owned Petroleos del Ecuador is poised to play a broader role as oil operator in Ecuador.

The new Petroecuador, consisting of several independent units and a central governing body, has expanded its involvement in all phases of the industry, from exploration and production to transportation, refining, and marketing.

Petroecuador hiked its initial budget for 1990 by about 30% from 1989, with added investment primarily earmarked for exploration and production.

However, owing to a need for a general cut in public spending, the state company's budget was trimmed about $80 million in second half 1990.

This reduction was not expected to affect exploration and production.

Current E&P

Petroecuador's activities in exploration and production are now carried out through Petroproduccion and Petroamazonas.

Petroproduccion operates fields in the northern sector of the Oriente basin: Bermejo, Libertador, Tetete, Shushuqui, Sansahuari, Cuyabeno, and Frontera.

Petroamazonas has taken over Texaco Inc.'s role as operator of the Shushufindi area. The two units will be merged in 1992, when the association with Texaco concludes.

In first half 1990, Petroproduccion spent about $35 million for seismic work—including acquisition, processing, reprocessing—one wildcat and 14 development wells, and production equipment and pipelines.

Fields operated by Petroduccion will sustain a daily production of about 65,000 for the next 2 or 3 years.

Petroamazonas has operated the Petroecuador-Texaco block since July 1990.

Since takeover, production has increased 15,000 b/d to about 225,000 b/d. This output will be maintained the next 2 years.

Petroamazonas drilling plans for 1991 and 1992 provide for wildcats close to production: Palanda and Conga, south of Yuca field, Anaconda, west of Yuca, Pindo, east of Auca, and Palo Rojo, north of the consortium area. Petroecuador will assume all costs for the wildcat program.

Preliminary steps for transfer of Texaco's activities in the

PETROAMAZONAS refinery processes 10,000 b/d in the Shushufindi area of Ecuador's Oriente jungle region. Petroamazonas is a unit of state oil company Petroleos del Ecuador.

LATIN AMERICA

consortium when the contract ends in 1992 have begun with some auditing on inventory and ancillary technical and financial matters.

These measures won't preclude Texaco's right to bid on the tender for a major enhanced oil recovery project in giant Shushufindi field, to have gone out early in 1991.

Crude produced by Petroproduccion and Petromazonas-Texaco will continue to be carried by the Lago-Agrio pipeline, operated now by Petrotransporte, another unit of Petroecuador.

Current refining-marketing

Petroecuador's role in Ecuador's refining sector expanded with its takeover of what had been the last privately owned refinery in the country: a small topping plant in Santa Elena Peninsula, close to the Anglo refinery.

With this acquisition, made in 1989, Petroecuador's refining capacity rose to 140,000 b/d. Petroecuador also operates two small refineries at Santa Elena, the 90,000 b/d Esmeraldas refinery, and the 25,000 b/d Shushufindi refinery. Petroecuador exports diesel, gasoline, and fuel oil.

After Iraq's invasion of Kuwait, Ecuador decided to increase production by 15,000 b/d to about 300,000 b/d.

GUYANA

CAPITAL: Georgetown
MONETARY UNIT: Dollar
REFINING CAPACITY: NA
PRODUCTION: NA
RESERVES: NA

EXPLORATION HAS PICKED UP IN GUYANA.

An affiliate of Hunt Oil Co. obtained a petroleum exploration license and production sharing agreement in the Takutu graben area of Guyana's dense jungle interior. The block contains Guyana's first oil discovery.

Total in 1990 took a farmout on the Essequibo permit off Guyana originally awarded to an affiliate of ECI Petroleum, Houston. Plans call for seismic surveys and a wildcat.

MEXICO

CAPITAL: Mexico City
MONETARY UNIT: Pesos
REFINING CAPACITY: 1,679,000 b/cd
PRODUCTION: 2.633 million b/d
RESERVES: 51.983 billion bbl

MEXICO'S RETURN TO THE WORLD OIL MARKET AS a major player is unlikely but should not be ruled out, says George Baker, Berkeley, Calif., oil consultant.

Mexico's ability to profit from rising oil prices is severely constrained by the cumulative effect of 5 years of underinvestment in exploration and production.

The crisis provoked by the Iraqi invasion and takeover of Kuwait brings to mind Mexico's basic argument since the late 1970s that it is the West's best source of crude oil outside the Organization of Petroleum Exporting Countries.

Initial benefits

In a more perfect world, Baker says, Mexico would benefit from the crisis in four ways.

Petroleos Mexicanos, the state oil company, benefits in the first instance because the export price formula in its long term contracts is sensitive to changes.

Pemex bills its crude export clients retroactively based on average daily prices for the billing period, according to an agreed price weighting formula.

Rising export prices, therefore, will bring up the average monthly price on existing, long term contracts.

Second, during the last oil boom, the first commandment of Mexico's crude export policy was: Thou shalt not sell on the spot market.

Along with other sweeping changes such as the shakeup of the oil union leadership in early 1989, the administration of Carlos Salinas de Gotari has dropped this commandment.

In 1989, a new subsidiary, Petroleos Mexicanos International (PMI) was formed for reasons that at the time were not

UNITED STATES
TEXAS
LOUISIANA
MISSISSIPPI
New Orleans
Houston
Corpus Christi
El Paso
Ciudad Juarez
Celulosa de Chihuahua
CHIHUAHUA
Chihuahua
Delicias
Cuidad Camargo
Jimènez
MEXICO
COAHUILA
Nueva Rosita
Sabinas
Nuevo Laredo
Laredo
Monclova
Gomez Palacio
Torreón
Parras
Saltillo
Monterrey
Apodaca
Cadereyta
NUEVO LEON
Reynosa
MacAllen
Brownsville
Matamoros
San Fernando
TAMAULIPAS
Soto la Marina
Ciudad Victoria
SINALOA
Culiacán
DURANGO
Durango
Mazatlán
ZACATECAS
Mateahula
SAN LUIS POTOSI
San Luis Potosi
Zacatecas
NAYARIT
San Blas
Chicura
Canuch
Jayabo
Aguascalientes
Lagos
Leon
Guanajuato
GUANAJUATO
Irapuato
QUERETARO
Queretaro
Salamanca
Celaya
La Barca
Guadalajara
Puerto Vallarta
JALISCO
Autlán
COLIMA
Manzanillo
Tecoman
Lago de Cuitzeo
Morelia
MICHOACAN
Ebano
Ciudad Madero
Tampico
Naranjos
Tuxpan
Tianchinal
Poza Rica
HIDALGO
Pachuca
Tula
Mexico City
Toluca
MEXICO
Cuernavaca
Azcapotzalco
Apizaco
Tlaxcala
Puebla
PUEBLA
Orizaba
Tierra Blanca
Veracruz
Alvarado
VERACRUZ
Coatzacoalcos
Minatitlan
Cardenas
Villahermosa
Pemex
Cactus
TABASCO
INSET SHOWN
Xicalango
Ciudad d
OAXACA
Oaxaca
Tehuantepec
Juchitán
Salina Cruz
CHIAPAS
GUERRERO
Chilpancingo
Acapulco
Rio Mexcala
Pacific Ocean
Gulf o
Golfo de Campeche
Golfo de Tehuantepec
Maloob
Bocab
Kutz
Ixtoc
Akal
Nohoch
Chac
Luch
Kanaab
Abkatun
Takin
36 in. 91 cm
Chiapas-Tabasco area
Gulf of Campeche
Xicalango
Dos Bocas
Cunduacan
Ojiacaque
El Golpe
TABASCO
Cardenas
Samaria
Villahermosa
Iride
Paredon
Ciudad Pemex
El Plan
Nispero
Cactus
Rio Nuevo
Sitio Grande
Agave
Sabancuy
Coatzacoalcos
Nanchital
Pajaritos
Cangrejera
La Venta
Minatitlán
Nuevo Teapa
VERACRUZ
CHIAPAS
C. Nanchital
Rio Teapa
25 Mi
40 Km.

clear, says Baker.

The current international crisis has brought forth what may have been the underlying reason: the administrative ability to sell crude oil on the spot market. If crude supplies are available, PMI could take advantage of spikes in oil prices.

Capital needs

In the third place, Baker says, Pemex's ability to raise money abroad for capital projects will be enhanced by a sustained increase in world oil prices.

The main item needing outside funding is the Cantarell Project, which needs a benefactor with about $800 million in his pocket.

The Cantarell Project is one that indirectly addresses the cumulative effect of a government policy of austerity that includes exploration and production.

Without the Cantarell Project, which was conceived as an investment package that might be attractive to funding agencies such as the Ex-Im Bank of Japan, Mexico's ability to hold onto its present share of the oil market is in jeopardy.

Increasing internal demand and falling crude production undermine Mexico's ability to assume its former prestigious role as a crude supplier to the U.S.

Soaring domestic demand

While Pemex's production of refined products in 1989 was only 4.29%

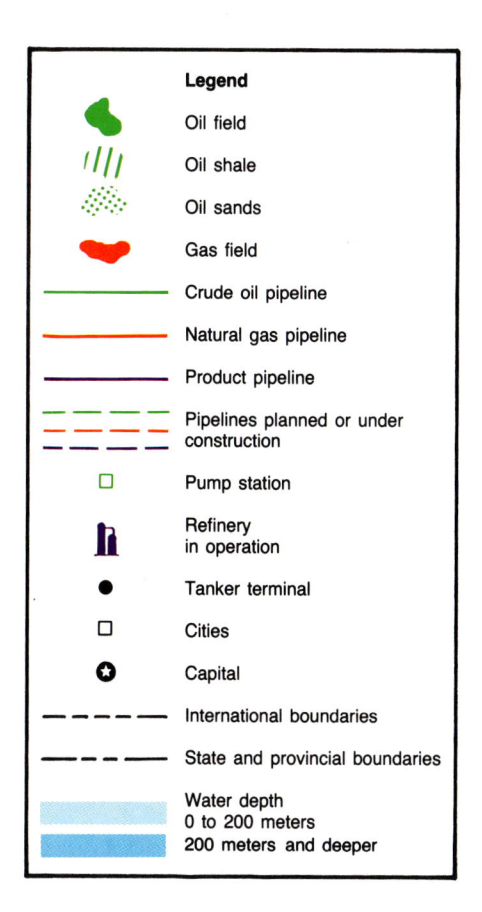

LATIN AMERICA

OCCIDENTAL Petroleum Corp. wells are under way on Block 1-AB in Peru's northern jungle. The block provides about half of Peru's crude production.

above that of 1988, Mexican imports told another story:
- Fuel oil imports rose 17.7% to 69,500 b/d in 1989 from the year before.
- Gasoline imports in 1989 shot up almost 1,500% to 28,200 b/d from 1988's 2,100 b/d.
- Natural gas imports in 1989 soared to an average 36.4 MMcfd from 6.3 MMcfd in 1988.
- Petrochemical imports jumped to 55,498 metric tons in 1989 from 34,433 metric tons in 1988.

In the absence of increased production, such increases in internal demand must come at the expense of exports. In 1989, Mexican gasoline exports fell to 394 b/d from 16,298 b/d in 1988. Overall petroleum product exports in 1989 were down 30.9% from the year before. Further, Pemex's petrochemical exports fell 14% in the year to year comparison.

Mexico's potential

Most U.S. observers, says Baker, believe Pemex is not developing the full potential of Mexico's undiscovered petroleum resources.

Two issues are at stake: the recovery factor for known fields and the question of undiscovered resources.

The word at Pemex is that the official 1981 figure of 70 billion bbl of oil equivalent (BOE) of proved reserves was first arrived at using a 40% recovery factor. Recent estimates have reduced this figure to 20-30%.

However, that original reserves number includes an estimate of reserves in the Chicontepec complex of fields of 10 billion bbl of oil, 26 tcf of gas, and 1.3 billion bbl of condensate.

Pemex first estimated a recovery factor for Chicontepec at 20% but has cut that to 5%. Chicontepec production is likely to average only about 100 b/d/well, and full development would involve 10,000 wells.

U.S. private investor groups have sounded out Pemex on the possibility of a risk contract agreement to help develop Chicontepec. Pemex has rebuffed those efforts so far, and Chicontepec still is not producing.

As for undiscovered resources, Pemex emphasizes that only a small percentage of hydrocarbon potential areas have been explored. Bernardo F. Grossling, a petroleum geologist with the U.S. Geological Survey, believes there is a minimum reserves potential of 100 billion BOE in the Reforma-Campeche region out of a potential resource base of 260 billion BOE at 80% probability, with an upper limit estimate of 700 billion BOE at 10% probability.

Pemex's published figures place onshore Reforma reserves at 13.7 billion BOE and offshore Campeche reserves at 30.7 billion BOE.

U.S.-Mexico relations

The fourth way Mexico stands to gain from higher oil prices stemming from an extended crisis in the Middle East lies in gaining leverage in U.S.-Mexican government relations.

Mexico might propose, for example, that for every $1 less of international debt servicing expense Pemex will invest an additional 50¢ in exploration/production. The other 50¢ would be set aside for other urgent programs such as public health, education, and agriculture. Mexico's proposal to increase crude production in the Bay of Campeche by 100,000 b/d for 60 days was an appreciated symbolic gesture, said U.S. Treasury Sec. Nicholas Brady.

In Mexico, however, government critics insisted that such output, if achieved, would be at a net loss to Mexico:

"The cost of the damage to oil fields from overproduction and the waste of flared natural gas will easily offset the additional revenues that Pemex may receive," one critic said.

PARAGUAY

CAPITAL: Asuncion
MONETARY UNIT: Guaranies
REFINING CAPACITY: 7,500 b/cd
PRODUCTION: NA
RESERVES: NA

PARAGUAY DOES NOT PRODUCE PETROLEUM, despite extensive exploration by 11 international companies.

There are, however, indications of hydrocarbons in the western and eastern regions of the country. A large part of Paraguay's territory, 406,752 sq km, shares sedimentary basins with neighbors Brazil, Bolivia, Argentina, and Uruguay. Argentina has oil production only a few miles from Paraguay's bordering Formosa province. In 1989 Paraguay had a stockpiled crude and products totaling an equivalent 1.1 million bbl of oil, enough for 90 days' consumption.

PERU

CAPITAL: Lima
MONETARY UNIT: Intis
REFINING CAPACITY: 188,820 b/cd
PRODUCTION: 132,000 b/d
RESERVES: 405.937 million bbl

REVIVAL OF PERU'S MORIBUND OIL AND GAS

LATIN AMERICA

industry in the 1990s hinges on whether the new administration of President Alberto Fujimori is successful in attracting foreign investment to Peru.

Fujimori's success would mean Peru pushing ahead into stepped up exploration and major development projects, such as the huge Camisea gas/condensate field discovered 2 years ago. His failure could mean Peru continuing to fall further behind in its already lagging low oil production.

Barriers to be lifted

The government was preparing to lift two barriers to investment: the lack of an international arbitration agreement to guarantee against nationalization, and Petroperu's monopoly on oil refining and distribution with the consequent price controls.

The Fujimori administration's declared policy is to seek foreign investment for exploration and development and possibly also for refining and marketing.

Opposition to foreign investment is expected to remain strong. Left wing political opposition, coupled with difficulties in raising credit, played a big part in stopping Camisea development in 1988.

Minister of Energy and Mines Fernando Sanchez, a member of Izquierda Socialista party, has accused the far left Izquierda Unida party of trying to undermine the government by blocking investment in Camisea through harassment of major oil companies operating in Peru.

The remote Camisea field, in the Amazon jungle out of reach of roads or rail lines, could cost about $2.2 billion to develop.

Near term outlook

Petroperu expects oil production in 1991 to remain close to the average 130,000 b/d produced in Peru the past 2 years.

The state oil company estimates domestic demand will average 110,000 b/d, leaving a 20,000 b/d surplus for export. Export revenue will be offset, however, by the cost of importing about 18,000 b/d of light crude to cover refinery runs.

Exports through the end of August 1990 were valued at $121.5 million. But oil imports during the period cost $172.2 million. Petroperu had expected a deficit of only $5 million during September-December 1990 following a fall in local demand from first half 1990 levels.

Domestic demand rebounded to 100,000 b/d the first week of November after falling to 40,000 b/d following August price hikes.

If the government succeeds in reviving the economy next year to 1986-87 levels, it would, however, take demand back up to 132,500 b/d, according to Petroperu estimates.

Crude oil production was down to 123,000 b/d by October 1990 vs. 142,000 b/d in October 1989. Output fell still further in November 1990 to 118,100 b/d.

Flow could also fall further in 1991, following the trend of the past 8 years.

Renewed exploration interest

Much of the revived interest is focused on or near the large Block 8, which Petroperu wants to carve up for operating contracts for private companies.

The company says it is negotiating with Occidental Petroleum Corp. for a new contract in modified Block 40—to include part of Block 43 outside Manu National Park—in the Madre de Dios basin, although Occidental earlier withdrew from Block 43 because of environmental concerns.

A local company, Cia. Petrolera San Juan SA, has renewed negotiations for Blocks Grau 1 and Grau 2 south of the Talara and Secura basins along the northern coast.

ER Operating Co., Dallas, is negotiating an exploration and development contract in the Lobitos area and the Block B area of the Talara basin.

Energy World Trade Ltd. is interested in exploration and development in the Maranon and Ucayali basins. It also has shown interest in participating in developing Chambira field in Block 8.

Santa Fe Energy Resource Inc. requested information on the Maranon basin apparently also in or near Petroperu's Block 8.

Chevron Corp. is interested in the Ucayali and Madre de Dios basins and has held preliminary talks with Petroperu.

Tripetrol requested information on the Maranon basin.

American International Group, insurers for the expropriated Belco oil field assets, seeks data on offshore Blocks Z1 and Z28, where it hopes to recoup compensation paid to Enron Corp., Belco's present owners.

Total CFP has requested information covering possible exploration and development work on offshore Blocks Z1, Z2A, and Z28.

Petroperu 1991 program

Petroperu scheduled a modest exploratory drilling program for 1991, subject to funding.

Along the northwest coast, the company plans to drill five wells at a cost of $3.5 million targeting postulated reserves potential of about 2 million bbl.

In the northern jungle, Petrperu plans a step-out on the Pavayacu structure at a cost of $3.5 million to prove additional postulated reserves estimated at 10 million barrels. The field currently has 7 million bbl of proven reserves.

In the central jungle, the state company will drill two wildcats at a cost of $3 million targeted structures with a combined postulated reserves of about 7 million bbl.

Mobil exploration

Mobil Oil Corp.'s plans to step up Peruvian jungle oil exploration after it moved into a fourth block in the central jungle suffered a setback from a guerrilla attack.

Mobil suspended its contract with Petroperu for exploration covering almost 9 million acres in the northern Peruvian jungle after guerrillas attacked a seismic crew camp at Barranca on the Biabao River.

The contract will be renewed once Mobil and contractor Sereal, a unit of Halliburton Co., replace seismic equipment, computers, and communications equipment damaged in the attack and receive government assurances the area is secure again. Damage is expected to run several million dollars.

Mobil's 6 year exploration program in the Huallaga basin, covering Blocks 28, 29, 30, and 53, entails conducting 3,600 line km of seismic surveys and spudding five wildcats.

If Mobil extends the exploration stage for a seventh year it must drill at least one more wildcat. Mobil is committed to spend a minimum $107.5 million in the exploration stage.

The first wildcat is scheduled for September 1991.

Oxy's dispute

Oxy could renew vigorous exploration activities if it reaches a new agreement with Petroperu.

Oxy had until March 1991 to decide whether to drill a third wildcat in Block 36 in the central jungle or move exploration to another block.

Meantime, the company still is trying to resolve a long running dispute with Petroperu over payment for crude received from existing operations.

Oxy has told Petroperu it could increase output from its fields by as much as 10,000 b/d if it receives regular payment for the oil it delivers.

LATIN AMERICA

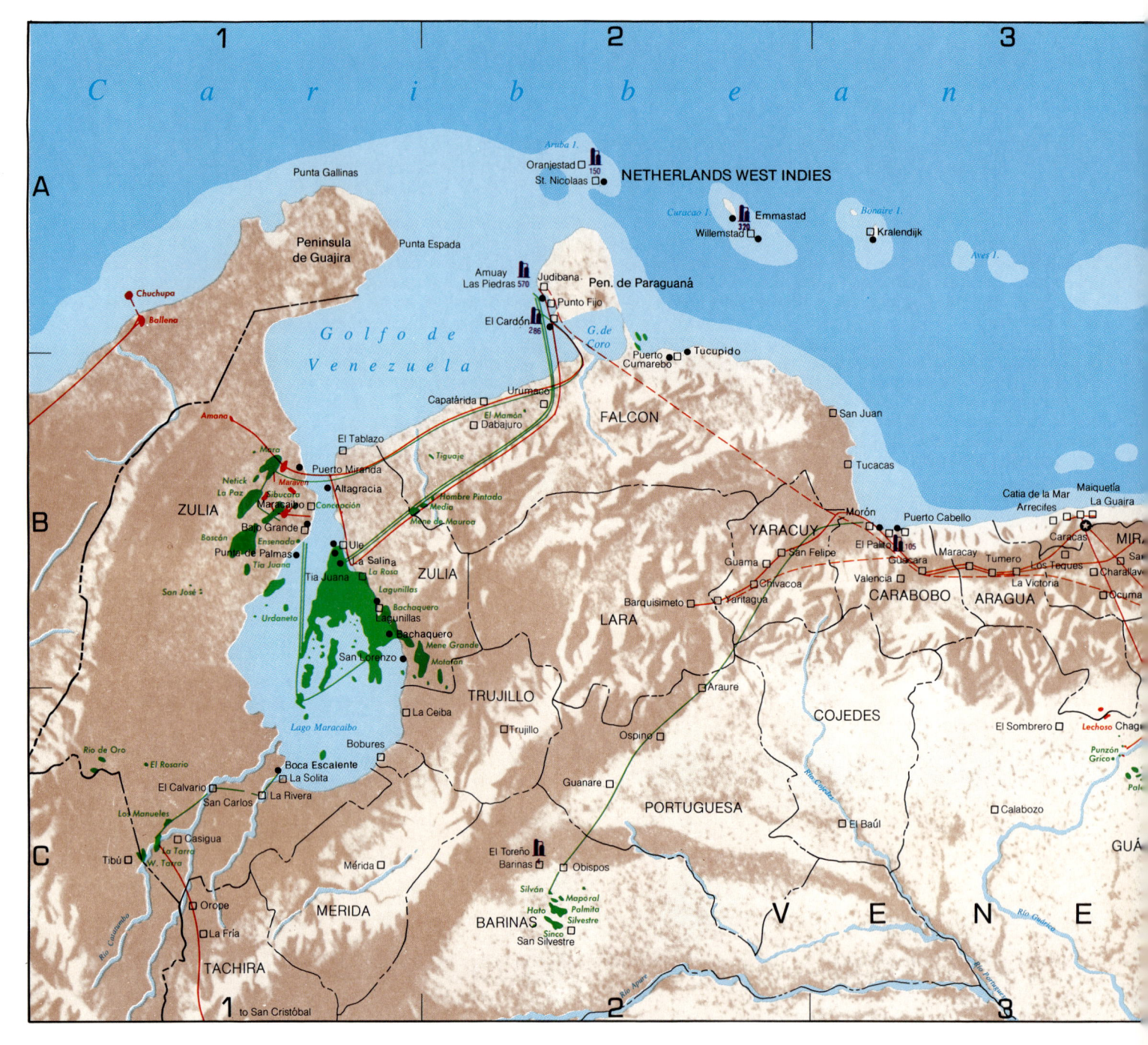

Oxy produces about half of Peru's 130,000 b/d output, mainly from its wells in the northern jungle.

Development projects

Petroperu also will bolster efforts in 1991 to find a partner for a joint venture to start up flow of 20,000 b/d from Chambira oil field. Petroperu discovered the 50 million bbl oil field in its northern jungle Block 8 in March 1989.

The Chambira discovery well, Petroperu's first jungle wildcat since 1986, flowed 5,680 b/d of 26.5° gravity oil.

Development of the Aguaytia gas/condensate field in the Pucallpa area of the central jungle is unlikely to get under way in 1991 unless the armed forces dislodge guerrillas and drug traffickers that occupy the area.

Production would reach 30 MMcfd and an undisclosed volume of condensate.

Cost is pegged at $65 million, including pipelines and gas processing facilities.

Production from fields off Peru's northern coast will increase in 1991 if the government's preliminary agreement with AIG goes through.

Most of the increase, however, will go towards reimbursing AIG for payments made to compensate Belco parent Enron for expropriation of Belco's Peruvian oil field assets under a political risk insurance policy.

Output from the former Belco offshore fields has fallen to about 20,000 b/d from 27,000 b/d before the 1985 takeover.

Camisea development

The biggest project in Peru, Camisea gas condensate development, has been stymied by financing and political problems.

Royal Dutch/Shell Group, which discovered gas/condensate in San Martin/Cashiriari fields in the Camisea area after a $175 million exploration campaign during 1982-88, shelved development plans after a dispute with Petroperu on financing.

Petroperu has revised its estimate of Camisea reserves,

LATIN AMERICA

VENEZUELA

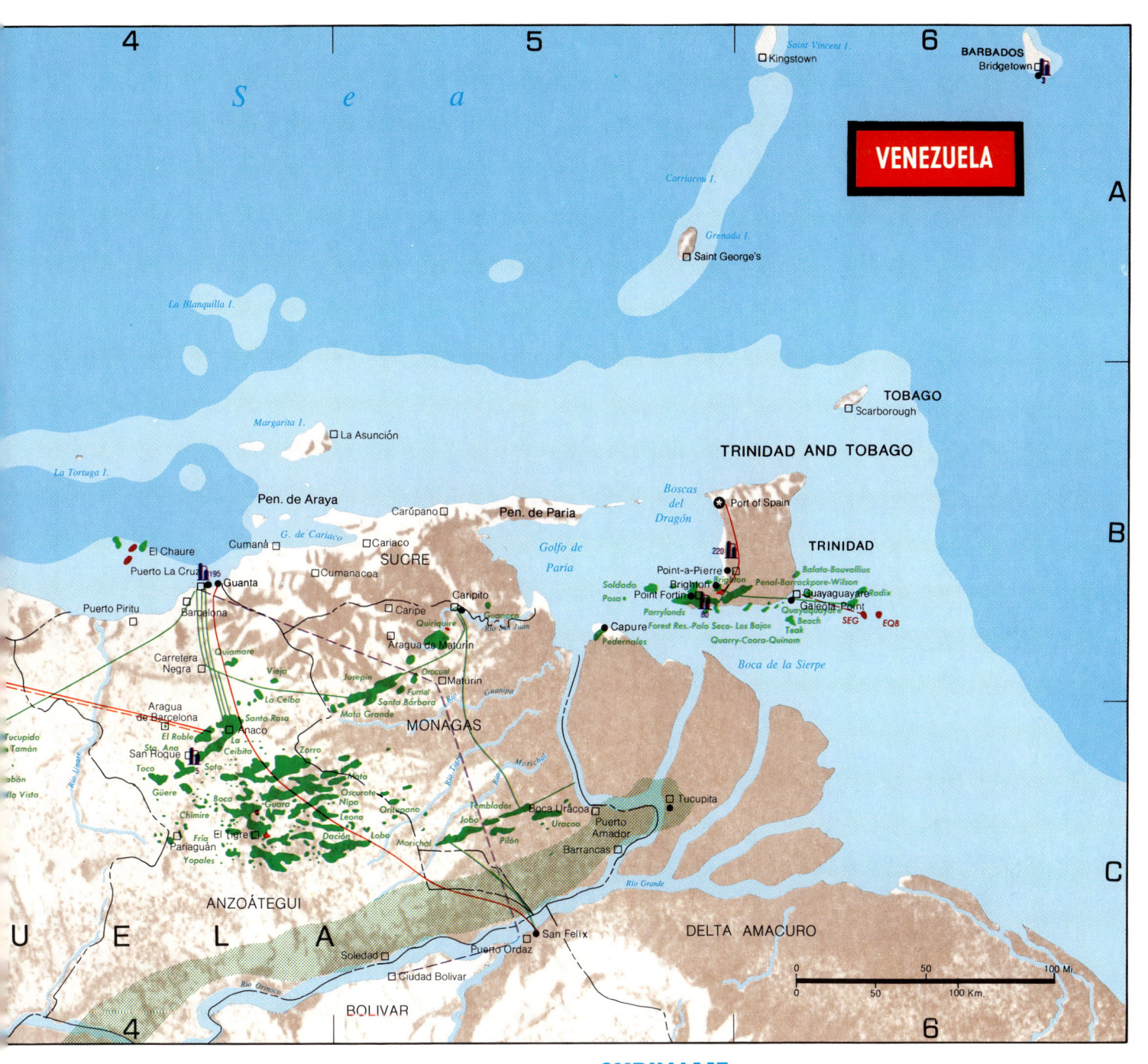

cutting the field's estimated gas reserves to 10.8 tcf and boosting condensate reserves to 725 million bbl.

Initial development would entail producing about 900 MMcfd of gas and as much as 50,000 b/d of liquids.

At first, some of the gas will be used to back out residual oil in power generation in Central and Southeast Peru and in industrial facilities near a gas grid installed in the Camisea area.

Refinery projects

Heading Petroperu's downstream projects is an upgrading project at its 93,900 b/d La Pampilla refinery at Ventanilla, near Lima.

Petroperu said it called an international tender for installation of three conversion units at the refinery.

The state company is trying to meet domestic demand for light products while reducing crude runs.

Petroperu also plans a $60 million project to hike yield of gasoline, kerosine, diesel, and LPG at its Talara refinery.

SURINAME

CAPITAL: Paramaribo
MONETARY UNIT: Guilder
REFINING CAPACITY: NA
PRODUCTION: 4,000 b/d
RESERVES: 27.3 million bbl

STATE OIL COMPANY SURINAME NV (STAATSOLIE), founded in December 1980, produced an average 4,000 b/d of heavy, sweet crude in 1989.

It estimates proved reserves in its sole producer, Tambaredjo, at 25 million bbl and probable reserves at 38 million bbl.

For 1990-92, Staatsolie plans capital spending of $150 million in onshore exploration and development, eyeing a production goal of 6,500 b/d.

Some of the funds will be earmarked for a small refinery with throughput capacity of 7,000 b/d.

LATIN AMERICA

OLD OIL PRODUCING REGIONS, such as depicted here in Venezuela's Lake Maracaibo, will be revitalized as South American oil producing nations lead an expansion of Latin America's petroleum sector in the 1990s (Photo courtesy Petroleos de Venezuela SA.).

Suriname's national assembly passed a law governing foreign investment in the oil and mining sectors, scrapping its existing petroleum legislation. It allows longer term concessions, tax incentives, and foreign exchange benefits for remitted profits.

TRINIDAD AND TOBAGO

CAPITAL: Port of Spain
MONETARY UNIT: Dollar
REFINING CAPACITY: 246,000 b/cd
PRODUCTION: 151,000 b/d
RESERVES: 536 million bbl

ON MAR. 31, 1985, THE GOVERNMENT OF Trinidad and Tobago signed an agreement with Texaco Inc. to transfer to the country the main assets of Texaco Trinidad Inc.

These assets include onshore and offshore oil and gas fields, a 30% interest in offshore Block L, and Texaco's interest in the refinery and petrochemical complex and ancillary facilities at Point-a-Pierre with all its auxiliary installations.

Total combined throughput capacity at the downstream facilities is equivalent to 220,000 b/d of oil.

Trinidad and Tobago, which has had oil production since the 19th Century, is grappling with a long term slide in crude production.

Oil production fell to 150,000 b/d at yearend 1989 from 229,000 b/d in 1978.

However, major recent discoveries in the country have ramped up reserves and production estimates for natural gas.

Gas reserves for Trinidad and Tobago are estimated at more than 10 tcf, and production has jumped to more than 650 MMcfd.

Despite another drop in crude production in 1989, the petroleum industry accounted for 60% of the country's exports and 24% of the government's revenue.

In addition to programs being implemented by state oil company Trintoc, the government has granted licenses for offshore oil operations to a joint venture of Trintoc and Pecten to operate the Lower Reverse L Block and to Trintoc-Mobil Oil Corp. to operate two blocks off the eastern coast of Trinidad coast.

At the same time, four other offshore blocks were put up for tender in early 1990.

Trintoc plans to spend about $65 million for oil and gas exploration in 1991, part of a 5 year plan to focus on deeper horizons and natural gas plays.

URUGUAY

CAPITAL: Montevideo
MONETARY UNIT: New Peso
REFINING CAPACITY: 33,000 b/cd
PRODUCTION: NA
RESERVES: NA

ALTHOUGH URUGUAY'S ADMINISTRACION NACIONAL de Combustibles, Alcohol, y Portland (Ancap) has been trying since 1932 to stimulate exploration in the country by foreign operators, it has no commercial hydrocarbon production.

Uruguay imports all of its oil and gas. Mexico supplies about 40% of Uruguay's oil.

Most of the rest of its oil supplies are bought on the spot market.

In 1988, Uruguay bought 20% of its crude from the United Arab Emirates, 11% from Iran, 10% from Saudi Arabia, 10% from the Soviet Union and 9% from Nigeria.

Ancap operates a refinery at La Teja, Montevideo, where it increased throughput capacity in 1990 to 50,000 b/d from 33,000 b/d.

The refinery is linked via a 150 km pipeline to a marine terminal that has a monobuoy for tankers of as much as 150,000 dwt.

Ancap also operates several products distribution and storage terminals around the country.

Ancap plans to upgrade and modernize the La Terra refinery and ancillary facilities at a cost of $56 million, with a World Bank loan providing about half that funding. Also on tap is a project to modernize the Ancap's lubricants plant with a projected investment of $4 million.

Ancap maintains some exploration in sedimentary basins that cover more than 200,000 sq km offshore and onshore.

LATIN AMERICA

VENEZUELA

CAPITAL: Caracas
MONETARY UNIT: Bolivares
REFINING CAPACITY: 1,167,000 b/cd
PRODUCTION: 2.118 million b/d
RESERVES: 59.04 billion bbl

VENEZUELA'S NATIONAL OIL COMPANY, PETROLEOS de Venezuela SA (Pdvsa) is moving ahead with an ambitious investment program designed to substantially expand its activities in oil and gas exploration and production, refining, petrochemicals, and coal in the 1990s.

The company, which has stakes in refining and marketing companies in the U.S., Europe, and the Caribbean, also is seeking new investment opportunities in U.S. and European markets as well as in the Far East.

Pdvsa officials expect the company by 2000 to have developed a much stronger presence in the global energy market, with:
- Crude productive capacity of more than 4 million b/d.
- Crude output averaging about 3.7-3.8 million b/d.
- Exports of 500,000 b/d-1 million b/d of Orimulsion—a new boiler fuel for electric power generation made from an emulsion of extra heavy crude, water, and surfactants—and synthetic crudes.
- Liquefied natural gas exports to the U.S. totaling 110,000 b/d of oil equivalent.
- One million b/d in additional refining capacity, consisting of a new 200,000 b/d high conversion unit, 400,000 b/d added to existing domestic refineries, and another 400,000 b/d consisting of new domestic capacity plus stakes in foreign refineries.
- More than a threefold increase in petrochemical output capacity to more than 14 million metric tons/year.
- An enlarged and renovated tanker fleet that will handle 30% of the company's international shipments.
- Coal exports of 10 million metric tons/year.

Medium term plans

Pdvsa's medium term development plan for 1991-96 calls for total investments of about $25 billion.

That breaks out as exploration $1.5 billion, production $9 billion, refining $6 billion, petrochemicals $6 billion, coal $1.5 billion, domestic marketing $800 million

One of Pdvsa's planning officials put the imported component at about 40% of total investments.

Foreign partners are also working with Pdvsa on coal and LNG projects and are expected to provide equity in these areas.

In addition to these areas, the company hopes to attract international investment in exploration and production. Up until now, these two sectors were off limits to foreign companies and were exclusively the domain of Pdvsa.

Exploration

Pdvsa plans to add 10 billion bbl of light crude to its proven reserves, which now stand at 59 billion bbl.

Pdvsa puts Venezuela's probable reserves at 85.5 billion bbl and possible reserves at 55.6 billion barrels, for a combined total of 200.1 billion bbl. That does not include 271 billion bbl of extra heavy crude and bitumen the company believes it can recover from the Orinoco Belt, which has an estimated 1.2 trillion bbl in place.

Much of the reserves to be added through 1996 will come from the northern zones of Monagas and Anzoategui states in eastern Venezuela, where Pdvsa operators Lagoven and Corpoven are developing large fields discovered in 1985.

Other producing areas that will provide new reserves are Apure in the southwest and Ceuta on Lake Maracaibo.

Pdvsa is carrying out exploratory and development work in both areas.

Other exploratory efforts target the Andean flank south of Lake Maracaibo, Perija in Zulia state, sites east of Maturin in Monagas, Guarumen, and offshore natural gas deposits.

Through 1996, Pdvsa plans about 50,000 line km of seismic surveys and more than 130 wildcats.

Proven natural gas reserves now stand at 105.7 tcf, with probable reserves estimated at another 27.6 tcf and possible at 93.9 tcf.

Oil production

Pdvsa plans to hike Venezuela's crude production potential to more than 3.6 million b/d by yearend 1996 from the present 2.75 million b/d.

During that time, average production is targeted to climb to 3.25 million b/d from about 2.3 million b/d in 1990.

Achieving this extra increment in crude production potential will involve drilling 7,000 exploratory and development wells, performing 10,000 workovers, adding 4.238 bcfd of gas compression capacity, building 1,000 km of oil and gas pipelines, and installing at least 4 gas processing plants.

Orimulsion, the product dubbed "liquid coal," is targeted for $1.5 billion in Pdvsa investments through 1996. Orimulsion production potential at that time is projected to be 750,000 b/d—of which 70-75% would be extra heavy crude—with average output of 500,000 b/d.

Pdvsa estimates it will need to drill 1,500 exploratory and development wells in the Orinoco belt, build 1,200 line km of oil and gas pipelines, set up 25 production stations, and lay 500 line km of electric power lines for the Orimulsion project alone.

Natural gas

Venezuela's natural gas production in 1989 was 3.67 bcfd, virtually all of which was associated gas.

Of the total, 37% was reinjected, 57% was sold directly on the domestic market or supplied to the petrochemical industry, and the rest used in Pdvsa operations.

Production capacity for gas now stands at 4.2 bcfd of gas per day. Plans call for that to jump by 1.6 bcfd by 1996. Capacity for extracting gas liquids will rise substantially from the current 108,000 b/d.

Pdvsa will be giving the gas sector a great deal of attention the next 6 years.

Refining plans

Pdvsa refining capacity will be raised by 400,000 b/d by 1996 through additions to its existing refineries.

That currently stands at 1.12 million b/d of atmospheric distillation capacity and 815,000 b/d of conversion capacity.

Pdvsa also plans to build a 200,000 b/d high conversion refinery in eastern Venezuela and to find additional partners for refining/marketing ventures overseas, with an eye toward increasing exports to the northern Atlantic basin.

Petrochemicals

Pdvsa's current expansion program in petrochemicals is an extension of policies that were established several years ago.

The company's strategy is to make greater use of Venezuela's natural gas reserves, reduce imports of petrochemicals and fertilizers, and establish new businesses with significant export potential. Most of the new projects being undertaken by Pequiven are joint ventures with Venezuelan and international investors.

Pequiven's updated development plan calls for increasing gross productive capacity to 14 million metric tons/year by 1996 from 4 million metric tons/year in 1991.

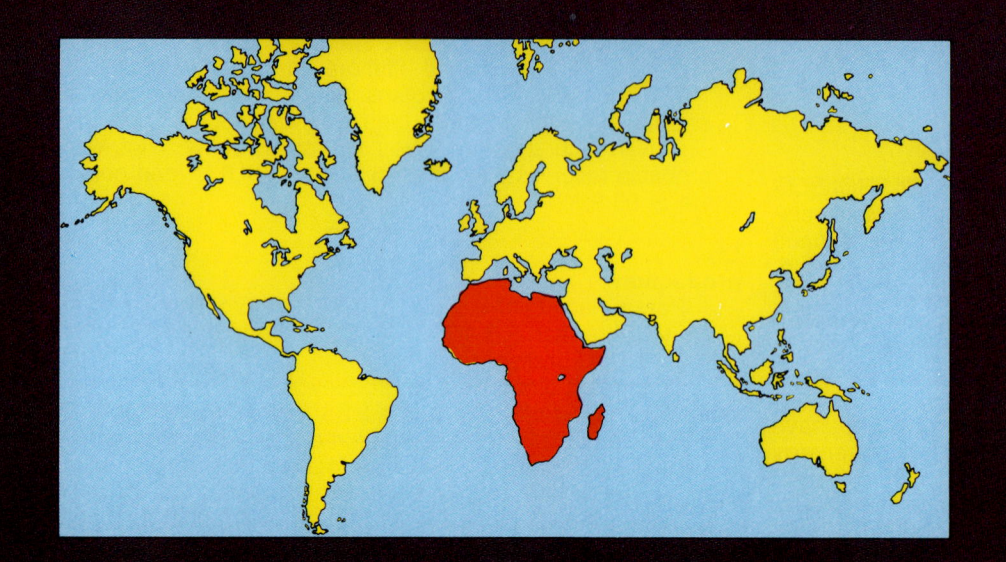

IPE ATLAS

AFRICA

ALGERIA

CAPITAL: Algiers
MONETARY UNIT: Dinars
REFINING CAPACITY: 464,700 b/cd
PRODUCTION: 797 Mb/d
RESERVES: 9,200,000 Mbbl

OPENING OF ALGERIA'S OIL AND GAS INDUSTRY to international companies continued to gather momentum in 1990.

After setting out first to attract foreign companies to the exploration scene and then into the crucial LNG business, Algeria's state owned Sonatrach turned its attention to LPG.

Sonatrach disclosed outline proposals to double gas liquids production capacity to more than 8 million tons/year.

Algeria, like many established producers and exporters, finds it needs to invest heavily in exploration and production to prepare for the expected upturn in oil demand in the 1990s.

For smaller producers such as Algeria, with pressing demands for revenue, tapping the financial resources of foreign companies and the reservoir of technology at their disposal is becoming increasingly attractive.

Sonatrach has a 4 million ton/year LPG plant and export facilities at Arzew on the Mediterranean coast. The plant is linked to large oil and gas production complexes at Hassi R'Mel and Hassi Messaoud by two pipelines.

Sonatrach's plans call for a 500-625 mile pipeline to collect gas liquids from several fields in Southeast Algeria and link up with the existing pipeline at Hassi Messaoud.

Foreign companies may also have the opportunity for a joint venture to develop gas/condensate production from fields in the Hassi Messaoud area. Plans to produce as much as 50,000 b/d of condensate were under study.

Sonatrach also is feeling the benefits of improved prospects for the European gas market.

In late 1989, the Italian gas company, SNAM, facing increased demand for gas in the main consuming areas of Italy, started negotiations with Sonatrach to expand capacity of the trans-Mediterranean gas pipeline from Algeria through Tunisia to Italy.

The twin pipelines that make up the onshore part of the pipeline in Algeria and Tunisia have a capacity of 13-14 billion cu m/year.

But there is only 12.5 billion cu m/year capacity in the subsea section. Originally, the two sides were talking about adding a fourth subsea line to boost submarine capacity by 4-6 billion cu m/year.

Seismic activity

Anadarko Petroleum Corp. is upgrading Algeria's seismic interpretation capability as part of its agreement to explore highly prospective acreage in eastern Algeria.

Its Anadarko Algeria Corp. unit let contract to CogiSeis Development Inc. and Geophysical Development Corp., both of Houston, to install and operate a dedicated seismic processing center in Boumerdes, about 30 miles east of Algiers.

CogiSeis will install a Convex 210 computer system, and CDC will manage and operate the processing center. Project cost is not disclosed.

The center is part of Anadarko's plans to explore four blocks covering 5.3 million acres in the Ghadames and Illizi basins.

Anadarko and Sonatrach signed a production sharing agreement in October 1989 covering Birkine Block 404a, El Merk Block 208, Sidi Yedda Block 211, and Garet Tessegt Block 245.

Those blocks together cover about 8,000 sq miles of Sahara desert.

The seismic center will enable the companies to reprocess as much as 25,000 line km of existing data in addition to processing new data in the license area, said GDC Pres. Reg N. Neale.

An initial seismic acquisition and processing phase will cover a period of 3 years.

"We are basically providing technical services to Anadarko for 3 years, then it will be passed over to Sonatrach for them to operate," Neale said.

By combining a CogniSeis software package and CDC's customized programs, the companies hope to solve a wide range of geophysical problems caused by the Algerian terrain.

AFRICA

Those problems include similar primary and multiple velocities, long wavelength, low velocity layer thickness variations, and the implied depth conversion consideration, Neale said.

The center will also correct seismic data for surface static caused by sand dunes, thus eliminating phantom structures common in such terrain.

Some of the world's largest dunes, more than 1,100 ft high, are in Block 211.

In addition to the seismic center, Anadarko will establish office headquarters in Algiers and a field camp in Hassi Messaoud and build desert roads to remote drilling locations.

Anadarko planned to spend about $10 million in Algeria in 1990, mostly for mobilization, seismic, and drilling costs, and $100 million over 10 years, an Anadarko official said.

The company hired a rig for the first initial wells and planned to spud the first well by yearend 1990.

Advanced seismic processing capability is part of the deal that Anadarko hammered out with Sonatrach in exchange for the right to explore the four blocks, Anadarko said. It planned to shoot 500 line km of seismic in 1990.

The agreement would not have been possible if Algeria had not altered its hydrocarbon law in 1986 to provide foreign companies more control, the Anadarko official said.

Anadarko is the first U.S. company to sign an exploration contract under Algeria's new hydrocarbon law.

BHP Petroleum Pty. Ltd., Melbourne, signed an agreement with Sonatrach that has not been finalized.

Agip SpA of Italy was the first foreign firm to sign an exploration/development agreement under Algeria's new law. Agip's pact, signed in January 1988, covers Permits 403 and 407, where it tested a 5,000 b/d oil discovery in late 1986.

Spain's Cia. de Investigacion y Explotaciones Petroliferas also signed an exploration contract with Sonatrach in early 1988.

Total-CFP planned to shoot 4,500 line km of seismic surveys on a 5,897 sq km exploration tract in Algeria, about 560 miles southwest of Algiers and east of Bechar.

If the surveys are successful, Total and Sonatrach will sign an exploration agreement.

ANGOLA

CAPITAL: Luanda
MONETARY UNIT: Escudo
REFINING CAPACITY: 37,630 b/cd
PRODUCTION: 480 Mb/d
RESERVES: 2,074,000 Mbbl

TEXACO INC. WAS BUSY EXPLORING OFF ANGOLA IN 1990.

A group led by Texaco tested two Cretaceous dolomite discoveries 5 and 9 miles off Northwest Angola on Block 2, north of Luanda.

Its 1 Estrela flowed more than 2,100 b/d of 36.2° gravity oil from 7,794-7,824 ft. The 1 Morsa West wildcat flowed a combined 1,600 b/d of 36-37° gravity oil from three pay zones at 4,916-5,066 ft.

Interests are Total Angola and Petrobras Internacional SA 27.5% each, Soc. Nacional de Combustiveis de Angola (Sonangol) 25%, and Texaco 20%.

Texaco's 1 Savelha flowed at a combined rate of more than 6,000 b/d of 22-25° gravity oil from three Cretaceous dolomite zones at 7,515-8,010 ft on Block 2 off Northwest Angola.

The fourth oil discovery on the block for Texaco and partners, it was drilled to 10,620 ft in 69 ft of water.

Texaco Latin America/West Africa and partners tested another oil discovery off Angola.

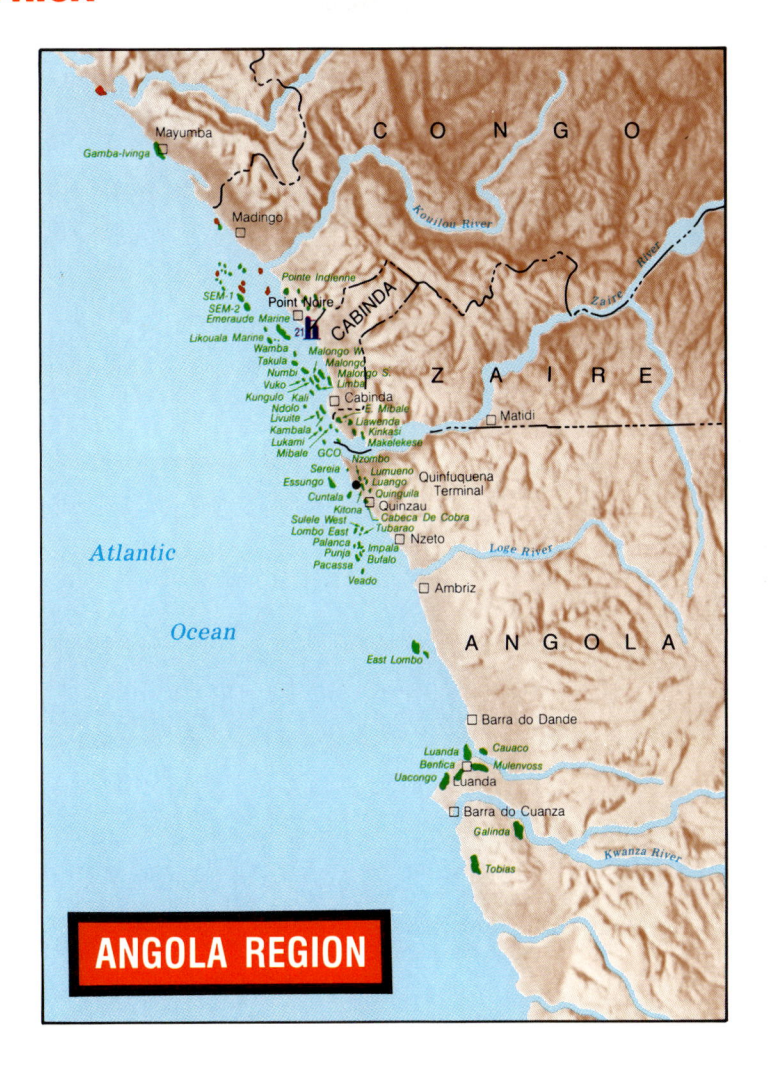

Texaco's 1 Bagre flowed 34.6-36.8° gravity oil at a combined rate of 13,488 b/d from five zones in Cretaceous dolomite at 9,130-9,640 ft.

The group was assessing the discovery's potential for development.

It was drilled to 11,086 ft measured depth in 72 ft of water on the 1 million acre Block 2 about 7 miles off Angola's northwest coast and 200 miles north of Luanda.

The well is 7½ miles south of Raia oil and gas field, discovered on Block 2 in June 1988 and under preliminary development.

The Texaco group operates other fields on Block 2 under a 1980 production sharing agreement with Angola's state owned Soc. Nacionale de Combustiveis de Angola (Sonangol).

Since the early 1980s, the Texaco group has operated Essungo and Cuntala fields, and both are on production in the northern portion of the block.

In late 1987, the group started production from East Lombo, Tubarao, West Sulele, and South Sulele fields in the southern corner of the block.

Texaco in 1989 reported a 6,100 b/d strike, 1 Tamboril, on the block.

In addition, Elf Aquitaine Angola gauged combined flows of 28,000 b/d of 30° gravity oil from its 1 Cobo wildcat on Block 3 off Angola.

The find is 5.6 miles south of Bufaloa field and 18.6 miles south of the Palanca field.

Partners in the block are Elf, Angola's Sonangol, Naftagas, Ina Naftalin, Ajex Petroleum, a Mitsubishi Development Corp. subsidiary, and units of Agip SpA, Repsol SA, and Svenska Petroleum.

AFRICA

Equatorial Guinea contract area

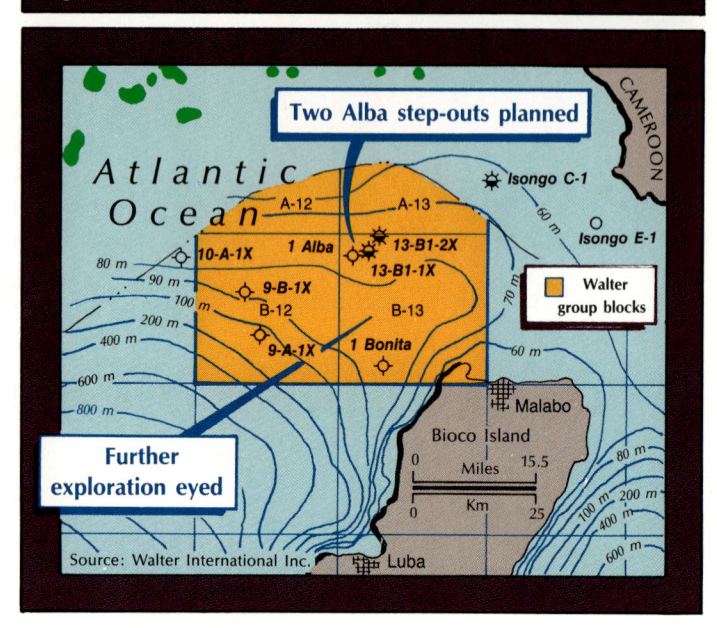

CAMEROON

CAPITAL: Yaounde
MONETARY UNIT: CFA francs
REFINING CAPACITY: 42,000 b/cd
PRODUCTION: 164 Mb/d
RESERVES: 400,000 Mbbl

CAMEROON LET AN 18 MONTH CONTRACT TO study proposed development of its gas resources to Gaz de France unit Sofregaz and Beicip, both of France.

The study, the first phase of Cameroon's national gas development plan, will cover potential and available gas and LPG reserves, projection of their potential demand, evaluation of various options for development, estimation of investment costs and infrastructure needs, and training.

CONGO

CAPITAL: Brazzaville
MONETARY UNIT: CFA francs
REFINING CAPACITY: 21,000 b/cd
PRODUCTION: 161 Mb/d
RESERVES: 830,000 Mbbl

IN THE WEST CENTRAL AFRICAN COUNTRY OF Congo, Amoco Congo Exploration Co. awarded a $12 million contract for the installation hookup and commissioning of the Yombo field floating oil storage unit to Rockwater International.

The firm earlier took delivery of four 375,000 kw steam turbines for the unit from Peter Brotherhood Ltd., Peterborough, U.K. The turbines will power the FPSO, which is the converted tanker Conkouoati.

Conoco Kayes (Congo) Ltd. let contract to Helmerich & Payne International Drilling Co., Tulsa, to drill three wildcats with options for as many as three more wells in Congo. Depths will range from 8,000 to 14,000 ft.

H&P planned to move Rig 117, an SCR land unit, from Guatemala to Congo. The rig was to drill the first well from a barge at a river location.

EGYPT

CAPITAL: Cairo
MONETARY UNIT: Pounds
REFINING CAPACITY: 523,153 b/cd
PRODUCTION: 873 Mb/d
RESERVES: 4,500,000 Mbbl

EGYPT INVITED BIDS FOR ACREAGE IN THE Gulf of Suez and the Mediterranean part of the Nile Delta and for onshore acreage in the Western Desert, the Nile Delta, and the northern part of the Eastern Desert.

Closing for bids was June 30, 1991.

Repsol SA agreed to buy Conoco Inc.'s production interests in Egypt's Western Desert, subject to Egyptian government approval and partner preemptions.

Conoco operates Khalda Petroleum and has a 50% interest in the Khalda concession, producing about 20,000 b/d. Conoco's partners are Phoenix Resources Co. 40% and South Korea's Samsung Group 10%. The deal does not include Conoco's Hurghada concession in Egypt's Eastern Desert.

Khalda Petroleum Co., Cairo, expected its 95 km, 10 in. Egyptian gas pipeline from Salam to the Mersa Matruh area to be on stream by yearend 1990. Khalda is a venture of Egyptian General Petroleum Corp. 50%, Conoco Inc. 25%, Phoenix Resources 20%, and Samsung Ltd. 5%.

Shell Egypt planned to develop an oil discovery on its East Gemsa concession in Zeit Bay off Egypt through SUCO's production facilities for Zeit Bay field.

The 25 million bbl reservoir likely will produce about 5,000 b/d.

Western Desert Operating Petroleum Co. let a turnkey contract to Technip Geoproduction, Paris, for construction of a 330 MMcfd onshore gas processing plant as part of an expansion of Abu Qir gas field in the Mediterranean Sea off Egypt.

ETHIOPIA

CAPITAL: Addis Ababa
MONETARY UNIT: Birr
REFINING CAPACITY: 18,000 b/cd
PRODUCTION: N.A.
RESERVES: N.A.

MAXUS ENERGY CORP., DALLAS, SIGNED A production sharing contract covering more than 27 million acres in eastern Ethiopia.

It agreed to spend at least $3.5 million, mostly for geological and geophysical studies, under the contract's 2 year first phase. There is no provision for a minimum number of wildcats.

Maxus, which holds a 100% interest in the area, can extend the agreement after assessing seismic results.

EQUATORIAL GUINEA

CAPITAL: Malabo
MONETARY UNIT: CFA francs
REFINING CAPACITY: N.A.
PRODUCTION: N.A.
RESERVES: N.A.

A GROUP OF U.S. INDEPENDENTS SIGNED A

AFRICA

AFRICA

Where Shell will explore

production sharing contract in Equatorial Guinea covering potential development of offshore Alba gas/condensate field and exploration of adjoining areas.

Operator Walter International Inc., Houston, 25%, McMoRan International Inc., New Orleans, 50%, and Samedan Oil Corp., Ardmore, Okla., 25%, signed the contract covering Blocks A-12, A-13, B-12, and B-13 northwest of Bioco Island.

The Walter group is committed to drill two wells on the Alba structure to assess the extent of reserves, estimated at 1.3 tcf of gas and 68 million bbl of condensate.

The first delineation well was to be spudded in late summer 1990 with a jack up in about 250 ft of water near 1 Alba about 22 miles off Bioco Island. It was targeted to 10,170 ft, with the top of the pay expected at 8,858 ft. The second delineation was to follow immediately.

Alba's pay is Miocene Isongo sandstone at 8,528-9,840 ft.

The group also has the option to commit to an exploratory well on Block B-13 by the end of the contract's second year. If it makes that commitment, the wildcat must be drilled within 18 months of commitment.

It is likely the well will be drilled in the first 2 years of the contract, said Will Frank, Walter managing director.

The group could commit to more exploratory wells, with provision for 25% acreage relinquishments per period, at 24 month intervals for 7 years.

If enough reserves can be established and a commercial development plan approved for Alba, the Walter group would then build a pipeline to shore, Frank said.

The production sharing contract carries a 30 year term in the event of an oil development, 50 years for gas development.

It isn't clear what market could be established for Alba gas or how the field would be developed.

Petroconsultants SA, Geneva, adviser to Equatorial Guinea, has recommended that Alba gas be exported to nearby fields off Cameroon for gas lift operations.

Guineo-Espanola de Petroleos SA (Gepsa), a 50-50 venture of Spain's Hispanica de Petroleos SA (now Repsol SA) and the government of Equatorial Guinea, drilled Alba discovery and confirmation wells in 1984-85.

Gepsa 13-B1-1X found pay at 12,635 ft, and 13-B1-2X step-out found pay at 13,851 ft. Repsol relinquished the acreage after failing to find a gas market.

Weather in the area is stable with no heavy seas. Seabed conditions are good for anchoring and pipelaying.

Equatorial Guinea has no history of oil production. Investment in infrastructure that might serve the petroleum industry has been minimal. The nearest supply bases are in Port Gentil, Gabon, and Douala, Cameroon. Malabo, the capital of Equatorial Guinea, has good facilities for offices.

GABON

CAPITAL: Libreville
MONETARY UNIT: CFA francs
REFINING CAPACITY: 24,000 b/cd
PRODUCTION: 269 Mb/d
RESERVES: 730,000 Mbbl

OIL PRODUCTION IN GABON RETURNED TO normal at midyear 1990 after a short slump triggered by antigovernment rioting.

The fall in oil flow followed a decision by France's Ste. Nationale Elf Aquitaine to shut down all its offshore operations in response to threats against Elf personnel and installations by rioters.

Production resumed when French troops stationed in Gabon moved to protect oil installations.

Sources said Elf's decision to resume production may also have been influenced by threats of retaliation against Elf by Gabonese President Omar Bongo. He reacted to the Elf shutdown by threatening to transfer the company's concessions to new partners.

Elf's decision to shut down production came after riots in the Gabonese oil town of Port Gentil in which seven Elf employees and three employees of a Royal Dutch/Shell Group unit were held hostage for several hours.

Rioting was set off by discovery of the body of opposition leader Joseph Rendjambe in a Libreville hotel room. His death left Marc Saturnin Nan Nguema, former Elf Gabon deputy general manager, as head of the Gabonese Progress party.

In addition to halting production, Elf evacuated most of its expatriate personnel and families.

Shell also evacuated the families of expatriate workers and a number of nonessential staff but kept production flowing from its onshore Rabi Kounga field at reduced levels.

Before the outbreak of trouble Gabon was producing about 270,000 b/d, of which about 110,000 b/d came from Elf fields. Most of the remainder came from Shell's Rabi Kounga field.

Other exploration/production activity

Amerada Hess Corp. and the government of Gabon formed a joint venture company that bought a 10% interest in Rabi-Kounga oil field.

Amerada spent $300 million for a 67% interest in the company.

Rabi-Kounga's proved and probable reserves are estimated at 425 million bbl, and gross production was expected to reach 140,000 b/d during 1991.

Sun Gabon Oil Co. let contract to Parker Drilling Co., Tulsa, for a multiwell exploration program in Gabon. Parker was to transfer a TBA 2000 rig from its Chad division and planned to spud the first well in late summer 1990.

Groups led by Ste. Nationale Elf Aquitaine drilled two oil discoveries in Gabon.

WEST AFRICA

AFRICA

The Lasmo group's blocks

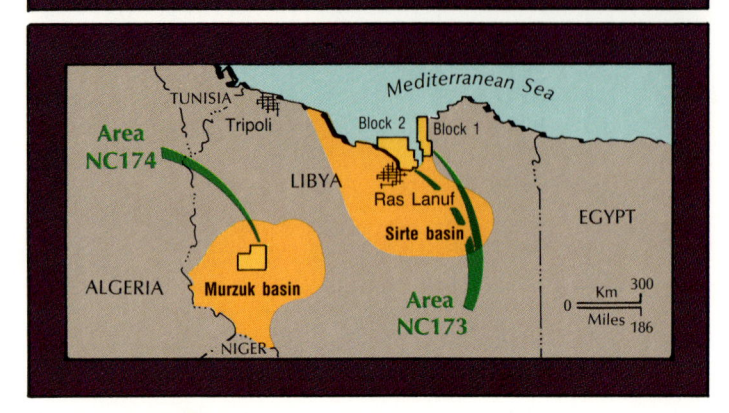

The Hylia Marine wildcat flowed 1,575 b/d of 24° gravity oil on the Eyena permit 17 miles northwest of Grondin field and 11 miles offshore.

Onshore, the Elf-Shell Gabon combine gauged 630 b/d at the 1 Avocette well on Ogooue Dianongo permit 16 miles north of Rabi-Kounga field.

KENYA

CAPITAL: Nairobi
MONETARY UNIT: Shillings
REFINING CAPACITY: 90,000 b/cd
PRODUCTION: N.A.
RESERVES: N.A.

SHELL EXPLORATION BV TOOK A FARMOUT FROM Amoco Kenya Petroleum Co. on a 42,250 sq km block in Northwest Kenya.

Shell will earn a 50% interest in Block 10 by conducting a 500 km seismic survey and drilling two wildcats in 1990-91.

The region, on Lake Turkana and surrounding the lake on three sides, has not been explored for oil and gas.

Shell and British Petroleum Co. plc pioneered onshore and offshore exploration in Kenya in the 1950s but withdrew because of lack of success.

LIBYA

CAPITAL: Tripoli
MONETARY UNIT: Dinars
REFINING CAPACITY: 347,600 b/cd
PRODUCTION: 1,369 Mb/d
RESERVES: 22,800,000 Mbbl

A GROUP LED BY LASMO PLC, LONDON, SIGNED agreements to explore three blocks in Libya.

The deal, in 50-50 partnership with a group of South Korean companies, covers two offshore blocks and one onshore block. Area NC173 involves the 7,602.68 sq km Block 1 and 16,030 sq km Block 2 in the Gulf of Sirte portion of the Sirte basin. Area NC174 covers 11,310 sq km in the onshore Murzuk basin.

Lasmo expected to start seismic surveys in 1991. In its first 5 year program the Lasmo group is committed to drill seven wells, four of them on the onshore block.

State companies from Romania and Bulgaria have made several significant discoveries in blocks adjoining the Lasmo acreage.

Detailed plans are under consideration for a 250 mile crude oil pipeline from the Romanian discovery into the pipeline transmission system.

Industry sources say Romania's Rompetrol has tapped a 2 billion bbl oil resource in the area. An initial development phase would produce about 75,000 b/d of light, low sulfur crude, with start-up set for 1993. Production is expected to rise ultimately to about 150,000 b/d.

The South Korean group is led by state owned Korean Petroleum Development Corp. and includes Daewoo, Hyundai, Majuko, and Daesung.

Gas projects

Libya also is pressing a campaign to develop and utilize more domestic gas reserves with gas processing and pipeline projects.

The country commissioned a new gas processing plant in Sahl gas field in the Sirte basin. Sahl produces 150 MMcfd of gas and 4,000 b/d of condensate from 15 wells.

Field output moves through a 30 in. gas pipeline that links Zelten field with the LNG plant and petrochemical complex at Marsa al-Brega.

State owned Sirte Oil Co. also is proceeding with plans to link Zueitina 103-D field and Bu Attifel field to the Sahl gas plant with a pipeline by yearend 1991. The project will tie in another 600 MMcfd of associated gas supply to the plant.

Tahaddi field development

Plans were advancing for development of Tahaddi field, at 9 tcf Libya's biggest dry gas reservoir.

The project's first stage calls for a 36 in. pipeline from Tahaddi to Marsa al-Brega. Initial throughput would be 250-300 MMcfd.

AFRICA

Additional volumes of gas will require new outlets. Libya's state owned National Oil Corp. plans a spur from the 400 mile, 34 in. coastal pipeline from Marsa al-Brega to Khoms.

Yugoslavia's Razvoj i Inzinjering will undertake a feasibility study for a 150 mile extension of the coastal line to Bukammash near the Tunisian border.

Mediterranean pipeline

Italy's plans to boost its gas consumption in the 1990s has triggered a Libyan project for direct gas exports through a new pipeline under the Mediterranean Sea.

The 350 mile line would cost $3 billion. Preliminary engineering studies have started, but so far there is no sign of a gas sales contract or a likely start-up date.

Libya is the second North African country to target the Italian gas market. Algeria was negotiating increased gas takes with Italy via the existing Transmediterranean line.

Italy wants gas to displace coal and nuclear power in the 1990s, pushing demand to 2.12 tcf/year by 2000 from the current 1.27 tcf/year. Agip SpA pegs Italian gas demand at 2.65 tcf/year in 2000.

Agip has started negotiations with Libya to include a gas clause in its production sharing agreements. Agip now has only the right to oil from development projects. Correspondingly, Agip is assessing gas potential of the offshore area around Bouri oil field. Bouri, operated by Agip, produces 75,000-80,000 b/d and flares big volumes of gas. Flaring could increase as production is boosted to more than 100,000 b/d. Agip first had planned 50 wells in initial Bouri

AFRICA

What the Shell Nigeria team plans in...

Spending (Million U.S. $, 1988–1994): Exploration, Production, Other

Exploration/appraisal drilling (Number of wells, 1988–1994)

Seismic surveys

Two dimensional (Party months): Offshore, Shallow water, Swamp, Land
- 1988: 60
- 1989: 59
- 1990: 48
- 1991: 26
- 1992: 29
- 1993: 30
- 1994: 28

Three dimensional (Party months):
- 1988: 22
- 1989: 33
- 1990: 56
- 1991: 63
- 1992: 65
- 1993: 61
- 1994: 64

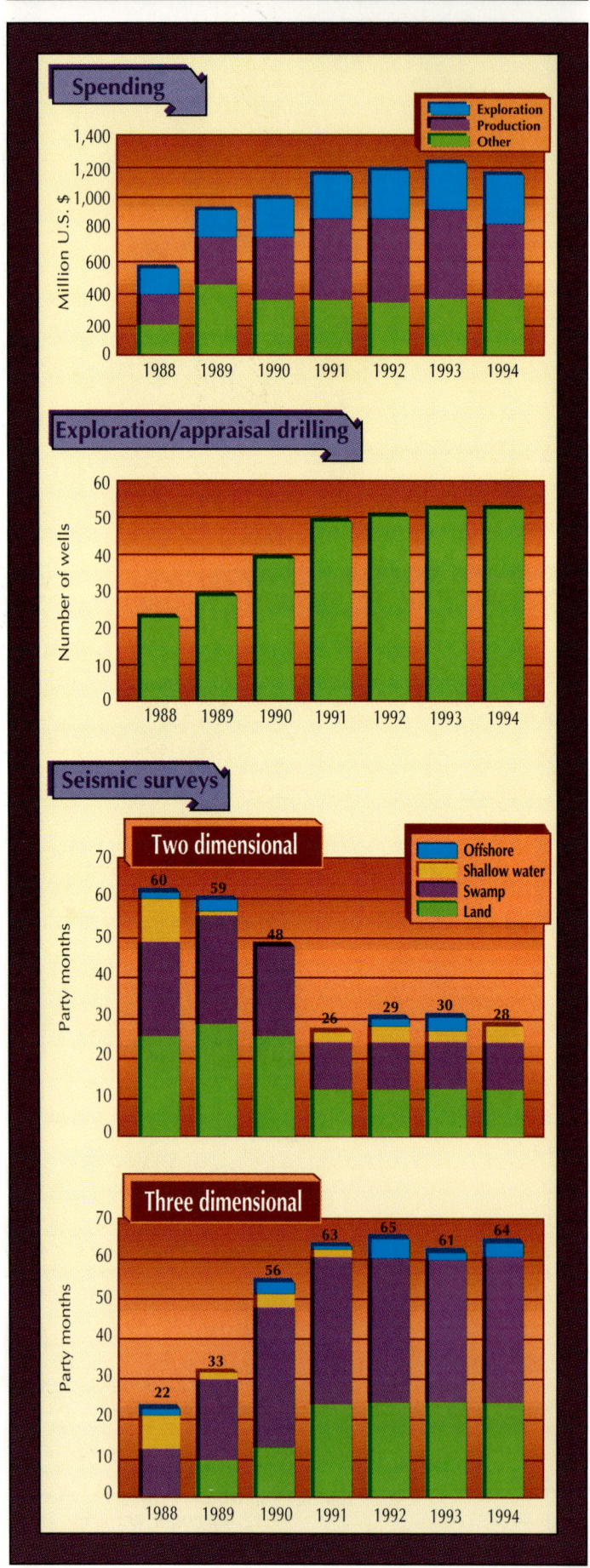

development but now is evaluating horizontal wells. The first horizontal well was a success, a second has been spudded, and another four could be approved.

The appraisal of offshore gas resources will cover Bouri and 10 other structures in the Agip operated Block NC-41, Sirte Oil Co. operated Block NC-35, and Ste. Nationale Elf Aquitaine operated Block NC-137.

MADAGASCAR

CAPITAL: Tananarive
MONETARY UNIT: Francs
REFINING CAPACITY: 16,350 b/cd
PRODUCTION: N.A.
RESERVES: N.A.

BHP PETROLEUM, MELBOURNE, AGREED WITH Omnis, Madagascar's state resources company, to explore a 25,000 sq km permit area off the northwest coast of Madagascar.

The first 3 years of an 8 year program will see geological studies, acquisition and processing of 5,000 line km of seismic surveys, and drilling of one well.

MOROCCO

CAPITAL: Rabat
MONETARY UNIT: Dirhams
REFINING CAPACITY: 154,600
PRODUCTION: 0.3 Mb/d
RESERVES: 2,000 Mbbl

ACTIVITY IN MOROCCO IN 1990 INCLUDED:
• Morocco's state owned Office National de Recherches et d'Exploitations Petrolieres awarding Agip SpA a production sharing contract covering three blocks in the Atlantic Ocean southwest of Agadir.

Royal Dutch/Shell Group and Texaco Inc. also signed agreements under the new terms.

• Maxus Energy Corp., Dallas, agreeing to a 1 year technical evaluation contract with the country calling for geological and geophysical studies, including seismic surveys, on more than 11 million acres in Central Morocco.

Maxus has the option to obtain exploration and development licenses after it completes the study.

• Soc. Cherifienne des Petroles of Morocco letting contract to Sofresid's OTP Engineering unit for engineering work on a 14 in. crude pipeline from the Atlantic port of Mohammedia to the Sidi Kacem refinery, 185 km northeast.

OTP began a feasibility study for the project in 1988.

NIGERIA

CAPITAL: Lagos
MONETARY UNIT: Naira
REFINING CAPACITY: 433,250 b/cd
PRODUCTION: 1,808 Mb/d
RESERVES: 17,100,000 Mbbl

NIGERIA, WEST AFRICA'S BIGGEST OIL producer, has started a major spending program to boost its oil flow in the mid-1990s.

The goal is to take advantage of an expected increase in demand for oil from members of the Organization of Petroleum Exporting Countries.

Money also is being plowed into gas development. Nige-

AFRICA

ria's biggest single investment in the early 1990s aims to turn the country into West Africa's first gas exporter.

By yearend 1990 the final go-ahead was expected for a $2.5 billion liquefied natural gas project at Bonny, Rivers State, to ship gas to Europe and the U.S.

Nigerian productive capacity is 1.8-1.9 million b/d. Production, reined by an OPEC quota, averaged 1.601 million b/d during the first 11 months of 1989, up 16% from the same period of 1988.

By the mid-1990s the government wants to raise sustainable capacity to 2.4-2.5 million b/d. Foreign operators, which run the Nigerian industry in partnership with Nigerian National Oil Corp. (NNPC), believe there are enough resources to meet that target.

With Nigeria's light, low sulfur crude highly prized by refiners in the U.S.—and to a lesser extent in Europe—operators are prepared to ensure that money is available for increased exploration and production.

Biggest program

Nigeria's biggest producer, a combine made up largely of Shell Petroleum Development Co. of Nigeria Ltd. and NNPC, plans to more than double capital spending in 1988-93.

This will involve an increase of more than 150% in exploration and appraisal drilling, start of the country's first deep drilling program, upgrading current production facilities, placing more fields on stream, and Nigeria's first heavy oil production.

The combine, which made a major hike in spending in 1989, budgeted $1 billion for 1990. Further increases are scheduled until spending reaches a peak of more than $1.2 billion in 1993.

Big investments also were under way by the Mobil Corp.-NNPC partnership. With three major offshore projects in progress, the combine plans an offshore wildcat program that will include evaluation of deeper plays.

Mobil produces about 200,000 b/d in Nigeria, mainly offshore.

Nigeria's other major producers are Chevron Nigeria, Texaco Nigeria, Elf, and Agip. Ashland Nigeria and Pan Ocean (Nigeria) have much smaller producing operations.

For all foreign operators, the big problem for the 1990s lies in the ability of NNPC to fund its majority interest in new investment programs. Even at the far lower investment levels of the late 1980s, NNPC usually met cash calls with crude oil rather than currency.

The government has reorganized NNPC to give the company a more commercial view of life. It also approved NNPC's sale of 20% of its interest in the Shell-NNPC partnership.

Shell took an additional 10%, increasing its interest to 30%. Units of Ste. Nationale Elf Aquitaine and Agip SpA took 5% each. The deal netted the state about $2 billion and reduced its commitment to fund new investment planned by the joint venture during the next 5 years.

Recovery plans

Nigeria probably felt the jolt more than any other OPEC member from the slump in oil demand in the 1980s, followed by the 1986 nosedive in oil prices.

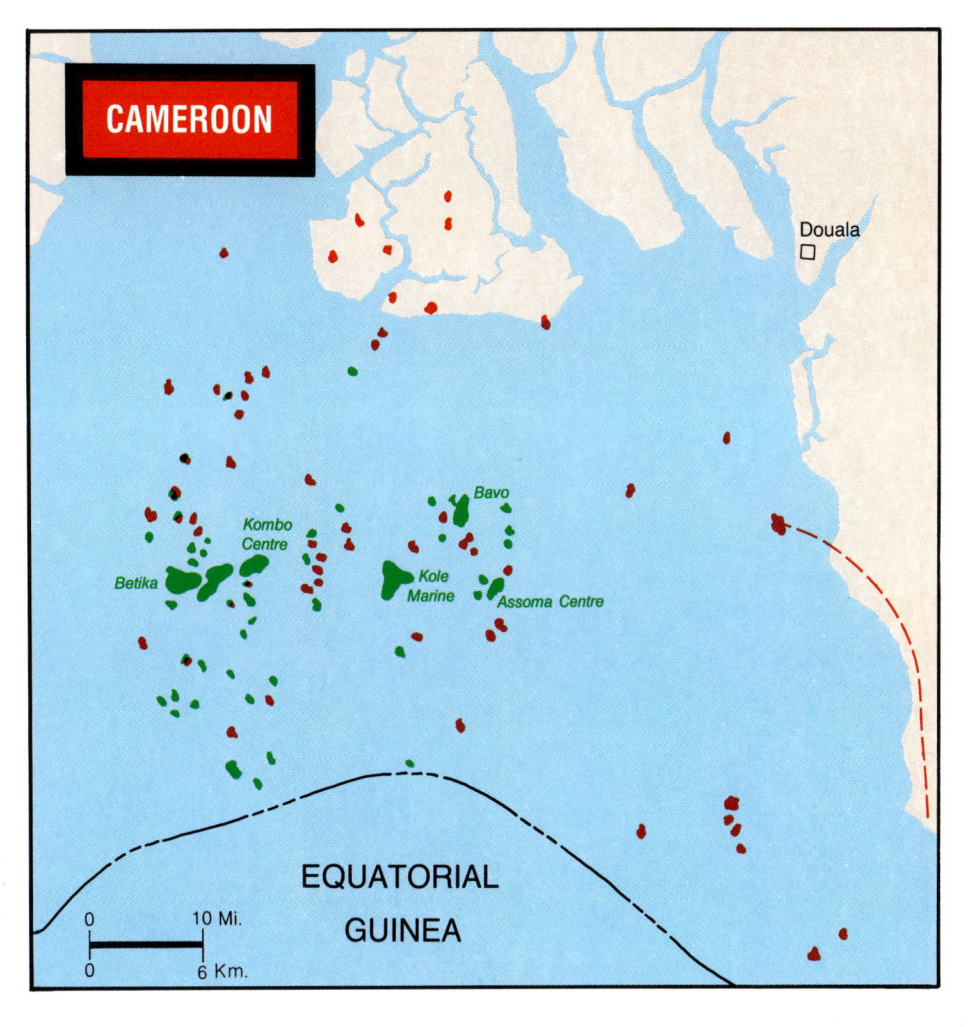

Throughout the second part of the 1970s Nigerian production averaged more than 2 million b/d, peaking at 2.3 million b/d in 1979. Exports crumbled in the early 1980s to a low of a little more than 1.2 million b/d in 1983.

Nigeria's revenues from crude oil exports peaked at a little less than $25 million in 1980 and slumped to less than $7 billion/year between 1986 and the start of 1989, when improved world prices boosted government take to about $8 billion/year.

As production collapsed in the 1980s, the Nigerian industry stagnated. With so many wells shut in, there was little requirement for exploration and development. Operators also were reluctant to invest under terms of an agreement for calculating profits.

Brian Lavers, chairman and managing director of Shell Nigeria, said the Shell combine's operation was mostly care and maintenance.

Nigeria is a land of small reservoirs. Productive capacity can decline very quickly. During this period of stagnation the combine's capacity dipped from its peak of 1.4 million b/d to well below 1 million b/d. Production fell to as low as 600,000-700,000 b/d.

The turnaround in the Nigerian industry started in 1986 when foreign companies and the government signed a new operating agreement. Operators were guaranteed a profit of $2/bbl in return for increased work programs to restore production levels.

The Shell partnership is working toward a target of 1.1 million b/d productive capacity by 1993-94, which would enable production to rise to 1 million b/d by then. But the group also has contingency plans to boost capacity to 1.3 million b/d in response to any request for higher production

AFRICA

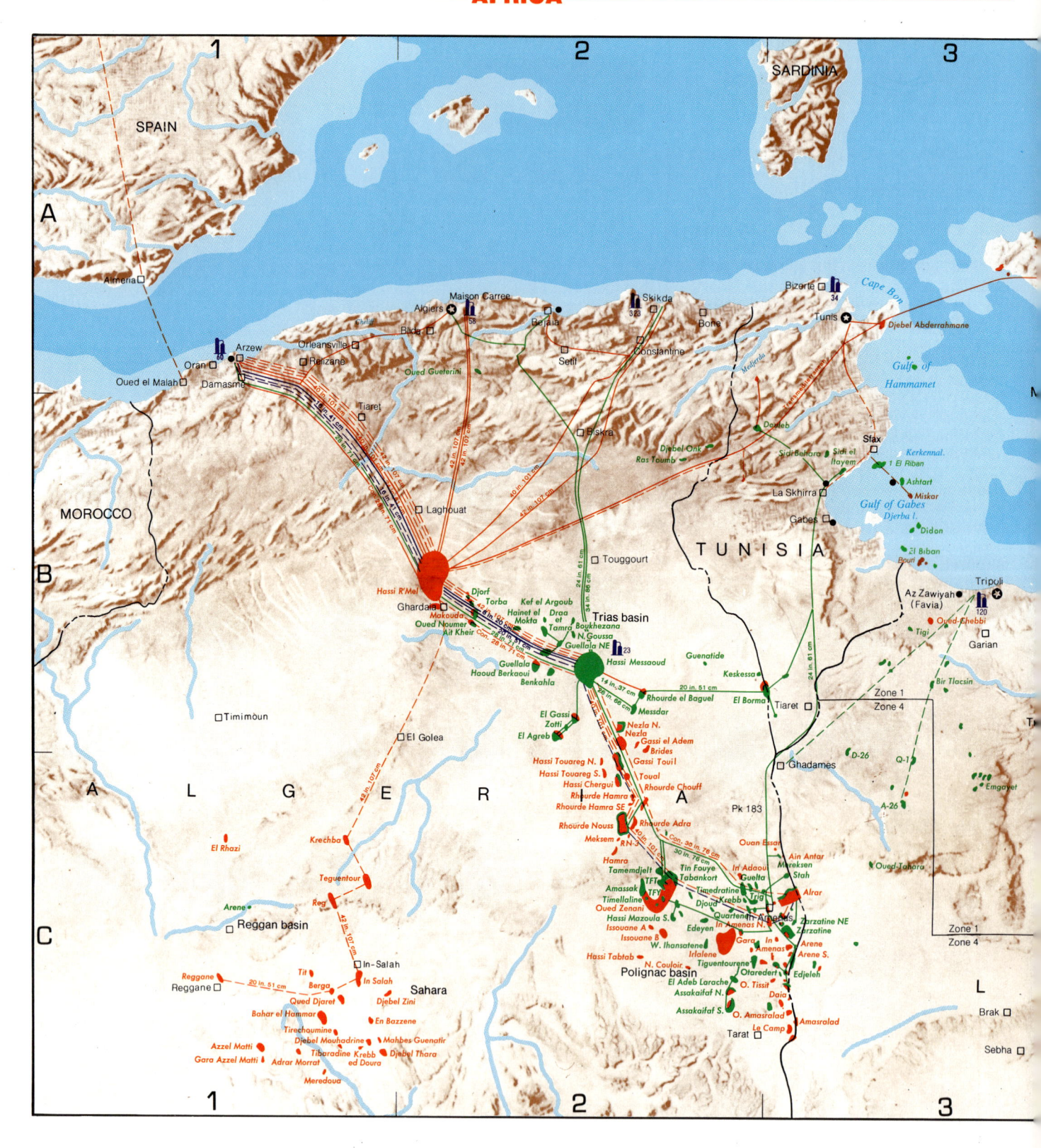

levels from the government.

During second half 1989 Nigeria participated in the OPEC production free-for-all. Production was pushed close to its technical potential of 1.8-1.9 million b/d with the Shell combine producing as much as 980,000 b/d. Those levels could not be sustained very long.

Spending plans

The scale of the Shell combine's spending plans was outlined by David Thomson, Shell Nigeria's general manager, commercial.

Spending, which climbed to $924 million in 1989 from slightly more than $550 million in 1988, was expected to be $1.012 billion in 1990.

The rise in spending will continue into the early 1990s. New investment is centered on increased exploration and upgraded and new facilities to boost oil flow. But the group continues to make substantial payments to the government as part of the renegotiation of its leases, which run for a further 19-29 years.

E.M. Daukoru, Shell Nigeria's exploration manager, outlined the transformation of Nigeria's exploration scene.

In 1981-82 only two wildcats were drilled. In 1986, the

AFRICA

NORTH AFRICA

Legend

- Oil field
- Oil shale
- Oil sands
- Gas field
- Crude oil pipeline
- Natural gas pipeline
- Product pipeline
- Pipelines planned or under construction
- Pump station
- Refinery in operation
- Tanker terminal
- Cities
- Capital
- International boundaries
- State and provincial boundaries
- Water depth 0 to 200 meters
- 200 meters and deeper

---

year in which the new deal on profits and work programs was announced, the number rose to seven as part of a 20 well exploration and appraisal program.

The 1990 exploration and appraisal program was expected to include 40 wells. The high level of drilling will continue through the early 1990s with a peak of 50-55 wells in 1993 and 1994.

The number of rigs working for Shell has risen to 22 from 10 in 1988, and more will be added during the next few years.

The increase in drilling is yielding results. The company discovered 511 million bbl of oil in 1988 and 475 million bbl in 1989.

Daukoru said the biggest aid to Nigerian exploration has been wide scale introduction of three dimensional seismic. After several limited programs in the early and mid-1980s, 3D crews began to appear regularly in 1986.

Total 3D use by all operators was about 20 party months in 1986.

NNPC's leases cover a variety of conditions from land to seasonal swamp, swamplands in the Niger Delta, shallow estuaries, and deeper acreage offshore.

Daukoru said the delta is still the main area of interest, but there is potential offshore in 650-1,300 ft of water. He reckons there are 300 undrilled prospects on the Shell combine's acreage, including 50 undrilled very deep prospects.

Most wildcats are drilled to about 14,000 ft. The exploration department plans to probe deep, high pressure prospects at or below 18,000 ft.

Initially Shell is acquiring a swamp barge with capacity to handle these depths and pressures but will need a large land rig and an offshore rig to complete the deep well program.

The current exploration program has produced finds of 10-100 million bbl, but Lavers said the deep prospects present the best chance of finding more substantial reservoirs.

Shell has formed divisions east and west of the Niger River, and both will undertake deep drilling.

Production plans

The Shell combine has an extensive network of pipelines through the eastern and western divisions. Most of the

AFRICA

Tunisian blocks

[Map showing Tunisian blocks El Franig and Baguel, with El Bourma-Ghannouche gas pipeline, near Algeria border]

steam injection projects on even more viscous crude deposits.

Gas developments

Shell describes Nigeria as a major gas province.

Official recoverable gas reserves are put at 87 tcf, but the potential is enormous. It should be possible to boost this figure to well over 100 tcf with relatively little additional drilling.

Most licensed acreage is gas prone. Even though companies have tended to avoid exploration in areas of known gas potential, they are still finding more natural gas than oil.

Developing Nigeria's gas resources has been a long and frustrating business.

Shell started to develop a market in the delta, but prices were poor and did not cover costs. However, the opening of the 220 mile Escravos to Lagos pipeline has improved prospects for increasing gas sales.

David Balogan, who heads Shell's domestic gas unit, says there is a need for greater infrastructure.

Industry in the northern part of the country wants gas.

The cost of an 18-24 in. pipeline to the northern city of Kaduna would be about $500 million. Domestic gas prices could not justify investment on this scale, and there are doubts the government has the resources to subsidize the project.

Nigeria is also assessing gas exports to other West African countries. There are outline plans for a pipeline to Nigeria's western neighbors, where gas could play a part in slowing deforestation of West Africa for firewood. A shortage of funds ensures this will be a long term project.

Mobil's projects

Mobil Producing Nigeria, which has liquids capacity of about 250,000 b/d from its offshore concessions, is involved in three development projects.

The first platform on the 850 million bbl Edop field, installed in 1987, holds five wells producing about 40,000 b/d.

Mobil has let construction contracts for a production platform with a capacity of as much as 250,000 b/d and a 24 in. pipeline to the shore to start up in 1991.

Full field development will require 36 wells from six platforms. More wells will be drilled and platforms installed as needed, depending on oil demand.

Mobil also is responsible for Nigeria's biggest offshore gas/condensate project.

Mobil agreed with NNPC and the Nigerian government to develop Oso gas/condensate field.

The two companies let a $400 million construction contract to MBJ Consortium, composed of McDermott Inc., New Orleans, Bouygues Offshore SA, Montigny-le-Bretonneux, France, and Japan Gasoline Corp.

MBJ will build eight platforms, a condensate pipeline from Oso to Mobil's Qua Iboe terminal, condensate storage tanks at the terminal, and dedicated export structures. It also will lay a low pressure gas gathering system between Oso and Mobil oil fields in Oil Mining Leases 67 and 70.

Oso field is in OML 70, 35 miles from Qua Iboe terminal. Mobil, as operator, holds a 40% interest in the lease, NNPC 60%. Reserves are about 500 million bbl.

Mobil expects production to begin in early 1993 and reach 100,000 b/d of condensate by yearend 1993.

It plans to inject the 300 MMcfd of gas Oso will produce and 200 MMcfd from OMLs 67 and 70 to maintain reservoir pressure.

This will use nearly 100% of associated gas. About 3.5 tcf will remain in Oso field after the condensate is produced. The project has loans for about 70% of its total required investment of about $1 billion.

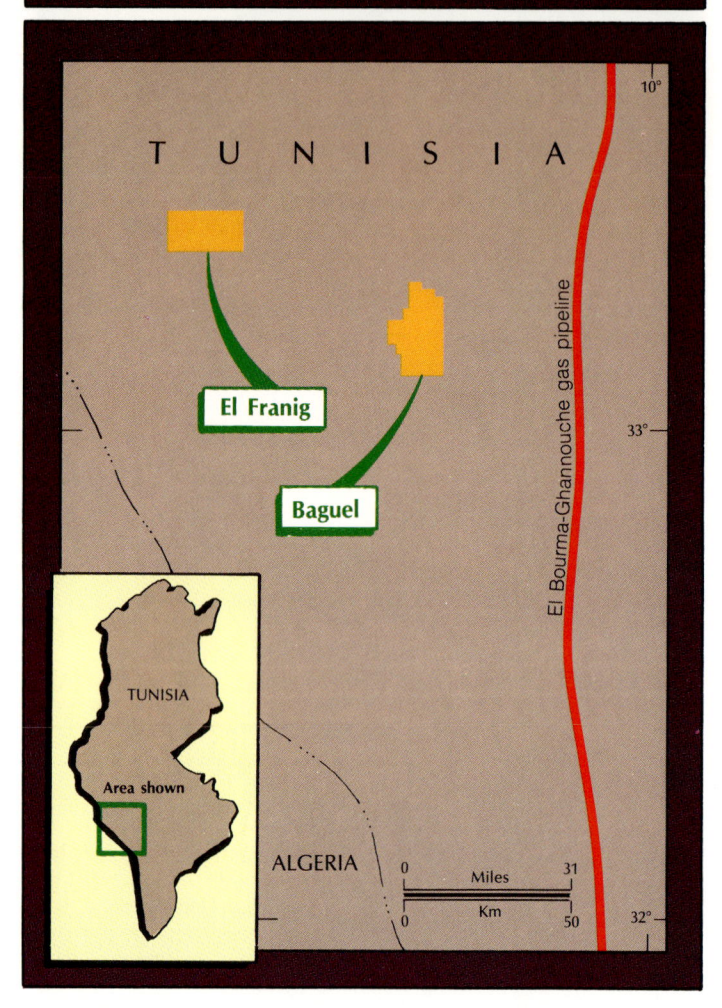

spending in this sector is to tie in small fields.

Most of the new flow stations will handle 10,000-30,000 b/d, but larger units will be built. For example, Tunu, to start up in 1992 in the western division, will have a capacity of 60,000 b/d.

Offshore, Shell's western division will develop EA field, expected to produce about 30,000 b/d of 30° gravity crude by 1993.

The combine is also introducing gas lift more widely and building more facilities to handle associated gas.

Improvements also will be made to terminal and transportation facilities, including a major extension of storage and treatment facilities at Forcados terminal that will involve jacking up existing tankage and installing new foundations.

The most ambitious production project the combine is undertaking is a condensate soak scheme designed to extract heavy, viscous crude from 700 million bbl Sapale field in the Niger Delta.

Shell plans to inject as much as 8,000 b/d of condensate into part of the heavy oil reservoir through four closely spaced wells.

The condensate should mix with the heavy crude in the reservoir, reducing viscosity and making production possible for the first time.

Condensate will be delivered from a nearby gas plant that lacks a liquids transportation system. With condensate readily available, the injection process looked more viable than steam injection.

If the four well pilot is a success, condensate soak could be extended to the rest of the field.

Outside the delta, NNPC is investigating possible pilot

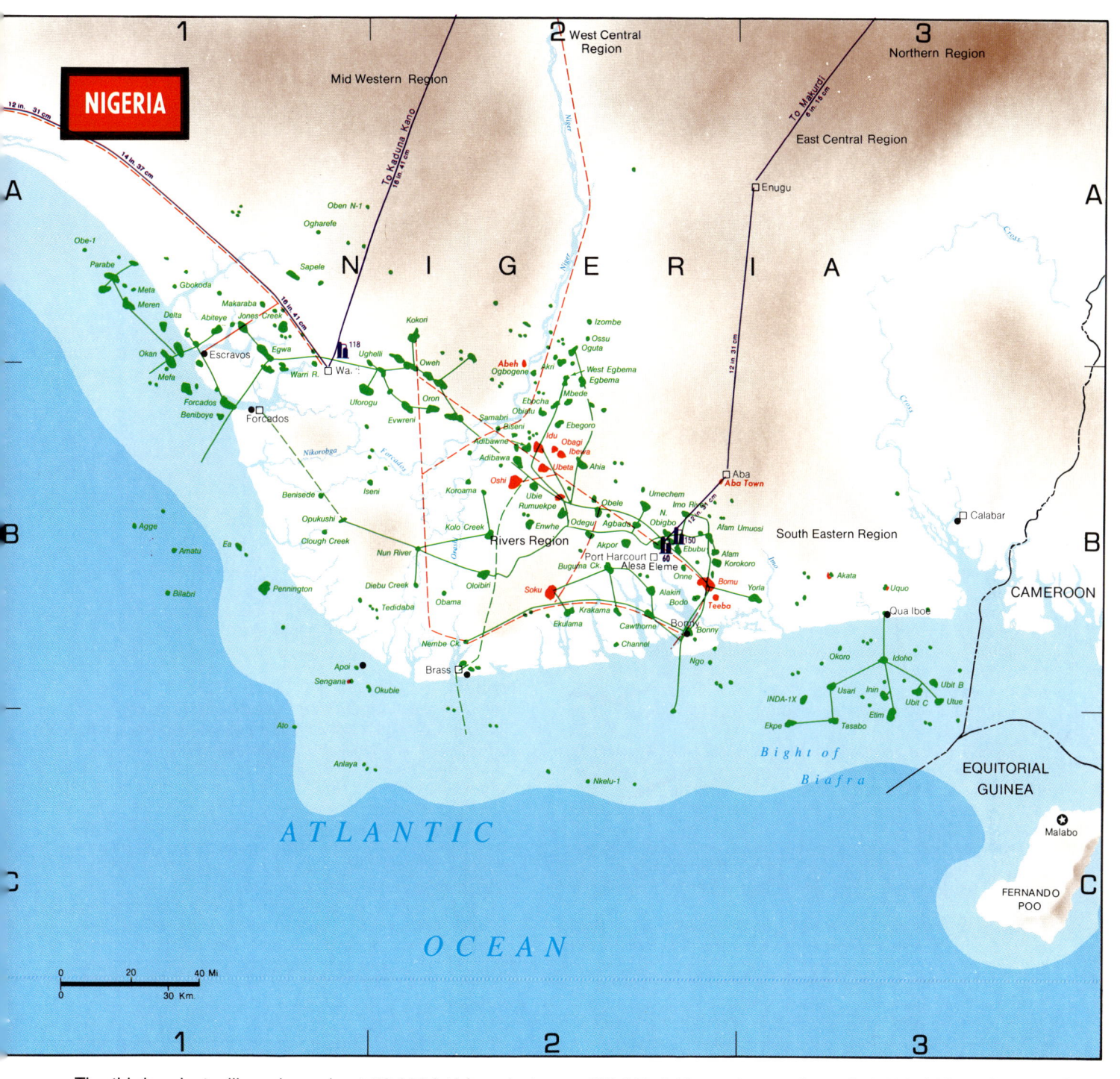

The third project will produce about 52,000 b/d from Iyak field in 1993. Mobil will drill 14 wells and install two platforms in the 146 million bbl field. First liquids are expected in 1991.

Mobil plans to increase its Nigerian productive capacity by an additional 100,000 b/d during the next 5 years.

In 1989, the company almost trebled its seismic activity, shooting 6,289 line-miles of 2D and 3D surveys. It drilled three successful wildcats.

Mobil also plans a wildcat drilling program for the next 3 years, a substantial part of which will target deeper plays at 10,000-15,000 ft.

Chevron program

The biggest of the other operators in Nigeria is the Chevron-NNPC joint venture. Chevron inherited a 3.5 million acre concession through its acquisition of Gulf Oil Corp.

The joint venture, with a productive capacity of about 300,000 b/d, produces about 270,000 b/d onshore and offshore.

Chevron, saluting the changed investment climate, is looking at expansion projects that would sustain production.

Since the increase in operators' margins in 1986, Chevron's exploration activity has averaged seven to nine wells/year and 10-19 seismic crew months/year.

Activity will remain at 1990 levels.

Moves also were under way to sell associated gas used for fuel. Chevron was studying a project to gather flared gas for LPG extraction and deliver the residue into the domestic distribution network.

Texaco, Elf

Chevron also is a partner with Texaco in two offshore concessions that produce about 60,000 b/d.

Texaco Chairman Alfred DeCrane Jr. said exploration is

AFRICA

at a relatively low level. Texaco drilled a dry hole in the outer offshore trend late in 1989 and planned to spud another offshore wildcat before yearend 1990.

Elf Nigeria, a major West African operator, has placed great emphasis on exploration and development in Nigeria. The company acquired new acreage offshore, became a partner in the LNG operation, and finally bought a 5% stake in the Shell combine's concession.

The company's onshore fields produce about 95,000 b/d. Basic engineering has started for development of four fields—Afia, Odudu, Ime, and Edikan—on offshore permit OPL95.

Onshore, Elf is developing Olo field and connecting facilities to the transportation network.

It continues a drilling program on its OML 57 and 58 permits.

Gas for the Bonny LNG project will not be required for several years, but preliminary studies have started on developing Ibewa field, which will supply Elf's initial share of the feedstock.

Elf's exploration program is aimed at maintaining production levels. The company requires two rigs/year offshore and about 1.5 rigs/year onshore.

Downstream activity

NNPC let major contracts for its proposed $1 billion Eleme petrochemical complex in Rivers state, Nigeria.

In planning since 1974, the project has been dogged by financing problems and doubts about economic feasibility. Production is to start up in 1993.

Nigeria's minister of petroleum resources, Jibril Aminu, said that the project had to cross some major hurdles in terms of feedstock supply, demonstrating project viability to the international community, and negotiating foreign financing.

Chiyoda Corp. will construct an olefins plant to produce 300,000 metric tons/year of ethylene and 90,000 tons/year of propylene and provide utilities and offsites.

Kobe Steel will build a 250,000 ton/year polyethylene/butene 1 complex.

Tecnimont and JGC Corp. will install an 80,000 ton/year polypropylene unit, expandable to 120,000 tons/year.

Spie Batignolles will provide infrastructure and site development.

NNPC negotiated a feedstock supply agreement for natural gas liquids with its joint venture partners Agip SpA and Phillips Petroleum Co. NGL will be supplied by pipeline from Obrikom, 50 miles away.

In addition, export agreement for finished products has been reached with Dupont of Canada for 150,000 tons/year of polyethylene and Techint International for 40,000 tons/year of polypropylene.

Foster Wheeler International Corp., Reading, England, project management consultant to NNPC since 1974, said it will continue to assist NNPC unit Eleme Petrochemicals Co. Ltd. in managing the current contracts.

SOMALIA

CAPITAL: Mogadishu
MONETARY UNIT: Shillings
REFINING CAPACITY: 10,000 b/cd
PRODUCTION: N.A.
RESERVES: N.A.

A JOINT VENTURE OF MOBIL EXPLORATION Somalia Inc. and Pecten Somalia Co. planned to spud the first wildcat on a 14.9 million block off Somalia early in 1991.

Pecten, awarded the concession in 1988, has conducted extensive seismic surveys of the concession, which extends along 1,200 km of Somalia's Indian Ocean coast.

The two will share 50-50 exploration risk and any resulting production.

TANZANIA

CAPITAL: Dar es Salaam
MONETARY UNIT: Shillings
REFINING CAPACITY: 17,000 b/cd
PRODUCTION: N.A.
RESERVES: N.A.

TANZANIA-ZAMBIA PIPELINE CO. REPAIRED 128 km of its crude pipeline system, finishing the second phase of a program started in 1987 with funding from Italy's government and European Investment Bank.

With 90% of capacity restored, pipeline throughput is up one third to 20,000 b/d.

Work on the third phase is to begin in Zambia.

TUNISIA

CAPITAL: Tunis
MONETARY UNIT: Dinars
REFINING CAPACITY: 34,000 b/cd
PRODUCTION: 93 Mb/d
RESERVES: 1,700,000 Mbbl

TUNISIA IS RENEWING ITS EFFORTS TO ATTRACT oil and gas companies to explore its territory.

About 400 exploratory wells have been drilled and about 100,000 line km of seismic data recorded.

Four of the five concession or seismic option agreements signed since Jan. 1, 1989, have covered areas near Ashtart oil field in the Gulf of Gabes.

The 1985 petroleum law, as amended in 1987, remains in effect. New legislation being considered is intended to promote deep drilling, gas development, and extension of existing exploration commitments.

It isn't known to what extent, if any, the deep drilling and gas legislation might apply to existing permit holders.

Most of the country's oil and gas fields are small, but several large areas are unexplored or lightly explored.

Exploration history

Exploration in Tunisia consisted of drilling about one well/year during 1909-42. First production came in 1948, when gas was produced from Lower Cretaceous in the Abderrahman structure on Cap-Bon peninsula.

The first commercial oil production appeared in 1964 from Middle Triassic sandstone on the large El Borma structure along the Algerian border. Oil also was discovered in Lower Cretaceous Aptian carbonates at Douleb in west central Tunisia. The country's deepest test, to 17,525 ft, was drilled in the mid-1960s on the Abderrahman structure. Exploration then slowed. Ashtart, Tunisia's second major oil field after El Borma, was discovered off Sfax in Lower Eocene El Gueria limestone in 1971, and Sidi El Itayem field was discovered in the plain of Sfax in southern Tunisia the same year in the same formation.

In 1975 Miskar gas field was discovered in Upper Cretaceous carbonates in the Gulf of Gabes. British Gas Tunisie (BGT) planned to decide in late 1990 whether to develop the field, which it estimates contains about 1 tcf of gas reserves.

Also in the 1970s, Birsa, Tazerka, and Yasmin oil fields

AFRICA

DRILLING in Tanzania's Ruaruke North-1. (Photo courtesy Shell World).

were found in Middle Miocene sands in the Gulf of Hammamet; and oil, gas, and condensate were discovered in Ordovician quartzite in the Chott area. Tunisian exploration peaked in 1981, when 32 wells were drilled. Activity later slowed markedly, especially after oil prices collapsed in 1986. Oil output may at least level off for several years, owing to a number of onshore discoveries by BGT.

BGT said that production from its Kerkennah concession near Sfax will rise to at least 10,000 b/d in 1991 from 3,200 b/d in 1990. BGT also plans to drill two wildcats.

Blocks acquired by Noble Affiliates

Noble Affiliates Inc., Ardmore, Okla., acquired a 25% interest in two onshore concessions in Southwest Tunisia from Walter International Inc., Houston, including one on which gas development is planned. Noble unit Samedan of Tunisia Inc. agreed to buy the interest in the 108,680 acre Baguel and 54,340 acre El Franig concessions, both formerly operated by Amoco Tunisia Oil Co.

The Baguel block lies about 30 miles west of the El Bourma-Ghannouche gas pipeline. Two gas wells have been drilled on it, one each on the Baguel and Tarfa prospects. The Baguel prospect well cut about 55 ft of Triassic pay and flowed 10.4 MMcfd through a 1 in. choke with 525 psi flowing tubing pressure. The first of two development wells on the prospect were set for yearend 1990. Once both are drilled and tested, the companies will lay a gas pipeline from the field to the El Bourma-Ghannouche line at a cost of about $6 million.

The Tarfa well, about 10 miles north of the Baguel drillsite, cut 33 ft of pay but was not tested.

The El Franig concession, about 30 miles northwest of the Baguel block, holds one well that flowed 12.6 MMcfd of gas and 1,843 b/d of condensate. Noble said there are no plans for further drilling on the block.

Ezzaouia field start-up

Marathon Oil Tunisia started production of 41° gravity oil from Ezzaouia field on the Zarziz permit in Southeast Tunisia. The 12,000 b/d capacity production facilities and a new 35,000 dwt capacity marine export terminal at Zarzus, about 5 miles away, cost $85 million. Marathon operates on behalf of state oil company ETAP, Ste. Nationale Elf Aquitaine, and Oranje-Nassau of Netherlands.

ZIMBABWE

CAPITAL: Salisbury
MONETARY UNIT: Dollars
REFINING CAPACITY: N.A.
PRODUCTION: N.A.
RESERVES: N.A.

ZIMBABWE HOPES TO INAUGURATE ITS CHEMICAL industry with a $110 million polyvinyl chloride plastics plant.

PVC Polymer Zimbabwe Pvt. Ltd. let contract to Foster Wheeler Energy Ltd., Reading, U.K., to complete by yearend 1990 a second phase feasibility study for the project, tentatively scheduled for completion in 1993.

If approved, the plant would produce 20,000 metric tons/year of PVC, as well as caustic soda, chlorine gas, sodium hypochlorite, and hydrochloric acid.

IPE
ATLAS

U.S.S.R.

RUSSIA

CAPITAL: Moscow
MONETARY UNIT: Ruble
REFINING CAPACITY: 12.3 million b/cd
PRODUCTION: 11.5 million b/d
RESERVES: 57 billion bbl

THE SOVIET UNION LAST YEAR UNINTENTIONALLY played a significant role in deepening and prolonging the world oil supply shortfall precipitated by Iraq's Aug. 2, 1990, invasion of Kuwait.

Official Moscow data indicate that if the U.S.S.R.'s liquid petroleum production and exports hadn't plunged in late summer 1990, the loss of Persian Gulf oil supplies would have had a milder and shorter effect on world markets. Increased output from other members of the Organization of Petroleum Exporting Countries last year in September and October thus could have steadied prices sooner.

Still the world's leading oil producer, the Soviet Union suffered a huge drop in crude and condensate flow during August-September 1990. Production those 2 months averaged only about 11.13 million b/d vs. 11.73 million b/d in first half 1990 and about 11.56 million b/d in July. It now appears that Soviet crude/condensate production will not rebound soon.

During 1991, oil flow will continue to fall, reports the deputy chairman of the U.S.S.R. State Planning Committee. He said this decline can only be prevented by considerably greater capital investment in crude production. Difficulties have mounted in developing new oil regions. Supergiant Tengiz field in Kazakhstan hasn't gone on production for a number of reasons, including irregular equipment deliveries and delays by construction organizations.

In other fields, he said, oil flow is less than expected because of planning errors. A simply calamitous situation prevails in providing material and technical resources for enterprises of the oil producing industry.

Oil production in the U.S.S.R. peaked at 12.48 million b/d in 1987. It fell to 12.45 million b/d in 1988 and to 12.14 million b/d in 1989. Moscow cut its oil production target to 12.04 million b/d for 1990, and authorities warned the nation's economy couldn't operate properly with less than that output. Shortly before production topped out in 1987, some Soviet oil industry personnel predicted flow would rise well above 13 million b/d in the 1990s.

Soviets pin big hopes on gas resources

Natural gas is today's and tomorrow's bright spot in the nation's energy picture. Boosting gas production will be easier than halting declining oil flow, but even natural gas growth faces political and technological challenges.

Corrosive well streams require special equipment and operating techniques. And in areas where new supplies will come from—western Siberia, for one—there is growing local resistance to development of oil and gas resources.

Environmental problems are acute in the U.S.S.R. During the last 5 years, the expense for environmental protection has doubled, Soviet officials say. And the cost will grow even faster in the future. Growing concern about improving the environment will have its effect on oil and gas supply.

There have been "some bloody battles" for the right to explore, one Soviet official said. Public opinion was "100% against" building a complex to handle high hydrogen sulfide gas from Astrakhan fields. Most wells there were drilled without problems, but the area is densely populated and a gas handling problem would be very serious.

Despite these hurdles, natural gas production and use will grow rapidly in the next decade, says the chief geologist for the U.S.S.R.'s Gas Association, successor to the Ministry of Gas Industry.

Natural gas accounts for 40% of the Soviet Union's primary energy production, and it's the only segment that has shown healthy growth in recent years. Gas exports have increased by 20 times during the past 15 years. Gas production will continue to grow in the 1990s, the chief geologist said.

Production will reach more than 1 trillion cu m/year by the mid-1990s. Production a year ago was an estimated 828 billion cu m, up from 727 billion cu m in 1987 and 587 billion cu m in 1984. Currently, western Siberia accounts for 58.5% of the Soviet Union's gas production. European regions provide 21.2%, and central Asian and Kazakhstan contribute 19.2%. The remaining 1.1% is from other areas.

Increased exports won't be the only growing market for natural gas, the association chief geologist predicts. He expects stricter rules covering the operation of nuclear power plants in the Soviet Union. That will limit the amount of electrical power from this source, and natural gas will take up some of the slack in power generation.

About 1,150 gas fields have been discovered in the U.S.S.R. Many await development, some because of special equipment needs. For example, an estimated 13% of the gas resource contains significant amounts of H_2S, in some concentrations of more than 20%.

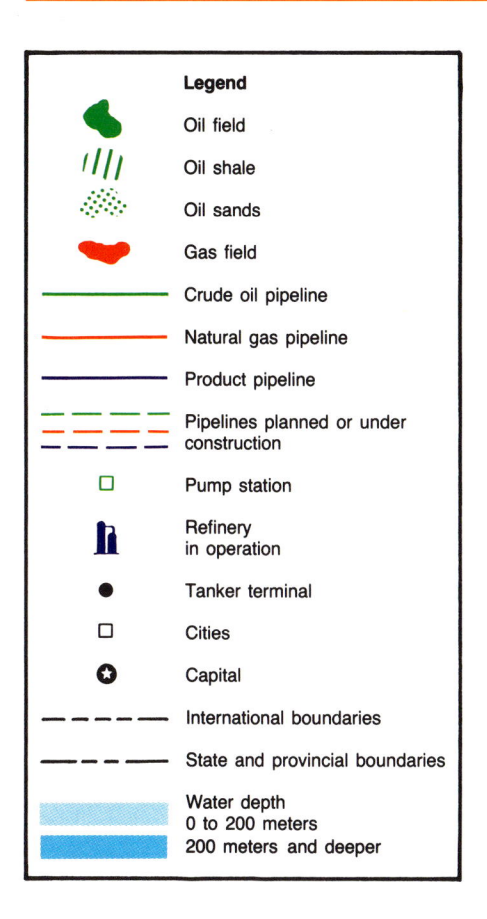

Gas with more than 3% ethane is considered a prime source of petrochemical feedstocks, and new plants are expected to be built to handle new supplies. More than 40% of the gas in western Siberia exceeds this ethane content, Soviet specialists say.

In addition to present reserves, there is a vast undeveloped gas resource in the remote areas of western Siberia alone. And there is potential in eastern Siberia. Those prospects are inferior to those of western Siberia, the Soviets say, and they are even farther from Soviet markets than western Siberia regions.

Development of eastern Siberia's gas resources wll be costly, and it may be an opportunity for joint ventures between the Soviets and firms from the U.S., Japan, and China. Because the region is closer to Japan than to big Soviet markets, a Soviet/Japanese venture may make sense.

Present fields will be the source of most gas until 2010, the chief geologist predicts. Exploration is still under way, but field development has higher priority.

Gas resources in western Siberia are enormous. But they also are 2,000-3,000 km from markets. More than 70% of the development investment in western Siberia is for main pipelines. Twenty pipelines from the area were finished during 1989-90. Huge Urengoi field now has about 1,200 wells producing about 250 billion cu m/year (8.83 tcf/year) form the Upper Cretaceous. Another 750 wells in the Lower Cretaceous contribute 21 billion cu m/year (0.74 tcf/year) of gas and 6 million tons/year (120,000 b/d) of condensate.

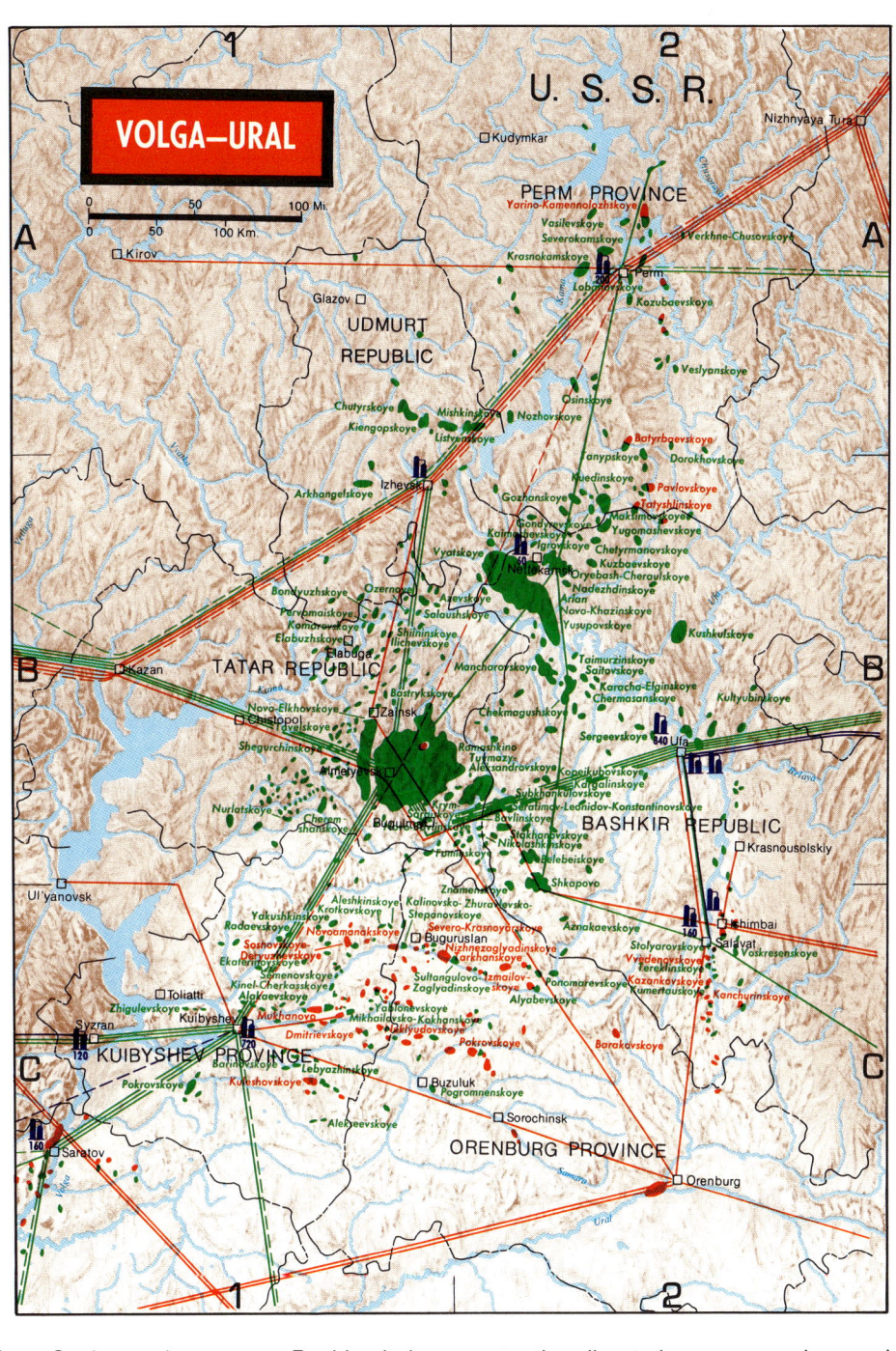

Besides being remote, the climate is severe, and ground water is lacking. Construction costs are 2-2½ times higher than in the European part of the country. And new technology will be needed to develop more reserves in the region. Recently, equipment modules weighing 300-1,000 tons for Yamburg gas field were built under factory conditions 1,500 km from the field and moved by a water route to the area during the summer.

Western Siberia's operations face a challenge in a rising volume of gas containing significant amounts of H_2S. That trend will require a change in design and operating techniques, the association official said.

The Soviets expect to place large volumes of Yamal Peninsula gas on stream in 1994 or 1995. Development drilling is under way, and preparations are being made to lay a pipeline from Yamal to the Central U.S.S.R.

Difficult geology and permafrost are characteristic of the area. To prevent permafrost melting, development drilling involves grouping directional wells in clusters of up to 20.

Much of the development experience gained in Yamburg

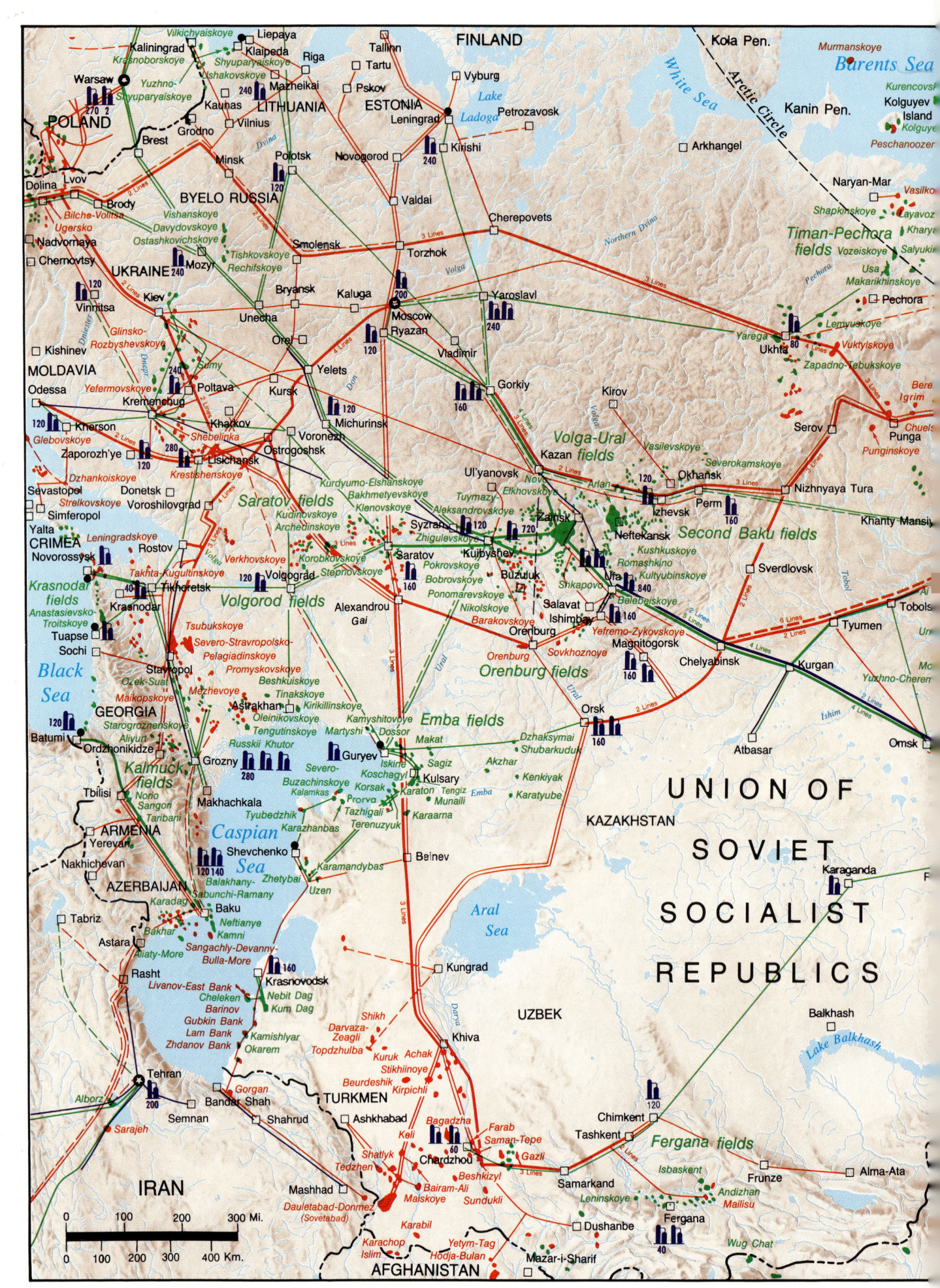

U.S.S.R.

U.S.S.R.

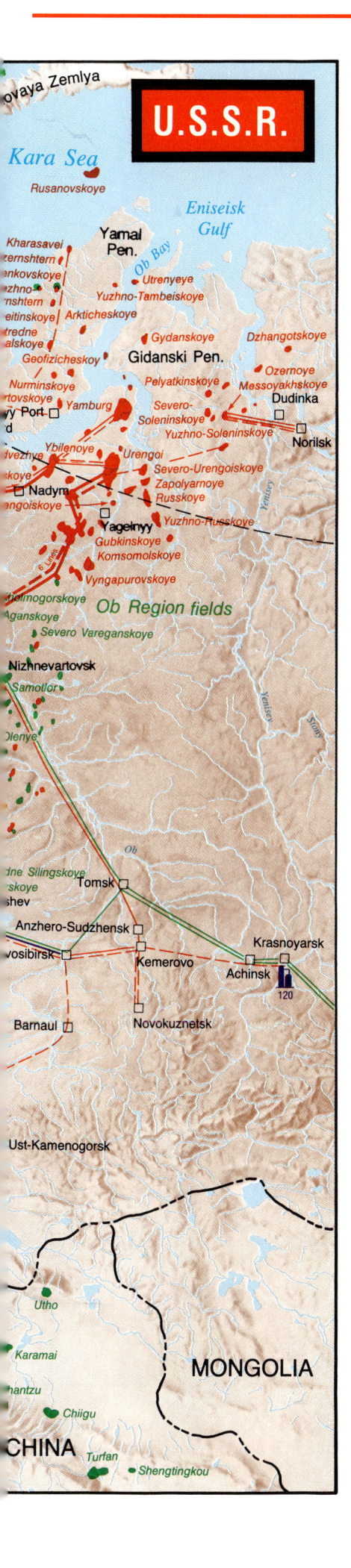

will be used in Yamal, Soviet specialists say. There is some oil in a few Yamal fields, too. Ten years ago, a lot of oil was expected in the northern Yamal Peninsula area, but not much has been found. The Gas Association chief geologist still thinks the oil potential in the region is significant, but for the time being it is much more economic to develop oil farther south.

The arctic offshore is "very prospective," he says. Some Yamal fields extend offshore, but development will be very expensive and likely will not begin until after 2000. There is oil in Urengoi, too—maybe 1 billion tons in place, 200 million tons recoverable. But it is distributed in many horizons, some of which are only 3 m thick at depths of 3,100-3,500 m.

A first for the U.S.S.R.: Competitive bids sought

The Soviet Union has taken the historic step of opening part of its territory to competitive bidding for oil and gas exploration and development by non-Soviet companies.

Two areas in the South Caspian and Amu-Daria petroleum provinces in Turkmen Soviet Socialist Republic are being offered by the U.S.S.R. Ministry of Geology and the government of the Turkmen republic. One area lies near Chardzhou, where there is 60,000 b/d of refining capacity. The areas cover a combined 34,749 sq miles and are highly prospective, said Wavetech Geophysical Inc., Denver. The areas have existing major hydrocarbon production, well developed infrastructure, and favorable climate.

Small tract competitive bidding will be used to select participating western companies. Size of the tracts was not disclosed. The Soviets and successful bidders are expected to sign exploration and production agreements that conform with existing Soviet legislation on joint ventures.

The bidding expected to result from the Soviet initiative isn't expected to be a one of a kind occurrence. The Soviets are considering a few additional territories for future rounds of bidding.

Soviet explorationists rate the resource potential of the offered territories as very high, even by Soviet standards. Among the large existing oil and gas/condensate fields in the area is 965 sq mile Dauletabad-Donmez field.

In spite of the large fields, the areas are largely undrilled and hold more than 100 major undrilled, prospective structures, Soviet data show. A big part of the areas contains widespread Jurassic salt deposition with major unexplored objectives and potential below it.

A well understood, fair, and uniform method of bidding for exploration rights will allow western oil companies of all sizes relatively easy access to large Soviet reserves, Wavetech said. At the same time, the Soviets can expect a more equitable overall return on their resources and significantly accelerated and more efficient development of their reserves and infrastructure.

Exploration success spreads from Barents into Kara Sea

Soviet offshore arctic exploration success has fanned out east of the Barents Sea into the even harsher environment of the neighboring Kara Sea. Discoveries and geophysical data disclosed by Moscow authorities and unofficially by Soviet petroleum industry personnel indicate that shallow waters of the Yamal Peninsula's Kara Sea shelf may hold one of the world's biggest concentrations of giant and supergiant offshore gas fields.

There is little doubt that gas reserves in the first Kara Sea field exceed the total for all developed Soviet offshore areas combined, including the Caspian and Black seas and the Sea of Okhotsk in the Far East.

That first Kara Sea field, Rusanovskoye, was discovered in 1989 less than 70 miles from the Yamal Peninsula's northwest tip. It could hold gas reserves of as much as 8 trillion cu m (282.4 tcf). If so, it rivals western Siberia's onshore Urengoi field, which has been tagged as the world's largest. It was found astride the Arctic Circle in 1966. Until Rusanovskoye appeared on the scene, Qatar's North field in the Persian Gulf was generally regarded as the world's largest gas reservoir lying entirely offshore. Its reserves are estimated at 150 tcf.

Soviet maps and sketchy data reaching the West provide evidence that besides Rusanovskoye the southwest arm of the Kara Sea may hold a second supergiant gas field on the Leningradskaya structure about 30 miles south of Rusanovskoye. Drilling at last report was under way on a third large, promising structure, Zapadno-Sharapovskaya, about 80 miles south of Leningradskaya.

Besides those prime targets, the U.S.S.R. has found and will test at least nine other structures in the Kara Sea's southwest arm. Moreover, two giant—possibly supergiant—Yamal Peninsula fields have major extensions offshore under the Kara Sea. At least one structure on Yamal Peninsula and another on Belyi Island

U.S.S.R.

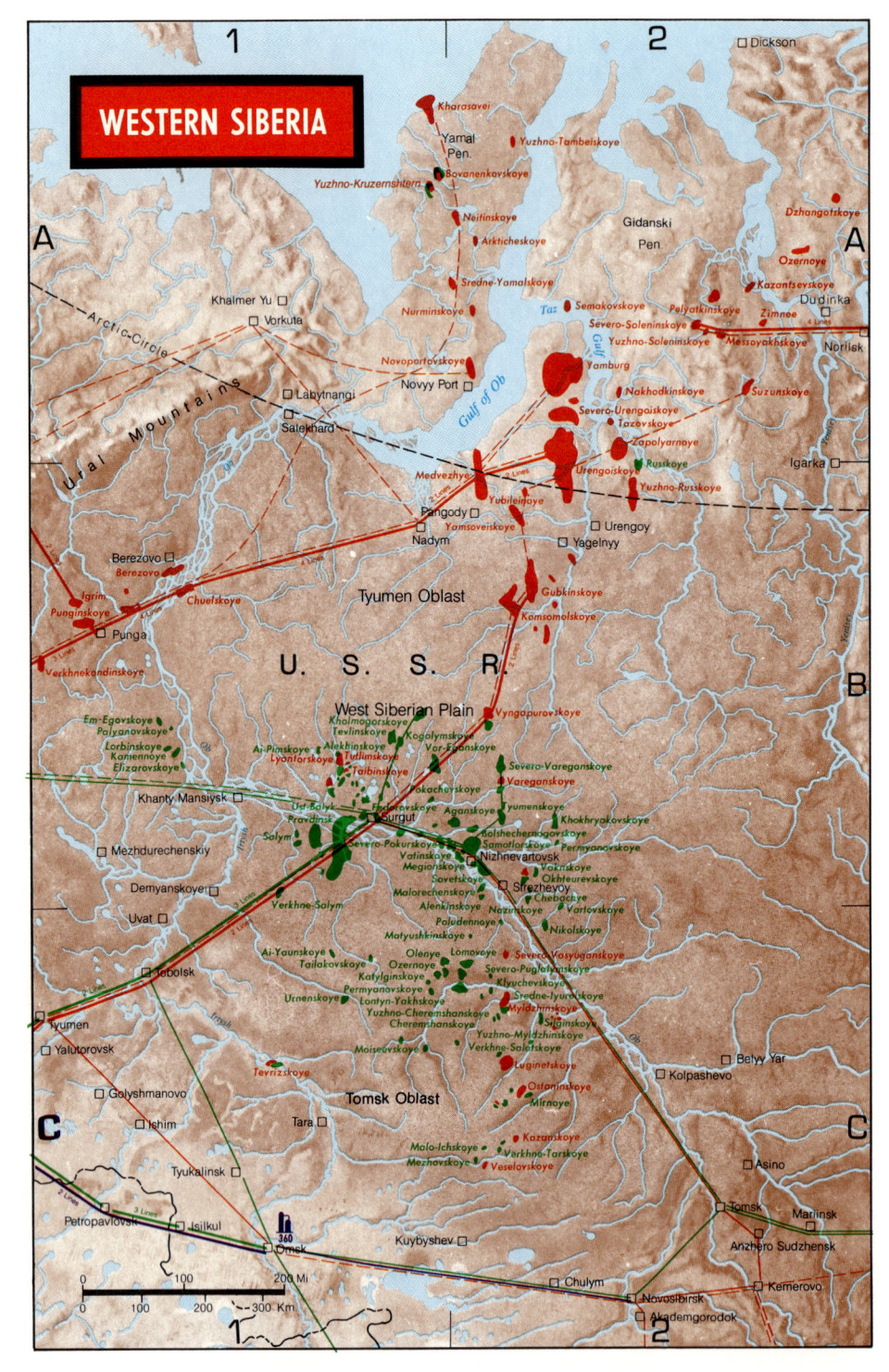

just north of the peninsula, which have not yet been designated as commercial fields, extend into the Kara Sea.

While the Soviets have revealed sparse reserve data on their offshore arctic fields, some western observers believe the Kara Sea's Rusanovskoye field may be twice as large as the more widely publicized Shtokmanovskoye field discovered in the Barents Sea in 1988. Last year, Moscow newspaper reports estimated Shtokmanovskoye's gas reserves at 4 trillion cu m (141.2 tcf). They said Shtokmanovskoye holds "at least four times as much fuel" as had been attributed to Norway's Troll field in the North Sea.

Shtokmanovskoye was found in a generally ice free area of the Barents Sea in water 200-300 m (656-984 ft) deep. It is believed that Rusanovskoye lies in water no more than 50 m (164 ft) deep but in a Kara Sea area that is navigable only 2-3 months/year without the aid of icebreakers. Whereas Rusanovskoye is less than 70 miles from the Yamal Peninsula, Shtokmanovskoye is about 200 miles west of huge Novaya Zemlya Island, which separates the Barents and Kara seas, and almost 300 miles north of the nearest point of the European Russian mainland. The field is about 350 miles northeast of the Soviet port of Murmansk.

Studies are under way to determine when development of

U.S.S.R.

Shtokmanovskoye can begin under a proposed joint venture agreement between the U.S.S.R.'s Ministry of the Oil and Gas Industry and a group that includes Norway's Norsk Hydro AS along with American and Finnish interests. It is believed Shtokmanovskoye won't be on production before 2000.

Odds are that the Soviet Union will continue to find more oil and gas fields in the Barents Sea than in the Kara Sea. Area of the Barents is about 542,000 sq miles with average water depth of 750 ft, while the Kara Sea covers 341,000 sq miles with average water depth of 387 ft.

There's a scattering of oil and gas fields entirely offshore in the Barents Sea and its southwest extension, the Pechora Sea. This doesn't include an oil field on the Pechora Sea's Kolguyev Island, where an oil and gas field extends offshore. The number of structures, drilled or which are still undrilled in the Barents Sea far outstrips those in the Kara Sea.

A spokesman with the U.S. Geological Survey's World Energy Resources Program believes there are good prospects for finding oil along the southern flank of the Kara Sea's submerged North Siberian ridge. The ridge apparently extends east from a point near the northern end of Novaya Zemlya Island to the big Taimyr Peninsula on the central Siberian mainland, which separates Kara and Kaptev seas.

Significantly, oil shows have been found on the most northerly Kara Sea structure designated on a new Soviet arctic exploration map prepared by the U.S.S.R.'s PA Arktikmorneftegazrazvedka (Arctic Offshore Oil and Gas Exploration Association). It is the Byeloostrovskaya structurer on Belyi Island's northwest coast and extending offshore.

Most discoveries in the Kara Sea's southwest arm are expected to be gas or gas/condensate fields with mainly Cretaceous pay. Their geological profiles are likely to be similar to those found in Yamal Peninsula fields along the Nurminsky megaarch. But deeper Kara Sea drilling may tap some Jurassic oil as in Nurminsky megaarch fields. No commercial hydrocarbons are likely to be found on the west side of the Kara Sea's southwest arm and probably not even in the middle sector beyond the Yamal shelf.

The huge Novaya Zemlya trough with water more than 500 m (1,640 ft) deep in some places is on the west side of the Kara Sea's southwest arm. This trough extends virtually the entire length of 500 mile long Novaya Zemlya Island, part of the boundary between the Barents Sea and Kara Sea basins.

Novaya Zemlya, where the U.S.S.R. has been conducting nuclear tests, has no hydrocarbon potential, Soviet authorities say.

The Yamal Peninsula's western shelf is only 250 miles long and 100 miles wide at best. It is in this comparatively small area of 25,000 sq miles or less—only about 7% of the Kara Sea's total expanse—that all of the sea's designated fields and structures are shown on the association's offshore arctic exploration map.

The onshore sector of the Yamal Peninsula's Nurminsky megaarch, which trends southeast to northwest, is about 130 miles long. It attains maximum width where it dips beneath the Kara Sea in the area of Kharasavei and Kruzenshtern fields and extends underwater in the direction of the big Leningradskaya structure and Rusanovskoye.

Besides giant—possibly supergiant—Kharasavei and Kruzenshtern gas/condensate fields with their offshore Kara Sea sectors, the Nurminsky megaarch includes Bovanenkovskoye gas/condensate field. Lying about 30 miles from the Yamal Peninsula's Kara Sea coast, Bovanenkovskoye has been credited with as much as 141.2 tcf easily making it a supergiant by Soviet standards (at least 35.3 tcf).

Giant Neitinskoye and Arkticheskoye fields together with more recently discovered Severo-Bovanenkovskoye gas fields are also on the Nurminsky megaarch.

Where Soviets will seek competitive bids

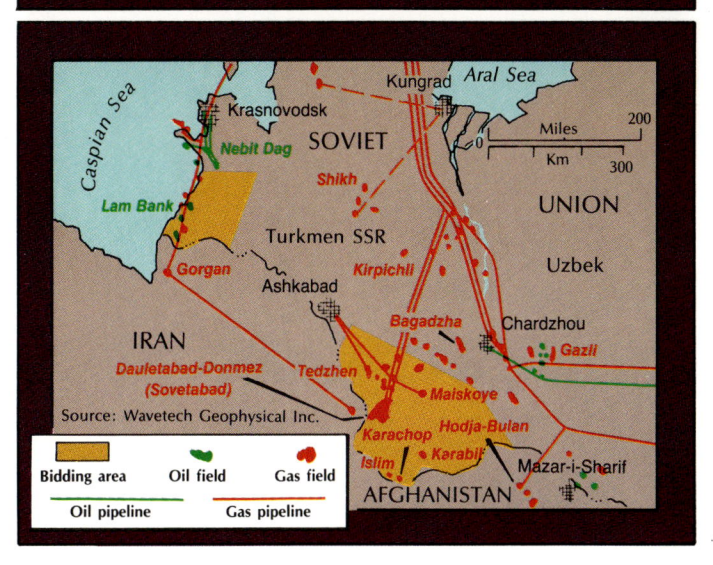

Source: Wavetech Geophysical Inc.

Both Bovanenkovskoye and Kharasavei have at least 16 pay zones from Jurassic to Upper Cretaceous (Cenomanian), Neitinskoye at least 11, Kruzenshtern at least 9, and Arkticheskoye at least 8.

The prolific Cenomanian gas zones in all Nurminsky megaarch fields are quite shallow and are generally topped between 1,700 and 2,650 ft. Small amounts of oil have been found in Kharasavei, Bovanenkovskoye, Arkticheskoye, and Neitinskoye fields.

Despite Rusanovskoye field's location in the Kara Sea's harsh environment, its location is favorable to feed gas into proposed Yamal Peninsula pipelines that likely will run south from Nurminsky megaarch fields to join with the enormous transmission system connecting Urengoi field with European Russia.

Construction of a railway form a point near the base of the Yamal Peninsula to Bovanenkovskoye field began several years ago. But work was suspended or drastically slowed following protests from environmentalists and the small local population of reindeer herdsmen and fishermen. In fact, the main reason for delaying the railroad's construction may have been the extremely high cost of building an embankment for the tracks across Yamal's swampy—sometimes submerged—terrain.

Tengiz holds promise, problems for Soviets

The U.S.S.R.'s Tengiz oil field near the northeast coast of the Caspian Sea in western Kazakhstan has posed the Soviet Union's most difficult problems in evaluating and developing a supergiant reservoir.

Tengiz probably won't be developed enough to permit an accurate determination of reserves until 1995, Soviet geologists say. However, it's possible the timetable could be advanced if a joint venture between Moscow and Chevron Corp. goes into operation and includes Tengiz. The proposed project, called the SovChevroil joint venture, would involve exploration and production in an area of about 8,900 sq miles, which includes Tengiz.

Preparing the pre-Caspian basin field for relatively small first stage production of 60,000 b/d has been enormously costly and technologically challenging. Construction of a plant to process Tengiz crude with its high hydrogen sulfide content in dissolved gas is running behind schedule, and wells remain shut in until the plant is complete. Inefficiency and shortages of equipment and materials are widespread, Soviet sources report.

Soviet industry is unable to manufacture high quality pipe

U.S.S.R.

Baltic region

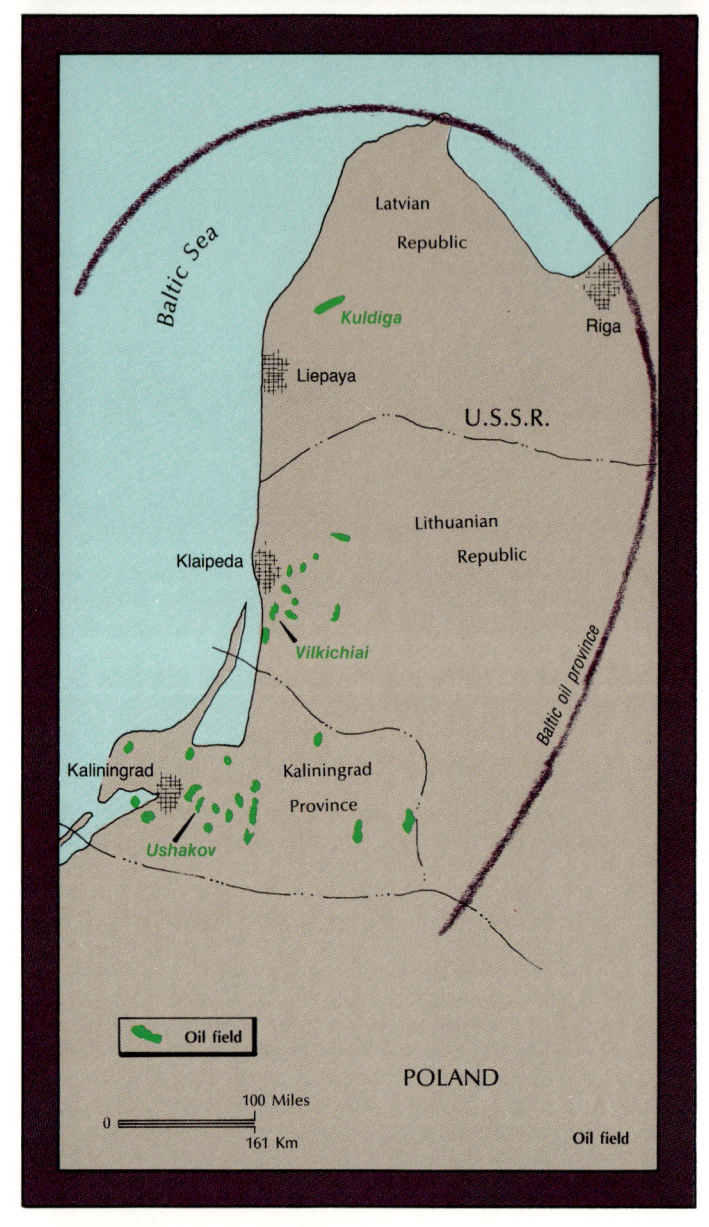

Northeast Caspian fields

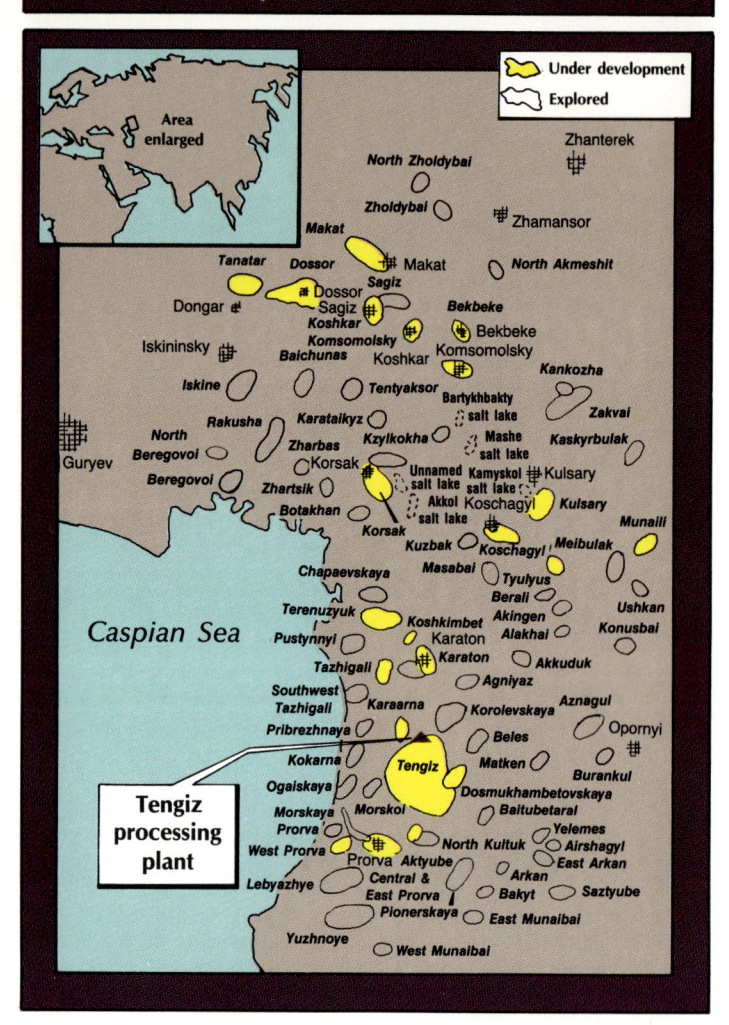

and equipment required to handly the extemely sour, corrosive Tengiz crude, forcing costly purchases from western firms at a time when Moscow is scraping the bottom of its hard currency reserves.

In 1987, Occidental Petroleum Corp. and Italy's Montedison SpA signed a memorandum of intent to form a joint venture to develop and operate a massive petrochemical complex in Tengiz field. Japan's Marubeni Corp. agreed to participate in the enterprise.

Compared with Tengiz, placing the Volga-Ural area's largest field—Romashkino—and western Siberia's supergiants—Samotlor and Urengoi—on large scale production was fast and easy. What's more, Romashkino, Samotlor, and Urengoi were developed with relatively little foreign assistance. Moscow has been forced to turn to Communist and western nations for extensive aid in bringing Tengiz on stream.

The U.S.S.R. discovered Romashkino in 1948 and began commercial production in 1952. By the early 1960s, it was the U.S.S.R.'s most prolific oil field.

Western Siberia's Samotlor is a 1965 discovery in a remote, roadless, swampy area plagued by extreme winter cold and summer flooding. Yet Samotlor by 1974 was producing more than 1.2 million b/d and became the U.S.S.R.'s leading oil field.

Urengoi, one of the world's largest gas fields, was discovered in 1966 astride the Arctic Circle, far from population centers and surface transportation. Developed in 1978, Urengoi produced more than 2 tcf in 1980 and the following year became the U.S.S.R.'s largest gas field by a wide margin.

In sharp contrast, Tengiz, although lying in a desert area, is close to a crude oil pipeline, a railroad, and towns. It is only 100 miles southeast of the Guryev refinery. Last year, the Moscow newspaper Rabochaya Tribuna (Worker's Tribune) reported that work delays on Tengiz facilities had cost the U.S.S.R. 3 million tons (21.9 million bbl) of crude oil, along with large volumes of natural gas, sulfur, ethane, and propane. It placed the value of the loss at 148 million rubles.

To speed construction of the Tengiz oil and gas processing plant, the U.S.S.R. brought in 80 brigades of Hungarian workers. They are paid three to four times as much as their Soviet counterparts, and Hungary will receive oil products in partial exchange for projects they complete. Kazakhstan officials have angrily objected to the higher pay granted to the Hungarians. But Rabochaya Tribuna pointed out that the Hungarians are highly skilled, whereas Soviet workers flown to Tengiz area construction projects for short tours of duty are likely to be former farmhands, miners, "and even artists."

The Soviets admit many deficiencies and "unresolved problems" in drilling deep pre-Caspian basin holes such as those in Tengiz field. They say the shortcomings have slowed drilling and reduced well quality.

The Soviet magazine Neftianik (Oil Worker) criticized chronic shortcomings of corrosion resistant blowout prevent-

U.S.S.R.

ETHNIC TURMOIL in Azerbaijan sharply reduced drilling by the U.S.S.R.'s Caspian Sea mobile rig fleet, which includes eight jack ups and this rig, one of four semisubmersibles.

ers and casing, suitable casing string equipment, ball cocks, and other items. It said development and introduction of modern drilling technology for conditions in the pre-Caspian basin are lagging badly.

Even so, it said, "The increase in oil reserves in Tengiz and adjacent fields was achieved despite failure to meet goals for exploratory drilling."

Last year, drilling of 18 Tengiz development wells was in progress. One well was projected to 19,685 ft, another to 21,352 ft. Deepest Tengiz hole went to 17,759 ft, but even it didn't reach the bottom of the carboniferous reservoir. No. 54 well in the salt dome sector of Tengiz at last report was in the planning stage and was projected to 22,966 ft. The Soviets believe holes that deep will establish a still greater pay thickness, originally estimated at almost 3,000 ft and later increased to more than 1,600 m (5,249 ft).

Despite the problems, the U.S.S.R. is stepping up drilling activity in the western Kazakhstan area, especially in the Tengiz district. A rig manufactured by the Uralmash plant in Sverdlovsk many be used for dilling in Tengiz and other pre-Caspian depression fields. The BU-8000 rig is designed to drill to 8,000 m (26,246 ft).

Fifty-six drilling brigades and 88 rigs were available to the Pre-Caspian Oil Drilling Association by the beginning of 1990. Since 1986, the number of pre-Caspian basin drilling brigades has increased by 90% and in western Kazakhstan, including Tengiz field, by 120%.

During 1986-89, the association started exploratory drilling in 19 areas, and the plan for increasing oil reserves was exceeded. New subsalt oil pools were found in Korolevskoye field just north of Tengiz and in other areas of western Kazakhstan.

The Soviets hope to complete geological exploration of the Korolevskoye structure in 1992.

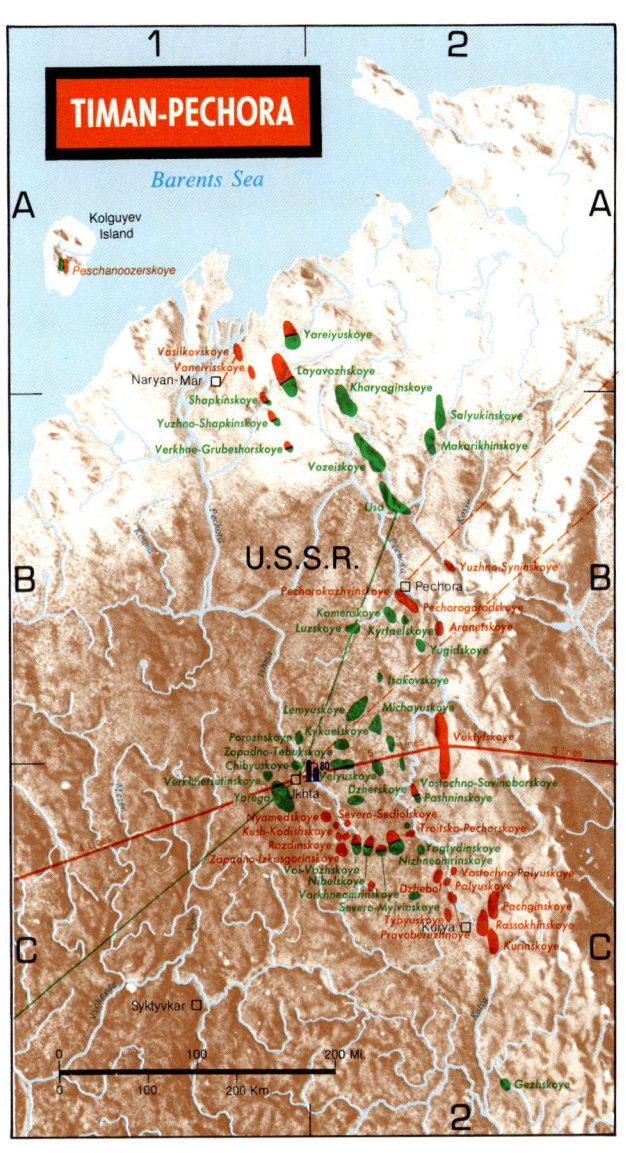

Moscow discloses Lithuanian oil production

The U.S.S.R. for the first time disclosed that the breakaway Lithuanian republic has commercial oil production. It has 20 oil wells capable of yielding 3,358 b/d, Rabochaya Tribuna reported. Lithuania's normal oil consumption is about 121,000 b/d, it said. The Soviets estimate Lithuania's "explored" oil reserves at 36.5 million bbl. Western nations have no intention of delivering crude, diesel, and gasoline to Lithuania, the newspaper declared. It added that Lithuania couldn't afford $13.79/bbl for foreign crude or $20.55-27.40/bbl for gasoline.

Meantime, Moscow newspaper Sovetskaya Rossiya (Soviet Russia) said Lithuania received 260,000 b/d of oil in 1989 and 262,000 b/d last year.

The newspaper placed Lithuania's 1988 natural gas imports from Russia at 180 bcf and 247 bcf last year. Lithuania in 1989 brought Russian crude for about 30 rubles/ton vs. the world market price of 64.5 rubles/ton and diesel for 68 rubles/ton vs. the world price of 78 rubles/ton, Sovetskaya Rossiya said. It reported that Lithuania, which produces no natural gas, bought that fuel from Russia at about half the world price.

The republic's territory accounts for the largest area of what Moscow calls the Baltic oil province. At least 10 oil fields have been discovered in Lithuania, most of them during the late 1960s and the 1970s. But indigenous oil supply is limited, and the western press reported Lithuania's sole refinery began shutting down in April last year for lack

U.S.S.R.

Soviet Union's oil, gas outlook

Oil production*
(Million b/d, 1975–2000)
- Urals-Volga
- Western Siberia
- Other

Gas production
(Billion cu m/day, 1975–2000)
- Central Asia
- Western Siberia and Far East
- Ukraine
- Other

Oil and gas capital spending
(Billion rubles, 1975–2000)

Drilling activity
(Million m; 1985-90, 1991-95, 1996-2000, 2001-05)
- Exploration
- Development

*Crude oil and condensate
Source: Scottish Development Agency

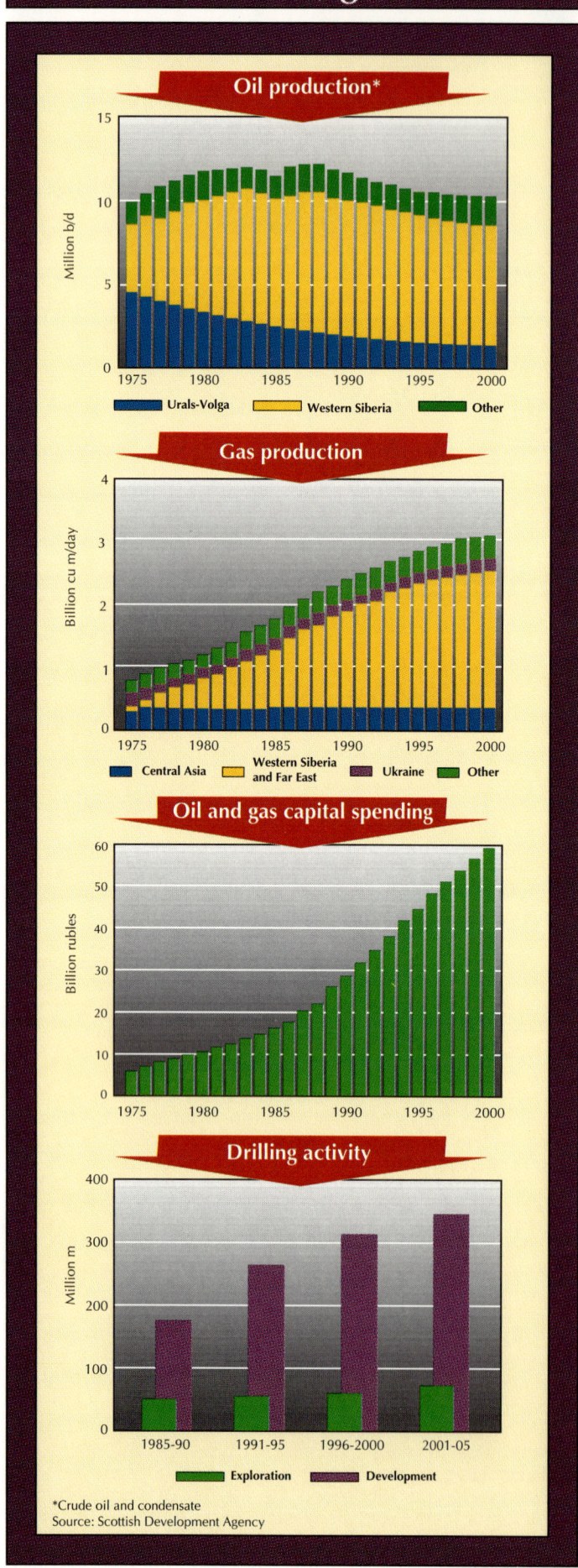

of feedstock 5 days after the U.S.S.R. cut shipments of key goods to the republic. Plant site is at Mazeikiai near the border with Latvia.

The Baltic oil province extends along the Baltic Sea coast from the Polish border to northern Latvia. It covers 38,600 sq miles. Total number of onshore oil province fields found by the early 1980s was at least 26. Largest oil flow reported for a Lithuanian discovery well was 904 b/d in Vilkichiai field near the Baltic Sea coast southeast of the port of Klaipeda. Found in 1969, Vilkichiai field was being prepared for development in the early 1980s.

No Lithuanian field was officially listed as having commercial production by the early 1980s. However, two fields were being "prepared for development." Just southwest of Lithuania, Kaliningrad Province, which is part of the Russian Soviet Republic, had 14 oil fields by the early 1980s. Latvia, immediately north of Lithuania, had one.

Kaliningrad Province had nine oil fields on production by 1980. Combined flow from 150 wells during the early 1980s was about 30,000 b/d. Almost all Baltic oil province discoveries are productive in the middle Cambrian. Very small volumes of oil were found in the Ordovician and Silurian.

Largest flows from oil discoveries have been in Kaliningrad Province. There, in Ushakov field near the city of Kaliningrad, initial test flows of 2,736 b/d and 2,194 b/d were reported from the north and south domes respectively in the middle Cambrian. Discovered in 1969, Ushakov went on production in 1972.

Pay depth in Baltic oil province onshore fields ranges from more than 8,000 ft in the middle Cambrian in Kaliningrad Province to a little more than 2,700 ft in the upper Ordovician in Latvia's Kuldiga field northeast of the port of Liepaya.

In the Baltic Sea, off Kaliningrad Province's Cape Taran, the U.S.S.R. during 1984 found commercial oil in 98 ft of water at about 3,000 m (9,842 ft). Several more wells were slated to be drilled from a fixed platform, but Moscow reports said plans for development were abandoned because of protests by environmentalists.

Caspian Sea drilling declines

Caspian Sea drilling early last year fell to its lowest point since the 1950s. Renewed ethnic unrest in Azerbaijan caused much of the sharp drop. Problems were compounded by failure of the big fabricating yard near Baku to provide jackets for platforms as scheduled.

The Soviets had hoped to increase Caspian Sea drilling by 40% in 1986-90 compared with 1981-85, the previous 5 year plan. Official figures show that the U.S.S.R.'s Caspian Sea drilling, which during recent years accounted for more than 90% of total Soviet offshore footage, was only 46.8% of plan in January last year. Only two wells were completed that month vs. the scheduled four in 28th of April field, by far the most productive in the Caspian.

During 1989, the U.S.S.R. completed 95 Caspian Sea wells instead of the planned 86. Fourteen of those were exploratory holes. Even drilling in the far southern Caspian was seriously affected by the turmoil in Azerbaijan that exploded in late January. The Soviet jack up Kaspii-2, which is drilling the first well on Iran's coastal shelf, was unable to operate normally following the Azerbaijan riots, the Baku newspaper Vyshka reported. In late December 1989, the hole off Iran reached 2,500 m (8,202 ft).

Especially poor performance was reported last year for the Caspian's fleet of about 12 mobile rigs—four semisubmersibles and eight jack ups. They drilled only 37% of scheduled footage during January. Vyshka explained that 48% of the "specialists" working on Caspian mobile rigs were Russian. As Azerbaijan's violence against "foreigners"—especially Russians and Armenians—grew, Russian personnel aboard mobile rigs quit work and returned to Baku, fearing for the safety of their families.

U.S.S.R.

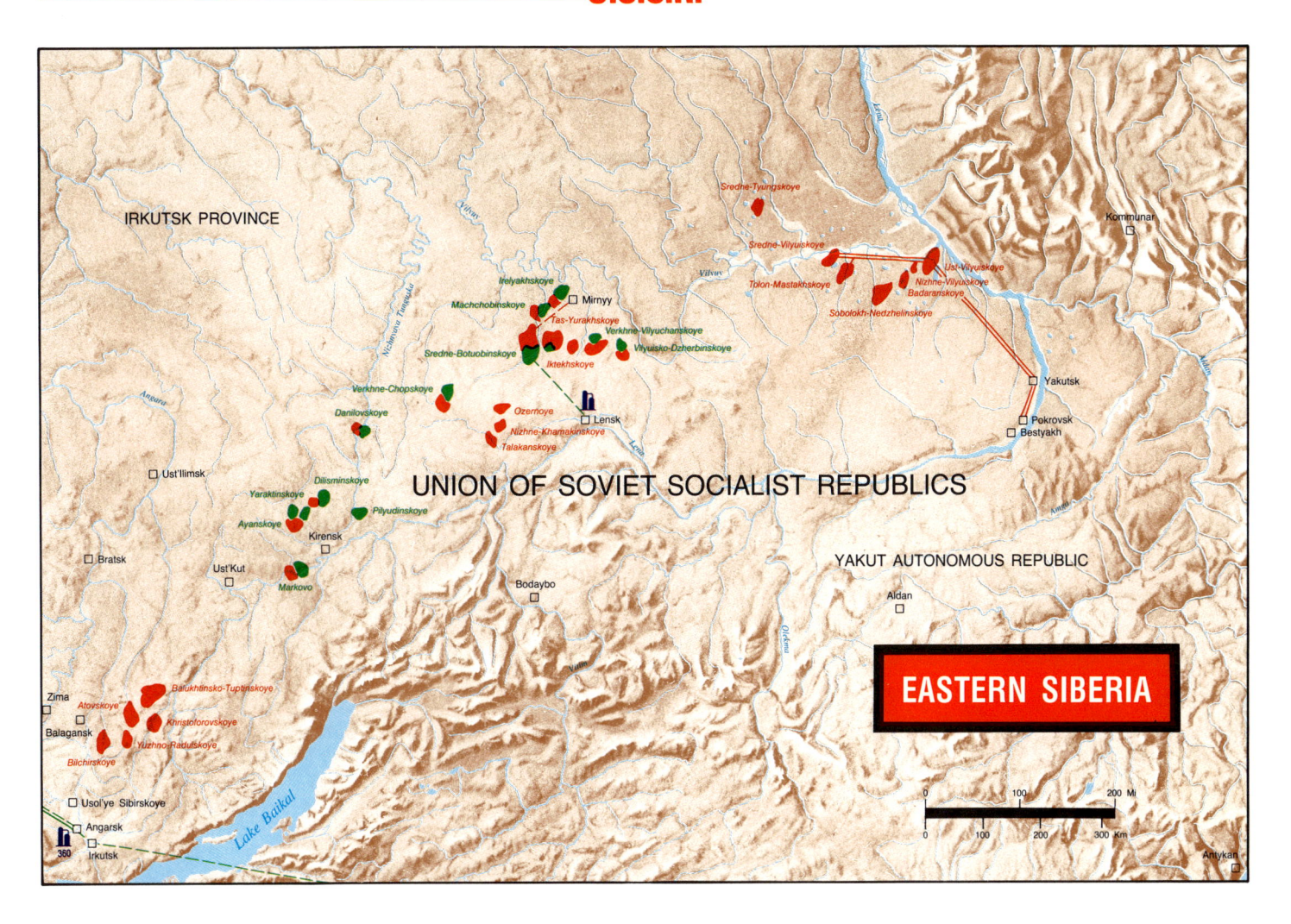

Caspian mobile rig workers remained at their jobs during 1989's periods of Azerbaijan unrest.

Vyshka reported that inexperienced replacement crews on supply vessels that normally deliver pipe, materials, and equipment to platforms in Caspian Sea fields refused to leave port during bad weather during winter 1989. As a result, even drinking water on the platforms had to be rationed. Morale of drilling personnel, already sagging because of onshore unrest, fell further.

Besides platforms built alongside the Caspian's 250 miles of shallow water trestles, there are about 30 free standing platforms. Most of them are in water less than 150 ft deep. The Soviets planned to install about 40 platforms in greater water depths—as much as 800 ft—in 1989-2005. This program is off to a poor start. Close to 1,000 wells were slated to be drilled from these platforms. Most of the structures were planned for installation in deepwater sectors of 28th of April field and other promising, undeveloped fields farther east along the Apsheron Sill.

Because of failure to build the expected platforms in 28th of April field, only seven of the nine available drilling crews were working early last year. Soon the number of active crews in 28th of April slumped to five, and footage fell correspondingly.

Meanwhile, the Soviets report some success in Caspian drilling off Turkmenia near the eastern end of the Apsheron Sill. In January 1990, the 28th of April jack up made a significant strike at Gubkin Bank off Turkmenia's Cheleken Peninsula. Each of the four zones tested flowed 1,825 b/d of "oil" (possibly condensate) and 11.65 MMcfd of gas. There are nine potential pay zones. Several appraisal wells are planned at Gubkin Bank.

Oil and gas have been found in at leat six offshore Turkmenian sites since the late 1960s. Total crude and condensate production from the Caspian's Turkmenian shelf remains small—apparently less than 10,000 b/d.

Natural gas line starts up

Despite sharp reductions in outlays for pipeline construction, the Soviet Union has started shipping gas through a major new line from Surgut in the Middle Ob district of western Siberia's Tyumen Province to Novosibirsk in the southern part of the region. Plans call for extending the line farther southeast to the Kuznetsk industrial area.

First 48 in. section of the pipeline, from Surgut to Omsk via the city of Tyumen, was completed in late 1988. The Omsk-Novosibirsk segment, finished far behind schedule, is believed to be of 40 in. diameter.

A severe shortage of funds forced the U.S.S.R. to curtail construction of big pipelines during 1989-90. Middle Ob gas has been delivered to Novosibirsk and the Kuznetsk region since the early 1980s via a 40 in. line from Nizhnevartovsk through Parabel and Tomsk. But even the new line won't meet the demand for gas in western Siberia's southern sector.

In other transportation action:

- The Urengoi-Petrovsk pipeline caught fire about midyear 1990 near Kuibyshev after gas leaking from a 1.5 m pipe ignited. Fire fighters closed off the line and brought the fire under control about a day after it was reported by nearby villagers who saw the smoke. Tass said officials blamed the leak on heavy rains that made surrounding ground sink.
- The U.S.S.R. increased its capacity to export liquefied petroleum gas. A new tanker built in Germany to transport LPG and chemicals last year made its first voyage from the

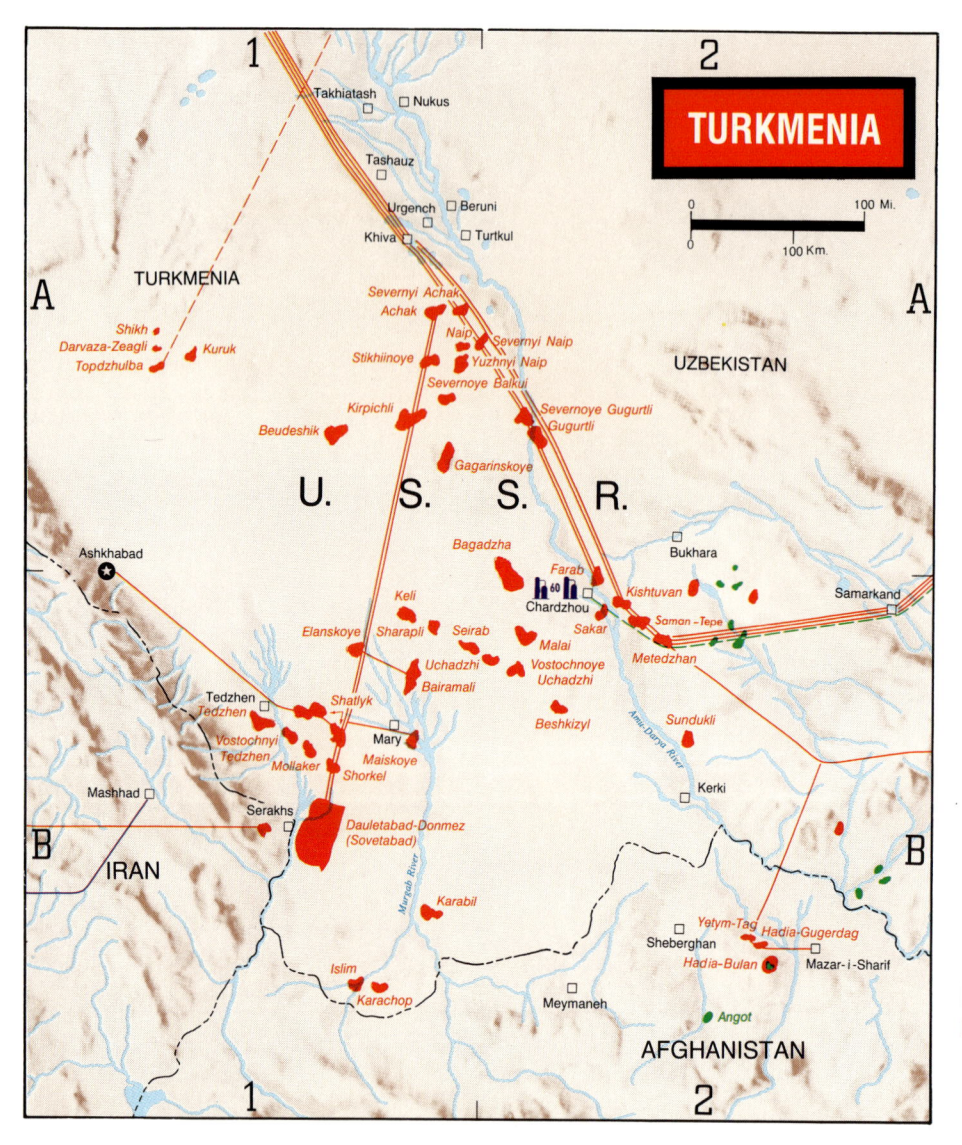

U.S.S.R.

local concerns about handling high H_2S gas from the field—H_2S concentrations as high as 25%.

Each processing train will handle full well streams from wells in the field. Each train includes separation and stabilization, gas purification, sulfur production, and gas liquids recovery. Stabilized oil and dry gas will be shipped from the plant by pipeline, and sulfur and LPG will move by rail.

Composition of the approximately 50 multiphase well streams that will converge at the inlet header will vary, making estimates of feed volume difficult. But expected annual output from Tengiz 1 plant's two trains is 3 milion tons of stabilized 0.805 density oil (about 70,300 b/sd), 77 million cu m (8.1 MMcfd) of dry gas, and about 500,000 tons of sulfur. Natural gas liquids also will be produced.

The first train was built by the consortium "LLL," made up of Lafarge Coppee Lavalin S.A., Litwin S.A., and Lurgi S.A. The consortium is also supervising construction of the second train, scheduled to begin operation in the fall of 1992.

The first additional capacity from plants now in the early design stage is planned for commissioning in 1993, according to a Soviet spokesman. They likely will be of the same configuration as Tengiz 1.

Moscow plans to bolster problem plagued oil industry

The U.S.S.R.'s problem plagued petroleum industry is likely to be given unprecedented leeway to rectify policies that have pushed it and the entire Soviet economy close to collapse. But the window of opportunity to improve horrendous conditions in oil and gas production, pipelining, refining, and petrochemical manufacture may be open for a relatively short time—less than 2 years.

If definite improvement isn't achieved in that period, the U.S.S.R. could reembrace much of the rigid, supercentralized, command system that has prevailed for so long. Under radical market based reforms endorsed before the Supreme Soviet by President Mikhail Gorbachev, the U.S.S.R. would attempt to stem the steepest oil production slump in its history and reinvigorate a natural gas industry that has recently experienced the slowest production growth since World War II. Aided by foreign loans, expansion of joint ventures, and perhaps increased purchases of equipment and advanced technology from the West, the U.S.S.R. can hope to at least patch its shoddily constructed pipelines and obsolete refineries and petrochemical plants that frequently break down or explode.

Opponents of perestroika (economic restructuring) have made it abundantly clear they will cooperate only reluctantly with Gorbachev's proposal to give more economic authority to the U.S.S.R.'s 15 constituent republics, regional and local entities, and industrial managers outside of Moscow. Even before Gorbachev outlined his reforms, old line Communist officials and bureaucrats, a vocal segment of disgruntled petroleum industry personnel, and a general population still inclined toward xenophobia were critical of the modest changes that have taken place.

Gorbachev implicitly pledged to reverse declining oil

Baltic port of Riga to western Europe and Mediterranean countries. France has been a leading buyer of Soviet LPG.

• The Soviet prosecutor's office arraigned seven pipeline officials involved in managing the West Siberia-Urals-Volga NGL pipeline, which leaked and sparked an explosion on the Chelyabinsk-Ufa railway June 3, 1989, killing 575 people and injuring 623 on two passing trains. Soviet prosecutors alleged gross violations of construction and operations rules that contributed to the accident. The pipeline was damaged by an excavator in October 1985, the damage ignored, and the damaged pipe buried and later leaked, Izvestia reported.

• The Soviet Union's Volgoneft-263 tanker spilled an estimated 850 metric tons of oil in the Baltic Sea after colliding with the Betty, a German dry cargo carrier, in heavy fog May 14, 1990, off Karlskrona, Sweden. Tass said seven Swedish vessels, including three equipped for spill cleanup, arrived at the collision site soon after the mishap. The German vessel headed for Karlskrona, while Volgoneft-135, another Soviet tanker, was dispatched to the site to take on Volgoneft-263's remaining cargo.

New processing plant nears start-up

The first of two trains of an oil and gas processing plant designed to handle high pressure corrosive well streams from giant Tengiz field was due to start up near the opening of 1991. Train two in this first plant to be built in the field is under way, and additional multitrain plants are planned. Construction of the plant was delayed 2 years because of

U.S.S.R.

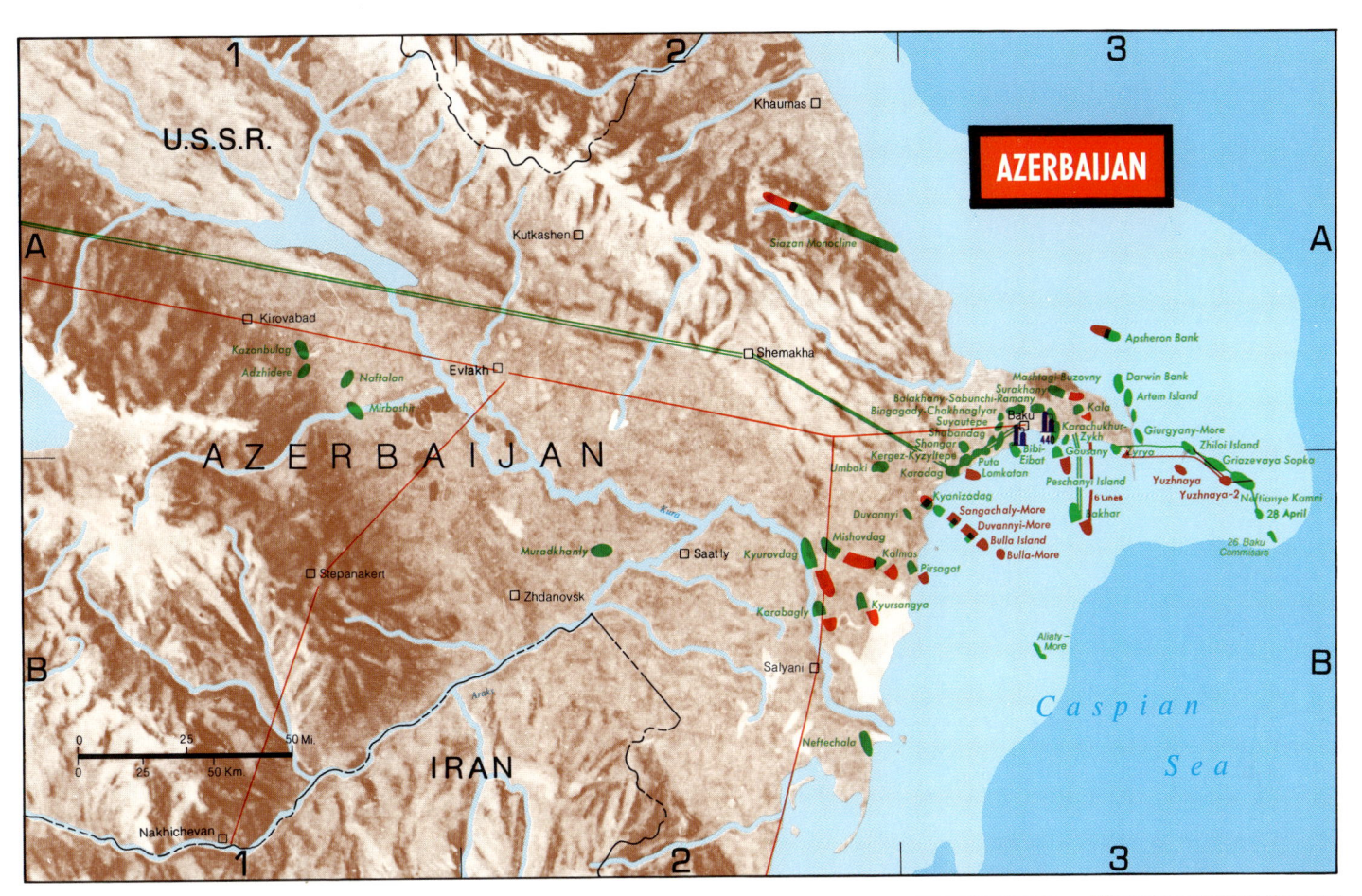

EQUIPMENT AIRFREIGHTED to the Soviet Union by Gripper Inc., Houston, for repair of a broken Caspian Sea crude oil pipeline is uncrated in a warehouse at the port of Baku, Azerbaijan.

exports to European countries able to pay for such deliveries with hard cash. Speaking at a Moscow press conference, he said the Soviet Union can't pull itself out of its economic depression without earning more of the currency that increased exports, especially of oil, would provide. "Our domestic needs and our domestic interests will push toward raising exports," Gorbachev said. "We have the capability to broaden and increase exports, and we will do so." His statement was in response to a question of whether deliveries of oil to eastern and western European countries will have to be reduced further because of the U.S.S.R.'s domestic economic problems.

While the Soviet president was specifically asked to comment on oil exports to European nations, his reply was directed more toward raising total sales abroad. Because the Soviet drive to gain markets for other Russian goods, especially machinery, has faltered badly, higher energy deliveries seem to be the only way for the U.S.S.R. to hike exports substantially in the near term at least.

Soviet oil exports tumbled from a peak of more than 4.1 million b/d to less than 3.7 million b/d in 1989. Foreign sales last year were expected to be the lowest since the 3.33 million b/d reported for 1985.

Foreign agency outlines U.S.S.R.'s equipment needs

The Soviet Union will need a wide range of imported equipment to meet crude oil and natural gas production goals to 2000. The Scottish Development Agency (SDA), Aberdeen, says Moscow's shopping list will include equipment for modernization, inspection, and repair of pipelines, meters and measuring equipment, services for the offshore industry, and equipment and services to maintain production in western Siberia, its No. 1 producing area.

U.S.S.R.

OIL PRODUCTION in Yakutsia, Eastern Siberia.

The Soviets will need equipment to improve environmental and safety performance, as well as for production in high pressure, high temperature, and corrosive environments and equipment to cope with severe weather that's encountered in frontier areas.

SDA said in hard currency terms, many of the skills and products that have been developed by western companies for use in the North Sea and frontier regions are relatively low cost and will provide the Soviet Union with a fast payback.

Romania is the Soviet Union's biggest oil field equipment supplier, delivering goods valued at 426 million rubles in 1988. In 1987 it provided 50% of all drilling equipment imported by the U.S.S.R.

The Soviets aim to prevent liquids production from falling to less than 10 million b/d in the second half of the 1990s and to boost gas production to about 105 bcfd by the turn of the century. "Perestroika and developments in eastern Europe place even greater emphasis on production," the Aberdeen agency said. It estimated achievement of planned production will require that 10% of all Soviet Union spending go to the crude oil and natural gas industry to 2000. As a result, drilling activity will increase significantly during the next three 5 year periods.

The Soviets assume their current fleet of drilling rigs is operating at capacity. So increased drilling rates can be achieved only by technological innovation. Areas in which improved technology and equipment are required include drill bits, automation, mud engineering, and mud pumps.

Biggest Soviet joint venture on track

Combustion Engineering Inc. and Neste Oy have formed the biggest joint venture in the U.S.S.R. to develop the first phase of a multibillion dollar petrochemical complex in Siberia. The project will entail construction of a $2 billion petrochemical plant in Tobolsk, western Siberia, to be complete in 1993. CE and Neste signed a joint venture agreement with Tobolsk Petrochemical Co., which is under the aegis of the Soviet Ministry of Chemical & Oil Refining Industries. McDermott International Inc., which together with CE signed letters of intent in spring 1988 with the ministry to participate in the Siberian petrochemical development, has pulled out of the group.

The Soviets earlier had envisioned a series of gas processing and petrochemical plants in western Siberia's Tyumen province that ultimately would cost a total of about $10 billion. At least five petrochemical plants were planned, including projects at Nizhnevartovsk and Novyi Urengoi. However, Moscow's campaign to reduce its huge budget deficit resulted in a major cutback in plans for Siberian petrochemical expansion. The Tobolsk plant is seen as the flagship of that refocused effort.

Rather than a grassroots project, as Nizhnevartovsk or Novyi Urengoi would be, the Soviets apparently are first emphasizing expansion and modernization of two existing major petrochemicals centers, Tobolsk and Tomsk, to western standards. Efforts are proceeding apace at Tomsk but have fallen behind at Tobolsk. The venture with CE-Neste emphasizes the reordered priority.

Under the agreement, the venture will build, own, and operate the Tobolsk complex. The plant will produce propylene, polypropylene, and thermoplastic elastomers for the domestic market and for export. The agreement completes almost 2 years of economic and technical feasibility studies and preliminary development work.

CE will be responsible for overall project management, including design, procurement, supply or process control instrumentation and automation systems, and foreign currency financing. Neste will provide global export marketing for the venture's products.

CE, Stamford, Conn., not long ago agreed to a merger with ABB Asea Brown Boveri Ltd., one of the world's biggest engineering, contractor/suppliers. Neste is Finland's state

U.S.S.R.

owned oil company.

In other activity involving opportunities for joint ventures:

- Moscow has offered 11 areas in five regions for E&P joint venture consideration. The areas encompass 21 fields and three exploratory areas. Further, the Soviets are offering vast stretches of the country to review by foreign oil companies for the companies to propose their own joint venture programs. The 12 areas on offer for joint venture review stretch from the western Siberian basin to the Black Sea and cover a total area about the size of the U.S.

The offerings were detailed by MDSeis, a joint venture of Professional Geophysics Inc. (PGI), Houston, and the Soviet Central Geophysical Expedition of the Ministry of Oil & Gas Industry. They include new opportunities for E&P in the western Siberian basin, Volga-Urals, North Caucasus, and Ukraine areas.

PGI estimated the identified oil reserves covered by the offering at 2 billion bbl. That brings total reserves covered by offerings through MDSeis to 10 billion bbl, PGI said.

- A joint venture of Fairfield Industries Inc., Houston, and the Yakut-Sakha Soviet Socialist Republic offered a data package covering a 4,100 sq km portion of the Chayadin-Botouba region in southwestern Yakut S.S.R. Fairfield said the offering will be followed the next 2 years by 16 more offerings of much larger areas, designated by the republic as available to joint venture development with non-Soviet partners. Fairfield said its joint venture with Yakut S.S.R. is the first between a western company and a sovereign Soviet republic.

- BP Exploration and Norway's state owned Den norske stats oljeselskap AS signed a protocol with the Soviet Union that could lead to an exploration and development joint venture in the Caspian Sea. They signed the protocol with Caspmorneftgas, the Caspian Sea operating arm of the Soviet Ministry of Oil and Gas.

BP says the protocol commits the two companies to study feasibility of deepwater drilling and possible development of Azeri field, formerly 26 Baku Commisars field. If approved and market conditions are right, a joint venture could undertake drilling in the underexplored areas of the Caspian region beyond Caspmorneftgas' onshore and shallow water fields that now produce about 200,000 b/d. Azeri, about 93 miles southeast of Baku, was discovered in 980 ft of water in 1987.

- Conoco Inc. signed a protocol covering a possible venture to develop oil and gas reserves in the Timan-Pechora basin of Soviet northern Europe. It is the third Soviet joint venture being considered by Conoco or parent DuPont. Conoco also is studying the feasibility of a joint venture with the Soviet Union's Tyumenneftegas, Tyumengeologia, and Noyabrskneftegas to develop hydrocarbons in the Tyumen, Purpa, and Noyabrsk regions in the western Siberian basin, east of the Ural Mountains.

Combined, the two Conoco study areas in the western Siberian and Timan-Pechora basins have potential recoverable reserves of 3-4 billion bbl, a company executive said.

- Phibro Energy Inc., Greenwich, Conn. reached an agreement with Anglo-Suisse Inc., Houston, to join the White Nights joint venture in the western Siberian basin. White Nights is a 50-50 venture with Varyegannefteqaz U.S.S.R. It plans to drill about 550 horizontal oil and gas wells and attempt to recomplete more than 100 wells currently shut in.

Phibro, Anglo-Suisse, and Varyegannefteqaz recently completed a joint engineering feasibility study, the first step in the Soviet Union's joint venture registration process. The firms were to complete registration as 1991 neared, begin equipment mobilization in first quarter 1991, and start field operations in the next quarter. White Nights expected to begin producing 800 million bbl of proved reserves in mid-1991. Under terms of the joint venture agreement, Phibro and Anglo-Suisse will export their production shares.

- The U.S.S.R. started its first gas marketing joint venture in Europe through a linkup with Germany's Wintershall AG, the oil and gas subsidiary of BASF Group. An agreement with Wintershall calls for Zarubeshgaz, a gas trading subsidiary of the Soviet Union's Gasprom group, to participate in joint marketing of gas from the Soviet Union and other sources in the former East Germany.

Wintershall's joint venture with Zarubeshgaz will become operational as soon as possible. It will plan, lay, and operate gas transmission lines and distribution networks. The venture could move into gas marketing in western Germany and other parts of western Europe once Wintershall's Midal and Stegal pipeline project starts up.

The joint venture will import about 8 billion cu m/year of Soviet gas at first. It is thought about 2 billion cu m/year could be used in BASF petrochemical operations.

- Global Natural Resources Inc., Houston, signed a joint venture agreement with the U.S.S.R.'s Tatneft, which operates the oil and gas fields of Tataria, west of the Ural Mountains on the Volga River. This venture, named Tatex, at first will install and operate crude oil vapor recovery units and compressors on most of Tatneft's production facilities, which currently handle about 800,000 b/d of oil. Tatex will exchange the recovered vapor for export crude oil.

The Tatex accord culminates a protocol signed in second half 1989. Tatex will be owned 50% by Tatneft and 50% by Texneft Inc., an 89% owned subsidiary of Global. The remaining interest in Texneft, which is an after payout interest, will be owned by Core Resources, a privately held company of San Antonio, which initiated and assisted in formation of the venture.

- Texaco Petroleum Development Co., a unit of Texaco Inc., signed an agreement on principles of cooperation with the Soviet Union's Ministry of Geology for a proposed upstream joint venture. The agreement covers exploration, development, and production in areas of the Timan-Pechora region. Texaco said the areas have high prospects for exploration and development of oil and gas fields with reserves estimated at "well in excess" of 5 billion bbl of oil.

The company is studying ministry data covering oil and gas resources of the areas and their infrastructure. Negotiations are to proceed toward details of a joint venture project for the region. Operations would be carried out mainly by Soviet personnel, supplemented by Texaco technical staff.

The agreement also calls for Texaco to organize seminars for Ministry of Geology personnel, dealing with technical and economic aspects of exploration/development.

- The presence of Soviet-foreign joint enterprise operating in the Soviet Union is beginning to be felt in the Soviet market, although in a very small way. About 1,300 joint ventures were registered with the Ministry of Finance through February 1990. Nearly 1,000 of them are said to have been formed during 1989.

According to the latest available State Statistical Committee data, 250 joint enterprises began to operate during the first 9 months of 1989. Of that, 184 produced goods valued at about 500 million rubles for Soviet consumers and 70 million rubles for export. Four hundred to 450 joint ventures would likely be operating as 1991 neared, a spokesman for the State Foreign Economic Commission of the U.S.S.R. Council of Ministers said last year.

Combined authorized capital of the joint enterprises has reached 3 billion rubles, of which 43% is foreign capital investment. Soviet joint enterprises with German companies number 151, Finland 121, the U.S. 110, Austria 76, Great Britain 75, and Italy 68. Joint venturers hail from 54 countries, including 23 western nations, 22 developing countries, and 9 Socialist nations.

The 1,800 Soviet partners in joint ventures included 564 state enterprises, 320 producer associations, 257 cooperatives, 120 scientific-research institutes and design bureaus, and 160 administrative agencies. [IPE]

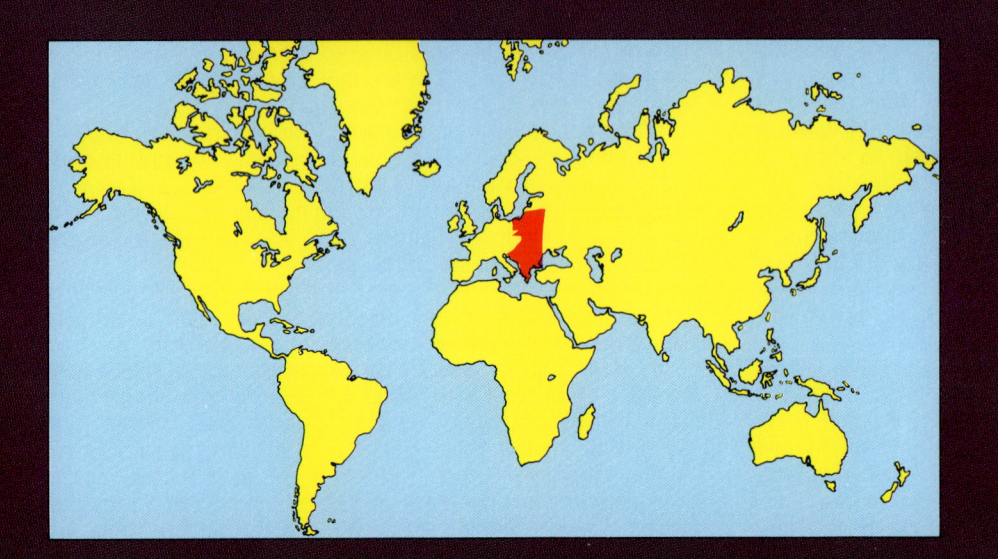

IPE ATLAS

EASTERN EUROPE

Political upheaval in the Soviet Union's East European satellites and the first steps toward a market economy herald changes for the oil and gas industries in the region. From Poland on the Baltic Sea to Romania on the Black Sea, state oil importing organizations face tougher commercial attitudes from their major supplier, the Soviet Union.

Oil for industrial products barter deals and internal crude price formulas are disappearing in favor of international crude oil prices paid in hard currency. In response, East European countries are turning to members of the Organization of Petroleum Exporting Countries and western producers for larger volumes of oil.

At the same time, international oil companies are eyeing the potential of eastern Europe, where the advent of market economies and the need to clean up the environment signal greater use of natural gas at the expense of locally mined coal. In addition, the growth of automobile ownership in the new market economies will require updating of refining and marketing operations.

East Germany, reunified in 1990 with West Germany, is setting the pace for progress among former Soviet satellites. Oil companies from western Germany are setting up joint ventures in gasoline retailing and hope to expand cooperation into refining. In the gas industry, the western German pipeline network is being linked into the eastern Germany system. Ruhrgas AG, the biggest western Germany gas company, and BEB Erdgas und Erdol GmbH, an established West German joint venture of the Royal Dutch/Shell Group and Exxon Corp., have acquired a 45% interest in the East German gas company VEB Verbundnetz Gas.

A study by Gaffney Cline Associates, Alton, Hampshire, England, of all aspects of the energy business in eastern Europe says that as various domestic oil industries develop, it is possible market prices will start to reflect the movements in prevailing spot market prices in western Europe. Refinery economics will be driven by demand for a particular oil product, which may result in increased supply of other refined products being sold at more competitive prices. On occasions this could make oil more economic than other forms of energy.

In the region's "highly inefficient" petrochemical industry, which generally uses outdated technology, GCA said the cost of feedstock and processing margins had played little part in determining the economics of plants. But they will become more important in determining whether plants remain in operation.

Gaffney Cline says ample supplies of gas from the Soviet Union and countries outside the former eastern bloc will ensure that gas provides stiff competition for coal and oil in all sectors with the possible exception of transportation.

Some East European countries manufacture town gas from coal—particularly lignite. That produces high levels of pollution as well as a product with a heating value 50% less than natural gas.

The prospects for meeting future energy demand from indigenous resources do not look bright, says Gaffney Cline. Coal reserves are relatively large, but supply prospects for oil and gas will be a totally different proposition. Oil demand has been constrained in favor of other fuels and production and import levels have shown only marginal increases. With an expanding economy the supply of lighter oil will be necessary, particularly in transportation.

Gaffney Cline says upgraded refining capacity is limited, and construction of more catalytic cracking units will be needed to enable eastern Europe to compete with refining in the West. Future increases in oil demand will soon stretch refining capacity in Bulgaria, Czechoslovakia, Hungary, and Poland. Crude processing capacity is likely to be further constrained if no further upgrading takes place.

GCA says the ability of the Soviets to export has reached its limit, and they are treading a fine line between demands of their East European dependents and the need for hard currency generated from oil exports to the West. The price of Soviet oil to its neighbors has in the past been calculated on a 5 year rolling average crude price. But the volatility of crude markets during the past 5 years exposed the drawbacks of this system.

Payment by bilateral bartering, the hallmark of eastern bloc trade, could well have outlived its usefulness. More conventional petrocurrencies look more attractive, said Gaff-

EASTERN EUROPE

EASTERN EUROPE

ney Cline.

An inefficient power generation system has made cuts in electricity part of the way of life in eastern Europe. Poor plant performance, inefficient transportation, miners strikes, and lack of foreign currency to buy oil have contributed to the situation.

The Chernobyl nuclear power plant disaster of 1986 dampened the enthusiasm of east Europeans for nuclear power. As a result, a number of projects have been shelved or canceled.

Assessments of eastern Europe's energy conditions, prospects

Gaffney Cline and Gustavson Associates, Boulder, Colo., offer these assessments of the energy situation, as well as geologic conditions and prospects, in the eastern European bloc:

Albania

Gustavson notes this country is rich in natural resources, especially chromite, oil, and gas. It has far greater petroleum reserves for its size than any other bloc country. Albania is the most politically isolated nation in Europe and one of the most xenophobic countries in the world. Long ruled by the repressive dictator Enver Hoxha, Albania is still a Stalinist Communist state. The nation consists of three main geologic provinces. Strongly folded late Paleozoic and Mesozoic strata comprise the subelagonian and Pelagonian massifs in Northeast Albania. Slightly folded Mesozoic and Cenozoic rocks form an extension of the Yugoslavian Dinarides (Albanides) in the center of Albania.

Coastal basins filled with Tertiary and Quaternary sediments are located in western Albania. These basins, including the Durres basin, produce hydrocarbons. The search for oil and gas in the future is likely to be concentrated in these coastal basins and in the relatively shallow waters of the Adriatic Sea. Hydrocarbons have been trapped in anticlines and tilted fault blocks, mainly in lenticular sandstones of the upper Miocene and in Helvetian (Tortonian) limestones.

Exploration for stratigraphic and other nonstructural traps is thought to be in its infancy and may represent the best potential for discoveries onshore. Albania's greatest oil and gas potential probably is in the Albanian shelf of the Durres basin, offshore.

Hydrocarbon exploration and production has continued since the early 20th century, with essentially all activity since 1948 under the control of the national government and its Soviet, later Chinese, allies.

Cumulative production figures aren't published officially, and western estimates of annual production rates vary widely, Gustavson pointed out. Albania has expressed little desire to open up for international exploration, especially by U.S. companies. European companies may have an advantage in this arena, due partly to established diplomatic and commercial relations with the Albanian government.

If Albania is opened to western investment in the petroleum sector, there is little doubt that modern seismic techniques and the deliberate search for subtle stratigraphic traps may be expected to lead to substantial reserve additions, Gustavson said.

Bulgaria

Gaffney Cline doesn't see large investments taking place in the energy sector. Because there is little prospect of discovering and developing sizable oil and gas reserves, most capital spending will go for continued development of the coal industry, where $3 billion could be spent, and power generation, which could claim $2-4 billion.

Projected spending on refining is estimated at $1 billion to upgrade existing units and build new capacity.

Since the Soviet decision to limit oil exports to its satellites in the mid-1980s, Bulgaria has turned to Algeria, Iran, Iraq, and Libya to meet the balance of requirements for its three refineries. They have a combined capacity of 290,000 b/d.

Bulgaria also is diversifying its sources of gas imports. A new pipeline to deliver Soviet gas to Greece and Yugoslavia runs through Bulgaria and will increase Soviet deliveries. In 1989, Bulgaria agreed to buy 95 MMcfd from Iran starting in 1990. However, the gas delivery comes from the Soviet Union under an agreement between Iran and the U.S.S.R.

GCA said Bulgaria has been running an energy conservation program during the 1980s, which has held its increase in demand to 1.8%/year during 1980-88. This has been largely a result of increased use of natural gas and nuclear power. Conservation has been enforced by gasoline rationing, electrical power cuts, and a 40-50% increase in the cost of electricity.

GCA says although the desired result was achieved, industrial production suffered, and Bulgaria came up short when the U.S.S.R. demanded larger volumes of higher quality goods in exchange for energy supplies.

Czechoslovakia

With relatively minor oil and gas reserves, it hasn't historically been a major producer of hydrocarbons, Gustavson said. The country relies heavily on imports of oil and gas from the Soviet Union.

It produces only about 3,000 b/d of oil and has reserves of about 19 million bbl. Most oil production comes from one field: Gbely in the Vienna basin. The country has seven state owned refineries with combined capacity of 455,000 b/d. Gas reserves are estimated to be about 330 bcf.

The Vienna basin, a Tertiary pull-apart basin in the Alpine-Carpathian overthrust belt, has been the site of most hydrocarbon production in Czechoslovakia. The largest oil field in the country is in this basin. Shallow oil has been produced in Gbely field since 1913 from Miocene sandstones in a locally faulted anticline. This hydrocarbon accumulation still accounts for about half of Czechoslovakia's oil production.

The opportunity for significant increases in oil and gas reserves is slight. In spite of the country's poor prospects for major new hydrocarbon discoveries, liberalization of joint venture laws is taking place. This move will make foreign investment in extraction industries more attractive.

Hungary

The country has maintained its oil production at about 40,000 b/d for the past 20 years and should be able to sustain 30,000-40,000 b/d for the rest of the decade, GCA said. Hungary's three refineries have a combined capacity of 220,000 b/d, with most feedstock coming from the Soviet Union.

Spending to develop natural gas resources could amount to $4 billion to 2005, said GCA, in addition to $2 billion for the upstream oil business. Hungary's refineries also need new investment, which is likely to run about $1 billion.

In East European terms, Hungary has one of the higher private car ownership rates with one vehicle/6½ persons in 1987.

Gas production has been a disappointment. GCA says despite hopes of an increase during the past decade, production slipped to 620 MMcfd in 1988 from 725 MMcfd in the mid-1980s and will decline further to 580 MMcfd by 2000. Domestic supplies were supplemented by 510 MMcfd of imports from the Soviet Union in 1988.

Gustavson Associates notes the nation's main geologic feature is the Pannonian basin, a large Neogene feature consisting of several small, deep subbasins separated by shallow basement blocks. Extensional and strike-dip deformation are characteristic. Rapid Miocene and Pliocene sedimentation deposited thick deltaic and lacustrine sandstone reservoirs.

Hydrocarbon source rocks and seals are thick Neogene

EASTERN EUROPE

shales that are relatively low in organic content. The geothermal gradient is unusually high, and most pre-Miocene strata are postmature.

Not unexpectedly, a recent deep drilling program with assistance from the U.S. Geological Survey had negative results, and the potential of any deep hydrocarbon reserves is remote, Gustavson said.

Poland

Only 3,000 b/d of the feedstock required for Poland's 385,000 b/d refining capacity comes from domestic resources, GCA reports. The rest is imported from the Soviet Union, with some of the imports from the West. Poland is one of the few East European countries with plans for grassroots refining capacity with help from the Japanese.

Gas production is on the decline. Volume fell to 505 MMcfd in 1988 from 630 MMcfd in 1985. GCA says further declines are forecast. Imports from the Soviet Union are increasing, in payment for work on the Progress pipeline and various industrial plants, and will climb to more than 900 MMcfd soon.

The country is starting to plan its future gas supplies. An official trade and industrial delegation from Norway was told the Polish gas industry is interested in supplementing its gas supplies from the Soviet Union with Norwegian gas.

Coal dominates Polish energy. New investment in coal to meet increased demand could amount to $40-50 billion by 2005, GCA says.

There is a severe shortage of power generating capacity, and spending of $20-45 billion could be required by 2005.

Elsewhere, what is known of the geology of Poland doesn't justify an overly optimistic outlook for oil and gas exploration in regard to large or giant discoveries, Gustavson points out.

Poland's oil and gas industry, years behind world standards, will benefit greatly from western technology and equipment, Gustavson says. Emphasis will be placed on improving petroleum exploration, development, and production as Poland's new leaders set priorities in an autonomous and non-Communist eastern Europe. Poland appears to be very receptive to the western world. Recent agreements should encourage exports of technology, money, and equipment to Poland.

Romania

This is the only East European country with significant oil reserves. In the 1970s it was a net exporter, but production is now on the decline from the 188,000 b/d recorded in 1988, GCA said. The country's 13 refineries have a combined capacity of 617,000 b/d. Most of its imports come from the Soviet Union, although crude is purchased from members of OPEC and other countries for processing. Products are exported to raise hard currency.

A concerted effort will be needed to arrest the decline in oil and gas production, GCA says. Required spending will be $13 billion in the oil sector and $25 billion in the gas industry. Refining will need further upgrading, likely to cost $6 billion, and coal exploration and development could require another $5 billion.

The country is reported to be considering opening its sector of the Black Sea beyond the 90 m water depth mark to foreign companies. There also is a possibility that companies may be invited into joint ventures in some of the shallower waters.

Petroconsultants SA, Geneva, which published a report on Offshore Romania, says the overview of this sector shows that at least 620 million bbl of undiscovered recoverable oil and condensate resources may be available in the four main structural units in the Romanian Black Sea. There could be 3 tcf of undiscovered gas.

Yugoslavia

This nation has been a producer of oil for more than 100 years. Gas production has been under way for nearly as long, Gustavson points out. Although crude production and reserves have declined in recent years, estimated 1988 reserves are about 211 million bbl. Gas reserves have increased since 1980 and are estimated at about 3 tcf. Yugoslavia imports about 75% of the oil it consumes and about 60% of the gas.

Gas holds the country's best hope for future production, Gustavson said. The largest gas field, Molve, should provide one third of all domestic gas production. Recent gas discoveries in the Adriatic Sea have helped bolster reserves.

Most hydrocarbons have come from the southern margin of the greater Pannonian depression. Several small, deep subbasins filled with Neogene deltaic and lacustrine sediments provide source rocks, reservoirs, and seals for dozens of oil and gas fields.

Gas fields have been discovered in the northern Adriatic Sea. Drilling farther south in the Adriatic has yielded hydrocarbon shows and some discoveries of questionable commercial value.

A favorable political and commercial climate exists in Yugoslavia, where joint venture systems are already in place.

Several U.S. oil companies and foreign groups have been evaluating offshore blocks for a few years.

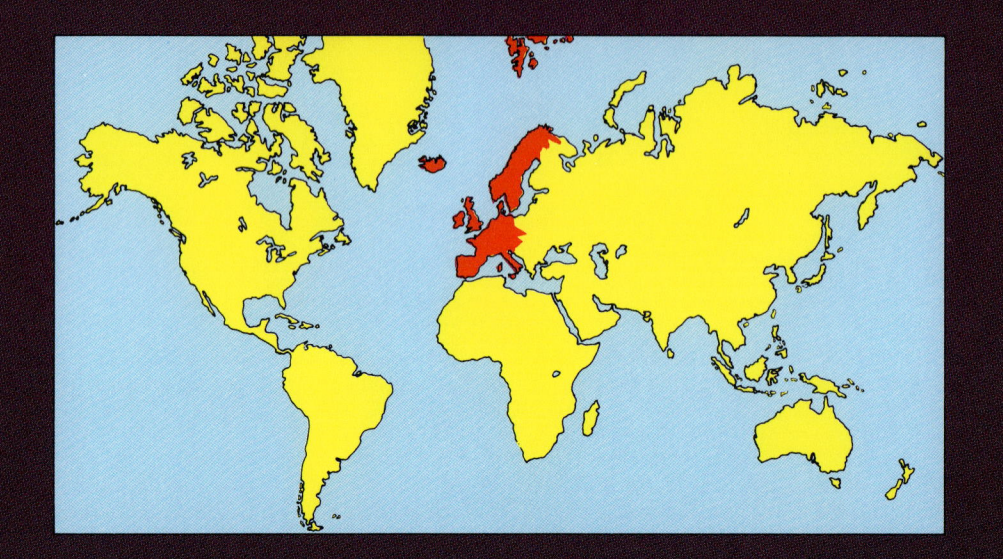

IPE ATLAS

EUROPE

AUSTRIA

CAPITAL: Vienna
MONETARY UNIT: Schilling
REFINING CAPACITY: 204,000 b/cd
PRODUCTION: 24,000 b/d
RESERVES: 81.443 million bbl

OMV, THE FULLY INTEGRATED STATE OIL COMPANY, has been active in efforts to expand its business into Central Europe.

The company plans to launch a network of at least 400 service stations in Austria, Yugoslavia, Hungary, and Czechlovakia. It was negotiating with the Soviet Union to develop an unidentified oil field, in exchange for a supply of 158,000 b/d of crude for its refineries at Schwechat, Austria, and Burghausen, Germany. OMV remained active in a project by Wintershall AG to build a gas pipeline to link Norwegian North Sea production with the Austrian distribution grid.

The 348 mile, 32 in. line from Emden to Ludwigshafen, Germany (on the Austrian border), will be rated at 290 MMcfd, expandable to 775 million. Wintershall has a two-thirds interest in the project and OMV a third.

In 1988 OMV contracted with the Norwegian state oil company, Den norske stats oljeselskap AS (Statoil), for 35 billion cu ft/year of gas for 20 years, starting in 1993.

The pipeline, known as Midal, will link up with the Stegal line near the former East German border. Wintershall and the Soviet Union's Gazprom are building Stegal from western Germany to Czechlovakia.

BELGIUM:

CAPITAL: Brussels
MONETARY UNIT: Franc
REFINING CAPACITY: 602,000 b/cd
PRODUCTION: 0
RESERVES: 0

BELGIUM'S DISTRIGAZ HAS NEGOTIATED TO TAKE 2 billion cu m/year of Norwegian North Sea gas beginning in 1996, with the option for a further 1.6 billion.

The gas will come from giant, 1.3 trillion cu m Troll field, which is due on stream in 1996. It will be moved through the Zeepipe system to Belgium.

In 1990 Norway's Statoil, operator of the Zeepipe system, awarded an $85 million contract for construction of a pipeline terminal at Zeebrugge, Belgium.

Distrigaz also receives Algerian liquefied natural gas at the rate of 3-4 billion cu m/year, but is due to increase that total to 4.5 billion cu m/year by 1992. It originally contracted for 5 billion cu m/year, but reduced it to 2.84 billion in a 1987 agreement.

Distrigaz operates a 600 MMcfd LNG terminal at Zeebrugge harbor to receive the Algerian gas.

Esso Belgium completed a $50 million alkylation/higher olefins plant at its 218,000 b/cd Antwerp refinery. It will use liquefied petroleum gas from the refinery's FCC unit and Esso's flexicoker in Rotterdam.

Solvay & Cie. SA was planning to build a 100,000 metric ton/year polyethylene and polypropylene plant at Lillo, near Antwerp.

It would be on stream in late 1991.

Ethyl SA was building a 200,000 metric ton/year linear alpha olefins plant at its complex in Feluy. Startup of the $100 million plant is due in late 1991.

DENMARK:

CAPITAL: Copenhagen
MONETARY UNIT: Krone
REFINING CAPACITY: 185,500 b/cd
PRODUCTION: 118,000 b/d
RESERVES: 799.435 million bbl

THE DANISH GOVERNMENT PLANS TO SELL ALL OR part of its state oil, gas, and pipeline companies when the

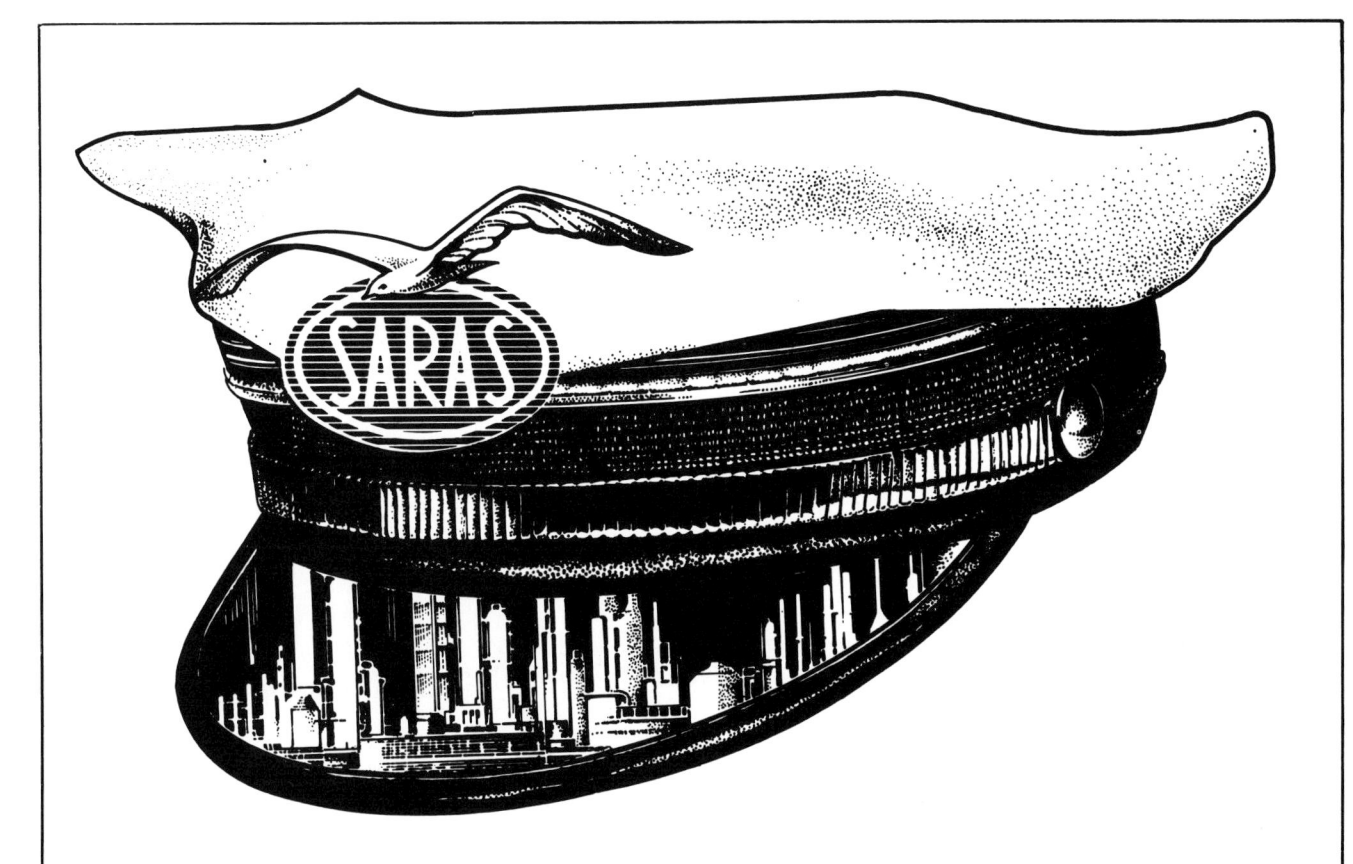

The largest oil refinery in the Mediterranean at your service

SARAS has always advanced technologically to process crude oil on behalf of third parties with the highest added value.

With a processing capacity of 18 million metric tons per year (360,000 BSD) and 5 million tons per year of conversion capacity, SARAS refinery can receive crude oil from tankers of up to 260,000 DWT, store into its huge tank farm (4 million cu.mt.) and deliver the entire range of refined products according to customer requirements.

SARAS S.p.A. RAFFINERIE SARDE

HEAD OFFICE - 20122 MILAN - GALLERIA DE CRISTOFORIS, 8 - TEL. (02) 77371 - TELEX 311273 - FAX (02) 76020640
REFINERY - 09018 SARROCH (CAGLIARI) - S.S. SULCITANA KM. 19 - TEL. (070) 90911 - FAX (070) 900209
BRANCH OFFICE - 00187 ROME - SALITA S. NICOLA DA TOLENTINO, 1-B - TEL. (06) 4820263 - FAX (06) 4871439

EUROPE/NORTH SEA

firms are econmically viable again.

Dansk Oile & Naturgas (DONG), the parent of the group of companies, was formed in the early 1980s to ensure Danish energy supplies, but has shown a small profit in only one of the past 5 years from oil and gas operations.

Meanwhile, Denmark's new energy plan links new energy investments to environmental goals, stressing the use of more natural gas and alternative fuels and less use of coal.

The government plans to convert existing coal-fired heating plants to gas in the 1990-94 period, completing it by 1998.

Efforts will be made to capture more surplus heat from manufacturing plants.

Coal will continue to be the main fuel for power generation, but new plants will be fueled by gas.

The energy plan predicted no increase in oil and gas production from the Danish sector of the North Sea. It predicted crude production would stay at 160,000 b/d until after 2000 while gas production would peak at about 6 billion cu m/year at that time.

Greenland exploration.

ARCO and Italy's Agip abruptly abandoned their joint Jameson Land concession in eastern Greenland, after completing seismic surveys but before any wildcats were drilled.

After the withdrawal, the Greenland Mineral Resources Administration ordered a new seismic search of likely oil areas off southwestern Greenland.

That effort will be separate from the Kanumas project, in which oil companies and Danish officials plan to shoot seismic off northeastern Greenland beginning in 1991.

Failure of the ARCO-Agip program, which cost the oil companies about $100 million, marks the end of a second attempt to find oil in Greenland.

The Danish government granted a number of exploration concessions off the west coast in the mid-1970s, but only dry holes were drilled.

Last year the Greenland Geological Survey said the reprocessing of seismic data from the mid-1970s, using modern equipment and methods, indicates the west Greenland continental shelf contains a number of sedimentary basins not previously reported.

FINLAND:

CAPITAL: Helsinki
MONETARY UNIT: Markka
REFINING CAPACITY: 241,000 b/cd
PRODUCTION: 0
RESERVES: 0

THE FINNISH GOVERNMENT ENDED NESTE OY'S monopoly of importing oil effective July 1, 1991, due to a new trade agreement with the Soviet Union which switches trade from one based on barter to a hard currency system.

The action was expected to reduce the volume of oil imports from the Soviet Union from 220,000 b/d to about 180,000 b/d.

That was not a problem, since Neste also was under pressure to increase imports of low sulfur crude. A new Finnish law limits the sulfur content of heavy fuel oil used in the southern part of the country to 1%, half the rate for the other parts of the country. Soviet oil supplied to Neste has a sulfur content of 1.6 to 3.5%.

Neste also was investigating the possibility of new import routes for Soviet crude, including barge shipments via the Barents Sea or a pipeline from Soviet oil fields to Finland's southeastern coast. Currently, Soviet crude moves via pipeline to a port and then is tankered to Finland.

Finland says the Soviet Union will remain its primary source of natural gas supplies.

But it also opened talks with Norwegian producers for possible transport of Norwegian gas to Finland about the turn of the century. Finland wants about 2.5 billion cu m/year, and the imports would require construction of a pipeline across Sweden and under the Gulf of Bothnia to Finland.

Finland and Sweden also have considered building a pipeline to move Russian gas to Sweden, either through Lapland or across the Barents Sea.

NEW dimethyl ether plant at Conoco Ltd.'s Humber refinery on the east coast of England.

FRANCE:

CAPITAL: Paris
MONETARY UNIT: Franc
REFINING CAPACITY: 1,815,600 b/cd
PRODUCTION: 62,000 b/d
RESERVES: 184.766 million bbl

THE FRENCH GOVERNMENT DISMANTLED THE state-owned petrochemical company Orken SA, formerly CDF Chimie, and distributed the assets between the two state oil groups, Ste. Nationale Elf Aquitaine and Total CFP.

Most of the assets went to Elf, including the petrochemicals, plastics, and fertilizers businesses. Total received Orkem's inks, adhesives, and paints businesses, plus Elf's paint operations.

The additions made Elf the fifth largest chemical producer in Europe.

Also, the government earmarked $25 million for research over 5 years into development and use of electric vehicles. And it planned to offer fiscal incentives for companies that

EUROPIPE

Stars and Pipes

The stars on the European flag symbolize new partnerships and new perspectives – perspectives which have just led to a new partnership between two internationally leading pipe manufacturers. The French Usinor Sacilor Group and the German Mannesmannröhren-Werke have fused their large-diameter pipe activities to create a new company: EUROPIPE.

EUROPIPE sets new standards in the international pipeline industry – with a truly comprehensive supply program that meets all demands for every type of pipeline. Backed up by unbeatable manufacturing capabilities and the superb quality standards of its six production locations in France and Germany, the company offers large-diameter pipe – both longitudinally welded and spiral-weld – in ODs up to 64", material grades up to API X 80, a full range of coatings and linings and – of course – perfectly matching pipe bends.

EUROPIPE's combined energies have created a source of excellence which can solve any problem where pipeline construction is concerned – no matter which media are to be transported, or whether the project is monumental or minor.

For detailed information please contact:

EUROPIPE GmbH
Postfach 40 55
D-4030 Ratingen
Phone (21 02) 8 57-0
Fax   (21 02) 8 57-285
Telex via Teletex
211 4550 = MRW

EUROPE/NORTH SEA

buy electric vehicles.

After Iraq's invasion of Kuwait, the government rolled back 4 years of oil price decontrol by imposing a temporary limit on refining-distribution margins.

Another government decree partially deregulated natural gas prices. The decree said publicly run distribution networks, rather than federal directives, will set gas prices.

Gaz de France exercised its option to take an extra 2 billion cu m/year of natural gas from Statoil under the Troll gas sales agreement, bringing it to 8 billion cu m/year in the year 2000.

Under the Troll agreement, gas deliveries will begin in 1993 and will be delivered through the Zeepipe system.

Gaz de France has formed a joint venture with the Geostock oil storage group to use seven of Geostock's 36 salt dome storage units at Manosque near Aix en Provence, north of Fos, to store 8.5 billion cu ft of gas beginning in 1992.

The storage will facilitate the import of liquefied natural gas from Algeria, which is landed at Gaz de France's Fos terminal in southern France.

The new storage will help meet gas requirements of southeastern France. Also, Gaz de France has 18 billion cu ft of gas storage at 12 sites, including two salt domes.

Production of methane from abandoned coal mines in northern France, a first in Europe, has begun. Under a 1988 agreement between Gaz de France and Nord Pas de Calais, the gas will be cleaned and dehydrated before delivery at a rate of 4 MMcfd to Gaz de France's network.

Their joint venture, Methamine, will produce 29 billion cu ft from abandoned mines during the next 12 years.

Ste. Nationale Elf Aquitaine acquired all of BP France's upstream assets, including 19 exploration and eight production permits in the Paris basin and one production and nine exploration permits in the Aquitaine area. Elf was BP's main partner in French exploration.

Petrole Saint-Honore acquired 14 permits in the Paris and Aquitaine basins owned by the British firm Clyde Expro. It gave PSH a 25% producing interest in Bagneaux field, operated by Triton, and 8.75% in the la Motte Noire field, operated by Total. Total CFP had a discovery at 1 Sancy-les-Provins, which flowed 160 b/d from the Dogger. The discovery, on the Montmirail permit in the Paris basin, is 11 miles southwest of Villeperdue field, which Total operates in a 50-50 partnership with Triton France.

Refining/Marketing

Ste. Nationale Elf Aquitaine and BP France plan to increase spending on their French refining and marketing operations.

Elf will spend $540 million in 1990-93 on refining and marketing, part of it for a 230,000 metric ton/year hydrotreater at its Grandpuits refinery east of Paris and a 145,000 ton/year alkylation unit at its Feyzin refinery located near Lyon.

Elf was negotiating to buy two large independent French gasoline retailers, which would increase its market share from 14.35% to 20%.

It owns 20% of Bianco, which is the second largest independent retailer in France, and Elf wanted to completely take it over.

And it was seeking an interest in the Tardy group, the largest independent, with 9% of the gasoline and heating oil markets.

BP France will spend $144 million on an isomerization complex at its Lavera refinery in Southeast France to meet growing demand for unleaded gasoline.

The $115 million unit was due to go on stream in early 1992.

Shell France will spend about $670 million to modernize its 158,000 b/d Petit-Couronne refinery near Rouen, Normandy, over the next 10 years.

The investment will be a third for environmental mitigation, a third for improving product quality, and a third to improve refinery performance.

Shell France said similar investments may be needed at its 128,000 b/d Berre l'Etang refinery in southeastern France.

Soc. due Pipeline Mediterranee-Rhone was building a 12 3/4 in., 150 km products pipeline linking La Mede refinery near Etang de Bare in southern France with storage facilities at Puget sur Argens.

EUROPE/NORTH SEA

WYTCH FARM drillsite on Fursey island, off the southern coast of England.

GREECE:
- **CAPITAL:** Athens
- **MONETARY UNIT:** Drachma
- **REFINING CAPACITY:** 395,300 b/cd
- **PRODUCTION:** 15,000 b/d
- **RESERVES:** 30 million bbl

GREECE IS ACCELERATING PLANS TO ESTABLISH A $1.3 billion national gas grid financed in large part by the European Community and fed by Soviet and Algerian gas imports.

Plans include construction of a pipeline from the border with Bulgaria to Thessalonika to be used for importing Soviet gas by 1993, and LNG storage/gasification terminals located near Athens.

Greece signed a 25-year supply deal with the Soviets in 1988 covering 35 billion cu ft/year and an LNG supply deal with Algeria in 1989.

In the first phase of the pipeline project, the line from Bulgaria will supply Thessaloniki, Volos, Larissa, and Athens.

In the second phase, a spur will go to the northern towns of Xanthi and Komitini.

Greece planned to end its state monopoly on oil and gas exploration, and invite participation by Greek and foreign private companies.

State owned Dep-Eky has held an exclusive right to explore in Greek territory.

The country's only oil production is from Prinos field in the northern Aegean Sea, where output was 18,000 b/d and declining sharply.

The field has a foreign operator and shareholders in partnership with the state company.

Also, the government was seeking international oil companies to invest in 50% of its state owned refineries, the 120,800 b/d plant at Aspropyrgos and the 108,000 b/d unit at Thessaloniki.

The Greek government, beset by economic problems, is seeking about $400 million in foreign investment in the refineries.

IRELAND:
- **CAPITAL:** Dublin
- **MONETARY UNIT:** Irish pound
- **REFINING CAPACITY:** 56,000 b/cd
- **PRODUCTION:** 0
- **RESERVES:** 0

MARATHON AGREED TO SELL production from Ballycotton gas field to Bord Gais Eireann, Ireland's gas board.

Marathon's plans for developing the small Block 48/20 field calls for a single subsea well tied back to the company's Kinsale Head Platform Bravo, 9 miles south. Production will flow through the pipeline from the Kinsale platform to Cork. Startup is due in the fourth quarter of 1991.

The field will be the first subsea development in the Celtic Sea, and costs will be about $35 million.

Kinsale Head field, Ireland's only domestic gas production, was flowing 220 MMcfd.

Meanwhile, Marathon had a discovery with its 48/24-3, which flowed 1,600 b/d of oil and 5.4 MMcfd of gas about 16 miles south of Kinsale Head gas field. It was the last of five wells that Marathon had agreed to drill to end a long dispute with the Irish government.

Local independent oil companies were responsible for an upsurge of exploration activity off Ireland in 1990.

A group led by Bula Oil Ltd., Dublin, acquired four exploration blocks in the Celtic Sea near Kinsale Head field. Bula will have 50%, Oliver Resources plc 40%, and Gaelic Oil plc 10% in Blocks 48/15, 48/18, and 48/19. Bula also acquired block 48/17 in an 80-20 partnership with Gaelic.

The companies made a firm commitment to one well and have options for several others. The Bula group also will shoot 200 line km of seismic survey on Block 48/18, where BP Petroleum Development Ireland tested a 13.7 MMcfd gas discovery in 1985.

Marathon has began drilling a block 48/24 hole on a structure that Bula said extends north into its block.

Aran Energy plc, Dublin, began a feasibility study of developing a 10-year old Atlantic Ocean discovery, Connemara field, in Block 26/28a, in the Porcupine basin. The study could lead to a long term production test of the field in 1,200 ft of water.

Connemara was discovered in 1979 by a BP group. The discovery well flowed 5,589 b/d of 32-38° gravity crude. But the field was estimated to contain only 200 million bbl of oil, and after crude prices dropped the non-Irish partners in the group withdrew, leaving Aran with a 63% interest and Bula with 37%.

Aran said if a study shows development is feasible using a floating production system in 1,200 ft of water, a production test would go ahead.

Aran funded two previous feasibility studies, in 1981 and 1984, neither of which favored development. But it said deepwater floating production vessel technology has advanced since then.

Atlantic Resources plc took a farmout on Marathon's Celtic Sea Block 50/3 in return for drilling a wildcat. New ownership will be Atlantic 50%, Marathon 25%, Enterprise Oil plc 10%, and Neste Oy and Oliver Petroleum Ltd. 7.5% each.

The Irish government has dropped plans to allow Nigerian

Location / No. Refineries

United Kingdom
- South Killingholme ... 1
- Fawley ... 1
- South Humberside ... 1
- Ellesmere Port ... 1
- Coryton ... 1
- Eastham, Cheshire ... 1
- Port Clarence ... 1
- Shell Haven ... 1
- Stanlow ... 1

Scotland
- Grangemouth ... 1
- Dundee ... 1

Wales
- Milford Haven ... 2
- Llandarcy ... 1
- Pembroke ... 1
- Dyfed ... 1

Belgium
- Antwerp ... 4

France
- L'Orcher ... 1
- La Mede ... 1
- Mardyck ... 1
- Reichstett-Vandenheim ... 1
- Donges ... 1
- Feyzin ... 1
- Grandpuits ... 1
- Fos sur Mer ... 1
- Port Jerome ... 1
- Notre Dame de Gravenchon ... 1
- Berre l'Etang ... 1
- Petit Couronne ... 1
- Dunkirk ... 1
- Lavera ... 1

Netherlands
- Rotterdam ... 2
- Pernis ... 2
- Amsterdam ... 1
- Vlissingen ... 1

Italy
- Livorno ... 1
- Porto Marghera ... 1
- Rho, Milan ... 1
- Sannazaro, Pavia ... 1
- Taranto ... 1
- Falconara, Marittima ... 1
- LaSpezia ... 1
- Augusta, Siracusa ... 1
- Frassino, Mantova ... 1
- Busalla ... 1
- Priolo Gargallo ... 1
- Naples ... 1
- Rome ... 1
- Milazzo ... 1
- Gela ... 1
- Sarroch ... 1
- Trecate, Novara ... 1
- Priolo ... 1
- Cremona ... 1

Germany
- Merseburg ... 1
- Hamburg ... 1
- Heide ... 1
- Wesseling ... 1
- Burghausen ... 1
- Godorf ... 1
- Harburg Grasbrook ... 1
- Neustadt ... 1
- Neustadt-Donau ... 1
- Karlsruhe ... 2
- Harburg ... 1
- Zeitz ... 1
- Leuna ... 1
- Woerth ... 1
- Schwedt ... 1
- Vohburg/Ingolstadt ... 1
- Geisenkirchen ... 1
- Lingen ... 1
- Salzbergen ... 1

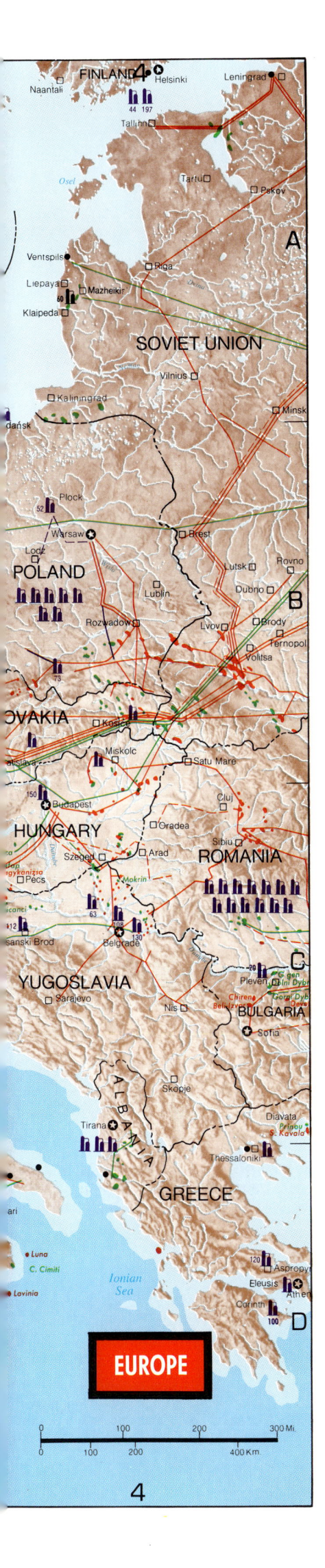

National Petroleum Corp. to take an interest in the 56,000 b/d Whitegate refinery near Cork.

NPC would have financed the modernization of the outdated refinery. The government said it would try to attract money and expertise for the modernization through a joint venture with international oil companies.

ITALY:

CAPITAL: Rome
MONETARY UNIT: Lira
REFINING CAPACITY: 2,385,358 b/cd
PRODUCTION: 97,000 b/d
RESERVES: 693.503 million bbl

SNAM, THE ITALIAN GAS COMpany, and Algeria's Sonatrach have agreed to higher gas sales which will require an expansion of the trans-Mediterranean gas pipeline from Algeria through Tunisia to Italy.

Sales will increase from the current 8 billion cu m/year to the 20 billion cu m/year level.

The Trans-Med pipeline has capacity of 13-14 billion cu m/year and SNAM was proposing that it be looped, and more compression added.

Algeria wants to use the pipeline as the conduit for expanded deliveries of gas to Central and Eastern Europe via Italy.

Italy wants gas to displace coal and nuclear power in the 1990s, pushing demand to 2.12 trillion cu ft/year by 2000 from the current 1.27 trillion. Agip SpA estimates Italian gas demand at 2.65 trillion cu ft/year in 2000.

Enimont, the Italian basic chemicals group, will return to 100% state ownership following sale of privately held Montedison SpA's 40% interest to state owned Ente Nazionale Idrocarburi for $1.27 billion. The sale followed more than a year of disputes and court actions over control of Enimont.

Meanwhile, Montedison formed a new holding company, Montecatini SpA for its chemical, energy, and pharmaceutical businesses following its merger with Ferruzzi Agricola Finanziaria. Montedison will hold 100% of Montecatini.

Production

Agip also is reassessing the gas potential of the offshore area around Bouri oil field. Agip operates the field, which produces 75,000 to 80,000 b/d and flares big volumes of gas. Flaring could increase as production grows to more than 100,000 b/d.

Agip initially had planned 50 wells for the field, but now is considering drilling another horizontal hole. The first horizontal well was a success.

Selm Petroleum, a unit of Italy's Montedison group, drilled two horizontal wells from Platform Vega off southern Sicily. The wells are believed to be the world's longest horizontal displacement holes with short radius.

They were plugged back from vertical wells, sidetracked to horizontal with a radius of 42 m, and drilled horizontally to lengths of 321 and 372 m. The reservoir lies at 2,600 m vertical depth.

Agip planned to start up a third gas production platform in Luna field, about 5 miles off Crotone in the Ionian Sea.

Refining

Union Petroliferia has proposed a $15 billion, 10-year investment program to upgrade the nation's refineries and reduce pollution.

It said refiners need to make cleaner fuel oils for power stations, cleaner gasoline and diesel fuel, and substantial investments in refinery and distribution network modernization.

Tamoil Italia SpA was building a 6,325 b/sd catalytic dewaxing unit at its 95,000 Cremona, refinery for more than $20 million. The unit will eliminate wax from atmospheric heavy gas oil to improve cloud and pour point in diesel fuel.

API Raffineria di Ancona SpA was considering installing a thermal cracking heater at its 78,900 b/sd refinery at Falconara. Initial capacity will be 10,950 b/sd, expandable to 16,425.

A Kuwait Petroleum Corp. affiliate agreed to buy Mobil Corp.'s refining/marketing assets in Italy for about $300 million.

Mobil Oil Italiana SpA operated a 100,000 b/d refinery and adjoining aromatics plant at Naples, and about 2,000 branded service station. Mobil planned to form an affiliate to continue lube sales in Italy.

MALTA:

CAPITAL: Valletta
MONETARY UNIT: Pound
REFINING CAPACITY: 0
PRODUCTION: 0
RESERVES: 0

EXPLORATION INTEREST WAS renewed in Malta's territorial waters last year, as the island country granted licenses.

Malta sits on a trend between the two largest producing fields in the Mediterranean: Sicily's Vega field

NETHERLANDS' Groningen onshore gas field has increased its estimate of reserves by 9.88 trillion cu ft. (Photo courtesy Shell World.)

and Libya's Bouri field, the largest offshore oil accumulation in the region.

Malta awarded offshore licenses in the 1970s, stimulating interest by international oil companies and launching the first stages of seismic surveys and wildcat drilling.

But a territorial dispute with Libya prevented exploration south of the island, and interest also waned in the area between Malta and Italy, with Agip SpA releasing the last of the acreage in 1985.

Recently, the International Court of Justice ended the jurisdictional dispute with Libya, and Malta renewed its quest for domestic production.

It awarded 4,218 sq km in Blocks 2 and 3 in Area 3 northeast of the island to a venture by Amoco Mediterranean Petroleum Co. (65%) and BHP Petroleum Mediterranean Inc. (35%).

With Amoco operating, work has started on magnetic and seismic surveys, to be followed by the drilling of two holes on a 4,150 sq km spread.

And Malta gave Texaco Exploration Malta Inc. an exploration license in two offshore blocks covering 760 sq miles in the strait between Malta and Sicily near Italy's Vega oil field.

The country plans to offer acreage south of Malta, in which oil companies are more interested than the northern acreage, during 1991.

NETHERLANDS:

CAPITAL: Amsterdam
MONETARY UNIT: Guilder
REFINING CAPACITY: 1,196,700 b/cd
PRODUCTION: 71,000 b/d
RESERVES: 157.2 million bbl

NEDERLANDSE AARDOLIE MIJ. BV, A SHELL-ESSO company, increased its estimate of reserves in giant Groningen onshore gas field in the Netherlands by 9.88 trillion cu ft after tests showed Groningen could produce at reservoir pressures of less than 290 psi, vs previous drawdown at 725 psi.

The new data raise the original proved reserves estimate to 94.63 trillion cu ft. Remaining reserves as of Jan. 1, 1990, were 51.9 trillion.

Meanwhile, Gasunie suspended plans to store natural gas in salt caverns to compensate for declining pressures in Groningen field.

Gasunie had planned to store gas in caverns by late 1993 so gas could be withdrawn quickly in the winter. The scheme was delayed due to discovery of a new gas field near Oude Pekela with reserves of 5-6 billion cu m, and the development of other small fields.

Conoco's Continental Netherlands Oil Co. BV agreed to take a farmout on the 47 sq km Haulerwijk block operated by Bula Oil Netherlands BV. Conoco will fund a seismic program to earn a 40% interest in the block, which offsets the concession containing giant Groningen gas field.

Dutch courts nullified an onshore concession awarded to Mobil in 1986, saying the government failed to clearly set forth environmental terms under which drilling may occur. The concession covered areas in north Holland and Utrecht.

Total Raffinaderij Nederland NV was revamping and expanding its Vlissingen refinery at a cost of $42 million. Work included a platformate fractionation unit.

Shell Nederland Chemie BV was planning a 20% capacity expansion of its Moerdijk, Netherlands, styrene propylene oxide plant. The plant currently produces 320,000 metric tons/year of styrene and 125,000 metric tons/year of propylene oxide.

Shell was considering modernizing its 348,000 b/d Pernis refinery at Rotterdam. The updated complex would come on stream in 1995.

Kuwait Petroleum Europoort BV commissioned a 6,200 b/sd unit at its Europoort refinery to increase the clear research octane number of light straight run naphtha.

NORWAY:

CAPITAL: Oslo
MONETARY UNIT: Krone
REFINING CAPACITY: 288,000 b/cd
PRODUCTION: 1.574 million b/d
RESERVES: 7,609,412,000 bbl

NORWAY OFFERED 52 BLOCKS IN THE 13TH ROUND

FRANCE

EUROPE/NORTH SEA

PARIS BASIN

Legend
- Oil field
- Oil shale
- Oil sands
- Gas field
- Crude oil pipeline
- Natural gas pipeline
- Product pipeline
- Pipelines planned or under construction
- Pump station
- Refinery in operation
- Tanker terminal
- Cities
- Capital
- International boundaries
- State and provincial boundaries
- Water depth 0 to 200 meters
- 200 meters and deeper

of licensing, and was expected to award licenses early in 1991.

The offshore licensing round was the largest since the first allocation in 1965. Of the blocks, 25 were in the Barents Sea, 22 in the North Sea, and 5 in the Norwegian Sea.

The North Sea blocks were concentrated on oil prospects that easily could be linked into the existing offshore platform and pipeline network.

Despite a lack of success in the Barents Sea so far, the Ministry of Petroleum and Energy offered blocks there to maintain exploration in the region.

Norway received applications for most of the 52 blocks, with most interest focusing on 22 blocks in the North Sea. Although Barents exploration has been a disappointment so far, companies showed a high level of interest in the 25 blocks offered there but less so in the Haltenbanken. Norway will announce awards in 1991.

Neste Oy of Finland purchased ARCO Norge SA from ARCO International Oil & Gas Co. ARCO Norge had about 135 million bbl of crude reserves and 800 billion cu ft of gas off Norway.

Onshore facilities

Den norske stats oljeselskap AS has agreed to join Conoco Inc. in construction of an 830,000 metric ton/year methanol plant in mid-Norway. Statoil will hold an 80% interest, will be operator, and will be responsible for marketing and sales.

The plant will use gas feedstock from Norske Conoco's Heidrun oil and gas field in the Haltenbanken areas of the Norwegian Sea.

Conoco needed an outlet for the Heidrun associated gas in order to win government approval for the project, which will use the world's first concrete hull tension leg platform.

The company wants to put Heidrun, which will produce 200,000 b/d of oil and 1.7 trilion cu ft of gas at its peak, on stream in 1995.

The methanol plant will require 72 MMcfd of gas and will be built at the landfall of a 20-24 in. gas pipeline from Heidrun.

The plant will meet 15-20% of western Europe's methanol demand, which currently depends on imports for about 60% of its supply. It also will be Norway's first large scale use of natural gas. All existing gas production is exported.

Statoil let initial contracts for a condensate reception terminal at Karsto, north of Stavanger, to receive liquids from Sleipner gas field after April 1993.

Sleipner condensate will be delivered at a peak rate of 75,000 b/d through a 140 mile, 20 in. pipeline to be laid in 1992. The terminal will be integrated into the existing gas processing unit for the Statpipe project.

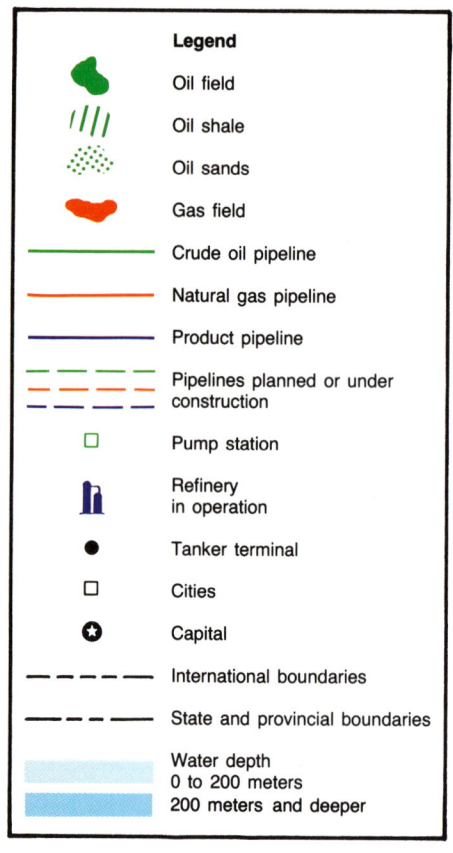

Statoil and the Dutch gas company Gasunie are cooperating on a study for a new gas pipeline from the North Sea. The line, which would land at Eemshaven in the Netherlands, would cost up to $920 million and would be operational by 1995.

Statoil was conducting a similar study which would use Emdem, West Germany, as the terminus. Either line would connect to either the Zeepipe or Statpipe lines in the Ekofisk area, running about 280 miles to shore.

Statoil was building a 130,000 ton/year propane plant at its Mongstad refnery.

PORTUGAL:

CAPITAL: Lisbon
MONETARY UNIT: Escudo
REFINING CAPACITY: 294,300 b/cd
PRODUCTION: 0
RESERVES: 0

PORTUGAL IS MOVING AHEAD WITH PLANS TO CREate a natural gas pipeline network.

Construction of the first phase of the program is to begin in 1991 and be completed in about 5 years.

The first stage calls for laying a line from Setubal, south of Lisbon, to the northern town of Braga near the Spanish border. There it will link to the European network through the Spanish system. The line will be the cornerstone of Portugal's drive to increase gas use. Bids for the job will be sought from domestic and foreign contractors, which would arrange for the gas supplies.

The project also calls for construction of a liquefied natural gas reception terminal at Setubal, a trunkline, and several distribution lines. It will require an $800 million investment. LNG imports would come from Algeria or Norway.

The Portuguese government wants the system to go on

EUROPE/NORTH SEA

stream in 1995. When fully operational, it will supply gas to about 2 million residential customers, 80,000 to 100,000 commercial users, and 4,000 to 5,000 industrial outlets in the western part of the country.

Applications for distribution rights in three of the areas covered by the grid—Setubal, the central western region, and Porto—will be issued later. The fourth distribution area, Lisbon, will be operated by Petroquimica e Gas de Portugal EP, the state company that operates the manufactured gas network in the area.

EUROPE/NORTH SEA

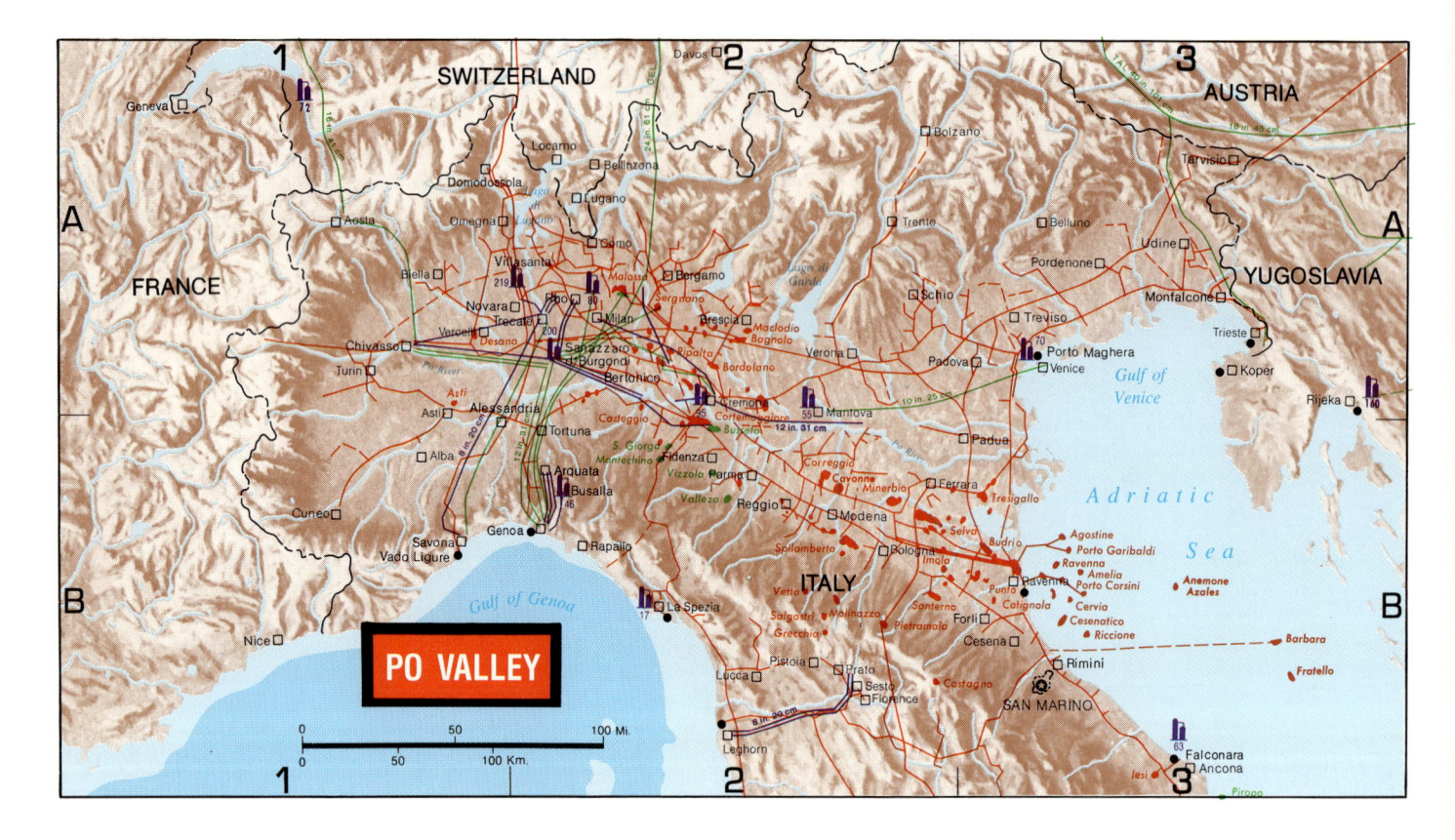

PO VALLEY

The government plans to give all the distribution companies open access to the transmission line that will transport gas as a common carrier.

The government said it would retain control over Petrogal, the state oil company, despite the shift to privatization.

Petrogal planned to add a 35,000 b/sd fluid catalytic cracking unit at its Sines refinery. It also planned to add a gas concentration unit and power recovery system.

Neste Productos Quimicos was considering revamping and expanding its Sines ethylene plant. It planned to expand high density polyethylene capacity to 120,000 tons/year from 80,000 tons/year and build a compounding unit on the site.

Neste also plans to raise capacity of Cia. Nacional Petrochimica's Sines ethylene cracker, which it operates under a 15 year lease, to 350,000 tons/year from 300,000 tons/year. Both projects are scheduled for completion in 1992.

Repsol Petroleo SA of Spain acquired full control of Enpetrol Portugal SARL, a Portuguese distributor of fuel oil and gas oil in which it had held a 50% interest.

SPAIN:

CAPITAL: Madrid
MONETARY UNIT: Pesceta
REFINING CAPACITY: 1,321,000 b/cd
PRODUCTION: 14,300 b/d
RESERVES: 20 million Mbbl

SPAIN CONTINUED ITS HEADLONG RUSH TOWARD EC membership in 1992, with more reorganizations, acquisitions, and construction plans being announced.

Government officials have estimated the country will need to invest $4.3 billion, including $370 million in the refining industry, to comply with European Community environmental standards over the next 10 years.

Under terms of Spain's entry into the European Community, the country's highly protected oil products market must be opened to competition in 1992

Spain also will need to increase production and imports of natural gas. Demand is expected to grow 25% to 221 billion cu ft in 1993 from 177 billion in 1989. And by the end of the decade, demand could rise to 259 billion.

Spain's Campsa SA, the oil products retail monopoly, says liberalization of the country's oil market could cost it about $700 million. Much of the expense would come from acquiring previously state owned gasoline stocks worth about $300 million.

Foreign companies continue to enter the Spanish market. Mobil Corp. entered a joint venture with Larious SA to operate 200 gasoline stations by 1995, and BP Espania has a venture with Petroleos de Mediterraneo SA for 400 stations.

Petroleos Mexicanos SA completed acquisition of a 2.88% share in Repsol SA in exchange for the 34.29% share it held in Spanish refinery Petroleos del Norte SA (Petronor). Pemex has a 2-year option to increase its interest in Repsol to 5%. The deal boosts Repsol's ownership of Petronor to 89.04%.

Ste. Nationale Elf Aquitaine received the Spanish government's approval to buy a 25% interest in Cia. Espanola de Petroleos SA (Cepsa) in a deal worth about $475 million. Elf initially bought a 20.5% interest in the refiner/marketer, and planned to buy the remaining 4.5% in 1991.

Meanwhile, Cepsa was seeking to buy an interest in Ertoil, the refining/petrochemical arm of Ercros, Spain's largest chemical group. Earlier in 1990, Ercros (controlled by the Kuwait Investment Office) agreed to sell a 25% interest in Ertoil to Nigerian National Petroleum Corp.

Exploration/Pipelines

Lack of significant discoveries in Spain in recent years has trimmed spending for exploration and production. Activity in the field by Spanish and foreign operators has fallen substantially in the last few years.

Chevron Corp. was seeking to sell its production and

EUROPE/NORTH SEA

undeveloped acreage. The company's net proved reserves in three fields are estimated at 6 million bbl of oil equivalent, 80% crude oil and 20% natural gas. Repsol, the Spanish state oil company, is operator of all three.

Chevron offered the properties for sale in May 1989 but determined the bids received were too low.

The sale offer includes Casablanca offshore oil field, where Chevron has an 8.92% working interest in almost 14,000 b/d. Repsol holds a 37.95% interest and five other companies the remainder.

Chevron holds a 25% working interest in 600 b/d from Ayoluengo offshore oil field. Repsol has a 50% interest and Floyd Oil (Spain) Inc. the balance. And it holds a 25% working interest in 10 MMcfd of gas production from Marismas offshore complex. Repsol holds 50% and Floyd 25%.

In addition, the Chevron offering involves six exploration permits and two exploitation concessions offshore covering about 460,000 acres in the Gulf of Valencia basin and 380,000 acres in the Sedano and Rio Guadalquivir basins.

Hideoelectrica del Cantabrico SA completed a natural gas distribution system serving about 30,000 residential and industrial customers in the Asturias region.

Empresa Nacional del Gas SA agreed to link its gas distribution system to the European network across the Pyrenees Mountains with a 44-mile, 24 in. line from Serrablo in northern Spain to Lacq, France. The project was due to be completed in 1992.

Cia. Arrendataria del Monopolio de Petroleos SA was laying two 10 in. oil products pipelines. A 94 mile line will run from Corunna to Vigo in northeastern Spain, and a 60 mile line will link Cartagena with Alicante on the Mediterranean coast. Both were due to be completed in late 1991.

It also laid a 134 mile, 8-12 in. oil products pipeline from Tarragona to Barcelona and Gerona in northeastern Spain.

Empresa Nacional del Gas SA (Enagas) planned to renew for 20 years its contract to buy LNG from Libya after the pact expires in 1991. Enagas also wanted to increase volumes from the current 33 billion cu ft/year.

Campsa is building storage capacity in preparation of the free market, adding tankage at Castellon, Corunna, and Valencia with combined capacity of 824 million bbl.

AMETHYST A2D PLATFORM, operated by British Petroleum Co. plc, adds to gas flow in the southern basin of the U.K. North Sea.

Refining

Repsol planned to spend $600-700 million to build a 400,000 metric ton/year steam cracker and associated petrochemical units at Cartagena on the Mediterranean coast, where it operates a 100,000 b/d refinery.

Cepsa planned to complete aromatics and unleaded gasoline units at its San Roque complex in 1991. Total cost was $70 million and capacity was 35,000 b/d.

Repsol completed a 500,000 ton/year desulfurizer at its Puertollano refinery in central Spain.

Repsol added 100,000 tons/year of chemical grade propylene capacity at its 130,000 b/d Corunna refinery with addition of propylene splitters to the plant's fluid catalytic cracker.

Petronor planned to build a 5,500 b/d alkylation unit at its Somorrostro, Bilbao, refinery in northern Spain. Completion was planned in mid-1992.

Petrochemicals.

Dow Chemical Iberia completed a $15 million upgrading of its styrenic products unit at Axpe-Erandio, Bilbao, northern Spain. It involved modernizing the polystyrene and styrene-butadiene latex plants and installing a recycling line in the polystyrene foam unit.

Dow launched a major petrochemical expansion at its Tarragona complex. The program, due completion by 1992, includes increasing ethylene capacity to 450,000 tons/year from 340,000, adding a 150,000 tons/year linear low density polyethylene plant, and increasing low density polyethylene capacity by 45,000 tons/year to 250,000.

Dow will raise high density polyethylene capacity by 40,000 tons/year to 90,000 and expand an existing linear polyethylene capacity from 40,000 to 60,000 tons/year. Products will go to Spain, southern Europe, and North Africa.

Hispavic Industrial SA, a joint venture between Belgium's Solvay & Cie. and Britain's ICI, planned to increase polyvinyl chloride capacity at its Martorell, Barcelona plant, to 175,000 tons/year to 130,000 tons/year.

E.I. Du Pont de Nemours was considering building a $1.2 billion complex at Asturias in northern Spain for the production of engineering polymers, special fibers, and other petrochemicals—mostly for export to other European Community nations. The entire complex won't be complete before 2000.

Atochem Espana SA increased polystyrene capacity at its El Prat de Llobregat, Barcelona, plant to 75,000 tons/year from 50,000.

Repsol Quimica raised styrene capacity at its Puertollano plant to 120,000 tons/year from 100,000 tons/year at a cost of $25 million.

Repsol is raising ethylene capacity at its Tarragona complex to 455,000 tons/year from 385,000 tons/year. Completion was due in 1991.

Hoechst AG of Germany plans to spend $450 million to increase capacities and widen the range of products of its Spanish affiliates in polyolefins, synthetic resins, paints, pharmaceuticals, and industrial gasses. The group's main petrochemicals and plastics subsidiary is Taqsa SA, Tarragona.

LOWLANDS

4
5
6
A
B
C
D
SWEDEN
D E N M A R K
Baltic Sea
GERMANY
POLAND
CZECHOSLOVAKIA
Arhus
Fredericia
Kalundborg
Copenhagen
Malmo
Korsor
Nyborg
Skaelskor
Flensburg
Kiel Bay
Fehmarn
Schwedeneck
Kiel
Preetz
Great Plon Lake
Schleswig
Heide
Warnau
Pion
Boostedt
Lubeck Bay
Bramstedt
Holstein
Lubeck
Stralsund
Rostock
Reinkenhagen
Wismar
Schwerin
Mölln
Hamburg
Harburg
Sinstorf
Reitbrook
Potrau
Dibbersen
Mechelfeld-S.
Harburg Grasbrook
Volkensen
Kolberg
Koslin
Wolin
Stettin
Walez
Schwedt
230
Achim
Lower
Saxony
Basin
Verden
Bomlitz
Wittenberge
Nienburg
Landsberg
20 in. 51 cm
NDO 28 in. 71 cm
24 in. 61 cm
20 in. 51 cm
Wittingen
Landesbergen
Lachendorf
Berlin
Rudersdorf
Bokeloh
Hannover
Lehrte
Wolfsburg
Roxforde
Brandenburg
Potsdam
Frankfurt-an-der Oder
Braunschweig
Helmstedt
Sarstedt
Buckeburg
Kemme
Magdeburg
Rybaki
Bad Salzdetfurth
Salzgitter
Guben
Fallstein
Staakow
Cottbus
Zary
Zagan
Schwarzheide
Kassel
Volkenroda
Merseburg
16
Leuna
100
Leipzig
Muhlh
Langensalza
Gorlitz
Eisenach
Erfurt
Jena
Zeitz
73
Dresden
Bad Hersfeld
Karl Marx Stadt
Zittau
Gera
20
Zaluzi
Hesse
Suhl
Kralupy
50
Prague
Plauen
Karlovy Vary

EUROPE/NORTH SEA

BASF AG of Germany will spend $280 million to expand plastics and synthetic resins capacities at its Tarragona site as well as paints capacity at Madrid and Barcelona. The main projects call for construction of 50,000 tons/year of polypropylene capacity and 10,000 tons/year of polyester fibers. Capacities for polyamides and glycols also will increase.

Bayer AG of Germany will spend $260 million at its Tarragona complex, including construction of an 18,000 tons/year formaldehyde plant and an increase to 60,000 from 24,000 tons/year in methyl diphenyl isocyanate output.

SWEDEN:

CAPITAL: Stockholm
MONETARY UNIT: Krona
REFINING CAPACITY: 427,500 b/cd
PRODUCTION: 0
RESERVES: 0

SWEDEN HAS DECIDED TO JOIN the European community. It will make formal application in 1991 but does not expect to be admitted before the end of 1994.

The country long has followed an independent, nonaligned foreign policy, and for that reason had decided against EC membership in the past.

Politicians argued that with the dissolution of the Iron Curtain, Sweden had less reason to remain neutral. Parliament voted 198 to 105 for EC membership, with 26 abstentions and 20 members absent.

Swedegas says it will look for one gas supplier—either Norway or the Soviet Union—for its sole source of natural gas in the future.

Swedish officials are due to clarify the country's energy policy in 1991, particularly regarding carbon dioxide limits, enabling gas talks to proceed.

SWITZERLAND:

CAPITAL: Bern
MONETARY UNIT: Franc
REFINING CAPACITY: 132,000 b/cd
PRODUCTION: 0
RESERVES: 0

THE FEDERAL COUNCIL, SWITZERLAND'S EXECUtive body, has a proposal to reduce carbon dioxide emissions which calls for higher gasoline taxes.

Before it goes into effect, both houses of parliament had to approve the plan and individual cantons could study it, delaying the effective date until 1994 at least.

The 72,000 b/d Collombey refinery was restarted in late 1990 by new owners, the Libyan controlled Tamoil SpA with 65% and Sase of Geneva with 35%. The refinery was closed in late 1988 when the previous owner, Gatoil, could not agree on the price of crude delivered from Italy.

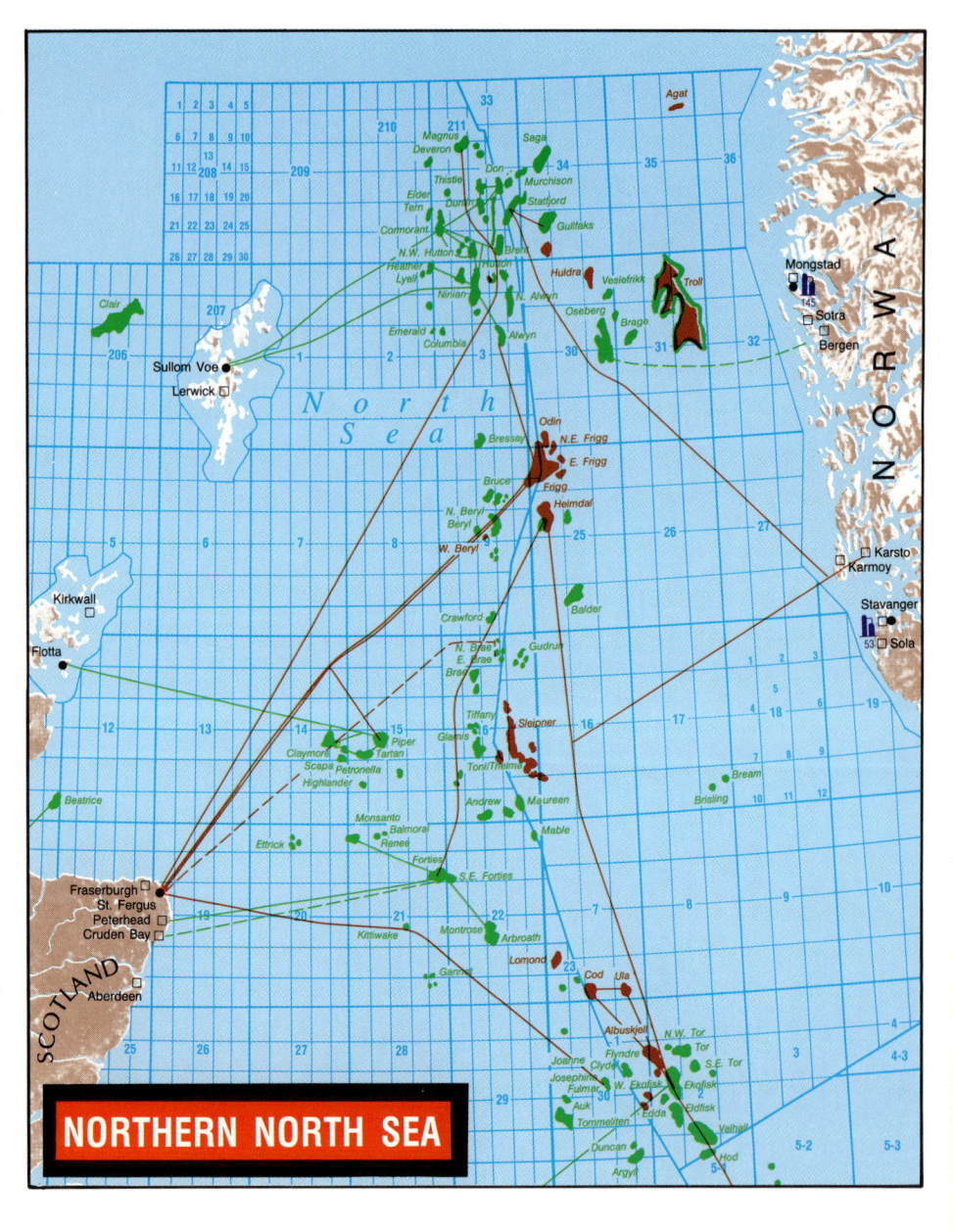

Court appointed administrators accepted Tamoil-Sase's $140 million offer, which included 350 service stations.

Late in 1990 the plant was processing about 25,000 b/d of Libyan crude and its owners were planning $80 million in improvements and to restore throughput to 70,000 b/d by late 1992.

Meanwhile, Shell planned to upgrade Switzerland's other refinery, a 60,000 b/d plant at Cressier.

Shell will spend $155 million in the 1990s to make the refinery more efficient, and to meet environmental regulations. Cressier recently added equipment at the refinery to reduce sulfur dioxide emissions and reduce emissions from storage tanks and products loading facilities.

UNITED KINGDOM:

CAPITAL: London
MONETARY UNIT: Pound
REFINING CAPACITY: 1,866,940 b/cd
PRODUCTION: 1.86 million b/d
RESERVES: 3.825 billion bbl

BRITAIN PLANNED TO OFFER 117 BLOCKS IN THE

EUROPE/NORTH SEA

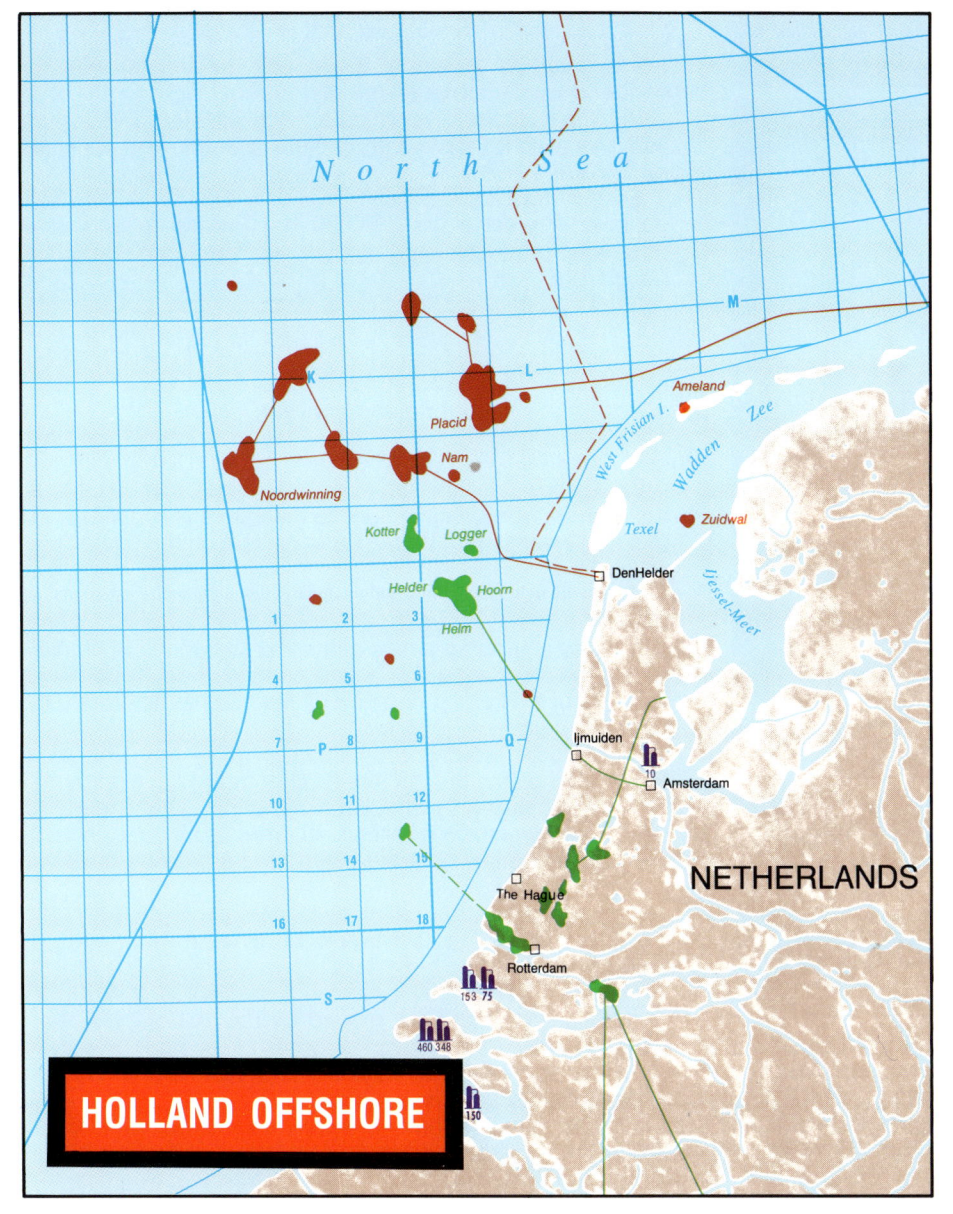

HOLLAND OFFSHORE

operations, effective Aug. 1, 1990.

Amoco had about 2% of the U.K. gasoline market and operated the 102,000 b/d Milford Haven, Wales, refinery, in which it had a 70% interest. Before the purchase, Elf had 450 gasoline stations in the U.K. and 2.3% of the market.

Not included in the sale were Amoco's exploration and production activities, and chemical and synthetic fiber operations in the UK.

Pipelines

The U.K. gas industry's evolution into a competitive business continued in 1990, as pipeline/power projects continued to jell.

Enron signed contracts for joint venture operations and power sales covering all 1.725 million kw of capacity at its proposed cogeneration power plant adjoining ICI's petrochemical complex at Teesside, U.K.

Four U.K. regional power companies, ICI, and an EnronICI joint venture marketing firm committed to buy power from the project, to be built by an Enron affiliate under a $1.1 billion turnkey contract. It will be fed by 300 MMcfd of gas for 15.5 years from U.K. North Sea fields proposed for development by an Amoco group via the Amoco group's proposed 1.4 bcfd pipeline system to Teesside.

Power sales are for 15 years beginning Apr. 1, 1993, with options to extend. The four power companies are expected to take a 50% equity state in the joint venture operating the project, and ICI may buy as much as 10% of the remaining equity from Enron after startup. Separate Enron-ICI joint ventures will have rights to transport another 300 MMcfd through CATS and certain rights to NGL in the CATS stream.

Kinetica Ltd., a joint venture of Conoco (U.K.) Ltd. and PowerGen plc, plans a major addition to its independent gas pipeline proposal in Britain.

It will lay a 175 mile line from the Conoco gas terminal at Theddlethorpe on the east coast of England to London and build a large gas fired power generation plant at Isle of Grain, east of the capital.

The joint venture earlier applied for its first major transmission system in Britain, a 31-mile link between the Theddlethorpe gas terminal and a cogeneration plant under construction for PowerGen at Killingholme.

Construction on the Midlands-London line is due to begin in the summer of 1993 with completion in late 1995. The 1.2 billion cu ft/day line will cost $106 million. It will be a direct competitor to the British Gas system.

The line will run from Theddlethorpe to Waddington, south of Lincoln.

A 36 in. line then will run east of London, under the Thames River, ending at the Isle of Grain, where PowerGen is converting a large oil fired power station to dual gas and oil fuel.

The line also could link into PowerGen's planned cogeneration plant at Rye House, north of London. And a spur could be laid from Waddington to a coal fired power plant at Cottam, Nottinghamshire, where PowerGen wants to install

lightly explored Atlantic Ocean off the west coast of Scotland in its first frontier licensing round in 1991.

The 117 blocks will be grouped in 11 two-part licenses. In the first 3 years the companies will shoot and interpret seismic.

Work programs will be negotiated on the basis of seismic results.

Also, the government was offering 120 blocks in the 12th licensing round, which will be held in 1991. Most of the blocks are in well-explored parts of the U.K. offshore.

As a sign of the changing U.K. gas market, regulatory officials told British Gas plc to rethink price increases to industrial customers designed to stop the deliberate wasting of gas.

British Gas had imposed a new tariff in an attempt to eliminate deliberate wasting, which enabled some customers' overall consumption to exceed the trigger point for the cheaper tariffs available to bigger consumers.

BP Exploration and Norway's Den norskje stats oljelskap AS have agreed to assess opportunities for several international joint ventures, including use of the existing and new North Sea facilities to move U.K. and Norwegian gas to markets in the U.K. and continental Europe and the joint marketing of gas in the U.K.

Ste. Nationale Elf Aquitaine of France completed the purchase of Amoco Corp.'s U.K. refining and marketing

EUROPE/NORTH SEA

CATS route

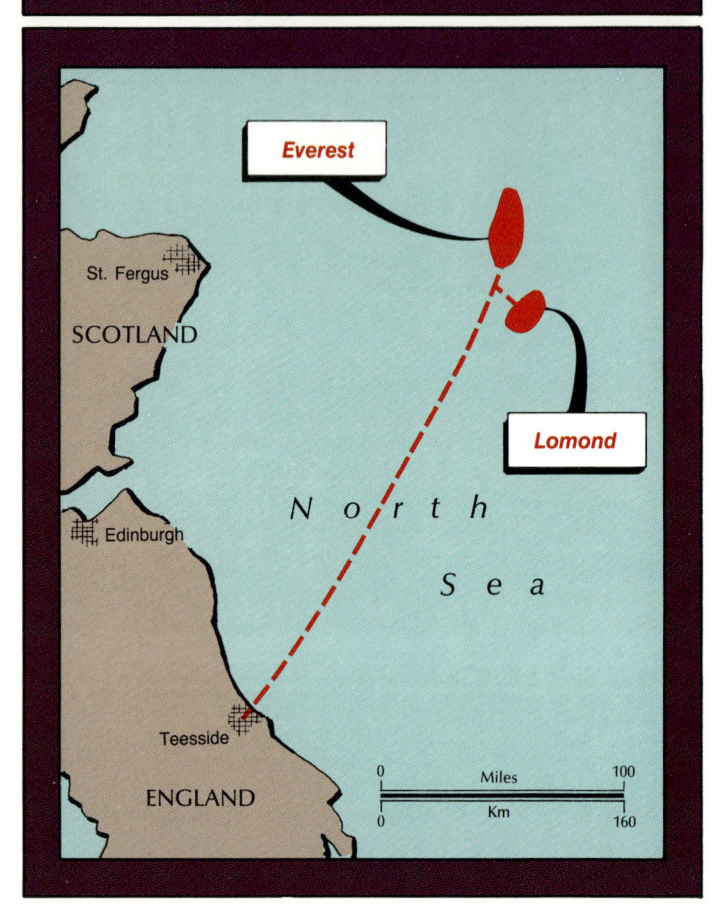

gas cycle topping.

Total Oil Marine let a $72 million contract for construction of reception facilities at St. Fergus gas terminal in northeast Scotland to handle gas from BP Exploration's Miller field in the North Sea. Startup is due in early 1992.

BP Exploration began work on an expansion of oil and gas liquids handling capacity at its Innneil terminal near Grangemouth.

The expansion is part of BP's project to boost onshore capacity of the Forties Pipeline in line with an offshore capacity increase to 900,000 b/d from 630,000 b/d.

Petrofina (U.K.) Ltd. laid a 143 mile, 10 in. products pipeline from its Lindsey refinery in South Humberside to a distribution terminal at Buncefield, Hertfordshire.

Refining

British Coal has begun a $25 million, 2.5 metric ton/day coal liquefaction pilot project at Point of Ayr in northern Wales to produce gasoline and diesel fuel.

The 3-year project, which uses a liquid solvent extraction process, is supported by the European Community, the U.K. Energy Department, Amoco Corp., and Ruhrkohle AG, Germany's largest hard coal producer.

The pilot plant is designed as an integrated unit capable of continuous operation. It also will test the ability to make products such as coke and special chemicals, and will be able to upgrade heavy residual oils either separately or in conjunction with coal.

British Coal hopes the project will pave the way for commercial scale plants producing 50,000 b/d of high grade gasoline and low sulfur diesel fuel.

Conoco Ltd. started up a 15,000 metric ton/year dimethyl ether (DME) unit at its 135,000 b/d Humber refinery on the east coast.

The unit will produce Dymel A being marketed by Du Pont as a partial replacement for fully halogenated chlorofluorocarbons.

The plant is a stand alone unit within the refinery, operated by Conoco personnel.

Du Pont chose the Humber site for the DME plant because of its location near markets in Britain and continental Europe.

Esso Petroleum Co. halted work on a $100 million resid upgrader project at its 300,000 b/d Fawley refinery near Southampton, following a dispute with the main contractor, Davy Corp.

Shell Chemicals U.K. Ltd. was building a $123 million, 100,000 metric tons/year low density polyethylene plant at its Carrington complex in northwestern England. It will replace two small trains and raise total LDPE capacity at the complex to 170,000 metric tons/year.

Wytch Farm

BP Exploration intends to submit plans in 1991 for an artificial island it wants to build to develop the 100 million bbl offshore extension of Wytch Farm oil field on the south coast of England.

Production startup from the island could begin as early as 1995.

BP also started up a new gathering station in the onshore part of the field that will enable production from Wytch Farm field, Europe's largest onshore oil field, to rise to 60,000 b/d from the current 11,000 b/d.

Wytch Farm started production in 1979 from the small Birdport reservoir, but in 1986-87 BP started development of the larger underlying Sherwood reservoir. That project required a new gathering station, 41 wells, a 56 mile, 16 in. oil line to Hamble near Southampton, and a pipeline to deliver gas into the U.K. grid.

First production from new wells, along with oil from older wells, began to flow to Hamble terminal in 1990 rising to 60,000 b/d.

That was followed by associated production of 4,000 b/d of gas liquids, which are moved by rail, and 12 MMcfd of gas.

The onshore part of Wytch Farm has recoverable reserves of abut 200 million bbl.

Most of the offshore extension cannot be reached by drilling from onshore sites.

BP decided on an offshore island because it would be less obtrusive and provide better protection in case of an oil spill. The 15-acre island would be 1.2 miles offshore in 20-23 ft of water.

It will stand 33 ft above low tide level and will be landscaped to screen facilities, except for the drilling mast.

BP plans to drill 30-40 holes from the site, which would be linked to the existing Wytch Farm facilities by pipeline.

If there are no delays, the island could start production in 1995 when the 60,000 b/d flow from the onshore part of the field will start to decline. BP said the offshore production would build slowly to maintain the 60,000 b/d level, and will peak at 30,000 to 40,000 b/d.

BP owns a 50% interest in Wytch Farm. Other partners are ARCO with 17.5%, Premier Oil Dorset Ltd. 12.5%, Kelt Exploration Ltd. and Clyde Petroleum (Dorset) Ltd. 7.5% each, and Goal Petroleum plc 5%.

Wytch Farm facilities also will handle production from a small nearby reservoir. BP is developing the 6 million bbl Wareham field, which is producing about 2,800 b/d.

Exploration

Hamilton Oil Corp. had a discovery in Morecambe Bay off Western England. Its 2-110/13 wildcat flowed 1,800 b/d of 34° gravity oil through a 36/64-in. choke.

The discovery is 5.5 miles southwest of Hamilton's 70 MMcfd 1-110/13 discovery on the same block.

Hamilton also extended the gas discovery to the north

EUROPE/NORTH SEA

with 3-110/13.

It planned more exploratory and appraisal drilling on the block in 1991.

Hamilton holds a 45% interest, Ultramar Exploration Ltd. 30%, and Monument Resources Ltd. 25%.

Kelt UK Ltd. was drilling a horizontal well in Horndean oil field near Portsmouth.

Horndean was producing about 200 b/d from two vertical wells.

The latest hole, with a displacement of 2,500 ft into an untapped part of the reservoir, is the second horizontal hole in the U.K.

East Midlands Oil & Gas, a unit of Blackland Oil plc, plans a $948,000 first phase development of the 900,000 to 2.5 million bbl Whisby oil field southwest of Lincoln. Later phases could boost total investment to $4.18 million.

East Midland is operator with 58%, British Gas plc had 25%, and Edinburgh Oil & Gas Plc 17%.

Kelt Energy plc had a discovery at the 2 Caythrope near Bridlington, Humberside.

It flowed 10.5 MMcfd from the Rotliegendes and 7.7 MMcdf from Kirkham Abbey.

Kelt operates several nearby gas fields.

GERMANY:

CAPITAL: Bonn
MONETARY UNIT: Deutsche Mark
REFINING CAPACITY: 2,065,400 b/cd
PRODUCTION: 78,000 b/d
RESERVES: 425 million bbl

RWE-DEA of Hamburg has developed a process which should reduce the costs of secondary recovery in Plon Ost field in the Schleswig-Holstein area.

The process uses a polyethylene glycol to reduce the adhesion of reservoir rock, so only a small volume of polymer remains in the pores during polymer flooding for enhanced oil recovery.

Plon Ost, which has produced 541 million bbl of light oil since 1958 and is believed to have 131 million bbl remaining in place, is a tight sandstone interbedded with clays with 15% saline water and 195° F. temperature.

Increased oil production is expected by mid-1991 from a Plon Ost pilot area where injection began in the summer of 1990.

Mobil Erdgas-Erdoel GmgH's Z-1 West Walsrode discovery flowed 27 MMcfd of gas from the Rotliegendes Sand at 15,000 ft on its Ahrensheide permit in northern Germany. Mobil, with 100% interest in the strike, planned more drilling.

BEB Erdgas und Erdoel and Mobil Erdgas-Erdoel plans to invest $177 million to expand capacity of their jointly-owned natural gas storage facility in salt caverns storage facility at Doetlingen in northwestern Germany.

They plan to increase storage capacity from 1.1 billion cu m to 2 billion, making it the country's largest storage facility.

Refining

Oberrheinische Mineralolswerke GmbH started up a 110,000 ton/year methyl tertiary butyl ether plant at its 142,000 b/d Karlsruhe refinery. Deutsche BP and Deutsche Agip planned a $73 million catalytic cracker expansion at their jointly owned 102,000 b/d Vohburg/Ingolstadt refinery in Bavaria. Current cat cracker capacity is 18,700 b/d. Completion was planned for 1993-94. Mobil Oil AG briefly shut down its 93,000 b/d refinery at Woerth near the French-German border to make improvements totaling $53 million.

The upgrading will increase gasoline production 3.5% and reduce heavy fuel oil output, as well as reduce nitrogen oxide emissions. Mobil plans to add a MTBE unit at the refinery in the future.

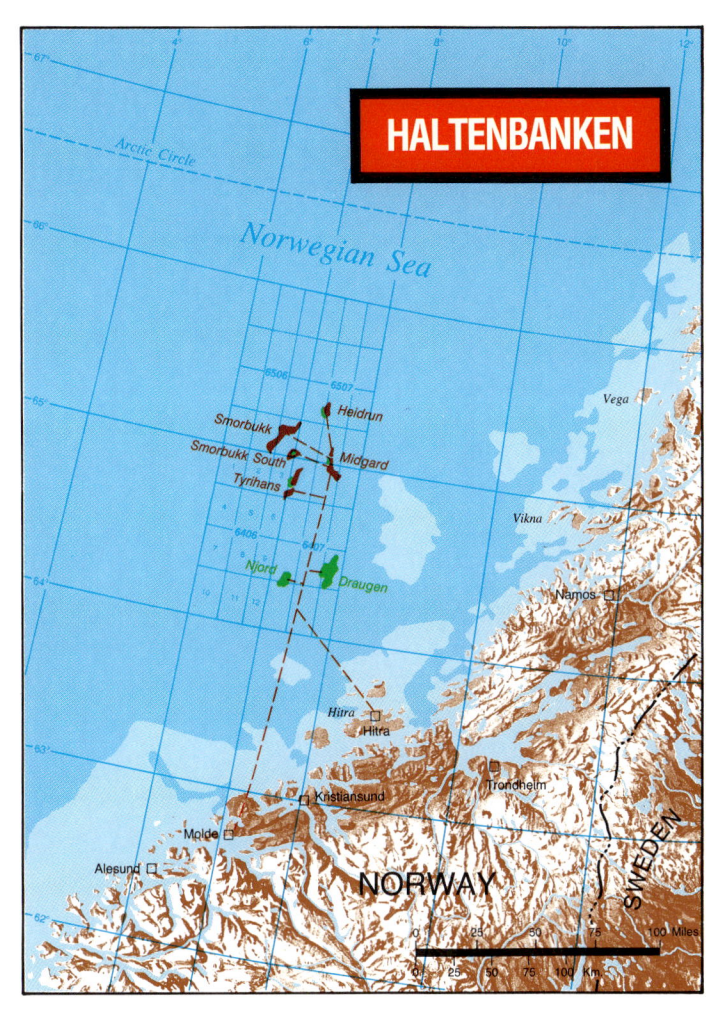

NORTH SEA

PROSPECTS FOR GAS PRODUCERS IN THE BRITISH, Norwegian, Dutch, and Danish portions of the North Sea continue to improve.

European demand for gas in the 1990s will be higher than first estimated. Despite competition from Soviet and African supplies, North Sea producers look set to corner much of the increased business.

In continental Europe, gas is benefiting from its relatively clean burning features at a time when government and consumers are looking for ways to reduce discharge of gases that might contribute to global warming.

The opening of eastern Europe's gas market, previously the unchallenged domain of Soviet supplies, will provide opportunities for North Sea producers as state gas companies try to diversify supply sources.

Poland's state gas authorities have sounded out Norway's Den norske stats oljelskap AS (Statoil) about the possibility of developing a supply network through the Baltic Sea to meet future Polish demand, which is expected to increase as pressure grows for environmental cleanup.

In the U.K., demand is growing for gas to burn in a new generation of power plants. The advent of competition in industrial markets also will stimulate offshore development.

U.K.
cogeneration

Sale of the state owned U.K. electrical power industry to private concerns allows new generating companies to enter the business.

Almost all the newcomers to power generation plan gas fired cogeneration plants. Detailed negotiations in progress

NORTH SEA

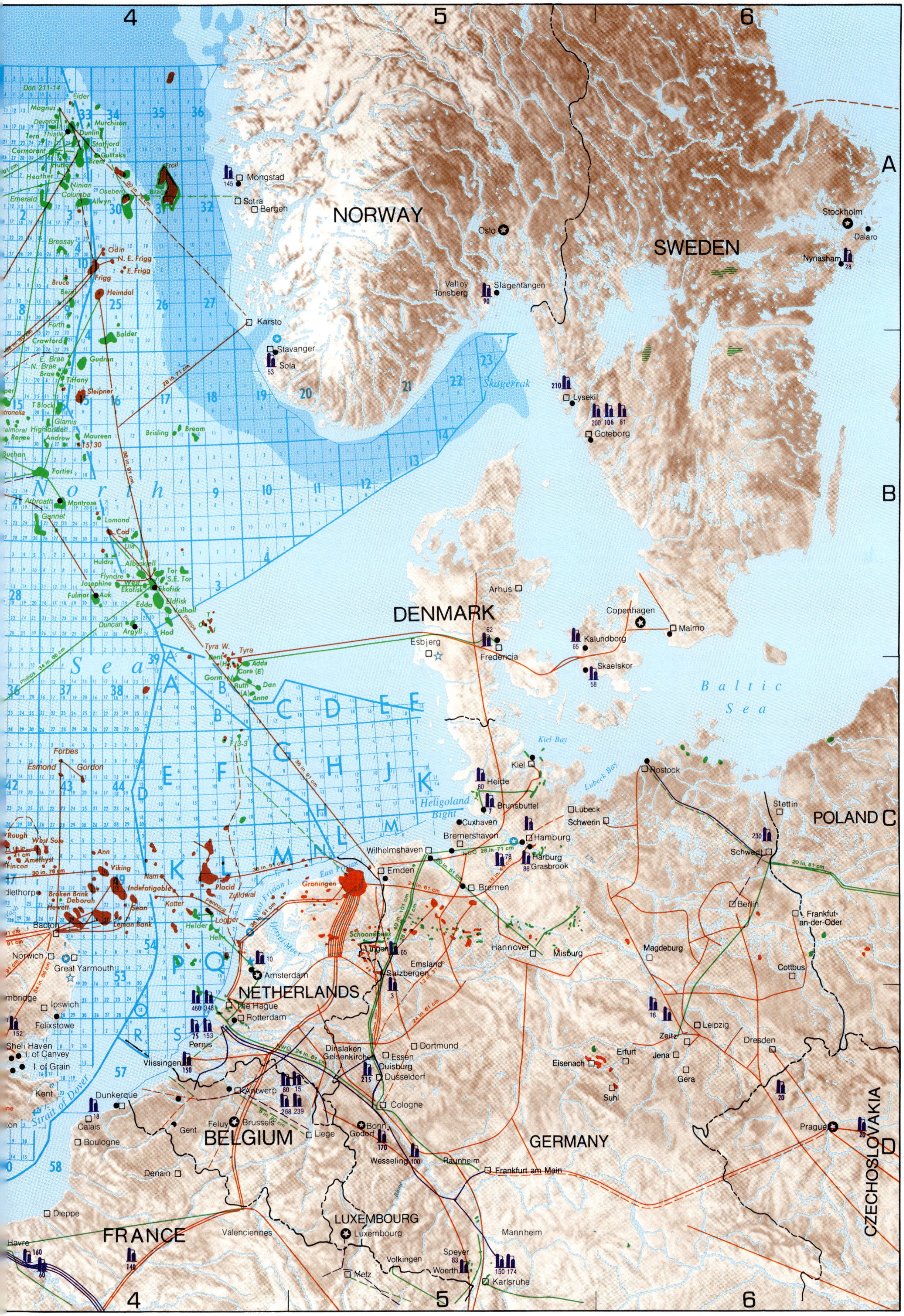
NORWAY
SWEDEN
DENMARK
NETHERLANDS
BELGIUM
GERMANY
FRANCE
LUXEMBOURG
POLAND
CZECHOSLOVAKIA
North Sea
Baltic Sea
Skagerrak
Heligoland Bight
Strait of Dover
Oslo
Stockholm
Copenhagen
Amsterdam
Brussels
Bergen
Stavanger
Goteborg
Hamburg
Bremen
Kiel
Rotterdam
The Hague
Antwerp
Cologne
Bonn
Prague
Berlin
Frankfurt am Main
Luxembourg
Karlsruhe
Mannheim
Groningen
Emden
Esbjerg
Malmo
Rostock
Stettin
Leipzig
Dresden
Ekofisk
Frigg
Brent
Statfjord
Troll
A
B
C
D
4
5
6

EUROPE/NORTH SEA

with many offshore producers aim to bring the first batch of new generating capacity on stream in 1993-94.

The power station fuel market is big enough for North Sea operators to develop new fields tied exclusively to electricity generating projects. ARCO British will sell all production from its Pickerill field in the southern North Sea to a large cogeneration unit on the east coast of England.

The most ambitious project is by the Amoco-Gas Council combine, which plans to develop two fields in the Central North Sea and build the first pipeline from that area into Northeast England to service a proposed power station at Teesside.

The industrial gas market also has been opened to competition.

British Gas plc has been forced to open its nationwide transmission system to third party customers. It is allowed to buy only 90% of the gas from any new source of supply. A producer may sell the remaining 10% to another buyer.

Currently, there is a shortage of gas for sale by the new breed of gas marketing companies. The first supply of new gas from a 90-10 deal will become available when ARCO starts up Welland field.

Although British Gas is no longer the monopoly buyer of all new supplies, it will remain the major force in the market. Offshore operators are concerned that the desire of British Gas to import supplies—probably from Norway—could restrict development of gas/condensate resources in the central North Sea.

While British Gas looks for cheap supplies outside the U.K., Norway is investigating the power generation market in Britain. Statoil reports that after talking to many of the proposed power station operators, it has started negotiations to sell gas to several projects.

The Troll shock

Norway, with estimated gas reserves of 82 tcf, was previously considered the most likely candidate to meet most new European gas demand through the 1990s.

Conventional wisdom saw 45 tcf Troll gas field as providing the base load for new contracts, with reserves on Haltenbanken off mid-Norway and in the arctic region of Tromsoflaket off the northern tip of Norway providing the basis for growth into the next century.

But the Norwegian gas industry suffered two shocks in the first half of 1990:

- Statoil's decision to shelve plans for an LNG export chain to the U.S. under an outline contract with Enron Corp. because of increased gas demand in Europe and possible problems bringing South Gullfaks field on stream.
- Norsk Hydro's successful horizontal drilling into oil zones in the western lobe of Troll field.

Norsk Hydro said Troll oil can be developed with a program of horizontal wells, which would preclude development of western lobe gas before 2005.

That leaves Statoil as operator of a huge volume of reserves in the eastern lobe but without the flexibility to take on new large supply contracts that existed when the only constraint was time required to develop reserves on the western lobe.

The upturn in demand could lead to earlier than expected development of gas from the Haltenbanken area. The decision to press ahead with a methanol plant using associated gas from Heidrun field will create a gas gathering infrastructure offshore.

If prices remain buoyant, a pipeline link between the Heidrun landfall and the gas infrastructure in the North Sea could emerge before the end of the decade.

Norwegian producers also are involved in long negotiations to sell gas to Sweden. Swedegas has agreed to buy an average 250 MMcfd, which Statoil claims would be enough to warrant laying a pipeline to Sweden from the North Sea. The two sides are negotiating prices.

Enthusiasm for gas

An indication of European enthusiasm for North Sea gas came from the decision of three German purchasers, Ruhrgas, BEB Erdgas und Erdol GmbH, and Thyssengas, to exercise an option to boost deliveries under their Troll gas purchase contract.

The first option will allow the German companies to increase the annual average to 996 MMcfd from 803 MMcfd.

The companies exercised the first option after surveys showed that gas demand in the second half of the 1990s and the early part of the next century could be much higher than expected.

The Ruhrgas and BEB link with the East German distribution network could increase volumes required even in the short term.

The Germans also are concerned that upheavals in the Soviet Union could have repercussions in the gas business.

The Soviets always have been reliable suppliers. Supply problems have been caused mainly by weather induced delays or technical problems rather than by political disinclination to meet full contract volumes.

Statoil is boosting gas sales to the Spanish state gas organization Enagas. The Spaniards were latecomers to Troll gas but were quick to exercise an option to start taking gas from Sleipner field in 1993 instead of waiting for Troll deliveries beginning in 1996. By exercising the option, Enagas will increase the plateau sales level to an average 205 MMcfd from 175 MMcfd. Total deliveries also will increase to 1.76 tcf from 1.34 tcf.

New Phillips deal

Phillips Petroleum Co. Norway, on behalf of the Ekofisk group, negotiated a new contract that will extend deliveries of gas from Ekofisk field to continental Europe from the original 1999 expiration date until 2011.

The new contract ends a disagreement between the Ekofisk group and its customers—Ruhrgas AG of Germany, Gasunie of Netherlands, Gaz de France, and Distrigaz of Belgium—over terms of the two original contracts of 1973 and 1974.

After major investments and upgrade projects, Phillips estimates there are at least 4.5 tcf of reserves remaining from the original 8.9 tcf. The Ekofisk partners felt the annual average offtake of about 1.2 bcfd was inadequate. Purchasers were unhappy with pricing.

Under the new contract, the gas price will continue to lag oil product prices at Wiesbaden, Germany, by 1 year. But they will take into account tax anomalies in German product pricing that will benefit the French, Belgian, and Dutch buyers.

County Natwest Woodmac, Edinburgh, Scotland, estimated the new gas price for Ruhrgas will be about $2.70/Mcf after taking into account the 12 month lag. On this basis, Ekofisk gas prices will remain significantly higher than Troll prices. The Ekofisk price can be renegotiated in 1999.

On the supply side, the contract is considerably more flexible. The old system of allocation of gas field by field is replaced by an annual contract volume covering existing and future development on the Ekofisk license.

Woodmac estimated gas sales from Ekofisk will remain at an average of about 1.1 bcfd in the early to mid-1980s, then decline gradually to 675-775 MMcfd in 1999.

Industry sources say there is an excellent chance of increasing the Ekofisk reserve base. Phillips is working on development proposals for South Eldfisk field, and other prospects may be drilled in the next few years.

Statoil also has broken new ground by buying 38 bcf of associated gas from four partners in Veslefrikk field, raising its share of reserves to 91%. Norsk Hydro owns the other 9%.

EUROPE/NORTH SEA

Troll field reservoirs

Oil province
- 22 billion cu m gas in place
- 121 million cu m oil in place
- 22-26 m oil pay

West gas province
- 182 billion cu m gas in place
- 141 million cu m oil in place
- 10-12 m oil pay
- Low productivity

West gas province
- 394 billion cu m gas in place
- 270 million cu m oil in place
- 10-12 m oil pay
- High productivity

East
- 1.072 trillion cu m gas in place
- 83 million cu m oil in place
- 0-3 m oil pay

Source: Norsk Hydro AS

The gas originally was sold for power generation, but the power station project was canceled, leaving as much as 25 MMcfd gas without a buyer when Veselfrikk reaches plateau production.

Veslefrikk gas will be stored in Statfjord or Heimdal field until a sales contract is arranged. Statoil's action could set a trend for other small parcels of gas that may become available.

Danish, Dutch gas

Danish offshore gas production, currently controlled exclusively by Dansk Undergrunds Consortium (DUC), could benefit from plans by the administration in Copenhagen to promote gas use in an effort to reduce pollution.

Unlike its neighbors in Netherlands and Norway, Denmark does not have access to massive untapped gas reserves. Remaining reserves at the beginning of 1990 were 4.45 tcf, all under the control of DUC.

Production from Tyra gas field and of associated gas from DUC's offshore oil fields has averaged about 265 MMcfd. All is sold to the state monopoly Dangas.

DUC has medium term plans to develop Harald and Roar fields under a new flexible gas contract that could help meet future requirements of a more environmentally oriented Danish energy policy.

Denmark is a relative newcomer to the use of gas. Outlets have been established in the industrial and domestic markets and for district heating projects. Only about 7.5% of total supply goes into power generation. Demand in 1990 was expected to be about 215 MMcfd.

To absorb its contract volumes from DUC, Dangas signed four export contracts with Swedegas and Ruhrgas.

The contracts expire in 2003-06. Dangas is exporting a little less than 100 MMcfd, but the contracts have a plateau level of 145 MMcfd. With sustained domestic growth and the export contracts, Dangas's requirements might rise to 375 MMcfd in 1995 from 310-315 MMcfd in 1990.

Dutch North Sea gas production averages about 1.7 bcfd. All gas is sold to Gasunie, which uses a price formula that allows very small discoveries to be linked into the extensive gas transportation system.

The Dutch government's policy is to maintain those favorable terms and keep a steady stream of new development proposals coming forward so reserves from giant onshore Groningen gas field can be conserved.

Other U.K. sector activity

BP Exploration's program of high angle wells on the periphery of Miller field in the U.K. North Sea resulted in the most demanding well undertaken by the company's drilling department.

The seven wells were being drilled from a template with the Santa Fe 135 semisubmersible prior to installation of the platform in 1991. Miller is to start up in early 1992 and produce about 113,000 b/d.

The fifth well in the program, 16/8b-AO5(37), a water injector, generated torque that stretched the rig's 38,000 ft/lb capacity top drive to its limit.

As a contingency, BP considered subsea completions for water injection, but they were not required because well AO5 demonstrated that the high angle wells were achieveable from the template.

In addition to operating at unprecedented torque levels, the drilling department had to deal with a number of unscheduled problems. A differentially stuck pipe forced a time consuming plugback and sidetrack, followed by a broken logging cable and a small well kick.

Field developments

Scott, with 450 million bbl of oil reserves, was at yearend 1990 the biggest oil field under development in the U.K. North Sea.

Associated gas production will start in early 1994 and peak at 90 MMcfd.

It will be transported through the Mobil-operated SAGE gas line to St. Fergus, Scotland. From St. Fergus, Mobil plans to distribute the gas using the BG distribution network.

The U.K. Department of Energy approved two new development projects in the central U.K. North Sea.

Marathon Oil U.K. Ltd. received the go-ahead for 1.5 tcf/300 million bbl East Brae gas/condensate field in Blocks 16/3a and 16/3b. Agip U.K. won the nod to develop 40 million bbl Toni oil field in Block 16/17 as a satellite to its Tiffany development.

Amoco (U.K.) Exploration Co. will use an innovative X-braced tower design for three platforms in its Everest, North Everest, and Lomond gas development projects.

The design concept, developed in house by Amoco, allows each of the three jackets to be installed in a single crane lift and provides a greater topsides area on which

EUROPE/NORTH SEA

integrated decks will be fitted. Parallel-sided towers are an unusual feature in the North Sea, and X bracing is rarely used outside Amoco operations.

The two jackets for Everest and Lomond production platforms will weigh 4,800 tons each, and the riser jacket to be installed in North Everest will weigh 3,800 tons. The two platform jackets will be installed over templates that are being fabricated at the Arnish Point yard on the Isle of Lewis off western Scotland.

Gas from the fields will move through the Central Area Transmission System (CATS), a 247 mile, 36 in. line to Teesside in Northeast England, where it will be used for electrical power generation. Condensate will move through a link into BP Exploration's Forties pipeline to Cruden Bay, Scotland.

Lasmo North Sea plc received U.K. government approval for its first field development as operator in the U.K. North Sea. Staffa field in Block 3/8b, 6.2 miles east of Ninian oil field, will produce about 8,000 b/d from two subsea wells tied back to Ninian's southern platform. Start-up is scheduled for yearend 1991.

Ranger Oil (U.K.) Ltd. won approval for development of its 235 bcf Anglia gas field in the southern U.K. North Sea.

In the northern North Sea, Conoco U.K. Ltd was preparing to modify Ninian oil field facilities to accommodate production from two satellite fields.

BP Exploration was seeking British government approval to develop Donan field in North Sea Block 15/20 using its new Seillean floating production vessel.

The £110 million ($170 million) ship took on its second cargo of crude from Cyrus field in North Sea Block 16/28.

Seillean, the world's first production vessel to hold station using only dynamic positioning, began operations in the spring of 1990.

One new oil field and five new gas developments went on stream in the U.K. North Sea in the fall of 1990.

The four gas fields, ARCO British Ltd.'s Welland field, Shell-Esso's Barque and Clipper fields, BP Exploration's Amethyst field, and Hamilton Bros. Oil & Gas's North Ravenspurn are in the southern basin of the sea.

The single oil field is Shell-Esso's Kittiwake in Block 21/18, 100 miles east of Aberdeen, which started production 5 months ahead of its original schedule.

Flow from the platform, linked to a tanker loading system, is to rise to a peak of 36,000 b/d during a 2 year drilling program.

The field, with reserves of 70 million bbl, was developed at a cost of £250 million ($475 million).

Shell-Esso's Barque and Clipper fields lie in the Sole Pit area of the southern basin. The three platform development program covering both fields cost £420 million ($798 million).

Initial production through a 46 mile, 24 in. pipeline to Bacton terminal on the east coast of England is about 130 MMcfd.

ARCO's Welland field is the only gas development among the five new gas fields in which 10% of reserves have been sold to buyers other than British Gas plc. AGAS Ltd. and the gas marketing operations of BP Exploration and Mobil North Sea will sell 10% of Welland gas on the U.K. industrial market.

The 240 billion cu ft of reserves will produce at a plateau rate of 100 MMcfd.

ARCO said for the first time in the U.K. North Sea, production flows through an unmanned satellite platform that combines platform and multiple subsea wells with total automation into a single project. Field control will be from ARCO's Thames complex 11 miles away.

Hamilton Bros.' North Ravenspurn development is another technical first. The company used the first shallow water concrete gravity base in the southern basin for the 1.3 tcf development.

The field, in Blocks 43/26a and 42/30, will produce an average 290 MMcfd through a 24 in. pipeline link into BP's Cleeton gas field, which has a 36 in. pipeline to Dimlington terminal on the east coast of England.

BP's Amethyst field has two platforms remotely controlled from Easington terminal. The two units will produce an average 150 MMcfd. Two more remote controlled platforms were to be installed in May 1991.

Partners in a group led by Chevron U.K. Ltd. approved development of Alba field.

The 300 million bbl field in Block 16/26 is expected to be on stream in late 1993 or early 1994 at 60,000-70,000 b/d, building to a peak of 100,000 b/d.

Partners approved a phased development plan with a steel drilling-production platform in the northern part of the field followed by a second platform in the southern part 5 years after start-up. Oil will be transported through a floating storage unit. Chevron said Phase 1 development will provide experience of production from shallow Eocene sands at 6,000 ft.

Cullen report ramifications

Some older, less profitable fields in the U.K. North Sea could be shut down prematurely when recommendations of the Cullen report into the Piper Alpha accident are implemented.

A study of implications of the Cullen report by the safety and reliability department of W.S. Atkins, Epsom, England, said it is not clear whether older fields will be abandoned because of increased costs, but it seems likely. The same situation could prevent development of marginal fields.

The cost of implementing the Cullen recommendations in modifications to installations, possible lost production, and additional manning will be considerable.

The Atkins study said introduction of formal safety assessments (FSAs) on all installations in the U.K. North Sea is unlikely to be a painful exercise and will bring significant benefits to operators.

It added that the recommendations will allow offshore safety to catch up and surpass requirements onshore.

The Cullen report identifies the failure of the permit to work system on Piper Alpha as one of the causes of the disaster. It recommends that operators and regulators should pay close attention to training and competence of contractors' supervisors who are required to operate the system.

While it is not necessary or practical to have a standard system throughout the industry, Cullen suggested companies should work toward harmonization in the colors used for different types of permits and the rules on the period in which a permit remains valid. All permit to work systems should have a procedure that involves locking off and tagging of isolation valves.

Because most of the equipment on the platform was not recovered from seabed wreckage and key witnesses did not survive, there are a number of possible explanations for the first condensate leak.

Cullen concluded the leak resulted from steps taken by night shift personnel after one of the two condensate injection pumps tripped.

The men tried to restart the other pump, which had been shut down for maintenance, and were unaware that a pressure safety valve had been removed from the relief line of that pump.

A blank flange assembly fitted to the side of the valve was not leak tight.

Lack of awareness of the removal of the valve resulted from failure in communication of information at shift handover earlier in the evening and failure in operation of the permit to work system in connection with the work that had entailed its removal.

A new vision of energy

From crude oil to fuel, from the oil well to the filling station. Energy is always on the move. So is ÖMV: Last year we started to set up our own network of filling stations. A move from merely producing and processing crude oil and natural gas to satisfying the needs of the end user. A move that is in line with our new corporate identity and has clearly set new standards in the market. We have chosen the direct approach to effectively cope with supply and demand.
We are determined to overcome all obstacles and to live up to our commitment: to understand man, to perceive our environment and to master technology.
ÖMV - power of competence.

ÖMV
Gruppe

EUROPE/NORTH SEA

How F3 field layout will look

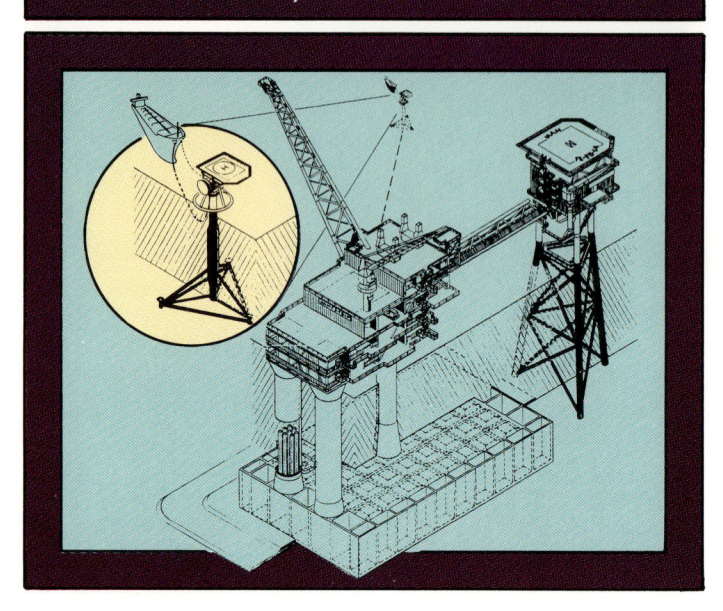

Norwegian sector activity

A Statoil survey shows there are 5.9 billion bbl of crude oil and 31.95 tcf of gas still to found in the Norwegian North Sea.

Those figures compare with 2.39 billion bbl of crude and 16.77 tcf of gas in Statoil's 1989 survey.

Statoil said it devoted more resources to its annual study of undiscovered resources off Norway so it could balance its exploration effort between Norwegian and foreign prospects.

The survey used new geological modeling techniques and included new seismic acquisition and interpretation.

The state company's figures are for internal use only. Official government estimates are based on an annual survey by the Norwegian Petroleum Directorate, which traditionally takes a more conservative view of resources.

Statoil's survey shows remaining proved reserves in the Norwegian North Sea are 11.43 billion bbl vs. 1989's 10.42 billion bbl. Proved remaining gas reserves are almost unchanged at 76.98 tcf, compared with 77.16 tcf in 1989.

While Statoil takes a more optimistic view of the North Sea's future, it is less optimistic about the Barents Sea and the Haltenbanken area.

In the Barents Sea, which only a few years ago was seen as source of oil discoveries for development in the next century, Statoil estimates only 3.18 billion bbl of oil will be found, down from its 1989 estimate of 3.71 billion bbl.

Gas potential of the area is also downgraded—to 22.95 tcf from 24.89 tcf.

Proved oil reserves in the Barents Sea are 220 million bbl. Proved gas reserves increased to 10.4 tcf from 10.06 tcf.

On Haltenbanken, where development of proved oil reserves of 2.04 billion bbl and 10.77 tcf of gas is starting to get under way, the Statoil survey says undiscovered resources are 5.67 billion bbl and 39.72 tcf, compared with 5.73 billion bbl and 47.67 tcf in 1989.

Proved reserves for Haltenbanken in 1989 were 1.98 billion bbl and 9.88 tcf.

Sleipner's Theta satellite commercial

In the Norwegian North Sea, the Sleipner group led by Statoil planned to declare Sleipner's Theta satellite commercial and bring the reservoir on stream with the main gas field development in 1993.

If the development plan is approved, Statoil will install a four well subsea template connected to Sleipner A platform by a 5.6 mile pipeline.

New gas pipeline

Statoil disclosed plans for a third major gas pipeline from the Norwegian North Sea to Northwest Europe.

The 403 mile, 40 in. line is to run from a riser platform on the Statpipe system in Block 16/11 to a landfall in Germany. Start-up date is set for Oct. 1, 1995.

Statoil said the line, with an initial capacity of 1.2 bcfd, will be required to handle increased demand for gas from European customers with existing purchase contracts covering Troll and Sleipner fields.

It also will transport associated gas.

Capacity of the new line could be increased considerably with compression.

Statoil, as head of the Norwegian gas sales committee, was in final stages of negotiations to sell as much as 480 MMcfd of offshore gas to Italy.

The new line, costing 8-9 billion kroner ($1.4-1.5 billion), will raise Norway's gas delivery capacity to 5.8 bcfd by 1995.

Currently there is a single link into the continent, the Statpipe/Norpipe system to Emden, Germany.

Design work has started on the Zeepipe system from Troll-Sleipner to Zeebrugge, Belgium, scheduled to open in autumn 1993.

Statoil is recommending that owners of the Zeepipe system finance and own the new pipeline.

By the middle of 1991 the project was to reach a stage at which a decision on financing is required.

Statoil hopes for final government approval by the end of 1991.

Mime field test production

Norsk Hydro, Oslo, started test production from Mime field in Norwegian North Sea Block 7/11.

Production from a single subsea well tied back to Phillips Petroleum Norway's Cod platform is about 3,000 b/d.

Partners in the project with Hydro 25% are Statoil 50%, Saga Petroleum 10%, Mobil Norway and Norske Conoco 7.5% each.

Tordis field development

Saga Petroleum, Oslo, and partners will develop Tordis oil field with a subsea production system (SPS) tied back to Gullfaks field 5.6 miles southeast.

Saga's proposals to use a floating production system (FPS) based on the successful Petrojarl design were rejected by its partners.

Saga was to present a development plan by yearend 1990 for the 100 million bbl field in Block 34/7 entailing an SPS with five producers and three water injectors.

If approved by summer 1991, Saga could have Tordis on stream in 1994.

Production from the $510 million project is expected to peak at about 56,700 b/d.

Crude oil and small quantities of natural gas will be processed on the Gullfaks C platform and transported through the Gullfaks tanker loading system.

Snorre development

Saga let two contracts for development of Snorre field in using a tension legged platform.

Rockwater AS will tow out the TLP and install it on Block 34/7 under a 140 million kroner ($23 million) contract.

Aker Drilling holds a $33 million, 5 year contract to drill production and water injection wells from the platform. Saipem SpA's Scarabeo 5 semisubmersible spudded the first of the Snorre wells that will be predrilled for production.

MANNESMANN DEMAG

Mannesmann Demag
Compressors and Pneumatic Equipment
P.O. Box 10 15 07, Wolfgang-Reuter-Platz
D-4100 Duisburg 1, Federal Republic of Germany
Tel.: Germany (2 03) 6 05-1, Fax: (02 03) 6 10 61-63

Compressors for the oil and gas industry

Process compressors are a traditional Mannesmann Demag field of activity. The Compressors and Pneumatic Equipment Division has contributed towards the dynamic development and the economics of chemical and petrochemical processes for a long time now. The product range includes single-stage and multi-stage compressors of centrifugal design, with horizontally or vertically split casings, geared turbocompressors, axial-flow and axial-centrifugal compressors. The degree of standardization we have selected ensures that specific designs meet the appropriate conditions. Ample proof of our international standing is demonstrated by the fact that we have supplied compressor installations all over the world, particularly to the oil and gas industry and the petrochemical industry, ranging from the tapping of energy resources through to the final further-processing stage.

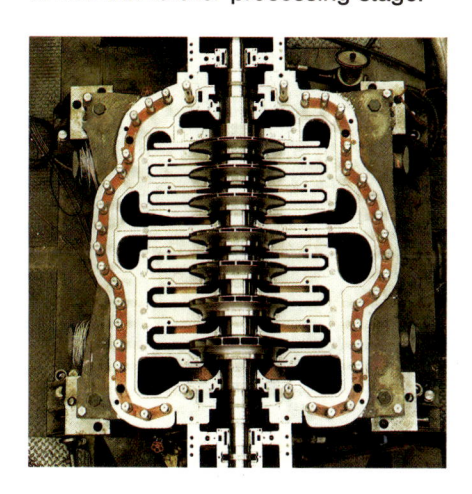

Compressors for gas processing

On offshore production platforms centrifugal compressors, installed as multicasing trains, gather natural gas from several wells and compress the separated product ready for transfer to the mainland. In onshore gas fields, where compressors are only required for a limited period at a particular location, mobile stations are used, i.e. single-stage blowers with complete peripherals or preassembled compact units with multi-stage compressors.

Compressors for gas liquefaction

In natural gas liquefaction plants, centrifugal compressors drive the refrigeration circuits. The compressors are multi-stage and, depending on the process involved, can have interstage sidestreams.

Compressors for pipelines

To compensate for pressure losses in pipelines, centrifugal compressors with fully automatic monitoring systems function with great reliability in booster stations no matter where they are located.

Compressors for petrochemical processing

The availability of an installation depends largely on the in-service behaviour of the compressors installed. The machines we have already supplied to all sectors of the petrochemical industry, in refineries, through to olefin and ethylene plants, demonstrate the reliability and soundness of Mannesmann Demag compressors.

We are represented by:
USA/Canada
Mannesmann Demag Corporation
Compression Equipment Division
1055 Parsippany Boulevard
Parsippany, New Jersey 07054
Tel.: (201) 402-5775
Fax: (201) 402-8452
Tlx.: Western Union 139006
Mannesmann Demag Corporation
1990 Post Oak Blvd., Suite 1800
Houston, Texas 77 056/USA
Tel.: (713) 960-1900
Tlx.: 77-5407

mannesmann technology

MANNESMANN ANLAGENBAU

Engineering + Construction

From Field to Product

Comprehensive and reliable solutions towards the full range of projects in the oil, gas, petrochemical and chemical industry is our business in many parts of the world.

Starting from the wellhead all the way downstream up to and including petrochemical and chemical plants, we offer our knowhow, experience and capabilities.

Our Spectrum

- Field gathering systems
- Oil and gas treatment
- Pump and compressor stations
- Pipeline systems
- Storage and loading facilities
- Gas treatment and extraction plants
- Refineries
- Ammonia plants
- Other chemical and petrochemical plants

For process plants, inhouse and acquired process knowhow are combined on a case-to-case basis, to provide the optimum solution for our clients' specific needs.

Engineering, procurement, construction and commissioning, provided by a single source, safeguard our clients' main concerns:

- safe, reliable and environmentally sound performance
- timely completion
- economical implementation and operation
- all encompassing guarantee

mannesmann *technology*

Mannesmann Anlagenbau AG
Theodorstraße 90
D-4000 Düsseldorf 30
Phone (211) 6 59-1
Fax (211) 6 59-23 72
Telex 8 586 677

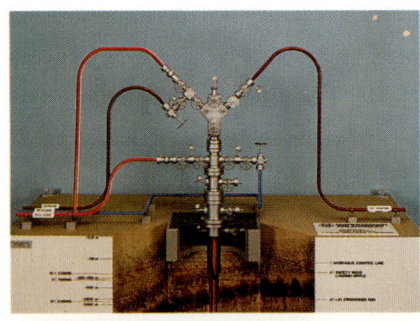

MANNESMANN RÖHRENWERKE

Mannesmannröhren-Werke AG
Postfach 1104, Mannesmannufer 3
D-4000 Düsseldorf 1, Fed. Rep. of Germany
Phone (211) 875-0, Fax (211) 875-3245
Telex 8 581 421

Mannesmann supplies a wide range of tubular products for the exploration, production, transmission, and processing of oil and natural gas, in compliance with all international standards and customers' special requirements.

Oil country tubular goods

Drill pipe, casing and tubing are available in all API sizes and grades as well as in special grades, e.g. arctic grades, special corrosion-resisting grades, or superhigh-strength grades.

Drill pipe
in OD's ranging from $2^{3}/_{8}$" to $6^{5}/_{8}$"
With weld-on tool joints according to API specifications or in special design

Casing
in OD's ranging from $4^{1}/_{2}$" to 26"
Thread types in accordance with API specifications or special gastight joints, such as BDS, MOS, HPC and MUST as well as: Omega, Mid Omega and Big Omega up to 26"

Tubing
in OD's ranging from $2^{3}/_{8}$" to $4^{1}/_{2}$"
Thread types in accordance with API specifications and special joints, such as MAT and gastight TDS, ST/C and ST/P

Our representative in the USA:
Mannesmann Oilfield Tubulars Corporation
1990 Post Oak Boulevard, 17th Floor
Houston, Texas 77056
Phone (713) 552-4069, Fax (713) 960-8631
Telex (3112) 27200144

Line pipe for oil and natural gas

Seamless line pipe
in OD's up to 28" in compliance with API Spec 5 L, ASTM, B.S., DIN, as well as other standards and special requirements

HFI-welded line pipe
high-frequency induction welded in OD's up to approx. 21" in compliance with API Spec 5 L, ASTM, B.S., DIN, as well as other standards and special requirements

All pipe is available PE coated with MAPEC®, the Mannesmann developed 3-layer corrosion protection system. Other types or corrosion protection and coating are also available, e.g. epoxy lining for gas lines.

Custom-bent steel pipe
made from seamless or welded steel pipe on a modern induction heat bending machine; in all customary steel pipe materials matching the high requirements of straight Mannesmann line pipe

Our representative in the USA:
Mannesmann Pipe & Steel Corporation
1990 Post Oak Boulevard, 18th Floor
Houston, Texas 77056
Phone (713) 960-1900, Fax (713) 960-1063
Telex (023) 166545

mannesmann technology

EUROPE/NORTH SEA

THE Claymore A Platform in the North Sea. Photo courtesy Smit International.

Ekofisk project

A group led by Phillips Petroleum Co. Norway started the second phase of a water injection project in Ekofisk oil field.

In addition, the group will step up waterflood operations with a third phase scheduled to begin in 1991.

The first two injection wells on Platform 2/4W in the main Ekofisk complex began injecting about 35,000 b/d.

Phillips plans to drill eight wells from the platform and inject a total of 120,000 b/d of water by early 1992.

Total cost of the second phase of injection is estimated at 2.2 billion kroner ($363 million). The investment is expected to contribute another 190 million bbl of production.

The second phase of the Ekofisk water injection project resulted from the success of the first phase, which involved installation of injection Platform 2/4K and drilling of 20 injection wells at a cost of 10 billion kroner ($1.65 billion).

The first phase is injecting 375,000 b/d of water. It is covering an estimated 65% of the main Ekofisk reservoir.

Oil production from the seven fields in the Ekofisk area had declined to 170,000 b/d in 1987 from a 624,000 b/d peak. August 1990 production was up to 240,000 b/d.

Phillips said about 40,000 b/d of the 70,000 b/d increase is the direct result of water injection. The rest came from workovers and new development wells. The third stage of water injection will boost Platform 2/4K's injection capacity to 500,000 b/d from 375,000 b/d of water. This will allow Phillips to inject water into the extreme southern portion of the reservoir and recover an additional 40 million bbl of oil at a cost of 950 million kroner ($156 million).

Gyda field start-up

BP Norway Ltd. UA started up Gyda oil and gas field in the summer of 1990, 9 months ahead of schedule.

Oil flow was expected to build to 60,000 b/d. Developing the 200 million bbl, 105 bcf Gyda reserve cost $1.29 billion. Oil and gas moves via pipeline into the Ekofisk complex and the Norpipe systems to Teesside, England, and Emden, West Germany.

Licensing round

Norway planned its biggest offshore licensing round since the first allocation in 1965.

The 13th round will cover 52 blocks—22 in the North Sea, 5 in the Norwegian Sea, and 25 in the Barents Sea. Applications were to close in the fall of 1990, and blocks awarded near the end of 1990.

The North Sea allocation was designed to make available prospects that, if productive, could be linked into the existing offshore platform and pipeline network.

Inclusion of the Barents Sea blocks demonstrates the Ministry of Petroleum and Energy's determination to maintain the exploration effort in this area despite lack of success so far.

Dutch North Sea

The first concrete gravity based structure in the Dutch North Sea will form part of a development program for F3 oil and gas field.

Nederlandse Aadolie Mij. BV found F3 in 1974. It extends under Blocks F2 and F6, bringing DSM-Energie, a Unocal group, and Elf Petroland into the development project.

The field is to be on production in July 1993.

Oil will move by pipeline to a tanker loading point, while gas will move to the Dutch mainland through the Northern Offshore Gas Transportation (Nogat) pipeline system. It is to be laid during the next 2 years.

The F3 production complex will consist of a three shaft, concrete gravity base, drilling and production platform linked by a bridge to a steel jacketed accommodation platform.

The base of the main platform will contain storage for 190,000 bbl of crude. One of the three shafts will contain utilities and facilities associated with oil storage. The other two will house a total of 16 well conductors.

A single process train will have a partial fractionation step in its 23,688 b/d oil stabilization section and a hydrocarbon dewpoint step in the 137 MMcfd gas treatment section.

The tanker loading structure will be positioned 1.24 miles from the main complex.

The Nogat pipeline, being developed by NAM and Petroland, will be laid and commissioned in two phases.

In the first phase a 93 mile, 36 in. line will be laid from Den Helder to the L2-FA/FB-1 platform and will be complete by October 1991.

The 66 mile, 24 in. second phase to the F3 complex is to be complete by October 1993.

Petroland is developing F15 field to be connected to the new line, and NAM will link the L12-L15 area and L2 field into Nogat.

Other fields will follow during the next few years.

One of the world's largest suppliers of quality petroleum products.

SOVIET FOREIGN ECONOMIC ASSOCIATION VVO "SOJUZNEFTEEXPORT" EXPORTS:

- Soviet export blend
- High-octane motor gasoline
- Straight-run gasoline
- Low-sulfur diesel fuel
- Straight-run fuel oil
- Wide range of lubricating oils and greases
- Petroleum coke and other petroleum products

VVO "Sojuznefteexport"

32/34, Smolenskaya sq., 121200, Moscow, USSR Tel. 253-94-88, 253-94-89
Telefax: 244-22-91 Telex: 411148A, 411148B, 411148C, 411148D, 411148E

Chronology of events for 1990

JANUARY
MARKETS: Average U.S. spot gas prices for the month jump to their highest level since July 1985 due to record cold throughout most of the U.S. in late December. Median price: $2.34/MMBTU vs. $1.69/MMBTU the previous month...Near-month futures price for light sweet crude reaches $23.27/bbl on the New York Mercantile Exchange (Nymex) at midmonth. The price a year earlier: $17.35/bbl. Prices subside as U.S. weather turns warmer in second half of the month.
RESTRUCTURING: Oryx Energy completes $1.1 billion purchase of oil and gas properties from British Petroleum...Roy M. Huffington Inc. transfers management of East Kalimantan, Indonesia, oil and gas joint venture to Virginia Indonesia Co., an affiliate of Union Texas Petroleum and Ultramar...Torch Energy Advisors completes purchase of Felmont Oil & Gas from Homestake Mining for $100 million...France breaks up the state owned petrochemical company, Orkem, and divides assets between Elf Aquitaine and Total.
LANDMARKS: Soviet Union resumes drilling, suspended in December 1983, of the world's deepest hole on the Kola Peninsula west of Murmansk. Tass reports rock sample recovery from 39,599 ft from the Kola SG-3, spudded in 1970 with a target depth of more than 49,000 ft.
PRODUCTION, RESERVES: API's yearend figures show 1989 U.S. oil production down 553,000 b/d—the biggest annual decline in history.

FEBRUARY
THE MARKET: Nymex price floats slightly above the $22/bbl mark most of the month. U.S. gas spot prices drop an average of 50¢/MMBTU from January.
PRODUCTION, RESERVES: OPEC Sec. Gen. Subroto urges Japan to help fund the exporter group's efforts to raise production capacity, saying OPEC needs $60 billion in foreign aid to hike capacity 6-7 million b/d by 1995...The Soviet Union reports crude and condensate production in 1989 of 12.14 million b/d, down 2.7% from 1988.
RESTRUCTURING: Imperial Oil receives regulatory approval for its $4.9 billion takeover of Texaco Canada.
SPILLS: The 82,300 dwt American Trader leaks 9,500 bbl of Alaska North Slope crude during lightering off Huntington Beach, Calif.
DRILLING: API reports U.S. oil and gas well completions dropped 12.8% in 1989 to 28,804.
GOVERNMENT: The U.S. DOE shifts oil research emphasis to maintenance of flow from wells most likely to be abandoned when prices drop...Venezuela says it will allow state owned Pdvsa to seek international participation in exploration and development for the first time since it nationalized the oil industry in 1976...Canada, as parts of a deficit reduction effort, plans to sell Petro-Canada and cancel federal support for the $4.1 billion OSLO oil sands development project in Alberta.
LANDMARKS: API reports that U.S. crude and product imports reached a record monthly average of 9.133 million b/d, 54% of demand.
ALTERNATE FUELS: Pacific Gas & Electric opens California's first public compressed natural gas auto fueling station.

MARCH
THE MARKET: OPEC crude production averages 23.7 million b/d—8% above the group's quota. An OPEC monitoring committee resists Iraqi demands for a higher price, leaving intact the 22.1 million b/d group ceiling and $18/bbl target price. Nymex price sinks below $20/bbl.
SPILL RESPONSE: Several companies move tanker routes 10 miles seaward of the Florida Keys...Exxon increases to $1.68 billion its 1989 writedown related to the March 1989 Exxon Valdez oil spill off Alaska...Exxon indefinitely suspends tanker and barge operations supporting its 130,000 b/d Bayway refinery at Linden, N.J., in response to three product spills since Jan. 1...BP America, accepting responsibility for the Feb. 7 spill off Huntington Beach, completes its cleanup after spending more than $19 million and collecting 9,000 tons of oil debris. The Coast Guard calls the 35-37% on-water oil recovery rate a world record.
ALTERNATE FUELS: Sun Refining & Marketing opens a retail methanol dispensing pump in Washington, D.C., first in the U.S. outside California.
LANDMARKS: Midland Cogeneration Venture, world's first conversion of a nuclear power plant to a gas fired cogeneration plant, starts commercial operation in Midland, Mich.

APRIL
MARKETS: With OPEC output reaching 24 million b/d, officials of Saudi Arabia, Kuwait, and Abu Dhabi meet to discuss falling oil prices. Kuwait and Abu Dhabi promise production cuts. Nymex price drifts below $18/bbl, but slide stops when OPEC calls for emergency meeting in May.
LANDMARKS: Iran resumes gas exports to the U.S.S.R. through a 1,300 km pipeline from southern fields to Astara on the Caspian Sea after an 11 year suspension...Placid Oil and partners shut down pioneering Green Canyon Block 29 floating/subsea drilling and production system in the Gulf of Mexico, citing disappointing flow rates.
ALTERNATE FUELS: Corpoven plans to install in Caracas Venezuela's first commercial pump to dispense compressed natural gas as a motor fuel...Pemex plans to invest $1 billion in an ecological program aimed at adding oxygen to and reducing pollution from vehicle fuels.
SPILL RESPONSE: U.S. Congress' Office of Technology Assessment says capability to respond to marine oil spills hasn't improved since the Exxon Valdez accident.
RESTRUCTURING: Gulf Canada and Home Oil agree to merge in a $492 million deal.

MAY
THE MARKET: OPEC members agree to reduce production by a total of 1.445 million b/d. U.K. North Sea output declines as operators begin their most extensive summer maintenance shutdowns ever. Nymex price climbs to nearly $20/bbl by midmonth.
GOVERNMENT: Argentina, privatizing state run YPF, plans to open major oil fields to participation by international companies.
LANDMARKS: U.S. Baker Hughes rig count in the month's last week tops 1,000 for the first time since January 1988...Lagoven claims Latin American drilling depth record at 1 Amarilis, drilled to 19,501 ft 20 km southeast of Maturin, Venezuela...Elf Aquitaine announces what it describes as the Soviet Union's first "classical oil development contract" signed with a foreign company, covering two onshore areas.
PRODUCTION, RESERVES: Chevron submits plans for development of Iagifu-Hedinia fields in Papua New Guinea, setting stage for the country's first oil production...Long term production test in Troll gas field in the Norwegian North Sea establishes significant oil production capacity.
ACCIDENTS: Explosion and fire kill one man and burn two others at Petro-Canada Products Inc.'s Edmonton, Alta., refinery.

JUNE
MARKETS: IEA reports OPEC overproduction continued in May

despite the OPEC agreement. Prices skid toward $15/bbl on Nymex.
LANDMARKS: Chevron signs protocol of intentions to add giant Tengiz oil field to a proposed Soviet joint venture covering the Northeast Caspian Sea region...Norsk Hydro claims North Sea record for most production from a single installation: 356,719 b/d from eight wells in Oseberg platform in a 24 hr test...Texaco says it drilled the first documented horizontal hole in the Gulf of Mexico—B-12 OCS G-094 in 180 ft of water drilled from East Cameron Block 265 Platform B.
PRODUCTION, RESERVES: A Statoil study doubles the estimate of Norwegian undiscovered oil and gas resources to 5.9 billion bbl and 31.95 tcf...Official figures show U.S.S.R.'s crude and condensate production averaged 11.5 million b/d in May, lowest level in 11 years.
ACCIDENTS: The Norwegian Mega Borg tanker explodes and burns 57 miles off Galveston in the Gulf of Mexico, killing two, leaving two missing and presumed dead, and injuring 17. The ship spills 71,400 bbl of Angolan crude.
SPILL RESPONSE: Elf, Shell, and others refuse to allow their tankers to use U.S. mainland ports for fear of unlimited oil spill liability.
GOVERNMENT: U.S. EPA further reduces allowable gasoline volatility—to 7.7 psi Reid vapor pressure in warm states and 9 psi elsewhere during summer...President Bush cancels eight sales of OCS leases off California and Florida.

JULY

MARKETS: Prices rise before a monthend OPEC meeting as overproducing members promise to trim output. The group agrees to a higher quota of 22.5 million b/d and a $21/bbl target price. Nymex price exceeds $18/bbl.
SPILL RESPONSE: Congress rejects U.S. participation in international tanker accords limiting tanker owners' liability for spills...Exxon returns the ill-fated Exxon Valdez tanker to service outside the U.S., renamed Exxon Mediterranean.
ACCIDENTS: ARCO Chemical Channelview, Tex., petrochemical complex explodes and burns, killing 17.
LANDMARKS: Mobil's High Island A-382 F-17 well off Texas flows 9,300 b/d of oil and 14 MMcfd of gas, which the operator calls the Gulf of Mexico's biggest flow from a single completion.

AUGUST

MARKETS: Crude oil prices leap after Iraq invades Kuwait. The U.S. sends troops to Saudi Arabia. An international embargo cuts off an estimated 4.2 million b/d of oil exports from Iraq and Kuwait. Nymex crude price leaps past $30/bbl.
SPILL RESPONSE: Congress passes an oil spill liability bill requiring double hulls on tankers and barges and allowing state liability laws to preempt federal statutes.
PRODUCTION, RESERVES: Saudi Arabia produces 2 million b/d of crude beyond its 5.38 million b/d quota to help make up for lost Iraqi, Kuwaiti supplies. Later, OPEC waives quotas.
GOVERNMENT: EPA requires an 80% reduction in sulfur content of U.S. diesel fuel...Australia replaces excise taxes with a resource rent tax for Bass Strait fields and takes other tax steps to boost exploration and production.
LANDMARKS: Soviet Union opens two onshore areas for the first competitive bidding of exploration and development rights in its history.

SEPTEMBER

MARKETS: Nymex price rises on rumors of war in the Middle East, falls on rumors of peace, fluctuating around the $30/bbl level and climbing at month's end.
SPILL RESPONSE: U.S. oil companies launch Marine Spill Response Corp., successor to the Petroleum Industry Response Organization, to set up a spill prevention and cleanup system.
LANDMARKS: NAM plans the Dutch North Sea's first concrete gravity-based structure, a production complex for F3 oil and gas field.
GOVERNMENT: President Bush announces a 5 million bbl test drawdown of the Strategic Petroleum Reserve.
PRODUCTION, RESERVES: EIA reports that U.S. crude reserves declined by 324 million bbl to 26.501 billion bbl at yearend 1989. Gas reserves declined by 908 bcf to 167.116 tcf, NGL reserves by 469 million bbl to 7.769 billion bbl.

OCTOBER

MARKETS: Nymex price exceeds $40/bbl early in the month, when war rumors dominate trading pit gossip. Later, conciliatory gestures by Iraq's Saddam Hussein send the price below $30/bbl...IEA says there's no shortage but points out the oil supply system is working at capacity.
GOVERNMENT: Leaders of the U.S.S.R.'s former European satellites request more oil from the Soviet Union, which cut subsidized exports in order to raise hard currency and ease fuel shortages of its own...Congressional conferees agree on Clean Air Act revisions that set reformulated gasoline specifications and oxygen content standards...Congress passes a budget bill raising U.S. gasoline and diesel taxes by 5¢/gal each and providing modest exploration and production incentives.
LANDMARKS: Myung & Associates, Tulsa, becomes the first U.S. company to sign an onshore production sharing contract in China...DOE delivers its first crude from SPR, part of the 5 million bbl test sale.
PRODUCTION, RESERVES: Prudhoe Bay interest owners approve a $1.1 billion project to expand gas handling capacity, which will boost field's ultimate liquids recovery by 330-450 million bbl.

NOVEMBER

MARKETS: War fears give way to concern about price collapse once Kuwait and Iraq return to market. Nymex price drops $4/bbl during month to $30.50-31.50/bbl.
PRODUCTION, RESERVES: Saudi oil minister says his country's production exceeds 8.2 million b/d and will reach 8.5 million b/d by early 1991...NAM adds 9.88 tcf to its estimate of reserves in Groningen gas field in Netherlands, pushing total remaining to 51.9 tcf...Chevron and partners approve midyear 1991 start-up of Point Arguello oil field off California, suspended 3 years by environmental controversies, at a limited rate of 20,000 b/d...Venezuela says it's studying a production capacity increase to 4.2 million b/d by 1996, up from a 3.5-3.65 million b/d target announced earlier.
GOVERNMENT: After a 13 month inquiry undertaken in response to the July 1988 Piper Alpha explosion and fire, a report by a Scottish judge calls for sweeping changes in oil and gas operations in the U.K. North Sea...FERC approves the 370 mile Iroquois pipeline to carry gas from Canada to the U.S. Northeast.
RESTRUCTURING: British Gas wins final approvals for its $1.1 billion (Canadian) takeover of Consumers Gas of Toronto...Amax Oil & Gas agrees to buy Ladd Petroleum from General Electric for $515 million...The Home-Gulf Canada merger stalls as stockholders of Home parent Interhome consider spinoff of the subsidiary...Arkla and Diversified Energies Inc. (DEI) shareholders approve a $630 million merger of DEI into Arkla...Italy's Montedison sells its 40% stake in Enimont to ENI for $1.27 billion, returning Enimont to full state control.

DECEMBER

MARKETS: Nymex price falls below $30/bbl as the market's shaky balance shows staying power...IEA cuts by 1.2 million b/d its estimate of fourth quarter demand in market economies...Despite frigid North American weather, gas spot and futures prices drop below $2/MMBTU.
PRODUCTION, RESERVES: Esso Resources Canada plans to reactivate part of its Cold Lake, Alta., heavy oil project mothballed in 1988 due to low oil prices.
LANDMARKS: Iran extinguishes the 5 Dehloran blowout in Ilam field, which had burned out of control for 3 years.
PROJECTS: Coastal suspends plans to lay a 670 mile gas pipeline from Wyoming to California.
RESTRUCTURING: RWE-DEA of Germany makes cash tender offer worth $590 million for Vista Chemical.

Indonesia, Japan key players in LNG activity

INDONESIA AND JAPAN WILL REMAIN THE KEY PLAYers amid a rising volume of world trade in liquefied natural gas.

As 1990 drew to a close, Indonesia was pushing its campaign to step up shipments and retain its title as the world's No. 1 exporter of LNG.

Heavily industrialized Japan must depend on tanker shipments of LNG for nearly all of its supply of gas, an environmentally friendly fuel. Japan was destined to receive more LNG from Alaska and other regions.

Among new and planned supply sources, Australia was increasing its LNG shipments to Japan, Venezuela was moving to launch an ambitious LNG export project with foreign partners, and Nigeria was homing in on a similar operation.

The result will be a growing number of LNG cargoes on the oceans.

Outlook for world LNG, LPG trade

Among those predicting strong growth in seaborne shipments of LNG—and liquefied petroleum gas, as well—is Drewry Shipping Consultants Ltd.

The outlook for liquefied gas trading has rarely been better, the London firm said in a June 1990 report.

Drewry estimated world seaborne trade in LNG would rise to 52 million metric tons in 1990 from 45 million in 1989 and only 40.9 million in 1987. This rapid growth will continue with shipments totaling 62.5-73.5 million metric tons in 1995 (Table 1).

Growth in LPG seaborne trade was not seen as brisk.

Drewry reckoned LPG shipments would reach 31.1 million metric tons in 1990, based on results for the first quarter of the year. Shipments had built up from 26.2 million metric tons in 1987 and should reach 35 million in 1995.

Drewry's report acknowledged that demand for LNG has been stimulated by increasing concern about the environment, which has highlighted the clean burning properties of gas.

In Japan, Drewry's figures showed, LNG imports have risen from 15 million metric tons in 1983 to 32 million metric tons in 1989. Projections are for imports of 40-50 million metric tons/year by the turn of the century.

Other Far East growth markets are South Korea, where imports will grow from 1.5 million tons in 1987 to 4-6 illion tons by the end of the century.

Taiwan, which began importing LNG in March 1990, could be importing 3-5 million metric tons/year by 2000.

Drewry said in the Atlantic basin the bright outlook for LNG trade was tempered by realization that it faced stiff competition from other energy sources, notably pipeline gas.

In the U.S., Drewry said, the gas surplus is a thing of the past. Gas utilities are becoming increasingly concerned about securing supplies, which could lead to opportunities for LNG imports.

Nigeria's LNG export project is scheduled to get under way in 1995, but few other projects in the planning stage are likely to be operational before the late 1990s, Drewry predicted.

In the meantime, deliveries under existing contracts are likely to rise—if shipping capacity is available. But Drewry said examination of voyages completed by LNG carriers in 1989 revealed that under most contracts there is very little room to increase deliveries, given the size of the current LNG tanker fleet.

Employing a small number of idle LNG tankers would provide some added tonnage, but the report said in the long

BEKALANG TANKER takes on a cargo of LNG at Lumut terminal in Brunei. Photo courtesy of Royal Dutch/Shell Group.

LNG

Indonesia's oil, gas balance of payments*

	Fiscal 1987-88 at $17.50/bbl	Fiscal 1988-89 at 16/bbl$	Fiscal 1988-89 at 15/bbl	Fiscal 1988-89 at $14/bbl$
	\multicolumn{4}{c}{Billion U. S. $}			
Net oil	2.334	2.229	1.532	1.431
Net LNG	1.426	1.345	1.483	1.298
TOTAL	**3.760**	**3.574**	**3.015**	**2.729**

*Exports less imports and srevice payments for oil, srevice for gas. †Actual oil price.
$Budgeted oil price. Source: Bank Indonesia

Fig. 1

run newbuildings will be required if trade is to expand at the forecast rates.

Drewry cited 1989 as a golden year for LPG shipping companies with freight rates at record highs, reflecting, in part, continued growth in seaborne trade.

Other positive factors for carriers were a static fleet at 7.8 million cu m capacity and the increasing concentration of vessel ownership in the hands of a small number of companies.

The report said if the volume of tanker orders in 1989 is not repeated in 1990, projected increases in LPG trade will absorb capacity on order.

As a result, freight rates for most sizes of LPG tankers appeared set to remain comparatively buoyant for the next 1-2 years.

Indonesia drives to hike shipments

Faisal Abdaoe, president of state owned Pertamina, late in 1990 told newsmen in Jakarta Indonesia's LNG exports will total 20.7 million tons for the year, then rise to 21.7 million tons in 1991 and 24 million tons/year by 1994.

The occasion was a signing ceremony for a contract in which Pertamina agreed starting in 1994 to ship another 2 million tons/year of LNG for 20 years to Osaka Gas, Tokyo Gas, and Toho Gas, three of Japan's major LNG importers.

Under earlier contracts, Indonesia shipped 16.5 million tons of LNG to Japan, 2 million tons to South Korea, and 1.5 million tons to Taiwan in 1998.

The additional LNG to be supplied to Japan will come from Bontang, East Kalimantan, where Pertamina will add a production train with design capacity of 2.3 million tons/year. In cooperation with foreign oil and gas companies, Pertamina operated five LNG trains in Bontang and six in Arun, Aceh.

The new train at Bontang will cost $637 million, to be obtained in the form of loans. Eight companies from Japan, the U.S., and France have shown interest in providing required credits, Abdaoe said.

Pertamina also was negotiating a short term and a long term contract to sell Korea Gas Co. 44 million tons of LNG during 21 years.

Talks also were under way on an agreement to ship Singapore 140 MMcfd of LNG for 15-20 years from Block D in Natuna field.

There also were plans to increase Indonesia's LNG shipments to Taiwan by at least 700,000 tons/year, Abdaoe said.

Among Indonesia's newest projects, PT Badak NGL Co. in summer 1990 started up the fifth 2.3 million ton/year gas liquefaction train at the Badak LNG plant on the east coast of East Kalimantan at a cost of about $294 million. PT Badak capacity is 11.5 million tons/year of LNG and 400,000 tons/year of LPG.

Chiyoda Corp. handled engineering design and construction, and Union Carbide Co. provided its solvent gas treating technology for the plant.

Prior to the gyrations in oil prices set off by the Aug. 2, 1990, seizure of Kuwait by Iraq, Indonesia expected LNG and LPG exports to surpass oil exports as the biggest source of its foreign exchange in the 1990s.

As Indonesia's crude oil reserves dwindle, foreign demand for its LNG is expected to accelerate. The nation was stepping up the search for gas to expand its share of world LNG exports.

Indonesia accounted for a little more than 40% of world LNG exports in 1988. That's the last full year for which data were available when the American Gas Association published its November 1990 Gas Energy Review (Table 2).

In an April 1990 report, Bank Indonesia said although LNG accounted for less than one third of petroleum export revenues in fiscal 1988-89, almost half the petroleum sector's total contribution to net foreign exchange earnings came from LNG. The government forecast that relationship would stay the same in fiscal 1989-90, based on a very conservative budgeted oil price of $14/bbl (Fig. 1).

"However, with additional LNG sales being negotiated and facilities being expanded, LNG export earnings may surpass those of oil on a net basis next fiscal year and in gross export revenues within 5 years," the bank said.

Indonesia also will continue to increase its use of indigenous gas.

Total domestic use of Indonesian gas is expected to jump to 1.7 bcfd by 2000 from 1990's 850 MMcfd.

Notable in that scenario are the country's efforts to develop a domestic petrochemical industry to trim surging imports. A group of Indonesian banks, along with the Bank of Japan, agreed to fund construction of a $200 million plant at Cilegon, West Java, to produce 200,000 metric tons/year of polypropylene. That will help back out imports costing about $800 million/year.

More Indonesian exploration is targeted toward gas, notably near Natuna Island in the South China Sea, where about half of Indonesia's 90 tcf of gas reserves lies.

During 1982-88, Indonesian gas production climbed to almost 5 bcfd from about 3 bcfd.

Indonesia's LPG exports to grow

In 1988, Indonesia positioned itself to become a major exporter of LPG with start-up of new Arun and Bontang fractionators at a cost of more than $800 million.

The two plants, with a combined capacity of 2.25 million metric tons/year, will provide 1.95 million tons/year to Japan under a 10 year contract.

Indonesian LPG production jumped by 61% to 1.25 million tons in 1988. During first half 1989, Indonesian LPG exports totaled 1.24 million tons. With the first full year of expanded production, LPG exports could reach 2.7 million tons.

Before the Arun and Bontang LPG plants were completed, Indonesian LPG production was limited to four LPG recovery units in oil fields with associated gas and four LPG recovery units at refineries operated by Pertamina.

To upgrade the refinery LPG for export, Pertamina and a group of Japanese companies installed a $68 million, 450,000 ton/year fractionator in 1986 at Tanjung Uban, Bintan Island. The Tanjung Uban unit was last reported operating below capacity because of limited refinery LPG feedstock.

Indonesia had more LPG capacity under construction and planned further capacity additions. Its LPG production capacity was expected to increase to 3 million tons/year in the early 1990s, according to government forecasts.

Indonesia's government also was pressing a campaign for increased domestic consumption of LPG to back out kerosine use. Domestic consumption of LPG had increased fivefold in the preceding 10 years.

A World Bank study estimated that by 2000 LPG will replace 20% of Indonesia's kerosine demand, currently about 123,000 b/d. Government estimates peg domestic LPG demand at 350,000 tons in fiscal 1990 and 464,00 tons in fiscal 1993.

Japan to keep role as No. 1 LNG importer

Japan will retain its title as the world's top consumer of LNG for the next 20 years, the July 1990 issue of Geopolitics of Energy predicted.

Meanwhile, said the monthly publication of Conant & Associates Ltd., Washington, D.C., the U.S. LNG market will remain stagnant through 2010, and Europe's may even decline.

Both of those regions will depend more on domestic gas resources and on pipelines.

Overall, the volumes of internationally traded gas moved through pipelines will far exceed those shipped as LNG by tanker because the U.S. and Europe hold reserves of their own or have access to transborder gas pipelines.

Japan, however, receives about 70% of the world's LNG exports and uses half of the world's LNG tankers. Geopolitics of Energy said it is unlikely Japan will greatly reduce its share because there are no major gas reserves in East Asia.

In fact, Korea, Taiwan, and China also are likely to increase their shares of global LNG consumption, although to a lesser extent. By 2000, Asian demand is expected to total 56.2 million tons/year, World Gas Intelligence reported, but only 38.2 million tons/year of production capacity were on order or planned.

Japan is forecast to require 57 million tons/year of LNG by 2010 (Table 3).

Korea, Taiwan, and China could require a total 75 million tons/year by then.

Although Indonesia, Malaysia, and Brunei are likely to remain Japan's largest sources for LNG, Australia is making a big push to step up its exports.

Australia has spent $14 billion (Australian) on a project to make gas exports the country's major hard currency source in the 1990s.

It will have seven LNG carriers shipping to eight Japanese utilities, with intent to meet 15% of those companies' requirements by the mid-1990s.

LNG from the Yukon Pacific project proposed to tap Alaska's North Slope gas is deemed a comparatively expensive source, and supply was still uncertain, Geopolitics of Energy said. Japan may "reluctantly" agree, however, to accept deliveries of North Slope LNG out of deference to U.S. concerns about its trade deficit.

Other Japanese supplies will come from more distant sources such as Abu Dhabi and Qatar.

Tokyo's persistent, wide ranging search for gas sources stems from the costliness of the LNG industry.

It is a risky business, Geopolitics of Energy pointed out, and could become much riskier if LNG evolves into another spot trade.

"But the risks are less with Japan, which is likely to calculate its needs carefully and maintain competition between sources."

It is doubtful that LNG imports by the U.S. will play a much more important role in the next 2 decades, Geopolitics of Energy said. If supply and demand forecasts are not off by more than 1-2 tcf, the U.S. will need no more LNG than its present terminal capacity of 800 MMcfd, or about 3-4% of total gas demand. The reason, Geopolitics of Energy said, is that the U.S. gets at least 80% of its gas supplies from domestic sources and almost all of the rest from Canada via transborder pipelines.

Europe, with large gas reserves of its own, is close to huge Soviet reserves and to potential supplies in the Middle East, also deliverable by pipeline.

In addition, seabed pipelines developed by Algeria, Italy, and France will increase the efficiency of gas supplies from North Africa and eventually the Middle East. Seabed pipelines, said Geopolitics of Energy, may have permanently changed the logistics of gas trade in Europe.

"It is conceivable that after the next several decades, LNG to Europe may be only a memory, as pipelines connect the continent to oil suppliers, especially to those in the Persian Gulf."

Australian exports to Japan

Shipments of Australian LNG to Japan were scheduled to increase with the maiden voyage late in 1990 of Northwest Snipe, the fourth ship in the North West Shelf Project's planned seven tanker fleet. Gas production to feed the project moves by pipeline from Platform A in North Rankin field off Western Australia.

Northwest Snipe was built at Chiba shipyard of Mitsui Engineering & Shipbuilding. It is the newest in the world fleet of LNG tankers (Table 4).

Partners in the project's LNG phase, each holding a 16⅔% interest, are operator Woodside Petroleum Ltd., BHP Petroleum (North West Shelf) plc, BP Developments Austra-

Table 1

What's ahead for seaborne trade in...

	1987	1988	1989	1990	1995
...LNG...					
	—————— Million metric tons ——————				
Total tanker shipments	40.9	44.2	45.0	52.0	62.5-73.5
Imports					
Japan	29.1	31.0	32.4	34.5	37.5-42.5
U.S.	0.1	0.4	0.9	2.4	5.0-7.0
Europe	10.2	10.9	9.8	12.0	15.0-18.0
Average number of idle tankers	16	14	9	7	—
	————————— Million U.S. $ —————————				
Newbuilding price*	120.0	145.0	175.0	220.0	NA
...and LPG					
	—————— Million metric tons ——————				
Total tanker shipments	26.2	26.9	28.1	31.1	35.0
Imports:					
Japan	12.7	12.7	14.0	15.2	16.4
Western Europe	7.5	7.0	7.1	7.3	7.9
U.S.	2.8	3.0	2.5	3.2	4.8
	———————— Million cu m ————————				
Tanker supply†	7.6	7.7	7.8	9.2	9.6
Tanker demand†	6.6	6.9	7.4	8.0	8.9
Average time-charter rate					
	——————— $1,000 (U.S.)/month ———————				
75,000 cu m	580.0	695.0	1,130.0	1,800.0	NA
12,000 cu m	330.0	450.0	550.0	530.0	NA
Average new building price					
	————————— Million U.S. $ —————————				
75,000 cu m	51.0	58.0	66.0	71.0	NA
12,000 cu m	23.0	27.0	34.0	37.0	NA

*125,000 cu m capacity tanker. †Ships of more than 1,000 cu m capacity.

Source: Drewry Shipping Consultants Ltd.

LNG

TELLIER tanker, rated at 40,000 cu m capacity, unloads the 2,000th cargo of Algerian LNG at Gaz de France's Fos sur Mer terminal in southern France in April 1990. Inset photo at upper left is a view of the terminal from the port side of the ship. Deliveries from Algeria's Skikda liquefaction plant started 18 years previously. Three LNG tankers shuttling between the North African plant and Fos sur Mer met more than 10% of France's gas requirements in 1989.

lia Ltd., Chevron Asiatic Ltd., Japan Australia LNG (MIMI) plc, and Shell Development (Australia) plc.

Its present fleet consists of the Northwest Swallow, which took on its first LNG cargo from Burrup Peninsula in December 1989, Northwest Sanderling, and Northwest Swift. As of May 1990, the three ships had transported a total 32 cargoes of LNG.

A fifth vessel, Northwest Shearwater, under construction at the Sakaide yard of Kawasaki Heavy Industries, was scheduled for delivery in September 1991. The sixth and seventh ships, Northwest Sandpiper and Northwest Seaeagle, are planned for launch in first quarter 1993.

The LNG phase was moving an average of one tanker/week to eight utilities in Japan following start-up in 1998 of its liquefaction plant.

LNG Trains 1 and 2 were producing 5,000-6,000 tons/day, depending on shipping needs.

Attention of the North West Shelf Project had shifted to the $2.7 billion Phase 3, which includes construction of $1.1 billion LNG Train 3.

Train 3 is scheduled for completion in first half 1993. When finished, it will boost capacity of the LNG plant to 6 million tons/year.

In April 1990, the North West Shelf Project let four contracts totaling $228 million (Australian) for construction of Goodwyn A production platform. The platform is to be the centerpiece of a $1.6 billion development program in Goodwyn field, also off Western Australia.

Abu Dhabi slates expansion

Elsewhere among LNG suppliers, Abu Dhabi Gas Liquefaction Co's plans for a $1.5 billion expansion of its Das Island LNG chain drew approval by the Abu Dhabi Supreme Petroleum Council in July 1990.

A new train will produce 2.3 million tons/year of LNG and 250,000 tons/year of LPG. Capacity of the two existing trains is 2.3 million tons/year of LNG, 750,000 tons/year of LPG and 300,000 tons/year of condensate.

The expansion is due for completion in 1993-94. Talks for added sales were in progress with Tokyo Electric Co., which buys the entire LNG and LPG output from Das Island.

Cost of the expansion was estimated at $1.5 billion. About half of the total was to go for four new LNG tankers.

Abu Dhabi National Oil Co. has a 51% interest in the Das Island project in partnership with Mitsui & Co., British Petroleum Co. plc, and Total Cie. Francaise des Petroles.

More LNG due from Algeria

Algeria's state owned Sonatrach late in 1990 said a proposed contract to sell LNG to the U.S. through a joint venture between Shell Oil Co. and Columbia Gas System Inc. was in an advanced stage of negotiations.

The statement from Sonatrach followed a meeting in Algiers with senior executives from Shell International Gas Co. (SIG) and the two partners in the proposed new Cove Point Trading Co. joint venture, Shell Oil Co. and Columbia Gas.

Guidelines had been given to Sonatrach and SIG for speedy completion of a contract.

The meeting was scheduled before settlement of a long dispute involving ownership and sale of three tankers earmarked for use in U.S. LNG imports.

Earlier in the year, Sonatrach's deputy general manager, Mustapha Faid, said Algeria expects to have a gas export capacity of 2.1-2.8 tcf/year by the end of the century.

He said by 1992 LNG export capacity would be increased to 1.16 tcf/year from the current 883 bcf/year. In 1989 Algeria exported 607 bcf.

By the end of the century another 176 bcf/year of LNG capacity is to be in place. LPG capacity, currently 4 million metric tons/year, will be increased to 7 million tons/year.

In 1990 Sonatrach let three major contracts for refurbishment and expansion of its three largest gas liquefaction plants.

Contracts awarded to M.W. Kellogg Co., Bechtel International Inc., and Gaz de France (GdF) engineering affiliate Sofregaz carried a value of about $600 million.

Their initial aim is to restore the plants to their original combined capacity of 1.02 tcf/year. The contracts also provide for capacity expansion at the three plants to 1.23 tcf/year in the mid-1990s.

Poor maintenance procedures cut operations of the three plants to about 50-60% of the original design capacity.

Sonatrach has 21 trains of production capacity in four complexes at Skikda, Arzew, and Bethioua.

Algeria is GdF's largest gas supplier.

Table 2

World LNG exporting/importing countries*

EXPORTERS	1986 Volume (bcf)	1986 % of world trade	1987 Volume (bcf)	1987 % of world trade	1988 Volume (bcf)	1988 % of world trade
Indonesia	716.0	39.6	781.2	39.4	868.0	40.6
Algeria	423.8	23.4	495.2	25.0	523.7	24.5
Malaysia	242.3	13.4	282.9	14.3	291.7	13.7
Brunei	246.5	13.6	247.6	12.5	256.0	12.0
Abu Dhabi	103.5	5.7	101.4	5.1	112.3	5.3
U.S.	45.9	2.5	44.5	2.2	45.9	2.2
Libya	30.4	1.7	28.3	1.4	37.4	1.8
Total	**1,808.5**	**100.0**	**1,980.8**	**100.0**	**2,135.1**	**100.0**
IMPORTERS						
Japan	1,346.9	74.5	1,388.2	70.1	1,479.7	69.3
France	270.9	15.0	330.5	16.7	316.1	14.8
Spain	87.6	4.8	87.9	4.4	114.8	5.4
Belgium	91.5	5.1	100.6	5.1	101.7	4.8
South Korea	5.3	0.3	69.2	3.5	94.3	4.4
U.S.	2.1	0.1	—	—	20.1	0.9
Italy	—	—	—	—	6.7	0.3
U.K.	—	—	—	—	1.8	0.1
West Germany	4.2	0.2	4.2	0.2	—	—
Total	**1,808.5**	**100.0**	**1,980.8**	**100.0**	**2,135.1**	**100.0**

*Totals may not add due to rounding.
Source: American Gas Association from Cedigaz data

The three phase GdF contract, worth $176.6 million, covered the Skikda GL1-K LNG complex in eastern Algeria.

In the first phase Sofregaz will conduct a technical assessment of what needs to be done. In the second phase Sofregaz will return the unit to its original 275.4 bcf/year capacity during 1991-92. And in the final stage of the contract, capacity will be increased to 350 bcf/year in 1994.

The Bechtel contract covered a 1 year technical audit and subsequent engineering/construction management for the 370.8 bcf/year, six train, GL1-Z plant at Bethioua.

Bechtel estimated the project could require 4 years, but the time will depend on how much renovation will be needed.

Bechtel built the plant in the late 1970s.

Kellogg will revamp Arzew GL2-Z, which also has a design capacity of 370.8 bcf/year. The three phase project called for a detailed audit in the first phase to determine the scope of work to be included in the second and third phases, which are renovation and production enhancement with related training and improvement of systems, procedures, and organization for operation.

Kellogg's contract required multinational financing, which was being developed jointly by Sonatrach and Kellogg.

GdF Pres. Francis Gutmann said the Sonatrach contract was in line with the trend of developing partnerships between gas suppliers and customers worldwide. The French company offered 100% financing for its engineering contract.

In addition, Algeria was consulting GdF on the first stage of its proposed Algeria-Morocco-Spain gas pipeline in which France was not involved.

GdF also had 12 joint venture agreements under negotiation with the Soviet Union, its second largest gas supplier, involving gas transmission and distribution systems, as well as manufacture of equipment.

Gutmann said the Soviet Union might not be able to maintain its export volumes long term because of rapidly growing internal demand.

Nigerian LNG is the best prospect for future supply, he said. GdF is lined up to take 17.6 bcf/year from the first phase of the Nigerian project. Increased liftings will depend on prices.

Nigeria's LNG project advances

Nigeria's plans for an LNG export project took a big step forward in September 1990 when the U.S. Maritime Administration (Marad) and several companies agreed to settle a dispute centering on ownership and sale of three LNG tankers.

The aim of the project, to cost more than $2.5 billion, is to export about 4 million metric tons/year of LNG.

Marad's agreement covering the Arzew, Southern, and Gamma vessels was struck with units of Royal Dutch/Shell, Argent group, and Cabot Corp. Marad acquired the tankers by repossession from subsidiaries of El Paso Corp.

Here are key provisions of the settlement:

• All litigation among the parties was to be dismissed with prejudice.

• Gamma was to be sold to Cabot, which planned to charter it to Nigeria LNG Ltd. on a half-yearly basis after start-up of Nigerian LNG exports. Cabot was to have free use of the tanker at all other times. It planned to use the vessel for LNG shipments to the U.S. from Algeria and Nigeria.

• Arzew and Southern were to be sold to Argent, which planned to charter them to Shell.

• Total purchase price paid to Marad for each vessel was to be $18.1 million, including option fees previously paid to Marad.

• The vessels were to undergo refurbishment in U.S. shipyards and receive maintenance and repairs in the U.S.

• The vessels were to be dedicated to LNG trade between the U.S. and other countries.

The following October, Energy Transportation Group Inc. (ETG), New York, filed a federal lawsuit to block Marad's sale of the tankers. But U.S. Transportation Sec. Samuel Skinner denied ETG's request to review the sale agreement, and a federal district court judge declined to block the deal and dismissed ETG's suit.

The suit claimed Marad's sale agreement violated terms of the agency's original bid solicitation. The ETG suit also alleged several improprieties by Marad in reaching its accord with Argent, Cabot, and Shell.

LNG

Table 3
Japan's LNG sources

	Japan's imports 1988	Suppliers' export capacity 1988	1995	2000
		Million tons/year		
Alaska, Cook Inlet	1.0	1.0	1.0	1.0
Alaska, North Slope*	—	—	8.0	8-14
Australia	1.6	1.6	6.0	6.0
Abu Dhabi	2.2	2.2	4.3	4.3
Brunei	5.1	9.1	5.1	4.0
Malaysia	6.0	10.0	10.0	10.0
Indonesia	15.0	18.5	28-29	28-29
Total†	**31**	**42**	**62-63**	**61-68**
Forecast for 1995	38			
Forecast for 2000	46			
Forecast for 2010	57			

*Yukon Pacific project. †Total and forecasts rounded.

Source: Geopolitics of Energy

ETG's suit charged that terms under which Argent and Cabot agreed to buy the vessels violated Marad's original bid solicitation in part because Argent, although a separate entity from Shell, was simply acting as Shell's agent for the deal.

Royal Dutch/Shell in 1988 purchased an exclusive option to buy the LNG carriers, but federal law required the vessels to be sold to a U.S. citizen. Shell nominated Argent to be the U.S. citizen buyer in the deal.

The project's drive to assemble a fleet of LNG tankers got a boost in June 1990 when Nigeria LNG disclosed the purchase of two laidup LNG vessels. But until LNG shipments to Europe and North America start in 1995 the French built vessels Gastor and Nestor, each of 122,000 cu m capacity, were to be time chartered for use in the trans-Atlantic LNG trade.

The vessels, laid up since they were built in 1977, were purchased by Enellengee Ltd., a subsidiary of Nigeria LNG Ltd., using a loan under which it also had acquired two Swedish built LNG carriers in January 1990.

Nigeria LNG Ltd. was a partnership of project sponsors Nigerian National Petroleum Corp. 60%, Shell Gas 20%, and Agip SpA and Ste. Nationale Elf Aquitaine 10% each.

Nigeria LNG in 1989 let a contract to a combine of Technip and M.W. Kellogg Co. for project specification preparation for a liquefaction plant in Rivers State.

Norway again a possibility

Norway's state owned Den norske stats oljeselskap AS (Statoil) disclosed in October 1990 that negotiations were about to begin for the sale of LNG from offshore arctic Norway to ENEL, Italy's national electrical power company.

Statoil, which earlier in the year dropped its plans for LNG exports to the U.S. based on supplies from the North Sea, said ENEL was interested in LNG purchases based on Snohvit field on the Tromsoflaket.

The field, on the western edge of the Barents Sea in 900-1,110 ft of water, has 3.5 tcf of reserves and is capable of supplying an 8 billion cu m/year LNG export project. Negotiations with ENEL were to center on a project of 4-5 billion cu m/year.

ENEL was the second Italian state company negotiating for Norwegian gas. SNAM, part of the ENI group, was still talking to Statoil as head of a Norwegian gas negotiating group for as much as 5 billion cu m/year from the North Sea delivered to Northwest Europe through existing pipelines.

Talks for increased volumes of North Sea gas coincided with approaches from Algeria and Libya to boost deliveries from North Africa to Italy through proposed increased pipeline capacity under the Mediterranean.

Portugal's Petrogal also voiced interest in Norwegian LNG as supply for a natural gas grid proposed to link Lisbon with Oporto, Braga, and Coimbra.

Petrogal, in partnership with the Portuguese tanker company Soponata, Enagas of Spain, and SNAM of Italy, planned to build an LNG import terminal near Lisbon.

Petrogal said Algeria and Norway are under consideration as the source for 2-3 billion cu m required for the initial import supply.

In April 1990, Statoil disclosed it had dropped the proposal for a 2.5 billion cu m/year LNG export project based on gas from the Norwegian North Sea. The gas would have been sold to the U.S. through an agreement between Enron Corp. and Statoil.

Statoil said geological complexities in South Gullfaks field, the proposed supply source, and improved prospects for selling North Sea gas in Europe were responsible for the decision.

Venezuela's plans taking shape

Venezuela disclosed in June 1990 it had chosen Exxon Corp., Mitsubishi Heavy Industries Ltd., and Royal Dutch/Shell Group to participate in its $3 billion LNG export project.

It marked the first time international oil companies had been involved in an oil or gas project in Venezuela since the country nationalized its oil industry in 1976. Units of Exxon and the Shell group were nationalized.

The three companies were to be partners with Lagoven SA, a unit of state owned Petroleos de Venezuela SA. Tentatively, the interests broke out as Lagoven 32%, Shell 31%, Exxon 29%, and Mitsubishi 8%.

The Cristobal Colon LNG export project called for development of gas discoveries in the Gulf of Paria dating to 1978, installation of a 45 km subsea pipeline, and construction of a processing/liquefaction/tanker loading complex at Mapire Bay between the towns of Guiria and Macuro on Paria Peninsula.

Construction was scheduled to begin in 1992, with exports of 4.4 million metric tons/year of LNG, valued at about $500 million/year, to begin in 1996. Most LNG exports were expected to go to the U.S.

Plans called for Lagoven to drill 55 wells from four to eight platforms. From them, gas will move through the subsea pipeline to the complex on Mapire Bay

Pdvsa broke out project costs as about $1 billion for the platforms, wells, and pipeline, $1.3 billion for the processing/LNG complex, and $600 million for three LNG tankers. It was not immediately clear whether LNG carriers will be purchased or chartered for the project.

With wells producing an average 13 MMcfd, Pdvsa said it expected the Gulf of Paria fields to produce for 20 years. Pdvsa also said believed the project will be economic with oil prices at $20/bbl.

Because of the nationalization law, the project required approval by Venezuela's Congress.

LNG supplies from Alaska

Two events in 1990 heralded a buildup in LNG shipments from Alaska:

• In November, Yukon Pacific Corp., Anchorage, Alas., let contract to Bechtel Corp., Houston, for design of the first phase of its proposed gas liquefaction plant and marine terminal at Anderson Bay, Alas., near Valdez. It is to be part of Yukon Pacific's Trans-Alaska Gas System (TAGS) project to export LNG to Japan, South Korea, and China. Yukon Pacific said the plant will be the largest of its kind when all

LNG

Current chains in world LNG trade*

No.	Chain	Loading ports	Unloading ports	Distance (miles)	Tankers in operation
1	Algeria-U.K.	Arzew	Canvey Island	1,556	Methane Princess
2	Algeria-France	Arzew	Le Havre	11,428	Hassi R'Mel
		Skikda	Le Havre	1,780	
3	Algeria-France	Arzew	Fos sur Mer	526	Hassi R'Mel
		Skikda	Fos sur Mer	398	Tellier
					Descartes
					Snam Palmaria
4	Libya-Italy	Marsa el Brega	La Spezia	969	Snam Palmaria
5	Libya-Spain	Marsa el Brega	Barcelona	1,068	Laieta
					Snam Palmaria
					Snam Elba
					Havfru
					Isabella
6a	Algeria-Spain	Arzew	Barcelona	355	Isabella
		Skikda	Barcelona	349	Annabella
					Snam Elba
					Havfru
					Laieta
					Descartes
6b	Algeria-Spain	Arzew	Huelva	373	Annabella
		Skikda	Huelva	718	Isabella
					Cinderella
					Havfru
7	Alaska-Japan	Kenai	Negishi	3,240	Polar Alaska
					Arctic Tokyo
8	Brunei-Japan	Lumut	Negishi	2,380	Bebatik
			Sodegaura		Bekalang
			Senboku		Bekulan
					Belais
					Bilis
					Belanak
					Bubuk
9	Abu Dhabi-Japan	Das Island	Futtsu	6,425	Norman Lady
			Sodegaura		Hilli
					Gimi
					Khannur
					Golar Freeze
10	Indonesia-Japan	Bontang	Chita	2,400	LNG Aquarius
			Senboku		LNG Aries
			Himeji		LNG Capricorn
			Tobata		LNG Gemini
		Blang Lancang	Chita	3,200	LNG Leo
			Senboku		LNG Libra
			Himeji		LNG Taurus
			Tobata		LNG Virgo

phases are complete.

• In July, Phillips 66 Natural Gas Co. and Marathon Oil Co. signed a contract with Ishikawajima-Harima Heavy Industries Co. Ltd. (IHI), Tokyo, to build two LNG tankers. The new vessels will replace Phillips-Marathon's Arctic Tokyo and Polar Alaska LNG tankers, currently in service to contract customers in Japan.

In March 1990, Yukon Pacific Corp. disclosed it had signed a memorandum of intent to sell Korea Gas Corp. 2 million tons/year of Alaskan North Slope gas beginning in 1997.

The preceding November the U.S. Department of Energy approved Yukon Pacific's proposal to export the gas to Pacific Rim nations. DOE later affirmed that decision when it denied a rehearing request by Alaskan Northwest Natural Gas Transportation Co. and Foothills Pipe Lines (Yukon) Ltd.

They were sponsors of the Alaskan Natural Gas Transportation System (Angts) project to move Alaskan North Slope gas to the Lower 48 states.

Angts sponsors have built the southern third of their pipeline system but not a trunk line to move gas from the North Slope to Alberta, where the present system starts.

Yukon Pacific's $11 billion TAGS project included a chilled 800 mile pipeline paralleling the trans-Alaska crude oil line from the North Slope to Alaska's southern coast, a terminal and liquefaction plant at Anderson Bay, and LNG tankers. Completion target is 1996.

Alaska and the U.S. Interior Department granted rights-of-way for the pipeline, and a final environmental impact statement was been prepared.

William McHugh, Yukon Pacific president and CEO, said, "The leadership position taken by South Korea provides the TAGS project with the momentum we need to move forward in conducting negotiations with other market countries.

"South Korea is the second largest market we targeted. The memorandum is a tremendous endorsement of our proposal.

"We are working to obtain similar expressions from Japan and Taiwan."

McHugh said the project will provide the U.S. with an estimated $80 billion in trade revenues during 25 years.

He said, "Far East demand for LNG is expected to increase more than 60% by 2000. TAGS is one of the few

LNG

Table 4

No.	Chain	Loading ports	Unloading ports	Distance (miles)	Tankers in operation
11	Algeria-U.S.	Skikda	Everett	3,634	Mostefa Ben Boulaid Bachir Chihanij Larbi Ben M'Hidi Mourad Didouche Louisiana
12	Algeria-France	Arzew Skikda	Montoir de Bretagne Montoir de Bretagne	1,257 1,619	Ramdane Abane Edouard L.D. Tellier Hassi R'Mel
13	Malaysia-Japan	Bintulu	Sodegaura Higashi-Oghishima Futtsu	2,480	Tenaga Satu Tenaga Dua Tenaga Tiga Tenaga Empat Tenaga Lima
14	Indonesia-Japan	Blang Lancang	Niigata Higashi-Ohgishima Futtsu	3,530	Echigo Maru Kotowaka Maru Dewa Maru Wakaba Maru
15	Indonesia-Japan	Bontang	Chita Senboku Himeji Yokkaichi	2,400	Bishu Maru Banshu Maru Senshu Maru
16	Indonesia-Korea	Blang Lancang	Pyeong Taek	3,100	Golar Spirit Hoegh Gandria
17	Indonesia-Japan	Bontang Blang Lancang	Yokkaichi Yokkaichi	2,400 3,200	Asake Maru
18	Algeria-Belgium	Arzew	Zeebrugge	1,550	Methania Mourad Didouche
19	Australia-Japan	Withnell Bay	Sodegaura Chita Himeji Yanai Oita Senboku	3,670	NW Sanderling NW Swift NW Swallow
20	Algeria-Spain	Arzew Skikda	Cartagena Cartagena	113 387	Annabella Isabella Cinderella Havfru
21	Algeria-U.S.	Arzew	Lake Charles	4,962	Louisiana Mostefa Ben Boulaid
22	Algeria-Japan	Arzew	Sodegaura Senboku	9,179	Bachir Chihani

*As of May 1990.

Source: Institute of Gas Technology from data compiled by Groupe International des Importateurs de Gas Natural Liquefie

projects in the world able to satisfy such a large demand and the only U.S. project capable of achieving such dramatic reductions in our trade deficit."

The new Phillips-Marathon tankers will use IHI's SPB product containment system, which is a self-supporting prismatic tank built of aluminum.

Among the advantages of the design are greater safety, easier maintenance, and better maneuvering ability, Phillips said.

Each of the carriers is to have a capacity of 87,500 cu m of LNG, compared with only 71,500 cu m each for the Arctic Tokyo and Polar Alaska tankers they will replace. The two present tankers have been carrying LNG from Kenai, Alas., to Tokyo since 1969.

The first tanker is to be delivered in mid-1993, the second at yearend 1993.

Phillips-Marathon customers are Tokyo Electric Power Co. Inc. and Tokyo Gas Co. Ltd.

Phillips operates the liquefaction plant on Alaska's Kenai Peninsula, while Marathon operates the tankers. Phillips holds a 70% interest in the sales and facilities, Marathon a 30% interest.

U.S. terminal scheduled to reopen

Along the Chesapeake Bay in southern Maryland is a multimillion dollar investment Columbia Gas System wants to turn profitable.

A partnership of Columbia LNG and Shell Oil Co. plans to reopen Columbia's mothballed Cove Point, Md., LNG terminal. The $300 million, 1 billion cfd terminal is linked to a Columbia trunkline in northern Virginia by an 87 mile, $400 million pipeline with the same capacity.

Columbia and Consolidated Natural Gas Co. received the first shipment at the 1 billion cfd plant in March 1978, but shipments were suspended in April 1980 in a dispute over gas price. Columbia became sole owner in April 1988 after CNG walked away from its part of the investment.

The new Cove Point Trading partnership expected to sign purchase agreements for Algerian and Nigerian LNG before yearend 1990, enabling Columbia to reopen Cove Point by early 1993 if regulatory permits are issued.

A major stumbling block was removed when Marad agreed to the sale of three LNG tankers.

After LNG supply contracts were signed, Columbia LNG

LNG

TANKER LOUISIANA, owned by Panhandle Eastern Corp., offloads cargo of Algerian LNG at Trunkline LNG Co.'s regasification terminal in Lake Charles, La. Deliveries of Algerian LNG to Trunkline, a Panhandle Eastern subsidiary, resumed in December 1989. Trunkline began importing Algerian LNG at Lake Charles in September 1982 and halted the program in December 1983 because of the high cost of LNG vs. U.S. gas prices. Photo courtesy of Panhandle Eastern.

Pipeline in northern Virginia.

Columbia LNG said it hoped to have regulatory approvals by January 1992, then begin recommissioning the terminal and hiring staff.

That will take another year.

The plant is to be ready for its first shipment of Algerian LNG in late 1992 or early 1993.

Shell International Gas Ltd. was negotiating a 15 year Algerian LNG supply agreement, extendable to 20 years, for 220 MMcfd.

Nigerian LNG planned to ship 160 MMcfd to Cove Point after Nigeria completes a 600 MMcfd liquefaction plant in 1995 at Finima.

The Algerian deal was to involve 32 cargoes/year, the Nigerian contract 22 cargoes/year. Both were to be based on netback, giving Sonatrach 65%, tanker companies 15%, and Cove Point Trading 20% of the sale price of the regasified LNG.

Even though it was used only 2 years, construction of the Cove Point facility was a success story.

Operators had to reach major agreements with environmentalists to build the terminal, including construction of a park, protection of a freshwater swamp, and construction of a 1.2 mile tunnel, rather than a pier, to a tanker terminal in Chesapeake Bay.

Taiwan begins LNG imports

Taiwan's state owned Chinese Petroleum Corp. (CPC) started up its long-delayed $1.26 billion Yung-an LNG reception terminal in 1990. The first tanker carrying Indonesian LNG arrived Mar. 26 at the terminal on the island's southwest coast.

Until then, Taiwan had not imported LNG. The country planned to convert several oil or coal fired power plants to burn gas because of environmental concerns. Under an agreement signed in 1987, Indonesia will supply 1.5 million metric tons/year of LNG to the new terminal for 20 years.

The terminal was originally scheduled for completion in 1988. But a series of delays, including almost 140 work days lost because of blockades by fishermen protesting out of fear of damage to their fishing grounds, resulted in the project falling more than 1 year behind schedule.

CPC estimated Taiwan's demand for LNG will increase to 5.5 million metric tons/year by 1998. Taiwan currently requires 3 million metric tons/year for residential gas use and power generation, CPC said.

To accommodate the expected increase, CPC was drawing up plans to expand the terminal. That project, details of which were not disclosed, was scheduled for completion in 1995.

World oil supply/demand changes due to...

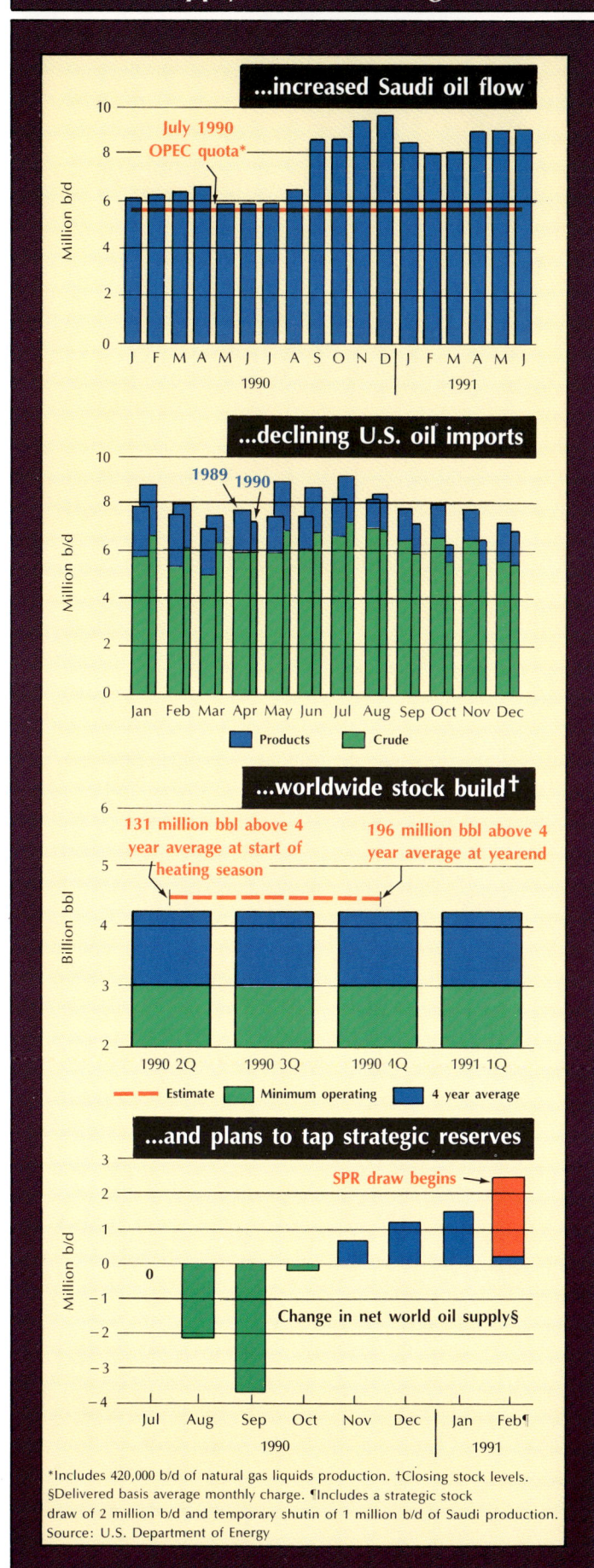

...increased Saudi oil flow
(July 1990 OPEC quota*)
Million b/d, monthly 1990–1991

...declining U.S. oil imports
Million b/d, 1989 vs 1990, by month
Products / Crude

...worldwide stock build†
- 131 million bbl above 4 year average at start of heating season
- 196 million bbl above 4 year average at yearend

Billion bbl, 1990 2Q, 1990 3Q, 1990 4Q, 1991 1Q
Estimate / Minimum operating / 4 year average

...and plans to tap strategic reserves
SPR draw begins
Change in net world oil supply§
Million b/d, Jul 1990 – Feb 1991

*Includes 420,000 b/d of natural gas liquids production. †Closing stock levels. §Delivered basis average monthly charge. ¶Includes a strategic stock draw of 2 million b/d and temporary shutin of 1 million b/d of Saudi production.
Source: U.S. Department of Energy

Mideast erupts in first war over oil

IRAQ'S 1990 INVASION OF KUWAIT TRIGGERED HISTO-ry's first war that began over oil.

Oil was a root cause for the invasion: Iraq violently disagreed with Kuwait's oil price and production policies and disputed the ownership of an oil field that straddled their common boundary.

Foreign oil workers became Iraq's hostages after the invasion but later were released.

When United Nations forces attacked Iraq Jan. 16, 1991, among the main targets for allied aircraft were Iraqi oil refineries, and Iraq shelled and launched rockets at Saudi Arabian oil facilities.

Oil was even made a battlefield weapon, as Iraq intentionally spilled crude in the Persian Gulf, placed oil in trenches as defenses, and blew up wellheads to create smokescreens.

The irony of the oil war was that the international oil market was little affected, even though the pre-war world oil market had been somewhat tight before losing a combined 3.5 million b/d in exports from the two countries.

For years, pundits had predicted another war in the Middle East would result in the third global oil shock—similar to the 1973-74 Arab oil embargo and the 1979 Iranian revolution.

That did not occur.

Oil prices rose sharply and then fell, without a major disruption.

The world oil market took the disruption in stride because members of the Organization of Petroleum Exporting Countries (OPEC) brought additional productive capacity on line and governments of consuming nations allowed the law of supply and demand, not bureaucratic decrees, to balance the market.

The invasion

Iraq attacked suddenly on Aug. 2, 1990, after Kuwait had rejected Iraqi demands in talks at Jeddah on Aug. 25.

The Iraqis demanded $2.5 billion compensation for crude it claimed Kuwait had taken from Iraq's South Rumaila field on the disputed border, demanded Kuwait forgive $10-12 billion in loans made to support Iraq's war with Iran, and

KUWAIT

AMERICAN F4G Weasels patrol over Bahrain on the Persian Gulf prior to the Jan. 16 launch of Operation Desert Storm, the allied attack designed to drive Iraq out of conquered Kuwait. Photo courtesy of U.S. Department of Defense.

demanded annexation of two Kuwaiti islands at the head of the Persian Gulf.

An underlying reason was the low price of crude on world markets, and Kuwait's reported overproduction of its OPEC quota.

Iraq's President Saddam Hussein wanted a $25/bbl marker price for OPEC crude, although OPEC had been unable to sustain an $18/bbl marker for most of the year.

Demand was soft and world oil stocks were high, offering little hope for higher prices for the final months of 1990.

Iraq's takeover of Kuwait, if unchallenged, would have won it a dominant role in OPEC. Iraq would have been a victorious and strong military presence, and able to influence two of the largest Middle East producers—Saudi Arabia and Iran—because of its strategic position between them.

Quick reaction

The United Nations was quick to condemn the invasion and place an embargo on trade with Iraq, enforced through a U.S.-led naval blockade.

Iraq's export pipelines through Turkey and Saudi Arabia were closed. Many nations froze Iraqi assets.

Markets reacted sharply to the loss of Iraqi and Kuwaiti production.

Just before the invasion, oil prices had risen on news of growing political instability in the Persian Gulf amid OPEC efforts to pursue a $21/bbl marker in a soft market.

North Sea Brent crude closed at $19.93/bbl Aug. 1 but spurted to $23.50/bbl the following day. On Aug. 2, Dubai crude soared to $21.15/bbl from a pre-invasion $17.83/bbl. West Texas Intermediate for September had closed at $21.54/bbl Aug. 1, up almost $1 on reports of Iraqi troops massing on Kuwait's border. At midday Aug. 2, WTI approached $24/bbl.

The International Energy Agency (IEA) met soon after the invasion and decided no physical shortage of oil existed and there was no need to activate the emergency response oil sharing system.

IEA urged oil companies to "avoid abnormal spot purchases" and instead draw down stocks—at near record levels. It warned consumers against unnecessary fuel purchases.

Japan, West Germany, and the U.S. stopped buying oil for their strategic reserves, further reducing the call on oil supplies.

Although world oil demand was relatively soft and oil stocks were brimming, oil futures prices rocketed to levels not seen since before the 1986 oil price collapse because of the possibility of Iraq continuing south to capture or damage the supergiant Saudi oil fields, throwing the world into a deep shortage.

That fear was quelled as United Nations members rushed troops, airplanes, and warships to Saudi Arabia. After oil prices climbed as much as $8/bbl in 5 days, the dispatch of U.S. military forces for Saudi Arabia sliced $2/bbl off U.S. futures prices Aug. 8.

Yet oil prices continued high through 1990, on the speculation that the outbreak of a shooting war could damage Saudi production/refining facilities. Some analysts even predicted prices reaching $45-50/bbl.

Shortfall seen

Oil traders knew the world was better prepared to accommodate oil supply disruptions than in 1973-74 and 1979. There was some spare global productive capacity, strategic oil stockpiles had been built, and world crude and products stocks were relatively high.

IEA pegged demand for OPEC oil at 22.3 million b/d in the third quarter of 1990 and 24.6 million b/d in the fourth. The projected fourth quarter call on OPEC oil was thought to be well beyond OPEC capacity available with the loss of Kuwait's and Iraq's production.

In July 1990, Iraq produced 3.1 million b/d and Kuwait 1.6 million b/d.

Combined internal demand for the two countries was 400,000-450,000 b/d. The loss of supply to the international market was about 4.2 million b/d.

The U.S. Department of Energy (DOE) estimated the world shortfall could be reduced to 1.45 million b/d if Iran, Saudi Arabia, and Venezuela produced at capacity, and that was in doubt at the time.

The Iraqi invasion highlighted the sharply decreased spare productive capacity within OPEC compared with the

KUWAIT

mid-1980s.

Earlier, First Boston had estimated OPEC production was at 89% of its capacity of 25 million b/d. Even with potential OPEC capacity additions totaling almost 5 million b/d during 1990-94, First Boston estimated OPEC capacity utilization rates at 85-90% during that period.

First Boston estimated OPEC's maximum sustainable capacity at 24.7-25.7 million b/d, putting current output close to 90% of capacity, a higher utilization than at any time in the 1980s.

OPEC's response

At first it appeared that OPEC would not react to the challenge.

Iraqi Foreign Minister Tareq Aziz quickly warned OPEC countries might have to "pay hard in the future" if they increased their production.

Several OPEC countries—including Indonesia, Iran, Nigeria, Venezuela, and the United Arab Emirates—maintained the industrialized countries should be forced to draw down their high stocks before OPEC members agreed to exceed their existing quotas.

Iranian President Ali Akbar Hashemi Rafsanjani warned OPEC countries exceeded their production quotas would be committing treachery against their own people in favor of "the global oil devourers."

However, Saudi Arabia—the country most threatened by Iraq's action and the country with the largest spare capacity—was not cowered.

After other OPEC nations declined to call an emergency meeting, the Saudis threw OPEC into disarray by proceeding unilaterally with a 2 million b/d increase in crude oil production to 7.3-7.5 million b/d from its second half 1990 OPEC quota level of 5.38 million b/d.

Some analysts predicted Saudi Arabia's action was the first step in the breakup of OPEC's quota system, saying it would lead to a market share free-for-all when the Middle East crisis had passed.

They said an OPEC collapse could set off an oil price plunge even more severe than in 1986, because OPEC would be unable to regain control of the falling market.

The Saudis also indicated they intended to produce more oil for a long time.

Saudi Aramco let a 5 year contract to John Brown company for Hawtah field development in central Saudi Arabia, where Aramco made significant discoveries of superlight crude.

Brown's contract included engineering and construction management for a gas/oil separation plant, a 350 km pipeline, and field infrastructure facilities.

And the Centre for Global Energy Studies, a London think tank headed by former Saudi Oil Minister Ahmed Zaki Yamani, said Saudi Arabia planned to accelerate plans to expand productive capacity to 10 million b/d, reaching that level in 1992 vs. 1995 planned earlier.

Higher production

The Saudi action forced an Aug. 27 emergency OPEC ministerial meeting, which resulted in an agreement supported by 10 of the group's 13 members.

That pact essentially gave OPEC members a free hand to increase crude production to offset the shortfall caused by the United Nations embargo.

Oil prices plummeted as much as $4/bbl after the meeting.

Saudi Arabia and Venezuela pushed the deal through in the face of opposition from Iran and concern from some other members that OPEC's tradition of consensus politics was being abandoned.

Iraq and Libya declined to attend the meeting. Iran said it would support the concept of higher production levels only if the industrialized nations agreed to draw down stocks at the same time.

By allowing a majority decision, OPEC avoided the oblivion some analysts were forecasting for it, and lost the benefit of prolonged higher prices.

The United Arab Emirates boosted output to 2 million b/d from 1.55 million b/d, with the increase coming from Abu Dhabi.

Qatar pushed output to 400,000 b/d, just 29,000 b/d above its quota.

Iran stayed at its 3.14 million b/d quota.

Venezuela climbed from 1.95 million b/d to near 2.45 million b/d. Libya, with a quota of 1.233 million b/d, went to about 1.5 million b/d. And Nigeria jumped from 1.611 million b/d to 1.85 million. Gabon had a quota of 197,000 b/d and was pushing its new onshore fields to produce about 270,000 b/d.

Western oil companies reacted cautiously to higher crude prices.

BP Exploration Chief Executive Officer John Browne warned industry not to increase spending plans, because the higher prices "can melt like snow in the desert, leaving overoptimistic plans and their authors sinking into the sands."

Markets shuffled

Saudi Arabia allocated its additional oil production to countries and companies directly suffering from the loss of supplies from Iraq and Kuwait.

It drew up a priority list headed by developing countries such as India, Pakistan, and the Philippines, that formerly relied on Iraq or Kuwait for a big share of their oil imports. But it was unable to help with product deliveries.

The Saudi state marketing company Samarec declared force majeure on deliveries of middle distillates to Far Eastern destinations. That enabled about 55,000 b/d from the Jubail and Ras Tanura refineries to be used to fuel the military buildup.

Saudi Arabia also became a major importer of petroleum products because of the need to acquire more diesel fuel for vehicles and jet fuel for warplanes.

IEA said refiners around the world faced the prospect of producing more hard-to-sell heavy fuel oil as a result of the heavier crudes coming onto the market to replace supplies from Iraq and Kuwait.

Most of the Middle Eastern crudes replacing Kuwaiti and Iraqi oils, which averaged 32.5° gravity and 2% sulfur, were less than 30° gravity but slightly lower in sulfur content.

The loss of Kuwait's refining capacity took about 600,000 b/d of exported products off the world market, a sharp loss since global refining capacity was already running at a very high utilization rate.

But it was counterbalanced by Japan's restarting of 250,000 b/d of surplus refining capacity and a sizable jump in throughput at the Rabigh refinery on Saudi Arabia's Red Sea coast.

KPC status

Kuwait's state owned Kuwaiti Petroleum Co., a significant European refiner/marketer, continued operations in exile at the London offices of international affiliate KPI.

Kuwait Oil Tanker Co. also moved its headquarters to London.

It operated 30 crude and products tankers and gas carriers with a combined capacity of 2.5 million dwt.

Under its Q8 brand, KPI had about 24% of the Danish market, 12% in Sweden, 10% in Italy, 7% in Belgium, and a little more than 2% in the U.K.

KPI also operated a 100,000 b/d refinery in Naples, a 75,000 b/d refinery in Rotterdam, and a 56,000 b/d refinery

KUWAIT

How the industry's petroleum stocks have varied

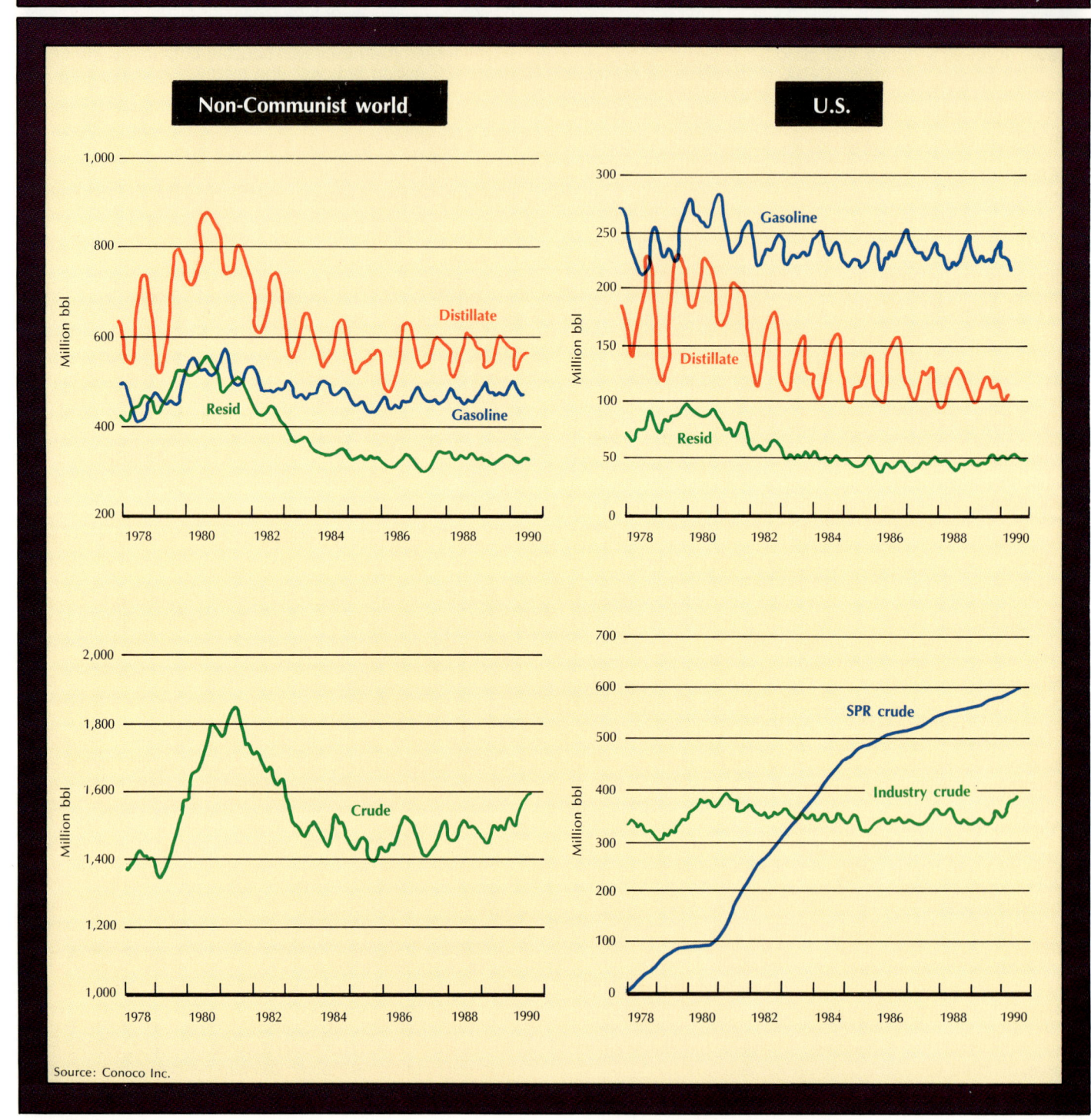

Non-Communist world | **U.S.**

Source: Conoco Inc.

in Denmark.

KPI initially met its crude supply needs with the sizable volumes of crude and products that left Kuwait before the Iraqi invasion, but then had look to other sources to obtain the 400,000 b/d of oil it had received from Kuwait. Saudi Arabia became a major supplier.

Many European countries froze Kuwaiti assets to prevent them from falling into the hands of the Iraqis, causing brief operational problems for KPI.

U.S. reaction

The U.S. imported an average 610,000 b/d of oil from Iraq during January-May 1990, about 8% of U.S. net oil imports and 3.6% of total U.S. demand.

Kuwait supplied a little more than 1% of net U.S. oil imports in the period, about 120,000 b/d.

At the time of the invasion, U.S. crude and product stocks were at their highest level in 8 years at a combined 1.123 billion bbl. And the U.S. had almost 590 million bbl in the Strategic Petroleum Reserve (SPR), equal to 95 days of imports.

Higher domestic products prices resulted in a political backlash slapping the U.S. petroleum industry. Consumer groups alleged price-gouging by oil companies.

Several congressional subcommittees held short investigations.

The U.S. Justice Department pledged to prosecute any

KUWAIT

antitrust violations it found related to the jump in petroleum product prices.

President George Bush urged oil companies to "show restraint" in gasoline pricing.

DOE proposed a variety of oil conservation and production measures to soften the effect of lost Iraqi and Kuwaiti supplies.

Energy Sec. James Watkins said the initiatives could replace 607,500 b/d of imports, or the equivalent, by the end of 1990 and another 530,000 b/d by end of 1991.

The actions included an array of conservation measures and programs to increase production, including faster permitting for two Alaska fields and construction of gas pipelines to enable more California heavy oil production.

In early December DOE completed a "test" drawdown and sale of 3.9 million bbl of SPR crude to 11 high bidders. It injected more crude into the supply system while removing doubts that the SPR could be tapped effectively and expeditiously.

Allies attack

Tension in oil markets relaxed after Jan. 16, when Allied forces launched a massive air attack on Iraqi military bases and the Iraqi army in Kuwait.

Within a few hours, U.S. spot crude prices plunged because traders realized massive U.S. led air strikes had eliminated Iraq's capability to retaliate against Saudi Arabia and sever critical oil supply lines in the Persian Gulf.

Oil prices dropped Jan. 18 to their lowest level since early July 1990—before an OPEC agreement to adhere to quotas.

Contributing to market softness were IEA announcements that strategic oil stocks would be released.

The price of Brent blend for immediate delivery, which fell to less than $20/bbl within 5 days of the start of allied air strikes, climbed $2.90 to $22.15/bbl in reaction to subsequent Iraqi missile attacks on Israel and Saudi Arabia and on reports Kuwaiti oil fields were afire.

Light sweet crude futures on the New York Mercantile Exchange for February delivery jumped about $3 on the week to close Jan. 22 at $24.18/bbl. Nymex crude for February delivery had slumped to $19.25/bbl on closing Jan. 18, the lowest since July 20.

Contributing to the price drop was announcements by U.S. and European refiners that they would freeze product prices and ensure adequate supplies.

IEA action

Upon the outbreak of the air war, IEA notified member governments to activate a 2.5 million b/d contingency plan. All IEA members plus Finland, France, and Ireland participated.

Of the total, 2 million b/d was to have been made available by drawing down stocks.

IEA estimated demand restraint would account for a further 400,000 b/d, and 100,000 b/d from fuel switching and surge capacity.

IEA recommended oil companies continue to draw on commercial stocks and that companies and consumers exercise restraint in purchases.

It reiterated world supplies and refinery were adequate, but said the outbreak of hostilities could lead to heightened uncertainty and volatility in the market, requiring additional oil to meet any possible temporary shortfalls.

The 2.5 million b/d was to come from a 1.125 million b/d stockdraw in the U.S., 187,000 b/d from Germany of which about 80% was from stocks, 350,000 b/d from Japanese stocks, and 120,000 b/d from U.K. stocks.

In France, oil companies agreed to a freeze on oil prices, and the government asked companies to reduce commercial stocks.

Government strategic stocks were to be used only as a last resort.

Sadek Boussena, Algeria's oil minister and OPEC president, protested IEA's decision to put 2.5 million b/d on the market when there was no shortage of supplies. He said it would undermine world oil prices.

"This would have a disastrous effect on the economic growth and social stability of countries strongly dependent on their oil resources," Boussena said.

U.S. SPR drawdown

U.S. refiner/marketers froze products prices and took emergency supply allocation measures in response to the outbreak of war.

Those price freezes became ceilings when the price of crude unexpectedly plunged after early reports of astonishing allied military successes.

As part of the international drawdown, President George Bush ordered the sale of 33.75 million bbl from the 585 million bbl Strategic Petroleum Reserve on the U.S. Gulf Coast. But the sale was trimmed to 17.3 million bbl.

Bush also ordered the Treasury secretary to waive Jones Act provisions requiring use of U.S. flag vessels to move crude between U.S. ports if they are moving SPR oil.

The SPR sale offered about one third sweet crude and two thirds sour.

But industry bid for twice as much sweet as it did for sour and offered less for the sour than DOE expected. So DOE chose to sell more sweet and much less sour than originally offered.

The crude was drawn down from four of the six SPR sites: Bryan Mound near Texas City, Tex., West Hackberry near Lake Charles, La., Bayou Choctaw near Baton Rouge, and Weeks Island near New Orleans.

Saudis cut back

A surplus of crude oil on the market soon forced the Saudis to cut production by 2.5 million b/d to 6 million b/d.

Storage facilities at Ras Tanura and Juaymah on the gulf coast and Yanbu on the Red Sea coast were full, forcing Saudi Aramco to throttle back production.

Fewer tanker liftings also were a factor.

The Middle East Economic Survey (MEES) estimated January production in Saudi Arabia was unchanged at 8.3 million b/d with total OPEC production at 23.1 million b/d, down from 23.86 million b/d in December.

MEES said Saudi Aramco's sustainable productive capacity was close to 9 million b/d, resulting from demothballing and construction work on production and associated handling facilities.

In December capacity was 8.6-8.7 million b/d.

That high level of capacity was causing second thoughts in the Saudi administration about an accelerated program to expand capacity to 10 million b/d by 1992-94, given the potential oil surplus in the medium term.

Saudi sources told MEES once the war is over the government will begin a thorough review of future productive capacity needs in light of projected oil supply/demand trends.

MEES said second thoughts about the investment program do not necessarily mean the 10 million b/d target will be changed.

But there could be a different time frame and investment levels.

Saudi planners will review future demand for crude with particular emphasis on demand for different qualities, the extent and timing of the resumed flow from Iraq and Kuwait, supply trends in the Soviet Union, and productive capacities of other OPEC members, it said.

KUWAIT

IEA/OECD stocks on land

	Aug. 1, 1990 Total stocks	Nov. 1, 1990 Company stocks	Government stocks	Total stocks
Canada	15.8	15.2	—	15.2
U.S.	213.7	127.8	79.5	207.2
North American	**229.5**	**143.0**	**79.5**	**222.4**
Australia	5.0	4.1	—	4.1
Japan	77.1	49.3	28.2	77.5
New Zealand	1.0	0.9	—	0.9
Pacific	**83.1**	**54.4**	**28.2**	**82.6**
Austria	2.7	2.4	0.3	2.7
Belgium	4.7	4.3	—	4.3
Denmark	4.8	2.6	1.9	4.5
Germany	39.4	12.4	25.9	38.4
Greece	4.2	3.9	—	3.9
Ireland	1.1	1.2	—	1.2
Italy	21.4	21.5	0.8	22.3
Luxembourg	0.3	0.3	—	0.3
Netherlands	9.9	7.1	2.3	9.4
Norway	4.0	4.0	—	4.0
Portugal	2.8	3.0	—	3.0
Spain	9.8	9.3	—	9.3
Sweden	6.6	6.1	0.2	6.4
Switzerland	5.7	5.8	—	5.8
Turkey	3.4	3.3	—	3.3
U.K.	17.4	17.1	—	17.1
IEA Europe	**138.2**	**104.5**	**31.4**	**135.9**
IEA	**450.9**	**301.8**	**139.1**	***440.9**
Iceland	—	—	—	—
Finland	3.4	—	—	3.4
France	20.5	—	—	20.2
Total OECD	**474.9**	—	—	**464.5**

*Total stocks of 440.9 million metric tons are the equivalent of 150 days of net imports.

Effect on Japan

The embargo touched off a scramble for supplies among oil importing nations.

Japan was the biggest lifter of Iraqi crude, but initially Japan's government and oil companies were more concerned about the possible effect of higher crude and product prices on inflation than a supply shortage.

That confidence was based on high levels of strategic crude stocks in Japan. After the supply crises of the 1970s, the Japanese government pledged to introduce its own strategic reserves and started stockpiling in 1978 through Japan National Oil Co.

In August 1990 Japanese strategic stocks were 520 million bbl, some 197 million bbl government held and 322 million bbl in private storage. Japan's stocks were equal to 142 days of consumption and compare with stocks of only 50 days' consumption in the mid-1970s.

Japanese companies had been lifting about 220,000 b/d from Iraq and had contracts for about 150,000 b/d from Kuwait, together totaling about 12% of its oil supplies.

Buyers tried to replace some of the shortfall from Iran. Early in 1990 it had bought 290,000-300,000 b/d from Iran, but it had fallen to 230,000 b/d at the time of the invasion.

Abu Dhabi withdrew an earlier decision to cut Japanese liftings and supplied Japan 220,000 b/d. The earlier cut was spurred by U.A.E.'s efforts to cut its production to 1.5 million b/d in line with OPEC's July agreement on production and pricing.

Japan decided to release its national strategic oil reserve for the first time. It had stocks totaling about 142 days' consumption but the Japan Petroleum Federation said a 60 day reserve was adequate.

Early in 1991 Japan's Ministry of Trade and Industry reduced local oil companies' requirements to hold stocks to 78 days' consumption from 82 days to meet its IEA requirement for a 350,000 b/d release of crude.

Eastern Europe

Iraq's invasion of Kuwait has transformed eastern Europe's critical liquid fuel shortage into a pervasive financial catastrophe.

Higher crude prices cut deeply into Polish, Czechoslovakian, and Bulgarian export revenues.

With Soviet oil deliveries to eastern European nations drastically reduced, those countries had turned to Iraq for crude. But the embargo against Iraq prevented either delivery of oil already contracted for or payments on the $5 billion debt that Baghdad owed eastern European countries.

Bulgaria was hardest hit. Iraq owed Bulgaria $1.2 billion, was receiving no Iraqi oil, and lost $160 million in embargoed exports to Iraq and Kuwait.

Iraq owed Yugoslavia $1.5 billion, with Belgrade expecting repayment in crude. Not only did Iraq crude deliveries stop, but Yugoslavia was unable to export $1 billion in goods

KUWAIT

What U.S. refiner/marketers did after Mideast war started*

Company	Supply action	Price action	Time frame
American Petrofina	None	Reduced wholesale gasoline 18-13¢/gal	Jan. 18 until further notice
Amoco	None	Froze wholesale gasoline and posted jet fuel and distillates, rescinded planned Jan. 17 price hike	Jan. 17 until further notice
ARCO	None	Froze gasoline, diesel and jet fuel	Jan. 16, with Jan. 23 review
Ashland	Supplying at least 100% of product needs; continuing "equitable" allocation system	None	NA
BP America	On West Coast only, 50% of January 1990 allocations	None	Jan. 15-31
Chevron	None	Capped wholesale prices for all branded products and froze pump prices at all 680 company owned stations, then cut wholesale branded prices by as much as 5¢/gal for gasoline and as much as 9¢/gal for diesel	Freezes Jan. 16, cuts Jan. 18, freezes lifted Jan. 23
Citgo	None	None	NA
Coastal	None	None	NA
Conoco	Allocating 100% of historic supply to wholesale customers	Reduced gasoline 6-8¢/gal, diesel fuel 8-10¢/gal	Jan. 17 until further notice
Crown Central	Allocation program if necessary	Reduced wholesale gasoline, heating oil at Crown distribution terminals in eastern U.S., at La Gloria division terminals in Tyler, Tex., and U.S. Midwest	NA
Exxon	None	Reduced wholesale prices in U.S. by 5¢/gal for branded gasoline and 10¢/gal for distillates	Jan. 17
Marathon	For branded products, allocating 100% of contract monthly and 150% of daily limits; other wholesale products, 150% of daily limits	None	Jan. 16 until further notice
Mobil	Started ratable lifting program for gasoline and heating oil; liftings restricted to 100% of contract entitlements	Froze all petroleum products prices	Jan. 16 until further notice.
Phillips	Distributing motor fuel and distillates each day in amount equal to montly contracts divided by business days in the month	Increased wholesale supply management surcharge to 25¢/gal from 10¢/gal for motor fuel and distillates	Jan. 14
Shell	Supplying 100% or normal requirements	Froze prices for gasoline, aviation fuel, heating oil, lubricants, and other light products	Jan. 16 until further notice
Sun	None	Reduced unbranded product prices "substantially;" froze Sunoco and Atlantic gasoline prices to all classes of customers	Jan. 17 until further notice
Tesoro	None	None	NA
Texaco	None	Froze branded wholesale	Jan. 17 until further notice
Unocal	None	Capped all product prices in West and Southeast U.S. and at all U.S. truck stops	Jan. 16 until further notice

* As of Jan. 23. Frozen prices in some instances are considered ceiling prices.

ordered by Iraq.

Faced with unexpected cuts in Soviet oil deliveries, Hungary also turned to Iraq for oil as payment on Baghdad's $150 million debt to Budapest.

Hungary, Bulgaria, Czechoslovakia, Poland, and Romania were operating their refineries far below capacity because of crude shortages.

Kuwaiti refineries damaged

In January 1991 Iraqi artillery rounds set a storage tank afire at the Japanese owned Arabian Oil Co.'s Ras al Khafji 30,000 b/d refinery in the Saudi sector of the Neutral Zone.

The refinery at Khafji apparently not damaged by the fighting around the town because Iraqi forces never reached the southern side of the city, where the oil installations are located.

The unit had been shut down as a precautionary measure. The smallest refinery bordering the Persian Gulf, it was not a major contributor to exports or the war effort.

An Iraqi strike into Saudi territory and the battle at Khafji forced Saudi Aramco to halt production from Safaniyah, Zaluf, and Marjan offshore fields, cutting total output by about 25% to 6 million b/d.

The platforms serving the three fields remained fully manned during the battle and were undamaged. Production was resumed a few days later. The shutdown did not affect Ras Tanura or Juaymah liftings because of huge Saudi crude stocks.

Iraqi forces also set fire to storage tanks at the 190,000 b/d Mina Abdulla and 187,000 Shuaiba refineries on the coast. Kuwait's three extremely sophisticated refineries at

KUWAIT

Ahmadi, Mina Abdulla, and Shuaiba, with a combined capacity of 670,000 b/d, were not operational before the war started.

Iraq had stripped key pieces of processing equipment from those units, including advanced computer control equipment, but industry sources say the major processing facilities at all three refineries are still intact.

Wells or storage tanks also were briefly set afire in Wafra and South Fuwaris fields in the Neutral Zone and South Umm Gudair oil field in Kuwait.

Combined production of those three fields was 135,000 b/d in normal times.

Months before the allied invasion began, Kuwaiti government officials were planning a massive postwar reconstruction in response to earlier reports Iraq had planted explosives in Kuwaiti oil facilities.

Before withdrawing, the Iraqi army blew up virtually all wellheads in Kuwait, plus other oil field facilities, setting hundreds of fires.

An American general described that action as "a scorched earth policy."

Iraq's facilities pounded

Iraq had to ration gasoline early in the embargo because the country's 320,000 b/d of refining capacity could not get sufficient chemicals and additives from normal suppliers.

After fighting began in earnest, it halted all sales of gasoline, diesel, and fuel oil because allied air assaults had damaged refining and distribution facilities.

Iraq's refining capacity was mauled after the air war began and it had to shut in much of the 350,000-400,000 b/d of crude it was producing to meet domestic demand.

The 70,000 b/d Daura unit near Baghdad was hit several times during the first week of allied attacks, with a substantial fire observed there. The 300,000 b/d Baiji refinery between Baghdad and the 150,000 b/d Basra refinery were damaged.

Iraq has five other small refineries with capacities of 12,000-30,000 b/d near main producing areas.

James Schlesinger, a former U.S. Secretary of Defense, said "The Iraqi refineries are gone, courtesy of the U.S. Air Force. They will take some time to come back."

However, allied commander Gen. Norman Schwarzkopf said the allies did not attempt to damage Iraq's oil producing ability although it was "something we could have done very easily." He said they only concentrated on destroying Iraq's ability to refine the oil into gasoline for military vehicles.

Air attacks on tank trucks moving products and some crude slashed Iraqi exports to Jordan to about 5,000 b/d, forcing Jordan to seek supplies elsewhere. Iraq previously supplied about 80% of the country's 60,000 b/d requirements.

Shipping threats

The war, and a sharp drop in Persian Gulf business forced the international tanker industry to adjust to reduced oil shipments, soaring bunker costs, and rising insurance premiums.

Although allied forces established air superiority in the Persian Gulf and neutralized the Iraqi missile threat, the threat of Iraqi mines continued to threaten shipping operations.

The Maritime Liaison Office in Bahrain reported several mines had been destroyed about 90 miles north of Bahrain and four more were eliminated farther south in the gulf.

Early in the war, 80 vessels were reportedly waiting off the Indian Ocean coast of Oman and Fujairah, reminiscent of the worst days of the Iran-Iraq war when tankers anchored there while waiting for orders to load in the gulf.

The sharp rise in war risk insurance for tankers—up to 5% of hull value for vessels trading with northern gulf terminals—plus war bonuses for crews persuaded many owners to hold vessels outside the gulf.

After the air war began, the allied naval command advised tanker owners it was safe to use the major gulf terminals in Saudi Arabia, Qatar, and the United Arab Emirates—leaving only Iran's Kharg Island still in the danger zone.

But owners of Japanese-flagged vessels agreed with the seamen's union not to enter terminals in Saudi Arabia, Bahrain, at Kharg Island, and the onshore Qatar terminal at Umm Said. Offshore loadings in Qatar were unaffected.

That prompted exporters to revive shuttle tankering, which proved so effective during the Iran-Iraq war.

Iran and Qatar chartered tankers to move crude from

The OPEC oil situation

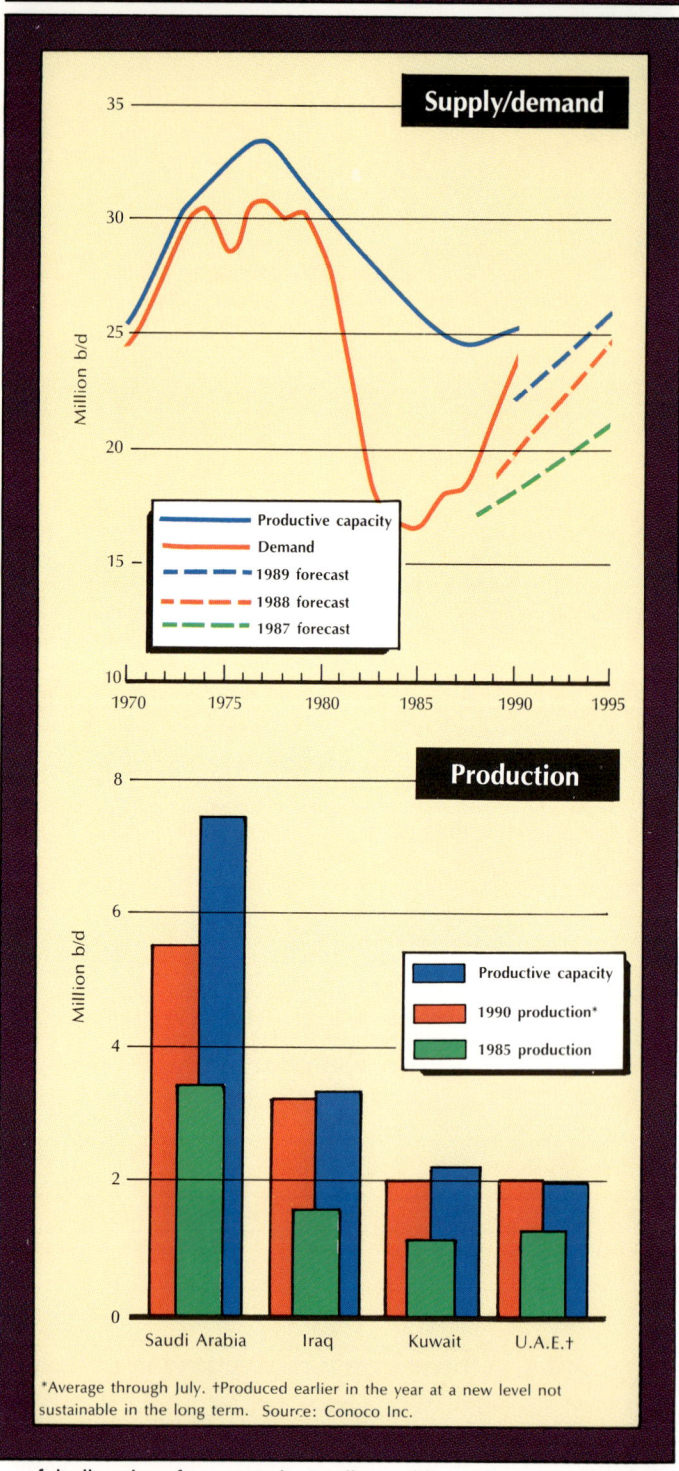

Supply/demand chart (1970-1995): Productive capacity, Demand, 1989 forecast, 1988 forecast, 1987 forecast (Million b/d)

Production chart: Productive capacity, 1990 production, 1985 production for Saudi Arabia, Iraq, Kuwait, U.A.E.† (Million b/d)*

*Average through July. †Produced earlier in the year at a new level not sustainable in the long term. Source: Conoco Inc.

KUWAIT

export terminals within the theoretical range of Iraqi strikes by warplanes and missiles, and offload it into customers' tankers in safer areas. Some tankers continued to load at Kharg.

Japan's crude oil shippers agreed to a fee of 50¢/bbl for crude shuttled in Iranian ships from Iran's Kharg Island terminal to Lavan Island for transshipment to Japanese tankers.

Saudi Arabia was operating a shuttle from its terminals in the gulf to a transshipment point in the Indian Ocean, but the shuttle was used by only a few of Saudi Aramco Oil Co.'s Third World customers.

Later, lack of casualties among all types of shipping spurred a 50% cut in war risk insurance rates.

Oil spill

Iraq used oil as a weapon against Saudi Arabia, spilling oil from Kuwait's Mina al Ahmadi terminal. The resulting slick initially was estimated at 10.7 million bbl but later was revised to about 1.5 million.

U.S. officials called the intentional spill "an act of environmental terrorism."

Smart bombs delivered by U.S. aircraft hit two onshore tank farm manifold stations, cutting off the terminal's source of oil flow Jan. 26.

The floating oil threatened the Saudi industrial city of Jubail, which has refining and desalination plants and a deep canal that utilizes seawater for use as cooling water in petrochemical and other industrial plants.

Not all the oil came from the Ahmadi sea island terminal. A small amount of the crude came from two tankers bombed by allied aircraft in January.

Another slick was in the head of the gulf, where Iraq released a small volume of oil from its Mina al Bakr terminal east of Bubiyan Island. Some oil also was dumped into the gulf after Iraqi guns hit a storage tank at the Khafji refinery in the early days of the war.

That slick polluted a large part of the northern Saudi Arabian coastline. Oil was reported around offshore Safaniya oil field and the coastal settlement at Ras Tanajib, where oil from Saudi offshore fields moves ashore by pipeline.

In Saudi Arabia, desalination plants, refineries, and export terminals were well protected by booms and work crews.

Saudi Arabia chartered the Norwegian owned Al Waasit cleanup vessel in an effort to protect the water desalination plant at Jubail. Saudi Arabian Oil Co. had sizable stocks of booms and skimmers and other equipment, along with the ships and personnel needed to deploy them. The equipment buildup started during the Iran-Iraq war when an Iraqi attack on an Iranian offshore platform resulted in oil spilling into the gulf for more than a year.

The oil spread into the central part of the gulf, headed for a number of islands that are the breeding ground for turtles and threatening the environmentally sensitive salt marshes along the coast of the United Arab Emirates.

Oil price collapse?

Yearend 1990 stocks on hand in Organisation for Economic Cooperation and Development countries were the highest since yearend 1982 at about 3.5 billion bbl, equal to 96 days of consumption, near the 1981 peak, and 122 million bbl higher than a year ago. Company stocks were the highest since yearend 1985.

The oil industry was expected to retain some additional stocks as insurance against supply disruptions but with a drop in demand likely in 1991, companies had little need to carry additional stocks for very long.

Among oil traders, there was concern that a brief war would be followed by a collapse in world oil prices.

Former Saudi Oil Minister Ahmed Zaki Yamani said crude oil prices could slump to $12/bbl if OPEC supplies remain high. He said OPEC production of 23.5 million b/d would soon swamp the market, adding that finding an agreement within OPEC to tackle the problem could be difficult because such countries as Saudi Arabia, which previously had exercised production restraint, face heavy financial demands resulting from the fighting.

Bank of America sees the price of oil dropping to the teens if a short (less than 8 weeks) war leaves Saudi production facilities unscathed, Kuwaiti facilities are severely damaged, and Iraqi facilities are moderately damaged.

It said Iraqi and Kuwaiti oil production could soon return to normal and Iraq's defeat would eliminate its influence in OPEC, permitting to continued production over quota by other OPEC members.

Then oil prices could drop to $16/bbl, but stabilize at $20/bbl by late 1991, averaging $18.71/bbl for the year.

Merrill Lynch said even systematic destruction of Kuwaiti oil facilities, putting Kuwait's production out of commission for a long time, would have little effect on markets if there is still a surplus of capacity in Saudi Arabia.

It said after the war the overhang of surplus crude on the market will put downward pressure on prices, but overhang may not be severe. It said oil prices might drop to about $20/bbl or lower after the war, but not as low as $15.

OPEC's future

The Iraqi invasion of Kuwait came just a month before OPEC's 30th anniversary, and it appeared certain to affect the organization's future.

Merrill Lynch said if Iraq withdraws from Kuwait in defeat, "OPEC may be weakened as a price administering institution.

"It is conceivable, however, that Saudi Arabia may step in to fill the leadership vacuum in OPEC by acquiescing to act as a swing supplier in order to support prices at reasonable levels."

Schlesinger said "It is apparent the the real winner of the war is Iran. It is likely to become the dominent power in the region."

Salomon Bros. said the Middle East crisis will cast a pall over the Persian Gulf for many years, noting the Middle East was to have accounted for 40% of all new refinery capacity in the world the next 3-5 years.

"This region now looks far less hospitable to such massive capital investment—perhaps totaling $20 billion, had all this Middle Eastern capacity been built. The result of the possible postponement of these facilities could be greater downstream bottlenecks for a longer time than most observers...would have thought likely before this crisis erupted."

Salomon said a new geopolitical order for the Persian Gulf may be in the making. "A sort of 'Middle Eastern NATO' is a conceivable security arrangement in which U.S., European, Soviet, and Japanese partners might guarantee the external security of regional treaty partners."

Salomon said this could pave the way for crude exporters to invest more in consumers' downstream industries, and consuming countries might become equity investors in crude exporting nations' domestic oil and gas E&D.

Melvin Conant, president of Conant & Associates Ltd. and editor of Geopolitics of Energy, said OPEC could face changes.

He said OPEC cohesion will further weaken if the group's members increase downstream investments in major consuming markets, noting OPEC is being split between members focusing on crude oil prices—notably Iran—and those focusing on products prices—notably Saudi Arabia.

Conant said, "The split is almost bound to grow as cartel members look for 'guaranteed' market shares through their overseas downstream refineries and marketing outlets, especially in anticipation of a long term world crude surplus."

Mideast crisis casts cloud on tanker building

THE MIDDLE EAST CRISIS THAT BEGAN WITH THE Aug. 2, 1990, seizure of Kuwait by Iraq cast a cloud on a surge in construction of large crude oil tankers that was well under way.

Typical of the tanker industry's fears was this warning from shipping analyst Marsoft Inc., Cambridge, Mass.:

"If the crisis is drawn out or fighting erupts, Marsoft foresees a rapid decline in shipping rates at least through mid-1991 as oil shipments dry up, oil stocks are drawn down, and economic growth slows."

When the crisis is resolved, however, the tanker market at least may be set to recover rapidly as oil stocks are replenished, Marsoft said in a report it issued just 12 days after the Iraqi invasion.

The Middle East eruption could spell trouble for investors who have committed tremendous amounts of money to shipping investments. They were under pressure to develop strategies to help them navigate through the 1990 crisis and possible future ones in the volatile Middle East, a vast storehouse of oil.

It holds two thirds of the world's total oil reserves. In normal times, it accounts for 13.389 million b/d of crude and products exports, or a little more than 44% of the world's main interregional oil trade (Fig. 1 and Table 1).

Marsoft's and shipping owners' fears were based on the possibility of a long Middle East war with serious damage to oil installations and loss of a significant part of the region's oil flow.

"The resulting high oil prices could swing the U.S. into a recession by yearend, significantly reducing oil demand," Marsoft said. "The impact would be a sharp drop in shipping demand just as the large number of ships ordered during the industry's boom years are entering the market."

Further clouding the outlook for ship owners was U.S. comprehensive oil spill legislation signed into law Aug. 18, 1990, by President Bush.

Among other things, the law requires double hulls for all new tankers operating within the 200 mile U.S. exclusive economic zone. There is a phaseout for all existing single skin tankers scheduled during 1995-2010.

As a result, any ship owner thinking about building a tanker will have to consider specifying double hull construction at added cost with no solid assurance such construction will prevent oil spills or reduce their volume.

The American Bureau of Shipping said, "Marine industry opinion—business and technical—is fairly evenly divided as to whether double hulls will do much to protect the environment in the event of a high energy grounding or collision."

The new U.S. law, which stiffened spill cleanup liability provisions, was a direct response to the Mar. 24, 1989, Exxon Corp.'s Exxon Valdez tanker spill that dumped 258,182 bbl of North Slope crude oil in Prince William Sound and the Gulf of Alaska off Alaska. It was North America's biggest tanker spill.

The last vessels and crews from Exxon Co. U.S.A.'s 1990 Alaskan cleanup campaign off Alaska were demobilizing in September 1990 at Seward and Valdez.

That wound up what the company called "a successful summer cleanup program" in the wake of the Exxon Valdez spill. Cleanup efforts ended Sept. 15 to protect personnel from winter storms.

Main interregional oil trade—1989*

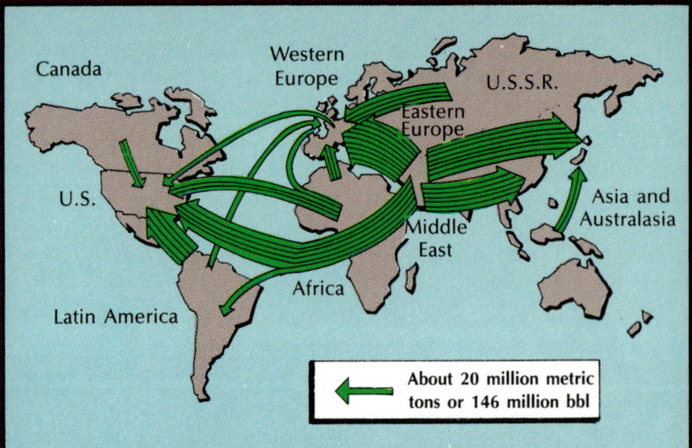

Fig. 1

← About 20 million metric tons or 146 million bbl

* Excludes bunkers and intraregional shipments—between individual countries in western Europe and between Alaska and the U.S. Lower 48, for example. Source: British Petroleum Co. plc

Exxon the previous May had disclosed a plan to increase by about $8 million its payments to some Alaska commercial fishermen for 1989 income losses caused by the Exxon Valdez oil spill. The increased payments under Exxon's voluntary claim program were to go to Kodiak and Chignik salmon fishermen.

Kodiak fishermen were to get about $5 million more, Chignik fishermen about $3 million more.

Of the more than $200 million Exxon had disbursed under the program, Kodiak permit holders and their crewmen were paid more than $50 million and those from Chignik more than $4 million.

A burst of tanker orders

The surge in tanker orders hit full stride during the first 6 months of 1990, when shipyards' order books for vessels of more than 200,000 dwt nearly doubled while orders for tankers of 100,000-200,000 dwt climbed by more than 25%.

The burst of orders was a response by tanker owners to the belief that growing world dependence on crude oil imports from the Middle East in the 1990s will bolster freight rates.

One of the prime growth markets predicted for Middle East crude is the U.S., easily the world's biggest customer for crude oil and petroleum products exports. But an increasing number of tanker owner/operators were withdrawing from that trade rather than face the financial jolt of unlimited liability for oil spills in U.S. waters.

Because of the impending spill liability law, by midyear 1990 at least five tanker owner/operators—three oil companies and two independent shippers—had vowed not to trade with mainland ports in the U.S.

A midyear report by the American Petroleum Institute underscored the potential seriousness of a widespread U.S. boycott by tanker owners.

TANKERS

Outlook for VLCC, ULCC construction (Fig. 2)

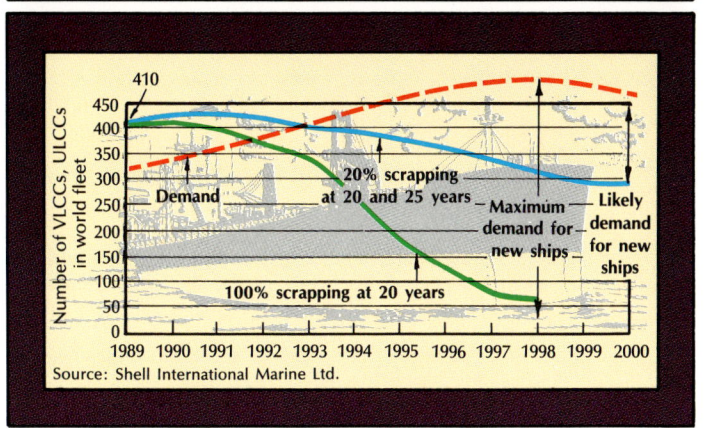

Source: Shell International Marine Ltd.

The non-Communist world's tanker fleet (Fig. 3)

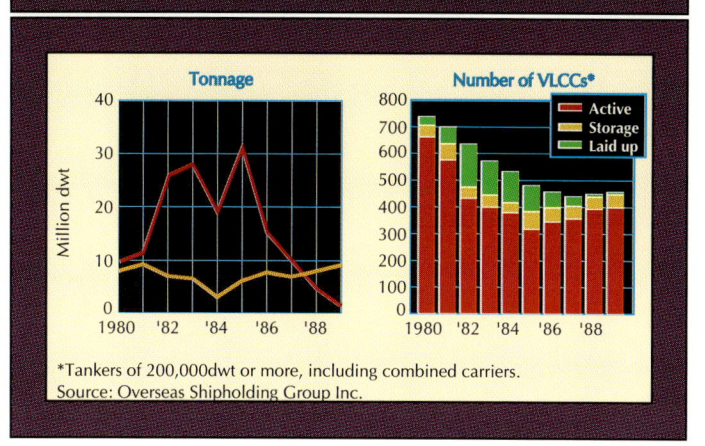

Tankers of 200,000dwt or more, including combined carriers.
Source: Overseas Shipholding Group Inc.

U.S. dependence on imported oil climbed to a record high in the first 6 months of 1990, API said.

Imports, driven by a continued decline in domestic crude production and a strong buildup in inventories, met 49.9% of the nation's oil needs in January-June, breaking the 48.8% high mark set in first half 1987. What's more, about 90% of U.S. imports—all but those from Canada—arrived in tankers.

Order books growing fatter

A tally by the International Association of Independent Tanker Owners (Intertanko), Oslo, showed that on July 1, 1990, there were 338 tankers and combined carriers on order around the world representing 40 million dwt. At the start of 1990 the total order book was 255 vessels of 25.4 million dwt.

At the lower end of the market for vessels of 15,000-50,000 dwt the numbers declined slightly from 106 vessels at the start of the year to 102 July 1. Tonnage was down slightly from 3.5 million dwt to 3.45 million dwt.

As tanker sizes increase so does the order book.

In the 50,000-100,000 dwt class the number of vessels on order jumped to 98 with 8.5 million dwt from 57 and 4.8 million dwt at the start of the year. Two years previously only 37 vessels of 2.8 million dwt were on order.

Orders for vessels of 100,000-200,000 dwt rose to 73 on July 1 from 58 at the start of 1990, and the tonnage on order jumped to 10.3 million dwt from 8.2 million dwt. At midsummer 1988, only 40 vessels of this class were on order with a total tonnage of 5.3 million dwt.

The biggest increase occurred in orders for the largest tankers.

At the start of 1990, order books for vessels of more than 200,000 dwt stood at 34 tankers with 8.9 million dwt. The buying boom by owners in the first 6 months of the year increased the number on order to 67 with 17.8 million dwt.

Two years previously the order book for such tankers stood at 29 vessels of 7.4 million dwt.

The prospect of better times also put a stop to the sale of vessels for scrap. In the first 6 months of 1990 no vessels of more than 200,000 dwt were sent to the breakers, while there were only 17 tankers in the 10,000-200,000 dwt class sold for scrap. These had a total weight of 830,000 dwt.

At the depth of the tanker depression, 30.5 million dwt of capacity was scrapped in 1985. Total scrappings throughout the 1980s amounted to 142.8 million dwt.

Intertanko said less than 6 million dwt has been scrapped since the start of 1988. Almost 13 million dwt/year has to be replaced, assuming that the average life of the fleet is extended to 22 years.

The upturn in tanker construction started several yeas ago, and the number of new tankers being delivered reflects the earlier revival. Thirty-nine vessels of 4.3 million dwt were delivered in first half 1990, compared with only 27 tankers of 2.7 million dwt in first half 1989.

Lloyds Shipping Economist, a publication of Lloyds of London, said the surge of supertanker orders in the early months of 1990 inevitably raised fears of a repetition of the unbridled tanker fleet expansion of the 1970s.

The results of that period of speculative excess were so disastrous that the tanker industry remains sensitive to any sign that the supply of new ships, particularly very large crude carriers (VLCCs), may again be threatening to get out of control.

The publication said this deep seated neurosis is increasingly counterbalanced by a conviction that this time around things are different.

Intertanko Chairman Seigo Suzuki acknowledged that demand for tanker transportation grew steadily during the preceding couple of years, recovering from a long depression. Recovery of the tanker market and optimism about the future call for caution, he said, because the most serious mistakes are made during good times.

World fleet growing older

Another factor that fueled the rush of orders was the increasing age of the world tanker fleet.

In the category between 255,000-319,000 dwt, 150 out of the 187 available vessels are 13-16 years old. In some parts of the industry, vessels more than 20 years old are deemed to be approaching the end of their working lives.

Data on tanker age compiled by Clarkson Research Studies, London, show a large number of ships—about 150—in the 255,000-320,000 dwt ton range that are 13-16 years old. They have a combined tonnage of 41.1 million dwt, amounting to 17% of the total world tanker fleet of 240 million dwt.

If all very VLCCs of at least 200,000 dwt and more than 13 years old are included, the total rises to 308 ships of 82.8 million dwt, or 34.5% of world tanker tonnage.

Clarkson's figures covered all oil tankers of 200,000 dwt or more as of July 1, 1989 (Table 2).

Owners also were concerned that the new U.S. oil spill law and pressure for introduction of double hull tankers in other countries would restrict the operating parameters of older tankers. That could lead to a two tier freight market with modern tankers commanding a premium over older ones.

Construction costs move up sharply

Intertanko charted the dramatic growth in tanker construction costs at midyear 1990.

Three years previously a tanker could be built in Japan or Korea for $42-43 million. During the great ordering boom of first half 1990 owners were committing to pay $85-90 million/

TANKERS

U.S. single hull phaseout calendar*
Fig. 4

	Tanker size (gross tons)			
	Less than 5,000	5,000-14,999	15,000-29,999	More than 30,000
		Number of years or older		
1995		40	40	28
1996		39	38	27
1997		38	36	26
1998		37	34	25
1999		36	32	24
2000		35	30	23
2001	Exempt	35	29	23
2002	until	35	28	23
2003	2015	35	27	23
2004		35	26	23
2005		25	25	23
2006		25	25	23
2007		25	25	23
2008		25	25	23
2009		25	25	23
2010		Must have double hull		
2011				
2012				
2013				
2014				
2015	Must have double containment system			

* In all cases, vessels with double bottoms or double sides are allowed an extra 5 years to comply. Source: American Bureau of Shipping

---

vessel. And reports from Japan said shipbuilders wanted to boost prices even further with the target of $100 million/vessel.

Strongest growth in orders for new tankers occurred during the first 3 months of the year. But the momentum was not fully maintained into the second quarter, and the June order book was slightly thinner than in the previous month. Industry sources said it was too early to attribute a small decline in June orders to high construction prices.

Of greater interest to the industry was the attitude of tanker owners who had not yet joined the ordering spree but will need to replace part of their fleet during the 1990s.

The Royal Dutch/Shell Group, one of the largest oil company owners, made its position clear on tanker construction.

It said there is no case on economic grounds for a VLCC replacement program. And in the short term, shipyard capacity does not exist to handle a major replacement program that could flow from likely demand for 150-200 new VLCCs and ultralarge crude carriers (ULCCs) by 2000 (Fig. 2).

Shell figured that to earn even a modest 8% return on a $90 million investment in construction of a new tanker, a charter rate of about $35,000/day would be needed. That compared with averages of $10,000-15,000/day in recent years. Shell expected rates to average about $20,000/day during 1990.

Instead of newbuildings, Shell planned to extend the life of its VLCCs. The company contended the age of a vessel is not relevant in an assessment of its reliability—if effective maintenance and repair programs are followed.

In making the case for extending the life of VLCCs, Shell said closure of many shipyards in the previous decade reduced construction capacity to 15 million dwt/year. Not every yard can build VLCCs, so the upper limit of larger tanker production probably is nearer 10 million dwt/year.

Replacing 80 million dwt of existing VLCC capacity would take 8 years if all yards capable of doing so built nothing but VLCCs.

But in practice, there are other demands on shipbuilding capacity from buyers of bulk carriers, container ships, and cruise ships. And because it is impossible to replace all older VLCCs with new ships quickly, their continued use for many years is inevitable.

A background of rising rates

The surge in tanker orders took place against a background of rising short term freight rates.

The tanker industry benefited from the Organization of Petroleum Exporting Countries' unofficial policy of maintaining high production levels throughout the second quarter of the year when liftings traditionally decline.

The effect on freight rates for VLCCs was startling. Rates shot up to $21,000/day on average for first half 1990, a 233% gain from the $9,000/day in first half 1989.

Increases for medium sized vessels were substantial but less dramatic.

The first half 1989 average of $13,000/day grew to $17,000/day in the same period of 1990, an increase of 31%.

Rates for product carriers in the 30,000-35,000 dwt range were unchanged at $10,000/day, while there was a small increase in rates for large product carriers from $14,000-15,000/day in first half 1989 to $16,000-17,000/day during the first 6 months of 1990.

OPEC's production policies also were reflected in the amount of tonnage used for floating storage.

At the start of 1990 there were only 46 vessels with a combined 8.5 million dwt operating as floating storage, the lowest level since the early 1980s. But by June 1 the number of storage tankers had grown to 68 with a combined 15.5 million dwt, the highest level since the middle of 1987.

Tanker demand to surpass newbuilding

Long term growth in tanker demand will outstrip newbuilding, Overseas Shipholding Group Inc. (OSG), New York, predicted in mid-1990.

With current orderbooks full, OSG's 1989 annual report projected delivery delays of as much as 3 years on orders for new tankers. "In this environment," it said, "trading conditions for modern, efficient tankers should improve over the next few years."

Considering the large tanker surpluses of the mid-1970s, soft freight rates for very large crude carriers, and increased building activity, a tanker supply crunch may seem unlikely.

But the OSG report outlined four factors in support of its forecast: increased production by members of the Organization of Petroleum Exporting Countries, increased non-Communist oil consumption, decreased worldwide shipyard capacity, and an aging international tanker fleet.

Because of falling scrappings rates and high construction and resale prices, OSG said it believed two thirds of the world tanker fleet will be at least 15 years old by early 1995 if the present newbuildings rate remained constant and scrapping rate was zero.

There is no reason to believe there will be any short term increase in the number of vessels scrapped as long as building and resale prices remain high, OSG said. In fact, the 1989 scrappings rate fell to less than 2 million dwt, the lowest since 1974 (Fig. 3).

In the long run, however, scrappings should begin to surge after 1995, OSG said. Older vessels will begin to need frequent, costly repairs and renewals to comply with insurance and safety regulations. Such demands will make them more expensive to maintain and operate than more modern tankers.

At some point, OSG said, economic considerations require older tankers to be scrapped. Growing worldwide concerns for tanker safety and environmental protection are

TANKERS

likely to speed retirement of substandard tonnage.

Tanker newbuilding figures appeared, at first glance, to be encouraging to shippers. Deliveries in 1989 were 9 million dwt, the highest rate since 1978, and another 9 million dwt were on line for delivery in 1990.

In addition, non-Communist tanker fleet capacity rose by 7 million dwt to 239 million dwt in 1989. But that barely kept up with demand.

Non-Communist oil consumption was up 2.2%, and oil supply, bolstered by steady increases in OPEC production, increased 2.6% in 1989.

U.S. oil spill liability law

The legislation Bush signed increased shipowners' liability for spills in U.S. waters from the former $150/vessel ton to $1,200/vessel ton. Owners of the crude are not jointly liable for spills.

Minimum liabilities are $10 million for vessels larger than 3,000 gross tons and $2 million for smaller vessels. Previous federal laws capped liability at $14 million.

Liability for offshore facilities was set at $75 million plus unlimited removal costs. Onshore facilities or deepwater ports would be liable for $350 million in damages. There are no liability limits if a spill results from a spiller's gross negligence, willful misconduct, or violation of a federal operating or safety standard.

The law uses an existing 5¢/bbl fee on oil to build a $1 billion oil spill cleanup fund, half of which can be used for a single spill. Oil spill victims can draw from the fund when the spiller's liability limit has been reached, the spiller is unknown, or settlement is delayed.

Table 1
Breakout of 1989 interregional trading*

	Imports Crude	Imports products	Exports Crude	Exports products
		1,000 b/d		
U.S.	5,778	2,241	130	687
Canada	358	160	625	319
Latin America	756	448	2,285	1,226
Western Europe	7,475	1,909	493	837
Middle East	10	48	11,678	1,711
North Africa	1	120	1,878	537
West Africa	10	77	2,144	175
East, South Africa	416	42	29	5
South Asia	430	269	33	42
Other Asia†	1,833	869	1,005	678
Japan	3,582	967	—	59
Australasia	190	138	128	90
U.S.S.R., Eastern Europe, China	1,106	186	1,728	1,750
Destination unknown§	211	642	—	—
Total	**22,156**	**8,116**	**22,156**	**8,116**

*Excludes bunkers and intraregional shipments—between individual countries in western Europe and between Alaska and the U.S. Lower 48, for example. †Excludes China and Japan. §Includes things like transit losses, minor shipments not otherwise listed, and unidentified military use.

Source: British Petroleum Co. plc

The new law requires double hulls on most oil tankers operating in U.S. waters under a timetable based on the vessel's age and gross tonnage (Fig. 4).

Exceptions included the delivering vessel in a lightering operation and tankers calling at Louisiana Offshore Oil Port (LOOP), both until 2015.

The law also requires improved spill response and preparedness. The federal government will direct cleanup of all major spills.

Congress voted not to have the U.S. join a pair of international oil spill liability and compensation agreements because they would preempt more stringent federal and

Table 2
Breakout of world VLCC, ULCC fleet by age and size

	200,000–254,999		255,000–319,000		320,000+		Total	
Year built	No. of vessels	dwt	No. of vessels	dwt	No. of vessels	dwt	No. of vessels	dwt
1969	1	206,972	...	...	...	...	1	206,972
1970	3	700,504	1	280,420	...	...	4	980,924
1971	7	1,563,645	3	795,062	...	...	10	2,358,707
1972	14	3,322,494	7	1,824,463	...	...	21	5,146,957
1973	22	5,131,703	17	4,610,579	...	...	39	9,742,282
1974	26	6,191,366	48	13,071,005	3	1,018,270	77	20,280,641
1975	29	6,877,935	45	12,410,869	9	3,400,242	83	22,689,046
1976	16	3,777,487	40	11,017,444	17	6,624,112	73	21,419,043
1977	3	694,705	12	3,379,663	14	5,584,239	29	9,658,607
1978	2	451,469	3	796,797	7	2,774,287	12	4,022,553
1979	...	...	4	1,133,059	4	1,898,316	8	3,031,375
1980	...	...	1	275,271	1	355,020	2	630,291
1981	2	474,998	1	267,672	1	355,020	4	1,097,690
1982	...	...	1	290,084	...	...	1	290,084
1983	...	...	2	603,313	...	...	2	603,313
1984	1	234,733	...	...	...	...	1	234,733
1985	2	490,006	...	...	...	...	2	490,006
1986	9	2,150,286	1	317,353	...	...	10	2,467,639
1987	5	1,170,334	1	259,992	...	...	6	1,430,326
1988	10	2,458,104	...	...	...	...	10	2,458,104
1989	3	748,903	...	...	...	...	3	748,903
Total	**155**	**36,645,644**	**187**	**51,333,046**	**56**	**22,009,506**	**398**	**109,988,196**

Source: Clarkson Research Studies, London

CHEVRON SHIPPING CO. tanker plies its trade against a backdrop of the San Francisco skyline. Photo courtesy of Chevron Corp.

state liability laws.

Congressional committees had considered oil spill liability bills since 1975, but a major hurdle was a controversy over whether legislation should preempt state liability laws. The new law allows states to impose stricter liability laws and set up their own oil spill compensation funds.

The American Petroleum Institute warned that the absence of federal preemption opens the door for unlimited liability at the state level.

The new law allows oil companies to establish a planned Petroleum Industry Response Organization to combat spills. PIRO will hold personnel and equipment in readiness at five regional centers in the event of spills.

A U.S. Coast Guard report and recommendations related to a Feb. 7, 1990, crude oil tanker spill off Huntington Beach, Calif., stirred controversy and could lead to tough new U.S. marine terminal restrictions.

The report, issued in June 1990, found the spill occurred because tanker terminal operator Golden West Refining Co., Santa Fe Springs, Calif., and the mooring master or his employer failed to maintain accurate water depth readings.

The Coast Guard found the American Trader tanker, with a 43 ft draft, was punctured twice by its anchor in 50-51 ft of water during mooring operations in a sea berth near the tanker terminal. The mooring master mistakenly believed the water depth in the sea berth was about 56 ft and thus assumed an underkeel clearance of 13 ft, the Coast Guard said. There were 4-6 ft sea swells at the time.

As a result of the accident and the investigation, the Coast Guard recommended new guidelines to ensure minimum safe keel clearance for vessels in ports. The agency also called for tighter standards for mooring masters' pilotage requirements.

The tanker captain and crew and the crude's owner, BP Oil Shipping Co., were found free of blame in the Coast Guard investigation.

Second deepwater port studied

A group led by Phillips Petroleum Co. disclosed plans in September 1990 to begin the second phase of a feasibility study for a supertanker oil terminal off Texas.

Eighteen oil and oil related companies took part in the first phase of the study that began earlier in the year. The study could lead to construction of a project designated Texport about 27 miles off Freeport in 110 ft of water.

The facility, which could cost $600 million to $1.3 billion, would be designed to handle 1-2 million b/d.

The country's only supertanker port at present is LOOP.

The second phase of the study aimed to explore the possibility of seeking government approvals needed for construction, a market study dealing with things such as rate structure, and government and public relations aspects of the project.

The first phase focused on three key things in determining whether to build a deepwater port: technical feasibility of the project, economics of construction and operation, and federal and state regulatory factors.

Phillips expected a decision on whether to pursue licensing for Texport in 1991. It could take more than 5 years to complete construction, including time required to obtain permits and licenses.

The possible site off Freeport, relatively close to shore, would provide ready access to an existing pipeline system. The pipeline system interconnects with refineries that can process as much as 2 million b/d of oil.

Texport also would be located at its proposed site because the world's largest supertankers need as much as 110 ft of water depth to load and offload.

Phillips said a major factor interesting coventurers in the

TANKERS

Texport study is that an oil port seems to be an effective method of handling imported oil in light of the growing number of vessels using U.S. ports.

It also said the increasing number of port calls raises questions about the ability of U.S. ports to handle vessel traffic in the future if imports rise as predicted.

For example, using the Energy Information Administration's 1989 Annual Outlook for Oil and Gas, an Interior Department study estimates that port calls by tankers will nearly double from 4,000 in 1988 to about 7,600 in 2000 if oil imports grow to about 12 million b/d.

The volume late in 1990 was more than 8.8 million b/d—6.532 million b/d of crude oil and 2.309 million b/d of products.

Phillips declined to disclose its partners in the study group because membership could change. However, it said, participants included major oil and petroleum related companies, U.S. and non-U.S.

Members of the Texport group asked Phillips to be lead coordinator because of the company's experience in similar offshore port studies.

Phillips, Seaway Pipeline Co., Dow Chemical U.S.A., and Continental Pipe Line Co. received a U.S. Department of Transportation license in September 1981 for construction of a project designated Texas Offshore Port about 12 miles off Freeport.

The port, planned as a 500,000 b/d capacity project with a 56 in. pipeline to shore, fell victim to things such as high interest rates and declining oil imports.

Excluding volumes for the Stategic Petroleum Reserve, U.S. crude oil imports hovered at a little more than 3 million b/d in 1982-85 after falling from 4.144 million b/d in 1981. Imports slid because U.S. production began rising in response to higher oil prices.

Oil imports began a steady climb in 1986 as domestic production slumped due to a price collapse brought on by flood of cheap oil on world markets.

LOOP, 18 miles off Louisiana in 155 ft of water, has an average throughput of 930,000 b/d in 1990, up from 840,000 b/d in 1989. It was completed in 1978 at a cost of $830 million with a design capacity of 1.4 million b/d.

Owners are Marathon Pipe Line Co. 32.1%, Texaco Inc. 26.6%, Shell Oil Co. 19.5%, Ashland Oil Inc. 18.6%, and Murphy Oil Corp. 3.2%.

A new generation of offshore loading systems

A NEW GENERATION OF OFFSHORE LOADING SYStems is set to make an appearance in 1991.

Imodco, a member of the IHC/Calland group, will install two catenary anchor leg moorings (CALMs) in Petrobras' Marlim field off Brazil in more than 1,300 ft of water.

Brazil has set the pace in deepwater production throughout the 1980s with a handful of mooring system in more than 400 ft of water. In 1986, Petrobras achieved a deepwater record with a system installed in 754 ft of water in Albacora field.

The two new Imodco CALMs will permanently moor storage tankers of 120,000 dwt and load crude through fluid swivels rated at 1,140 psi. Shuttle tankers, also of 120,000 dwt, will offload the crude by tandem transfer.

Operators of offshore fields are turning increasingly to tanker storage systems that are equipped to transfer crude. There is an increasing trend toward turret moorings for those units. In addition, Imodco was to moor a newly built 125,000 dwt storage barge in Dulang field off Malaysia.

The barge, which will have a working life of 25 years, will be moored with an Imodco single point turret system that will be externally mounted. The crude oil swivel unit will be piggable, and the stack will include a gas swivel and a 350 amp electrical power slipring.

Turret mooring systems increasingly used

Single Buoy Moorings Inc. (SBM), Monaco, expects turret mooring systems to be used for most floating storage and offloading (FSO) projects and floating production storage and offloading (FPSO) projects worldwide in the future.

The company devoted considerable effort to developing

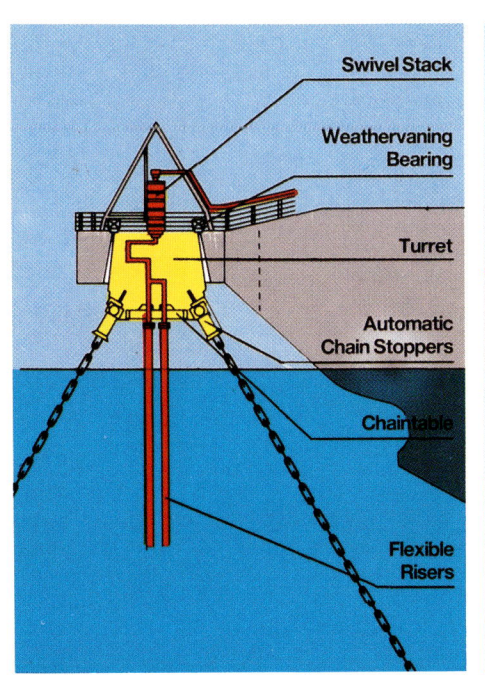

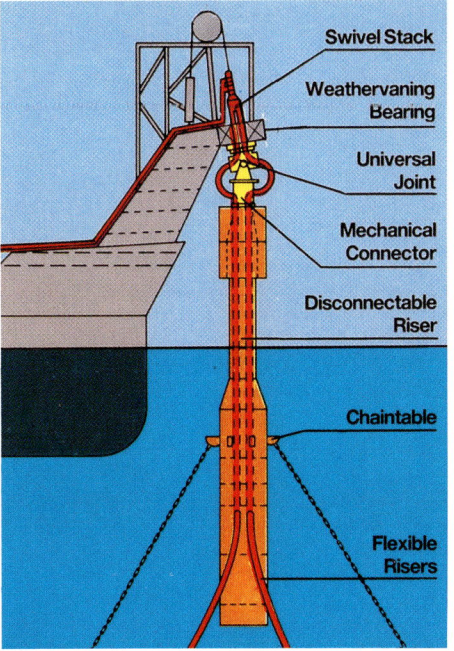

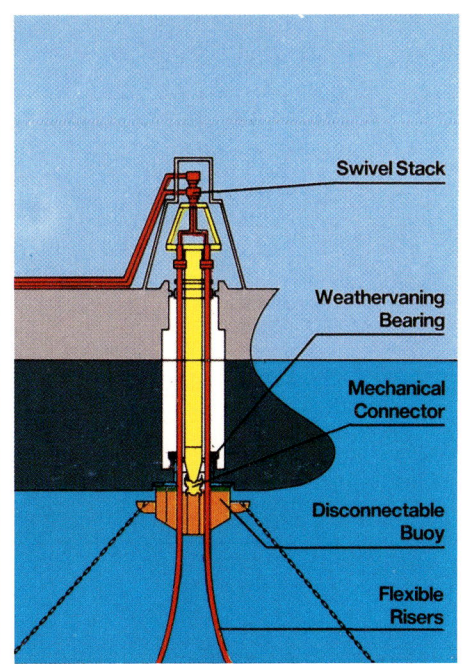

THREE types of turret mooring systems from SBM. At left, permanent turret mooring. In the center, disconnectable riser turret mooring. At right is shown disconnectable buoyant turret mooring (Fig. 1).

TANKERS

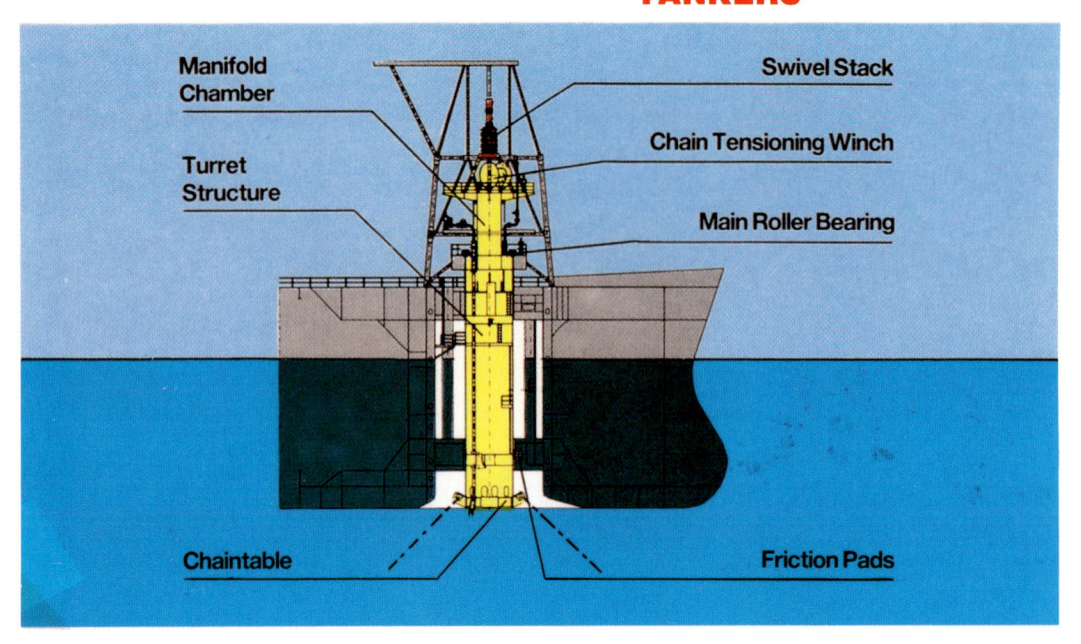

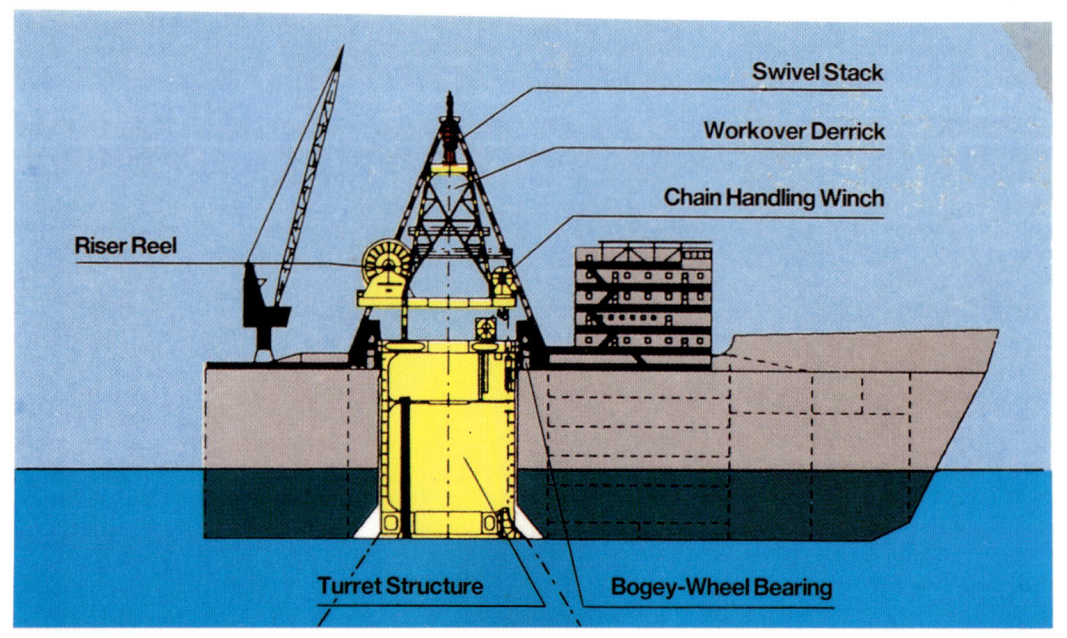

SMALL-DIAMETER internal turret from SBM (top photo). SBM's large-diameter internal turret (bottom photo). (Fig. 2).

SBM has patented the RTM. In areas where hurricanes or typhoons are a feature of seasonal production, the system allows a tanker to disconnect quickly and sail to safer waters when hurricanes or typhoons are forecast.

The interface between the riser and the mooring arm is a bearing, a universal joint and a mechanical connector. The riser is moored by CALM, and the disconnection is carried out by remote control. Reconnection is performed by equipment on the tanker without external assistance. SBM installed an RTM in BHP Petroleum's Jabiru field in the Timor Sea off Australia. The system design allows for connection of four wells.

The 140,000 dwt storage and production tanker is moored in 400 ft of water about 400 miles west of Darwin in an area subject to severe cyclones. The system, which has been handling about 27,000 b/d, can be disconnected in less than 2 min. The BTM also was developed for typhoon areas where fast disconnection is required.

The disconnectable part of the turret is a submerged buoy that supports the catenary anchor legs. In operations, the mooring buoy is linked to the turret in the bow of the ship and supported on a weathervaning bearing.

The turret extends through the tankers with the reconnection winch and fluid swivels located above the main deck. Disconnection and reconnection are carried out from the tanker without external help.

SBM develops internal turrets

SBM developed internal turrets that are located inside a vessel's hull when harsh environments and a larger number of risers have to be accommodated. Different types of systems have been developed using small and large diameter turrets (Fig. 2). SBM said a small diameter internal turret would typically be used in North Sea conditions for applications involving multiple risers. It is suitable for tankers converted either to FSOs or FPSOs. The slender turret is supported vertically by a roller bearing at deck level. Friction pads near the keel provide horizontal support. Special structures are used for the connection of the turrets's rigid bearing support to the relatively flexible vessel deck.

such systems and has three applications: a permanent turret mooring (TM), a disconnectable riser turret mooring (RTM), and a disconnectable buoyant turret mooring (BTM) (Fig. 1).

SBM said during the relatively short period turret systems have been in use, they have proved reliable and suitable for a wide range of applications. Turrets are particularly attractive for mooring vessels in deep water. Designs for as much as 1,100 ft of water are readily available, while systems for deeper water are feasible. By carefully designing the anchoring lay and using a balanced combination of chain and wire, SBM said, safe turret mooring systems can be adapted from shallow waters for all types of vessels.

The TM is a steel box type structure mounted directly on a tanker's hull. It provides a foundation for the weathervaning bearing arrangement carrying a chain table to which mooring chains as well as fluid transfer hoses are attached.

Product and electrical lines run from the seabed via swivels to the tanker. SBM said one of the advantages is that the complete structure is above water level, providing easy access for inspection and maintenance.

TANKERS

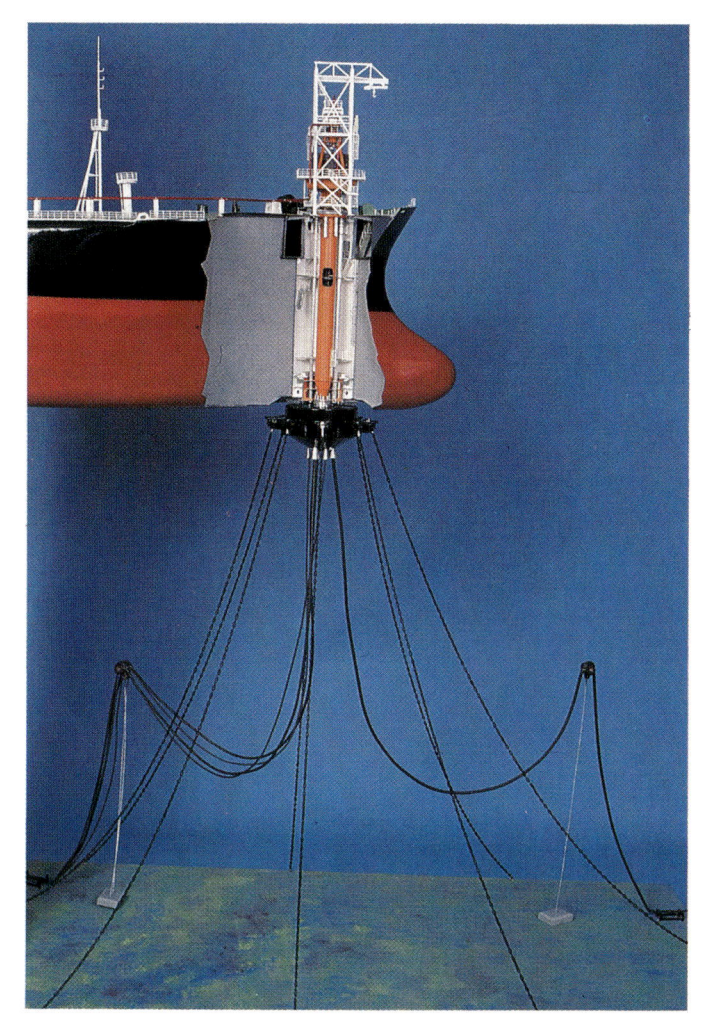

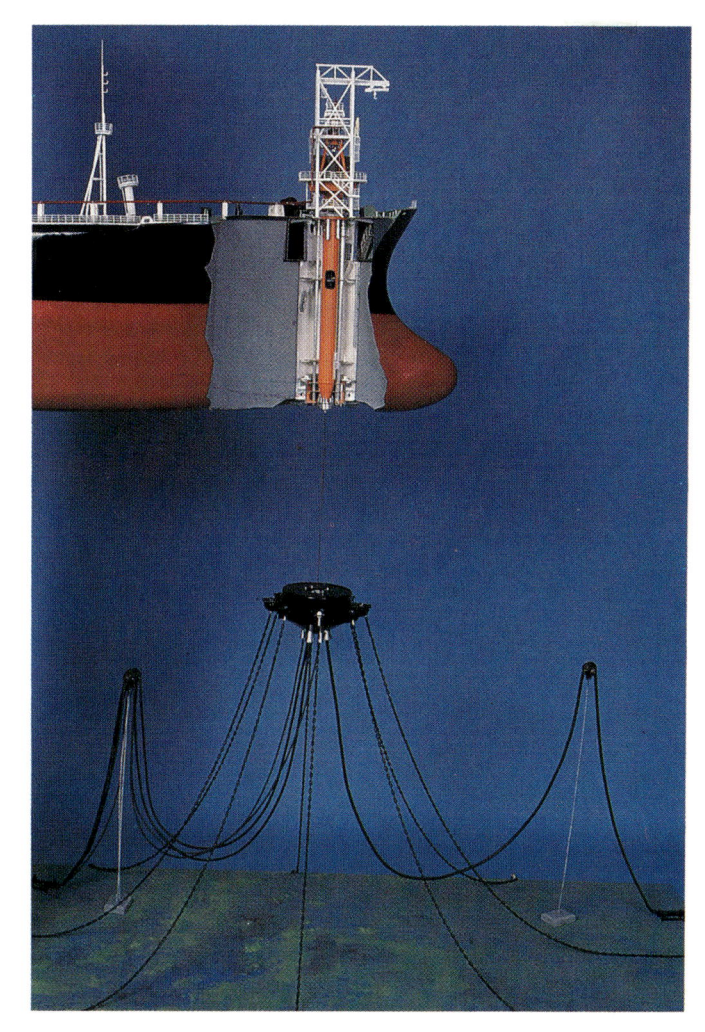

ACT OPERATORS group's disconnectable buoyant turret mooring, in Huizhou field, South China Sea, offshore China. At left is shown the system in operation. At right, after disconnection, the mooring buoy has sunk to a depth of 115 ft below the surface (Fig. 3).

Design is arranged so that turret connection and tanker conversion can be performed in different yards.

SBM also has a large diameter internal turret developed for harsh environments in which a very large number of risers must be accommodated. This system allows well workovers to be carried out by repositioning the system in the field through individual anchor leg adjustment.

The turret, which can be integrated into newly built vessels, is connected to the ship through a system of bogey wheels at deck level. The bogeys are arranged on a large diameter pitch circle to transfer the vessel mooring loading and the overturning movement. A typical diameter of such a turret could be 20 m.

How Rospo Mare's system works

One of the first turret moored FSOs was installed in Rospo Mare field in the Italian part of the Adriatic Sea in 1987. A contract to design, supply, and install the turret mooring system for the 140,000 dwt tanker Alba Marina was placed with SBM by Elf Italiana SpA. The turret system was integrated into the stern of the tanker, which had been converted to a storage unit. The turret is moored on six catenary mooring chains anchored by driven piles.

The fluid transfer system, transporting crude oil from the nearby production platform, is formed by two 8 in. hoses in a steep "S" configuration. A swivel stack arrangement consists of three electric swivels and one fluid swivel. The electric swivels allow transfer of electrical power from the FSO to the production platform along subsea power cables.

Stored crude is offloaded into tandem moored shuttle tankers as large as 100,000 dwt. The terminal, in 233 ft of water, is designed to withstand 100 year storm conditions: a 53 ft maximum wave height, 93 knot wind, and 2½ knot current.

Huizhou field system off China

ACT Operators Group, a combine of Agip SpA, Chevron Inc., Texaco Inc., and Nanhai East Oil Corp., placed an order for a BTM for installation in Huizhou field in the South China Sea (Fig. 3). Huizhou lies about 100 miles southwest of Hong Kong in 380 ft of water. SBM installed an internal turret in the forepeak structure of a converted 250,000 dwt tanker. The mooring is achieved by an array of eight equally spaced catenary anchor legs connected to the turret through the submerged mooring buoy.

The mooring buoy is connected to the turret structure by a collet-type structural connector. During a disconnect, the mooring buoy will fall away from the keel of the tanker and stabilize at a depth of 115 ft below sea level.

To reconnect, the mooring buoy is lifted up under the tanker using a wire rope hauled in a drum winch. The mooring system for the ACT tanker is designed to remain attached in sea states with waves up to 26 ft. Reconnection can be made in waves of as much as 11½.

The FPSO for Huizhou is fitted with a plant to process partly stabilized crude from the HZ-21 and HZ-26 platforms flowing to the vessel through 8 in. and 12 in. flow lines and risers. The mooring system also accommodates a 6 in. fuel gas line, a 6,800 kw power cable and an 8 in. water injection line.

Gulf of Mexico activity inches upward slowly

PLATFORM in the Gulf of Mexico's South Pass area produces natural gas. (Photo courtesy Panhandle Eastern Corp.)

GULF OF MEXICO EXPLORATION AND DEVELOPMENT during 1990 continued inching upward despite weak oil and gas prices in the first half of the year.

Indicators of the pace of future gulf oil and gas activity were mixed.

Participation in U.S. Department of the Interior Minerals Management Service's area-wide lease sales in the central and western gulf diminished during 1990, as price incentives deteriorated.

Less leasing activity translates into fewer opportunities in the gulf for future exploration.

At the same time, federal and state officials declared moratoriums on leasing in all U.S. waters but the central and western gulf and the Beaufort Sea off Alaska, eliminating development in important oil and gas frontiers and infusing new urgency in efforts by gulf operators to replace dwindling U.S. reserves.

In terms of drilling and well completions, 1990 gulf activity was more similar to 1989 than any other year in the past decade. But the action moved to a still higher plateau.

Well completions, active rigs

American Petroleum Institute figures showed that by the end of November 1990 gulf oil well completions had surpassed 1989's total of 265 and seemed certain to top 300 by yearend. Similarly, gulf gas well completions at the end of November were on track to eclipse 1989's total of 475, itself a 10 year high.

API data also showed that activity off Louisiana paced 1990 oil and gas well completions, registering 243 and 304 wells, respectively, by the end of November. Oil well completions off Texas during 1990 totaled 28, gas well completions 136.

As compiled by Baker Hughes Inc., the average count of active rigs in the gulf during 1990 remained flat compared with 1989, with 76 rigs in the central gulf and 22 rigs in the western gulf. During 1989, respective average weekly rig counts in the gulf were 72 and 22.

Mobile drilling units engaged in drilling, workovers, or completions in state and federal water off Texas averaged 35 during 1990, up four units from 1989, according to the biweekly rig count of Grasso Marine Service Centers

But as the year progressed, Grasso's average rig count fell. During the first 6 months of 1990, the count averaged 38.5 units. During the last 6 months it averaged 31.6, just above 1989's yearlong average of 31.

Despite only a slight gain in gulf drilling activity, by the end of November 1990 API had recorded 2,858,095 ft in oil well footage drilled in the gulf, surpassing 1989's 12 month total by more than 365,000 ft.

Gas well footage, 4,151,734 ft by the end of November, was the only measure of gulf drilling activity that didn't outpace the previous year. But gas well footage in 1989, 4,669,135 ft, was the highest compiled by API in a decade and nearly 37.5% greater than the closest 12 month total of 3,396,797 ft in 1984.

Solid foundation for a rebound

Factors that made the gulf ripe for a rebound in upstream work in 1989 endured into 1990.

Although participation was off, MMS area-wide lease sales provided explorationists with access to a broad range of prospects in the central and western gulf. High quality seismic data on offshore blocks and growing portfolios of prospects provided a firm foundation on which to base 1990 gulf exploration and development (E&D) programs.

Because of the high cost of 3D seismic surveys, most companies operating in the gulf limited their use to delineation on producible leases.

But several seismic contractors during 1990 were offering

GULF OF MEXICO

speculative 3D seismic surveys in shallow and deep waters.

An infrastructure of pipelines and drilling and production platforms, as well as ready availability of offshore rigs and related equipment, allowed operators to implement E&D plans on schedule.

Those factors instilled offshore players with enough confidence to plan gulf projects with more ambitious technological requirements.

For example:
• Texaco U.S.A. drilled the first two documented horizontal wells in the gulf.
• Exxon U.S.A. disclosed details of a development plan in Mobile Bay that required a combination of technological achievements.
• Mobil Exploration & Producing U.S. Inc. completed one of the world's longest extended reach offshore wells.

Ultradeepwater exploration continued at a lower level, again paced by Shell Offshore Inc. BP Exploration Western Hemisphere moved ahead with plans during the year to plunge into ultradeepwater plays.

But all was not automatic in ultradeepwater plays, as witnessed by the decision of Placid Oil Co. and partners to shut down their floating/subsea drilling and production system in the ultradeep water of Green Canyon Block 29, citing sustainable flow rates too low to justify continuing the project.

A variety of prospects

Still, the variety of gulf prospects was adequate to continue attracting U.S. independents and non-U.S. production companies.

Miocene gas plays in water as much as 400-600 ft deep, Pliocene and Pleistocene oil and gas prospects in water deeper than 400 ft, and deep Jurassic Norphlet gas in shallow water off Alabama offered an array of risks and rewards that fit a host of E&D strategies.

While prospects in ultradeepwater and off Alabama garnered much attention—and a considerable share of investment dollars—most gulf drilling took place in more mature, traditional areas of activity in shallower water in the central and western gulf.

The number of bids decreased for ultradeepwater acerage during gulf area-wide lease sales, partly as a result of a reduced numer of blocks that were available.

Of the 538 central gulf tracts that received bids in Outer Continental Shelf Sale 123, conducted in March 1990 by MMS, 370 were in water less than 600 ft.

In OCS Sale 125 the following August, competition was light for ultradeepwater acreage. Only 48 bids were offered for 46 tracts in the Garden Banks area off Texas, three bids were offered on tracts in Keathley Canyon, and two on Alaminos Canyon tracts.

Technological successes in deeper water

Gulf drilling still is trending toward deeper water. The trend is being led by the search for fields with greater reserves.

Gulf operators are seeing constant innovations with significance in acquisition and processing of seismic data. That trend will continue, particularly the ability to analyze amplitude anomalies.

Innovation also is likely to continue in the way wells are drilled in the gulf, based on improved understanding of developmental and structural concepts.

Several gulf projects during 1990 recorded technological advances.

Amoco Production Co. revealed a world record water depth flow test in the gulf. During a 4 day test, the company's 1 Sidetrack 2 well in 3,492 ft of water on Viosca Knoll Block 957 flowed 22.5 MMcfd of gas and 2,670 b/d of condensate through a $^{48}/_{64}$ in. choke with 2,800 psi average flowing tubing pressure.

Baker Hughes active rig count

	Central gulf	Western gulf
1980	139	72
1981	154	74
1982	137	78
1983	124	48
1984	132	55
1985	129	48
1986	69	19
1987	65	19
1988	88	22
1989	72	22
1990	76	22

At midyear, Texaco claimed the first documented horizontal well in the gulf, a province where highly deviated holes are common.

Texaco's B-12 OCS G-0974 horizontal well on East Cameron Block 265 flowed at a rate of 12.1 MMcfd of gas from a sand at 1,700 ft. The well had a horizontal reach of 670 ft.

Texaco completed a second horizontal well, B-12, from the same platform. That well produced gas at a rate of 7 MMcfd, also from a sand at 1,700 ft.

At year's end, both wells were shut in, awaiting four point tests.

Texaco planned no other horizontal wells from B platform on the block but was working on designs of horizontal programs for other areas of the gulf.

Among other operators, Unocal Corp. was considering drilling horizontal wells in the gulf to improve drainage of old fields in which sands with low resistivity looked to be a promising source of substantial unrecovered reserves.

Other Texaco successes

Early in fourth quarter 1990 Texaco discovered a field about 14 miles off Louisiana in 23 ft of water on South Marsh Island Block 240-241, where it logged about 230 ft of gas pay in 10 sands and 90 ft of oil pay in five sands of Upper and Middle Miocene age.

The deepest pay was about 12,200 ft deep, the shallowest about 7,100 ft.

At yearend, the company had drilled the discovery well and two development wells on South Marsh Island 240-241 and had plans to drill another six to 10 development wells on the acreage by the end of third quarter 1991.

Texaco planned to develop the acreage by setting caissons in each block and laying flow lines to a production platform on South Marsh Island 239.

Texaco had production on about 20 blocks in the area, which provided a solid base of knowledge on which to plan development. Many reservoirs delineated on South Marsh Island 240-241 produced on adjacent Texaco acreage, eliminating the need for an extensive test program.

Also during 1990, Texaco drilled a series of 10 exploratory and development wells from a platform ins'alled in August 1989 in 622 ft of water about 90 miles off Louisiana on Green Canyon Block 6.

Before it set the platform, Texaco said, exploratory drilling had identified six shallow Pleistocene gas pay zones. After setting the platform and commencing development drilling, the company found five more, generally deeper Pleistocene oil pays ranging from 3,000 to 8,000 ft deep.

Altogether, pay columns of Green Canyon Block 6 wells average 495 ft of gas and 320 ft of oil.

At the end of 1990, Texaco was producing as much as

GULF OF MEXICO

8,000 b/d of oil, with gas production marginal while it brought oil zones on stream. When Green Canyon 6 reaches full production during 1991, Texaco expects gas production to climb to 60 MMcfd.

Texaco expected to drill another six wells on Green Canyon 6 by August 1991, completing the first development phase that project.

In May 1990, operator Unocal and Texaco installed a production platform on West Cameron Block 196 and laid flow lines back to existing wells in 57 ft of water, about 30 miles off Louisiana.

The first new well drilled from the platform—drilled to deeper strata found five new Middle Miocene reservoirs at 13,000-14,000 ft with 102 ft of net gas pay.

Original discovery date was June 1988.

Gulf-wide in 1990, Texaco participated in 38 gross wildcat wells, about a 47% increase from 1989, and in 55 development wells, about the same as in 1989.

Texaco was operator on about 50% of exploratory and development wells.

Most development drilling targeted Middle and Upper Miocene strata in water less than 600 ft and Pleistocene reservoirs in water 600-1,500 ft deep.

Texaco said its Gulf of Mexico exploration and development program of 1991 will be similar to 1990 in terms of trends and numbers.

Mobil operations in Mobile Bay

Mobil expanded its Mary Ann field development in Mobile Bay during 1990, increasing production to 100 MMcfd of gas. Also, the company's 1990 activity on Mobile Block 823 raised expectations it will become the first producer in federal water off Alabama.

Mobil's large production increase in Mary Ann field was the result of several factors. The company completed another Norphlet well at more than 22,000 ft on Mobile Bay Block 95, performed a workover of a well on Mobile Bay Block 77, and expanded capacity of Mary Ann gas processing plant in southern Mobile County, Ala., to more than 100 MMcfd.

At yearend, Mobil was working on additional Mary Ann field wells with expected production capacities of more than 30 MMcfd.

In June, Mobil nearly doubled the previous best gas test in federal water off Alabama with a well on Mobile Bay Block 823 in 44 feet of water 4 miles south of Dauphin Island. The company's A-3 development well flowed 62.3 MMcfd of gas through a $^{48}/_{64}$ in. choke with 5,067 psi flowing tubing pressure from 322 ft of Norphlet pay. The previous record gas flow off Alabama was 36 MMcfd.

Mobil also drilled the A-4 development well on Mobile Bay Block 823, the last of four initial wells planned for the field.

Earlier in the year, Mobil disclosed plans to build a gas treatment plant for Mobile Bay Block 823 field at the site of its Mary Ann field facility near Coden, Ala. Production from Block 823 was to begin in third quarter 1991.

Other Mobil activity

During 1990, Mobil's U.S. exploration and production operations were reorganized into 16 multidisciplined teams, each of which focuses on a regional portfolio of assets. Eleven of the 16 are considered strategic teams, and five of those 11 focus on the gulf.

Mobil's five gulf teams drilled 28 development and two exploratory wells during 1990, while managing workover and completion programs. Despite low gas prices during the year, the company's average gross working interest production was 691 MMcfd of gas and 39,000 b/d of condensate.

Gulf-wide during 1990, Mobil added reserves equal to 112% of its production there.

In July, the company's Lake Charles Deep asset team claimed the gulf record for a single completion oil flow when the A-17 well on High Island Block A-382 flowed 9,300 b/d of oil through a ¾ in. choke with 2,744 psi flowing tubing pressure. Production came from 485 ft of Lower Pleistocene pay, 319 ft of which was a continuous oil sand.

This well is 110 miles southeast of Galveston, Tex., in 243 feet of water.

High Island Block A-382 is part of High Island A-573 field, which began producing in 1979. More than 100 wells had been drilled in the field at the time of Mobil's record oil flow.

During April 1990, Mobil's Morgan City West asset team installed a production platform on South Marsh Island Block 205 and the same month began producing 120 MMcfd of gas from four wells. The area held Mobil's largest gas production on the gulf coast.

Mobil's Morgan City East asset team attained a 90% success rate with 10 wells in 1990. At the end of the year the team had five wells on Green Canyon Block 18 flowing at a combined rate of 8,400 b/d of oil and 7 MMcfd of gas.

A sixth Green Canyon Block 18 well was to be placed on production.

The wells on Green Canyon Block 18 represent Mobil's largest gulf oil flow. But with an average depth of 15,000 ft and average vertical deviation of 30%, the wells also were a challenge to the company's drilling capabilities. One Green Canyon Block 18 well exceeded 19,000 ft. Deviation of some wells exceeded 50%.

Mobil also conducted a successful test of its of its shunt gravel pack system in a Green Canyon Block 18 completion.

Exxon development in Mobile Bay

Exxon disclosed plans for a long term development program in Mobile Bay during 1990 that will deal with the same set of technical requirements faced by Mobil.

By late 1993, when Exxon begins moving 300 MMcfd of sour gas from 12 Jurassic Norphlet wells to an onshore treating plant in southern Mobile County, Ala., it will have spent $2 billion.

Delineation wells will cost $15-20 million each. As many as 26 wells could be drilled during the 35-40 year life of the project.

Exxon will produce gas from Bon Secour Bay field in Mobile Bay and Northwest Gulf and North Central Gulf fields in the gulf.

Gas will be produced through subsea well templates at a production platform in each field and through six remote subsea well templates up to 4 miles away from the production platforms. Twenty-three miles of subsea flow lines will deliver gas to the platforms.

At each platform gas will be metered and dehydrated before being delivered by gathering line to the gas processing plant onshore.

The onshore treatment plant is designed to process 300 MMcfd of inlet gas with an average H_2S content of about 3%. Exxon also expects to recover 150 long tons/day of sulfur and small volumes of carbon dioxide.

Exxon's areas of development virtually surround Mobil's Mary Ann field development, the first production off Alabama.

Both companies had to design their projects to avoid disturbing the environment of Mobile Bay.

Circulation of Mobile Bay was one of the first environmental variables that had to be considered. Water movement within the bay is so restricted it flushes itself only about once a month. Yet the bay is the site of commercial and recreational fishing, and its beaches are tourist attractions.

The bay's estuary environment includes wetlands and beds of sea grass and oysters, all of which must be protected from drilling and production operations.

Mobile Bay also is ringed with a world class botanical garden and historical sites dating from the Civil War, which

Alabama's Mobile Bay

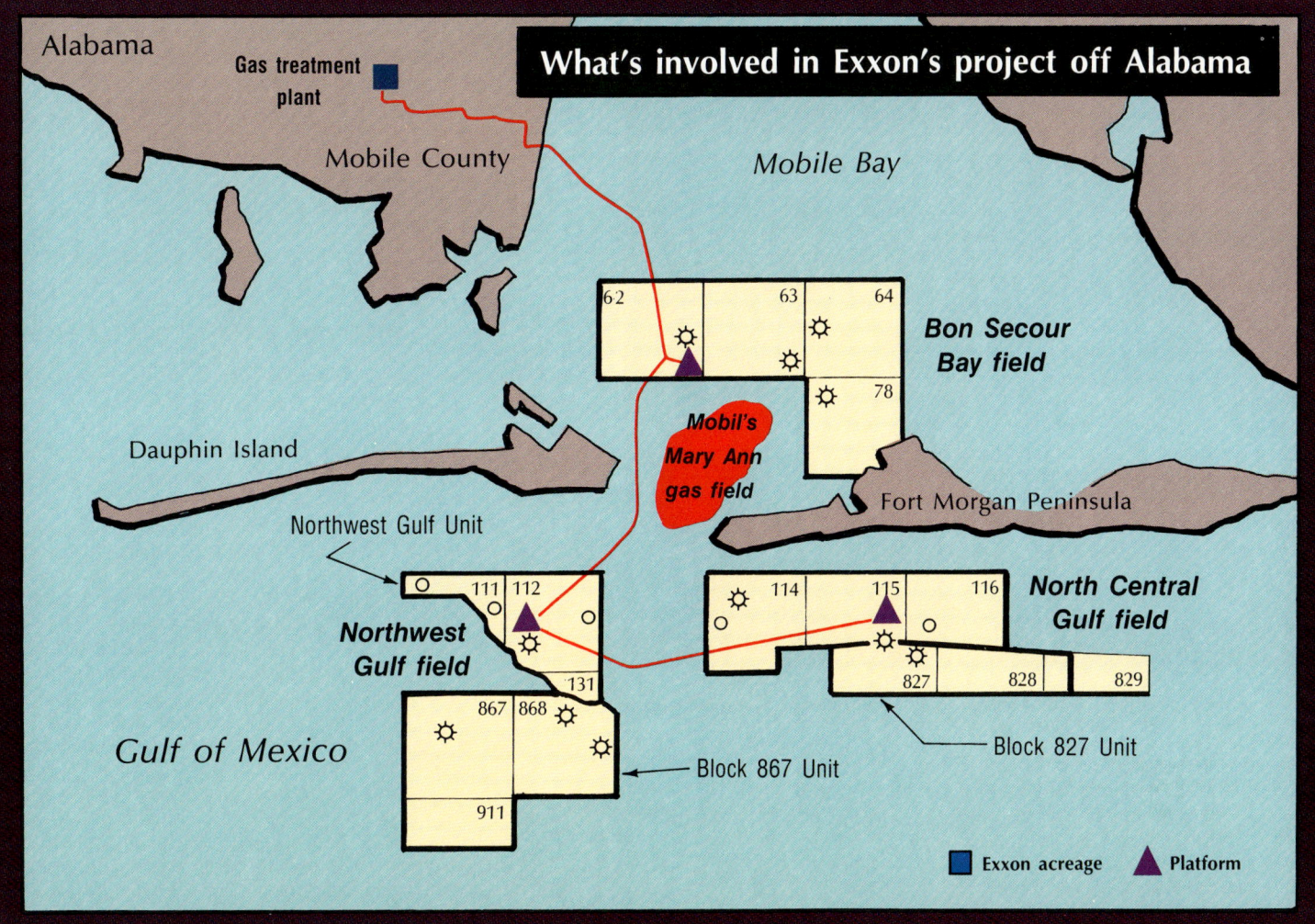

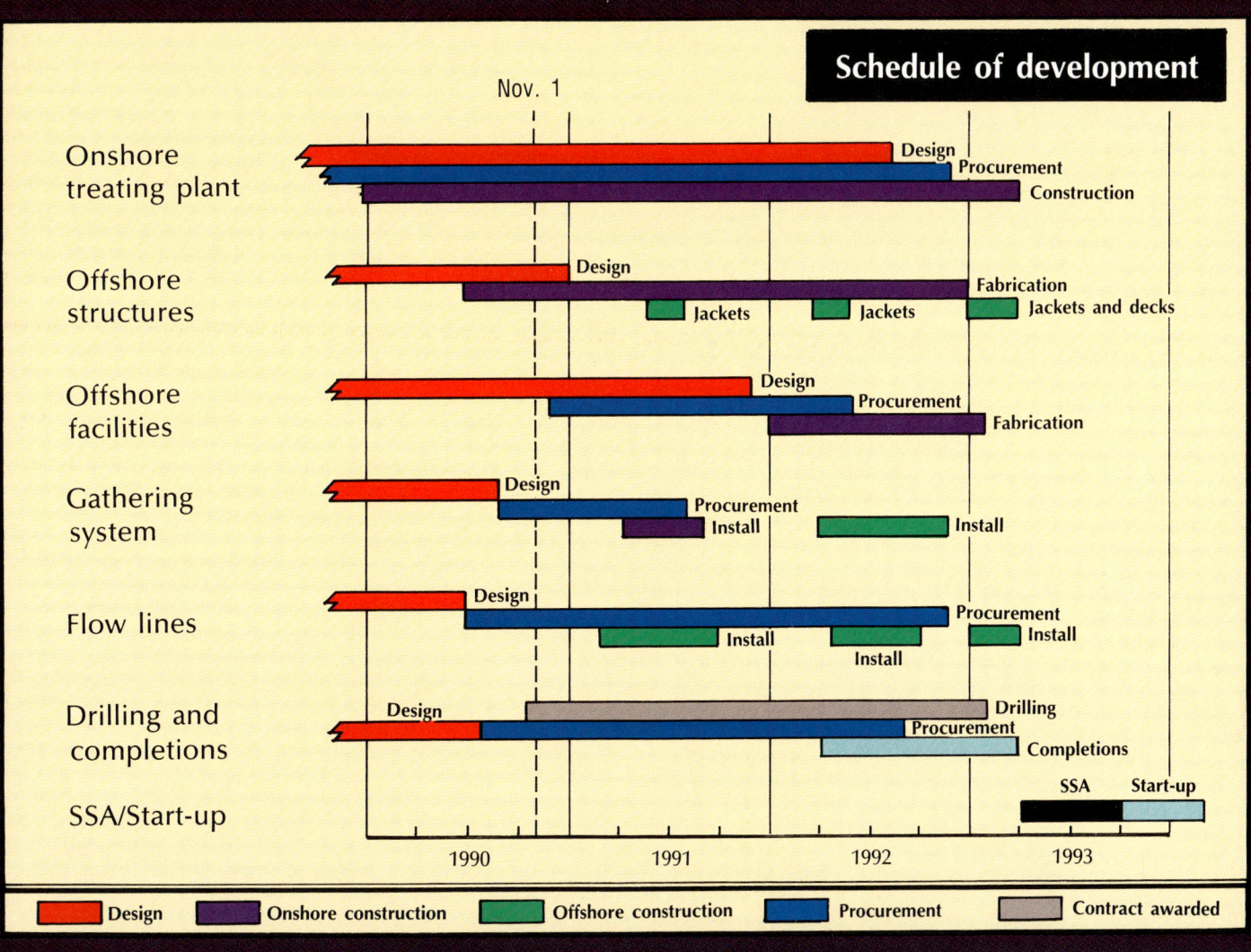

GULF OF MEXICO

CNG Producing Co. platform on East Cameron Block 346 in the Gulf of Mexico. (Photo courtesy Consolidated Natural Gas Co.)

had to be protected within terms acceptable to the industry and the public. As part of obtaining permits for the project, Exxon agreed not to discharge drilling mud and cuttings into the water.

Mobile Bay technical hurdles

The challenge of protecting Mobile Bay's environment paled when compared to the technical hurdles Exxon had to clear as a result of the hostile Norphlet downhole environment and its highly corrosive production stream.

To begin with, gas bearing Norphlet sands are overlain with more than 20,000 ft of sediment, rendering seismic analysis difficult at best because of loss of signal strength.

Although Mobil's nearby Mary Ann unit produces from the same deep formation, production data are only about 2 years old and don't reflect a significant percentage of depletion.

Because Exxon found no similar producing reservoirs elsewhere to use as models, predictions of reservoir performance were difficult.

To compound problems, the Norphlet reservoir pressures range upward to 13,500 psi with temperatures of 420° F. Concentrations of H_2S occur up to 10% and of CO_2 up to 4%, making produced fluids extremely corrosive.

Not only does H_2S have to be removed from the production stream for sales, its presence means much of the produced fluid is extremely toxic, raising the need to protect workers.

Additionally, solids carried by fluid in the production stream—in some instances in concentrations greater than 300 ppm—can plug production equipment. And gaseous hydrocarbons in the production stream crystalize at ambient temperatures.

Corrosion resistant alloys

To deal with the combination of high temperatures, high pressures, and the extremely corrosive production stream, Exxon decided to fabricate all production equipment and components of various grades of corrosion resistant alloys (CRAs) composed of nickle, chrome, lithium, tungsten, and iron.

Carbon steel would last only a few months under those conditions, and chemical inhibition would not be reliable enough to mitigate safety and environmental risks.

To permit full scale testing of the unusual CRA materials under service conditions, Exxon had to built prototype tubulars, connections, wellhead valves, elastomer seal assemblies, and other completion hardware.

High strength CRAs, used in flow lines to keep wall thickness manageable, typically lose strengh when welded. So Exxon developed a proprietary field welding procedure that avoids the problem.

Flow line properties also present unique thermodynamic problems.

Seawater would cause solid hydrocarbon compounds and hydrates to precipitate in uninsulated flow lines. Exxon had to insulate submarine lines to minimze cooling.

Gas coolers then had to be installed in well templates to keep insulation within acceptable temperature ranges and avoid prohibitive thermal stressing.

Exxon expects flowing wellhead pressures to be about 9,000 psi, up to 12,000 psi when shut in. So flow lines and wellhead equipment were designed to withstand pressures of 10,000-12,000 psi to avoid the need of costly well template safety and emergency relief systems.

Onshore, Exxon had to design the gas treatment plant to handle H_2S concentrations that will vary by a factor of 2,000 and still meet stringent air emission regulations.

The treatment plant also had to accomodate a 50:1 swing in inlet CO_2 content, a ratio that exceeds standard industry practice by a factor of four.

Exxon believes its Mobile Bay installation is the only one of its kind in the world.

The company planned to resume drilling operations in Mobile Bay during 1991 and to begin laying gathering lines for the project about midyear.

Ultradeepwater projects

Also during 1990, Exxon disclosed plans to spend about $500 million to develop two ultradeepwater gulf fields 50 miles south of Grand Isle, La.

Exxon planned to set a conventional platform, dubbed Alabaster, in 468 ft of water on Misissippi Canyon Block 397 to receive production from Mississippi Canyon Block 354 via a multiwell subsea system, named Zinc, in 1,500 ft of water.

Exxon expected to start production by early 1992 from 22 wells 7,000-15,000 ft deep. Production during the 20 year life of the project was expected to peak at 115 MMcfd of gas and 2,500 b/d of condensate.

Exxon officials said the Zinc production and gathering system will be the company's first gulf application of deepwater subsea technology proved by its submerged production system more than a decade ago.

Designed for diverless operation, the Zinc production system will be the oil and gas industry's largest multiwell subsea satellite application in the gulf.

Zinc will surpass the Lena guyed tower as Exxon's deepest water development in the gulf. The Lena tower was installed in 1,000 ft of water in 1983.

Shell Offshore also advanced several significant ultradeepwater projects in the gulf during 1990.

Last summer, the company let contracts valued at $475 million for construction and installation of components of its Auger development on Garden Banks Block 426 off Louisiana, where it plans to install a tension leg platform (TLP) in 2,860 ft of water.

Shell Offshore's Auger project, about 214 miles southwest of New Orleans, includes Garden Banks Blocks 426, 427, 470, and 471. Development costs are estimated to total about $1.3 billion.

McDermott Inc. will fabricate the TLP deck, mate the deck and hull, and install the TLP, lateral mooring system, and deepwater pipelines. Belleli SpA wil fabricate the hull.

GULF OF MEXICO

Oil and gas completions and footage for Gulf ores

	\- Oil well completions \-				\- Gas well completions \-			
	Alabama offshore	Gulf of Mexico northern federal waters	Louisiana offshore	Texas offshore	Total Gulf of Mexico	Alabama offshore	Gulf of Mexico northern federal waters	Louisiana offshore	Texas offshore	Total Gulf of Mexico

Year	Alabama offshore	GoM northern federal waters	Louisiana offshore	Texas offshore	Total Gulf of Mexico	Alabama offshore	GoM northern federal waters	Louisiana offshore	Texas offshore	Total Gulf of Mexico
1980	NA	14	126	6	146	NA	0	206	44	250
1981	NA	4	150	9	163	NA	0	207	72	279
1982	NA	17	113	17	147	NA	10	196	94	300
1983	NA	13	231	21	265	NA	8	124	129	261
1984	0	6	310	2	318	1	1	278	75	355
1985	0	7	139	22	168	3	3	150	90	246
1986	0	16	159	10	185	1	2	117	90	210
1987	0	24	163	9	196	1	2	106	35	144
1988	0	13	141	1	155	1	12	113	57	183
1989	0	47	205	13	265	0	28	293	154	475
1990*	0	16	243	28	287	2	10	304	136	452

Year	Alabama offshore	GoM northern federal waters	Louisiana offshore	Texas offshore	Total Gulf of Mexico	Alabama offshore	GoM northern federal waters	Louisiana offshore	Texas offshore	Total Gulf of Mexico
			Oil well footage					Gas well footage		
1980	NA	168,728	1,314,357	43,136	1,526,221	NA	0	2,076,886	415,276	2,492,162
1981	NA	47,329	1,412,245	80,713	1,540,287	NA	0	2,112,923	679,994	2,792,917
1982	NA	183,832	1,117,360	181,734	1,482,296	NA	123,187	1,991,511	824,041	2,939,369
1983	NA	158,276	2,264,523	187,287	2,610,086	NA	91,322	1,430,347	1,189,822	2,711,491
1984	0	56,310	2,952,490	15,953	3,024,753	21,000	10,275	2,658,371	707,151	3,396,797
1985	0	75,900	1,402,400	195,600	1,673,900	64,400	27,400	1,540,400	907,800	2,540,000
1986	0	176,320	1,457,400	96,602	1,730,322	22,390	15,481	1,298,367	998,752	2,334,990
1987	0	260,845	1,534,846	100,091	1,895,782	21,710	21,128	1,168,353	376,961	1,588,152
1988	0	131,995	1,306,788	10,200	1,448,983	22,950	92,953	1,171,555	593,957	1,881,415
1989	0	504,356	1,861,900	126,060	2,492,316	0	254,838	2,915,083	1,499,214	4,669,135
1990*	0	169,057	2,405,389	283,649	2,858,095	5,024	103,962	2,868,221	1,174,527	4,151,734

*1990 well completions and footage through November.

Sources: Rig count — Baker Hughes Inc.
Well completions — American Petroleum Institute

Data provided by Oil & Gas Journal Energy Database.

Plans call for McDermott to install the Auger TLP in 1993.

Sonat Inc.'s George Richardson semisubmersible rig began drilling Auger wells about midyear 1990.

Shell Offshore expects to begin Auger production in late 1993, peaking at 40,000 b/d of oil and 150 MMcfd of gas. Early estimates place ultimate recovery at about 220 million bbl of oil equivalent.

Also in 1990, Shell Offshore completed development drilling from Platform Bullwinkle on Green Canyon Block 65 in 1,350 ft of water. The company expected to start production through permanent facilities late in 1991. Bullwinkle is the world's tallest fixed platform.

Shell Offshore also had a redevelopment program under way on Cognac platform in 1,025 ft of water on Mississippi Canyon Block 194.

Until Bullwinkle was installed in 1988, Cognac held the gulf deepwater platform record.

In 1987, Shell Offshore set the world water depth drilling record with an exploratory well in 7,520 ft of water on Mississippi Canyon Block 657.

Long a leader in ultradeep water in the gulf, in 1987 Shell Offshore drilled no wells in water less than 1,500 ft deep.

More recently, the company has been active at shallow and deepwater sites throughout the gulf. In 1990, it drilled 13 exploratory wells and 75 development wells.

Shell Offshore produces about 123,000 b/d of oil and 987 MMcfd of gas in the gulf. The company holds about 15% of all gulf acreage, and about 37% of all deepwater gulf acreage.

Shell Offshore intended to start production from Fairway field in Mobile Bay during second half 1991. The company made significant progress with development drilling there in 1990.

Shell Offshore said its Yellowhammer gas treatment plant in southern Mobile County will be ready for start-up in time to receive gas from Fairway field.

Placid pullout

Not all news in 1990 involving ultradeepwater Gulf of Mexico activity was positive.

Placid Oil Co. and its partners in April ended operations on the $400 million floating subsea drilling and production project on Green Canyon Block 29. The decision came when the system's 31-4 satellite well in 2,243 ft of water on Green Canyon Block 31 was lost because of downhole equipment failure.

Placid said sustainable flow rates from the few wells drilled at the site weren't enough to justify continuing the program.

When Placid's Green Canyon Block 29 project went on stream Nov. 14, 1988, it launched an era of ultradeepwater development in the gulf. With a flow rate of 2.5 MMcfd of gas, the 31-4 satellite well established the water depth record for gulf production.

The well's rate of production was gradually increased to

GULF OF MEXICO

8.15 MMcfd and 1.066 b/d of condensate. But the flow fell into a steady decline, ceasing in late 1989.

Two other subsea wells and one template well had been drilled in the project, which was intended to involve as many as eight satellite and 12 template wells. However, all three wells were beset with problems.

Placid's 999-1 satellite well on Ewing Bank Block 999 began sanding up early in 1989 when a gravel pack in the lower of two producible zones failed.

Following failure of the lower gravel pack, the lower zone was plugged and the upper zone was perforated and completed.

That zone still was producing when the project was shut in, but Placid said the flow rate had declined.

The 31-6 satellite well, in the same water depth as the 31-4 well on Green Canyon Block 31, and the lone template well brought on stream never performed as expected.

Coming at a time when other gulf operators were stepping up ultradeepwater exploration and production, the failure of Placid and its partners on Green Canyon Block 29 was a reminder of the risks in such projects.

Activity by ARCO

Among other operators, ARCO in 1990 drilled 38 wells in the gulf, 15 of which were successful.

The company has a diverse portfolio of exploration and development prospects in the gulf and doesn't specialize in a particular type of play. But during 1990 most of its success involved gas prospects, mainly stratigraphic Miocene plays based on seismic anomalies scattered all over the gulf at different geologic levels.

Some were purely stratigraphic traps, and others were fault assisted.

Its most significant successes were concentrated in a swath cutting across northern High Island, West Cameron, and East Cameron areas.

Off Louisiana, ARCO had projects under way on East Cameron Block 60 and in the West Cameron area, where it drilled seven wells in water depths of less than 200 ft during 1990, most of which were extensions of Miocene and Pliocene projects.

ARCO also drilled in the Upper Miocene play in West Cameron Block 241.

Along the Texas coast, just outside state waters, ARCO found commercial gas reserves along a highly faulted Miocene ridge below 12,000 ft on Matagorda Island Block 591, the only block in the area that had not yielded a commercial field.

At the end of 1990 ARCO was shooting 3D seismic surveys in the area, delineating fault blocks before continuing its drilling program.

ARCO has extensive 3D seismic data on a spread of acreage in the Mustang Island area where it was drilling Miocene fault traps about 7,000 ft deep. In 1990, the company drilled Mustang Island Blocks 800 and 792 and will be developing additional plays over time.

The company also pursued Miocene prospects on Galveston Block 239 and High Island 177.

The lone ultradeepwater well ARCO drilled during 1990 was in 2,600 ft of water on Mississippi Canyon Block 118, where it had hoped to find pay in turbidite sands on the northwestern and southern flanks of a large salt dome. After finding the sands—but not commercial pay—ARCO abandoned the well as a dry hole.

ARCO has significant acreage in the Mississippi Canyon area, but in the near term won't be concentrating its efforts there.

Instead, the company's deepwater exploration will focus on prospects in water less than 2,000 ft deep.

Still, ARCO says it has been ramping up its gulf exploration and development program since 1986.

Chevron's gulf activity

In terms of number of wells drilled during 1990, Chevron was perhaps the most active operator in the gulf.

The company drilled 62 wells in the gulf during 1990 in shallow and deep water, stretching the breadth of the gulf. Only six of the wells were wildcats.

Most of Chevron's wells were drilled to Miocene objectives. Next largest group could be classified as Pliocene prospects, followed by Pleistocene.

Chevron's lone ultradeepwater prospect was a delineation well with multiple sidetracks in 2,800 ft of water on Green Canyon Block 161.

That well was a considered a success, and at yearend Chevron was studying the area, with plans for more delineation drilling before embarking on an extensive development program.

Chevron drilled prospects mainly on the basis of 2D seismic information.

But it deepened an existing well on Garden Banks Block 147 seeking Pleistocene sands based on 3D seismic information.

Using 2D seismic data, Chevron deepened a well to 17,000 ft on High Island Block 71 to test the Miocene. The company explored Middle to Lower Miocene with wells on West Cameron Block 362 off Louisiana and High Island 153 off Texas.

Chevron drilled a Miocene-Pliocene well on Vermilion Block 176, and tested cap rock with a well on Grand Isle Block 172.

In a Lower Cretaceous carbonate play off Alabama, the company drilled an exploratory well on Viosca Knoll Block 169.

Chevron's exploratory program previously focused on opportunities in mature areas with Miocene or Pliocene reservoirs. But about midyear 1990, Chevron began to redirect its exploratory program toward prospects in which there were cost effective opportunities to find larger reserves.

That meant more thoroughly evaluating the potential of its producing properties to include deeper horizons or reservoirs that in the past had not been objectives. More attention also was given to opportunities in frontier areas of the gulf.

Spending by BP Exploration

Among other operators, BP Exploration significantly increased its investment in Gulf of Mexico exploration and development during 1990.

At the central gulf Sale 123 in the spring, BP led all bidders with 79 apparent high bids, exposing a total of $86,824,700. BP led all bidders on deepwater tracts, winning 60. Shell was second with bids on 50 deepwater tracts.

In western gulf Sale 125 of August 1990, BP again was the most active bidder, with 28 apparent high bids and gross exposure of $11 million.

BP also disclosed ambitious E&D plans last year.

The company plans to use a platform with a four leg jacket 1,060 ft tall in 1,032 ft of water to develop Mississippi Canyon Block 109. When it is installed about mid-1991, it will be the first conventional four leg jacket platform to be sited in water more than 1,000 ft of water and the second tallest jacket in the world.

Shell Offshore's Bullwinkle jacket stands 1,365 ft tall in 1,350 ft of water. The eight leg jacket weighs 49,375 tons.

BP's Mississippi Canyon Block 109 jacket will weigh 21,500 tons. The platform will accomodate 35 well slots. Peak production of 20,000 b/d of oil and 20 MMcfd of gas is expected.

Total cost of platform fabrication and installation and development drilling is projected at $250 million.

BP also planned to begin production during 1990 from its

GULF OF MEXICO

WORKERS check underside of Chevron Corp. platform in the Gulf of Mexico. (Photo courtesy Dennis Harding, Chevron)

first subsea well in the gulf, in 460 ft of water on High Island Block A-587.

The company planned to run two 4 in. flow lines to its Snapper platform on East Breaks Block 165, 5.7 miles away in 860 ft of water, where it expected to produce as much as 25 MMcfd of gas.

Meantime, a new entrant to gulf operations, British Gas plc, disclosed plans to spend $80 million to participate in 20 wells during 1990-91, mostly in the gulf and Gulf Coast area. Expenditures during its last fiscal year totaled some $250 million, $140 million for exploration.

Most wells in the company's U.S. E&D program were to be joint ventures with ARCO off Louisiana and Alabama, and with BHP Petroleum off Texas and Louisiana.

By yearend 1990, British Gas hoped to have its first U.S. gas production, and build U.S. gas reserves of 500 bcf during the next 5 years.

Deepwater pipelining

Plans to extend gulf pipelines into deep and ultradeep water have pushed offshore contractors against technological barriers.

In another Auger TLP contract let last year by Shell Offshore, McDermott International Inc. will set a water depth record for gulf pipelines with the first commercial application of the vertical J pipelaying technique.

Plans call for McDermott to start laying pipe in 2,860 ft of water on Garden Banks Block 426 and finish with connections in 1,200 ft of water. Installation is scheduled for early 1993.

Until then, two McDermott pipelay installations in 1989 will be the gulf's deepest.

McDermott used the J tubing pull method to lay 7½ miles of 12 in. line from Shell's Bullwinkle platform, in 1,354 ft of water on Green Canyon Block 65, to Shell's Boxer plaform in 750 ft of water on Green Canyon Block 19.

For Conoco's tension leg well platform (TLWP) in Jolliet field on Green Canyon Block 184, McDermott used a barge to lay 13 miles of rigid steel pipe in 1,390 ft of water. Coflexip used a reel ship to install flexible pipe to a connection with Conoco's TLWP in 1,760 ft of water.

All told, more than 53 miles of pipe were installed across Joliet field.

Near yearend 1990, Leviathon Gas Pipeline Co. signed a letter of intent that calls for Offshore Pipelines Inc. (OPI) to lay the first deepwater gathering system in the gulf. For that project, OPI was to lay 80 miles of 12 and 16 in. pipe in about 600 ft of water from East Cameron South Addition Block 338 to Eugene Island South Addition Block 330.

Access to offshore acreage

At yearend 1990, many companies operating in the gulf were concerned that policies of federal, state, or local governments are curtailing opportunities to work on the U.S. Outer Continental Shelf.

In June 1990, President Bush eliminated the possibility of drilling until after 2000 in eight offshore U.S. areas by canceling federal lease sales outside the central and western gulf, off Alaska, in the mid- and South Atlantic, and a small area off southern California. But there was little prospect of sales in those areas occurring as originally schedule by MMS, anyway, because of lawsuits, state opposition, and congressional spending moratoriums.

In the waning hours before adjournment in autumn 1990 the U.S. House passed—and the Senate accepted—an Interior Department appropriations bill that continued bans on leasing off the coasts of California, Washington, Oregon, the mid-Atlantic states, and Florida.

Congress also approved coastal zone management legislation that surrendered to affected states more control over operations on the OCS.

Raw materials · Energy · Environment

Take Oil and associated Natural Gas:
Whatever the process needed,
Lurgi has the right answer. **LURGI**

... die Anlagen baut Lurgi

Lurgi Gas- und Mineralöltechnik GmbH · Lurgi-Allee 5 · Postfach 11 12 31
D-6000 Frankfurt am Main · Tel. (0 69) 58 08-0 · Tx 41236-0 lg d · Fax (0 69) 58 08-38 88

CO_2, HC injection lead EOR production hike

OIL PRODUCTION FROM CARBON DIOXIDE AND LOW-molecular-weight hydrocarbon miscible/immiscible (HC) gas injection projects is rapidly increasing in the U.S.

Since Oil & Gas Journal's previous worldwide biennial enhanced oil recovery survey in 1988, covering the situation at the beginning of 1988, enhanced oil production has more than doubled from HC projects and increased by 49% from CO_2 projects (Table 1).

Total enhanced oil production in the U.S. at the start of 1990 was 656,700 b/d or a 6.2% increase from the same period in 1988 (Fig. 1).

Worldwide there has been little increase in EOR production, although the survey shows an increase of 11.5% to about 1.2 million b/d.

This increase is misleading because the data for two large projects, the Duri field steamflood in Indonesia and the Intisar 103D field HC miscible project in Libya, were not included in the 1988 survey.

The former came in too late and the latter was left out by mistake.

After taking into account the 115,000 b/d these two projects were producing at the start of 1988, the worldwide enhanced oil production has remained virtually the same with only a 0.1% increase.

The U.S.S.R.'s All-Union Scientific Research Oil & Gas Institute in Moscow says the Soviet Union has a large number of EOR projects scattered from Sakhalin Island in the east to the Ukraine in the west and from the Arctic Circle in the north to Azerbaijan in the south.

The U.S.S.R. has thermal, gaseous, chemical, and microbiological projects. A total EOR production figure was also not available.

The 1990 survey reveals that the more expensive enhanced oil recovery processes that involve chemical injection have experienced a dramatic decrease in the U.S. The number of projects with chemical EOR at the beginning of 1990 had dropped to 50 from the 124 projects at the beginning of 1988 (Table 2).

Likewise, production from chemical projects had been halved to 11,856 b/d from 22,501 b/d in 1988. Observers say that the high cost of chemicals in most cases put these projects in jeopardy. Also the demise of the federal "windfall profits" tax took away the incentive for maintaining some such projects.

The 1990 survey includes a look at heavy oil projects.

The definition OGJ used for heavy oil fields are those that could not be produced through conventional primary or

PUMPING UNIT in Duri steamflood on Sumatra, Indonesia. In the background are 50 MMBTU steam generators. (Photo courtesy of PT Caltex Pacific Indonesia)

U. S. EOR production climbs steadily

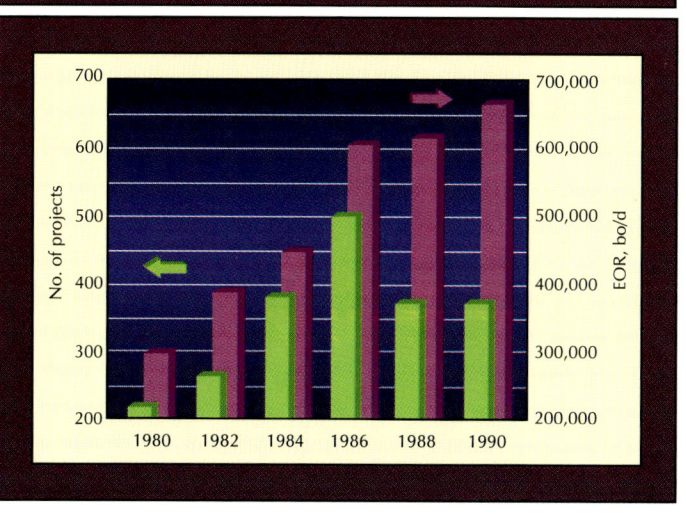

WORLD'S LARGEST EOR PROJECT, Duri field steamflood, is located on the island of Sumatra in Indonesia. (Photo courtesy of PT Caltex Pacific Indonesia)

Abbreviations

Formation type
Si: Sandstone
LS: Limestone
Dolo.: Dolomite
Congl.: Conglomerate
Tripol.: Tripolite
US: Unconsolidated sand

Project maturity
JS: Just started
HF: Half finished
NC: Nearing completion
C: Completed
PP: Postponed
Term: Terminated

Previous production
Prim.: Primary
WF: Waterflood
GI: Gas injection
C: Cyclic steam
HW: Hot water
SS: Steam soak
S: Steam
SF: Steam flood
HC: Hydrocarbon

Project evaluation
TETT: Too early to tell
Prom.: Promising
Succ.: Successful
Disc.: Discouraging

Project scope
P: Pilot project
FW: Field wide
LW: Lease wide
RW: Reservoir wide
Exp. L: Expansion likely
Exp. UL: Expansion unlikely

secondary means and were not mining projects.

All responses on active projects came from Canada and included 19 fields producing a total of 115,007 b/d. The largest by far was Esso Resources Canada Ltd.'s Cold Lake cyclic steam stimulation.

The project started in 1964 and involves 1,200 wells. Current production is 88,060 b/d.

For 1990, eight projects were slated for start-up. Two were in the U.S., the others Canada. Planned investment was $51 million.

Duri field:
international pacesetter

The most noteworthy international project has been expansion of the Duri, Indonesia, steamflood. PT Caltex Pacific Indonesia operates the project under a production sharing contract with the Indonesian state oil company, Pertamina.

Duri in 1990 became the largest EOR project in terms of daily production in the world, surpassing the production rate from Kern River field in California. Steam injection in Duri started in 1985.

As 1990 progressed, the production rate in Duri was about 160,000 b/d, or 15,000 b/d more than shown in the OGJ tabulation. The cutoff date for the OGJ survey was the end of 1989.

A peak production of 330,000 b/d was expected in Duri by 1993.

About 20% of this oil will be burned to generate steam for injection.

Ultimate tertiary recovery from this $1.8 billion project was estimated at 2 billion bbl of oil.

In Duri, development of four of the planned 12 areas was complete.

The total area developed eventually will grow to about 15,100 acres from the 4,000 acres shown in the OGJ tabulation.

The number of wells is scheduled to peak at 3,202 producers and 1,408 steam injectors.

At the cutoff date for the OGJ survey, 952 wells were producing and steam was being injected into another 288 wells.

The dominant process for Canadian EOR remained HC miscible/immiscible injection, accounting for 82.8% of the

If you've got the drive, Chaser can multiply your profits right now.

Chaser products can double or triple base oil production in EOR programs. These chemical enhancers require no re-fitting or risk to equipment. Inject them, and they go right to work.

The Chaser line includes formulations for diverse EOR operations in all parts of the globe. Production information pouring in since 1985 solidly demonstrates Chaser's ability to bring up oil from the most stubborn reserves:

San Joaquin Valley Cyclic Steam Wells:
26,135 BBL IN SIX MONTHS

Midway-Sunset Section 26c Steam Drive:
39,500 BBL IN TEN MONTHS

Athabasca Region Thermal Operation:
75-100% increase over baseline

The Chaser Technical Service Team can customize exact formulations and application procedures for individual sites and conditions to maximize EOR returns in any formation. Our research and hands-on experience make us the leading authority on EOR technology today.

But all the expertise and documentation won't bring in one additional barrel of oil until you put it to work. So if you want Chaser to start bringing up your profits, give us a call and let us know what kind of operation you have.

CHEVRON CHASER™ EOR Chemicals

Chevron Chemical Company
Enhanced Oil Recovery Chemicals
6001 Bollinger Canyon Road
Post Office Box 5047
San Ramon, CA 94583-0947

1-800-553-1196

ENHANCED OIL RECOVERY

152,953 b/d Canadian EOR production. Because of the increase in HC projects to 51 from 42 in 1988, Canadian EOR production increased by 5.8% from the previous survey.

Venezuelan enhanced oil production decreased. OGJ's previous survey showed 44 projects producing 216,360 b/d of enhanced production.

Although the number of projects dropped only by two to 42, EOR production decreased 45% to 118,788 b/d. Three of the largest drops were Bare (F.O.) field from 28,000 to 8,280 b/d, East Tia Juana field L.L. zone from 18,000 to 6,800 b/d, and Bachaquero field from 41,600 to 4,000 b/d.

The response for the Bachaquero field noted that the steam soak activity was being constrained.

Steam projects in the U.S.

Oil production from steam EOR in the U.S. continued a slow decline from the peak reached in 1986 of 468,692 b/d. A 2.7% decline occurred between the 1986 and 1988 surveys, followed by a further decline of 2.5% from 1988 to 1990. Although declining, steam still accounted for 67.8% of all U.S. enhanced oil recovery. The latest rate was 444,137 b/d.

The six reported steam projects that were planned in the U.S. in 1990-91, because of their limited size of 336 acres, will not greatly influence the production decline from steam EOR projects.

Environmental laws inhibited the expansion of steam projects in California.

Crude burned to generate steam has a relatively high sulfur content.

But one positive development that could spur added enhanced oil production was proposed pipelines to move more gas into the steam EOR areas of California. As long as gas remains cheaper than oil, gas should replace much of the crude oil burned in generating steam.

Better steam EOR economics may spur new or expanded development.

Chemical and gas

The hardest hit by low crude oil prices were chemical projects. Of the 91 terminated or completed projects in the

U.S. EOR production

Table 1

	1980	1982	1984	1986	1988	1990	Change from 1988, %
Thermal							
Steam	243,477	288,396	358,115	468,692	455,484	444,137	−2.5
Combustion in situ	12,133	10,228	6,445	10,272	6,525	6,090	−6.7
Hot water	—	—	—	705	2,896	3,985	37.6
Total thermal	255,610	298,624	364,560	479,669	464,905	454,212	−2.3
Chemical							
Micellar-polymer	930	902	2,832	1,403	1,509	617	−59.1
Polymer	924	2,927	10,232	15,313	20,992	11,219	−46.6
Caustic/alkaline	550	580	334	185	0	0	—
Surfactant	—	—	—	—	—	20	—
Total chemical	2,404	4,409	13,398	16,901	22,501	11,856	−47.3
Gas							
Hydrocarbon miscible/immiscible	—	—	14,439	33,767	25,935	55,386	113.6
CO_2 miscible	21,532	21,953	31,300	28,440	64,192	95,591	48.9
CO_2 immiscible	—	—	702	1,349	420	95	−77.4
Nitrogen	—	—	7,170	18,510	19,050	22,260	16.8
Flue gas (miscible and immiscible)	—	—	29,400	26,150	21,400*	17,300	−19.2
Total gases	74,807	71,915	83,011	108,216	130,997*	190,632	45.5
Other							
Carbonated waterflood	—	—	—	—	—	—	—
Grand total	332,821	374,948	460,969	604,786	618,403*	656,700	6.2

*Corrected

Active U.S. EOR projects

Table 2

	1971	1974	1976	1978	1980	1982	1984	1986	1988	1990	Change from 1988, %
Thermal											
Steam	53	64	85	99	133	118	133	181	133	137	3.0
Combustion in situ	38	19	21	16	17	21	18	17	9	8	−11.1
Hot water	—	—	—	—	—	—	3	10	9	−10.0
Total thermal	91	83	106	115	150	139	151	201	152	154	1.3
Chemical											
Micellar-polymer	5	7	13	22	14	20	21	20	9	5	−44.4
Polymer	14	9	14	21	22	55	106	178	111	42	−62.2
Caustic/alkaline	—	2	1	3	6	10	11	8	4	2	−50.0
Surfactant	—	—	—	—	—	—	—	—	—	1	—
Total chemical	19	18	28	46	42	85	138	206	124	50	−59.6
Gas											
Hydrocarbon miscible-immiscible	21	12	15	15	9	12	16	26	22	23	4.6
CO_2 miscible	1	6	9	14	17	28	40	38	49	52	6.1
CO_2 immiscible	—	—	—	—	—	—	18	28	8	4	−50.0
Nitrogen	—	—	—	—	—	—	7	9	9	9	0.0
Flue gas (miscible and immiscible)	—	—	—	—	8	10	3	3	2	3	50.0
Total gases	22	18	24	29	34	50	84	104	90	91	1.11
Other											
Carbonated waterflood	—	—	—	—	—	—	—	—	1	—	—
Grand total	132	119	158	190	226	274	373	512	366	295	−19.4

ENHANCED OIL RECOVERY

U.S., 54 of them were chemical.

Three of those were micellar/polymer, while the rest were polymer. In addition, operators asked that an additional 21 polymer projects be deleted from the OGJ survey.

Because of the infrastructure in the West Texas area, CO_2 EOR remained strong and was expected to grow, judging from the number of planned projects. The infrastructure makes expansion of CO_2 into nearby fields relatively economical. Half of the almost 30,000 b/d increase in enhanced oil production from HC injection projects came from Prudhoe Bay. HC injection is the only EOR process active on the Outer Continental shelf.

Future projects

In the U.S., 34 projects were in planning stages late 1989 through 1992. More than half, 18, were CO_2 projects.

The largest in area, 23,000 acres, was Mobil Oil's planned project in Postle field in Oklahoma's Texas County.

One large project in West Texas was Shell Western E&P Inc.'s expansion of the CO_2 flood in Wasson field's 28,758 acre Denver Unit.

Guide to EOR project tabulations

A. U.S. planned EOR projects by type
B. Planned projects outside U.S. by country
C. Producing Canadian EOR projects
D. Completed, terminated Canadian projects
E. Producing EOR projects outside U.S. and Canada
F. Completed, terminated projects outside U.S. and Canada
G. Producing thermal EOR in U.S.
H. Producing CO_2, gas EOR in U.S.
I. Producing chemical EOR projects in U.S.
J. Completed, terminated U.S. projects
K. Producing Canadian heavy oil EOR projects

Table A

U.S. planned EOR projects by type

Project type	Operator	Field	Location	Pay zone	Size, acres	Depth, ft	Gravity, °API	Start date
CO_2 immiscible	Chevron	Bay Marchand	Offshore Louisiana	Pliocene-Miocene	3 wells	8,500	15-20	6/90
CO_2 miscible	Citronelle Unit	Citronelle	Mobile County, Ala.	Rodessa	1,680	11,000	42	1/90
CO_2 miscible	Conoco	South Huntley	Garza County, Tex.	San Andres	450	3,100	37	1/90
CO_2 miscible	Conoco	Huntley East	Garza County, Tex.	San Andres	700	3,100	37	1/92
CO_2 miscible	Mobil	Postle	Texas County, Okla.	Morrow	23,000	6,100	40	6/91
CO_2 miscible	Mobil	Slaughter	Cochran, Hockley, Tex.	San Andres	4,800	5,000	32	1990
CO_2 miscible	Mobil	Salt Creek	Kent County, Tex.	Canyon	12,100	6,300	39	1990
CO_2 miscible	Mobil	Slaughter	Cochran County, Tex.	San Andres	1,280	4,900	32	1990
CO_2 miscible	Mobil	Levelland	Cochran, Hockley, Tex.	San Andres	12,100	4,800	32	1990
CO_2 miscible	Mobil	Russell	Gaines County, Tex.	Lower Clearfork	4,820	7,500		1992
CO_2 miscible	Phillips	Aneth	San Juan County, Utah	Desert Creek	3,250	5,500	40	1/91
CO_2 miscible	Phillips	Jess Burnes	Reeves County, Tex.	Delaware	100	3,300	36	1/90
CO_2 miscible	Phillips	Leamex	Lea County, N.M.	Paddock	120	6,200	39	2/90
CO_2 miscible	Phillips	Ranger Lake	Lea County, N.M.	Bough C	320	9,000	40	1/90
CO_2 miscible	Phillips	South Cowden	Ector County, Tex.	Grayburg/San Andres	800	4,700	35	1/91
CO_2 miscible	Phillips	Vacuum	Lea County, N.M.	San Andres	320	4,500	37	1/91
CO_2 miscible	Shell	Wasson	Yoakum Co., Tex.	San Andres	9,586	5,200	33	1992
CO_2 miscible	Texaco	Paradis	St. Charles, La.	10,000 feet	102	11,400	38.5	1/90
CO_2 miscible	Texaco	Paradis	St. Charles, La.	Main Pay	89	10,500	37.5	1/90
HC miscible	ARCO	Kuparuk River Unit	North Slope, Alaska	Kuparuk "A" & "C"	40,000	6,200	24	6/90
HC miscible	ARCO	Prudhoe Bay	Alaska		3,200			1990-91
HC miscible	Marathon Oil	Sand Dunes	Converse County, Wyo.	Muddy S.S.	6,400	12,500	40	12/90
HC miscible	Marathon Oil	So. Coles Levee Unit	Kern County, Calif.	Stevens (Fl)	40	9,700	34	4/90
Micellar	ARCO	Sabine Tram SW	Newton County, Tex.	Wilcox	2,010	11,525	49	12/89
Microbial	Phillips	N. Burbank	Osage County, Okla.	Burbank	40	2,900	39	6/90
Polymer	Mitchell Energy	Alba	Wood County, Tex.	Sub Clarksville	1,132	4,000	15.5	1990
Polymer	Phillips	N. Burbank	Osage County, Okla.	Burbank	160	2,900	39	3/90
Steam	Bechtel	Shallow Oil Zone	Kern, Calif.	Sub-Scalez	20	2,800	27	1991
Steam	Mobil	Bayou Bleu	Iberville Parish, La.	Miocene	16	11-1,600	16	1/91
Steam	Mobil	Lost Hills	Kern County, Calif.	Etchegoin	80	400	15	1/90
Steam	Mobil	Lost Hills	Kern County, Calif.	Etchegoin	100	400	15	10/90
Steam	Mobil	Saratoga	Hardin County, Tex.	Miocene	20	850	20	1/90
Steam/polymer	Unocal	Brea-Olinda	Orange Co., Calif.	1st-3rd Pliocene	100	3,500	16	2/90
Surfactant	Marathon Oil	Yates	Pecos County, Tex.	San Andres	80	1,500	30	11/89

Table B

Planned projects outside U.S. by country

Type project	Operator	Field	Country	Pay zone	Size, acres	Depth, ft	Gravity, °API	Start date
CO_2 immiscible	Trintoc	Oropouche	Trinidad	Retrench Sand	175	2,160	29	3/90
CO_2 miscible	Pan Canadian	Countess	Canada/Alta.	Glauconite	80	4,000	30	10/90
Combustion	Pan Canadian	Countess	Canada/Alta.	Glauconite	80	4,000	30	7/90
HC miscible	Chevron Canada	Acheson	Canada/Alta.	Leduc D-3	3,810	5,000	39	1991
HC miscible	Esso Res. Canada Ltd.	Boundary Lake Unit #1	Canada/B.C.	Triassic Bound	320	4,300	34	10/90
HC miscible	Mobil Oil Canada	Rainbow Twp 110,R6,W6M	Canada/Alta.	Keg River QQ	200	5,300	38	1/91
N_2 miscible	Mobil Oil Canada	Carson Creek North	Canada/Alta.	BHL	4,000	8,799	45	1/91
Polymer	OMV	Matzen	Austria	9th Tortonian	66	4,330	21	7/90
Steam	BEB	Georgsdorf III	West Germany	Valanginian	80	2,100-2,800	27	10/90

ENHANCED OIL RECOVERY

Two thirds of the unit was under continuous CO_2 injection. This area was to be converted to a water-alternating-gas (WAG) injection, and the remaining one third of the unit was to be put on continuous CO_2 injection followed by WAG at a later date. Shell's Denver Unit is the largest CO_2 EOR project in the world in terms of production. At the end of 1989 the unit was producing 41,100 b/d, of which 25,300 b/d were considered enhanced recovery.

Internationally only nine projects were planned, six of them in Canada. The largest in area, 4,000 acres, was Mobil Oil Canada's miscible N_2 project in Carson Creek North field of Alberta.

Total planned expenditures through 1992 on these projects is estimated to be $502 million.

Producing Canadian EOR projects

Operator	Field	Province	Start date	Area, acres	Number wells Prod.	Number wells Inj.	Pay zone	Formation type	Porosity, %	Permeability, md
Alkaline										
Amoco Canada*	Cessford	Alta.	7/84	680	19	7	Basal Colorado	S	24.0	350
Amoco Canada*	David	Alta.	3/86	245	22	7	Lloydminster	S	29.0	1,480
CO₂ miscible										
Home Oil	Silverdale	Sask.	12/85	160	9	3	Sparky	S	29.0	2,000
Shell Canada	Harmattan East	Alta.	8/88	1,600	20	8	Mississippian Turner Val	Dolo.	14.0	10-300
Vikor	Joffre Viking	Alta.	12/82	320	4	4	Viking	S	13.0	500
Vikor	Joffre Viking	Alta.	8/85	480	6	3	Viking	S	13.0	500
Vikor	Joffre Viking	Alta.	6/88	560	5	1	Viking	S	13.0	500
Combustion										
Mobil Oil Canada	Battrum	Sask.	10/66	4,920	76	15	Battrum/Roseray	S	26.0	1,265
Mobil Oil Canada	Battrum	Sask.	8/67	2,400	35	7	Battrum/Roseray	S	25.0	930
Mobil Oil Canada	Battrum	Sask.	11/65	680	22	3	Roseray	S	27.0	930
Mobil Oil Canada	Fosterton Northwest	Sask.	1/70	200	1	0	Roseray	S	29.0	958
Pancanadian	Countess Upper Mannville B Pool	Alta.	1/83	40	4	1	Glauconite S.S.	S	25.0	800
Murphy	Evehill	Sask.	6/80	180	21	9	Cummings Waseca	S	34.0	6,000
Foam flood										
Signalta Resources	Pembina	Alta.	10/86	2,240	13	1	Ostracod	S	12.0	70
Hydrocarbon miscible										
Amoco Canada	Ante Creek Unit 1	Alta.	1968	6,080	15	10	Beaverhill Lake	LS	6.0	9
Amoco Canada	Bigoray Nisku B	Alta.	2/80	251	5	1	Nisku D-2	LS	5.0	1,130
Amoco Canada*	Kaybob South	Alta.	7/84	8,000	40	9	Triassic	Dolo.	11.5	92
Amoco Canada	Nipisi	Alta.	2/84	28,480	168	68	Gilwood	S	18.0	200
Amoco Canada	Rainbow S.	Alta.	8/72	490	5	2	Keg River	Dolo.	6.0	40
Amoco Canada	South Swan Hills	Alta.	1973	28,480	163	51	Beaverhill Lake	LS	8.0	49
Amoco Canada*	Willesden Green	Alta.	5/75	26,000	107	55	Cardium	S	11.0	5
Canadian Hunter Expl.	Brassey	B.C.	1/89	6,043	10	6	Artex	S	16.6	137
Chevron Canada Resources	Acheson D-3A	Alta.	7/87	610	15	1	Leduc D3	LS	12.0	5,000
Chevron Canada Resources	Bigoray Nisku F	Alta.	3/87	150	2	1	Nisku F	LS	10.0	250
Chevron Canada Resources	Kaybob North BHL Unit #1	Alta.	1/88	5,400	34	9	Beaverhill Lake	LS	7.5	50
Chevron Canada Resources	Mitsue, Stage 1 & 2	Alta.	5/85	17,280	140	55	Gilwood	S	13.4	282
Chevron Canada Resources	Pembina Nisku A	Alta.	2/84	400	3	1	Nisku A	LS	8.0	250
Chevron Canada Resources	Pembina Nisku D	Alta.	11/85	400	5	1	Nisku F	LS	12.0	750
Chevron Canada Resources	Pembina Nisku F	Alta.	6/86	200	3	1	Nisku F	LS	12.7	660
Chevron Canada Resources	W. Pembina Nisku A	Alta.	8/81	200	2	1	Nisku A	LS	10.0	750
Chevron Canada Resources	W. Pembina Nisku C	Alta.	4/84	130	2	1	Nisku C	LS	11.0	1,100
Chevron Canada Resources	W. Pembina Nisku D	Alta.	5/81	320	2	1	Nisku D	LS	10.0	300
Esso Resources Canada†	Pembina 'G' Pool	Alta.	9/89	328	4	1	Nisku	Dolo.	8.0	900
Esso Resources Canada†	Pembina 'K' Pool	Alta.	1984	126	2	1	Nisku	Dolo.	12.7	2,020
Esso Resources Canada†	Pembina 'L' Pool	Alta.	1985	625	5	2	Nisku	Dolo.	10.5	1,060
Esso Resources Canada†	Pembina 'M' Pool	Alta.	1983	192	2	1	Nisku	Dolo.	9.0	540
Esso Resources Canada†	Pembina 'O' Pool	Alta.	11/83	346	2	1	Nisku	Dolo.	11.8	3,100
Esso Resources Canada†	Pembina 'P' Pool	Alta.	10/83	420	4	1	Nisku	Dolo.	10.3	2,400
Esso Resources Canada†	Pembina 'Q' Pool	Alta.	2/85	301	2	1	Nisku	Dolo.	9.8	1,970
Esso Resources Canada†	Wizard Lake	Alta.	1969	2,725	49	14	Leduc D-3A	Dolo.	10.5	1,375
Esso Resources Canada Ltd.	Judy Creek	Alta.	5/85	29,050	186	65	Beaverhill Lake	LS	9.0	43
Esso Resources Canada Ltd.	Judy Creek	Alta.	1/87	8,290	52	22	Beaverhill Lake	LS	9.2	41
Esso Resources Canada Ltd.	Rainbow Keg River "FF" Pool	Alta.	5/72	102	2	1	344 ft	Dolo.	8.5	
Esso Resources Canada Ltd.	Rainbow Keg River "T" Pool	Alta.	5/69	222	3	1	394 ft	Dolo.	8.6	
Esso Resources Canada Ltd.	Rainbow Keg River "Z" Pool	Alta.	2/71	221	5	1	265 ft	Dolo.	4.25	
Esso Resources Canada Ltd.	Redwater	Alta.	1/85	1,000	25	4	Leduc D3	LS	5.0	10-5,000
Gulf Canada	Fenn-Big Valley	Alta.	4/83	1,268	26	7	Nisku	Dolo.	8.0	400
Gulf Canada	Goose River	Alta.	10/86	2,847	14	5	Beaverhill Lake	LS	8.0	240
Home Oil	Swan Hills	Alta.	10/85	19,440	196	39	Beaverhill Lake	LS	8.5	54
Husky Oil§	Rainbow B Pool	Alta.	6/84	2,500	28	9	Keg River	LS	8.0	300
Husky Oil§	Rainbow E Pool	Alta.	6/72	173	5	1	Keg River	LS	12.0	300
Husky Oil§	Rainbow EEE Pool	Alta.	4/70	150	2	1	Keg River	LS	16.8	500
Husky Oil§	Rainbow G Pool	Alta.	10/72	163	3	1	Keg River	LS	8.0	300
Husky Oil§	Rainbow H Pool	Alta.	6/73	165	3	1	Keg River	LS	9.4	200
Husky Oil§	Rainbow KRA Pool	Alta.	12/68	633	17	2	Keg River	LS	10.1	100
Husky Oil§	Rainbow KRD Pool	Alta.	3/76	85	1	1	Keg River	LS	10.0	200
Husky Oil§	Rainbow O Pool	Alta.	2/70	703	9	1	Keg River	LS	6.0	150
Mobil Oil Canada	Rainbow	Alta.	7/83	320	5	1	Keg River	Dolo./LS	8.5	100-1,000
Mobil Oil Canada	Rainbow	Alta.	9/72	800	14	2	Keg River	Dolo./LS	8.6	500
Petro-Canada	Brazeau River	Alta.	9/81	640	3	1	Nisku	LS	7.0	50
Petro-Canada	Brazeau River	Alta.	9/80	640	2	1	Nisku	LS	10.0	396
Petro-Canada	Brazeau River	Alta.	11/80	640	2	1	Nisku	LS	10.0	35
Petro-Canada	Caroline	Alta.	9/84	12,620	27	12	Cardium	S	10.0	12
Shell Canada	Simonette	Alta.	11/86	150	4	1	Leduc D-3	Dolo.	7.0	10-50
Shell Canada	Virginia Hills	Alta.	11/89	4,575	50	14	Beaverhill Lake	LS	9.0	1-500

ENHANCED OIL RECOVERY

Texas tax incentives

To spur EOR projects, the Texas legislature in 1989 reduced the oil severance tax to 2.7% from 4.6% for new tertiary recovery projects that started fluid injection after Sept. 1, 1989.

This reduced rate is to be in effect for a 10 year period as long as the project remains active. The rate applies to all oil from the project, not just the increase from enhanced recovery processes.

Not covered by this rate reduction are expansions of projects producing prior to Sept. 1, 1989, projects that change from one tertiary method to another, and pressure maintenance projects. [IPE]

Table C

Depth, ft	Gravity, °API	Reservoir oil cp	°F.	Previous prod.	Satur. % start	Satur. % end	Project maturity	Total prod., b/d	Enh. prod., b/d	Project eval.	Profit	Project scope
3,000	24.0	24.0	84	Prim./WF	62.0		NC	340	100	Prom.	Yes	P(Exp.L)
2,450	22.0	34.0	75	Prim./WF	60.0		HF			TETT		FW
1,837	16.0	5,000	60	Prim.	82.0	74.6	HF	130	130	Succ.	Yes	P(Exp.L)
8,500	39.0	0.3	198	WF	40.0		JS	190	140	TETT		Exp. L
4,000	42.0	1.14	133	WF	40.0	28.0	HF			Succ.	No	P(Exp.L)
5,140	42.0	1.14	133	WF	38.0	27.0	HF			Prom.		P(Exp.L)
5,075	42.0	1.14	133	WF	36.0		JS			Prom.		P(Exp. L)
2,900	18.0	70.3	110	Prim.	66.0		HF	3,330	3,330	Succ.	Yes	FW
2,900	18.0	70.0	110	Prim.	62.0		HF	1,635	1,635	Succ.	Yes	FW
2,900	18.0	70.0	110	Prim.	70.0		HF	1,400	1,400	Succ.	Yes	FW
3,100	24.0	16.0	125	WF	62.0	40.0	NC	50	50	Prom.	No	P(Exp. UL)
3,550	28.0	6.0	100	WF	45.0	25.0	NC			Prom.		Exp. L
2,450	14.3	2,750	70	Prim.	100.0	80.0	HF	350	350	Prom.	Yes	P(Exp. UL)
5,659	42.0	0.35	135	G	78.0	31.0	NC	168	168	Prom.	Yes	FW
11,000	40.0	0.14	233	Prim.	90.0	60.0	NC	750	750	Succ.	Yes	FW
7,500	34.0	1.4	169	Prim.	70.0	30.0	HF	800	800	Succ.	Yes	FW
6,981	40.0	0.419	190	WF	46.0	31.0	NC	4,500	2,460	Prom.	Yes	FW
5,500	41.0	0.84	131	WF	30.0	5.0	HF	24,000	10,000	Succ.	Yes	FW
6,200	40.0	0.3	183	Prim.			NC	1,250	1,250	Succ.	Yes	FW
8,500	38.0	0.4	225	Prim./WF	35.0	13.0	NC	13,200	13,200	Succ.	Yes	FW
6,400	38.0	0.55	136	Prim.	87.0	73.0	NC	1,600	1,600	Disc.	No	FW
9,850	57.0	0.097	210		98.0	33.0	JS	3,800		Prom.		FW:(Exp.L)
5,000	39.0	0.85	139	Prim.	90.0	29.0	JS	2,400	1,400	Succ.	Yes	FW
10,500	37.0	0.8	175	WF	90.0	5.0	JS	4,000	1,100	Succ.	Yes	FW
9,780	42.0	0.19	234	WF	30.0	5.0	JS	5,600		TETT	No	FW
5,000	41.0	0.65	147	WF	30.0	19.0	JS	14,300	9,000	Succ.	Yes	FW
10,000	41.0	0.3	210	WF	80.0	5.0	JS	4,450	1,700	Succ.	Yes	FW
9,000	41.0	0.7	176	WF	90.0	5.0	HF	6,400	1,700	Succ.	Yes	FW
8,000	40.0	0.5	208	WF	75.0	43.3	JS	2,271	705	TETT		FW
10,000	42.0	0.3	228	Prim.	90.0	5.0	HF	4,466	2,766	Succ.	Yes	FW
10,300	42.0	0.3	218	WF	75.0	9.0	HF	6,000	2,500	Succ.	Yes	FW
10,000	42.0	0.2	218	Prim.	90.0	5.0	HF	3,800	2,300	Succ.	Yes	FW
9,541	43.2	0.33	204	Prim.	80.0	5.0	HF	3,500	2,100	Prom.	Yes	FW
9,469	43.6	0.37	198	Prim	82.0	5.0	HF	2,900	1,450	Prom.	Yes	FW
9,415	40.9	0.42	199	WF	88.0	5.0	HF	7,633	2,979	Prom.	Yes	FW
9,333	41.1	0.14	198	Prim.	93.0	5.0	HF	3,131	1,461	Prom.	Yes	FW
9,332	43.4	0.32	190	Prim.	84.0	5.0	JS	2,200	1,490	Prom.	Yes	FW
9,531	45.4	0.36	200	Prim.	87.0	5.0	JS	5,800	3,610	Prom.	Yes	FW
9,421	41.3	0.42	196	Prim.	91.0	5.0	HF	3,280	1,710	Prom.	Yes	FW
6,500	38.0	0.54	167	Prim.	93.0	5.0	NC	9,600	9,600	Succ.	Yes	FW
8,850	41.6	0.38	206	WF	84.0	42.0	HF	14,750	8,200	Succ.		FW
9,150	42.1	42.0	206	WF	83.0	43.0	JS	4,720	1,760	Succ.		FW
4,160	37.0	0.59	188	Prim.	90.0		NC			Succ.		RW
4,330	40.0	0.69	188	Prim.	88.0		NC			Succ.		RW
4,040	38.0	0.547	190	Prim.	73.5		HF			Succ.		RW
3,280	35.0	2.85	94	Prim.	75.0	20-25.0	HF	685	550	Prom.	No	P
5,249	32.8	1.34	136	Prim.	33.0		HF	590	590	Succ.	Yes	FW
9,200	41.0	0.4	234	WF			HF	2,831	1,573	Prom.		Exp. L.
8,300	41.0	0.4	225	WF	30.0	5.0	JS	20,300	7,100	Succ.	Yes	FW
6,000	39.0	0.83	180	WF	52.0	23.0	JS	4,635	4,635	Prom.	Yes	FW
5,932	39.0	0.439	175	Prim.	91.0	27.0	HF	1,900	1,900	Succ.	Yes	FW
6,085	37.0	0.542	180	Prim.	95.0	29.0	NC	70	70	Succ.	Yes	FW
6,151	39.0	0.485	185	Prim.	92.0	28.0	HF	1,528	1,528	Succ.	Yes	FW
6,210	39.0	0.596	187	Prim.	92.0	28.0	HF	572	572	Succ.	Yes	FW
6,380	43.0	0.29	195	Prim.	90.0	30.0	HF	2,610	2,610	Succ.	Yes	FW
6,310	40.0	0.426	195	Prim.	90.0	27.0	NC	141	141	Succ.	Yes	FW
6,053	42.0	0.28	180	Prim.	87.0	26.0	NC	650	650	Succ.	Yes	FW
5,500	39.0	0.46	190	WF	40.0	10.0	HF	600	600	Prom.	Yes	FW
5,500	39.0	0.47	184	GI/WF	80.0	10.0	HF	1,900	1,900	Succ.	Yes	FW
10,091	42.0	0.38	216	Prim.	90.0	15.0	HF	3,100	3,100	Succ.	Yes	FW
10,215	44.0	0.4	218	Prim.	90.0	10.0	HF	7,400	7,400	Succ.	Yes	FW
10,544	46.0	0.2	222	Prim.	90.0	15.0	HF	2,500	2,500	Succ.	Yes	FW
8,200	46.0	0.25	173	Prim.	85.0	45.0	HF	1,900	1,550	Succ.	Yes	FW
11,600	47.0	0.13		WF			NC	160	135	Disc.		Exp. UL
9,500	34.0	0.53	221	WF	56.0	36.0	JS	11,000		TETT		Exp. L

233

ENHANCED OIL RECOVERY

Producing Canadian EOR projects—continued

Operator	Field	Province	Start date	Area, acres	Number wells Prod.	Number wells Inj.	Pay zone	Formation type	Porosity, %	Permeability, md
Hot Water										
Alberta Energy	S. Jenner Up. Mannville J. Pool	Alta.	9/89	2	3	1	Glauconitic	S	25.0	1,560
Polymer										
BP Exploration	Chauvin South	Alta.	2/88	300	10	3	Sparky E	S	15-27	1,076
Chevron Canada Resources	Taber Mannville D Unit #1	Alta.	6/87	480	7	2	Mannville D	S	22.3	1,140
Encor Energy	Rapdan Unit	Sask.	1/86	440	12	5	Upper Shaunavon	S	17.0	85
Steam										
Amoco Canada	Elk Point	Alta.	4/89	33	9	4	Lower Cummings	S	30.0	1,500
Amoco Canada	Lindbergh	Alta.	11/85	120	12	12	Lower Cummings	US	32.0	4,000
Esso Resources Canada†	Lone Rock	Sask.	11/84	45	13	3	Sparky	S	34.4	2,800
Home Oil	Kitscoty	Alta.	6/81	160	12	15	Sparky B	S	27.6	905
Mobil Oil Canada	Celtic (Multi-zone)	Sask.	9/88	160	16		General Petro & Waseca	S	35.0	1-2,000
Norcen	Provost	Alta.	3/86	20	8	1	McLaren	S	27-30.0	4,000
Sceptre Resources	Tangleflags North	Sask.	12/87	17	1	3	Lloydminster Sand	US	32.5	5,000

*Former operator: * Dome, † Texaco, §Canterra Energy.

Completed, terminated Canadian projects

Operator	Field	Province	Start date	Area, acres	Number wells Prod.	Number wells Inj.	Pay zone	Formation type	Porosity, %	Permeability, md
Caustic-polymer										
Pancanadian	Horsefly Lake	Alta.	12/84	34	5	4	Lower Mannville	S	18.0	260
CO$_2$ miscible										
Shell Canada	Midale	Sask.	10/85	5	3	4	Midale	LS/Dolo.	16.0	3
Combustion										
Husky	Tangleflags	Sask.	1985	240	19	3	GP	US	33.0	4,000
Mobil Oil Canada	Kitscoty	Alta.	9/75	1,040	84	7	Sparky	S	34.0	2,000
Electromagnetic										
Canada Northwest Energy	Wildmere	Alta.	1/86	120	3		Lloydminster	S	30.0	800
Steam										
Canada Northwest Energy	Atlee Buffalo	Alta.	9/86	640	10		Glauconite	S	30.0	2,000
Husky	Frog Lake	Alta.	1985	20	4	1	Colony	US	33.0	2,500
Husky	Charlotte Lake	Alta.	1985	20	2	2	Rex	US	33.0	3,000
Mobil Oil Canada	Celtic	Sask.	4/85	50	5		General Petro.	S	35.0	2,000
Mobil Oil Canada	Celtic	Sask.	10/80	125	25		Waseca	S	35.0	1,000
Steam soak										
Phillips Petroleum	Coleville	Sask.	11/84	35	2	2	Bakken	S	25.0	400

Producing EOR projects outside U.S. and Canada

Country	State/area	Operator	Project type	Field	Start date	Area, acres	No. wells Prod.	No. wells Inj.	Pay zone	Formation	Porosity, %	Permeability, md
Colombia	Mid. Magdalena basin	Texas Petroleum	Steam	Cocorna	3/87	15	8	3	Oligocene B	S	28	700
Colombia	Mid. Magdalena basin	Texas Petroleum	Steam	Teca	2/84	3,000	200		Oligocene A & B	S	28	1,200
France	Paris	Elf Aquitaine	Polymer	Chateaurenard (Courtenay)	1/89	250	12	4	Neocomian	S	30	2,000
France		Elf Aquitaine	Steam	Lacq Superieur	1977	30	8	1	Senonien	LS	15	1
France	Paris	Elf Aquitaine	Steam	Lacq Superieur	1/89	30	8	1	Senonien	LS	15	10
Hungary		NKFV	CO$_2$	Nacylengyel	10/88	2,890	104	11	I-IV Rudistic	LS	1.06	1,000
Hungary		NKFV	CO$_2$	Nagylengyel	9/80	200	5	1	Triassic	Dolo.	2.5	1,000
Hungary		NKFV	CO$_2$	Budafa	1981	1,013	77	60	Zala Kerettye	S	22	110
Hungary		NKFV	Combustion	Demjen East	1986	14	8	3	Oligocene c,da,df,e	S	19	100
Indonesia		Total	Micellar polymer	Handil	1982	4	1	3		S	30	1,000
Indonesia	Pekanbaru	PT Caltex	Steam	Duri	4/85	4,000	952	288	Pertama - Kedua	US	32-36	1,550
Libya		Zueitina Oil	HC miscible	Intisar 103D	1/69	3,325	20	6	"D" Reef	LS	23.9	226
Trinidad	Forest Reserve	Trintoc	CO$_2$ immiscible	Forest Reserve (Area 2102)	6/76	58	6	3	Forest sands	S	30	175
Trinidad	Forest Reserve	Trintoc	CO$_2$ immiscible	Forest Reserve (Area 2121)	1/74	29	4	3	Forest sands	S	30	150

ENHANCED OIL RECOVERY

Table C

Depth, ft	Gravity, °API	Reservoir oil cp	°F.	Previous prod.	Satur. % start	Satur. % end	Project maturity	Total prod., b/d	Enh. prod., b/d	Project eval.	Profit	Project scope
2,950	14.0	170.0	90	Prim.	75.0		JS	95		TETT		P
2,100	21.5	43.1	78	WF	60.0		JS	440		Prom.		P
3,200	18.0	55.0	121	WF	92.0	7.0	JS	17,600	16,400	TETT	Yes	FW
4,500	23.0	10.0	132	WF	55.0	43.0	HF	880	660	Succ.	Yes	P (Exp. L)
2,000	12.0	25,000	73	Prim./C	80.0		JS			TETT		P
1,950	12.0	15,000	75	Prim.	72.0		HF			Prom.	Yes	P(Exp L)
1,900	16.0	800	70	Prim.	82.0		NC	200	200	Prom.	No	P(Exp. UL)
2,066	13.0	8,000	60	Prim.	80.0	52.0	NC	400	325	Succ.	Yes	P(Exp. UL)
1,600	13.0	2-5,000	60	Prim.	70.0		JS	800		TETT		P
2,430	12.0	2,000	80	Prim.	82.0	45.0	JS	420	420	Prom.	No	P(Exp.L)
1,476	12-13.0	13,000	66	Prim.	80.0	15.0	HF	600	550	Prom.	Yes	P

Table D

Depth, ft	Gravity, °API	Reservoir oil cp	°F.	Previous prod.	Satur. % start	Satur. % end	Project maturity	Total prod., b/d	Enh. prod., b/d	Project eval.	Profit	Project scope
3,200	23.0	13	94	WF	51	40	Term.					Exp. UL
4,600	28.0	3	150	Prim./WF	35		C					P
1,700	14.5	4,000	74	Prim.	77		Term.	550	550	Succ.	Yes	UL
1,800	12.2	15,000	70	Prim.	85	60-68	Term.	610	305	Disc.	No	P
2,100	12	8,000	75	Prim.	85		Term.	80	60	Prom.	Yes	P (Exp. L)
2,900	14.5	800	80	Prim.	80	1	Term.	0	0	Succ.	Yes	Exp. L
1,400	12	100,000	60		80	65	Term.	45	45	Disc.	No	UL
1,400	10.8	107,000	70		80		Term.			Disc.	No	UL
1,576	13	7,000	67	Prim.	69	46	Term.	290	190	Prom.		P (Exp. L)
1,500	13	5,380	61	Prim.	70	47	Term.	630	630	Prom.		P (Exp. L)
2,600	14	1,000	84	WF			C	60	22	Disc.	No	P

Table E

Depth, ft	Gravity, °API	Reservoir oil cp	°F.	Previous prod.	Satur. % start	Satur. % end	Project matur.	Total prod., b/d	Enh. prod., b/d	Project eval.	Profit	Project scope
2,100	12.8	2,965			57	19	HF	390	190	Prom.		P (Exp. L)
2,100	12.8	2,965		Prim.			HF	17,000	8,500	Succ.	Yes	FW
1,950	27	40	86	Prim.	45		JS	337	161	TETT	Yes	
1,950-2,100	22	20	150	Prim.	45		NC	1,000	400	Succ.	Yes	FW
1,950-2,100	22	20	150	Prim.	45		JS			Prom.		P
6,700	18	22	237	Prim.			JS	3,200	3,200	Prom.		
7,400	16	92	248	Prim.			NC	130	130	Succ.		
3,000	42	0.8	147	Prim.			NC	500	500	Succ.		
820	39	7	78	Prim.	55		HF	60	52	Prom.		Exp.L
4,500	33	0.4	115	WF	35			400	400			
625	22.7	157		Prim.	62	25	JS	145,000	115,000	Succ.	Yes	Exp. L
8,849	39.2	0.46	226		80	18	NC	40,000	40,000	Succ.	Yes	FW
3,000	29	3	120	Prim.	56		JS	25	25	Prom.		RW
2,600	17	66	120	Prim.	60		JS	10	10	Prom.		RW

ENHANCED OIL RECOVERY

Producing EOR projects outside U.S. and Canada—continued

Country	State/area	Operator	Project type	Field	Start date	Area, acres	No. wells Prod.	No. wells Inj.	Pay zone	Formation	Porosity, %	Permeability, md
Trinidad	Forest Reserve	Trintoc	CO_2 immiscible	Forest Reserve (Area 2124)	1/86	184	2	1	Forest sands	S	31.5	283-370
Trinidad		Trintoc	CO_2 cyclic	Forest Reserve	5/84		6		Forest sands	S	30	150-300
Trinidad	Point Fortin	Trintoc	Steam	P/F Cruse "E" Steam Proj.	2/86	66	28	3	Cruse E	S	31	95
Trinidad	Point Fortin	Trintoc	Steam	Parrylands	7/81	90	46	6	Forest sands	S	30	500
Trinidad	Forest Reserve	Trintoc	Steam	F/R Project III	7/65	135	53	14	Forest zone 5.1	S	33	340
Trinidad	Forest Reserve	Trintoc	Steam	F/R Phase I Reactivation	8/76	58	10	0	Forest sands	S	31	205
Trinidad	Forest Reserve	Trintoc	Steam	F/R Phase I West Expansion	12/88	50	29		Forest sands	S	30	430
Turkey	Ankara	TPAO	CO_2 immiscible	Bati Raman	3/86	10,709	96	23	Garzan	LS	18	58
Venezuela	Anzoategui	Corpoven S.A.	Gas	Oveja	8/61	5,467	31	4	I2L,3(YAC.OG-503)	S	37	2,500
Venezuela	Anzoategui	Corpoven S.A.	Gas	Oveja	1/67	1,325	4	1	L1L(YAC.OM-101)	S	31	2,940
Venezuela	Anzoategui	Corpoven S.A.	Gas	Oveja	1/64	2,422	14	2	J-3(OM-100)	S	28	10,000
Venezuela	Monagas	Corpoven S.A.	Gas	Oritupano	9/76	1,054	7	1	A-7(YAC.ORM-33)	S	30	1,000
Venezuela	Anzoategui	Corpoven S.A.	Gas	Oveja	5/64	1,686	9	2	L1L(YAC.OM-104)	S	30.4	2,642
Venezuela	Campo Cerro Negro	Lagoven S.A.	Steam	B.E.P.-Cerro Negro	1984	49,090	144		Ofic. Morichal-Memb.	S	32	7,000
Venezuela	Campo Jobo	Lagoven S.A.	Steam	Jobo	12/69	4,329	18		Ofic. Jobo Member	S	32	2,500
Venezuela	Campo Jobo	Lagoven S.A.	Steam	Jobo - P.E.T.C.	8/85	267	17		Oficina	S	30	8,000
Venezuela	Campo Jobo	Lagoven S.A.	Steam	Jobo	12/69	27,522	99		Ofic. Mor-Member	S	31	5,000
Venezuela	Campo Pilon	Lagoven S.A.	Steam	West Pilon	12/69	1,065	29		Oficina-1	S	29.1	5,000
Venezuela	Zulia	Lagoven S.A.	Steam	Bachaquero	12/80	343	2		Bachaquero Superior	S	23	1,500
Venezuela	Zulia	Lagoven S.A.	Steam	Lagunillas	2/71	9,343	308	2	Bachaquero	S	34	4,000
Venezuela	Zulia	Lagoven S.A.	Steam	Tia Juana	2/70	1,692	25		Lagunillas Inferior	S	31	1,250
Venezuela	Zulia	Maraven S.A.	Steam	East Tia Juana	3/69	1,773	145		L.L.	S	38.1	780
Venezuela	Zulia	Maraven S.A.	Steam	Lagunillas	4/65	420	59		U.L.H.	S	35	
Venezuela	Zulia	Maraven S.A.	Steam	East Tia Juana	4/59	339	33		L.L.	S	38.1	
Venezuela	Zulia	Maraven S.A.	Steam	East Tia Juana	8/69	415	24		L.L.	S	38.1	
Venezuela	Zulia	Maraven S.A.	Steam	Lagunillas	11/79	3,101	267		L.L.	S	33.7	
Venezuela	Zulia	Maraven S.A.	Steam	East Tia Juana	9/64	415	36		L.L.	S	38.1	3,000
Venezuela	Zulia	Maraven S.A.	Steam	Main Tia Juana	10/63	873	82		L.L.	S	38.1	1,400
Venezuela	Zulia	Maraven S.A.	Steam	East Tia Juana	8/68	2,629	224		L.L.	S	38.1	3,000
Venezuela	Zulia	Maraven S.A.	Steam	East Tia Juana	8/69	1,642	131		L.L.	S	38.1	1,300
Venezuela	Zulia	Maraven S.A.	Steam	Lagunillas	4/70	2,565	220		U.L.H.	S	35	
Venezuela	Zulia	Maraven S.A.	Steam	Main Tia Juana	7/67	1,271	114		L.L.	S	38.1	675
Venezuela	Zulia	Maraven S.A.	Steam	Main Tia Juana	7/67	1,500	134		L.L.	S	38.1	750
Venezuela	Zulia	Maraven S.A.	Steam	Main Tia Juana	6/68	141	15	7	L.L.	S	33	675
Venezuela	Zulia	Maraven S.A.	Steam	Main Tia Juana	10/67	2,221	197		L.L.	S	38.1	750
Venezuela	Zulia	Maraven S.A.	Steam	Lagunillas	1/70	3,025	147		L.L.	S	33.7	
Venezuela	Zulia	Maraven S.A.	Steam	East Tia Juana	1/74	1,830	135	21	L.L.	S	38.1	780
Venezuela	Zulia	Maraven S.A.	Steam	Bachaquero	10/84	7,795	539		Post-Eoceno	S	37.3	600
Venezuela	Zulia	Maraven S.A.	Steam	East Tia Juana	2/61	36	7		L.L.	S	38	1,300
Venezuela	Zulia	Maraven S.A.	Steam	Lagunillas	7/67	618	54		U.L.H.	S	35	
Venezuela	Zulia	Maraven S.A.	Steam	Main Tia Juana	6/66	141	15		L.L.	S	38.1	675
Venezuela	Zulia	Maraven S.A.	Steam	Lagunillas	11/80	3,565	299		U.L.H.	S	35	
Venezuela	Zulia	Maraven S.A.	Steam	Lagunillas	8/70	2,114	175		U.L.H.	S	35	
Venezuela	Zulia	Maraven S.A.	Steam	East Tia Juana	5/68	2,667	232		L.L.	S	38.1	
Venezuela	Zulia	Maraven S.A.	Steam	East Tia Juana	12/68	1,956	168		L.L.	S	38.1	
Venezuela	Zulia	Maraven S.A.	Steam	Lagunillas	1/70	76	7		L.L.	S	33.7	
Venezuela	Anzoategui	Corpoven S.A.	Steam	Bare (F.O.)	3/85	16,452	47	48	U1,3(YAC.MFB-53)	US	31.9	6,600
Venezuela	Anzoategui	Corpoven S.A.	Steam	Arecuna (F.O.)	2/85	1,668	45	5		US	30.7	5,800
Venezuela	Anzoategui	Corpoven S.A.	Steam	Bare (F.O.)	3/87	8,066	61	65	U2,3(YAC.MFB-23)	US	28.6	5,000
Venezuela	Anzoategui	Corpoven S.A.	Steam	Arecuna (F.O.)	12/83	1,544	15	15	T(YAC.MFA-52)	US	30.6	4,500
West Germany	Lower Saxony	Wintershall AG	Hot Water	Emlichheim 14	10/67	64	5	1	Valenginian	S	30	6,000
West Germany	Lower Saxony	Wintershall AG	Hot Water	Emlichheim 17	10/74	82	9	2	Valenginian	S	30	6,000
West Germany	Lower Saxony	Wintershall AG	Hot Water	Emlichheim 11	9/73	70	8	1	Valenginian	S	30	6,000
West Germany	Lower Saxony	Wintershall AG	Hot Water	Emlichheim 07	10/67	82	5	1	Valenginian	S	30	6,000
West Germany	Northwest	Preussag AG	Polymer	Edesse-Nord	11/85	4	3	2	Wealden	S	24	1,200
West Germany	Northwest	Preussag AG	Polymer	Vorhop-Knesebeck	11/89	345	3	1	Dogger-Beta	S	28	1,000
West Germany		RWE-DEA*	Polymer	Hankenbuettel Westblock 4	4/77	70	7	2	Upper Dogger-Beta	S	28	2,000-4,000
West Germany		RWE-DEA*	Polymer	Hankenbuettel Westblock 2	5/79	75	6	3	Upper Dogger-Beta	S	28	2,000-4,000
West Germany		RWE-DEA*	Polymer	Westblock	1980	410	13	3	Upper Dogger-Beta	S	28	2,000-4,000
West Germany		RWE-DEA*	Polymer	Hankenbuettel, South Block	1/84	15	2	1	Dogger-Beta	S	26	1,000-2,000
West Germany		RWE-DEA*	Polymer	Ploen OST	1/89	30	3	1	Dogger-Beta	S	19	200-1,500
West Germany		RWE-DEA*	Polymer	Hohne	1/84	73	4	3	Dogger-Beta	S	26	2,000
West Germany	Lower Saxony	Wintershall AG	Polymer	Bockstedt	9/84	57	4	1	Valenginian	S	23	4,000
West Germany	Emsland	BEB	Steam	Ruhlermoor II	8/86	360	40	4	Valenginian	S	27-31.0	600-1,000
West Germany	Lower Saxony	BEB	Steam	Georgsdorf I	1/75	165	19	2	Valanginian	S	25	460-1,150
West Germany	Lower Saxony	BEB	Steam	Georgsdorf II	12/79	144	23	3	Valanginian	S	25	1,000-1,300
West Germany	Lower Saxony	BEB	Steam	Ruhlermoor I	11/80	220	30	4	Valanginian	S	28-30.0	300-1,000
West Germany	Emsland	BEB	Steam	Ruhlermoor III	12/87	400	40	6	Valanginian	S	25-28	300-800
West Germany	Lower Saxony	Wintershall AG	Steam	Ruehlertwist	12/78	237	26	3	Valenginian	S	28	5,000
West Germany	Lower Saxony	Wintershall AG	Steam	Emlichheim 06	5/81	121	15	4	Valenginian	S	30	6,000

Former operator: *Texaco AG

Completed, terminated projects outside U.S. and Canada

Country	State/area	Operator	Type projects	Field	Start date	Area, acres	No. wells Prod.	No. wells Inj.	Pay zone	Formation	Porosity, %	Permeability, md
Austria		OMV	Caustic	Matzen/Scho	7/80	33	5	1	Torton 9th	S	23.5	300
Austria		OMV	Caustic	Matzen/Scho	7/80	33	5	1	Torton 9th	S	23.5	300
Austria		OMV	Steam	Maustrenk	7/80	123	19	3	Schlier		24	147
Congo		Elf Congo	Steam	Emeraude-Offshore	1/82	9	8	2	Senonien	S	25	80
England		BP	Micellar-polymer	Bothamsall	7/83	10	3	1	Sub-Alton	S	12	14
France		Elf Aquitaine	Micellar-polymer	Chateaurenard	1/84	71	9	4	Necomian	S	30	1,300
France		Elf Aquitaine	Polymer	Chateaurenard	1/85	12	4	1	Necomian	S	30	1,000
Venezuela	Monagas	Corpoven S.A.	Gas	Oritupano	2/79	1,100	6	1	A-9(YAC.ORM-27)	S	28	1,300
Venezuela	Campo Jobo	Lagoven S.A.	Steam	Jobo Picv	3/81	90	22	6	Morichal-1	S	32	10,000
Venezuela	Zulia	Lagoven S.A.	Steam	Lagunillas	10/66		2		Lagunillas Inferior	S	38	2,150
Venezuela	Zulia	Lagoven S.A.	Steam	Bachaquero	6/70	55			Bachaquero	S	30	1,761

ENHANCED OIL RECOVERY

Table E

Depth, ft	Gravity, °API	Reservoir oil cp	°F.	Previous prod.	Satur. % start	Satur. % end	Project matur.	Total prod., b/d	Enh. prod., b/d	Project eval.	Profit	Project scope
3,850	25	5.7	120	WF	44		JS	60	50	TETT		RW
2,500	20-25.0	66-90	120	Prim.	60		JS	10	10	Prom.		P (Exp. L)
1,200-1,600	16-18.0	150-200	110-115	Prim.	58	18	JS	270	270	Prom.		FW
1,000-1,200	11.5	5,000	110	Prim.	75	20	JS	640	640	Prom.		FW
1,000-1,300	15.7	148	110	Prim.	70	20	HF	1,300	1,300	Succ.	Yes	FW
1,200	19	32	105	Prim.	57	15	JS	125	125	Prom.		RW
1,500	12.3-17.8	160	105	Prim.	67	25	JS	75	75	TETT		P (Exp. L)
4,265	13	592	129		78		HF	7,400	6,800	Succ.		FW
3,150	15	38	150	Prim.	90	67	HF	5,000	5,000	Prom.	Yes	FW
3,430	21	5.5	154	WF	90	63	NC	1,300	1,300	Succ.	Yes	FW
3,300	20	52	149	WF	90	50	HF	2,250	2,250	Succ.	Yes	FW
5,150	20.5	4.9	146		90	72	HF	550	550	Succ.	Yes	FW
3,350	18	12.5	152	WF	90	65	NC	2,600	2,600	Succ.	Yes	FW
2,800	8.5	5,500		Prim.	80		HF	10,000	8,000	Succ.	Yes	P (Exp. L)
	12.6	80		Prim.	80		JS	2,500	180	Succ.	Yes	FW
4,025	8.5	1,850		Prim.	75		HF	2,500	2,000	Succ.	Yes	P (Exp. L)
3,600	8.5	1,850		Prim.	85		JS	27,100	1,430	Succ.	Yes	FW
3,200	9.5	1,850		Prim.	82.2		JS	4,350	450	Succ.	Yes	FW
3,400	14	185	135	Prim.	85		JS	9		TETT	No	P (Exp. L)
2,690	12	600	128	Prim.	84		HF	13,511	7,216	Succ.	Yes	FW
2,537	15	93	119		72		HF	1,870	870	Prom.	Yes	
1,700	11.9	2,000	111	Prim.	92.7	74.5	HF	4,320	3,400	Prom.	Yes	
2,100	11.4	3,500	117	Prim.	96.9	58.7	NC	2,510	1,150	Succ.	Yes	FW
1,500	12.2	1,000	108	Prim.	92.6	72.3	HF	490	150	Prom.	Yes	FW
850	10.5	10,000	95	Prim.	99.9	84.6	HF	810	810	Prom.	Yes	FW
2,615	15	580	126	Prim.	83.3	68.8	HF	3,526	1,690	Prom.	Yes	FW
1,250	12	3,000	104	Prim.	95.6	78.6	HF	500	330	Prom.	Yes	FW
1,750	13.1	750	113		86	74	NC	2,036	1,297	Prom.	Yes	FW
1,000	11.6	4,000	104	Prim.	94.6	79.6	HF	3,217	2,550	Prom.	Yes	FW
1,250	10.2	12,000	102	Prim.	99.9	89.9	HF	2,325	2,325	Prom.	Yes	FW
1,750	11.8	2,500-11,500	115	Prim.	97.3	89.4	HF	12,150	6,500	Prom.	Yes	FW
1,220	13.1	6,000	123	Prim.	94.3	80.8	HF	2,188	1,375	Prom.	Yes	FW
1,400	11.8	1,000	110	Prim.	91.1	76	HF	3,195	1,950	Prom.	Yes	FW
1,250	13.1	1,000	113	Prim.	88	68		3,100	2,000	Prom.	Yes	P
1,746	13.1	4,100	110	Prim.	95.1	80.9	HF	5,924	4,700	Prom.	Yes	FW
2,500	15.2	500-4,500	122	Prim.	86	74.3	HF	2,840	1,130	Prom.	Yes	FW
1,590	12	1,000	111		81	56	HF	7,500	6,800	Succ.	Yes	FW
2,000	13	300-800	117	Prim.	85	73	JS	8,800	4,000	Prom.	Yes	FW
1,000	9.5	12,000	105	Prim.	100	76.9	NC	140	140	Prom.	Yes	P
2,100	11.4	3,500	117	Prim.	9	7	NC	3,080	1,400	Succ.	Yes	FW
1,250	13.1	1,300	104	Prim.	92	78	NC	120	25	Prom.	Yes	FW
2,398	11	2,000-3,700	125	Prim.	90.6	77.9	HF	16,260	5,510	Prom.	Yes	FW
1,725	11.4	4,000-9,000	118	Prim.	97.8		HF	7,380	3,620	Prom.	Yes	FW
1,200	12	500	104		93.6	82.5	HF	4,510	3,600	Prom.	Yes	FW
1,250	11.7	7,500	106	Prim.	97.5	82.5	HF	4,174	3,450	Prom.	Yes	FW
2,600	15.2	250	125	Prim.	73.7	63.5	NC	270	220	Succ.	Yes	P
2,650	9.3	376	131	Prim.	88	75	HF	26,675	17,340	Succ.	Yes	RW (Exp. L)
2,850	10	370	135	Prim.	82	74	NC	505	50	Disc.	No	P (Exp. UL)
3,050	9.2	351	136	Prim.	83	75	HF	177,000	8,280	Succ.	Yes	RW (Exp. L)
3,150	9.8	560	140	Prim.	80	72	NC	2,875	1,150	Prom.	Yes	RW
2,550	24.5	175	95	WF	73	58	NC	230	230	Succ.	Yes	FW
2,300	24.5	175	95	WF	66	55	NC	210	200	Succ.	Yes	FW
2,550	24.5	175	95	WF	72	68	NC	85	80	Succ.	Yes	FW
2,400	24.5	175	95	WF	70	56	NC	210	210	Succ.	Yes	FW
1,200	36	8	73	WF	53	37	NC	13	5	Succ.	No	P (Exp. L)
3,600	33	4	133	WF	65.5	55.2	JS	300	180	TETT		RW
4,900	27	13	136	WF	69	29	NC	120	90	Succ.	Yes	Exp. L
4,900	28	13	136	WF	72	33	NC	320	250	Succ.	Yes	Exp. L
4,900	25	17	136	WF	60	50	NC	200	160	Succ.	Yes	Exp. L
4,800	32	6	136	WF	65	47	HF	130	90	Succ.	Yes	Exp. L
9,400	35	1.3	195	WF	34		JS	150	0	TETT		P
3,800	32.6	5.3	133	WF	67		HF	150	50	TETT		
4,100	34	11	130	WF	39	33	HF	125	50	Prom.		P
1,700-2,100	25	120		Prim.	78		HF	2,880	2,220	Succ.	Yes	FW
2,130-2,790	27	120	95	Prim.			HF	1,120	975	Succ.	Yes	
2,130-2,790	27	120	95	Prim.			HF	1,340	1,020	Succ.	Yes	
2,070-2,460	25	120	96	Prim.			HF	1,780	1,520	Succ.	Yes	
1,700-2,100	25	120	96	WF			JS	3,100	2,160	Succ.	Yes	FW
2,650	25	175	100		51	42	HF	1,000	935	Succ.	Yes	P (Exp. L)
2,400	24.5	175	95	HW	62	45	NC	1,210	1,205	Succ.	Yes	P (Exp. L)

Table F

Depth, ft	Gravity, °API	Reservoir oil cp	°F.	Prev. prod.	Satur. % start	Satur. % end	Proj. maturity	Total prod., b/d	Enh. prod., b/d	Proj. eval.	Profit	Proj. scope
4,330	21	12.7	120	Prim./WF	62	43	Term.	78	21	Succ.	Yes	P (Exp. L)
4,330	21	12.7	120	Prim.WF	53	35	Term.	52	4	Disc.	No	P (Exp. UL)
3,018	26	7.81	98	Prim./WF	51	46	Term.	84	54	Disc.	No.	
600	22	100	85	Prim.			Term	1,600	1,600	Succ.		P
3,140	41	3	110	WF	57	50	C	15	5	Disc.	No	P
1,950	27	40	86	Prim.	48	23	C	118	32	Succ.	No	P (Exp. UL)
1,950	27	40	86	Prim.	55	30	C	144	108	Succ.	Yes	P (Exp. L)
5,200	21	4.9	146	Prim.	84	69	Term.	800	800	Disc.	No	Exp. UL
3,600	8.5	1,800		Prim.	82		C			Succ.	Yes	P (Exp. UL)
	16		135	Prim.	90		Term.	30	20	Disc.		Exp. UL
4,325		57	137		84		Term.	5,178	2,848	Prom.	Yes	

ENHANCED OIL RECOVERY

Completed, terminated projects outside U.S. and Canada—continued

Country	State/area	Operator	Type projects	Field	Start date	Area, acres	No. wells Prod.	No. wells Inj.	Pay zone	Formation	Porosity, %	Permeability, md
Venezuela	Zulia	Lagoven S.A.	Steam	Cabinas	8/71	96	3		Lagunillas Inferior		31	900
Venezuela	Zulia	Lagoven S.A.	Steam	Tia Juana	7/65	2,066	17		Lagunillas Inferior	S	36	1,064
Venezuela	Anzoategui	Corpoven S.A.	Steam soak	BARE (F.O.)	9/84	1,547	7	7	U-1 (YAC.OS-721)	US	27.9	2,000
Venezuela	Anzoategui	Corpoven S.A.	Steam soak	Melones Central	4/83	35,809	25	26	U1,2,2(YAC.MEL-15)	US	34	1,500
West Germany	Lower Saxony	Wintershall AG	Hot water	Emlichheim 15	1/67	115	6	2	Valenginian	S	30	6,000
West Germany		Wintershall AG	Hot water	Ruehlertwist	12/86	53	8	2	Valenginian	S	30	400
West Germany	Lower Saxony	Wintershall AG	Hot water	Emlichheim 10	11/81	41.5	3	1	Valenginian	S	30	1,200
West Germany	Northwest	Preussag AG	Polymer	Edesse-Nord	11/85	4.2	3	2	Wealden	S	24	6,000
West Germany		RWE-DEA	Surfactant	Hankenbuettel Westblock	1/81	83	7	3	Lower Dogger-Beta	S	28	300-1,000

Producing thermal EOR in U.S.

Operator	Field	State	County	Start date	Area, acres	Number wells Prod.	Number wells Inj.	Pay zone	Formation type	Porosity, %
Steam										
Amoco	Winkleman Dome	Wyo.	Fremont	1964	160	19	13	Nugget	S	22.8
ARCO*	Kern River	Calif.	Kern	1970	80	112	33	Kern River series	S	31.0
ARCO*	Kern River	Calif.	Kern	1970	95	71	29	Kern River series	S	33.0
ARCO*	Kern River	Calif.	Kern	9/72	38	32	16	Kern River series	S	31.0
ARCO*	Kern River	Calif.	Kern	1972	50	41	20	Kern River series	S	31.0
ARCO	Midway-Sunset	Calif.	Kern	1/72	8	31	4	Potter	S	32.0
ARCO*	Midway-Sunset	Calif.	Kern	4/84	25	24	2	Potter	S	34.0
ARCO*	Midway-Sunset	Calif.	Kern	1983	34	13	2	L. Monarch	S	30.0
ARCO*	Midway-Sunset	Calif.	Kern	3/73	29	24	2	Potter	S	34.0
ARCO*	Midway-Sunset	Calif.	Kern	1985	22	17	5	U. Monarch	S	29.0
ARCO*	Midway-Sunset	Calif.	Kern	4/85	30	26	5	Potter	S	34.0
ARCO*	Midway-Sunset	Calif.	Kern	1973	40	70	10	Monarch	S	32.0
ARCO	Midway-Sunset	Calif.	Kern	6/89	24	42	6	Potter	S	34.0
ARCO	Midway-Sunset	Calif.	Kern	8/72	9	35	4	Monarch	S	32.0
ARCO*	Midway-Sunset	Calif.	Kern	1984	15	12	0	Sub Lakeview	S	30.0
ARCO*	Midway-Sunset	Calif.	Kern	1969	50	60	15	Metson	S	30.0
ARCO	Midway-Sunset	Calif.	Kern	2/72	200	50	3	Potter	S	32.0
ARCO	Midway-Sunset	Calif.	Kern	12/84	160	12	0	MYA TAR	S	32.0
ARCO*	Placerita	Calif.	Los Angeles	1986	95	70	17	L. Kraft	S	28.0
Bechtel	Elk Hills	Calif.	Kern	7/87	20	9	4	Sub-Scalez, Shallow oil zone	US	30.0
Chevron	Cymric 26W	Calif.	Kern	10/89	30	31	19	Tulare	S	33.0
Chevron	Cymric 36W	Calif.	Kern	5/75	110	98	47	Tulare/Amnicola	S	33.0
Chevron	Cymric 6Z	Calif.	Kern	2/86	30	31	19	Tulare	S	33.0
Chevron	Cymric Sec. 31X	Calif.	Kern	4/79	69	38	17	Amnicola	S	35.0
Chevron	Edison 27-RT	Calif.	Kern	7/77	30	65	29	Kern River	S	30.0
Chevron	Kern River KCL 39	Calif.	Kern	10/75	118	67	42	Kern River	S	33.0
Chevron	Kern River MCCII	Calif.	Kern	4/71	80	129	30	Kern River	S	34.0
Chevron	Kern River Sec. 3	Calif.	Kern	9/68	424	345	155	Kern River	S	34.0
Chevron	Kern River Sec. 4	Calif.	Kern	6/78	156	200	64	Kern River	S	34.0
Chevron	Kern River MC1	Calif.	Kern	4/76	132	120	76	Kern River	S	34.0
Chevron	Kern River—ANO	Calif.	Kern	5/74	160	177	82	Kern River	S	32.0
Chevron	Lost Hills Sec. 18	Calif.	Kern	7/83	38	39	11	Tulare	S	38.0
Chevron	Lost Hills Sec. 18	Calif.	Kern	8/85	14	8	0	Tulare	S	38.0
Chevron	Lost Hills Sec. 30	Calif.	Kern	1/80	190	37	0	Etchegoin A	S	40.0
Chevron	McKittrick	Calif.	Kern	6/77	131	37	15	Amnicola	S	37.0
Chevron	McKittrick	Calif.	Kern	8/83	17	21	8	Tulare	S	36.0
Chevron	McKittrick	Calif.	Kern	6/87	70	25	14	Amnicola	S	37.0
Chevron	McKittrick 6Z	Calif.		4/85	30	29	18	Tulare/Amnicola	S	36.0
Chevron	McKittrick 9Z	Calif.	Kern	8/78	80	27	11	Amnicola	S	38.0
Chevron	Midway-Sunset Sec. 15A	Calif.	Kern	5/78	23	36	5	Potter	S	36.0
Chevron	Midway-Sunset Sec. 25A	Calif.	Kern	6/87	35	14	4	Tulare	S	35.0
Chevron	Midway-Sunset Sec. 26C	Calif.	Kern	11/75	84	243	81	Monarch	S	32.0
Chevron	Midway-Sunset Sec. 2F	Calif.	Kern	10/83	84	44	10	Webster	S	32.0
Chevron	NE. McKittrick 16-Z	Calif.	Kern	9/75	87	42	27	Amnicola	S	35.0
Chevron	West Coalinga	Calif.	Fresno	3/89	100	70	21	Temblor	S	33.0
Chevron	West Coalinga	Calif.	Fresno	7/84	60	40	12	Temblor	S	33.0
Chevron	West Coalinga	Calif.	Fresno	2/82	43	38	10	Temblor	S	33.0
Chevron	West Coalinga 13-D	Calif.	Fresno	5/73	550	182	104	Temblor	S	35.0
Chevron	West Coalinga 25-D	Calif.	Fresno	5/80	36	131	95	Temblor	S	29.0
Enercap Corp.√	Camp Hill	Tex.	Anderson	4/89	25	23	14	Carrizo	S	37.0
Exxon	Edison	Calif.	Kern	1965	1,100			Kern River		28.0
Exxon†	Midway-Sunset	Calif.	Kern		200	100	100	Monarch Sand	S	30.0
Exxon†	South Belridge	Calif.	Kern		90	41	30	Tulare	S	38.0
John Brown E & C∞	Teapot Dome NPR-3	Wyo.	Natrona	10/85	120	68	22	Shannon	S	18.0
M.H. Whitter	Midway-Sunset	Calif.	Kern	1965	200	336		Potter	US	35.0
MacPherson Oil Co.	Mt. Poso -West area	Calif.	Kern	5/89	240	40	3	Vedder	US	35.0
Marathon Oil	Garland	Wyo.	Big Horn	5/86	35	6	1	Madison	LS	15.5
Mobil	Bayou Bleu	La.	Iberville	5/88	120	26	7	E.B. Schwing RA SUA	S	30.0
Mobil	Cymric McKittrick Fee	Calif.	Kern	7/65	180	40	0	Welport Amnicola	S	35.0
Mobil	Midway-Sunset	Calif.	Kern	10/70	400	280	0	Monarch	S	34.0
Mobil	North Midway-Sunset	Calif.	Kern	11/67	410	372		Potter	S	35.0
Mobil	San Ardo	Calif.	Monterey	6/68	312	78	16	Aurignac	S	34.5
Mobil	San Ardo	Calif.	Monterey	3/80	84	38	12	Lombardi	S	32.5
Mobil	Saratoga	Tex.	Hardin	6/87	54	10	5	Miocene	S	35.0
Mobil	Sho-Vel-Tum	Okla.	Carter	11/86	55	7	7	Deese Sand	S	26.5
Mobil	South Belridge	Calif.	Kern	1969	870	446	147	Tulare	US	36.8
Mobil	South Belridge	Calif.	Kern	1965	1,987	50		Tulare	US	36.8
Oryx	Cymric	Calif.	Kern	1981	135	55	3	Tulare	S	35.0
Oryx	Midway-Sunset	Calif.	Kern	7/84	49	76	0	Potter	S	28.0
Oryx	Midway-Sunset	Calif.	Kern	3/64	106	225	14	Potter/Tulare	S	35.0
Oryx	Midway-Sunset	Calif.	Kern	1/84		153	0	Potter	S	35.0

238

ENHANCED OIL RECOVERY

Table F

Depth, ft	Gravity, °API	Reservoir oil cp	°F.	Prev. prod.	Satur. % start	Satur. % end	Proj. maturity	Total prod., b/d	Enh. prod., b/d	Proj. eval.	Profit	Proj. scope
1,985	16	15	135		80		Term.	138	80	Disc.	P (Exp. UL)	
2,100	14	93	118	Prim.	70		Term.	646	628	Prom.	Yes	P (Exp. UL)
3,400	9.8	232	145	Prim.	78.8	70	C	1,195	540	Prom.	Yes	RW (Exp. UL)
3,800	11.5	200	160	Prim.	78	63	C	3,955	2,150	Succ.	Yes	P (Exp. L)
2,300	24.5	175	95	WF	72	54	C	100	100	Succ.	Yes	FW
2,460	25	175		WF	76	74	C	60	45	Disc.	No	P (Exp. L)
2,460	24.5	175	95	WF	70	68	C	25	20	Disc.	No	FW
1,200	36	8	73	WF	53	37	C	20	12	Succ.	No	P (Exp. L)
4,900	27	11	136	WF	58	39	Term.	0	0	Disc.	No	Exp. UL

Table G

Permeability, md	Depth, ft	Gravity, °API	Reservoir oil cp	°F.	Prev. prod.	Satur. % start	Satur. % end	Proj. matur.	Total prod., b/d	Enh. prod., b/d	Proj. eval.	Profit	Proj. scope
481	1,225	14.0	1,000	81	Prim.	71		NC	215	215	Succ.		FW
2,000	600	13.0	5,000	78	Prim.	63	15.0	NC	600	600	Succ.	Yes	FW
4,000	1,200	13.0	8,000	90	SS	55	20.0	NC	1,030	950	Succ.	Yes	FW
2,500	900	13.0	7,000	84	Prim.	70	15.0	HF	480	480	Succ.	Yes	FW
2,000	800	13.0	8,000	90	SS	60	15.0	NC	1,100	1,100	Succ.	Yes	FW
1,675	1,200	12.0	2,000	100	SS	65	15.0	HF	450	350	Succ.	Yes	LW
4,000	1,600	11.5	10,000	95	SS	67	15.0	HF	700	500	Succ.	Yes	LW
2,000	1,500	13.0	5,000	10	C	60	30.0	HF	300	200	Succ.	Yes	FW
4,100	1,600	11.5	7,000	95	SS	65	15.0	NC	950	900	Succ.	Yes	LW
2,000	1,300	13.0	5,000	100	SS	60	18.0	JS	600	500	Succ.	Yes	Exp. L
3,700	1,500	11.5	10,000	95	SS	62	15.0	HF	1,200	1,000	Succ.	Yes	LW
1,500	1,200	13.0	3,000	200	SS	50	10.0	HF	2,300	1,400	Succ.	Yes	LW
2,854	1,850	11.5	10,000	95	SS	65	15.0	JS	500	50	TETT		LW
1,500	1,000	13.0	1,500	100	SS	50	10.0	JS	320	75	Succ.	Yes	RW
4,500	1,300	13.0	4,000	90	Prim.	60	35.0	HF	120	100	Succ.	Yes	FW
3,000	1,100	11.4	9,500	130	Prim.	75	57.0	JS	800	250	TETT		Exp. L
2,500	1,200	14.0	1,500	100	Prim.	65		NC	1,150		Succ.	Yes	FW
3,030	1,050	14.0	1,500	95	Prim.	65		NC	70		Disc.	No.	P (Exp. L)
3,000	1,800	12.0	7,500	100	Prim.	62	15.0	JS	1,200	700	Prom.		FW
500	2,800	27.0	6	120	Prim.	55	5.0	HF	350	50	TETT		P
2,900	1,200	12.0	5,500	100	C	51	15.0	JS	1,850	1,650	Succ.	Yes	P
1,800	1,200	13.0	4,000	100		52		NC	3,800	3,800	Succ.	Yes	FW
2,900	1,200	12.0	5,500	100	C	51	15.0	JS	2,500	2,400	Succ.	Yes	Exp. L
2,700	1,300	12.0	3,000	100	Prim.	50		NC	1,600	1,600	Succ.	Yes	FW
	1,000	14.0	2,000	90	Prim.	50		NC	1,600	1,600	Succ.	Yes	FW
	1,400	14.0	2,000	90		65		HF	600	600	Succ.	Yes	FW
2,800	960	14.0	2,000	90	Prim.	59		HF	1,950	1,950	Succ.	Yes	FW
3,389	775	14.0	2,000	90	Prim.	55		HF	8,700	8,700	Succ.	Yes	FW
	850	14.0	2,000	90		60		HF	3,100	3,100	Succ.	Yes	FW
2,800	960	14.0	2,000	90		59		HF	3,900	3,900	Succ.	Yes	FW
	1,000	14.0	2,000	90	Prim.	55		HF	4,800	4,800	Succ.	Yes	FW
2,100	300	14.0	400	75	C	86		JS	410	410	Succ.	Yes	P (Exp. L)
2,000	350	12.0	4,600	90	C	55	20.0	JS	20	20	Disc.	No	P (Exp. UL)
800	550	18.0	300	90	Prim.			HF	280	280	Prom.		
1,000-2,000	1,100	11.3	500	170	SS	60	36.0	NC	610	425	Succ.	Yes	
1,000-2,000	850	11.4	1,000	140	SS	60	27.0	NC	650	545	Succ.	Yes	
1,500	1,100	12.0	300		SF	41	31.0	JS	290		TETT		FW
2,900	1,300	13.0	2,700	100	C	48	15.0	JS	2,600	1,250	Prom.	Yes	P
2,700	1,300	13.0	4,000	100	C	47		HF	410	410	Succ.	Yes	FW
3,900	1,400	14.0	900	120	C	61	20.0	NC	2,650	2,650	Succ.	Yes	P
1,000	1,000	11.0	10,000	90	C	53		JS	400	400	TETT		P
1,100	1,300	14.0	1,500	150	C	58	20.0	HF	7,900	7,900	Succ.	Yes	
2,000	1,800	12.0	7,000	90		60		HF	2,300	2,300	Succ.		RW
2,700	1,300	12.0	4,000	100	C	57		HF	510	510	Succ.	Yes	FW
2,300	1,700	13.0	6,000	100	C	44	15.0	JS	1,800	300	TETT		P
2,400	1,500	13.0	5,700	100	C	49	15.0	HF	1,200	1,200	Succ.	Yes	LW
2,400	1,600	13.0	6,000	100	C	50	15.0	HF	800	800	Succ.	Yes	LW
1,000	1,200	13.0	1,000	90	Prim.	46	15.0	NC	3,400	3,400	Succ.	Yes	FW
1,000	2,500	12.0	2,300	100	Prim.	50	12.0	HF	5,800	5,800	Succ.	Yes	FW
2,500	400	18.4	1,200	75	Prim.	50		JS	300	300	Succ.	Yes	LW
1,100	1,100	12.0	8,000	95.0	Prim.			HF	450	450	Succ.	Yes	
2,500	1,500	12.0	11,000	95					600	600	Succ.	Yes	FW
4,000	1,250	13.0	450						1,500	1,500	Succ.	Yes	FW
63	325	33.0	10.00	65	Prim.	50	15.0	NC	775	734	Succ.	Yes	RW (Exp. L)
2,500	1,000-1,500	13.0	2,200	100	Prim.	60	30.0	HF	5,100	5,100	Succ.	Yes	FW
2,000	2,500	17.0	350	105	Prim.	50	20.0	JS	220	100	TETT		LW (Exp. L)
10	4,250	22.0	29	140	Prim.			NC	396	65	Disc.	No	P (Exp. L)
2,000-4,000	1,550	16.0	30	100	Prim.	70		JS	500	400	TETT		RW
3,000	500	13.0	1,500	95	Prim.	87.5	57.7	NC	300	300	Succ.	Yes	FW
4,000	950	13.0	800	85	Prim.	64	53.0	HF	1,600	1,600	Succ.	Yes	FW
1,000	1,000	12.0	2,000	100	Prim.	65	30.0	HF	5,200	5,200	Succ.	Yes	FW
3,000	2,300	12.0	300	130	SS	55	27.0	HF	1,628	1,628	Succ.	Yes	FW
5,000	2,100	11.0	3,000	125	SS	55	27.0	HF	3,129	3,129	Succ.	Yes	P (Exp. L)
3,000	500	15.0	500	74	Prim.	50	20.0	JS	380	380	Prom.	Yes	Exp. L
875	1,500	16.0	939	85	WF	80		JS	200	200	Prom.		P
3,000	1,000	14.0	1,600	95	SS	55.7	18.1	HF	33,500	33,500	Succ.	Yes	Exp. L
3,000	1,000	14.0	1,600	95	Prim.	57	53.6	NC	800	800	Succ.	Yes	Exp. L
1,500	600-1,500	12.0		95	Prim.			HF	1,016	1,016	Succ.	Yes	FW
3,000		12.0	2,300	80	Prim.	80	40.0	HF	648	648	Succ.	Yes	FW
3,000	1,200	12.0			Prim.			HF	7,036	7,036	Succ.	Yes	FW
3,000	1,100	11.0			Prim.			HF	2,203	2,203	Succ.	Yes	FW

239

ENHANCED OIL RECOVERY

Producing thermal EOR in U.S.—continued

Operator	Field	State	County	Start date	Area, acres	Number wells Prod.	Number wells Inj.	Pay zone	Formation type	Porosity, %
Oryx	Midway-Sunset	Calif.	Kern	1/84	80	106	12	Potter/Marvic/Tulare	US	35.0
Oryx	Midway-Sunset	Calif.	Kern	7/84	70	56	6	Potter	US	35.0
Oryx	Midway-Sunset	Calif.	Kern	8/65	80	195	0	Potter	US	30.0
Oryx	Midway-Sunset	Calif.	Kern	7/84	120	271	0	Potter	US	30.0
Oryx	Midway-Sunset	Calif.	Kern	2/88	90	18	4	Marvic	US	34.0
Oryx	Midway-Sunset	Calif.	Kern	5/64	80	220	0	Potter	S	35.0
Oxy USA	Kern Front	Calif.	Kern	11/81	342	120	8	Etchegoin Chanac	S	33.0
Phillips	Smackover	Ark.	Ouachita	10/71	985	91	5	Nachatoch	S	35.0
Santa Fe Energy	Coalinga	Calif.	Fresno	3/79	160	142	54	Lower Temblor	S	30.0
Santa Fe Energy	Coalinga	Calif.	Fresno	10/89	20	14	6	Lower Temblor	S	30.0
Santa Fe Energy	Coalinga	Calif.	Fresno	10/89	20	23	4	Upper Temblor	S	30.0
Santa Fe Energy	Kern River	Calif.	Kern	9/89	20	31	8	Kern River Series	S	30.0
Santa Fe Energy	Midway	Calif.	Kern	1/85	100	160	25	Potter S.D.	S	30.0
Santa Fe Energy	Midway	Calif.	Kern	1964	900	1,000		Potter Cyc.	S	30.0
Santa Fe Energy	Midway	Calif.	Kern	1970	400	460		Spellacy	S	30.0
Santa Fe Energy	Midway	Calif.	Kern	6/82	800	240		Tulare	S	33.0
Shell	Arroyo Grande	Calif.	San Luis Obispo	1/87	20	35	13	M6	S	31.0
Shell	Belridge	Calif.	Kern	1961	3,489	2,982	683	Tulare	S	36.0
Shell	Cat Canyon	Calif.	Santa Barbara	3/85	1,000			Basal Sisquoc	S	32.0
Shell	Coalinga	Calif.	Fresno	1981	700	150	29	Temblor Zone II	S	30.0
Shell	Coalinga	Calif.	Fresno	1980	550	196	39	Temblor Zone I	S	32.0
Shell	Coalinga	Calif.	Fresno	1984	300	60	8	Temblor Zone II	S	27.0
Shell	Coalinga	Calif.	Fresno	11/87	245	80	30	Etchegoin	S	34.0
Shell	Cymric	Calif.	Kern	12/86	200	124	27	Tulare	S	36.0
Shell	Kern River	Calif.	Kern	1970	600	1,100	225	Kern River Series	S	30.0
Shell	Kern River	Calif.	Kern	9/83	85	35	35	Kern River	S	31.0
Shell	McKittrick	Calif.	Kern	3/88	226	114	30	Tulare	S	36.0
Shell	Midway-Sunset	Calif.	Kern	1/89	5	8	2	San Joaquin	S	33.0
Shell	Midway-Sunset	Calif.	Kern	1979	95	73	0	Upper Spellacy	S	28.0
Shell	Midway-Sunset	Calif.	Kern	1/89	68	76	10	Lower	S	25.0
Shell	Midway-Sunset	Calif.	Kern	1983	143	205	27	Sub-Hoyt	S	31.0
Shell	Midway-Sunset	Calif.	Kern	1971	400	575	35	Potter	US	22-32
Shell	Midway-Sunset	Calif.	Kern	1980	85	130	15	Monarch	S	25.0
Shell	Mount Poso	Calif.	Kern	1971	2,800	253	46	Vedder	S	33.0
Shell	Poso Creek	Calif.	Kern	6/80	360	70	0	Etchegoin	S	33.0
Shell	White Castle	La.	Iberville	8/76	64	30	10	V	S	35.0
Shell	White Castle	La.	Iberville	8/76	31	12	3	U	S	35.0
Shell§	Yorba Linda	Calif.	Kern	7/62	80	76	0	Shallow Tar	S	27.0
Shell	Yorba Linda	Calif.	Orange	1971	310	300	16	Upper Congolomerate	S	22.0
Texaco	Cat Canyon	Calif.	Santa Barbara	1964	1,700	12		S1B	S	31.0
Texaco	Cat Canyon	Calif.	Santa Barbara	1977	390	91		S1A-S6	S	31.0
Texaco	Kern River	Calif.	Kern	8/62	5,070	4,324	1,642	Kern River Series	S	31.0
Texaco	Lost Hills	Calif.	Kern	1977	35	46	11	Etchegoin	S	40.0
Texaco	Lost Hills	Calif.	Kern	1977	49	33	17	Tulare	S	38.0
Texaco	McKittrick	Calif.	Kern	1984	18	15	6	Tulare	S	35.0
Texaco	Midway-Sec 35	Calif.	Kern	1977	27	51	7	Potter	US	35.0
Texaco	Midway-Sec 36	Calif.	Kern	1/89	38	62	11	Potter	US	36.0
Texaco	Midway-North Midway	Calif.	Kern	11/81	77	100	23	Potter	US	30.0
Texaco	Midway-Sec 35	Calif.	Kern	1984	22	34	5	Potter	US	35.0
Texaco	Midway-Security	Calif.	Kern	11/77	36	72	15	Potter	US	33.0
Texaco	San Ardo	Calif.	Monterey	7/87	51	24	7	Lombardi	S	32.0
Texaco	San Ardo	Calif.	Monterey	1965	1,108	53	0	Aurignac	S	39.0
Texaco	Sour Lake	Tex.	Hardin	11/81	26	18	3	Miocene OB-1	S	35.5
Texaco	Sour Lake	Tex.	Hardin	6/84	7	3	2	Miocene OB-4	S	35.0
Texaco	Sour Lake	Tex.	Hardin	3/85	9	2	1	Miocene OB-2	S	33.3
Texaco	Sour Lake	Tex.	Hardin	1/86	6	4	2	Miocene OB-2	S	33.0
Texaco	Sour Lake	Tex.	Hardin	7/89	74	13	2	Miocene OB-3	S	33.0
Texaco	Sour Lake	Tex.	Hardin	10/87	4	6	2	Miocene 700 ft	S	33.0
Union Pacific	Wilmington	Calif.	Los Angeles	4/89	85	38	27	Tar	S	30.0
Unocal	Cymric	Calif.	Kern	1964	300	259	0	Tulare	S	35.0
Unocal	Guadalupe	Calif.	San Luis	1955	2,970	150	0	Sisquoc	S	35.0
Unocal	Midway-Sunset	Calif.	Kern	1974	81	57	17	Potter	S	34.5
Unocal	Midway-Sunset	Calif.	Kern	1983	81	61	0	Tulare	S	31.0
Unocal	Midway-Sunset	Calif.	Kern	6/77	39	60	17	Potter	S	24.0
Unocal	Midway-Sunset	Calif.	Kern	5/69	161	173	0	Potter	S	24.0
Unocal	North Belridge	Calif.	Kern	11/65	160	14	0	Tulare	S	38.0
Unocal	Tensleep	Wyo.	Natrona	5/81	25	9	2	Tensleep	S	18.0
Hot water										
ARCO*	Kern River	Calif.	Bakersfield	1986	50	45	10	Kern River Series	S	31.0
ARCO*	Kern River	Calif.	Bakersfield	1986	28	42	11	Kern River Series	S	31.0
ARCO	Midway-Sunset	Calif.	Kern	8/85	6	6	2	Tulare	S	33.0
Mobil	South Belridge	Calif.	Kern	11/82	242	54	8	Tulare	US	36.8
Santa Fe Energy	Kern River	Calif.	Kern	1/80	150	101	57	Kern River	S	30.0
Texaco	Lost Hills	Calif.	Kern	1/86	11	13	4	Tulare	S	38.0
Texaco	Lost Hills	Calif.	Kern	1/86	38	31	13	Etchegoin	S	40.0
Texaco	McKittrick	Calif.	Kern	1981	12	50	14	Potter	S	35.0
Texaco	San Ardo	Calif.	Monterey	9/86	240	62	24	Aurignac	S	33.0
Combustion										
Bayou State	Bellevue	La.	Bossier	1970	200	85	15	Nacatoch	S	32.0
Chevron	W. Heidelberg	Miss.	Jasper	12/71	362	9	3	Cotton Valley 4 & 5	S	14.0
Greenwich Oil	Forest Hill	Tex.	Wood	9/76	1,900	100	21	Harris Sand	S	28.0
Mobil	Lost Hills	Calif.	Kern	4/61	164	45	7	Tulare	S	42.6
Mobil	Midway-Sunset	Calif.	Kern	1/60	150	32	6	Moco	S	36.0
Mobil	West Newport	Calif.	Orange	1958	300	139	36	Miocene	S	37.0
Santa Fe Energy	Midway-Sunset	Calif.	Kern	1982	24	40	10	Potter	S	32.5
Texaco	Bellevue	La.	Bossier	9/63	385	200	30	Nacatoch	S	33.0

240

ENHANCED OIL RECOVERY

Table G

Permeability, md	Depth, ft	Gravity, °API	Reservoir oil cp	°F.	Prev. prod.	Satur. % start	Satur. % end	Proj. matur.	Total prod., b/d	Enh. prod., b/d	Proj. eval.	Profit	Proj. scope
3,000	1,100	11.0	2,300	95	Prim.			HF	914	914	Succ.	Yes	FW
3,000		12.0	2,300	95	Prim.			HF	603	603	Succ.	Yes	FW
3,000	600	12.0	3,607	110	Prim.	80	20.0	HF	4,657	4,657	Succ.	Yes	FW
3,000	700	12.0	3,607	110	Prim.	80	80.0	HF	5,613	5,613	Succ.	Yes	FW
1,000	1,200	13.0	3,000	104	S	70	15.0	JS	245	245	TETT		P
3,000	1,200	12.0			Prim.			HF	3,734	3,734	Succ.	Yes	FW
3,250	1,300-1,500	14.0	1,525	95	Prim.	50	15.0	JS	1,400	1,400	TETT	Yes	FW
2,000	1,920	20.0	75	110	Prim.	71	50.0	NC	600		Succ.	Yes	FW
100-3,000	850-1,800	14.5	1,900	75	Prim.	35	15.0	NC	1,800	1,500	Succ.	Yes	FW
100-3,000	600-1,300	14.5	1,900	75	Prim.	40	15.0	JS	75	65	TETT		Exp. L
100-3,000	1,100-1,600	14.5	1,900	75	Prim.	40	15.0	JS	100	85	TETT		Exp. L
100-3,000	300-600	13.0	4,000	85	Prim.	45	15.0	JS	250	200	TETT		EXP. L
2,500	1,300	12.0	4,000	100	Prim./C	60	20.0		3,200	1,500	TETT		Exp. L
2,500	1,300	12.0	4,000	100	Prim.	60	33.0		15,200	13,400	Succ.	Yes	FW
2,250	900	11.5	6,500	90	Prim.	60	33.0		4,200	3,300	Succ.	Yes	FW
1,300	1,200	11.0	5,000	100	Prim.	50	43.0		1,200	700	Succ.	Yes	FW
500-800	200-2,000	13-15	2,500-3,000	80-100	Prim.	56		JS	700	550	Prom.	Yes	FW
2,400	400-1,400	13.0	1,900	95	Prim.			HF	88,831	88,831	Succ.	Yes	FW
500-1,000	2,500-4,500	8-12		90-150	Prim.	55		PP			Succ.	Yes	FW
100-500	1,400-1,800	15-20	100-300	80-120	SS			NC	1,300	1,300	Succ.	Yes	FW
200-2,500	825-1,650	11-14	2,000-10,000	84-98	Prim.	60	10.0	HF	3,200	3,100	Succ.	Yes	FW
100-500	1,500-2,000	15-20	100-300	80-120	Prim			JS	900	720	Succ.	Yes	P (Exp. L)
800-1,000	650-1,000	7-12		80-90	Prim.	55		JS	3,300	3,300	Succ.		FW
1,000-2,000	1,000	11-14	1,000-2,000	95-105	Prim.			JS	2,700	2,700	Prom.		LW
500-2,500	150-1,500	11-15		78-85	Prim.			HF	12,200	11,840	Succ.	Yes	FW
2,000	1,200	12.0		81	Prim.	55		PP	400	400	TETT		LW
1,000-2,000	600	10-12		90-100	Prim.			JS	100	100	TETT		LW
	1,200	12.0	2,900	95	SS			JS			TETT		LW
	900	13.0	2,235	90	Prim.			HF	1,050	950	Succ.	Yes	FW
	800	11.6	5,000	90				JS	850	750	Prom.		LW
2,000-3,000	1,080	13.0	3,000	100	Prim.			HF	6,000	5,800	Succ.	Yes	FW
	500-1,400	8.5-14		95	Prim.			HF	11,400	11,400	Succ.	Yes	FW
900	1,300	13.0	2,700	95	SS			HF	6,400	6,300	Succ.	Yes	FW
2,000-10,000	1,800-2,000	15-18	100-500	90-130	Prim.	55	10	NC	14,800	14,800	Succ.	Yes	FW
2,000-4,000	2,300-2,600	10-14	400-800	80-90	Prim			JS	800	500	Succ.	Yes	FW
3,000	1,400	16.0	312	92	Prim.	90		HF	1,490	1,490	Succ.	Yes	RW
3,000	1,100	16.0	145	88	Prim.	90		HF	210	210	Succ.	Yes	RW
1,000	250-1,650	13.0		97	Prim.	95		NC	600	100	Succ.	Yes	FW
500-3,000	500-1,000	10-14		70-100	SS	66		NC	2,300	2,300	Succ.	Yes	FW
3,000	2,300	8.0	100,000	100				NC	30	30	Succ.	Yes	RW
1,400	2,500	7.0	50-10,000	110				NC	1,320	1,320	Succ.	Yes	RW
4,000	1,000	13.0	4,060	90	Prim.	70	30.0	HF	84,600	84,600	Succ.	Yes	FW
2,000	400	13.0	10	250	S	63	30.0	NC	1,100	1,100	Succ.	Yes	FW
2,000	200	13.0	20	220	S	70	30.0	NC	725	725	Succ.	Yes	FW
2,800	1,000	13.0	35	220	S	60	30.0	NC	120	120	Disc.	No	P (Exp. UL)
2,000	1,600	12.0	6,500	230	SS	55	20.0	JS	670	565	TETT	Yes	FW
3,000	1,600	14.0		90	SS	61	21.0	JS	1,560	795	TETT	Yes	P (Exp. L)
3,400	1,000	13.0	1,500	220	S	60	20.0	HF	2,335	1,422	Succ.	Yes	FW
2,000	1,600	12.0	1,500	240	SS	55	20.0	HF	580	440	Succ.	Yes	FW
5,000	950	12.0	1,500	280	S	60	20.0	NC	2,000	1,200	Succ.	Yes	FW
5,500	1,900	11.0	1,200	135	C	58	15.0	JS	1,446	1,446	TETT	No	P (Exp. L)
2,200	2,200	13.0	4,500	100		58	24.0	PP	772	772	Succ.	Yes	RW
1,436	1,000	14.7	665	90	Prim.	53	17.0	NC	163	163	Succ.	Yes	RW
900	1,600	16.2	107	100	Prim.	50	20.0	NC	45	45	Succ.	Yes	RW
1,000	1,200	16.2	100	95	Prim.	50	20.0	NC	35	35	Succ.	Yes	RW
1,000	1,900	13.6	200	106	Prim.	50	20.0	NC	30	30	Disc.	Yes	RW
5,181	1,350	17.0	180	100	Prim.	50	20.0	JS	100		TETT		RW
1,000	1,425	14.5		100	Prim.	50	20.0	JS	32	32	TETT	Yes	RW
2,780	2,563	14.0	283	123	WF	55	20.0	NC	400	400	Succ.		P (Exp. L)
1,000	1,100	13.0			Prim.	65	45.0	NC	8,365	7,070	Succ.	Yes	FW
1,550	3,000	9.0	580	135	Prim.	80	60.0	HF	1,200	700	Succ.	Yes	FW
3,700	1,100	11.4	10,000	110	SS	57.2	32.3	HF	2,530	2,250	Succ.	Yes	FW
1,146	700	11.4	10,000	100	Prim.	54	37.8	HF	420	210	Succ.	Yes	FW
1,500	1,000	11.3	100	220	SS	55	34.0	NC	1,800	1,500	Succ.	Yes	FW
1,500	1,000	11.3	500	170	Prim.	57	35.0	NC	3,300	2,435	Succ.	Yes	FW
2,500	800	13.6	450	130	Prim.	63	49.0	HF	166	66	Prom.	No	FW
500	2,500	14.0	580	90	Prim.	67	30.0	HF	490	300	Prom.	Yes	P
3,000	900	13.5	8,000	90	S	30	20.0	HF	200	130	Succ.	Yes	FW
2,500	600	13.5	5,000	78	S	35	20.0	HF	200	50	Prom.	Yes	FW
2,500	1,000	11.0	24,000	100	SF	20	10.0	NC			Disc.	No	P (Exp. UL)
3,000	1,000	14.0	1,600	95	SF	37.1	28.2	HF	2,200	2,200	Prom.	Yes	P
100-3,000	700	13.0	4,000	85	Prim.	45	15.0	NC	1,600	1,200	Succ.	Yes	FW
2,000	200	13.0	20	250	SF			NC	125	125	Succ.	Yes	P (Exp. L)
2,000	400	13.0	10	250	SF			HF	150	150	Succ.	Yes	P (Exp. L)
4,000	1,100	13.0	40	210	SF	50	30.0	HF	300	30	Disc.	No	P (Exp. UL)
3,000	2,200	13.0	4,000	100	S	38	31.0	JS	650	100	TETT	Yes	RW
650	400	19.0	660		Prim.	94	49.0	HF	420	420	Succ.	Yes	FW
85	11,300	18.0-27.0	6.00	221	Prim.	80		NC	470	470	Succ.	Yes	FW
950	5,000	10.0	1,006	185	Prim.	63	32.0	JS	1,050	1,050	Prom.	No	FW
1,790	300	15.0	410	95	Prim.	63	22.0	NC	520	520	Succ.	Yes	LW
1,500	2,700	14.0	110	125		74	37.0	HF	1,100	1,100	Succ.	Yes	FW
500-1,000	1,600	13.0	750	100	Prim.	86	37.0	HF	980	980	Succ.	Yes	FW
500-4,000	1,700	11.5	2,770	110	Prim.	70		HF	800	700	Prom.	Yes	Exp. L
960	350	19.0	675	130	Prim.	60	28.5	NC	850	850	Succ.	Yes	FW

ENHANCED OIL RECOVERY

Producing CO₂, gas EOR in U.S.

Operator	Field	State	County	Start date	Area, acres	No. wells Prod.	No. wells Inj.	Pay zone	Formation	Porosity, %	Permeability, md
CO₂ miscible											
Amoco	Levelland	Tex.	Hockley	3/73	13	2	6	San Andres	Dolo.	11.5	4
Amoco	Lost Soldier	Wyo.	Sweetwater	5/89	1,345	49	66	Tensleep	S	9.9	31
Amoco	Lost Soldier	Wyo.	Sweetwater	5/89	790	31	33	Darwin-Madison	S/LS-Dolo.	10.3	4
Amoco	Slaughter	Tex.	Hockley	1984	6,412	230	79	San Andres	Dolo./LS	10.8	2
Amoco	Slaughter	Tex.	Hockley	12/84	5,700	201	121	San Andres	Dolo./LS	12.0	5
Amoco	Slaughter	Tex.	Hockley	12/84	1,600	70	52	San Andres	Dolo./LS	10.0	4
Amoco	Wasson	Tex.	Yoakum	11/84	7,800	321	282	San Andres	Dolo./LS	9.0	5
Amoco	Wertz	Wyo.	Carbon, Sweetwater	10/86	1,400	58	61	Tensleep	S	10.0	20
ARCO	Sable	Tex.	Yoakum	3/84	825	43	21	San Andres	Dolo.	8.4	2
ARCO	Sho-Vel-Tum	Okla.	Stephens	9/82	1,100	70	41	Sims	S	16.0	70
ARCO	Wasson-Willard	Tex.	Yoakum	1/86	8,000	286	225	San Andres	Dolo.	9.0	1
Chevron	Kingdom Abo	Tex.	Terry	5/85	2,240	59	48	Abo Reef	Dolo.	4.5	8
Chevron	North Ward Estes	Tex.	Ward	3/89	3,840	190	194	Yates	S	16.0	37
Chevron	Pittsburg	Tex.	Camp	6/85	43	4	1	Pittsburg	LS	11.0	2
Chevron	Quarantine Bay	La.	Plaquemines	10/81	57	2	1	4 Sand Reservoir	S	30.0	100-1,000
Chevron	Rangely Weber Sand	Colo.	Rio Blanco	10/86	15,000	130	111	Weber SS	S	12.0	5-50
Chevron	SACROC Unit	Tex.	Scurry	1/72	49,900	732	387	Canyon Reef	LS	3.9	19
Chevron	Timbalier Bay	La.	Lafourche	4/84	37	2	1	S-2B	S	30.0	9,244
Citronelle	Citronelle	Alas.	Mobile	5/81	1,680	42	10	Rodessa	S	10-16	5-75
Conoco	Ford Geraldine Unit	Tex.	Reeves & Culberson	2/81	3,850	91	69	Delaware	S	23.0	64
Conoco	Maljamar	N.M.	Lea and Eddy	1/89	1,200	70	15	Grayburg-San Andres	Dolo/S	10.8	5
Enron Oil & Gas	Twofreds	Tex.	Loving, Ward, Reeves	1/74	4,392	45	36	Delaware	S	19.5	32
Exxon	Cordona Lake	Tex.	Crane	12/85	2,084	29	25	Devonian	Tripol.	22.0	4
Exxon	Means (San Andres)	Tex.	Andrews	11/83	8,500	364	184	San Andres	Dolo	9.0	20
Exxon	Slaughter	Tex.	Hockley	5/85	569	19	11	San Andres	Dolo.	12.5	6
Exxon	Wasson (Cornell Unit)	Tex.	Yoakum	7/85	1,923	58	51	San Andres	Dolo.	8.6	2
George R. Brown	Rose City North	Tex.	Orange	4/81	800	3	5	Hackberry	S	37.0	4,500
George R. Brown	Rose City South	Tex.	Orange	1/83	900	8	5	Hackberry	S	37.0	4,500
Mitchell Energy	Alvord South	Tex.	Wise	1980	2,291	17	17	Caddo	Congl.	12.8	55
Mobil	GMK South	Tex.	Gaines	1982	1,143	31	29	San Andres	LS	9.8	3
Mobil	McElmo Creek Unit	Utah	San Juan	2/85	13,440	155	110	Ismay Desert Creek	LS	14.0	5
Mobil	Slaughter	Tex.	Hockley	6/89	2,495	79	46	San Andres	Dolo	10.3	3
Mobil	Wasson	Tex.	Yoakum	10/85	640	30	26	San Andres	Dolo.	13.0	6
Pennzoil	Tinsley	Miss.	Yazoo	11/81	1,338	15	15	Perry	S	26.4	49
Phillips	Bridger Lake	Utah	Summit	4/70	3,800	7	1	Dakota	S	12.8	79
Phillips	Vacuum	N.M.	Lea	2/81	4,900	192	100	San Andres	Dolo.	11.7	11
Phillips	Maljamar	N.M.	Lea	11/89	40	4	1	Grayburg	S/Dolo.	6.0	0
Santa Fe Energy	Raymond	Mont.	Sheridan	8/83	685	2	1	Nisku	LS	8.2	13
Shell	Crossett (N. Cross)	Tex.	Crane & Upton	4/72	1,500	27	11	Devonian	Tripol.	22.0	5
Shell	Crossett (S. Cross)	Tex.	Crockett	6/88	800	21	10	Devonian	Tripol	21.0	4
Shell	Little Creek	Miss.	Lincoln & Pike	12/85	6,200	56	20	Lower Tuscaloosa	S	23.0	33
Shell	Olive	Miss.	Pike	10/87	1,280	16	6	Lower Tuscaloosa	S	26.0	50
Shell	Wasson (Denver)	Tex.	Yoakum & Gaines	4/83	20,000	780	380	San Andres	Dolo.	12.0	8
Shell	Wasson (South)	Tex.	Gaines	1/86	3,500	100	60	Clearfork	Dolo.	6.0	2
Shell	West Mallalieu	Miss.	Lincoln	11/86	5,760	14	4	Lower Tuscaloosa	S	25.0	20
Stanberry Oil	Farnsworth, North	Tex.	Ochiltree	6/80	1,431	9	5	Marmaton B	LS	11.5	140
Stanberry Oil	Hansford Marmaton	Tex.	Hansford	6/80	2,010	9	10	Marmaton	S	18.1	48
Texaco	Paradis	La.	St. Charles	2/82	347	7	6	Lower 9,000-Ft	S	26.0	770
Texaco	Paradis	La.	St. Charles	2/82	320	1	0	No. 8	S	27.0	795
Texaco	Paradis	La.	St. Charles	1/88	44	1	1	9,500 Ft	S	24.0	252
Union Texas	Wellman	Tex.	Terry	7/83	1,400	32	7	Wolfcamp	LS	9.2	>100.0
Unocal	Dollarhide	Tex.	Andrews	5/85	6,183	83	66	Devonian	Dolo./Tripol	13.5	9
CO₂ immiscible											
Chevron	Timbalier Bay	La.	Lafourche	12/85	422	10		4,900 ft Sand	S	32.0	500-2,500
Marathon Oil	Yates	Tex.	Pecos & Crockett	11/85	14,300	862	23	Grayburg/San Andres	Dolo.	17.0	175
Shell Western E&P	Weeks Island	La.	New Iberia	1/88	480	18	6	S RES A	S	23.0	150-4,000
Shell Western E&P	Weeks Island	La.	New Iberia	1/88	908	8	2	R RES A	S	23.0	100-1,500
Hydrocarbon miscible											
Amoco	Levelland	Tex.	Hockley	5/71	1,422	81	50	San Andres	Dolo.	10.2	2
ARCO	Prudhoe Bay	Alas.		12/82	3,650	42	11	Sadlerochit	S	23.0	300
ARCO	South Pass Block 61	OCS		8/83	55	1	1	1 RK2	S	30.0	25
ARCO	South Pass Block 61	OCS		8/85	54	2	1	LL RBB1, 2	S	30.0	175
ARCO	South Pass Block 61	OCS		5/81	320	8	3	UM RAAO, 2, 3	S	30.0	180
ARCO	South Pass Block 61	OCS		5/81	112	5	2	UM RBB	S	30.0	300
ARCO	South Pass Block 61	OCS		8/83	26	1	1	LL RG	S	33.0	10
ARCO	South Pass Block 61	OCS		8/83	39	1	1	UJ RL	S	33.0	50
ARCO	South Pass Block 61	OCS		5/81	54	4	2	MM RBB	S	30.0	300
ARCO	South Pass Block 61	OCS		10/84	68	3	3	LK2 RU	S	29.0	100
ARCO	South Pass Block 61	OCS		2/84	37	2	1	LL ART	S	30.0	50
ARCO	South Pass Block 61	OCS		5/81	179	7	2	MM RAAO, 2, 3	S	30.0	180
ARCO	West Devil's Pocket	Tex.	Newton	7/87	1,726	6	1	Wilcox 11 500 ft	S	12.3	1
ARCO	Kuparuk River	Alas.		6/88	5,120	30	20	Kuparuk "A" & "C" Sands		25.0	50-1,000
ARCO	Prudhoe Bay	Alas.		2/87	17,200	190	56	Sadlerochit	S	22.0	500
ARCO	Sabine Tram. SW (Wilcox 11500)	Tex.	Newton	12/89	2,010	7	1	Wilcox 11,500 ft	S	14.0	27
Exxon	South Pass Block 89	Tex.		12/83	204	10	10	X and Y Series	S	29.0	100-1,500
Hunt	Fairway	Tex.	Anderson/Henderson	3/66	22,618	95	65	James	LS	12.6	11
Kerr-McGee	North Buck Draw	Wyo.	Campbell	12/88	5,700	11	8	Dakota	S	9.3	3

ENHANCED OIL RECOVERY

Table H

Depth, ft	Gravity, °API	Reservoir oil cp	°F.	Prev. prod.	Satur. % start	Satur. % end	Proj. maturity	Total prod., b/d	Enh. prod., b/d	Proj. eval.	Profit	Proj. scope
4,900	30.0	2.30	105	WF	74		NC	28	28	Succ.		P (Exp. L)
5,000	35.0	1.30	178	WF			JS	6,500	4,570	Succ.	Yes	FW
5,400	35.0	1.40	181	WF			JS	2,700	764	Prom.	Yes	FW
4,900	31.0	1.40	105	WF			JS	3,850	1,100	Prom.	Yes	FW
4,950	31.0	1.40	105	WF			JS	9,095	4,015	Succ.	Yes	FW
4,950	31.0	1.40	105	WF			JS	3,900	1,900	Succ.	Yes	FW
5,100	32.0	1.30	110	WF			JS	16,000	8,000	Succ.	Yes	FW
6,000	35.0	1.16	163	WF			HF	10,000	7,200	Succ.	Yes	FW
5,200	32.0	1.46	107	WF			JS	680	380	Succ.	Yes	FW
6,200	25.0	3.30	115	WF	59	42.0	HF	2,140	1,330	Succ.	Yes	LW
5,100	32.0	2.01	110	WF			JS	4,800	1,600	Succ.	Yes	FW
7,500	28.5	3.40	120	WF	58	40.0	JS	2,790		TETT		FW
2,600	35.0	1.40	83	WF	25	10.0	JS	1,730	480	TETT		
8,000	41.0	0.80	205	WF	59		NC	130	100	Succ.	Yes	P (Exp. L)
8,120	32.0	0.99	183	Prim.	33	15.0	NC	125	125	Prom.	No	P (Exp. UL)
5,500-6,500	35.0	1.70	160	Prim./WF	38	29.0	JS	30,700	7,150	Prom.		FW
6,700	41.0	0.35	130	Prim./WF	63.3	46.8	NC	25,540	13,780	Succ.	Yes	FW
7,400	39.0	0.39	180	Prim.	29		HF	250	250	TETT	No	P (Exp. L)
11,000	42.0	46	210	Prim./WF	63		PP	3,300	350	Prom.		Expl. L:
2,680	40.0	1.40	83	Prim./WF	41	35.0	HF	1,500	1,500	Succ.	No	FW
3,650-4,200	37.0	1.00	95		55.6	40.0	JS	1,000		TETT		Exp. L
4,900	36.0	1.50	105	WF			NC	680	680	Succ.	Yes	FW
5,500	40.0	0.50	101	WF			JS	2,100	360	Prom.	Prom.	LW
4,300	29.0	6.00	97	WF			JS			Succ.	Succ.	
4,900	32.0	1.30	110	WF			JS	650	90	TETT		LW
4,500	33.0	1.00	106	WF			JS	2,000	100	TETT		LW
8,200	37.0	2.00	180	WF	50	35.0	NC	160	160	Prom.		FW
8,200	37.0	2.00	180	WF	50	35.0	HF	800	600	Prom.		FW
5,700	44.0	0.39	154	WF	60	52.0	HF	297	80	Prom.	No	FW
5,400	30.0	2.57	101	WF	55	28.0	JS	1,800	300	Succ.	Yes	LW
5,600	41.0	0.50	125	Prim./WF	50		JS	5,450	1,600	Prom.	Yes	FW
5,000	32.0	1.60	107	WF	45	8.0	JS	1,400		TETT	Yes	LW
5,100	33.0	0.97	110	WF	54.4	39.2	NC	1,700	300	Prom.	Yes	FW
4,800	39.0	1.50	175	WF	65	38.0	HF	260	260	Prom.	No	FW
15,600	40.0	0.36	225	Prim.	70	61.0	NC	650	250	Succ.	Yes	FW
4,500	38.0	1.00	101	Prim.	70	50.0	JS	9,000		Succ.	Yes	FW
4,600	36.0	1.10	101	Prim.	60	29.0	JS	40		TETT		P
7,900	40.0	0.40	178	Prim.			HF	85	60	Prom.	Yes	
5,300	44.0	0.36	106		49	21.0	HF	1,934	1,934	Succ.	Yes	FW
5,200	43.0	0.60	104		43	24.0	JS	565		TETT		Exp. L
10,640	39.0	0.40	248	WF	21	2.0	JS	3,900	3,900	Prom.		FW
10,500	39.0	0.34	250	WF	17	2.0	JS	950	950	Succ.	Yes	FW
5,200	33.0	1.30	105	WF	51	30.5	JS	41,100	25,300	Succ.	Yes	FW
6,700	33.0	1.10	105	WF	60		JS	5,800	400	Prom.		FW
10,365	38.0	0.50	245	Prim.	15	1.0	JS	1,500	1,500	TETT		FW
6,400	44.0	1.61	131	WF	57		HF	300	300	Succ.	Yes	FW
6,500	44.0	1.56	142	Prim.	43		HF	420	420	Succ.	Yes	FW
10,400	37.0	0.50	205	Prim.	62	48.0	HF	400	400	Prom.	No	RW
8,600	39.0	0.40	192	Prim.	30	20.0	NC	125	125	Prom.	No	RW
9,800	38.0	35	192	Prim.	45	33.0	HF	100	100	Prom.	No	RW
9,800	43.5	0.54	151	WF	35	10.0	JS	3,600	400	TETT		FW
8,000	40.0	0.44	122	Prim./WF	35	22.0	JS	2,200	400	Prom.		
4,878	26.0	1.70	138	Prim.	60		NC	124	95	Prom.	Yes	RW
1,100-1,700	30.0	5.50	82	GI			HF	57,923		Prom.	Yes	FW
14,100	33.0	0.42	242	Prim.	28	5.0	JS			TETT	Yes	RW
13,200	34.0	0.41	241	Prim.	28	5.0	JS			TETT	Yes	RW
4,900	30.0	2.30	105	Prim.	90		HF	2,185	2,185	Succ.		P (Exp. UL)
8,800	27.0	0.90	200	Prim.	65	25.0	JS	50,000	10,000	Prom.		P (Exp. L)
4,900	39.0	0.70	150	Prim.	65	25.0	NC	57	6	Succ.	Yes	FW
5,750	29.0	1.70	158	Prim.	66	28.0	NC	130	55	Succ.	Yes	FW
9,000	41.0	0.38	200	WF	78	25.0	HF	2,800	2,800	Succ.	Yes	FW
6,300	37.0	0.70	165	WF	78	27.0	HF	1,100	400	Succ.	Yes	FW
6,800	43.0	0.40	170	Prim.	73	31.0	NC	37	10	Succ.	Yes	FW
5,500	40.0	0.40	150	Prim.	77	26.0	NC	237	24	Succ.	Yes	FW
6,400	37.0	0.70	165	WF	78	27.0	HF	2,080	1,960	Succ.	Yes	FW
7,900	36.0	0.50	180	Prim.	67	27.0	HF	420	150	Succ.	Yes	FW
8,500	33.0	0.86	190	Prim.	70	30.0	HF	162	41	Succ.	Yes	FW
9,100	38.0	0.30	190	WF	76	26.0	HF	1,650	1,650	Succ.	Yes	FW
11,525	45.9	0.13	266	Prim.	53	24.9	JS/HF	354	167	Prom.		FW (Exp. UL)
5,900		2.00			40-50	25-35		24,000				
8,800	27.0	0.90	200	WF	50	25.0	JS	300,000	10,000	Prom.		FW
11,525	49.0	0.10	271	Prim.	55	15.0	JS	550		TETT		FW (Exp. UL)
11,270	38-40.0	0.40-0.60	165	Prim.	89	73.0	HF	14,000	10,000	Prom.	Yes	RW
9,900	48.0	0.15	260	Prim.	73	36.0	NC	5,110	5,110	Succ.	Yes	FW
12,450	46.0	0.12	282	Prim.	85	35.0	JS	11,000	8,000	Succ.		RW

ENHANCED OIL RECOVERY

Producing CO₂, gas EOR in U.S.—continued

Operator	Field	State	County	Start date	Area, acres	No. wells Prod.	No. wells Inj.	Pay zone	Form-ation	Porosity, %	Permea-bility, md
Oryx	Fordoche W-12	La.	Pt. Coupee	5/80	3,400	6	0	Wilcox	S	19.0	5
Oryx	Fordoche W-8	La.	Pt. Coupee	5/80	3,300	8	3	Wilcox	S	20.0	9
Oxy USA	Northeast Purdy	Okla.	Garvin	9/82	8,320	106	102	Springer	S	13.0	44
True Oil	Red Wing Creek	N. Dak.	McKenzie	1/82	640	8	1	Mission Canyon	LS	10.0	0.1
Hydrocarbon immiscible											
ARCO	Kuparuk River	Alas.		2/86	30,000	100	85	Kuparuk A & C Sands	S	22.0	50-500
Nitrogen miscible											
Exxon	Blackjack Creek	Fla.	Santa Rosa	2/81	4,600	8	8	Smackover	LS	16.0	112
Nitrogen immiscible											
Chevron	East Painter	Wyo.	Uinta	11/83	1,500	17	7	Nugget	S	11.0	3
Chevron	Painter	Wyo.	Uinta	6/80	1,360	33	13	Nugget	S	11.9	4
Chevron	Stonebluff	Okla.	Wagoner	3/82	540	18	15	Dutcher	S	16.0	200
Exxon	Fanny Church	Ala.	Escambia	1985	600	3	0	Smackover	LS	12.0	5
Exxon	Jay-Little Escambia Crk	Fla/Ala.	Santa Rosa/Escambia	1/81	14,415	50	30	Smackover	LS	14.0	35
Phillips	Andector	Tex.	Ector	11/81	1,931	20	0	Ellenburger	Dolo.	3.0	100
Phillips	Binger	Okla.	Caddo	1977	12,960	62	30	Marchand	S	7.5	0.2
Unocal	Chunchula Fieldwide Unit	Ala.	Mobile	4/82	23,279	31	8	Smackover Carbonate	Dolo.	12.4	10
Flue gas											
ARCO	Block 31	Tex.	Crane	6/49	7,840	195	98	Devonian	LS	12.0	5
Exxon	Hawkins	Tex.	Wood	4/77	10,590	491	41	Woodbine	S	28.0	3,400
Flue gas immiscible											
Exxon	Hawkins	Tex.	Wood	8/87	2,800	100	6	Woodbine	S	28.0	3,400

Producing chemical EOR projects in U.S.

Operator	Field	State	County	Start date	Area, acres	Number wells Prod.	Number wells Inj.	Pay zone	Form-ation type	Porosity, %	Permea-bility, md
Polymer											
Anderson Oil	North Glo	Wyo.	Campbell	10/87	292	4	2	Minnelusa	S	15.9	150
ARCO	Hamilton Dome	Wyo.	Hot Springs	9/89	80	7	1	Tensleep	S	14.0	58
Cenex	Sage Spring Crk-Unit A	Wyo.	Natrona	8/78	1,456	8	5	Dakota	S	13.0	50
Chevron	N. Stanley	Okla.	Osage	2/76	810	18	11	Burbank	S	18.0	300
Chevron	S. Stanley	Okla.	Osage	6/83	1,280	10	16	Burbank	S	18.0	300
Chevron	W. Bay	La.	Plaquemines	10/81	64	3	2	3A Sand	S	30.0	40
Chevron	W. Bay	La.	Plaquemines	10/81	42	4	1	3A Sand	S	34.0	4,600
Conoco	Vacuum-State #35 Lease	N.M.	Lea	3/83	240	5	10	Grayburg/San Andres	S/LS	10.9	14
Enron Oil & Gas	Isenhour	Wyo.	Sublette	7/80	173	2	3	Almy	S	15.0	1.0-25.0
Enron Oil & Gas	Long Island-Star Corral	Wyo.	Sublette	8/69	173	3	2	Almy	S	15.5	1.0-50.0
Enron Oil & Gas	McDonald Draw #1TB	Wyo.	Sublette	1/72	435	1	1	Almy	S	16.7	2.0-59.0
Enron Oil & Gas	McDonald Draw #4TB	Wyo.	Sublette	7/72	420	10	3	Almy	S	18.0	1.0-200.0
Enron Oil & Gas	Ruben	Wyo.	Sublette	12/70	800	5	4	Almy	S	14.1	1.0-56.0
Enron Oil & Gas	Saddle Ridge Unit	Wyo.	Sublette	12/74	376	8	5	Mesaverde	S	22.0	10.0-95.0
Fina Oil	South Cowden	Tex.	Ector	6/84	666	25	15	Grayburg	Dolo.	10.1	4
Gallagher Drilling	Stewart East Minnelusa Unit	Wyo.	Campbell	7/82	200	4	1	Minnelusa	S	15.6	250
Hunt	East Texas	Tex.	Rusk	8/82	508	76	11	Woodbine	S	22.0	300
Lamamco Drilling¶	Naval Reserve	Okla	Osage	2/82	3,360	121	33	Bartlesville	S	15.8	12
Marathon Oil	Grass Creek	Wyo.	Hot Springs	10/85	740	40	18	Phosphoria/Tensleep	LS/S	21.6	20.0/112.0
Marathon Oil	Ore.Basin-No.Embar Tensleep	Wyo.	Park	11/83	3,000	154	97	Embar/Tensleep	LS/S	20.2	68.0/193.0
Marathon Oil	Oregon Basin-SETPA	Wyo.	Park	1/85	6,108	180	33	Embar/Tensleep	LS/S	15.9	9.0/140.0
Marathon Oil	Kinney Coastal	Wyo.	Park/Big Horn	11/84	1,500	36	23	Tensleep	S	11.2	50
Marathon Oil	Main Consolidated	Ill.	Crawford	7/81	130	11	11	Robinson	S	18.0	115
Marathon Oil	Yates	Tex.	Pecos	12/83	6,900	323	256	Queen Grayburg/San Andres	Dolo.	15.0	62
Mitchell Energy	Alba Northeast Unit	Tex.	Wood	7/80	410	5	1	Sub-Clarksville	S	21.4	471
Mitchell Energy	Alba SEFB Unit	Tex.	Wood	2/72	731	21	7	Sub-Clarksville	S	23.3	525
Mitchell Energy	Jacksboro S.	Tex.	Jack	5/82	270	5	8	Strawn	S	19.9	86
Natl. Coop. Refinery	Stewart Ranch	Wyo.	Campbell	2/72	1,200	14	14	Minnelusa	S	16.5	92
Oryx	Hitts Lake	Tex.	Smith	10/80		19	7	Paluxy	S	18.9	300
Oxy USA	Harmony Hill	Kan.	Rooks	10/84	130	5	3	Lansing/Kansas City	LS	12.5	
Oxy USA	Hoof	Kan.	Graham	6/82	960	18	8	Lansing/Kansas City	LS	12.0	74.0-113.0
Phillips	North Burbank	Okla.	Osage	9/80	1,440	65	33	Burbank	S	15.5	50
Phillips	South Cowden	Tex.	Ector	9/83	2,050	54	23	Grayburg	Dolo.	11.9	4
Santa Fe Energy	Candy Draw	Wyo.	Campbell	5/87	482	12	1	Minnelusa	S	16.0	
Unocal	Dos Cuadras	Calif.	Santa Barbara	6/83	320	37	10	EP/FP	S	27.0	260.0-680.0
Unocal	Farnsworth	Tex.	Ochiltree	9/83	13,005	37	40	Morrow	S	14.5	22
Unocal	Healdton	Okla.	Carter	12/82	2,010	174	107	Healdton Sands	S	26.2	346
Unocal	Osage-Hominy	Okla.	Osage	4/83	2,880	96	36	Miss. Chat	LS	30.0	27
Western Production	Clareton	Wyo.	Weston	10/85	240	6	1	Newcastle	S	15-18	1-15
Western Production	Mush Creek	Wyo.	Weston	1969	880	4	3	Newcastle	S	15-19	1-15
Micellar-polymer											
Chevron	Berry Hill (Kiefer)	Okla.	Creek	9/82	82	12	21	Glenn	S	20.0	150

ENHANCED OIL RECOVERY

Table H

Depth, ft	Gravity, °API	Reservoir oil cp	°F.	Prev. prod.	Satur. % start	Satur. % end	Proj. maturity	Total prod., b/d	Enh. prod., b/d	Proj. eval.	Profit	Proj. scope
13,650	45.0	0.13	274				NC	338	338	Succ.	Yes	FW
13,200	44.0	0.13	267				HF	490	490	Succ.	Yes	FW
9,400	38.0	1.20	148	WF	46	40.0	HF	3,400	2,000	Succ.		FW
9,000	40.0	0.25	241	Prim.	60	20.0	JS	1,200		Succ.	Yes	FW
6,000	24.0	2.20	158	WF	57	54.0	JS	100,000		Prom.	Yes	
15,700	48.0	0.30	286	WF	59		NC	1,280	580	Prom.	Yes	FW
12,000	46.0	0.20	185	HC	98.5	52.0	HF	9,700	3,700	TETT		FW
11,500	46.0	0.20	174	HC	47		HF	3,500	1,450	TETT		FW
1,200	28.0	5.00	85	Prim.	67		NC	48	30			FW
15,565	51.0	0.15	288		76		NC	1,000	500	Disc.		FW
15,400	51.0	0.20	285	WF	59	46.0	HF	12,800	11,000	Prom.	Yes	FW
8,835	44.0	0.60	132	Prim.	43.8	37.0	NC	4,420		Succ.	Yes	LW
10,000	38.0	0.30	190	Prim.	76	59.0	HF	2,000	2,000	Succ.	Yes	FW
18,500	54.0	0.18	325	Prim.	80	45.0	JS	8,100	3,000	Succ.	Yes	FW
8,600	46.0	0.30	130	Prim.			NC	7,700	6,300	Succ.	Yes	FW
4,530	24.0	3.70	168	Prim.	88	12.0	NC		11,000	Succ.	Yes	FW
4,600	24.0	3.70	168		35	12.0	JS			TETT	Yes	Exp. L

Table I

Depth, ft	Gravity, °API	Reservoir oil cp	°F.	Previous prod.	Satur. % start	Satur. % end	Project maturity	Total prod., b/d	Enh. prod., b/d	Project eval.	Profit	Project scope
7,900	20.2	29.70	130	Prim	74.4	33.3	HF	726	613	Succ.		FW
2,750	20.0	39	135	WF			JS	350		TETT	Yes	Exp. L.
7,400	36.0	1.40	160	WF	45	21.0	NC	265	130	Succ.	Yes	FW
2,900	36.0	2.36	107	WF	51		NC	300	200	Succ.		FW
2,900	32.0	2.50	165	Prim.	54		NC	190				FW
7,100	41.0	2.30	120	WF	36	32.0	HF					
7,100	30.0	2.40	165	Prim.	65		HF				No	
4,500	37.0	1.30	101	Prim./WF	52	43.0	HF	425		Succ.	Yes	FW
3,600	41.5	1.20	100	Prim.	38.1	27.0	NC	40	10	Succ.	Yes	FW
3,550	44.0	1.20	100	Prim.	43.6	23.9	NC	30	1	Succ.	Yes	FW
3,050	42.0	1.40	90	WF	40	36.7	NC	18	1	Succ.	Yes	FW
3,200	41.8	1.45	95	WF	43.8	29.1	NC	125	7	Succ.	Yes	FW
3,300	43.0	0.95	95	Prim.	42	25.4	NC	120	8	Succ.	Yes	FW
1,800	44.0	2.33	85	Prim.	37.5	27.3	NC	155	10	Succ.	Yes	FW
4,500	35.0	1.90	96-100	WF	57	53.0	HF	1,600	100	Succ.	Yes	LW
8,000	20.0	20.00	124	Prim.	65	37.7	NC	126	100	Succ.	Yes	FW
3,500	37.0	1.90	142	Prim.	57	33.0	NC	530	330	Succ.	Yes	LW
2,650	38.0	3.50	90	WF	27.2	26.7	NC	549	15	Succ.	Yes	FW
4,300/4,500	24.0/24.0	15.00/15.00	105/110	Prim./WF			NC	2,485	920	Succ.	Yes	FW
3,370/3,520	22.5/22.1	8.5/11.0	108/110	Prim./WF	52	40.0	NC	10,900	2,500	Succ.	Yes	FW
3,600/3,840	21.0/21.0	25/15	100	Prim./WF	54	39.0	NC	9,200	3,000	Succ.	Yes	FW
4,240	22.1	11.30	140	Prim./WF			HF	1,581	431	Succ.	Yes	FW
1,100	30.0	18	70	Prim.	51		NC	18	18	Disc.	No	P (Exp. UL)
1,500	30.0	5.50	82	Prim.			NC	12,159		Succ.	Yes	FW
4,100	15.1	75	150	WF	62.6	52.1	NC	38	38	Disc.	Yes	FW
4,200	15.5	75	150	Prim.	69.5	51.8	NC	117	117	Succ.	Yes	FW
1,975	40.0	1.73	100	WF	50	22.0	NC	85	85	Succ.	Yes	FW
8,100	20.0	25	136	WF	68	49.0	NC	950	1,013	Succ.	Yes	FW
7,250	26.0	2.71	210	WF			NC	950		Succ.	Yes	FW
3,130	38.6	3.70	105	Prim.	49	34.0	NC	81	81	Succ.	Yes	FW
3,700	41.0	2.30	120	WF	36		NC	75	33		No	FW
2,900	39.0	3.00	120	WF	53	47.0	NC	459	304	Succ.	Yes	FW
4,750	35.0	3.40	96	WF	67	59.0	NC	1,000		Disc.	No	LW
7,300	26.0			Prim.			JS	1,000	1,000	Succ.		
1,280-1,650	25.0-26.0	5-24	95-103	WF	54	52-57	NC	2,300	0	Disc.	Yes	FW
7,400	38.0	1	141	WF	43	40.0	HF	573	9	Disc.	Yes	FW
800	30.3	10	80	WF	56	47.0	HF	1,270	72	Succ.	Yes	FW
2,880	38.7	3	100	WF	51	48.0	HF	975	50	Succ.	Yes	FW
6,000	40.0		160	Prim.			NC	16	16	Disc.	No	P
4,400	40.0		115	Prim.			NC	7	7	Succ.	Yes	LW
1,300	39.0	4.40	100	WF	50		NC	350	350	Succ.		LW

245

ENHANCED OIL RECOVERY

Producing chemical EOR projects in U.S.—continued

Operator	Field	State	County	Start date	Area, acres	Number wells Prod.	Number wells Inj.	Pay zone	Formation type	Porosity, %	Permeability, md
Marathon Oil	Lawrence Robins 102 B Maraflood	Ill.	Lawrence	8/82	25	12	5	Bridgeport	S	21.0	398
Marathon Oil	Lawrence-Robins 202-B Project	Ill.	Lawrence	10/85	1	4	1	Bridgeport Main	S	20.6	352
Marathon Oil	Main Consolidated M-1 Project	Ill.	Crawford	2/77	407	34	27	Robinson	S	19.0	102
Marathon Oil	Main Consolidated M-2 Project	Ill.	Crawford	7/80	331	72	53	Robinson	S	19.0	150
Alkaline											
Conoco	San Miguelito	Calif.	Ventura	11/82	316	16	11	Third Grubb	S	15.1	36
Shell	White Castle	La.	Iberville	9/87	1	2	1	Q RA SU	S	31.0	1,000
Surfactant											
South Fork Inc.*	S. Johnson	Ill.	Clark	6/81	49	36	22	Partlow	S	18.3	302

Tables G, H, I, former operator: *Tenneco †Celeron §Chevron ‡Texaco 'Gary √DCR ∞Lawrence Allison

Completed, terminated U.S. projects*

Operator	Field	State	County	Start date	Area, acres	No. wells prod.	No. wells Inj.	Pay zone	Formation	Porosity, %	Permeability, md
Alkaline											
Unocal	Carroll Unit	Tex.	Van Zandt	11/81	212	4	0	Lewisville	S	26.0	100.0
Unocal	South Van Unit	Tex.	Van Zandt	9/80	800	9	4	Lewisville	S	26.0	100.0
CO$_2$ Immiscible											
Chevron	Heidelberg	Miss.	Jasper	12/83	40	1	0	Eutaw	S	25.0	74.0
Chevron	Pittsburg	Tex.	Camp	11/83	40	1	0	Sub-Clarksville	S	23.0	460.0
Exxon	Pewitt Ranch	Tex.	Titus	6/83		1		Paluxy	S	24.0	1,000
Shell	Weeks Island	La.	New Iberia	1978	8	2	1	S RES B	S	26.0	1,800.0
Union Pacific	Wilmington	Calif.	Los Angeles	3/81	41	3	4	Tar	S	24.0	465.0
Union Pacific	Wilmington	Calif.	Los Angeles	2/84	156	10	12	Tar	S	24.0	465.0
CO$_2$ miscible											
Amoco	Levelland	Tex.	Hockley	8/78	15	1	4	San Andres	Dolo.	11.8	3.8
Amoco	North Cowden Unit	Tex.	Ector	6/85	12	2	6	Grayburg	Dolo. LS	10.0	5.4
ARCO	North Coles Levee	Calif.	Kern	6/81	70	7	1	Stevens	S	19.5	9.0
Chevron	Kurten	Tex.	Brazos	8/81	672	5	4	Woodbine	S	13.0	0.8
Oxy USA	Welch	Tex.	Dawson	2/82	2,675	140	119	San Andres	LS	9.3	9.0
Pennzoil	Richardson	W.V.	Calhoun, Roane	8/86	78	8	4	Berea	S	14.0	40.0
Texaco	Bay St. Elaine	La.	Terrebonne	1/81	9	0	0	8,000 ft	S	32.9	1,480.0
Combustion											
Oxy USA	Bellevue	La.	Bossier	1971	180	186	48	Nacatoch	S	33.9	700.0
HC miscible											
Callon Offshore*	Chandeleur Sound Block 25	La.	St. Bernard	1/83	342	8	3	4,800 ft sand	S	21.3	1,980
Callon Offshore*	Chandeleur Sound Block 25	La.	St. Bernard	1/83	941	40	16	BB	S	33.0	1,680
Hot water											
Texaco	Midway-Sec 35	Calif.	Kern	3/87	10	16	6	Potter	S	35.0	2,000.0
Micellar/polymer											
Oxy USA	Madison	Kan.	Greenwood	8/82	40	15	16	Bartlesville	S	18.9	42.0
Texaco	Salem Consolidated	Ill.	Marion	8/81	60	20	12	Benoist	S	15.4	154.0
Texaco	Sho-Vel-Tum	Okla.	Stephens	8/83	9	3	3	Sims		17.0	83.0
Microbial											
John Brown E&C	Nav Pet. Res. 3-Teapot Dome	Wyo.	Natrona	8/88	188	24	2	Shannon	S	18.0	4-650
Polymer											
American Expl.†	Elaine (San Miguel)	Tex.	Dimmit	12/81	1,260	11	0	San Miguel	S	21.7	6
American Expl.†	Hospah Sand Unit	N.M.	McKinley	1/83	1,061	31	16	Upper Hospah	S	24.7	467
American Expl.†	S. Hospah Lower Oil Pool	N.M.	McKinley	10/83	595	50	13	Lower Hospah	S	27.3	1,282
ARCO	West Nelson Rozet	Wyo.	Campbell	3/84	440	2	2	Minnelusa	S	17.0	170.0
ARCO	West Rozet	Wyo.	Campbell	9/84	120	3	2	Minnelusa	S	17.0	170.0
Beard Oil Co.§	Sleepy Hollow	Neb.	Red Willow	2/85	1,600	41	10	Reagan Sand	S	24.0	2,900
Chevron	C-Bar	Tex.	Crane	6/83	2,600	56	32	San Andres	Dolo.	10.0	6.0
Chevron	Dune	Tex.	Crane	3/82	410	26	14	San Andres	Dolo.	14.0	28.0
Chevron	Goldsmith 5600	Tex.	Ector	9/83	3,840	139	51	Clearfork	Dolo.	15.0	28.0
Chevron	McElroy	Tex.	Crane	10/82	3,200	242	56	Grayburg	Dolo.	13.0	6.0
Chevron	N. Ward Estes	Tex.	Ward	12/81	17,986	785	666	Yates	S	18.0	10.0
Exxon	Camrick	Okla.	Beaver/Texas	1984		43	27	Upper Morrow	S	15.0	
Exxon	Conroe	Tex.	Montgomery	1985				2B	S		5,500
Exxon	Hewitt	Okla.	Carter	3/81	886	85	52	Hewitt	S	21.0	175.0
Exxon	Loudon	Ill.	Fayette	1985		760	554	Chester	S	19.0	
Exxon	Omaha	Ill.	Gallatin	1984		15	4	Chester	S	18.0	20.0-387
Exxon	Roland	Ill.	White/Gallatin	1985		12	8	Chester	S	19.5	13-279
Exxon	Roland	Ill.	White	1985		16	14	Chester	S	15.7	16-626
Exxon	Trapp	Kan.	Russell	1984		15	9	Lansing/Kansas City	LS	18.4	
Exxon	W. Yellow Creek	Miss.	Wayne	9/76	4,813	51	24	Eutaw	S	24.0	112.0
Exxon	Webster	Tex.	Harris	1985		7	3	Frio 1A	S	28.0	206.0

246

ENHANCED OIL RECOVERY

Table I

Depth, ft	Gravity, °API	Reservoir oil cp	°F.	Previous prod.	Satur. % start	Satur. % end	Project maturity	Total prod., b/d	Enh. prod., b/d	Project eval.	Profit	Project scope
950	29.0	18	70	Prim./WF	45		HF	140	140	Prom.	No	P (Exp. UL)
900	29.0	18	70	WF	45		NC	12	12	Prom.	No	P (Exp. UL)
1,000	36.0	6	70	Prim./WF	40		NC	50	50	Prom.	No	FW
1,000	35.0	7	70	Prim./WF	40		HF	65	65	TETT		FW
8,011	31.0	0.70	205	Prim.	70	55.0	HF	1,025		TETT		FW
5,600-5,800	25.0	3.00	135		22	11.0	NC			Succ.		P
475	30.0	20.00	65	WF	46	28.0	PP	36	20	Prom.	Yes	Exp. L

Table J

Depth, ft	Gravity, °API	Reservoir Oil cp	°F.	Prev. prod.	Satur. % start	Satur. % end	Project maturity	Total prod., b/d	Enh. prod., b/d	Proj. eval.	Profit	Project scope
2,600	33.0	2.30	130.0	WF			Term.	5	0	Disc.	Yes	FW
2,600	35.0	2.30	130.0	WF			Term.	50	0	Disc.	Yes	FW
5,060	20.0	15.00	150.0	Prim.			C	24	0	Disc.	No	P (Exp. UL)
3,800	14.0	2,200.00	120.0	Prim.	65.0		C	15	3	Succ.	Yes	P
4,500	19.0	30.00	160.0	Prim.			C			Disc.	No	P
12,760	32.0	0.50	225.0	WF	22.0	2.0	C	0	0	Succ.	No	P.
2,500	14.0	283.00	123.0	WF	51.0	30.0	C	50	50	Succ.	No	P
2,500	14.0	283.00	123.0	WF	51.0	30.0	Term.	250	250	Succ.	No	FW
4,900	30.0	2.30	105.0	WF	43.0		C	0	0	Succ.		P (Exp. L)
4,300	34.0	1.67	94.0	WF			Term.	20	20	Disc.		P
9,000	36.0	0.45	235.0	WF	34.0	25.8	Term.	70	0	Disc.	No	P (Exp. UL)
8,300	38.0	0.40	230.0	Prim.	40.0		C	118	0	Succ.		P
4,890	34.0	2.15	96.0	WF	30.0	18.0	Term.	1,990	100	TETT	No	
2,300	46.0	3.00	80.0	WF	48.0		Term.	80	40	Prom.	No	P
7,400	36.0	0.67	170.0	Prim.	20.0	5.0	C	10	10	Disc.	No	RW
	19.0	676.00	75.0	Prim.	72.0	32.0	C	0	0	Succ.	Yes	FW
4,800	27.0	3.50	120	Prim.	45	30.0	C	379	0	Succ.	Yes	FW
5,450	27.0	3.45	132	Prim.	45	30.0	C	2,104	0	Succ.	Yes	FW
1,600	12.0	38.00	200.0	SF			Term.	400	220	TETT	Yes	P (Exp. L)
1,900	38.0	2.90	97.0	Prim.	34.0		Term.			Disc.	No	P (Exp. UL)
1,750	36.0	36.00	72.0	WF	30.5	15.3	C	110	110	Prom.	No	P (Exp. UL)
7,000	28.0	12.00	135.0	WF	50.0	45.0	C	30	30	Disc.	No	P (Exp. UL)
350	35.0	10.00	65.0	Prim.	45.0	45.0	Term.	391	0	Disc.		P
4,450	36.0	3.00	140		80.2	67.0	Term.	96	0	Succ.	No	FW
1,600	30.0	13.00	95	WF	69	68.2	Term.	148	0	Succ.	No	FW
1,600	24.0	35.00	95	WF	82.5	65.9	Term.	440	0	Succ.	No	FW
8,700	21.0	12.00	197.0	WF			C	100		Disc.		P
8,700	22.0	24.00	170.0	WF			C	70		Disc.		P
3,500	31.0	20.00	95	WF			C	1,600	800	Succ.	Yes	FW
3,500	36.0	5.00	107.0	WF	39.0		C	220	0	Succ.	Yes	FW
3,350	32.0	3.50	95.0	WF	64.0		C	176	0	Succ.	Yes	FW
5,600	32.0	3.50	100.0	WF	49.0		C	1,180	350	Succ.	Yes	FW
2,800	32.0	2.70	88.0	WF	58.0		C	4,600	0	Succ.	Yes	FW
2,600	33.0	1.50	95.0	WF	42.0		C	5,364	0	Succ.	Yes	FW
7,530	39.0	1.50	142.0	WF			C			Disc.	No	
4,900			170.0	Prim.			Term.			Disc.	No	P
2,400	35.0	8.70	96.0	WF	43.0		C			Succ.	Yes	FW
1,800	36.0	6.00	75.0	WF			C			Disc.		FW
1,950	28.0	17.00	78	WF			Term.			Disc.		
3,000	32.0	10.00	83	WF			Term.			Disc.		
3,100	36.0	10.00	87	WF			Term.			Disc.		
3,215	38.0	1.40	97.0	WF			C			Disc.	Yes	LW
4,950	20.0	20.00	150.0	WF	84.0	75.0	C			Succ.		FW
5,900	29.0	1.30	165.0	WF			C			Succ.	No	RW

247

ENHANCED OIL RECOVERY

Completed, terminated U.S. projects—continued

Operator	Field	State	County	Start date	Area, acres	No. wells prod.	No. wells Inj.	Pay zone	Formation	Porosity, %	Permeability, md
Fina	Garza	Tex.	Garza	10/82	120	10	5	San Andres	LS	19.8	4.1
Fina	Westbrook	Tex.	Mitchell	12/80	4,740	149	43	Clearfork	Dolo.	7.4	6.3
Graham Resources	Carthage	Okla.	Texas	12/82	1,640	14	9	Morrow	S	19.0	200.0
Lamamco	Naval Reserve	Okla.	Osage	2/82	3,360	90	28	Bartlesville	S	15.8	12.0
Marathon Oil	Byron	Wyo.	Big Horn	12/82	1,500	41	36	Embar/Tensleep	LS/S	13.9	4.5/78
Mitchell Energy	Lucy N.	Tex.	Borden	10/83	550	8	4	Pennsylvanian	LS	9.7	30.0
Mobil	Salt Creek	Tex.	Kent	12/84	12,100	185	92	Canyon Reef	LS	12.0	20.0
Mobil	Vacuum	N.M.	Lea	5/84	320	10	4	San Andres	Dolo.	10.4	3-100.0
Oryx	Fitts	Okla.	Pontotoc	1982	2,030	103	112	Viola	LS	13.6	1-36
Oryx	Stephens County Regular	Tex	Stephens	2/80	3,900	73	73	Caddo	LS	14.2	11.0
Oryx	Stephens County Regular	Tex.	Stephens	2/80	700	10	15	Caddo	LS	14.2	11.0
Oryx	Stephens County Regular	Tex.	Stephens	3/86	3,026	62	78	Caddo	LS	14.0	10.0
Phillips	Vacuum	N.M.	Lea	8/83	320	14	8	San Andres	Dolo.	11.5	17.0
Prima Exploration	C-H	Wyo.	Campbell	8/84	40	3	1	Minnelusa	S	20.6	80
Shell	Big Mineral Creek-Barnes Sand	Tex.	Grayson	12/84	1,065	14	18	Barnes Sand	S	18.5	59.0
Shell	Big Mineral Creek-S Sand Unit	Tex.	Grayson	5/85	1,600	22	14	S Sand	S	13.0	28.0
Texaco	Carthage N.E.	Okla.	Texas	4/83	1,024	7	4	Morrow A	S	18.6	92.0
Texaco	Hewitt	Okla.	Carter	8/82	300	6	5	Hoxbar	S	19.0	270.0
Texaco	Hewitt	Okla.	Carter	4/84	220	8	5	Hoxbar	S	21.0	84.0
Texaco	Hewitt	Okla.	Carter	11/85	100	6	3	Lone Grove	S	18.0	50.0
Texaco	Sho-Vel-Tum	Okla.	Carter	5/85	570	32	22	Deese	S	17.0	70.0
Texaco	Sho-Vel-Tum	Okla.	Carter	2/83	225	14	5	Deese	S	17.0	59.0
Texaco	Sho-Vel-Tum	Okla.	Carter	1/83	1,390	73	34	Hoxbar Deese	S	20.3	200.0
Texaco	Sho-Vel-Tum	Okla.	Stephens	8/81	190	7	4	Deese	S	17.4	216.0
Texaco	Slaughter	Tex.	Cochran	2/84	1,887	33	15	San Andres	Dolo.	11.2	6.0
Texaco	Slaughter	Tex.	Cochran	12/81	7,406	275	90	San Andres	Dolo.	10.7	3.0
Texaco	Vacuum Grayburg-San Andres	N.M.	Lea	7/82	1,486	37	29	Grayburg/San Andres	Dolo.	11.0	18.0
Texaco	Black Diamond	Tex.	Jim Hogg	4/83	200	4	3	Pettus II	S	25.0	212.0
Unocal	Howard Glasscock	Tex.	Howard	1/83	169	12	11	Queen	S	22.0	37.0
Unocal	Southwest Van Unit	Tex.	Van Zandt	9/84	529	18	20	Lewisville	S	26.0	100.0
Steam											
Chevron	Cymric	Calif.	Kern	5/66	26	18	0	Tar	S	32.0	1,200
Chevron	Midway-Sunset. Sec. IF	Calif.	Kern	2/78	19	23	8	Upper 10-10	S	32.0	2,200
Chevron	Tejon-Grapevine	Calif.	Kern	12/78	40	8		Basal Chanac		30.0	2,160
Mobil	Edison	Calif.	Kern	12/81	100	11	0	Kern River	S	30.0	1,000.0
Mobil	Kern Front	Calif.	Kern	5/74	480	43	0	Etchegoin	S	37.0	1,000.0
Texaco	Midway-Reward	Calif.	Kern	2/77	10	36	0	Potter	US	33.0	3,000.0
Texaco	Midway-Santiago Creek	Calif.	Kern	12/83	10	10	4	Tulare	US	34.0	2,500.0
Union Pacific	Wilmington	Calif.	Los Angeles	7/82	20	9	4	Tar	S	30.1	2,780.0
Unocal	Gato Ridge	Calif.	Santa Barbara	1970	130	16	0	Sisquoc Sib & S2	S	32.0	2,500
Unocal	Midway-Sunset	Calif.	Kern	1/85	5	9	4	Potter	S	34.5	3,700
Unocal	South Belridge	Calif.	Kern	3/85	167	12	0	Tulare	S	38.0	4,000.0
Chevron	Cymric 62Z	Calif.	Kern	2/86	30	31	19	Tulare	S	33.0	2,900
Chevron	Lost Hills Sec. 32	Calif.	Kern	2/79	150	29		Etchegoin A	S	36.0	
Chevron	Lost Hills Sec. 32	Calif.	Kern	3/85	16	11	4	Etchegoin A	S	40.0	2,500
Shell	Brea Olinda	Calfi.	Orange	6/85	19	20	0	Middle A Sand	S	20.0	50-200
Chevron	Kern Bluff	Calif.	Kern	7/66	214	25		Santa Margarita	S	29.0	2,000
Shell	Midway-Sunset	Calif.	Kern	6/86	110	37	0	San Joaquin	S	33.0	

Former operator:
    * ARCO
    † Tesoro
    § Amoco

Producing Canadian heavy oil EOR projects

Operator	Field	Province	Start date	Area, acres	Number wells Prod.	Number wells Inj.	Pay zone	Formation type	Porosity, %
Combustion									
Amoco Canada	Lindbergh Section 18 02 Pilot	Alta.	12/82	10	6	1	Lower Cummings B&C	US	29.0
Amoco Canada	Morgan	Alta.	1/81		35	9	Lloydminster	US	31.5
Electromagnetic									
Mazzei Oil & Gas Ltd.	Frog Lake	Alta.	1/89	640	4		Cummings, McLarin	S	33.0
Steam									
Amoco Canada	Cold Lake	Alta.	1/85	50	16	16	20 meters	US	33.0
Amoco Canada	Soars Lake	Alta.	6/88		16		Sparky	US	33.0
Aostra	Athabasca Oil Sands	Alta.	7/84	2	3	3	McMurray	US	30-35
BP Resources Canada	Wolf Lake	Alta.	4/85	10,130	396	396	Clearwater FM	S	31-33
BP Resources Canada	Wolf Lake	Alta.	1/88	4-6	22	22	L. Grand Rapids FM (B)	S	35.0
Husky Oil	Pikes Peak	Sask.	1/81	240	75	9	Wasaca	US	33.0
Koch Exploration Canada	Fort Kent	Alta.	12/85	160	13		Upper Grand Rapids	S	35.0
Mobil Oil Canada	Iron River	Alta.	3/88	160	20	20	Sparky	S	30-35
Mobil Oil Canada	Iron River/Wolf Lake	Alta.	6/89		1	1	Grand Rapids-Gen. Petr	S	35.0
Murphy Oil	Lindbergh	Alta.	1/74	212	68	68	Lower Grand Rapids	US	33.0
Petro-Canada	Athabasca	Alta.	1/87		1	1	McMurray	US	32.0
Petro-Canada	Athabasca	Alta.	1/85		1	1	McMurray	US	32.0
Shell Canada Ltd.	Peace River	Alta.	10/79	49	24	7	Cretaceous-Bullhead	S	28.0
Shell Canada Ltd.	Peace River	Alta.	10/86	450	163	53	Cretaceous-Bullhead	S	28.0
Amoco Canada	Gregoire (Athabasca)	Alta.	9/85		3	1	McMurray	US	36.4
Esso Resources Canada	Cold Lake	Alta.	1964		1,200	1,200	Clearwater	US	35.0

ENHANCED OIL RECOVERY

Table J

Depth, ft	Gravity, °API	Reservoir Oil cp	°F.	Prev. prod.	Satur. % start	Satur. % end	Proj. matur.	Total prod., b/d	Enh. prod., b/d	Proj. eval.	Profit	Project scope
2,900	36.0	2.50	90.0	WF	40.0		C	680		Disc.	Yes	P (Exp. UL)
3,000	26.0	9.10	90.0	WF	50.0		C	15			Yes	
4,500	39.0		119.0	Prim./WF	50.0	40.0	C	600	450	Succ.	Yes	FW
2,650	38.0	3.50	90.0	WF	27.2	26.7	C	463	15	Succ.	Yes	FW
5,600	23.0	17.00	121.0	Prim./WF	53.9	50.7	C	2,450	0	Succ.	Yes	FW
7,640	40.0	0.44	140.0	Prim.	52.6	35.7	Term.	95	15	Disc.	No	FW
6,300	39.2	0.85	129.0	WF	60.0	58.0	C	29,000		Succ.	Yes	FW
4,700	37.0	1.02	101.0	Prim./WF	38.0	30.0	C	240			Yes	P (Exp. UL)
3,900	39.0	3.20	119.0	WF			C	2,024		Succ.	Yes	FW
3,300	39.0	1.80	131.0	WF			C	1,555		Succ.	Yes	FW
3,300	39.0	1.80	131.0	WF			C	117		Succ.	Yes	FW
3,200	39.0	3.00	130.0	WF			C	1,031		TETT		FW
4,500	37.0	1.20	101.0	Prim.	74.9	58.2	C	1,000		Succ.	Yes	LW
7,500	20.7	26	135	WF			Term.	94		Prom.		P (Exp. UL)
5,000	37.0	1.50	120.0	WF	50.0	47.0	Term.	300	0	Disc.	No	P (Exp. UL)
5,200	37.9	1.67	124.0	WF	44.0	37.0	Term.	570	0	Disc.	No	FW
4,400	40.0	2.00	113.0	WF	45.0	44.5	C	209	60	Succ.	Yes	FW
2,350	35.0	6.00	98.0	WF	40.0	39.8	C	41	10	Succ.	Yes	FW
2,400	33.0	10.00	94.0	WF	39.0	37.0	C	102	46	Succ.	Yes	FW
4,000	30.0	4.30	115.0	WF	39.0	37.0	C	135	31	Succ.	Yers	FW
3,500	34.0	10.00	110.0	Prim.	57.0	54.6	C	1,059	320	Succ.	Yes	FW
4,400	20-34	4.30	102.0	WF	38.0	37.2	C	94	8	Disc.	Yes	FW
3,100	25.0	20.00	95.0	WF	49.0	47.8	C	1,520	200	Succ.	Yes	FW
5,500	22.0	60.00	124.0	Prim.	64.0	56.0	C	88	29	Succ.	Yes	FW
5,000	31.0	1.47	110.0	WF	53.0	50.0	C	291	0	Disc.	Yes	FW
5,000	31.0	1.47	107.0	WF	49.4	45.6	C	6,150	900	Succ.	Yes	FW
4,720	38.0	1.47	105.0	WF	45.0	44.0	C	3,650	182	Succ.	Yes	FW
4,550	48.0	0.40	164.0	Prim.	68.0	41.0	C	160	178	Succ.	Yes	RW
1,600	32.0	10.00	90.0	WF	50.7	43.4	C	62	25	Succ.	Yes	FW
2,600	35.0	2.30	130.0	WF			Term.	475	0	Disc.	Yes	FW
700	11.0	6,500.00	81.0	Prim.	60.0	20.0	C	332	273	Succ.	Yes	FW
1,600	14.0	1,500.00	105.0	C	55.0	15.0	C	100	0	Succ.	Yes	FW
2,600	16.0	167.0	115.0	Prim.			Term.	0	0			
1,100	17.0	800.00	100.0	Prim.	87.0	33.0	Term.	30	20	Succ.	Yes	FW
1,800	14.0	800.00	100.0	Prim.	86.0	67.0	Term.	350	350	Succ.	Yes	FW
1,300	13.0	1,500.00	220.0	S	60.0	20.0	Term.	450	220	Succ.	Yes	FW
1,300	12.0	1,500.00	90.0	S	60.0	20.0	Term.	300	300	Prom.	Yes	Exp. L
2,563	14.0	283.00	123.0	WF	55.0	20.0	C	400	400	Succ.	Yes	P (Exp. L)
1,900	10.5	1,450	130	Prim.	65	49.0	Term.	60	5	Disc.	No	FW
1,100	11.4	10,000	110	SF	52	28.0	C	719	109	Succ.	Yes	P (Exp. L)
1,250	13.2	450.00	130.0	Prim.	66.0	52.0	C	103	0	Disc.	No	FW
1,200	12.0	5,500.00	100.0	C	51.0	15.0	Term.	130	0	Succ.	Yes	FW
1,150	18.0	200.00	90.0	Prim.	72.0		Term.	130	0	Succ.	Yes	
850	14.0	1,400.00	100.0	C	50.0	20.0	Term.	100	0	Succ.	Yes	FW
3,000	13-16	200-600	100-140	Steam soak	49.0		Term.	220	280	Succ.	Yes	FW
950	15.0	1,000.00	84.0	Prim.	74.0		Term.	0	0	Succ.	Yes	
1,200	12.0	2,900.00	95.0	Prim.			Term.	280		Disc.		

Table K

Permeability, md	Depth, ft	Gravity, °API	Reservoir oil cp	°F.	Satur. % start	Satur. % end	Project maturity	Total prod., b/d	Enh. prod., b/d	Project eval.	Profit	Project scope
4,000	1,900	11.7	6,500	81	70		HF			Prom.		P
4,400	1,940	12.0	8,100	75	80		HF			Succ.		P FW
	1,800	9-12	10,000	80			HF	220	80	Prom.		Exp. L
1,500	1,500	10.5	500,000	55	80		HF			TETT		P
5,000	1,480	11-13	5-10,000	70	78		JS		11	Disc.	No	P(Exp. UL)
0.1-10	500	8	10,000,000	50	85	15	NC			Succ.	Yes	P
1,000-3,000	1,600	10-12	30-230,000	60-61	54-66		JS	8,175	8,175	Succ.		FW
3,000	380	10	250-1,400,000	61	75		JS			Disc.		P
5,000-10,000	1,640	12	25,000	70	85		HF	4,400	4,400	Succ.	Yes	Exp. L
	1,000	11	21,300		85		HF			Prom.		P
500-2,000	1,150	11.7	40,000	52	65	49	JS	850	850	TETT		P
1,000-4,000	1,250	11	50,000	15	70		JS			TETT		P
2,500	1,600	11	102,500	70	82	65	PP	2,500	2,500	Prom.	Yes	Exp. L
1,000-5,000	1,200	8-9	1,000,000	50	85		NC			Prom.		P
1,000-5,000	1,200	8-9	1,000,000	50	85		NC			Prom.		P
100-2,000	1,800	7.5	100,000	60	80		NC	1,000		Succ.		P
100-2,000	1,800	7.5	100,000	60	80		JS	10,000		Succ.		Exp. L
3,000	615	10	850,000	55	70		HF			Succ.	No	P(Exp. UL)
1,500	1,509	10.2	100,000	55			HF	88,060		Succ.	Yes	FW

Oil, the environment, and the public

THE PETROLEUM INDUSTRY IS ARGUABLY THE world's most important. The Automobile Age that propelled the world's growth in the 20th Century would not have been possible without the inexpensive, widely available range of products that fueled it.

In its relatively short history, the petroleum industry has almost always managed to deliver its vital commodities to the consumer safely, efficiently, and inexpensively.

A similar case can be made for how the industry conducts itself in day to day business. The public relations and media professionals within the industry generally are no exception to this truism.

Industry's image

If that is the case, then why does the petroleum industry have such a poor image with the public and the media?

Oil companies routinely seem to be regarded by the public at large as ruthless, greedy, and indifferent to social concerns. Despite its contemporary connotation, this phenomenon is nothing new. It certainly predates Exxon Valdez, or Santa Barbara, or even the Seven Sisters era.

The roots of the petroleum's industry's poor image are in its monopolistic beginnings. There is a direct line from the editorial cartoons of the John D. Rockefeller/Standard Trust days depicting the bloated, sneering, greedy plutocrats dripping with oil to the scathing editorial cartoons that appeared after the Exxon Valdez, only the latter came with a "raped Mother Nature" twist.

After the Teapot Dome scandal scarred the Harding presidency, muckraking novelist Upton Sinclair wrote **Oil!**, a popular novel wherein the son of a wealthy oil tycoon comes to learn that politicians and oilmen are an unscrupulous lot; he turns to socialism.

The years after World War II proved a respite for industry's negative image, if only because it rode the coattails of the general go-go pro-business spirit of the era. The good, grey eminence of the Seven Sisters presided over an era of boom and bust and generally sheap energy prices, and the industry was personified by the Humble tiger and the singing service station attendants behind Uncle Miltie.

The petroleum industry is no more entitled to immunity from media and political criticism than any other. But it seems to have been especially effective at picking the worst possible times to hand the media just that opportunity. Recall the emerging environmental consciousness of 1968 and its resurgence in 1988 just before the major spills off California and Alaska. That trend continued in 1990. The Mega Borg tanker exploded in the Gulf of Mexico, threatening a 1 million bbl oil spill for a time just when Congress was trying to deal with spill legislation and President Bush was about to announce his decision on federal oil and gas lease sales off California and Florida.

Sometimes, it seems as if the petroleum industry can do no right. Exxon must be feeling particularly set upon these days, although to a significant degree that has been self-inflicted. A rash of spills in 1990 in New York Harbor led the company to suspend marine operations at its Bayway marine terminal at Linden, N.J., and its Bayonne, N.J.,

BP AMERICA SENIOR CONSULTANT Roger Herrera illustrates the oil potential of the Arctic National Wildlife Refuge Coastal Plain during a media tour of the area. Such initiatives to explain the industry's stance on key issues are crucial to the survival of the petroleum industry during the 1990s. Photo by Bob Williams.

refinery and set up a task force to study and make recommendations on the situation. Exxon in early June 1990 released the results of that study, resuming marine operations and announcing plans to spend $10 million to upgrade safety and environmental performance at the facilities. A few days later, there was another sizable spill in the harbor—not related to Exxon, but the linkage certainly was there in the public's mind.

About that same time, Exxon released a cursory report by three leading British scientists with expertise on the long term effects of oil spills in cold marine environments, who pronounced Prince William Sound recovering from the Exxon Valdez spill. At that same moment, singer John Denver and National Wilderness Society President Jay Hair were holding up oily rocks on TV cameras in the sound with Denver angrily claiming that he does not trust anything Exxon says.

The industry's image suffers when it is held up in the media as a fount of greed and unscrupulousness. It suffers even more when it is depicted as a despoiler of the earth, because a threat to the environment brings the problem a little closer to home. But when does industry's image suffer the most, even to the point of overriding the other two images?

When it hits John Q. Public in the pocketbook. There is a

ENVIRONMENT

sizable body of evidence in industry surveys that indicate the public's ire is manageable when it is perceived—rightly or wrongly—as corrupt or environmentally negligent when those issues do not affect them economically. In a sense this may seem cynical, but it is really just a common sense reading of how people generally react to issues of the day most strongly when they are directly affected economically.

Opinion Research Corp., Princeton, N.J., has tracked public attitudes toward the petroleum industry from 1965 to 1989. It is clear that the most significant increases in an unfavorable view of the industry and parallel drops in a favorable view of the industry occurred after major spikes in the price of crude oil and then gasoline. Although the industry's skid in public favorability started with the environmental boom, it gained its greatest momentum after the price shocks of 1971-73 and 1979. Conversely, the highest favorable ratings since the Arab oil embargo have come in the wake of the oil price slide of the mid-1980s.

ORC's research indicates industry has suffered substantially since the Exxon Valdez spill and the reflowering of environmentalism. About three out of five people surveyed in March 1990 viewed the operations of the chamical, and oil/gasoline industries as very harmful to the environment, faring the worst of the eleven industries studied.

ORC's surveys are based on telephone interviews with a random, scientifically selected nationwide sample of 1,046 adults split almost evenly between men and women.

Anger at the oil industry was especially marked soon after the Exxon Valdez spill. Residents there wore black armbands, telling reporters they felt as if there had been a death in the family. They spoke of being in shock, mourning a loved one. Residents of fishing communities were especially angry, fearful of a loss of their livelihood. At one point, the radical environmental group Greenpeace attempted to spur the fishermen in the Price William Sound area to blockade the Valdez Narrows with their fishing boats to keep oil tanker traffic from the port. Despite their anger and grief, the fishermen turned down the Greenpeace request. Such a blockade would effectively shut down almost 25% of the nation's oil supplies, eliminate all but a fraction of the revenue stream that accounts for almost 90% of Alaska's revenues, and by pinching North Slope producers' cash flow perhaps complicate the fishermen's efforts to be reimbursed for losses incurred by the spill.

No one had more of a right to be angry over the Exxon Valdez spill than Alaskans. That anger was certainly in evidence for weeks and months to come. A poll conducted by the Alaska Oil & Gas Association in October 1988, five months before the spill, indicated that 69% of the Alaskans had a positive view of the petroleum industry, versus 6% with a negative view. Shortly after the spill, another AOGA survey showed the positive rating fell to 50% and the negative rating jumped to 25%. By September 1989, another survey showed the approval rating had rebounded to 66.5% while the disapproval rating slipped to 21.5%. The last survey occurred as controversy erupted anew over Exxon's winter plans for cleanup activity in Prince William Sound.

Ironically, one possible factor in all of this is that the spill actually boosted Alaskan revenues for the year because of the short-lived spike in oil prices that followed. It would be too superficial an assessment to conclude that Alaskans' anger subsided solely because the spill was not damaging them economically, but it is conceivably a factor.

It was not a case of short memories, because the spill dominated much of the national news and certainly the state news as controversy over the cleanup bubbled all summer long. At yearend 1989, UPI editors placed the spill as the fourth top story of the year, following only the California earthquake, reform in eastern Europe, and Tienanmen Square and beating Hurricane Hugo, the abortion and flag-burning debates, Jim Baker and Oliver North convictions, and the Sioux City, Iowa plane crash.

How oil industry's image has changed

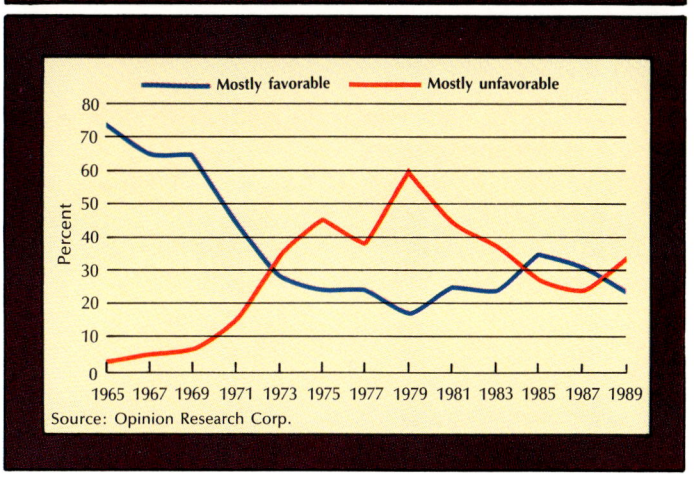

Source: Opinion Research Corp.

Time and again, Americans' attitudes toward the oil and gas industry have been linked to the price of energy. Much of the outrage following the Exxon Valdez spill was inflamed further by the spike in gasoline prices. Even more outrage was stirred by the revelation that Exxon would be able to deduct the costs of cleanup from its taxes, leading to legislation being introduced to prevent that sort of thing, even though this has been an acceptable practice for many years.

What of the flip side, the more favorable view the industry enjoyed after the oil price collapse of 1986? Certainly, some of that can be attributed to the sympathy factor for the small indepents and unemployed oil field workers and the economic devastation to oil state communities. But where was the gush of sympathy that might have produced an "Oil Aid" concert? Perhaps it was buried by the memory of $2/gal gasoline.

No doubt those were considerations. But it was likely simpler than that. The average consumer was simply delighted that the cost of a vital commodity was going down after a decade of often sharp increases; even more so at the sight of the price of any commodity going down in these inflationary times. What if the OPEC nations had declared an all-out production war in 1989? What if the price of crude oil had plunged to $10/bbl again? Is it too much to suggest that the anger over the Exxon Valdez would have been mitigated somewhat by the sight of 50¢/gal gasoline again?

Changing industry's image

There has been little study of whether the drastic cuts stemming from the price collapse contributed to the 1988-90 string of accidents that so marred its image. A top priority for industry should be putting together a task force to study that question realistically and unflinchingly and make appropriate recommendations.

It should not fear the critic who accuses the industry of putting profits before safety and environmental concern. Oil executives should not underestimate the public's understanding of how a corporate economy works; each consumer has his or her own economic agenda as well.

They recognize the exposure to added risk through belt tightening, for example, in buying a thrifty subcompact instead of a gas guzzler—which in turn increases one's chance of dying in a car crash. In any event, the industry must answer that question, because it already is being forced to spend more money on environmental concerns. Finding the answer to that question will help direct it, finding solutions to recurring environmental and safety problems instead of just throwing money at them.

ENVIRONMENT

The petroleum industry must reeducate itself to not conduct its operations as if in a vacuum—a holdover from the Seven Sisters era. This attitude tends to be smug and patronizing and more than a little arrogant by assuming that "We can do our job just fine, if we're just left alone, thank you, and you need not bother about the details."

Certainly, the industry is doing a much better job of staying away from that sort of attitude, but there remains too much of its residue. Companies ought to take a close look at themselves and decide whether they still have traces of that attitude lingering about.

Rooting out those outdated attitudes are critical to industry's efforts at managing public relations correctly

Dealing with the media

Management at some petroleum companies seems to regard media relations as something of a necessary evil. They would just as soon not have to do it, but sometimes, grudgingly, they go through the motions: a press release after the fact, a designated public relations person instructed to hew to a lawyer-approved script to field inquiries, a "No comment," or vague fluff that really answers nothing.

Other companies employ big P.R. staffs whose designated function seems to be: please submit your questions in writing and we'll get back to you on that, which is usually in the form of useless generic answers—and all too often after deadline.

Then there are those few antediluvian outfits who have no media relations whatsoever and will fire any employee who communicates with the press.

If some petroleum companies put a fraction of the effort into their media relations that they did their dog-and-pony shows for financial relations, they would be better off with both communities.

The companies that do well at media and public relations are those that require their executives in management and operations to be accessible to the press, train them accordingly, and are straightforward and responsive. Too many times, companies employ P.R. professionals as buffers against the outside world, and they simply earn a reputation for stonewalling that serves them ill in terms of a crisis.

This approach comes with a high price tag. In the early days following the Exxon Valdez spill, when Exxon's handling of media inquiries contributed to an overall negative image of how it was handling the crisis, that company's stock value fell by more than $1 billion. Bad P.R. may not have been the sole cause of that loss, but it certainly was a contributing factor.

The company learned something of a lesson later. When it prepared for shutting down intensive cleanup operations during the winter, it placed a seasoned P.R. professional, familiar with the Alaska scene at Valdez to coordinate a media tour of cleanup operations. The upshot was that instead of the newsmagazine doomsday covers that featured the spill in the spring, there was a newsmagazine cover that referred to "the disaster that wasn't."

Much of industry's difficulties in dealing with the media are not entirely of its own making. Many print and especially electronic media are not equipped to provide the kind of technical training reporters need to cover some of the crisis stories that arise within the petroleum industry. Often, media outlets that have someone assigned to an energy beat generally just plug in reporters from some other business beat to cover energy just as they would have covered financial institutions or aerospace. When large metropolitan media outlets outside the oil patch states (California a notable exception to the latter) cover something like offshore lease sales or tanker spills, the coverage is by an environment beat reporter (certainly valid, if a business side reporter is also assigned to the same story to cover industry's angle on the story).

Consequently, petroleum industry news coverage by the daily media often tends to be superficial and inaccurate. When oil companies are reluctant to be fully accessible and to take the pains to be patient in telling its side of the story, it is only natural for a reporter to turn to another source who can tell his readers just what the implications of an event are for the average consumer. The politicians with an ax to grind or career to promote and the environmental and consumer group lobbies with an agenda to promote or subscription rolls to fill are only too happy to oblige by filling that void.

It becomes a vicious circle. With this situation resulting in poor reporting, petroleum executives dealing with the next newsmaking situation come to situations involving the press with a chip on their shoulder. They've been burned by superficial, sensationalistic reporting by hostile or ignorant reporters, so they put up another stone wall—not realizing that their stone wall in the past contributed to that kind of reporting.

Every petroleum company has an obligation to help the daily media achieve a better understanding of how that company and the industry works, how it goes about its daily business and what that means for the average consumer. It should also duplicate that effort at the community level as well. These companies should not wait for a crisis to occur to implement such programs. It should not be necessary for companies to wait for community leaders to bang on the doors of their executive suites seeking answers.

There is absolutely no priority more important in the 1990s than industry's obligation to communicate its role on environmental issues specifically and generally to the public and the media.

Currently, the oil and gas industry has a fierce uphill struggle in this regard. If stronger initiatives are not taken soon, it will be a losing struggle.

A few recommendations

The first thing petroleum industry management must do in communicating a company's views on environmental issues of the 1990s is to ask itself: how best do we get our message across?

It is enough simply to leave that assignment to the public relations department? Here, many companies automatically have a roadblock to overcome. Many of the P.R. staffs at oil companies are already overburdened and understaffed, a result of the industry doldrums of the latter 1980s.

Another problem is management philosophy. Too many companies go to the trouble of hiring highly qualified, talented, experienced P.R. professionals and then just do not let them do their jobs. P.R. is this distasteful chore that senior management would rather not do—the corporate analog of janitorial work—so it hires specialists to take care of the "problem." THe real problem, however, is that a P.R. professional is doing his or her job best when it links a manager in operations or vice president with a reporter and helps facilitate that communication.

Just as companies are restructuring to accommodate environmental concerns in operations, they should restructure at the public relations level by creating an environmental affairs (EA) office. Staff in an EA office should not have their efforts diluted by other duties that corporate, media, or financial relations staffers could handle as easily. This office should be the contact for all media on environmental issues and breaking stories. It should be this office's responsibility to maintain a working awareness of what the chain of command is in a crisis, to know which technical people and which executives would be the appropriate people to field detailed media inquiries, and how it should implement the company's messages and underlying philosophy on environmental issues.

An especially important measure is to see to it that the EA officer has not just access but regular input to the highest

ENVIRONMENT

CLEANUP of the Exxon Valdez oil spill off Alaska continued in 1991. The spill, worst in North American waters, plunged the public's approval of the petroleum industry to its lowest level in years.

levels of management and the company's board of directors. The kinds of decisions oil and gas companies will be making in the 1990s on environmental issues will be make-or-break ones, and that should be reflected in corporate philosophies. A good approach would be to designate a special environmental communications committee consisting of board members, senior management, and EA staff.

The words "no comment" should be stricken from the vocabulary of anyone dealing with the media. If a company spokesman cannot answer a question because he does not know, then he should say so and offer to find the answer and then get back to the reporter. All too often, though, "I'll get back to you" carries the unspoken subtext of "after your deadline." It is important to be honest in that regard. If an executive can't be sure about answering the question truthfully and accurately within a set time, that should be made known at the outset.

With that in mind, a company should familiarize itself with the media it deals with, getting to know deadline constraints, need for photographs or other visual aids, what a reporter's strengths and shortcomings are on environmental and general industry issues.

Even an experienced energy reporter in the daily media cannot be expected to have background to deal with some of the issues that come up without someone with expertise in the industry to provide guidance. Even the most seasoned among the oil trade press will seek guidance and understanding. It is a tough and complicated business.

Companies should accommodate different media needs differently and not play favorites. Most print reporters recognize that television and radio reporters have more urgent deadlines and less time in which to tell a story. TV reporters in particular also have special demands because of their equipment and the need for certain controlled conditions. But calling a press conference that allows for a few quick sound bites for the cameras and then walking away from the more detailed questions after the cameras have packed up and left is the quickest way to invite a print reporter's hostility. That in itself is often seen as a sort of stonewalling. Press kits and backgrounders can be a godsend in such situations, for all types of media. But they should not be seen as panaceas. The personal touch is still required.

How a company communicates is of critical importance as well. It is a good idea to search for the key management people who have the authority to speak on behalf of the company in most matters and who are adept before a camera or a roomful of eager reporters. Training is essential, especially in programs that simulate press encounters, whether one on one or in crowded press conferences during a crisis. Workshops and training seminars should be built around such simulations.

Such programs are especially helpful in rooting out some of the most common problems reporters encounter in reporting on technical or otherwise complicated issues. Jargon is the death of clear communication. It is not patronizing to express technical issues in everyday language. The average newspaper reader or television viewer is not an engineer. They can tell, however, when someone hides information behind a fog of technical terms.

Most important, a company should take the initiative. If a reporter has been on a tour of a refinery and gone back to his office with photos, graphs, charts, background sheets, and conversations with the people likely to be his contact on a breaking story, it follows that he will do a better job of reporting a story involving that refinery when a story does break.

Companies can keep reporters abreast by issuing lists of contacts for possible stories and their areas of expertise. They can call reporters and ask if they might be interested in a tour or an interview. A company can offer its expertise on a general issues story, even on a story that may not involve it or its facility directly. Some of the best stories in the trade press are generated by public relations or even operating personnel within the companies.

Challenge of crisis management

Is there a petroleum company around that still believes a crisis management organization is not essential for operating effectively in the 1990s?

Unfortunately, there are—too many. According to industrial communications specialist, Alex Stanton, President of

253

ENVIRONMENT

Dorf & Stanton Communications Inc., New York, only about half of the Fortune 1000 companies in the U.S. had a formal crisis communcations plan in early 1989. The situation is even more dismal among small companies, he said.

Stanton maintains that most companies are totally unprepared to cope with a full-fledged corporate calamity, thereby raising the potential for a crisis to turn into a disaster when all the company can do is improvise.

Stanton says the first step a company should take is to designate a crisis management team, ideally a small group of quick decisionmakers who represent all key functions. This team should develop a comprehensive crisis plan and playbook in advance and "live it, minute by minute, hour by hour," he says.

"These should be people who operate well under pressure and are fully familiar with the crisis plan and playbook. Crisis team members should be accessible to one another on a 24 hour basis via a wallet sized telephone list or beepers."

Stanton contends the first priority of the crisis team in a crisis is to define the scope and severity of the crisis. There should be no barriers to information for the team.

This is best accomplished, says Stanton, when the crisis team meets to review the situation in half hour or hour sessions during the crisis.

It also is critical to identify key target audiences affected by the crisis and then decide who needs to know what. What Dorf & Stanton recommends is a cascading notification system in which the most important audiences hear first and the least important last. This system should be spelled out in detail and rely on key people within an organization for passing information along. The crisis team itself should not be responsible for all the communications while it also is managing the crisis, Stanton says. At the same time, a system should be in place for monitoring reactions to the crisis and new developments.

In communicating on a crisis, companies must be quick, open, and honest. Stanton also counsels against speculation or unverified information or placing blame. The emphasis instead should be on solutions to the problem at hand.

All too often, a company's employees are the last to know what is happening during a crisis.

"It also is important to recognize that every employee who is properly informed and armed with the correct messages can be a significant soldier in your communications army," Stanton says.

"So while you may expose middle management and rank and file employees to different levels of information, all must be told promptly and told the same story. Lack of a party line can be fatal to a crisis management effort."

Stanton also cautions against relying too heavily on legal counsel during the crisis. While a legal adviser may be an important member of a crisis team and consulted throughout a crisis, management should not be afraid to move against legal counsel when warranted.

"Ultimately, management has the responsibility to preserve the reputation of the company and its special relationship with customers.

"Attorneys often will seek to limit liability by releasing as little information as possible and moving more slowly than the situation demands.

"That's not always the right decision for the company or product involved."

Stanton also advises the need to keep a log of events during the crisis, so a company can tell its side of a story later.

It is just as important, he contends, to issue a final statement after the crisis that specifies what steps are being taken to avoid a repeat of the crisis.

As Stanton emphasizes, the three key words in effectively managing any crisis are communicate, communicate, communicate!

Mobil's challenge

A good example of how to respond to a public relations crisis and the effectiveness and wisdom of using outside public relations consultants is the campaign Mobil launched in early 1990 to combat Measure A in Torrance, Calif.

Facing a negative image and an uphill scrap in defeating the measure after a series of accidents at the refinery, Mobil hired the one of the top public relations firms, Braun & Co.

Measure A— limiting hydrofluoric acid storage to an impossibly low level by any business in the city—received national attention with parallels to Bhopal raised. Early polls indicated the measure would pass 3-1.

Braun talked to the Torrance refinery manager and conducted in depth ascertainment interviews with about 50 community leaders. These one on one interviews identify the client but don't attribute the comments solicited to identify the client and possible solutions. Braun uses that as the basis for a P.R. campaign.

Mobil created a risk management and prevention program for hydrofluoric acid use at Torrance, one of the few such programs in the state. It took out newspaper ads profiling employees and detailing safety programs at the refinery. At the same time, it set up a community advisory panel.

Mobil then started up a bimonthly publication and produced an audiovisual program and brochures, all providing detailed information about the refinery. The company also invited more than 2,000 people to tour the refinery.

Just before the election, Mobil's campaign kicked into high gear. It told voters what the measure really entails: an increased volume of hazardous chemicals used at and trucked into the plant, plus an $8 million lawsuit the company planned to file to overturn the measure. Mobil employees held coffee hours in their homes, and the company bought more advertising, direct mailings, and a 30 min local cable television program. Measure A lost by 3 to 1 margin. Braun Executive Vice Pres. Doug Jeffe contents that although city leaders were concerned about the costs of a lawsuit, "If we had not successfully addressed the safety issue, we wouldn't have won."

The challenge Mobil and other southern California refiners face now on the hydrofluoric acid issue is expanding that approach throughout southern California to combat a ban proposed by the South Coast Air Quality Management District.

BP's paradigm

The best way to illustrate the importance of effectively managing an environmental crisis and at the same time keeping it from turning into a P.R. nightmare is to offer as a paradigm the example established by BP America in tackling the Feb. 7, 1990, American Trader spill off Huntington Beach, Calif.

BP's crisis response was outlined by Chuck Webster, the company's manager of crisis management, in his words, "How we managed the crisis instead of reacting to it."

"We have two other full time professionals on staff here. Our responsibility is to develop plans in the event of a crisis to go beyond what a local facility or business might be capable of effectively managing.

"The concept is that businesses within a company like BP America are designed to be lean and mean. We recognize that when you deal with the products that we do, whether crude oil, petroleum products, or even animal food, you face the potential for getting involved in a crisis that requires many more resources than you have in your business or facility.

"That was the starting point for developing what we refer to as a sort of 'National Guard' of people with specific functions and skills to be called upon should you face a crisis that goes beyond what would normally be expected of that business or facility."

ENVIRONMENT

Industries' environmental images*

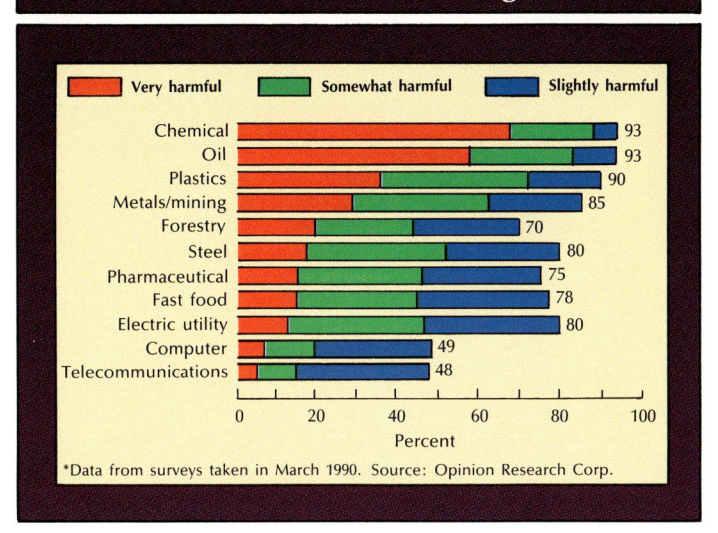

Industry	Percent
Chemical	93
Oil	93
Plastics	90
Metals/mining	85
Forestry	70
Steel	80
Pharmaceutical	75
Fast food	78
Electric utility	80
Computer	49
Telecommunications	48

Legend: Very harmful, Somewhat harmful, Slightly harmful

*Data from surveys taken in March 1990. Source: Opinion Research Corp.

BP was training for a crisis such as an oil spill in the weeks before the spill. It had put its new crisis management team through several drills, including a major crisis drill in December 1989 that simulated a alky-feed butane leak from a pipeline that involved a flashback and secondary explosion with civilian casualties. In January, Webster's team went through incident command system (ICS) training in Alaska involving an Alyeska pipeline drill in Valdez. ICS was adapted from an emergency response management approach originated by California firefighters.

"ICS allows you to go, in fire parlance, from a one company response to a multicommunity response while maintaining an organizational structure that allows you to manage a situation. It has an incident commander at the top; an operations section; a planning section; a logistics section; a financial section; and subsets below that to allow you to grow as much as you need to grow and cover everything from air operations to who's going to make sure there is food in the command center to lodging to people who can work skimmers—every phase of the operation."

It helped that the cities most immediately affected by the spill also happened to have crisis management systems based on ICS, helping effective communications.

"What is critically important," said Webster, "is to have local people from the mayor to the police chief to the fire chief to city leaders understand what you are doing to protect the environment. There have been times in the past that that aspect of communications liaison has not worked effectively. In this case, it worked very effectively."

BP sought input from the communities on priorities for beach cleaning, thereby advancing the opening of some beaches.

"This was appreciated by them, and it allowed us to be more sensitive to local community interests."

BP continued the emphasis on local input.

"For instance, in addressing the major beach contamination, there were many options to get oil off the beaches, from manual labor to front loaders. Before making a decision, we talked not only to state and federal officials, we went all the way down to lifeguards who lived on the beach and asked them to fill our heads with as much information about the makeup of the beaches and their history before we made any decisions. One of the key things that we discovered is that there is an erosion history to those beaches. So one of the things we decided to do is go the more manpower, to stay away from heavy machinery to reduce the removal of sand. It caused us to choose technologies such as pompoms and absorbent sausages that got the oil up without removing a lot of the sand in the process. It was absolutely the right kind of decision for that environment, but one that would have been much more difficult to arrive at had we not taken the organizational approach that we did."

Timing was critical in the response. The BP crisis team was on a plane headed west from Cleveland within 2 hr of notification, which came 30 min after the spill occurred.

"It is BP's policy that whenever there is oil in the water, we send a team with the view that, until we know for sure what caused the oil to be in the water, we will assume there is a hull fracture or something of an equally serious nature," Webster said.

"As a result, we have sent teams out on spills of less than 1 bbl, with the view that we would much rather come back for dinner than to find out that we should have been there sooner."

Before dawn the following morning, top BP officials were being briefed by the Coast Guard and talking to reporters—including an appearance on the Today television program. BP began calling in more than 100 professionals that comprise its crisis response team.

"We also made decision early on that were going to be absolutely available to the media, that we were not going to dodge questions, and that we were not going to try to put any 'spin' on the story," Webster said. "And that meant that when things didn't go well, we put them out with the same gusto that we tried to report successes.

"There was even a point where there was a system set up with state and federal officials to do surveys of the beaches to determine which areas were clean and which areas required additional work. We reached a point where we were finding buried oil in areas where the survey teams declared the beaches clean. So we put out a news relrease that we would like to see a more comprehensive survey. We weren't satsfied that all the oil was being found. It helped to develop credibility with those who were covering the story. And it just served the purpose of presenting ourselves as people who had a good measure of pride and wanted to see the job well done. I don't think that we made any promises that we couldn't keep."

To sum up the philosophy behind BP's crisis response, Webster said, "Information is your biggest ally and your biggest enemy. The more people understand about what you are doing and why you are doing it, the better shape you're in.

"We found in our team just a bunch of very, very skilled people who were within an organizational structure, given a task, and allowed to do it—and the results were very gratifying. There was a lot of pride of ownership in this job. We won some respect from local communities."

BP found a gauge of that respect when it returned a month later to Huntington Beach to throw a dinner for about 100 locals to thank them and tell them it will carry on the process of "lessons learned."

"I couldn't believe it," Webster recounted. "Here we were, in southern California, no less, and here were all these people coming up to us and thanking us for what a wonderful job we did."

The upshot of this effort was that polls of Californians shortly after the spill suggested a majority did not feel especially concerned about an immediate, pressing environmental crisis affecting California. Most today could not even name the American Trader.

The tanker that ran aground at Bligh Reef in Prince William Sound, however, has had to undergo a name change—Exxon Mediterranean, to show its new venue. And the name of the Exxon Valdez, like the Santa Barbara oil spill, will resonate forever to industry's chagrin. [IPE]

This article was adapted from a chapter of *U.S. Petroleum Strategies in the Decade of the Environment* (PennWell Books, 1991) by Bob Williams.

Natural gas: fuel of the future

WHATEVER THE ENORMOUS CHALLENGES AND pitfalls awaiting the petroleum industry in the Decade of the Environment, there is little doubt about one thing: natural gas has a bright future in the 1990s.

Growing demand for natural gas—in large part due to a rapidly accelerating shift from other energy sources on environmental grounds—together with the removal of regulatory constraints that have long hamstrung this abundant U.S. resource may be the main source of the U.S. petroleum industry's optimism in the 1990s.

With market forces and consumer preferences determining the role of natural gas in the U.S. energy mix for the first time in 50 years, U.S. petroleum companies are shifting their focus increasingly to gas in favor of oil.

That emphasis on gas will continue, even as natural gas prices rise—because those price hikes are not likely to be sudden or steep over a sustained period of time, and because environmental and energy security concerns together dictate that natural gas is an energy source whose time has come.

There will be environmental concerns for the natural gas industry, but they pale in significance in comparison with those for the rest of the petroleum industry. For example, the gas industry accounts for only 5-8% of world methane emissions, in turn making it a contributor to no more than 1-1.5% of the postulated greenhouse effect. These figures disregard the reductive influence the gas industry has on the greenhouse effect through utilization of coalbed or landfill methane.

There are some modest inroads natural gas will make in traditional oil markets, such as transportation fuels. Natural gas vehicles can cut emissions of carbon monoxide by as much as 99%, nitrogen oxides 65%, and reactive hydrocarbons 85%. Fleets that have shifted to natural gas have realized a significant savings, not only in fuel costs, but also in vehicle maintenance and repair. However, because it is likely to be limited to fleet vehicles, compressed natural gas as a motor fuel is not likely to amount to more than 5% of the overall gas market in the 1990s.

And natural gas demand will continue to grow in industrial and space heating and cooling sectors—albeit modestly as efficiencies improve.

But the spectacular growth in natural gas demand in the U.S. will occur in the electric power industry.

And the gas industry has the environmental movement to thank for that.

Brownouts and BTUs

Demand for electric power in the U.S. is expected to outstrip the overall rate of energy demand in the country in the 1990s.

The U.S. Department of Energy's Energy Information Administration in 1986 projected that total U.S. energy consumption will grow at an annual rate of 1.3% between 1987 and 2000. By contrast, EIA forecast, U.S. demand for electricity will jump 2.4%/year during the same period.

Growth in electricity demand will account for more than half the increase in total end use demand for energy in the U.S. in the 1990s. Electricity generation is expected to grow from 36% of primary energy consumption in 1987 to 42% in

Outlook for gas power market share

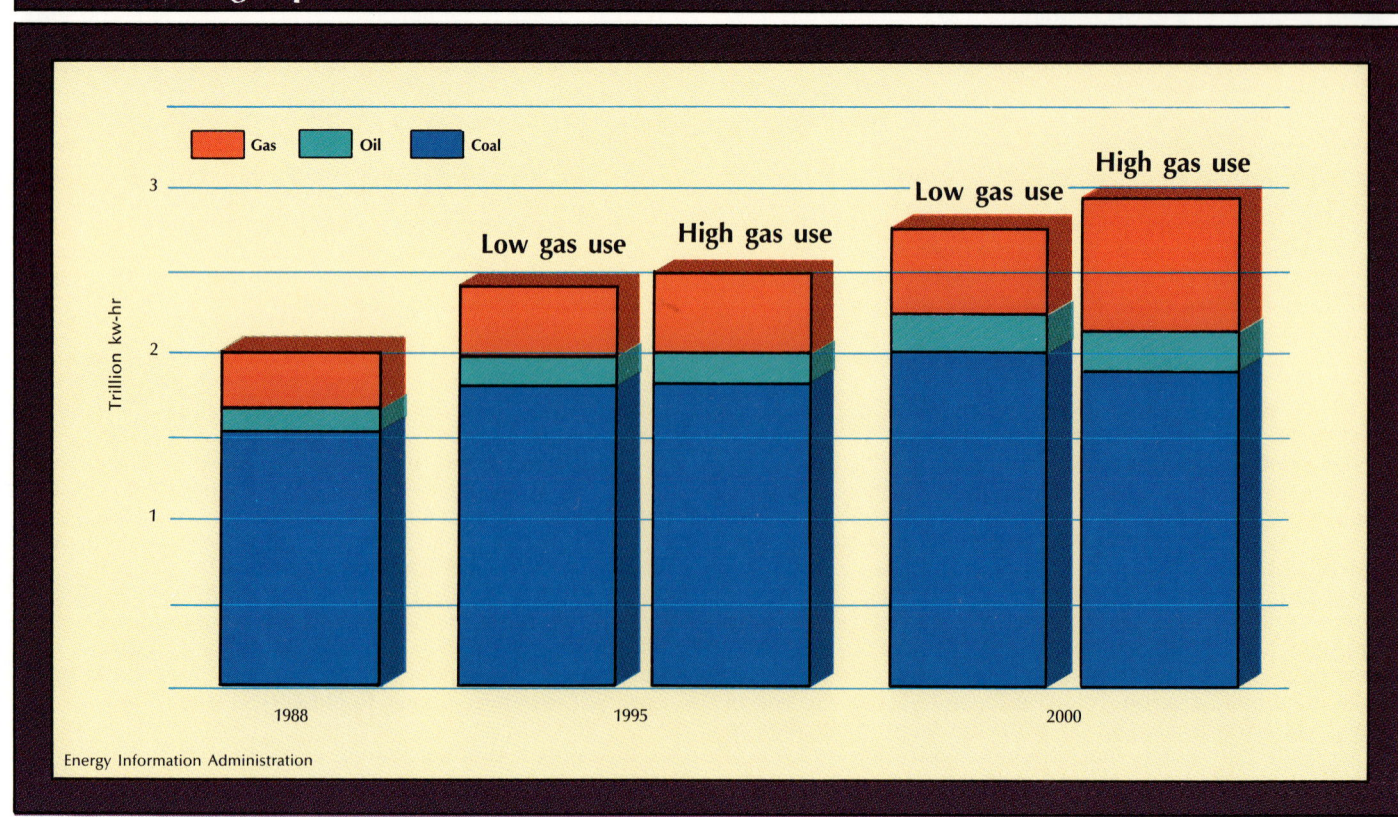

Energy Information Administration

FEDERAL ENERGY REGULATORY COMMISSION officials take testimony from gas producers at a 1990 conference on producer demand charges. FERC is in the forefront of regulatory change in the U.S. designed to boost use of and access to more gas supplies. Photo by Patrick Crow.

2000, according to EIA.

Electric utilities are expected to meet the growth in power demand by completing more than 44 gigawatts (GW) of capacity under construction or announced. Most of that construction will occur in 1991-95, projected EIA. Beyond that, utilities must construct an additional 32-73 GW (base case of 53 GW) of capacity that is currently unplanned. Most unplanned construction, accordingly, will occur in the latter half of the decade.

EIA figured in 1987 that most of the unplanned generating capacity will come from either pulverized coal steam plants (15 GW) or gas-fired/combined-cycle plants (22 GW) built as peak load turbines. Cogeneration power and independent power producers are projected to add almost 30 GW of capacity in the 1987-2000 time frame.

EIA in 1987 figured electric power's fuel mix will break out by 2000 as coal 53%, nuclear 17%, hydro 9%, and oil/natural gas the remainder. The agency estimates oil and gas fired plants' market share will rise from 15% in 1987 to 21% by 2000. By 2000, the combined percentage of generation from oil and gas is expected to exceed that from nuclear power. EIA pegged the total consumption of oil and gas by electric utilities at 3.4 million b/d of fuel oil equivalent from about 1.8 million b/d in 1987.

Bear in mind, these estimates were made shortly after the oil price collapse and before the revival of environmental concerns in 1988-90.

Even in 1988, concerns were arising among the utilities that a capacity shortfall might be in the offing.

In October 1988, the North American Electric Reliability Council warned that "several forces creating risks to the future electricity supply in the United States and Canada must be carefully managed."

The alternative? Soaring costs, draconian conservation measures, controlled brownouts.

"These forces that affect the reliability of electrical supply include a higher than expected growth in the demand for electricity, an imbalance between economy and reliability considerations, public policy disincentives, transmission access and deregulation issues, and environmental considerations..." NERC said.

"Several geographic areas where electric system reliability is at risk are identified...Emergency measures, including controlled interruptions of customers' supply, are possible in these areas."

Underpinning those projections of electric power demand growth are expectations of a healthy economy continuing.

In a 1989 study on North American electric power trends, Arthur Andersen & Co. and Cambridge Energy Associates cited the linkage between economic growth and electric power demand growth.

The study noted that from the end of World War II until the early 1970s, every added U.S. GNP dollar, in constant dollars, inevitably came in tandem with more than 1 kw-hr of electric power sales. That changed dramatically in the 1970s, when the ratio of added kilowatt-hours to increased GNP dollars fell by about half.

But the oil price collapse changed all that, restoring the benefits of low cost energy to the U.S. economy. In 1987 and 1988, electricity sales outstripped GNP growth by about 1.5%/year, according to Andersen/Cambridge.

"If this is evidence of another inflection point in the relationship between electricity sales and economic activity, it has extremely important implications for the utility industry," the study concluded.

"If the pattern that persisted from World War II until the mid-1970s has reappeared, then significant new power plant construction will be needed in the 1990s to avert regional power shortages."

More importantly for power planners, the linkage between economy and power demand remains unbroken, even after energy crises.

Although lagging electric power demand from the mid-1970's to the mid-1980s has been blamed on soaring electricity costs, there is reason to believe that a more significant factor in the falling rate of growth in electricity demand is the stagnant economy that persisted during the same period. That begs a chicken-or-egg proposition, of course, since the same spiraling oil and gas costs that contributed so much to the spike in electricity costs also contributed to depressing other areas of the economy.

If the economy remains healthy and summers prove to be hot and dry in the 1990s, the need to build new power generating capacity in the U.S. will take on a much greater urgency.

Although there was much surplus generating capacity in the 1970s and early 1980s, that won't be the case in the 1990s. Already, annual capacity margins are at their lowest point since the early 1970s. In some parts of the U.S., annual capacity margins are at their lowest levels since

U.S. incremental gas supply potential

	Within 1 year	Within 3 years
	(Bcf)	
Nonproducing reserves	250-750	500-1,000
Canadian gas*	400	500
Accelerated infill drilling	50-100	100-400
LNG†	40-100	200
Total	**740-1,350**	**1,300-2,100**

*Volumes are in addition to 1.283 tcf of imports estimated in 1988.
†Volumes are well below the physical limits of existing terminals.

Source: American Gas Association

NATURAL GAS

World War II.

A study by the American Gas Association in 1989, using NEC data, found that at least 50 million kw will be needed by 1997 in addition to the capacity currently under construction if peak electricity demand grows at 1.9%/year. If peak demand grows at a rate of 2.9%/year, which is a conservative estimate considering current trends, the U.S. electric power industry would have to construct almost 105 million kw of capacity in addition to plants already under construction.

In 1988, the electric utility industry had 73 million kw of capacity additions on the drawing board for the following 10 year planning period. That compares with 250 million kw of planned capacity additions by utilities for the 10 year period beginning in 1979.

For the 1988-98 planning period, planned coal capacity additions are down 86% compared with the previous planning period. Planned nuclear capacity additions are down 85% in the same comparison.

In large part, that situation can be blamed on the environmental movement.

Certainly conservation has played a role in downscaling planned capacity additions. And the overbuilt capacity of the late 1970s and early 1980s has understandably made utilities much more cautious in planning capacity additions.

But barring some draconian new conservation push nationwide or economic collapse, there simply will not be enough electric generation capacity to meet peak electricity demand in the mid-1990s unless more capacity additions make it onto the drawing boards.

Some see widespread power shortages and blackouts more commonplace as electricity demand continues to grow in the 1990s. The areas most likely to be affected are Florida and the New England, Mid-Atlantic, and North Central regions. A study by Salomon Bros. in 1989 pegged the shortfall at more than 100 million kw, perhaps as much as 400 million kw. Some mostly completed nuclear power plants are being held up by intervenor actions, exacerbating the situation.

The shortfall could come even sooner than any had expected a few years ago, according to the Salomon study. The analysts looked at electric utilities' efforts to maintain a reserve margin—committed total capability capacity minus peak demand—of about 20%, which is derived from a probability analysis done by each utility to minimize the risk of a blackout to a 1% chance in ten years.

In calculating utilities' aggregate capability, Salomon factored in planned nonutility power sources, including cogenerators, resource recovery plants, and independent power producers. The analysts used four scenarios of demand growth, assuming no demand growth in 1989, followed by, respectively, 2% growth each year thereafter, 3%, 4%, or 5%. Under the 2% growth case, reserve margins will fall below 20% in 1993; under 3%, 1992; under 4%, 1991; under 5%, early 1991. Even under the most conservative demand growth scenario, Salomon Brothers estimates utilities will have to add 107 million kw to their capability in the 1990s. Correspondingly, that calls for 196.6 million kw needed with 3% growth, 294.7 million kw with 4% growth, and 402.6 million kw with 5% growth, the study concluded. It also cautioned that the shortfall will be greater and sooner than under these projections if the nuclear plants under construction never start up.

Another development likely to further aggravate the situation is the fundamental change that is occurring in the structure of the electric utility industry. The Investor Responsibility Research Center in 1988 issued a report that concluded that the basic conditions were in place for the largest wave of mergers and financial restructuring activity the electric utility industry has seen since the 1930s.

The culprit?

"Financial problems associated with nuclear power plants are proving to be the single most powerful catalyst for industry reorganization," the IRRC report said.

There are other contributing factors as well, but the real dilemma for electric utilities is how to add capacity at a controlled cost. Yet if anything the lead times for construction and permitting of big baseload coal and nuclear power plants are getting even longer.

Building a large coal or nuclear power plant today brings into play concerns over environmental and safety regulations, long construction lead times, and public utility commissions disallowing certain outlays deemed imprudently incurred.

Three Mile Island set the Not-in-my-backyard syndrome in concrete for the nuclear power industry. And the tough new version of the Clean Air Act to be implemented in the coming years—could ensure a similar scenario for coal fired power plants. Oil might have been an alternative, had it not been for the Exxon Valdez, energy security worries in the wake of the Persian Gulf war stemming from Iraq's invasion of Kuwait, and to some extent, air quality concerns.

At least two-thirds of state utility regulatory bodies now require a least-cost approach to planning future load growth, in order to rein long term electricity costs.

As the industry undergoes a wave of restructuring and consolidation, utilities will become even more conservative in controlling costs of capacity additions. Survivors of the shakeout will be held more accountable for financial performance, and will be less likely to be bailed out by public utility commissions on cost passthroughs of "imprudent" costs.

The changing regulatory climate

Changes sweeping across the regulatory horizon of the electric power and gas industries generally will have a salutary effect on gas sales' incursions into the power market.

Many of those changes will provide a definite boost to natural gas, and would have even without the rebirth of environmentalism.

The jitters electric utility planners have developed in recent years are justified. Utilities and their investors can no longer presume they will be allowed to recover their full investment in power plants. An increasing trend among regulators has been to disallow returns on investments in what was later deemed to be excess capacity.

When planning big baseload plants, where the economics of nuclear and coal would hold sway, it used to be acceptable to sin on the side of largesse. No longer. Utilities now must use existing capacity, nonutility purchases, conservation, load management initiatives, and power imports to meet their needs.

That may work—until the mid-1990s. But the long lead times for nuclear—and increasingly, for coal—tend to preclude those fuels for near term planning purposes.

So it follows that as utilities seek to match planning with incremental growth in demand, they will turn increasingly to smaller, modular power plants and purchases from cogenerators and independent power producers (IPPs). Most of those will be fired by gas. Gas fired plants tend to be smaller, have much shorter lead times, and still burn a fuel whose price is relatively low.

Gas can be used in a variety of power plant schemes—steam plants, combustion turbines, combined cycle plants, and cogeneration plants. Moreover, gas fired turbines are seldom larger than 100,000 kw, and combined cycle gas fired plants are usually smaller than 200,000 kw. That kind of scale enables utilities to add new capacity when they need it, rather than commit massive outlays to a big baseload plant years in the future that could become a white elephant if demand falls short of projections. The modular approach eliminates the need for costly custom designed engineering and labor intensive construction programs.

NATURAL GAS

How electric power demand for U.S. gas will grow

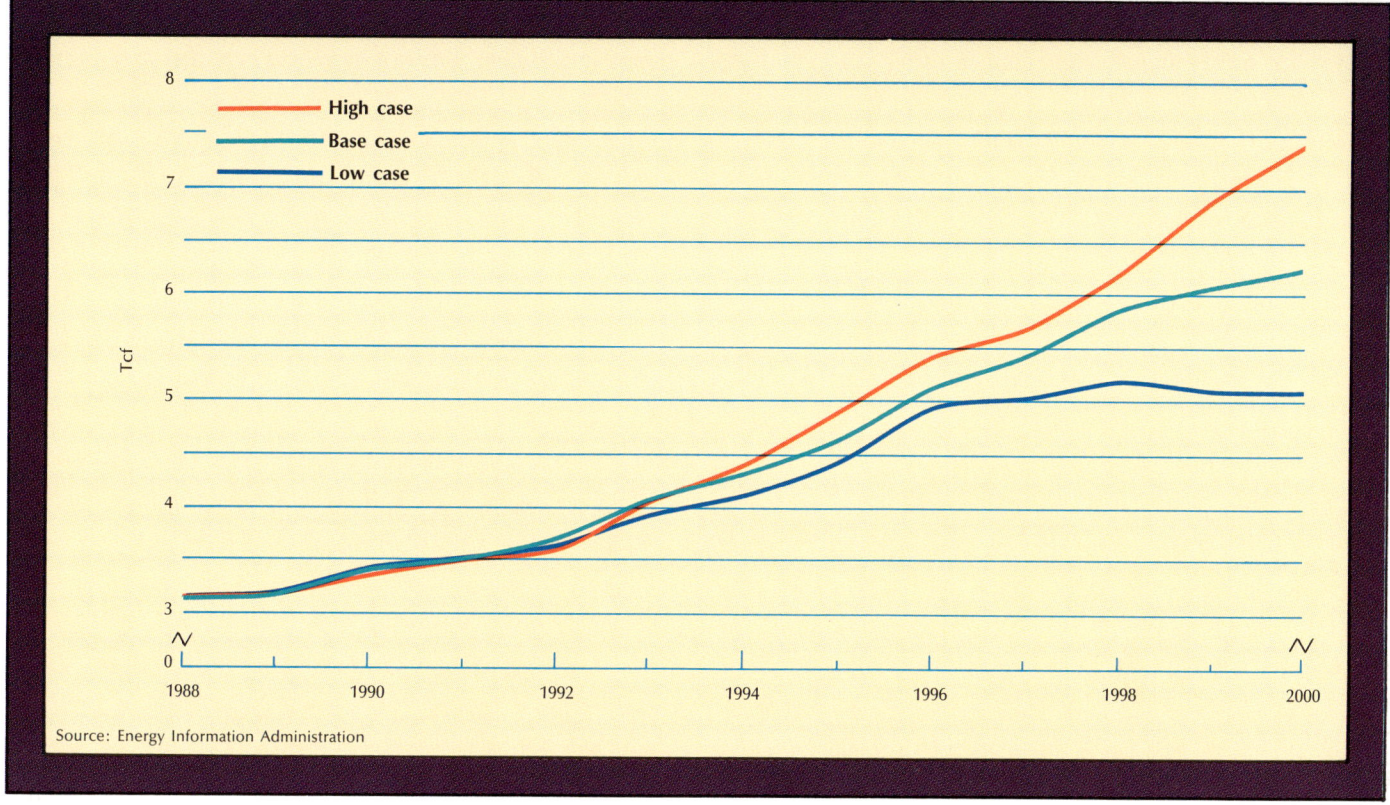

Chart showing Tcf (trillion cubic feet) projections from 1988 to 2000 for High case, Base case, and Low case scenarios.

Source: Energy Information Administration

Regulatory changes within the gas industry are paving the way for a greater market share for gas in power generation. In 1987, Congress amended the Powerplant and Industrial Fuel Use Act to relax restrictions on using natural gas for nonpeak electricity generation.

One of the biggest incremental new markets for gas in the U.S. has been the rapid rise of cogeneration. Although Congress in 1978 passed the Public Utility Regulatory Policies Act, it was not until the mid-1980s that its desired effect was seen. Court challenges and hindrances by public utility commissions stalled Purpa's intent, which was to give nonutility power generators a secure market by requiring electric utilities to buy power from them at the avoided cost rate. How one defined avoided cost made all the difference for the feasibility of cogeneration. The early, loose interpretation by some regulators essentially involved the avoided cost of having to build the same capacity from the ground up.

With the projected costs of power plants skyrocketing, that pushed avoided cost schemes up to premium levels and triggered a cogeneration boom that was first spawned in California, where the public utilities commission and the major utilities encouraged cogeneration growth. That boom spread, notably to the Texas Gulf Coast, where the vast refining and petrochemical complexes found the double benefits of producing huge volumes of process steam locally as well as providing a secondary profit center in power sales.

However, the cogeneration projects proliferated to the point where the utilities were faced with being forced to buy more power than they could use, perhaps finding themselves eventually in the bind of having to shut down existing baseload capacity in order to buy the cogeneration power. Regulators rescued the utilities by tightening the parameters for avoided cost sales, focusing on the avoided cost of burning competing fuels. Consequently, the economics of cogeneration projects shifted from sales of overpriced power sales to their thermal value. That has resulted in a shakeout of cogeneration projects and developers.

IPPs are not without some of their own economic hurdles as well. Such projects generally depend upon project financing—debt secured solely by revenues from a single plant—which tends to be costly and complicated. Some projections estimate that 40% of proposed IPP capacity will be delayed and 20% will never be built, according to engineering consultants Burns & McConnel, Kansas City, Mo.

And IPPs are not always the most economic choice from an investment standpoint, contends Burns & McConnell. The consultants cite their project for Old Dominion Electric for two new 400,000 kw coal fired units. They received fifteen proposals, from conventional contractors involving utility ownership and from IPPs. The conventional utility ownership alternative was 15-20% more economic on a life cycle basis than its nearest IPP competitor, according to Burns & McConnell.

The consultants also raise the question of reliability.

"It may not be politically popular, but public commissions should allow utilities to consider the reliability of IPPs in their long term plans," Burns & McConnell said in a 1989 report.

"Can we really depend on these plants being financed, permitted, and built on schedule, and to operate reliably for 40 years?"

Still, the deciding factor is likely to come down to cost, and IPPs can add coal fired electric capacity for a fraction that of conventional baseload utility plants.

And, the economics of cogeneration are very healthy today, partly because of its inherent energy efficiency and the low cost of its most attractive fuel: natural gas.

Because nonutility power generators were exempt from FUA restrictions on burning natural gas prior to 1987, more than one third of nonutility power generation is already fueled by gas. Low prices ensure that significantly more nonutility power generation will be fueled by gas in the future. Not only is the cost of gas as fuel low, the capital cost of gas fired plants is low compared with the alternatives. The price of natural gas delivered to electric utilities has fallen in

NATURAL GAS

constant dollars since 1982. A slow decline in that price occurred in 1983-85, followed by a steep drop in 1986 before continuing a slow slide again.

The fall in gas prices has as much to do with regulatory reform, particularly by the Federal Energy Regulatory Commission, as it has with the collapse of oil prices in 1986. Decontrol of gas supplies, gas wellhead price decontrol, open access to transportation on pipelines, unbundling of pipeline services, market-indexed gas pricing contracts, improved ratemaking flexibility, and trimmed regulatory constraints, all helped to foster gas on gas competition, which has reined prices.

Deregulation initiatives in the gas and power industries have helped fueled spectacular growth—with more to come—in independent power generation, notably fired by gas in cogeneration and conventional turbine plants.

Jean-Louis Poirer, in a 1989 study for RCG/Hagler, Bailly Inc., Washington, D.C., estimates that a total of 30.5 million kw of IPP capacity will come on line by 1995. That will mushroom to as much as 90 million kw by the end of the century, he projects.

Of the 42.76 million kw of IPP capacity under construction or in development in early 1989, cogeneration accounted for 71% of the active capacity and almost half the active projects, Poirer found. Further, natural gas accounted for almost 29 million kw of the active IPP capacity, of which 40% was on line at the time.

Some IPP developers have projected U.S. IPP capacity reaching 200 million kw by 2000, split 50-50 between load growth and replacement of existing baseload capacity.

Environmental constraints on nuclear, coal, and oil fired electric power will ensure a built-in premium for natural gas continuing even as surging demand growth and declining deliverability boost natural gas prices again in the 1990s. Gas prices might double in the 1990s, as some predict, but the added costs of scrubbing equipment and uncertain economics for nuclear will likely offset higher gas prices. To the extent that nuclear power can make its case for safety and public acceptability and coal can make its case for economically viable clean technologies, that will determine the strength of natural gas in the power generation market in the U.S.

Clean coal

For utilities, the preferred alternative to scrubbing equipment is an accelerated program of clean coal technology development. Clean coal is evolving into a healthy cottage industry funded by the Electric Power Research Institute and a $5 billion, 5 year research and demonstration program funded by the Department of Energy. Utilities maintain that, over the long term—say, the year 2010—clean coal technologies will produce an equal or greater amount of emissions reductions at a cost of about 20% less than that for scrubbers under proposed acid rain legislation.

Repowering with new, cleaner coal based technologies offer great promise. The hitch is in timing. In the near term, the emphasis on clean coal approaches likely will be on scrubbing, pressurized fluidized bed combustion, atmospheric fluidized bed combustion, hot gas cleanup, direct coal liquefaction, and underground coal gasification. Longer term, the emphasis will shift to surface coal gasification, advanced combustors, fuel cells, and indirect coal liquefaction.

By early 1991, DOE had chosen 40 projects under three rounds of solicitation for clean coal technology demonstration projects. Under the program, DOE plans by 1992 to help finance 50-75 demonstration projects. Early in the program DOE was blistered by environmentalists for weighting the program heavily with repowering projects at the expense of retrofit/scrubber technologies. However, the last round, at yearend 1989, had more than half the selections in the retrofit camp. Environmentalists want the mandatory installation of scrubbers to achieve emissions reductions more quickly.

The question of timing is critical to the controversy. The most appealing technologies, fluidized bed combustion and coal gasification, may not be widely available on a commercial basis until the latter part of the 1990s. Environmentalists insist that the future costs associated with delaying repair of environmental damage and health care costs will be far greater than the short term costs of scrubbers. Utilities contend that the situation for acid rain has stabilized and that there is time until the new repowering technologies and reduced cost scrubbing approaches are commercialized.

Cofiring

However, there appears to be an interim solution that could form a happy compromise and provide a wonderful marketing opportunity for the gas industry: cofiring.

Cofiring, or the burning of limited volumes of natural gas in combination with coal or fuel oil in utility and industrial boilers, reduces emissions, improves operations, and boosts economic performance. Plants can operate at full capacity and stay within air quality standards. At the same time, cofiring enables a plant operator to switch to lower quality coals, rein downtime, improve combustion efficiency, and cut maintenance of downstream equipment.

The potential for cofiring can be significant. Pioneer in the field Consolidated Natural Gas Co. estimates cofiring will mean an additional 40-45 bcf/year of new business for the company by the end of 1992. CNG's approach calls for convincing a plant operator of the economic and technological benefits of cofiring and then finding an easy way to introduce gas into the plant, such as to ignite the coal, followed by a gradual buildup in the gas feed to as much as 20% of the total fuel load.

The environmental results can be dramatic. A demonstration project in Pittsburgh in 1989 sponsored by the Gas Research Institute and Duquesne Light Co. produced reductions in sulfur dioxide emissions of 12% and nitrogen oxide of 25% by burning as little 4-9% with the coal.

A newer cofiring process, reburn, goes much farther, using the chemical properties of natural gas to cut nitrogen oxide emissions by as much as 60%. There are technologies involving coal/natural gas combustion that could result in reductions of as much 95% of both key acid rain pollutants.

Getting access to gas won't prove a problem for most power producers that need it for cofiring. Interstate Natural Gas Association of America (Ingaa) notes that the top 100 emitters of sulfur dioxide in the U.S. are an average 5 miles from one or more gas pipelines and that 13 have natural gas hookups.

Cofiring is just one example of where the natural gas industry can make significant headway in securing more of the surging electric power market. Other approaches include repowering boilers to burn natural gas, switching to gas on a seasonal basis, and installing gas fired combustion turbines.

Altogether, additional demand for gas created by electricity consumption growth could total as much as 2 tcf/year—based on new generating capacity already on the drawing boards, according to American Gas Association (AGA).

Gas advantages over coal and oil are manifold in power plant use. It takes about 3-4 years to site and build a gas fired power plant. A coal plant takes about 4-7 years.

Gas is the winner in operating economics as well. Figuring in operating costs and dispatch reliability, a gas fired, combined cycle power plant is more economic than an oil fired plant even if the delivered price of the gas is as much as 20% higher than fuel oil. The payout for an IPP or cogeneration power plant can be as short as 5 years for one that is gas fired compared with one that is coal fired. Capital costs for gas turbine generated power range $400-700/kw.

NATURAL GAS

GAS FIRED COGENERATION plant in California's San Joaquin Valley generates steam and electricity for thermal enhanced oil recovery operations. The boom in cogeneration growth during the 1980s has fueled one of the fastest growing markets for natural gas. Photo by Bob Williams.

For a coal fired facility, the cost jumps to as much as $1,400/kw. Longer term, however, the comparisons of economics tend to even out. A 50 year coal contract for an IPP can be had for as little as $1.50/MMBTU. Gas contracts usually do not run for more than 5 years, owing to the historic volatility of gas prices.

That leads to the biggest single stumbling block for the gas industry in making significant inroads in the electric power market in the 1990s: developing supply contracts for 15-20 years that are acceptable to gas producers and power plant operators alike.

Power plant operators generally want a gas price that is competitive with coal, because they base the price escalators in their contracts with IPPs on the price of coal. Gas producers, however, want escalators tied to finding costs or indexed to fuel oil or an alternate gas supply. In some areas of the country, gas producers will be hard pressed to meet long term contracts. In the Gulf of Mexico, for instance, fields tend to play out in as little as five years. So to fulfill a long term contract, a gulf producer will have to assume that he can replace his initial reserves three more times at a competitive cost near the same pipeline as under the original contract.

Alternate motor fuels

Gas producers, pipelines, and gas processors have a major stake in ensuring that they are not left behind in the rush to embrace alternate fuels.

While much of the government emphasis and political lobbying has been on alternate liquid fuels, proponents of compressed natural gas and liquefied petroleum gases have been slow to capitalize on their potential in the energy/environment debate.

That is due mainly to the search for a solution that will meet the needs of the average motorist. But there remains a sizable potential for CNG and LPG in the nation's commercial and private fleets of cars and trucks. It took the perceived tilt of the Bush administration toward alcohol fuels to awaken the gas fuels lobbies to more aggressively touting the virtues of their products.

According to the National Association of Fleet Administrators, fleet vehicles account for about 5% of all U.S. vehicles. Converting all such vehicles to LPG or CNG would amount to a big new market. There already is a sizable infrastructure for LPG vehicles in the U.S. CNG is making inroads as well, notably among local gas distribution utility fleets and municipal buses.

LPG proponents can point to their product's environmental benefits along with advantages over alcohol fuels. EPA tests of a propane fueled vehicle showed propane to be as clean as methanol while providing better mileage. The EPA tests showed propane emissions vs. gasoline were 93% lower in CO, 39% lower in NO_x, and 73% lower in net reactive exhaust hydrocarbons.

LPG associations contend that propane is the alternate clean fuel closest to gasoline in costs per gallon and in miles per gallon. Propane delivers about 85% the mileage of gasoline vs. methanol's 54% and ethanol's 70%, based on BTU energy content per gallon, according to the Western Liquid Gas Association.

Perhaps the strongest argument in favor of LPG is its track record, with more than 350,000 vehicles on the road in the U.S., mainly in commercial fleets. A clear economic advantage made this infrastructure possible. Gasoline engines can easily be converted to propane or a dual gasoline/LPG use, for only about $750. For vehicles that run about 30,000 miles per year, that yields a payout in less than 3 years. LPG fleet operators have noted a sharply reduced maintenance for propane vehicles because the fuel burns more cleanly than gasoline.

Drawbacks to LPG are the weight of the pressurized tank it requires and public perceptions of its safety. A 20 gal propane tank weighs about 150 lb more than a comparably sized gasoline tank. That may not prove daunting for a large van or truck, but it might for an owner or prospective buyer of a subcompact passenger vehicle.

Consumers tend to get skittish abut any fuel that is stored in a pressured tank. The rash of publicity over fires and injuries related to exploding disposable lighters in recent years has not helped mitigate that perception. In fact, propane is no more dangerous to handle than gasoline. And it is less susceptible to explosion or fire in an accident because the thick-walled pressurized tank on an LPG vehicle could withstand an impact better than a thin walled steel gasoline tank.

LPG advocates must face up to the task of winning consumer confidence in LPG safety and convenience. A broad U.S. network is in place to manufacture, distribute, and sell LPG. The price is right at about 70¢/gal, considering

NATURAL GAS

U.S. gas supply outlook

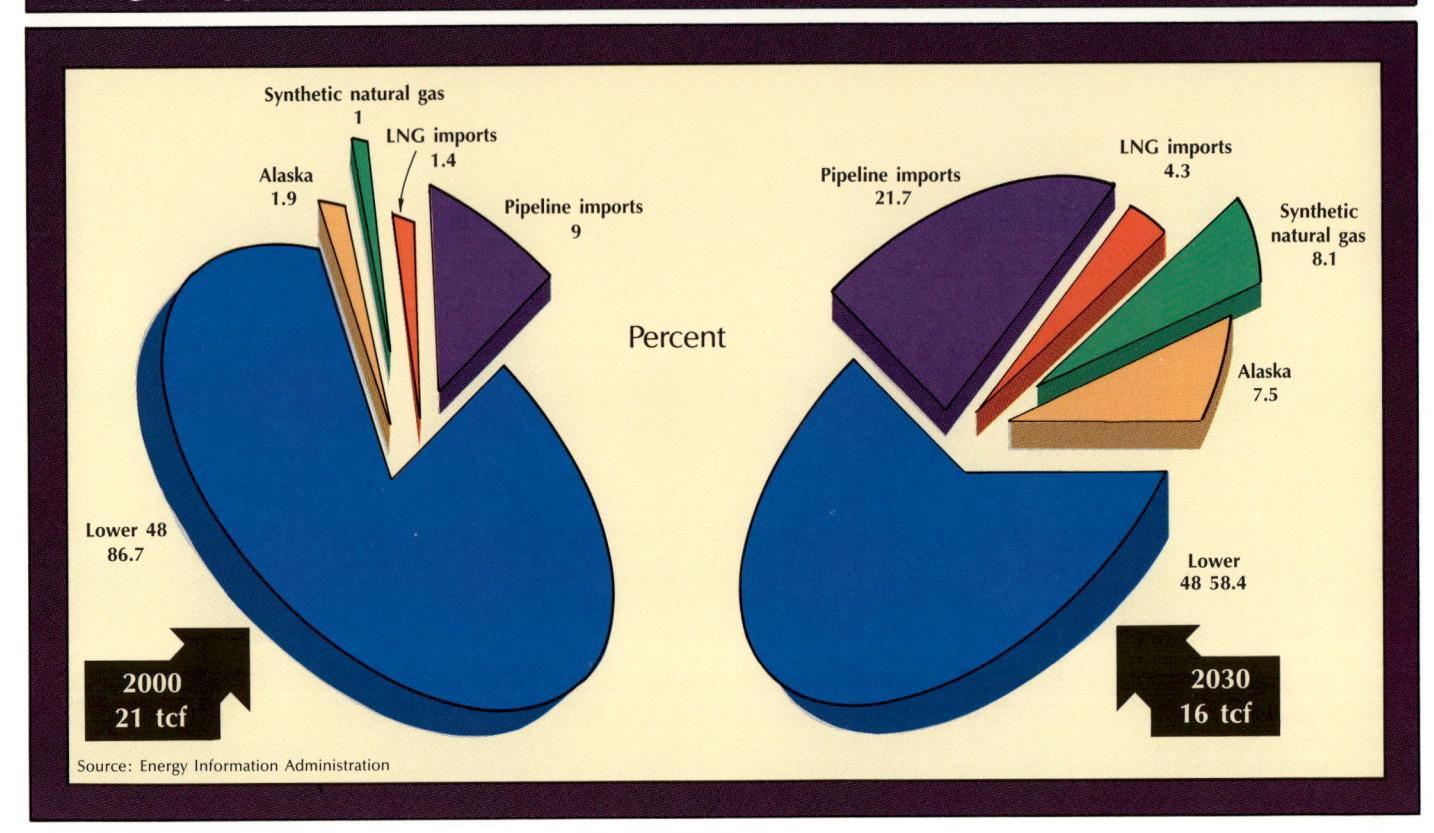

2000 — 21 tcf
- Lower 48: 86.7
- Alaska: 1.9
- Synthetic natural gas: 1
- LNG imports: 1.4
- Pipeline imports: 9

2030 — 16 tcf
- Lower 48: 58.4
- Pipeline imports: 21.7
- LNG imports: 4.3
- Synthetic natural gas: 8.1
- Alaska: 7.5

Percent

Source: Energy Information Administration

---

the high octane and mileage features.

Currently, only about 5-6% of U.S. propane supplies goes toward use as a motor fuel, according to the Gas Processors Association. More than one third of U.S. propane supplies is used as petrochemical feedstock, mainly because an LPG surplus has kept a lid on propane prices in recent years. Hiking propane use as a motor fuel would place more of a premium on the fuel, raising its price and therefore driving much of the supply out of the petrochemical market and into the motor fuels market.

Surplus gas processing capacity could be put to use as well, doubling U.S. output of propane to 1.7 million b/d. It would not require a massive effort to expand the nation's retail propane marketing network to accomodate nonfleet motorists in the major nonattainment urban areas, as it would for the alcohol fuels.

Infrastructure is a strong argument on behalf of CNG as well. A natural gas pipeline network already extends 1 million miles across the U.S. CNG also has a strong track record, with more than 500,000 CNG vehicles in operation worldwide.

Natural gas may be the best alternative as a motor fuel from an environmental standpoint, say advocates of CNG. Converting existing gasoline or diesel vehicles to natural gas can slash emissions of CO by as much as 99%, NO_x by as much as 85%.

According to the American Gas Association, converting half of the nation's 16 million fleet vehicles could cut U.S. oil consumption by 500,000 b/d, or 5% of total consumption by the transportation sector.

But the natural gas industry has come under some criticism by oil products marketers after AGA stepped up its lobbying efforts in 1989 after passage of the Alternative Motor Fuels Act. The Petroleum Marketers Association of America blasted AGA's efforts to portray natural gas as a "panacea" for U.S. energy and environmental needs. PMAA also claimed that increases in U.S. consumption of natural gas to meet growing motor fuel uses would result in increasing reliance on imports of that fuel, notably in the form of LNG from the Middle East. At the same time, said PMAA, it would take 25% of U.S. gas production to meet 20% of U.S. motor fuels consumption.

That sort of internecine squabbling among different sectors of the petroleum isn't necessarily a harmful thing, if it is simply a case of competing forces battling it out for market share on a level playing field. However, when one segment of the industry seeks preferential treatment, it automatically skews the economics of energy use and just invites more government intervention.

There can be room for all competing petroleum fuels, if they are placed on equal footing. Some competitors may even be of benefit to each other, such as the case with diesel and natural gas. Under EPA strictures, NO_x emissions in new diesel engines must be halved from current levels by 1991. Bus and heavy truck engines will have to cut particulates emissions to one sixth of today's levels by 1994.

Correspondingly, the Gas Research Institute is working on a dual fuel conversion kit for diesel engines that involves using diesel as a sort of pilot light, ignited by the heat of compression and in turn firing the natural gas. At idle and low speeds this kind of diesel engine would run on diesel fuel and at higher speeds on natural gas. That would provide some progress toward meeting future emissions reductions goals without eliminating diesel from the market and mandating costly and inconvenient single fuel alternate fuel engine designs.

Supply concerns

The disappearing gas "bubble," or surplus deliverability, is a cause of concern for electric power producers. Many in the industry surmise that the gas bubble is already effectively gone, as witnessed by the supply curtailments in recent winters in California and elsewhere in the nation.

It is not a question of the potential resource. Although the reserve life of U.S. gas totals more than 20 years at the

NATURAL GAS

current rate of consumption, spiraling demand is shrinking that number.

In a 1989 study, the U.S. Energy Information Administration projected that the U.S. may be using twice as much natural gas to generate electricity by the year 2000 as its does now. EIA sees the U.S. gas resource as adequate to meet that extra demand. Assuming an economic life of power plants of about 30 years, natural gas use to generate electricity will jump from 3.2 tcf/year to about 6.2 tcf/year in the year 2000 and beyond, as the gas generating plants are retired by the year 2030.

However—and this is a big if—since it is much cheaper to completely revamp an existing gas plant than it is to build any kind of new grassroots plant, utilities may opt to go that route instead. If that happens, according to EIA, the life cycle of power plants will be greatly extended, perhaps even doubled, eliminating a post 2000 slide in gas demand by the electric power industry.

"This implies a continuation of the 6.2 tcf of annual gas demand to the year 2030," EIA said. "The resource base now generally believed to exist in the U.S. would not be able to handle such a massive demand for so long."

The U.S. will have used up about 80% of its existing natural gas resources by the year 2030, EIA estimated. EIA pegged the U.S. gas resource in 2030 at about 180 tcf, compared with 830 tcf in 1988. By 2030, the U.S. will have to depend on imports for about 26% of its gas supply needs, from about 7% in 1988. In addition, about 10% of all pipeline quality gaseous fuel used in the U.S. in 2030 is likely to be synthetic, produced mainly from coal and liquid hydrocarbons, EIA projected.

Because of the tightening supply/demand balance for natural gas, the average wellhead price for gas is likely to double (in real dollars) by the year 2000. Consequently, utilities are likely to retire most gas burning plants at the end of their economic lives, without any effort to extend their lives or convert them to other, more economic fuels, EIA said.

The gas industry was quick to dispute EIA's conclusions, citing a study a year earlier by DOE to assess the potential for U.S. natural gas. The DOE study concluded that the recoverable resources of conventional gas totaled 1.059 quadrillion cu ft in the lower 48. Of that total, 800 tcf lies in conventional reservoirs and can be extracted at a cost of less than $3/MMBTU. The rest is economic at $3-5/MMBTU.

There is also the massive Alaskan North Slope gas resource to consider, at more than 30 tcf. Combined with frontier gas supplies in Canada, the North American Arctic's gas reserve potential is more than 60 tcf. Of course, price is the key. It is not likely that arctic gas supplies will be economic until almost 2000.

In the meantime, there are two other sources of potential gas supplies to help the U.S. bridge the gap to the 21st Century: coalbed methane and LNG.

Coalbed methane proved a surprise to the industry in 1988-90, generating frenzied drilling activity in areas like Alabama's Black Warrior basin and the San Juan basin of Colorado and New Mexico. Relatively low costs, aided by a federal tax credit that was renewed at yearend 1990, have led to a significant addition to the U.S. gas resource base. Recognition of this incremental increase is likely to result in the tax credit being extended through the 1990s.

Worldwide, the demand for LNG is accelerating as the economics of gas improve and more nations—notably those crude exporters seeking additional markets for their hydrocarbon resources—are developing low cost sources of gas for LNG projects. Countries like Algeria—previously a major exporter of LNG to the U.S.—are selling LNG at scarcely breakeven prices to ensure contracts that will certainly look better in the future. As a result, the U.S. again is importing substantial volumes of LNG after a period of negligible LNG imports in the latter 1980s. Joining Algeria in the 1990s in competition for the U.S. LNG import market will be Norway, Nigeria, Qatar, Venezuela, Indonesia, Australia, and Papua New Guinea.

In short, the gas supplies are likely to be available to meet expected demand in the U.S. electric power sector in the early part of the next century. Will those future incremental supplies be cost-competitive?

There are three factors to consider here. One is the uncertainty over future costs of generating power with coal. If scrubbing is mandated and not directly subsidized, gas almost certainly will boost its market share in power generation at the expense of coal in the 1990s. Studies show gas at even a sharply higher price can compete effectively with scrubber/coal power generation and provide more environmental benefits.

Secondly, there is the near term price-depressing effect of dual fuel boiler capability. Almost 4.9 tcf/year of gas in the combined power plant and industrial sectors is burned in such dual-capable facilities. The switches can be made in hours or even minutes and often on price differences of a few cents per million BTUs. About one-third of total gas consumption is subject to fuel switching—about 95% in power generation and 30% in industrial plants.

Gas now must compete with residual fuel oil to avoid big market losses. So it follows that as oil prices rise in the 1990s—most likely modest increases averaged out over the decade—gas prices will keep pace, rising enough to stimulate drilling from current depressed levels but not so sharply as to switch to fuel oil or improve the relative economics of scrubbed coal. As drilling accelerates, the resource base in the U.S., Canada, and overseas also will expand. This follows the same pattern established for oil in the late 1970s and early 1980s, and there is no reason to doubt that gas will do the same. Of course, natural gas could run into the same fate that oil did in the 1980s: Too much drilling for any kind of prospect inflating its cost and a skyrocketing price adding too much supply at the same time it spurs conservation and fuel switching.

There are critical differences for natural gas, however. Unlike with oil, a reasonable cost resource of natural gas has yet to be fully tapped—and there are the natural price constraints of a deregulated market. The U.S. just does not face the same kind of energy security or trade balance threat with gas that it does now with oil.

AGA contends that the U.S. surplus of gas deliverability is gone and that there will come several years of balance in U.S. gas supply and demand.

AGA figures that incremental supplies of as much as 1.35-2.1 tcf/year of gas could be made available within three years if needed.

Gas well completions are rising as well. AGA estimated U.S. producers would complete 10,040 gas wells in 1990, compared with 8,840 in 1989 and less than 8,000/year in 1986-88.

The winter of 1989-90 gave the gas industry a healthy taste of just how tight the situation can get. A sudden, severe and widespread cold snap spiked demand for gas and froze off significant production volumes when wells froze up.

That shot the spot price of gas up to levels not seen in almost five years.

Those are signs of health that is certain to turn more robust in the next few years. The gas industry's long awaited cure will come in large part because of the environmental ethos that is seeping into the national consciousness and permeating every part of the American fabric. It is abundantly clear that natural gas is the environmentally preferred fuel for the 1990s.

*This article was adapted from a chapter of **U.S. Petroleum Strategies in the Decade of the Environment** (PennWell Books, 1991) by Bob Williams.*

U.S. honing technology to burn coal cleanly

COAL IS THE UNITED STATES' MOST ABUNDANT ENergy source, but in this age of environmental awareness one of the most difficult to use.

The U.S. has one fourth of the world's coal reserves, and at current consumption rates its 300 billion tons could last well into the 22nd century.

Since 1973, coal has provided more new energy for the U.S. than any other source. One fourth of all primary energy consumed by the U.S. comes from coal. More than half of the electricity used by American consumers is produced by coal burning power plants.

Per capita, Americans use 19 lb/day of coal, mainly in the form of electricity. Greater coal use by U.S. electrical power plants has saved the equivalent of nearly 3.2 million b/d of oil since 1973.

To help the nation use more coal without harming the environment the Department of Energy has undertaken a $5 billion program to demonstrate full scale projects that burn coal more cleanly and lower power generation costs.

Clean Coal Technology program

Congress launched the Clean Coal Technology (CCT) program in 1986, and President Reagan expanded it in 1987. Under the program, the federal government provides $2.5 billion for coal projects, selected in five rounds of competition, that will be more than matched by private funding.

DOE selected 35 projects to receive federal matching funds in the first three rounds of competition. The projects are valued at more than $3.4 billion.

DOE plans to help finance 50-75 showcase projects by 1992, each demonstrating an advanced method for burning coal cleanly and efficiently. Project costs could total more than $5 billion.

By conducting several rounds of competition, the government hopes to attract the newest and best technologies that will become available as the program proceeds.

The solicitations will demonstrate technology options consistent with demands of the energy market and responsive to environmental considerations.

The projects will provide data on technical, environmental, economic, and operational performance to reduce the uncertainties of later commercial scale use of the technologies.

The technologies to be demonstrated must be capable of being applied to any segment of the U.S. coal resource base and involve new, retrofit, and repowering applications.

The first round focused on new and retrofit applications that, through increased efficiency, can increase the role of coal as an energy option.

The second round concentrated specifically on demonstrating technologies that can overcome barriers to the use of coal created by the issue of acid rain.

The third round targeted technologies capable of achieving significant reductions in emissions of sulfur dioxide and/or nitrogen oxide from existing facilities to minimize environmental impacts such as transboundary and interstate pollution and/or provide for future energy needs in an environmentally acceptable manner.

What's planned in the fourth round

DOE planned to announce the CCT fourth round demonstration projects in the fall of 1991. The fifth and final round will be held in 1992.

The fourth round solicitation will focus on innovative technologies that will result in lower greenhouse gas emissions.

Several features have been added to the solicitation that will encourage technologies to help solve U.S. future energy needs.

Another new criterion will rate projects on how the proposed technology can improve the thermal efficiency of utility or industrial facilities in a cost effective way.

DOE said, "More efficient technologies mean lower emissions of pollutants that will have to be offset after a permanent cap on sulfur emissions goes into effect after 2000." It added that increased thermal efficiency also will lower carbon dioxide emissions.

Another change in the solicitation asks sponsors to describe how commercial use of their technology will improve the nation's trade position, create new jobs, or generate revenue.

It permits DOE to protect trade secrets or confidential information for as long as 5 years after a project is completed. If a CCT process were licensed, DOE would receive 5% of the licensing fees or 0.5% of the value of equipment sales.

Table 1

U.S. clean coal technology funding

Funding round	1986	1987	1988	1989	1990	1991	1992
				(Million $)			
1	99.4	149.1	149.1	—	—	—	—
2	—	—	50	190	135	200	—
3	—	—	—	—	419	156	—
4	—	—	—	—	—	568	—
5	—	—	—	—	—	—	600
Total	99.4	149.1	199.1	190	554	924	600

Source: Department of Energy

COAL

Building on progress in reduced emissions

J. Allen Wampler, DOE assistant secretary for fossil energy, says U.S. sulfur emissions have dropped significantly in the last 15 years, while coal use has steadily increased, and the CCT program will build on that progress.

"We can fashion a program that accelerates emission reductions in the near term. Gas cofiring, for example, may be an important way to achieve some immediate reductions at relatively low cost.

"Obviously, there will be some projects in this program that will succeed and others that will fail. That is inherent in a program of this type. In fact, knowing what won't succeed in the marketplace, in many ways, is as important as learning what will succeed. That's why this is a demonstration program and the government is shouldering a portion of the risks."

Energy Sec. James Watkins has said, "The CCT program offers us a way to break the link between coal and concerns about the environment and over time to begin replacing oil tankers moving out of the Persian Gulf with increasing numbers of tugs and towboats moving coal along the Ohio River, colliers leaving Norfolk, or unit trains moving out of the Powder River basin in Wyoming and Montana.

"By the time this program is complete, the U.S. could have 50 or more clean coal projects up and running in this country, each showcasing a new, high tech concept for using coal cleanly and efficiently. Out of this partnership with industry will come state of the art systems that can sustain and accelerate the downward trend in air pollutants.

"This nation stands second to none in its commitment to the environment. It also must stand second to none in its commitment to technological excellence, such as that which has produced a wide spectrum of advanced pollution control equipment—innovative technology that can replace the 1960 vintage wet scrubber with techniques such as duct injection, simultaneous sulfur and nitrogen oxide removal, and in the future perhaps, the use of electron beams for advanced flue gas cleanup.

"It is that commitment that today is giving us new concepts in precombustion cleaning—ultrafine grinding, microbubble flotation, and the list could go on."

Benefits of demonstration period

When projects are complete, industrial participants will be in a position to use the information and experience gained in the demonstration period to promote, commercialize, and market the technology.

The detailed data and experience will be vital to firms deciding to invest in and build, retrofit, or repower plants using clean coal technologies.

DOE said building a new, large coal fired power plant would cost about $1.5 billion today. Of that, environmental controls will account for more than one third.

Many of those controls must be added as separate facilities, raising the cost and complexity of the power plant and reducing its efficiency because some of the plant's power must be used to operate the controls.

CCT represents a fundamental change in coal fired power plant technology because in many cases the emissions reductions and cost improvements are achieved concurrently, rather than being pitted against each other.

DOE said in terms of sulfur and nitrogen emissions CCT has the potential to make a coal fired plant as clean as an oil fired plant and, in some cases, as clean as a plant that burns natural gas, the cleanest burning fossil fuel.

It said unlike the use of scrubbers, CCT will not achieve a high environmental performance at the expense of efficiency. In many cases they boost a plant's performance at the same time they reduce pollution.

Clean coal technologies can be installed at any of three stages in the fuel chain of a power plant or in a fourth manner that departs from the traditional method of burning coal.

The three steps are:
- Precombustion, in which sulfur and other impurities in coal are removed before it reaches the boiler.
- Combustion, in which pollutants inside the combustor or boiler are removed while the coal burns.
- Postcombustion, in which flue gases released from coal boilers are cleaned in ductwork leading to the smokestack or in advanced versions of today's scrubbers.

The fourth method, conversion, bypasses the combustion process.

It changes coal into a gas or liquid that can be cleaned and used as fuel.

Precombustion cleaning could cut emissions

With the advent of mechanical mining and loading of coal, impurities in coal also have increased. Also, many coals contain high levels of undesirable minerals.

The Electric Power Research Institute has estimated wider use of coal cleaning processes could reduce total sulfur dioxide emissions by 10% nationwide.

DOE said about 40% of the coal bound for U.S. utility burners is cleaned before it is burned. Most commercial coal cleaning is done on eastern and midwestern bituminous coals, usually at the mine mouth.

When coal from the mine is crushed and then washed,

Cofiring controls pollutants from coal burning

NATURAL GAS INCREASINGLY IS being used as a method to control pollutants from the burning of coal.

Cofiring, a commercially available technique, burns gas and coal simultaneously in the same boiler.

Typically, the two fuels are not physically mixed. Different burners are used and often positioned at different heights within the boiler.

The amount of sulfur dioxide reduction is directly proportional to the amount of gas fired in place of coal. So if 10% of the fuel is natural gas, sulfur emissions will be about 10% less than if only coal were burned.

Gas cofiring also reduces nitrogen oxide emissions and can mitigate ash fouling in the burner.

A second technique, gas reburning, is used mainly to control nitrogen oxide emissions. Coal is fired in the lower regions of a boiler to provide 80-90% of the total heat released.

Natural gas is fired in the "reburn" region above the main combustion zone. Within the fuel rich reburn region, hydrocarbon fragments from the gas will react with nitric oxide produced in the main flame to form molecular nitrogen, the same form of nitrogen that exists naturally in the air.

Secondary air is added above the reburn region to finish the combustion at a lower temperature, preventing nitrogen oxide from forming.

Nitrogen oxide emissions from a gas reburning system are expected to be about 40% less than those from a unit firing only coal.

The U.S. Department of Energy said the capital cost of retrofitting a natural gas reburn system on a 500,000 kw boiler is about $12/kw, competitive with low nitrogen oxide burners. Depending on the boiler configuration and design requirements, costs could range from about $5 to $30/kw.

COAL

heavier impurities are separated.

But physical cleaning can remove only matter that is physically distinct from the coal, such as small dirt particles and rocks.

It cannot remove sulfur that is chemically combined with the coal, or nitrogen from the coal.

The advanced CCT techniques will grind the coal into much smaller sizes than is done commercially today, enabling up to 90% of impurities to be removed.

A second method, which may be tested through CCT projects, is through chemical or biological cleaning, such as the molten caustic leaching method.

In that method, coal is exposed to a hot, sodium, or potassium based chemical which leaches sulfur and mineral matter from the coal.

Biological cleaning is being tested in the laboratory, using naturally occurring bacteria or fungi to desulfurize coal.

DOE said chemical or biological coal cleaning appears to be capable of removing as much as 90% of the sulfur and 99% of the ash in coal.

Combustion methods of coal cleaning

Coal also can be cleaned while it is burning, an advantage because additional sulfur or nitrogen removal equipment is not required.

In most conventional coal combustion plants, raw coal is pulverized into particles small enough to form a combustible cloud and injected with hot air into burners along the lower portion of a hollow rectangular box called a steam boiler.

As the coal burns, the heat is transferred to water filled tubes in the sides of the boiler, which create steam that turn a turbine generator.

One of the newer technologies, fluidized bed combustors, uses crushed coal mixed with limestone and suspended on jets of air.

The "bed" of coal and limestone floats and tumbles inside the boiler.

The limestone soaks up the sulfur in the coal, capturing 90% of it.

Combustion temperatures can be held to less than 1,600° F., or almost half the temperature of a conventional boiler, and below the threshold at which nitrogen pollutants form. That allows fludized bed combustors to meet sulfur dioxide and nitrogen oxide standards without additional pollution control equipment.

The CCT program is demonstrating two types of fluidized bed combustors: atmospheric and pressurized methods.

Another advanced technology is the cyclone combustor, which burns coal in a separate chamber outside the furnace

U.S. clean coal projects

Sponsor/description	Project site	Total cost Million	DOE cost $	DOE share (%)	Status
American El. Power Service Co. Pressurized fluidized bed combustion	Brilliant, Ohio	167.5	60.2	36	1991 start
Babcock & Wilcox Co. Limestone injection multistage burner	Lorain, Ohio	19.4	7.6	39	Operation
Coal Tech Corp. Slagging combustor and sorbent injection	Williamsport, Pa.	980.4	490.2	50	Completed
Energy & Environ. Res. Corp. Gas reburn or sorbent injection	Springfield, Ill.	30	15	50	Construct.
Ohio Clean Fuels Inc. Coal-oil coprocessing liquefaction	Undetermined	225	45	20	Design
Colo.-Ute Electric Assoc. Circulating fluidized bed combustion	Nucla, Colo.	54.1	19.9	37	Operation
Western Energy Co. Western coal cleaning process	Colstrip, Mont.	69	34.5	50	Construct.
Combustion Engineering Coal cleaning	various sites	17.4	8.7	50	Testing
City of Tallahassee, Fla. circulating atmospheric fluidized bed	Tallahassee, Fla.	277	75	27	Design
Amer. El. Pwr. Services Corp. Pressurized fluidized bed combustor	New Haven, W.Va.	660	185	28	Design
Combustion Engineering Inc. Flue gas cleaning	Niles, Ohio	31	16.5	50	Construct.
Babcox & Wilcox Co. Blue gas cleaning	Dilles Bottom, Ohio	10.6	4.8	46	Construct.
Southern Co. Services Sulfur dioxide particulate control	Newman, Ga.	35.8	17.6	49	Construct.
Pure Air Advanced flue gas desfulurization	Chesterton, Ind.	151	63	42	Construct.
Passamaquoddy Tribe Cement kiln flue gas recovery	Thomaston, Me.	12.5	5.9	47	Operation
Bethlehem Steel Corp. Coke oven gas cleaning	Baltimore, Md.	45	13.5	30	Construct.
Southern Co. Services Nitrogen oxide control for wall-fired boiler	Coosa, Ga.	11.7	5.3	45	Operation
Babcox & Wilcox Co. Coal reburning for cyclone boilers	Cassville, Wisc.	10.6	5	48	Construct.

COAL

cavity.

The hot combustion gases then pass into the boiler where the heat exchange takes place.

Cyclone combustors keep the ash out of the furnace cavity, where it tends to collect on boiler tubes and lower heat transfer efficiency.

To keep ash from being blown into the furnace, combustion temperature is kept so hot that mineral impurities melt and form slag.

A vortex of air—the "cyclone"—forces slag to the outer walls of the combustor where it can be removed.

But the high temperature required to melt the ash produces unacceptable levels of nitrogen oxide. The CCT program is demonstrating cyclone combustors that overcome that drawback.

DOE said by positioning air injection ports so coal is burned in stages, nitrogen oxide emissions can be reduced 50-70%.

The flue gas scrubber

Until clean coal technologies emerged, the flue gas scrubber, developed in the 1960s, was the only commercial technology capable of achieving the 70-90% sulfur dioxide reduction required under the 1977 Clean Air Act amendments.

Scrubbers are complex chemical plants—separate gas processing facilities installed at the "back end" of a power plant leading to the smokestack.

As of 1988, 146 scrubbers had been installed at 82 of the 370 currently operable coal fired power plants in the U.S. Combined installation and operational costs for the scrubbers exceeded more than $17 billion.

In wet scrubbers, flue gases from combustion of coal are sprayed with a slurry made of water and an alkaline reagent, usually limestone.

The sulfur dioxide in the flue gas reacts chemically with the reagent to form calcium sulfite and calcium sulfate in the form of a wet sludge.

During its lifetime a 500,000 kw, coal fired power plant will produce enough wet sludge to fill a 500 acre disposal pond to a depth of 40 ft, often creating a waste disposal problem.

Wet scrubbers can remove 90% or more of the sulfur dioxide but are expensive to install. They often cost as much as $300/kw of capacity, or $150 million for a typical 500,000 kw plant.

They consume 5-8% of a power plant's thermal energy to run pumps, fans, and a flue gas reheat system, thereby reducing electricity output 1-2%.

They also require large amounts of water, typically 500-

Table 2

Sponsor/description	Project site	Total cost Million	DOE cost $	DOE share (%)	Status
Southern Co. Services Catalytic reduction of nitrogen oxide	Pensacola, Fla.	15.5	7.5	48	Construct.
Trans-Alta Resources Burner retrofit for cyclone boilers	Marion, Ill.	15.3	6.8	44	Operation
Southern Co. Services Tangentially fired combustion techniques	Lynn Haven, Fla.	8.5	4.1	49	Design
Combustion Engineering Inc. Airblown, entrained flow gasification	Springfield, Ill.	270	129	48	Design
AirPol Inc. Gas suspension absorption system	Paducah, Ky.	6.9	2	29	Design
Air Products & Chemicals Inc. Liquid phase methanol process	Beulah, N.D.	213	93	43	Negotiation
Alaska Ind'l Dev. & Export Auth. Advanced slagging coal combustor	Healy, Alas.	193	94	48	Design
Babcock & Wilcox Co. Low nitrogen oxide cell burner	Aberdeen, Ohio	9.8	4.7	48	Construct.
Bechtel Corp. Utility flue gas sulfur removal	Indiana Cty, Pa.	9.2	4.6	50	Construct.
Bethlehem Steel Corp. Blast furnace granulated coal injection	Burns Harbor, Ind.	143	31	21	Design
Clean Power Cogeneration Inc. Gasification combined cycle	Tallahassee, Fla.	242	121	50	Negotiation
Dairyland Power Cooperative Pressurized circulating fluid bed combustor	Alma, Wisc.	184	88	48	Negotiation
Encoal Corp. Mild gasification to clean coal	Gillette, Wyo.	72	36	50	Construct.
Energy & Env. Research Corp. Combined gas reburning	Denver, Colo.	14.4	7.2	50	Design
Lifac-North America Limestone injection for desulfurization	Richmond, Ind.	17	8.5	50	Construct.
MK-Ferguson Co. Flue gas cleanup system	Niles, Ohio	66.2	33.1	50	Design
Public Service Co. of Colo. Low nitrogen oxide burner demonstration	Denver, Colo.	26	13	50	Design

Source: Department of Energy

COAL

Clean coal technology commercial readiness

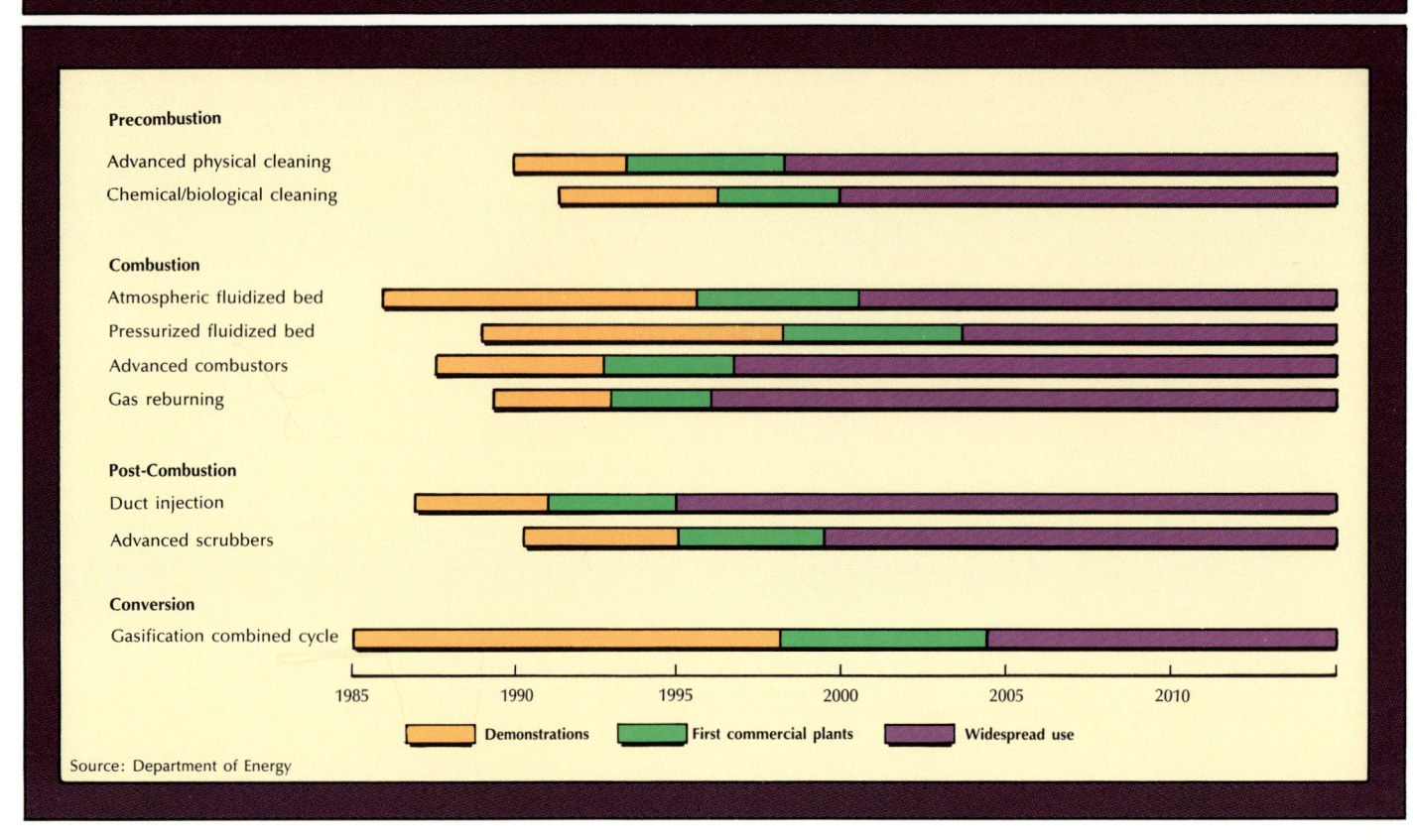

Technology	Demonstrations	First commercial plants	Widespread use
Precombustion			
Advanced physical cleaning	~1991–1994	~1994–1998	1998–
Chemical/biological cleaning	~1993–1998	~1998–2001	2001–
Combustion			
Atmospheric fluidized bed	1985–1993	1993–1998	1998–
Pressurized fluidized bed	~1991–1999	1999–2004	2004–
Advanced combustors	~1988–1994	1994–1997	1997–
Gas reburning	~1991–1995	1995–1998	1998–
Post-Combustion			
Duct injection	~1988–1994	1994–1998	1998–
Advanced scrubbers	~1990–1996	1996–1999	1999–
Conversion			
Gasification combined cycle	1986–1998	1998–2005	2005–

Source: Department of Energy

2,500 gal/min for a unit of 500,000 kw.

In a dry scrubber, the reagent slurry—usually lime—is injected in a finely atomized form.

The droplets evaporate in the hot gas, leaving only dry particles for collection as waste.

Although simpler in concept than the wet scrubber, the dry scrubber has not been as successful on high sulfur coal due to the increased amounts of expensive reagents required to reduce sulfur dioxide by 90%.

Postcombustion innovations

A number of CCT projects are demonstrating innovations in postcombustion cleaning.

In-duct sorbent injection works inside ductwork leading from the boiler to the smokestack.

Sulfur absorbers such as lime are sprayed into the center of the duct.

By controlling the humidity of the flue gas and the spray pattern of the sorbent, 50-70% of the sulfur dioxide can be removed.

The reaction produces dry particles that can be collected downstream, and extensive new construction is not needed because the plant's existing ductwork is used.

Advanced scrubbers, like their predecessors, place the flue gas processing facilities outside the main power plant.

These devices can regenerate the sulfur absorbing chemical, making the system more economical; remove both sulfur dioxide and nitrogen oxide; produce an environmentally benign, dry waste product; or streamline operations by reducing or eliminating the need for reheating or backup modules.

Other innovations in treating flue gases make it possible to reduce nitrogen oxide in flue gases leaving the boiler instead of modifying the combustor.

The selective catalytic reduction technique, which may be demonstrated in the CCT program, mixes ammonia with flue gas and passes it through a reaction chamber separate from the scrubber vessel.

In the presence of a catalyst nitrogen oxide is converted by the ammonia into molecular nitrogen and water, but problems arise with high sulfur coals. The technique can reduce nitrogen emissions by 50-80%.

Other techniques

Some techniques would convert coal into another fuel form and bypass the conventional combustion process. Coal would be converted into a gaseous or liquid form or a combination of gases, liquids, and solids.

The gasification combined cycle process breaks coal into gaseous molecules by bringing it into contact with high temperature steam and oxygen, purifies those gases, and then burns them, using the exhaust to generate electricity and any residual heat to boil water for a conventional steam turbine generator.

Such combined cycle systems also can be powered by a pressurized fluidized bed combustor.

DOE said gasification combined cycle systems are among the cleanest of the emerging clean coal technologies. Sulfur, nitrogen compounds, and particulates are removed before the fuel is burned in the gas turbine, before combustion air is added.

For that reason, there is a much lower volume of gas to be treated than in a postcombustion scrubber.

In a coal gasifier, unlike coal combustion processes, the sulfur in coal is released in the form of hydrogen sulfide rather than sulfur dioxide. And several commercial processes are capable of removing more than 99% of the hydrogen sulfide in gas.

Several commercial scale gasification combined cycle demonstration plants are being built in the U.S. Current research is concentrating on using techniques to clean the gas while it still is hot, rather than cooling it first.

Gas production (bcf)

	1990
WESTERN HEMISPHERE	
Argintina	816.1
Barbados	0.7
Boliva	108.1
Brazil	113.2
Canada	4,267.0
Chile	39.2
Colombia	140.5
Ecuador	3.6
Guatemala	0.3
Mexico	1,332.7
Peru	46.0
Trinidad	177.7
United States	18,358.0
Venezuela	862.0
Total	**26,265.1**
EUROPE	
Austria	41.6
Denmark	101.4
France	106.0
Germany, Fed. Rep.	565.7
Greece	3.1
Italy	736.8
Netherlands	2,618.1
Norway	896.5

	1990
Spain	46.3
Turkey	7.5
United Kingdom	1,700.7
Total	**6,823.7**
MIDDLE EAST	
Bahrain	209.9
Iran	854.4
Iraq	140.7
Israel	1.2
Jordan	7.5
Kuwait	147.8
Oman	88.2
Qatar	265.1
Saudi Arabia	1,401.2
Syria	25.8
United Arab Emirates	1,092.1
Total	**4,233.9**
ASIA-PACIFIC	
Afghanistan	10.5
Australia	749.9
Bangladesh	162.4
Brunei	278.6
India	437.7
Indonesia	1,383.2

	1990
Japan	57.8
Malaysia	569.7
Myanmar (Burma)	36.5
New Zealand	164.3
Pakistan	451.1
Taiwan, China	38.3
Thailand	212.0
Total	**4,552.0**
AFRICA	
Algeria	1,588.0
Angola	19.0
Egypt	240.0
Gabon	3.6
Libya	202.9
Morocco	1.1
Nigeria	133.9
Tunisia	13.1
Total	**2,201.6**
COMMUNIST	
China	517.7
Romania	1,181.1
USSR	28,800.0
Other	1,032.7
Total	**31,531.5**
WORLD TOTAL	**75,607.8**

World well completions—1989*

Country	Total wells	Total wildcats	Country	Total wells	Total wildcats
North America			Jordan	-	8
U.S.	†31,417	5,687	Kuwait	10	2
Western Canada	§5,513	2,525	North Yemen	32	12
			Oman	-	44
Europe			Qatar	11	3
Austria	19	12	Saudi Arabia	47	3
Denmark	18	4	South Yemen	65	21
France	73	47	Syria	50	20
Greece	4	4	Turkey	87	39
Ireland	5	5			
Italy	87	38	**Southwest Pacific**		
Netherlands	73	45	Australia	228	63
Norway	87	23	New Zealand	13	5
Spain	11	10	Papua New Guinea	23	-
Svalbard	1	1			
Sweden	17	17	**North Africa**		
Switzerland	1	1	Algeria	78	18
U.K.	212	184	Egypt	114	43
West Germany	-	12	Libya	87	26
Yugoslavia	135	23	Sudan	4	4
			Tunisia	11	8
Latin America					
Argentina	801	120	**Central, South Africa**		
Aruba	2	2	Angola	63	19
Barbados	9	2	Benin	2	1
Bolivia	31	13	Cameroon	3	1
Brazil	591	115	Chad	5	5
Chile	53	17	Congo	10	7
Colombia	208	72	Cote d'Ivoire	1	1
Costa Rica	1	1	Guinea Bissau	1	1
Cuba	2	2	Gabon	52	26
Ecuador	29	11	Ghana	3	3
Guatemala	-	4	Kenya	4	4
Mexico	82	24	Mauritania	1	1
Panama	3	3	Mozambique	1	0
Paraguay	1	1	Nigeria	91	47
Peru	86	7	Senegambia	4	4
Puerto Rico	1	1	Somalia	1	1
Suriname	59	8	South Africa	19	19
Trinidad & Tobago	85	8	Zaire	6	1
Venezuela	181	23			
			Far East		
Middle East			Bangladesh	4	-
Abu Dhabi	28	2	Indonesia	599	102
Bahrain	20	0	Japan	11	9
Dubai	24	1	Malaysia	27	17
Iraq	-	2	Pakistan	53	20
Israel	4	3	Philippines	8	6

*AAPG data; includes stratigraphic and service wells. †Includes 20,851 producers. §Includes 4,094 producers.

International Active Rig Count

Region	1983	1984	1985	1986	1987	1988 Land	1988 Off.	1988 Total	1989 Land	1989 Off.	1989 Total	1990 Land	1990 Off.	1990 Total
NORTH AMERICA														
Canada	201	259	311	178	181	194	2	196	129	1	130	138	0	138
United States*	2,229	2,429	1,976	970	936	784	152	936	747	123	870	900	108	1,008
Subtotal	**2,430**	**2,688**	**2,287**	**1,148**	**1,117**	**978**	**154**	**1,132**	**876**	**124**	**1,000**	**1,038**	**108**	**1,146**
LATIN AMERICA														
Argentina	73	82	81	47	61	63	0	63	54	1	55	63	1	64
Bolivia	8	6	5	4	6	5	0	5	6	0	6	6	0	6
Brazil	83	70	76	77	60	28	11	39	21	7	28	16	7	23
Chile	6	6	6	6	7	2	3	5	3	2	5	3	3	6
Colombia	18	16	21	17	14	19	0	19	15	0	15	11	0	11
Costa Rica	1	1	0	0	1	0	0	0	0	0	0	0	0	0
Ecuador	5	4	2	3	3	5	1	6	5	0	5	2	0	2
Mexico	187	196	196	163	143	121	34	155	69	34	103	57	33	90
Peru	16	17	12	13	15	6	5	11	6	4	10	4	2	6
Trinidad	10	9	10	11	9	3	4	7	2	4	6	3	5	8
Venezuela	41	30	33	29	18	15	10	25	18	10	28	20	12	32
Other	3	2	1	1	1	0	0	0	0	0	0	0	0	0
Subtotal	**451**	**439**	**443**	**371**	**338**	**267**	**68**	**335**	**199**	**62**	**261**	**185**	**63**	**248**
FAR EAST														
Australia	25	34	31	15	16	14	5	19	7	7	14	8	6	14
Bangladesh	2	4	5	5	5	2	0	2	4	1	5	4	0	4
Brunei	10	9	7	5	3	1	3	4	1	3	4	1	4	5
China-offshore	10	9	14	7	4	0	8	8	0	4	4	0	3	3
India	56	57	62	70	116	106	25	131	110	25	135	110	27	137
Indonesia	88	82	80	62	37	33	11	44	36	10	46	36	16	52
Japan	20	18	16	15	10	11	0	11	6	0	6	6	1	7
Malaysia	10	9	8	8	8	0	9	9	1	11	12	0	13	13
Myanmar (Burma)	36	33	33	33	29	26	0	26	23	0	23	19	0	19
New Zealand	2	2	4	4	3	1	1	2	2	0	2	0	1	1
Pakistan	17	17	18	17	14	13	0	13	13	0	13	11	0	11
Papua New Guinea	1	1	2	1	2	3	0	3	2	1	3	2	1	3
Philippines	8	3	2	1	2	3	1	4	6	1	7	6	0	6
Taiwan	5	6	7	7	4	2	2	4	4	1	5	2	1	3
Thailand	13	11	9	4	6	3	3	6	3	3	6	2	3	5
Other	0	0	0	0	0	0	1	1	0	1	1	0	1	1
Subtotal	**303**	**295**	**298**	**254**	**259**	**218**	**69**	**287**	**218**	**68**	**286**	**207**	**77**	**284**
AFRICA														
Algeria	54	27	35	41	40	32	0	32	24	0	24	35	0	35
Angola	12	7	11	7	10	0	10	10	1	7	8	1	7	8
Congo	4	3	3	2	3	0	2	2	0	1	1	0	2	2
Gabon	7	6	7	3	3	3	2	5	3	3	6	3	2	5
Kenya	0	0	0	0	1	1	0	1	3	0	3	1	0	1
Libya	24	26	30	20	12	14	2	16	13	4	17	12	3	15
Nigeria	17	11	10	10	11	7	6	13	10	4	14	12	6	18
South Africa	2	2	3	3	3	0	4	4	0	3	3	0	3	3
Tunisia	7	6	4	3	4	2	1	3	3	0	3	2	1	3
Other	20	17	15	8	3	3	2	5	2	2	4	2	0	2
Subtotal	**147**	**105**	**118**	**97**	**90**	**62**	**29**	**91**	**59**	**24**	**83**	**68**	**24**	**92**
EUROPE														
Austria	9	7	7	6	4	4	0	4	2	0	2	1	0	1
Denmark	5	3	4	3	3	0	2	2	0	2	2	0	2	2
France	17	20	22	15	8	8	0	8	7	0	7	6	0	6
Germany	25	19	20	19	12	9	0	9	8	0	8	8	1	9
Greece	3	4	1	0	2	2	0	2	1	0	1	0	0	0
Holland	17	18	24	15	10	4	10	14	4	8	12	3	11	14
Italy	26	26	40	33	26	17	7	24	17	5	22	16	5	21

International Active Rig Count—continued

Region	1983	1984	1985	1986	1987	1988 Land	1988 Off.	1988 Total	1989 Land	1989 Off.	1989 Total	1990 Land	1990 Off.	1990 Total
Norway	10	10	13	12	12	0	15	15	0	12	12	0	12	12
Spain	8	9	10	7	2	1	1	2	1	2	3	1	0	1
United Kingdom	42	60	63	43	44	5	52	57	4	42	46	2	46	48
Yugoslavia	21	25	27	30	31	28	3	31	29	3	32	30	1	31
Other	2	2	2	1	1	0	1	1	0	0	0	0	0	0
Subtotal	**185**	**203**	**233**	**184**	**155**	**78**	**91**	**169**	**73**	**74**	**147**	**67**	**78**	**145**
MIDDLE EAST														
Abu Dhabi	35	29	21	15	9	2	5	7	2	4	6	4	5	9
Dubai	7	7	5	3	2	0	3	3	0	3	3	0	3	3
Egypt	35	36	37	33	23	11	10	21	6	12	18	6	9	15
Iran	13	20	20	18	18	18	0	18	20	0	20	19	0	19
Iraq†	23	19	28	21	10	23	0	23	26	0	26	23	0	23
Jordan	3	3	2	2	2	2	0	2	3	0	3	2	0	2
Kuwait†	5	5	6	6	6	6	0	6	4	0	4	2	0	2
Oman	11	12	13	13	10	9	0	9	13	0	13	17	0	17
Qatar	3	2	3	1	1	2	1	3	1	2	3	1	1	2
Saudi Arabia	26	16	11	6	5	3	1	4	4	1	5	9	1	10
Syria	15	23	26	27	22	24	0	24	24	0	24	17	0	17
Turkey	26	24	25	27	26	20	1	21	20	0	20	17	0	17
Yemen	0	0	1	4	4	3	0	3	2	0	2	3	0	3
Other	4	1	3	2	1	0	0	0	1	0	1	1	0	1
Subtotal	**206**	**197**	**201**	**178**	**139**	**123**	**21**	**144**	**126**	**22**	**148**	**121**	**19**	**140**
Total	**3,722**	**3,927**	**3,580**	**2,232**	**2,098**	**1,726**	**432**	**2,158**	**1,551**	**374**	**1,925**	**1,686**	**369**	**2,055**

*U.S. offshore includes inland water rigs: 30 in 1988, 19 in 1989, 21 in 1990.

†1990 data for Iraq and Kuwait is average rigs January through August.

Source: Baker Hughes Inc.

International seismic crew count

	1985 Land	1985 Marine	1985 Total	1986 Land	1986 Marine	1986 Total	1987 Land	1987 Marine	1987 Total
				(Average crews operating)					
United States	319	45	364	161	22	183	154	24	178
Canada	86	4	90	53	1	54	58	1	59
Mexico	32	0	32	29	0	29	23	0	23
Central & South America	64	8	72	48	5	53	46	5	51
Europe	78	35	113	65	26	91	52	25	77
Middle East	40	2	42	33	4	37	35	3	38
Africa	55	6	61	44	4	48	37	5	42
Far East	92	9	101	74	9	83	87	9	96
Total	**766**	**109**	**875**	**507**	**71**	**578**	**492**	**72**	**564**

	1988 Land	1988 Marine	1988 Total	1989 Land	1989 Marine	1989 Total	1990 Land	1990 Marine	1990 Total
				(Average crews operating)					
United States	147	29	176	109	22	131	104	22	126
Canada	69	1	70	57	1	58	61	1	62
Mexico	21	0	21	17	0	17	17	0	17
Central & South America	53	4	57	53	5	58	53	7	60
Europe	49	26	75	42	25	67	39	24	63
Middle East	42	2	44	41	2	43	38	2	40
Africa	36	6	42	46	8	54	49	8	57
Far East	75	11	86	58	14	72	75	15	90
Total	**492**	**79**	**571**	**423**	**77**	**500**	**436**	**79**	**515**

*First half.
Source: Society of Exploration Geophysicists

Industry details spending schedule for E&P

SOLID INCREASES IN THE PETROLEUM INDUSTRY'S world exploration and production spending were in place as 1990 began.

More than 200 companies planned a combined 10.5% hike in E&P spending compared with 1989 outlays, an annual survey by Salomon Bros. Inc., New York, showed.

The firm's report, by James D. Crandell and Carol Y. Lau, revealed that surveyed companies planned to spend 12.6% more in 1990 outside North America and 12% more in Canada. In the U.S., 130 independents planned to spend 11.7% more than in 1989 and 19 majors 6% more.

Fueling increased spending in the U.S. were expectations of higher oil prices and a closer match between gas demand and deliverability.

The surveyed companies expected oil prices to average $18.37/bbl in 1990, 21% higher than they expected for 1989.

More than 70% of the companies said the U.S. gas surplus no longer affected their E&P spending. The companies' budgets, as estimated by Salomon Bros., showed they expected to receive $1.77/Mcf for their gas, compared with $1.78/Mcf in 1989.

Of the companies that were shifting budget emphasis between exploration and production, a majority was shifting toward exploration in 1990. In 1989 more companies shifted toward development.

More companies spent more of their budget on leases than spent less in 1989, a good sign for future exploration increases. A smaller majority planned to spend more in 1990.

About 71% of the companies surveyed in December 1989 said drilling economics were superior to those of purchasing reserves, up from 58% of those surveyed a year earlier.

About 65% of the companies were trying to purchase reserves in 1990, down slightly from 1989.

The survey excluded money from public drilling funds and, when possible, excluded money earmarked for large purchases of reserves or other companies.

A move toward exploration

Responses from most companies indicated a move toward exploration in 1990 for the first time in 8 years.

However, 63% of the companies surveyed said they would cut total E&P spending if oil prices averaged $15/bbl in 1990, and 46% would increase spending if oil prices averaged $21/bbl in 1990.

More companies responded that U.S. gas demand and supply would be in closer balance in the 1989-90 winter than any other period.

Companies that believed the U.S. gas surplus is limiting spending included 32% of surveyed U.S. independents, 21% of U.S. majors, 23% of Canadian companies, and 17% of international companies.

Oil and gas prices appeared to be less important factors in spending plans in 1990 than in the past. U.S. independents cited cash flow, gas prices, and development programs as the most important determinants of spending. Only 18% of U.S. independents listed gas prices as a major influence on 1990 spending, down from 44% in 1989.

Prospect availability, called a big influence on spending by 44% of the majors in 1989, was not mentioned by majors for 1990. It was a big influence for 13% of independents, down from 14% in 1989.

For the fourth straight year in 1990, a slight majority of major oil companies planned to spend less of its combined capital budget on E&P.

A majority of all companies planned to spend less than cash flow in 1990. This outlook was strongest among U.S. independents.

Of the companies that said financial leverage or lack of liquidity limited spending in 1989 and 1990, 21% said those factors would be more important in 1990, 46% said they would be about the same as in 1989, and 33% said they would be less important in 1990.

Spending results of 1989

More companies overspent what they thought they would spend at midyear in 1989. That was particularly true of companies operating in Canada and overseas.

Oil's 1989 price increase and the lack of a gas price hike had not halted the shift in spending toward gas from oil. The shift toward gas was not quite as dramatic as indicated for 1989 at the beginning of that year, but more than four times as many companies shifted spending toward gas in 1989 than shifted to oil. An even greater percentage planned to do so in 1990.

A slight majority of companies' E&P capital spending exceeded cash flow in 1989.

In 1989 about 57% of companies that explore onshore and offshore spent an increased percentage of their budget offshore.

The figure was substantially larger than originally indicated and largely reflected independent companies' increased emphasis offshore. The figure in 1990 was 41%.

More than half of all companies surveyed viewed drilling costs as stable in 1989, but more than twice as many companies thought costs had risen as those that believed they had fallen.

More Canadian companies believed costs had fallen than believed they had risen.

Exploration economics

A significant turnaround occurred in companies' views of gas exploration in the U.S.

Companies were asked to rate U.S. exploration economics for oil and gas as excellent, good, fair, or poor in 1988 and 1989.

Most companies judged the economics of U.S. oil exploration as poor and U.S. gas exploration fair in 1988, but a majority said U.S. gas exploration economics were good in

Breakout of exploration/production spending*

	1990	1989		1990	1989		1990	1989
	(Million U.S. $)			(Million U.S. $)			(Million U.S. $)	
MAJORS, U.S.†			Mesa Petroleum	50	48	Poco Petroleum	44	39
Agip SpA	62	43	Mitchell Energy	77	73	Renaissance Energy	84	75
Amerada Hess	200	200	Murphy	42	45	Sceptre Resources	48	46
Amoco	800	940	Nerco	50	25	Shell Canada	480	394
ARCO	1,455	1,325	Nicor	33	30	Suncor	50	50
BP	1,000	850	Noble Affiliates	59	59	Total Petroleum NA	33	31
Chevron	1,100	1,100	Nomeco	48	39	Union Pacific	33	33
Du Pont	610	580	Noranda	42	34	Unocal	70	70
Elf Aquitaine	80	100	Norcen Energy	51	53	Westcoast	56	52
Exxon	1,400	1,400	Odeco	68	55			
Mobil	1,160	1,080	Oneok	14	12	**OTHER INTERNATIONAL†**		
Occidental	250	270	Pacific Enterprises	160	140	Agip SpA	1,613	1,252
Oryx	330	332	PanCanadian	13	13	Amerada Hess	369	335
Pennzoil	136	186	Petrofina Delaware	35	35	Amoco	1,000	785
Royal Dutch/Shell	1,758	1,573	Pogo Producing	36	33	Ashland Oil	5	20
Texaco	1,000	750	Prairie Producing	50	32	ARCO	540	490
Total Minatome	70	75	Presidio	35	31	Bow Valley	103	139
Union Pacific	305	241	Questar	55	31	Bridge Oil	18	20
Unocal	400	400	Santa Fe Energy	75	75	British Gas	468	375
USX	495	450	Seagull Energy	38	24	Canadian Occidental	48	22
			Seneca Resources	27	19	Chevron	650	650
SELECTED INDEPENDENTS, U.S.†			Sonat	96	78	Clyde Petroleum	72	62
Adobe	76	47	Southwestern Energy	17	27	Du Pont	780	740
Amax	26	27	Taylor Energy	15	20	Elf Aquitaine	1,850	1,450
American Petrofina	97	75	Transco Exploration	35	20	Enterprise Oil	468	351
Anadarko	205	188	Ultramar	45	23	Exxon	2,200	2,200
Apache	125	80	Union Texas	90	90	Goal Petroleum	27	30
Arkla	60	60	Wolverine Exploration	30	42	Gulf Canada	60	34
Ashland	30	30				Hamilton Oil	150	68
Beard	35	15	**CANADA†**			Hardy Oil & Gas	35	35
Belden & Blake	24	20	Alberta Energy	96	80	Kerr-McGee	81	81
British Gas	50	10	Amerada Hess	68	65	Lasmo	400	250
Brooklyn Union	40	27	Amoco	375	300	Louisiana Land	72	47
Burlington Resources	240	260	Anderson	60	28	Maxus Energy	115	91
Cabot Corp.	30	53	Bow Valley	71	63	Mobil	805	740
Canadian Occidental	39	28	Canada Northwest	10	23	Murphy	30	17
Chieftain	24	14	Canadian Occidental	25	34	Norcen Energy	35	43
Coastal	75	75	Chevron	250	250	Norsk Hydro	585	585
Coho	33	21	Columbia Gas	30	28	Occidental	400	310
Columbia Gas	100	93	Co-Enerco	16	14	Odeco	35	28
Consolidated Natural	180	200	Dekalb Energy	25	25	Oryx	120	0
Dekalb Energy	25	25	Du Pont	40	50	Petrofina	322	343
Diversified Energies	44	30	Encor Energy	58	55	Premier Consolidated	57	57
Edisto Resources	55	50	Enron	25	20	Ranger Oil	70	50
Enron	165	170	Gulf Canada	240	214	Royal Dutch/Shell	3,166	3,166
Enserch	125	86	Imperial	450	428	Saga Petroleum	234	198
Equitable Resources	60	55	Mark Resources	46	46	Statoil	797	746
First Energy	15	25	Mobil	175	160	Sun Co.	100	100
Forest Oil	77	90	Murphy	30	25	Texaco	700	400
Freeport-McMoRan	150	150	Noranda	164	145	Total	491	491
Grace Energy	28	20	Norcen Energy	114	72	Triton Energy	54	54
Kelley Oil et al.	30	21	North Canadian	57	62	Ultramar	170	124
Kerr-McGee	130	130	Numac	23	20	Unimar	55	45
Lasmo	25	18	PanCanadian	240	222	Union Texas	150	107
Louisiana Land	110	100	PanContinental	23	17	Unocal	280	280
Maxus Energy	102	102	Placer CEGO	32	28	USX	308	280

* Estimates for both years. †Companies spending at least $20 million in either year.

Source: Salomon Bros. Inc.

1989. About 25% of companies said U.S. oil exploration opportunities were good in 1989, compared with 15% in 1988. But a 51% majority still rated oil exploration economics as fair in 1989.

The 71% of all companies that viewed the economics of drilling as more favorable than purchasing reserves was one of the highest percentages in recent years.

The 65% of companies seeking to buy reserves was down slightly from 1989 but still higher than in most recent years. Most companies, particularly U.S. and Canadian companies, preferred to buy gas reserves.

The companies collectively viewed seismic technology advances, most notably three dimensional seismic surveys, as the most important technological innovation in the oil service industry during the previous 2-3 years. Horizontal drilling was the only other advance mentioned by more than 10% of respondents.

Almost all expressing concern cited a shortage of qualified personnel, particularly among roughnecks and floorhands and including qualified reservoir and petroleum engineers, seismic and geophysical professionals, and drilling technicians.

Stratigraphic charts

STRATIGRAPHIC CORRELATORS ARE REPRODUCED here from six widely separated areas of the world.

The charts are from the onshore Burma basin of Myanmar, Midway-Sunset field in California, Brazil's Solimoes basin, the Arabian Gulf countries, eastern Venezuela, and the Northwest Palawan shelf off the Philippines.

Exploration and development were picking up during the year in Myanmar, formerly Burma, after a hiatus in 1989 due to civil unrest. The country was reopened to foreign operators in late 1989 for the first time since 1962.

Tertiary Burma basin, Myanmar

Age			Rock Units		Thickness m	Reservoir	Source rock
	Quaternary		Irrawaddy gp		3,000		
Neogene	Pliocene						
	Miocene	Upper	Pegu gp	Upper	As thick as 1,000 (Obogoun fm)		
		Middle			As thick as 1,500 (Kyoukkok ss)	☼ ●	
		Lower			As thick as 1,000 (Pyawbwe clays, ss)	☼ ●	
	Oligocene	Upper Middle		Lower	0-1,500 (Okhmintoung ss)	☼ ●	≡
		Lower			As thick as 750 (Padoung clay)	☼ ●	
					600-1,200 (Shwezelaw fm)	☼ ●	≡
Paleogene	Eocene	Upper			300-600 (Yaw fm)		
					As thick as 2,000 (Pondoung ss)		
		Middle			1,500 (Tabyin clay)		≡
		Lower			As thick as 1,500 (Tilin ss)		
					2,700-3,600 (Laungshe sh)		
	Paleocene				600-1,200 (Paunggyi congl.)		
Cretaceous and older			Metamorphic rocks, schist, phyllite, quartzite				

STRATIGRAPHIC CHARTS

Midway-Sunset field, California

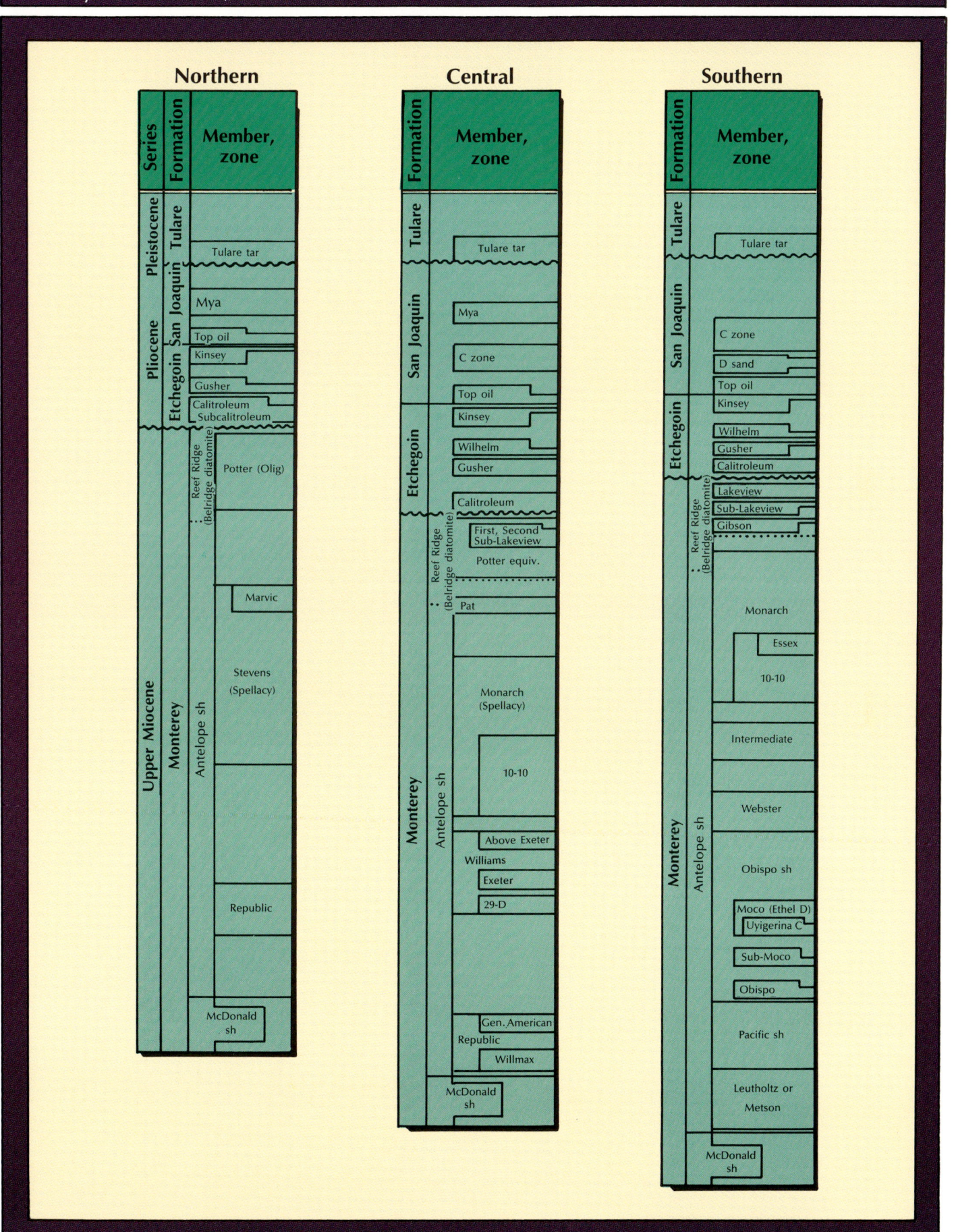

Northern

Series	Formation	Member, zone
Pleistocene	Tulare	Tulare tar
Pliocene	San Joaquin	Mya / Top oil / Kinsey / Gusher
	Etchegoin	Calitroleum / Subcalitroleum
Upper Miocene	Monterey — Reef Ridge (Belridge diatomite) / Antelope sh	Potter (Olig) / Marvic / Stevens (Spellacy) / Republic / McDonald sh

Central

Formation	Member, zone
Tulare	Tulare tar
San Joaquin	Mya / C zone / Top oil
Etchegoin	Kinsey / Wilhelm / Gusher / Calitroleum
Monterey — Reef Ridge (Belridge diatomite) / Antelope sh	First, Second Sub-Lakeview / Potter equiv. / Pat / Monarch (Spellacy) / 10-10 / Above Exeter / Williams / Exeter / 29-D / Gen. American / Republic / Willmax / McDonald sh

Southern

Formation	Member, zone
Tulare	Tulare tar
San Joaquin	C zone / D sand / Top oil
Etchegoin	Kinsey / Wilhelm / Gusher / Calitroleum
Monterey — Reef Ridge (Belridge diatomite) / Antelope sh	Lakeview / Sub-Lakeview / Gibson / Monarch / Essex / 10-10 / Intermediate / Webster / Obispo sh / Moco (Ethel D) / Uyigerina C / Sub-Moco / Obispo / Pacific sh / Leutholtz or Metson / McDonald sh

275

The Myanmar tertiary geosyncline covers about 140,000 sq miles, of which 111,000 sq miles could be considered to have hydrocarbon potential.

The main oil and gas producing formations are part of the Pegu group of Oligocene and Lower Miocene age. Crudes are often high in paraffin content with pour points of 80-100° F.

Exploration is characterized by strong structural elements, multiple pay sands, and shallow depths.

Many zones yield oil in Midway-Sunset field

Midway-Sunset field is in the southern San Joaquin Valley of California in Kern and San Luis Obispo counties.

The field has produced slightly less than 2 billion bbl of oil since discovery in 1894. The main producing formations are the Pleistocene Tulare, Pliocene San Joaquin and Etchegoin, and Upper Miocene Monterey.

Tulare lies at 200-1,400 ft and is 50-200 ft thick. It consists of fine to coarse grained sands and gravels with an average permeability of 3,000 md and average porosity of 36.9%. Oil averages 13° gravity.

San Joaquin consists of four producing zones and ranges from 150-750 ft thick. The individual zones are 20-150 ft thick. The most important, the Top Oil zone, averages 20-50 ft thick with average permeability ot 400 md and average porosity of 32%. San Joaquin oil averages 13° gravity, and Top Oil crude averages 15-23° gravity.

Etchegoin has five producing zones that contain lighter oils than the rest of the field. Thickness is 0-650 ft, with individual zone thicknesses of 0-75 ft. Oils are 14-26° gravity.

Monterey is divided into the Reef Ridge and Antelope shale members. Reef Ridge is 0-1,200 ft thick and has five producing zones each 0-500 ft thick.

Potter, the most important, is 200-2,500 ft thick in the field's northern and central areas. Potter permeability averages 3,000 md, porosity averages 35%, and oil averages

Solimoes basin, Brazil*

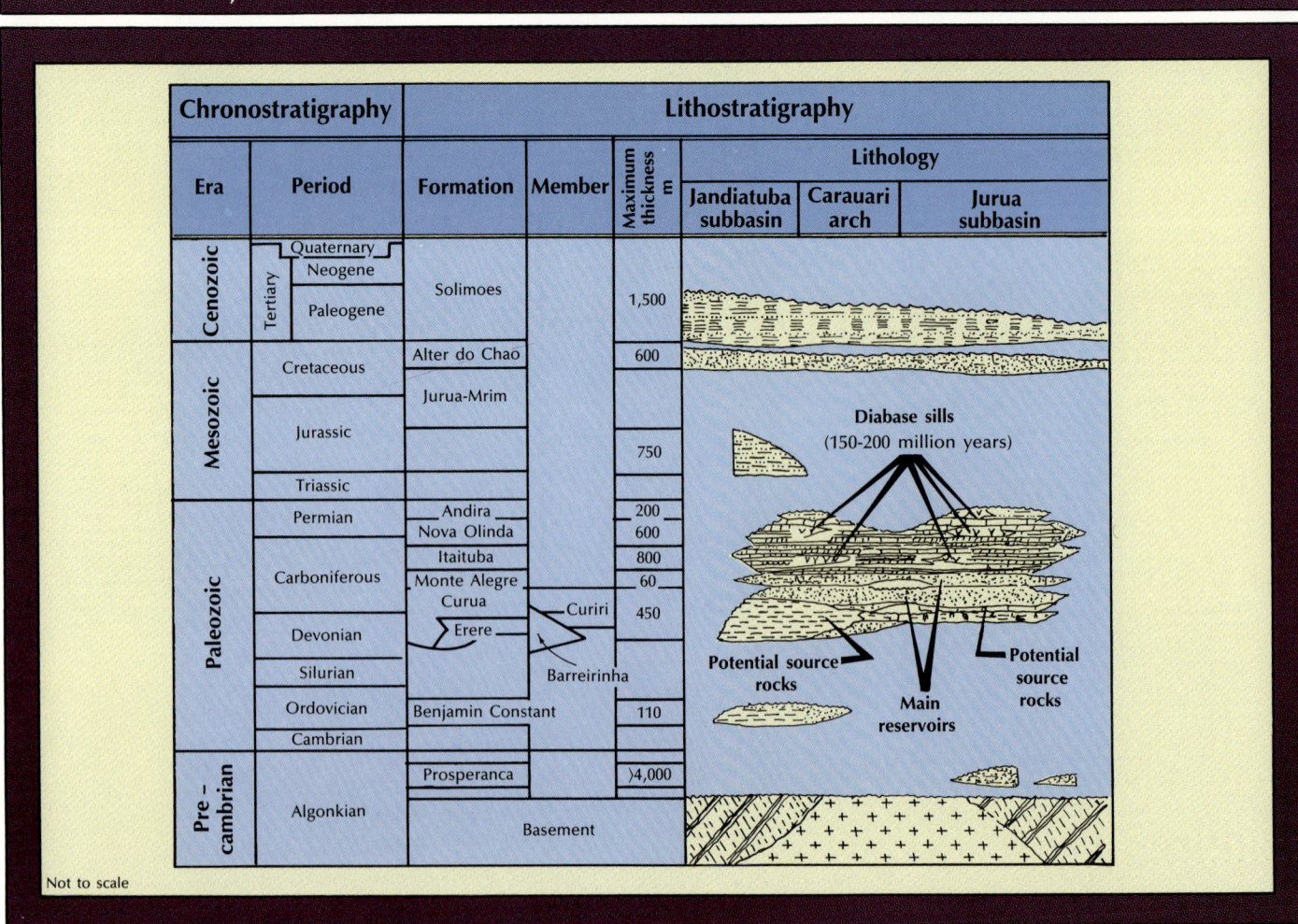

12° gravity.

The Reef Ridge, Lakeview, Sub-Lakeview, and Gibson permeabilities and porosities are similar to Potter's, but those zones contain 22° gravity oil.

Antelope shale has 12 zones and is as thick as 4,000 ft thick. Individual zones are 25-1,000 ft thick.

Monarch, the Potter equivalent, is the most important. Monarch is 50-400 ft thick with permeability of 3,000 md and porosity of 33%. It produces 13-17° gravity oils.

Brazil's Solimoes basin holds promise

Brazil's Upper Amazon/Solimoes basin is separated from the Middle Amazon basin on the east by the Purus arch and from the Acre basin on the east by the Iquitos arch.

In the Urucu area, features are distinguished as en echelon structures associated with thrust fault trends generated during the Jurassic-early Cretaceous compressional tectonic regime. Structures average 10 sq km in areal extent with closure of an average 120 m.

Pay in the Urucu area exclusively consists of sandstones in the Carboniferous Monte Alegre and Itaitube formations and Devonian Curua formation/Oriximina member.

Pay zones occur typically at about 2,400 m with pay thicknesses averaging 25 m. Porosity range is 15-22%, and permeability 60-2,100 md. Reservoir drive is a gas cap and water influx.

The Persian Gulf has produced since the discovery well of Masjid-i-Sulaiman field in Iran produced oil from Tertiary Asmari limestone in 1908.

Production originates mostly in Middle Cretaceous sandstones and Upper Jurassic limestone reservoirs. The Upper Jurassic Arab limestone is the region's most important carbonate formation. Pay zones are of Miocene, Oligocene, Eocene, Cretaceous, Jurassic, Triassic, and Permian age.

Petroleos de Venezuela is keeping up a brisk pace of exploration and development in the Eastern Venezuelan basin.

Persian Gulf stratigraphic correlations

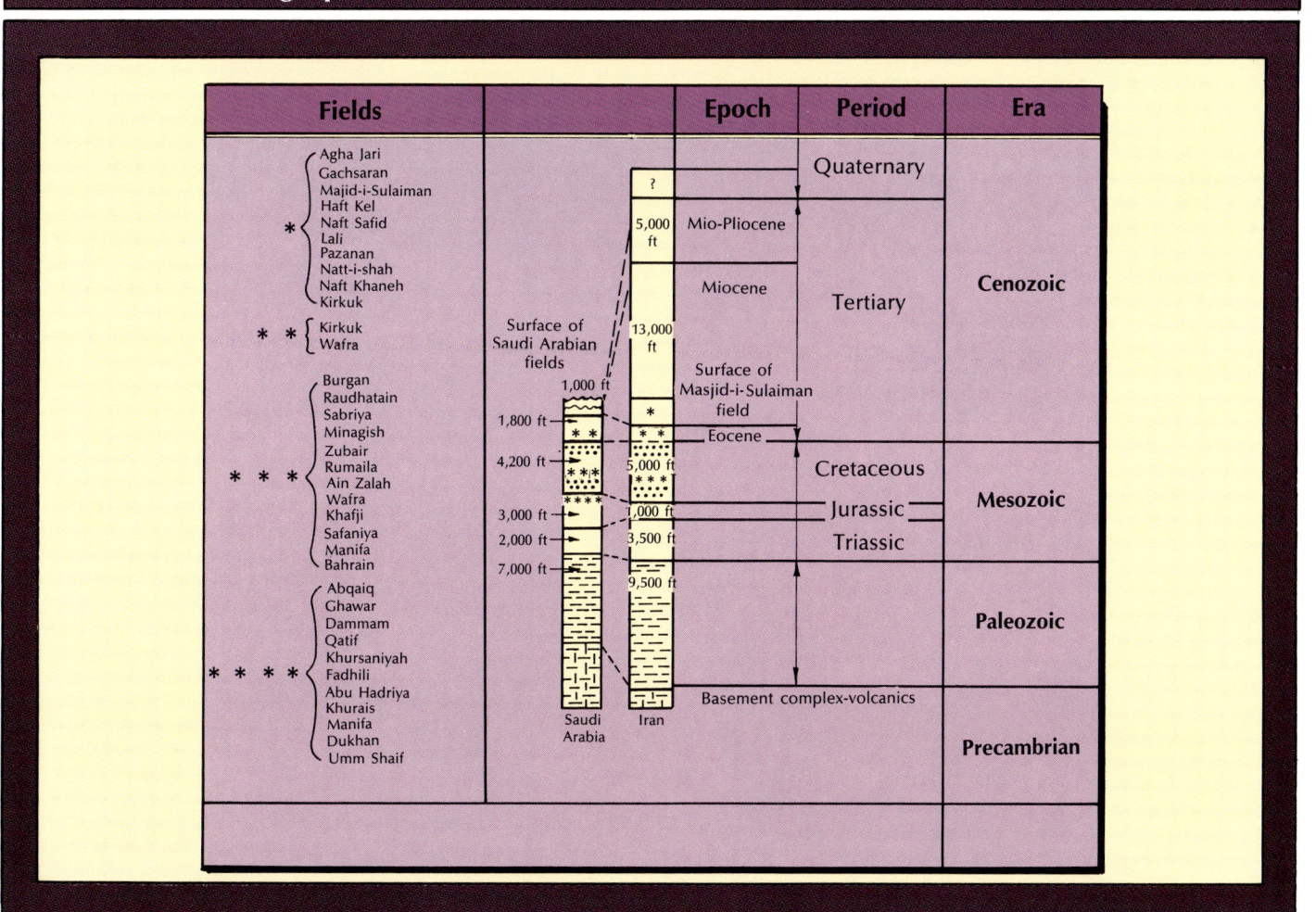

Fields		Epoch	Period	Era
* { Agha Jari, Gachsaran, Majid-i-Sulaiman, Haft Kel, Naft Safid, Lali, Pazanan, Natt-i-shah, Naft Khaneh, Kirkuk }		?	Quaternary	Cenozoic
		5,000 ft — Mio-Pliocene		
** { Kirkuk, Wafra }	Surface of Saudi Arabian fields — 1,000 ft	13,000 ft — Miocene	Tertiary	
		Surface of Masjid-i-Sulaiman field		
{ Burgan, Raudhatain, Sabriya, Minagish }	1,800 ft	* — Eocene		
*** { Zubair, Rumaila, Ain Zalah, Wafra }	4,200 ft	5,000 ft	Cretaceous	Mesozoic
{ Khafji }	3,000 ft	1,000 ft	Jurassic	
{ Safaniya, Manifa }	2,000 ft	3,500 ft	Triassic	
{ Bahrain }	7,000 ft	9,500 ft		Paleozoic
**** { Abqaiq, Ghawar, Dammam, Qatif, Khursaniyah, Fadhili, Abu Hadriya, Khurais, Manifa, Dukhan, Umm Shaif }	Saudi Arabia	Iran — Basement complex-volcanics		Precambrian

277

STRATIGRAPHIC CHARTS

Northwest Palawan shelf, Philippines

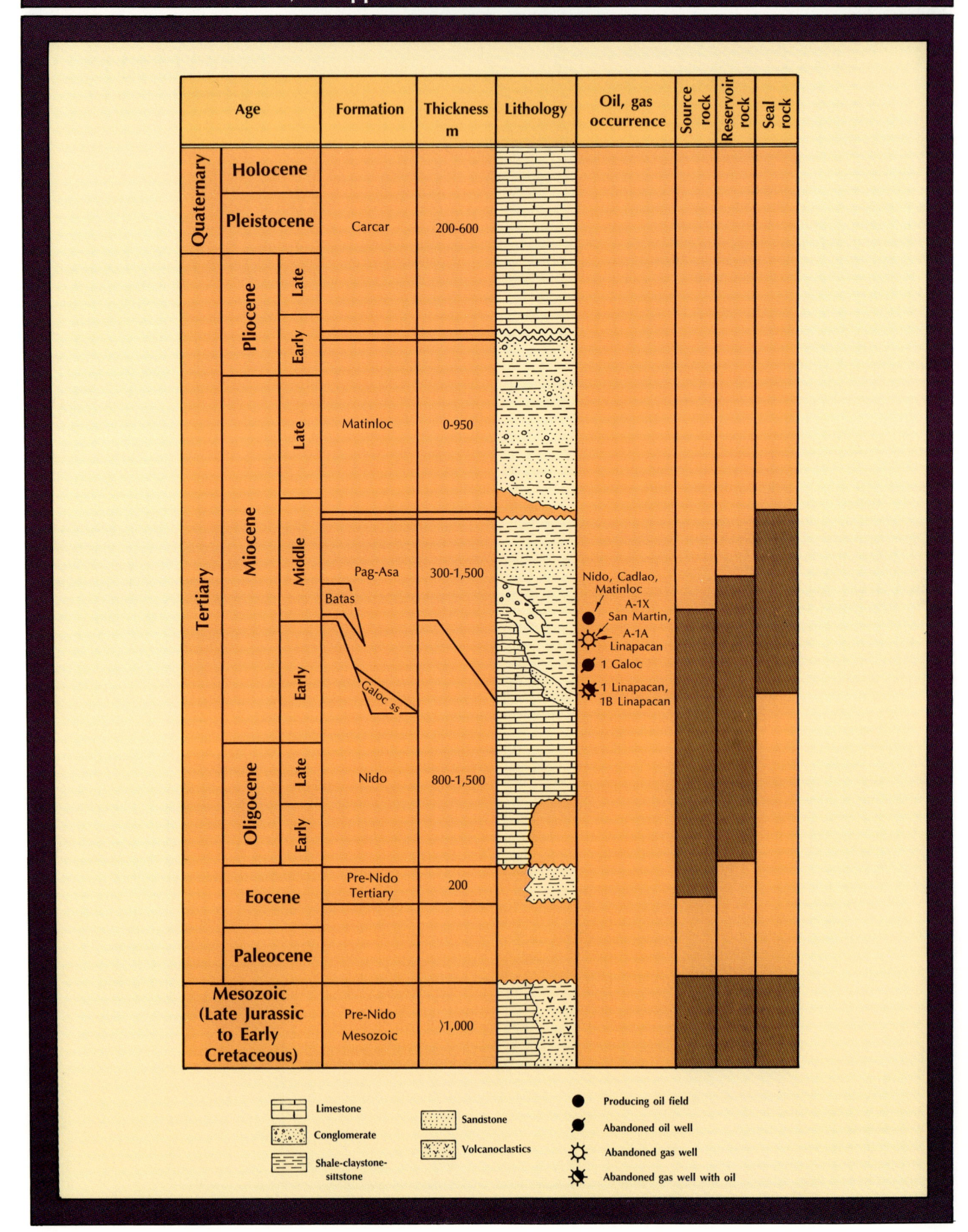

Age			Formation	Thickness m	Lithology	Oil, gas occurrence	Source rock	Reservoir rock	Seal rock
Quaternary	Holocene		Carcar	200-600					
	Pleistocene								
Tertiary	Pliocene	Late							
		Early	Matinloc	0-950					
	Miocene	Late							
		Middle	Pag-Asa	300-1,500		Nido, Cadlao, Matinloc A-1X ● San Martin, ☼ A-1A Linapacan ◐ 1 Galoc ✸ 1 Linapacan, 1B Linapacan			
		Early	Batas / Galoc ss						
	Oligocene	Late	Nido	800-1,500					
		Early							
	Eocene		Pre-Nido Tertiary	200					
	Paleocene								
Mesozoic (Late Jurassic to Early Cretaceous)			Pre-Nido Mesozoic	>1,000					

Legend:
- Limestone
- Conglomerate
- Shale-claystone-siltstone
- Sandstone
- Volcanoclastics
- ● Producing oil field
- ◐ Abandoned oil well
- ☼ Abandoned gas well
- ✸ Abandoned gas well with oil

STRATIGRAPHIC CHARTS

Eastern Venezuela

Age/period		Greater Mercedes	Greater Anaco	Greater Oficina	Temblador	Guarico mountain front	N.W. Anzoategui mountain front	Northern Anzoategui	Greater Jusepin
Pleistocene			Mesa	Mesa	Mesa	Terraces	Terraces	Terraces	Mesa
Pliocene				Algarrobo		Yucales	Caicaito		Quiriquire
			Las Piedras	Las Piedras	Las Piedras		Prespuntal		Las Piedras
							Las Piedras		
Miocene	Upper							San Mateo	La Pica
	Middle		Freites	Freites	Freites	Quiamare	Quiamare	Salomon	
	Lower						El Pilar C.	Revoltijo	
Oligocene	Upper	Chacuaramas	Blanco / Azul / Moreno / Naranja / Verde / Amarillo / Colorado	Oficina	Oficina	Chaguaramas	Capirical	Uchirito / Carapita	Carapita
	Middle							Capaya	Capaya
	Lower	Roblecito / La Pascua	Periquito			Quebradon	Naricual	Naricual	Naricual
						Batatal	Los Jabillos	Los Jabillos	Los Jabillos
Eocene	Upper					Tememure	Tinajitas	Tinajitas	Tinajitas
	Middle								
	Lower								
Paleocene						Guarico	Caratas	Caratas	Caratas
	Danian		Santa Anita			Vidoño	Vidoño	Vidoño	Vidoño
Cretaceous Upper	Maestrichtian					Arrayanes / Igneous		San Juan	San Juan
	Campanian								
	Santonian	Guavinita						San Antonio	
	Coniacian	Infante	Upper	Upper	Upper	Guayuta	Guayuta	Querecual	Guayuta
	Turonian	La Cruz							
	Cenomanian							Boqueron	
Middle	Albian	Temblador		Lower	Lower	Sucre	Sucre	Chimana / Borracha	Sucre
	Aptian	Mottled	Lower					Barranquin	
	Barremian								
	Neocomian					? ? ?			? ?
Precretaceous		Carrizal / Hato Viejo	?			Metamorphic series	?	?	?
		Basement	Basement	Basement	Basement				

Production is mainly from Oligocene Carapita and Basal Naricual, Eocene Los Jabillos, Paleocene Caratas, and Upper Cretaceous Vidono. Pdvsa has also tested the Upper Cretaceous San Juan and San Antonio formations.

One recent well, Lagoven SA's 1 Amarilis, in Monagas state 13 miles southeast of Maturin, flowed 1,500 b/d of light 40° gravity oil from Oligocene Carapita at 15,902 ft. Lagoven called the well, drilled in early 1990 to 19,501 ft, the deepest well drilled in Latin America.

Northwest Palawan shelf highly prospective

The Northwest Palawan shelf off the Philippines is in a class separate from the 12 other Philippine basins because its hydrocarbon potential is rated quite good.

It is regarded as a marginally explored basin with both oil and gas prospectivity. The shelf is part of the North Palawan-Mindoro basin, a 13,000 sq km area in the west central Philippines of which 95% is offshore.

The late Mesozoic and Tertiary marine clastic-carbonate section has a thickness of more than 10 km, of which the Tertiary fill probably attains as much as 4-5 km.

The basin is part of the Kalayaan-Calamian microplate that rifted from the South China continental margin and thus is considered a rift basin.

The first exploratory well, 1 San Teodoro, was drilled on Mindoro Island in 1959. Exploratory drilling moved offshore Northwest Palawan starting in 1971.

Philippine Cities Service Co. made the first discovery, 1 Nido, in 1976. Amoco International Oil Co. made another reef discovery, 1 Cadlao, in 1977.

Through the late 1980s, 51 wells have been drilled in the basin, almost all of them offshore. Fourteen are oil and five are gas discoveries.

Nine plays have been identified for the Northwest Palawan shelf and Mindoro-Cuyo platform regions. These include Oligocene/Miocene reef and platform carbonates, clastic and carbonate turbidite pinchouts, drapes and truncations, thrust folds, fault blocks, and anticlines. Stratigraphic and structural traps are present.

Worldwide oil and gas at a glance

COUNTRY	ESTIMATED PROVED RESERVES 1-1-91 Oil (1,000 bbl)	Gas (bcf)	OIL PRODUCTION Producing wells* 12-31-89	Estimated 1990 (1,000 b/d)	% change from 1989	No. of ref.	REFINING Capacity on 1-1-91 Crude	Thermal operations b/cd	Catalytic cracking	Catalytic reforming
ASIA-PACIFIC										
Australia	1,566,163	15,433	1,024	582.0	18.8	10	705,500	...	205,300	160,900
Bangladesh	500	12,700	10	0.9	...	1	31,200	...	...	1,650
Brunei	1,350,000	11,200	556	143.0	8.0	1	10,000	...	...	...
China	24,000,000	35,300	43,700	2,755.0	0.1	40	2,200,000	...	...	...
China, Taiwan	4,500	700	88	2.5	8.7	2	542,500	17,493	22,600	51,960
India	7,997,100	25,049	1,944	679.0	4.5	12	1,122,360	114,875	136,970	28,233
Indonesia	11,050,000	91,450	6,963	1,274.0	3.5	6	813,600	81,700	12,600	61,500
Japan	63,019	1,124	368	10.5	11.7	41	4,383,400	76,900	629,975	607,970
Korea, North	...	...	...	...	...	1	42,000	...	...	...
Korea, South	...	...	...	...	...	6	867,000	48,700	...	62,450
Malaysia	2,900,000	56,900	508	605.0	6.7	4	209,500	...	...	30,300
Myanmar (Burma)	51,000	9,400	450	13.0	8.3	2	32,000	5,200	...	...
New Zealand	209,000	4,106	51	39.0	1.8	1	95,100	...	...	26,600
Pakistan	162,087	19,449	98	60.0	13.6	3	120,975	...	...	4,800
Papua New Guinea	200,000	8,000	...	...	...	...	...	...	...	...
Philippines	38,688	1,800	12	5.0	−2.0	4	279,300	...	26,900	37,500
Singapore	...	...	...	...	...	5	878,000	190,700	...	63,600
Sri Lanka	...	...	...	...	...	1	50,000	12,500	...	3,750
Thailand	150,000	5,850	287	41.0	2.2	3	220,550	17,600	26,100	27,550
Viet Nam	500,000	100	NA	40.0	100.0	...	...	...	...	...
Total Asia-Pacific	**50,242,057**	**298,561**	**56,059**	**6,249.9**	**4.1**	**143**	**12,602,985**	**565,668**	**1,060,445**	**1,168,763**
WESTERN EUROPE										
Austria	81,443	403	1,070	24.0	9.6	1	204,000	23,426	24,000	32,000
Belgium	...	...	...	...	...	4	602,000	63,500	104,700	81,900
Cyprus	...	...	...	...	...	1	18,600	...	...	4,300
Denmark	799,435	4,488	95	118.0	5.6	3	185,500	73,670	...	31,100
Finland	...	...	...	...	...	2	241,000	46,300	41,800	42,900
France	184,766	1,324	632	62.0	−5.5	14	1,815,630	172,990	349,480	246,760
Germany	425,000	12,400	2,883	78.0	−3.7	20	2,065,400	386,962	240,250	376,500
Greece	30,000	47	13	15.0	−17.6	4	395,300	40,400	59,900	47,000
Ireland	...	1,700	...	...	...	1	56,000	...	...	11,000
Italy	693,503	11,624	198	97.0	16.9	21	2,385,358	390,450	306,300	315,050
Netherlands	157,200	60,900	188	71.0	4.4	7	1,196,700	160,060	128,000	182,600
Norway	7,609,412	60,674	318	1,574.0	6.8	3	288,000	68,000	41,000	47,000
Portugal	...	...	...	...	...	3	294,300	...	10,100	51,000
Spain	20,000	780	44	14.3	−20.6	10	1,321,000	179,500	157,300	191,000
Sweden	...	...	...	...	...	5	427,500	65,000	25,000	72,000
Switzerland	...	...	...	...	...	2	132,000	20,000	...	26,000
Turkey	650,000	1,150	653	70.0	25.0	5	728,644	4,177	40,575	66,557
United Kingdom	3,825,000	19,775	762	1,860.0	4.9	15	1,866,940	146,000	444,500	366,300
Total Western Europe	**14,475,759**	**175,265**	**6,856**	**3,983.3**	**5.6**	**121**	**14,223,872**	**1,840,435**	**1,972,905**	**2,190,967**
EASTERN EUROPE & U.S.S.R.										
Romania	1,170,000	4,700	NA	160.0	−11.1	13	617,000	NA	NA	NA
U.S.S.R.	57,000,000	1,600,000	145,000	11,500.0	−5.3	39	12,300,000	NA	NA	NA
Yugoslavia	240,000	2,900	1,671	65.0	−4.3	7	609,135	35,630	52,050	74,612
Other†	445,000	11,400	NA	81.0	−14.7	25	1,400,454	NA	NA	NA
Total Eastern Europe & U.S.S.R.	**58,855,000**	**1,619,000**	**146,671**	**11,806.0**	**−5.4**	**84**	**14,926,589**	**35,630**	**52,050**	**74,612**
MIDDLE EAST										
Abu Dhabi	92,200,000	182,800	1,001	1,587.0	12.9	2	192,500	...	...	30,100
Bahrain	97,460	6,250	286	42.0	7.7	1	243,000	20,000	39,000	18,000
Dubai	4,000,000	4,800	149	469.0	12.9	...	...	...	...	...
Iran	92,850,000	600,350	361	3,120.0	11.4	5	720,000	92,500	...	105,397
Iraq	100,000,000	95,000	820	2,083.0	−28.1	8	318,500	...	...	43,500
Israel	1,400	10	12	0.3	...	2	180,000	60,000	22,000	25,000
Jordan	20,000	400	4	0.4	...	1	100,000	...	4,410	8,640
Kuwait	94,525,000	48,600	363	1,080.0	−32.2	4	819,000	54,000	42,000	33,000
Lebanon	...	...	...	...	...	2	37,500	...	7,250	7,442
Neutral Zone	5,000,000	10,000	500	315.0	−20.7	...	...	...	...	...
Oman	4,300,000	7,200	1,240	658.0	2.8	1	76,932	...	...	15,386
Qatar	4,500,000	163,200	174	387.0	...	1	62,000	...	...	12,030
Ras al Khaimah	400,000	2,000	6	10.0	...	...	...	...	...	...
Saudi Arabia	257,504,000	180,355	858	6,215.0	24.4	8	1,862,500	71,900	87,541	183,922
Sharjah	1,500,000	10,800	20	35.0	...	...	...	...	...	...
Syria	1,700,000	5,500	963	385.0	13.2	2	237,394	37,285	...	31,352
Yemen	4,000,000	7,000	70	179.0	0.5	2	171,500	...	...	13,300
Total Middle East	**662,597,860**	**1,324,265**	**6,827**	**16,565.7**	**2.7**	**39**	**5,020,826**	**335,685**	**202,201**	**527,069**

COUNTRY	ESTIMATED PROVED RESERVES 1-1-91 Oil (1,000 bbl)	Gas (bcf)	OIL PRODUCTION Producing wells* 12-31-89	Estimated 1990 (1,000 b/d)	% change from 1989	No. of ref.	REFINING Capacity on 1-1-91 Crude	Thermal operations b/cd	Catalytic cracking	Catalytic reforming
AFRICA										
Algeria	9,200,000	114,700	840	797.0	11.2	4	464,700	...	...	55,600
Angola	2,074,000	1,800	439	480.0	6.1	1	37,630	...	...	1,714
Benin	100,000	...	6	4.0	...	...	...	...	...	...
Cameroon	400,000	3,880	175	164.0	−3.0	1	42,000	...	...	7,000
Congo	830,000	2,580	312	161.0	15.6	1	21,000	...	...	2,000
Egypt	4,500,000	12,400	896	873.0	2.6	8	523,153	16,470	...	33,540
Ethiopia	...	880	...	...	...	1	18,000	...	...	2,400
Gabon	730,000	490	310	269.0	27.5	1	24,000	7,200	...	1,400
Ghana	500	...	...	...	...	1	26,600	...	...	6,175
Ivory Coast	100,000	3,500	13	2.1	...	2	69,000	...	...	13,600
Kenya	...	...	...	...	...	1	90,000	...	...	9,000
Liberia	...	...	...	...	...	1	15,000	...	...	2,000
Libya	22,800,000	43,000	696	1,369.0	21.2	3	347,600	...	...	13,882
Madagascar	...	70	...	...	...	1	16,350	7,000	...	2,600
Morocco	2,000	53	5	0.3	...	2	154,600	...	5,600	26,800
Mozambique	...	2,290	...	...	...	...	...	...	...	...
Nigeria	17,100,000	87,400	1,432	1,808.0	8.0	4	433,250	...	82,700	70,070
Senegal	...	...	...	...	...	1	24,000	...	...	2,600
Sierra Leone	...	...	...	...	...	1	10,000	...	...	...
Somalia	...	210	...	...	...	1	10,000	...	...	...
South Africa	...	1,760	...	...	...	4	430,500	72,000	84,400	62,600
Sudan	300,000	3,000	...	...	...	1	21,700	...	...	1,900
Tanzania	...	4,100	...	...	...	1	17,000	...	...	6,700
Tunisia	1,700,000	3,000	147	93.0	−9.7	1	34,000	...	...	3,300
Zaire	55,640	30	110	28.0	−4.8	1	17,000	...	...	3,500
Zambia	...	...	...	...	...	1	24,500	...	...	5,600
Total Africa	**59,892,140**	**285,143**	**5,381**	**6,048.4**	**10.3**	**44**	**2,871,583**	**102,670**	**172,700**	**333,981**
WESTERN HEMISPHERE										
Argentina	2,280,000	27,000	10,161	473.0	3.3	11	696,285	164,100	171,300	41,300
Barbados	3,083	6	88	1.2	9.1	1	3,000	...	...	...
Bolivia	119,182	4,149	296	19.2	−3.5	3	45,250	...	...	11,300
Brazil	2,840,000	4,045	6,267	633.0	7.0	13	1,411,520	39,765	316,585	23,610
Canada	5,782,949	97,589	38,794	1,508.0	−3.2	28	1,882,060	88,978	387,870	367,060
Chile	300,000	4,100	366	20.3	−1.5	3	145,800	19,000	37,870	10,300
Colombia	2,000,000	4,500	3,771	445.0	10.1	4	247,400	60,000	91,000	6,000
Costa Rica	...	...	...	...	...	1	15,000	6,500	...	1,200
Cuba	100,000	100	NA	15.0	−6.3	4	280,000	...	...	...
Dominican Republic	...	...	...	...	...	2	48,000	...	...	9,100
Ecuador	1,420,000	3,950	947	287.0	2.9	5	141,800	25,200	16,000	2,780
El Salvador	...	...	...	...	...	1	16,300	...	...	3,000
Guatemala	36,205	10	14	4.0	2.6	1	16,000	...	...	3,000
Honduras	...	...	...	...	...	1	14,000	...	...	1,800
Jamaica	...	...	...	...	...	1	34,200	...	...	3,240
Martinique	...	...	...	...	...	1	12,000	...	...	3,000
Mexico	51,983,000	72,744	4,740	2,633.0	0.8	9	1,679,000	82,000	267,000	157,800
Netherlands Antilles	...	...	...	...	...	1	320,000	67,000	42,000	15,000
Nicaragua	...	...	...	...	...	1	16,000	...	...	3,000
Panama	...	...	...	...	...	1	100,000	...	...	7,500
Paraguay	...	...	...	...	...	1	7,500	...	...	...
Peru	405,937	7,082	3,536	132.0	...	6	188,820	...	24,600	1,760
Puerto Rico	...	...	...	...	...	2	125,000	...	12,000	25,500
Suriname	27,300	...	124	4.0	5.3	...	...	...	...	...
Trinidad & Tobago	536,000	8,910	3,199	151.0	1.1	2	246,000	8,000	26,000	14,400
Uruguay	...	...	...	...	...	1	33,000	...	4,100	2,500
Venezuela	59,040,000	105,688	12,752	2,118.0	22.4	6	1,167,000	134,100	139,000	6,000
Virgin Islands	...	...	...	...	...	1	545,000	80,000	...	125,000
United States	26,177,000	166,208	603,365	7,220.0	−5.2	190	§15,558,923	†1,972,400	§5,404,100	§3,930,470
Total Western Hemisphere	**153,050,656**	**506,081**	**688,420**	**15,663.7**	**0.5**	**301**	**24,994,858**	**2,747,043**	**6,939,425**	**4,775,620**
TOTAL WORLD	**999,113,472**	**4,208,315**	**910,214**	**60,317.0**	**1.4**	**732**	**74,640,713**	**5,627,131**	**10,399,726**	**9,071,012**

EDITOR'S NOTE: All reserves figures except those for the U.S.S.R. (and gas for Canada) are reported as proved reserves recoverable with present technology and prices. U.S.S.R. figures are "explored reserves" which include proved, probable, and some possible. Canadian gas figure, under criteria adopted by Canadian Petroleum Association in 1980, includes proved and some probable. *Does not include shut-in, injection, or service wells. †Includes Albania, Bulgaria, Czechoslovakia, Hungary, and Poland. §Estimates based on capacity as of Jan, 1. 1990.

281

Worldwide Production

- Offshore   (e) Estimated   (c) Condensate   (NA) Not available

WORLDWIDE PRODUCTION

		Name of field, discovery date	Depth, ft	No. of wells Producing	No. of wells Total	1989 average B/d	Production Cumulative to Dec. 31, 1989 Bbl	°API gravity
ABU DHABI(e)	ADCO, OTHERS	Asab, 1965	8,000- 8,700	106	248			40.8
		Bab, 1958	8,600- 9,300	100	232			40.6
		Bu Hasa, 1962	8,300- 9,200	201	371			39.0
		Sahil, 1967	9,500- 9,800	22	52			39.7
		• Umm Shaif, 1958	9,800-12,000	142	203			37.0
		• Zakum Lower, 1963	8,000-9,000	131	254			39.0
		• Zakum Upper, 1963	7,000-8,000	142	262			35.0
		• Umm Addalkh, 1969	7,700-8,500	32	54	1,405,600	9,391,306,800	32.5
		• Satah, 1975	8,550-9,500	10	21			39.8
		• Arzanah, 1973	11,000-11,300	14	34			44.6
		• Abu Al Bukhoosh, 1969	8,900-10,900	42	48			32.0
		• Mubarraz, 1969	8,300-10,000	28	28			37.0
		• Bunduq, 1964	9,400-10,900	23	40			40.0
		Total Abu Dhabi		**1,001**	**1,863**	**1,405,600**	**9,391,306,800**	
ALGERIA(e)	SONATRACH	Zarzaitine, 1957	4,700					42.2
		Edjeleh, 1956	1,500- 2,800					37.2
		Tiguentourine, 1966	2,160- 4,265					44.4
		El Adeb Lareche, 1958	4,100					43.3
		Assakaifaf N. & S., 1958	2,800					42.7
		Tan Emellel N. & S., 1960	2,800- 4,500					42.5
		Edeyen, 1964	5,580					41.5
		Hassi Mazoula S., 1963	4,430					40.2
		Hassi Mazoula B., 1965	4,920					42.0
		Acheb W. & Kreb, 1963 & 1960	7,360					45.6
		Tin Fouye N. +F6, 1962	4,265					39.6
		El Borma, 1967	8,200					42.9
		Keskassa, 1968	8,860					42.9
		Mereksen, 1974	9,420					45.8
		Gara, 1962	9,420					42.9
		Hassi Messaoud N., 1956	10,500					45.9
		Hassi Messaoud S., 1956	10,500					46.1
		Gassi-Touil & Gassi-Touil E., 1961	6,890					50.6-50.8
		Hassi Chergui	8,200					50.6
		Nezla N., 1965	8,530					51.5
		Rhourde El Baguel, 1962	9,850					40.2
		Mesdar E. & W., 1972	11,155					40.3
		El Gassi-El Agreb, 1959	10,170	840	840	714,000	8,032,996,000	47.9
		Rhourde Nouss & E., 1972	8,530					39.7
		Tin Foye Tabankort, 1966	4,600- 6,560					41.1-41.3
		Djoua W., 1967	5,580- 6,240					41.1
		Amassak, 1970	6,565					41.1
		Tamendjelt, 1970	5,585- 6,890					41.1
		Timedratine, 1964	8,155					42.7
		Ohanet N. & S., 1960	7,878					43.8
		Askarene, 1962	8,200					41.7
		Guelta, 1962	8,960					41.7
		Stah, 1971	9,515					45.8
		Ras Toumb, 1976	9,190					35.7
		Haoud Berkaoui-Ben Kahla, 1965	11,000					43.6
		Guellala-Guellala N.E., 1969	11,320					42.2-42.3
		N'Goussa, 1974	11,715					42.2
		Kef El Argoub, 1973	12,570					42.2
		Oued Noumer, 1969	8,860					48.3
		Djorf, 1974	9,025					48.3
		Djebel Onk, 1960	3,940					37.0
		Alrar FJT	...					...
		Hassi R'Mel(c)	...					...
		Gassi Touil(c)	...					...
		Total Algeria		**840**	**840**	**714,000**	**8,032,996,000**	
ANGOLA	FINA ANGOLA	Benfica, 1955	8,300	2	15	22	2,272,000	30.0
		Bento, 1972	7,350	1	5	225	5,075,000	36.0
		C. Cobra, 1969	4,000	11	15	1,056	3,795,000	41.0
		Cacuaco, 1958	7,870	1	12	38	279,000	30.0
		Galinda, 1959	7,220	...	8	...	2,245,000	30.0
		Ganda, 1975	5,000	8	9	4,096	8,497,000	36.0
		Legua, 1972	6,300	1	3	96	732,000	31.0
		Luango, 1977	3,800	3	6	955	2,555,000	38.0

PRODUCTION

WORLDWIDE PRODUCTION

	Name of field, discovery date	Depth, ft	No. of wells Producing	Total	1989 average B/d	Production Cumulative to Dec. 31, 1989 Bbl	°API gravity
	Lumueno, 1977	4,600	9	13	2,341	5,610,000	37.0
	Mulenvos, 1966	5,900- 6,600	7	16	267	5,750,000	25.0
	N'Zombo, 1973	5,000- 6,300	24	41	9,882	95,213,000	33.0
	Pambo, 1982	4,600	4	5	1,096	2,434,000	39.0
	Pinda, 1966	6,709	...	4	...	3,000	26.0
	Quenguela, 1968	6,000	26	68	1,164	38,058,000	31.0
	Quinfuquena, 1975	6,800- 7,100	14	19	4,223	27,428,000	32.0
	Quinguila, 1972	3,400	18	58	7,375	38,732,000	36.0
	Sereia, 1974	6,200	2	4	812	1,735,000	27.0
	Tobias, 1961	1,970	1	6	50	28,846,000	31.0
CHEVRON	● Kali, 1969	11,400-11,660	1	1	120	4,334,230	30.0
	● Kambala, 1971	10,500	4	5	5,430	21,778,134	32.8
	● Kungulo, 1975	4,000	16	17	9,846	50,196,610	32.2
	● Lifuma, 1984	7,800	...	1	15	205,368	31.4
	● Limba, 1969	2,660- 7,350	10	14	10,542	92,672,296	31.6
	● Livuite, 1979	8,000	3	7	406	2,796,577	31.7
	● Malongo North, 1966	1,170- 9,250	53	61	14,533	220,826,849	30.0
	● Malongo South, 1966	1,150- 1,440	26	29	7,514	78,253,055	28.0
	● Malongo West, 1969	1,680- 8,800	25	35	19,895	244,753,554	22.5
	● Numbi, 1980	3,446	24	26	46,044	36,341,483	34.0
	● Takula, 1971	3,070- 9,090	66	72	103,661	235,651,419	31.5
	● Vuko, 1983	3,700	3	6	7,049	12,574,990	35.0
	● Wamba, 1982	2,700	7	7	29,023	30,144,650	29.0
TEXACO-SONANGOL- TOTAL-BRASPETRO	● Cuntala, 1978	7,200	...	1	...	4,420,000	31.0
	● Essungo, 1975	6,500	8	10	3,575	39,795,000	35.0
	● Lombo Este, 1983	7,800	9	10	19,740	13,130,000	41.0
	● Sulele Oeste, 1985	7,600	6	6	2,986	2,630,000	41.0
	● Sulele South 1, 1986	7,600	1	3	1,712	1,085,000	39.0
	● Tubarão, 1984	7,600	6	7	11,164	7,800,000	41.0
ELF ANGOLA[e]	● Pacassa, 1982	11,500	39	39	125,677	145,391,043	38.0
	● Palanca, 1981	8,860					40.0
	Total Angola		**439**	**664**	**452,630**	**1,513,439,258**	

ARGENTINA

YPF[e]	Chubut, 1907	1,908- 9,168	2,261	2,325			21.1-26.0
	Formosa, 1984	11,466-11,490	12	14			42.0-42.9
	Jujuy, 1969	2,461-16,732	9	9			46.5-47.0
	La Pampa, 1968	3,609- 4,921	198	209			31.0
	Mendoza, 1932	2,178-16,778	1,156	1,187			23.7-32.0
	Neuquen, 1918	1,312-12,498	1,160	1,205	416,266	4,705,331,768	24.0-41.0
	Rio Negro, 1960	2,953- 7,546	897	937			22.5-41.0
	Salta, 1928	6,551-15,092	33	43			29.0-59.0
	Santa Cruz, 1946	1,468-17,060	3,129	3,274			21.1-30.0
	Tierra del Fuego, 1959	5,249- 7,824	206	206			43.4-46.0
AMOCO	Cerro Dragon, 1958	7,200	1,100	1,600	41,734	439,469,000	21.0
	Total Argentina		**10,161**	**11,009**	**458,000**	**5,144,800,768**	

AUSTRALIA

BARRACK ENERGY	Mount Horner, 1965	3,800- 5,800	7	12	360	219,165	38.0
BHP PETROLEUM PTY. LTD	● Jabiru, 1983	4,786	4	4	40,800	36,144,789	42.3
	● Challis/Cassini, 1984	5,000	7	7	24,348	535,666	39.5
HADSON ENERGY LTD.	● Harriet, 1983	6,230	10	10	11,200	14,253,000	38.2
AGL PETROLEUM	Alton, 1964	6,000	4	8	29	1,909,000	49.5
	Anabranch, 1965	4,188	...	...	...	13,718	44.0
	Bennett, 1965	5,960	1	1	4	118,592	45.0
	Cabawin, 1961	10,800	1	1	3	62,337	49.0
	Maffra, 1965	4,251	1	1	14	38,162	48.0
	Mereenie, 1964	4,000	24	37	1,737	4,605,023	48.9
	Merrit, #1, 1985	4,521	1	1	47	50,154	51.0
	Moonie, 1961	5,800	20	38	560	22,103,380	44.5
	Richmond, 1963	3,643	...	...	...	16,345	43.0
	Snake Creek, 1964	4,990	...	...	...	13,158	67.0
	Sunny Bank, 1962	6,139	...	...	...	706	40.0
	Trinidad 1 & 3, 1965	4,599	...	...	...	63,560	46.0
AMPOLEX (PPL) PTY. LTD.	Cranstoun, 1987	4,250	1	1	179	169,556	48.0
	Endeavour, 1989	5,085	1	1	202	9,494	49.7
	Ipundu, 1986	2,700	1	1	53	86,636	43.0
	Ipundu, North, 1989	4,354	1	1	25	1,055	49.7
	Monler, 1987	4,150	2	2	24	36,330	49.0
	Talgeberry, 1985	3,900	2	3	113	334,701	46.0
	Takyah, 1985	3,800	1	1	15	20,874	43.0
	Tarbat, 1988	2,700	1	1	7	13,648	43.0
	Tintaburra, 1984	3,500	5	6	271	643,194	43.0
	Toobunyah, 1985	3,500	4	5	241	403,172	43.0
EROMANGA ENERGY	Blina, 1981	4,300	5	5	236	1,263,829	37.0
	Lloyd, 1987	4,806	1	1	87	113,836	37.0
	Sundown, 1983	3,600	2	2	25	189,868	39.0
	West Terrace, 1985	3,800	2	2	35	135,648	33.0

PRODUCTION

WORLDWIDE PRODUCTION

	Name of field, discovery date	Depth, ft	No. of wells Producing	No. of wells Total	1989 average B/d	Cumulative to Dec. 31, 1989 Bbl	°API gravity
SANTOS LTD.	Alwyn, 1985	5,200	2	2	88	137,000	41.0
	Big Lake, 1984	6,600	2	3	133	419,000	45.0-48.0
	Bogala, 1984	6,900	2	2	210	381,000	40.0
	Bookabourdie, 1985	7,000	2	2	39	94,000	46.0
	Brolga, 1983	9,547	1	1	...	48,100	53.0
	Challum, 1983	...	...	1	...	21,000	...
	Chookoo, 1985	6,000	4	8	771	649,000	45.0
	Cook, 1985	...	...	1	...	29,000	...
	Corella, 1989	...	1	1	18	3,000	...
	Cooroo/Cooroo North, 1986	...	4	5	1,039	1,239,000	...
	Dingera, 1987	...	...	1	...	1,000	...
	Dirkala, 1986	7,000	1	1	108	572,000	56.0
	Dullingari, 1979	5,200	15	31	1,909	8,448,000	54.0
	Fly Lake, 1971	9,300	1	3	...	502,500	53.0
	Gidgealpa, 1984	6,150	13	13	2,276	3,770,000	45.0-50.0
	Gidgee, 1989	...	1	1	16	6,000	...
	Gunna, 1983	6,000	1	1	18	83,000	40.0
	James, 1989	...	1	1	12	4,000	...
	Jackson, 1981	4,750	35	36	9,030	27,774,000	40.0-48.0
	Jackson South, 1982	4,800	5	8	575	2,331,000	40.0
	Jena, 1985	5,200	4	4	220	162,000	41.0
	Kercummura, 1985	...	...	1	...	1,000	50.8
	Kerinna, 1984	4,300	...	1	3	18,000	41.0
	Limestone Creek/Biala, 1984	5,100	10	11	931	1,695,000	41.0
	Mawson, 1987	...	1	1	55	106,000	...
	McKinlay, 1985	5,100	1	2	6	33,000	41.0
	Meranji, 1985	5,660	3	3	142	433,000	1.0
	Merrimelia, 1983	5,250-7,070	10	11	1,958	4,073,000	5.5
	Mooliampah, 1985	6,000	2	2	110	117,000	...
	Moorari, 1971	7,705-9,400	6	6	...	1,598,000	6.9
	Munro, 1988	...	2	2	154	96,000	...
	Muteroo, 1985	6,000	5	5	1,479	2,507,000	5.0
	Naccowlah, 1989	...	1	1	25	9,000	...
	Naccowlah East, 1988	...	1	1	...	...	...
	Naccowlah South, 1983	6,000	9	10	831	2,608,000	0.0-46.0
	Naccowlah West, 1983	5,700	9	9	1,319	2,580,000	...
	Narcoonowie, 1983	6,000	2	2	106	193,000	2.0-53.0
	Natan	...	1	1	36	16,000	...
	Nungaroo, 1985	4,150	1	1	21	58,000	...
	Pinaroo, 1889	...	1	1	53	19,000	...
	Pintari North, 1988	...	1	1	35	33,000	...
	Pitchery, 1988	...	1	1	79	77,000	...
	Sigma, 1983	5,700	2	2	140	164,000	...
	Spencer, 1986	6,500	6	6	1,393	1,588,000	7.0
	Strzelecki, 1978	5,800	15	17	2,902	13,719,000	3.0-49.0
	Sturt, 1988	...	4	4	515	217,000	...
	Sturt East	...	4	4	1,862	687,000	...
	Taloola, 1988	...	2	2	535	317,000	...
	Tantanna, 1988	...	4	4	3,719	1,672,000	...
	Tickalara, 1984	5,600	5	5	1,123	830,000	...
	Tinpilla, 1983	6,000	1	1	22	87,000	...
	Tirrawarra, 1970	9,500	29	40	...	11,564,300	2.0
	Toby, 1987	...	...	1	...	25,000	...
	Ulandi, 1985	5,200	3	3	154	153,000	0.4
	Wandilo, 1989	...	2	2	106	39,000	...
	Wancoocha, 1984	6,000	3	6	344	1,400,000	4.0-54.0
	Watson, 1985	...	2	2	231	251,000	...
	Watson South, 1985	5,500	2	2	643	1,006,000	40.0
	Wilson, 1983	5,000	5	6	289	508,000	...
	Woolkina, 1982	9,700	1	2	...	101,900	1.5
	Yanda, 1984	8,000	1	1	18	63,000	...
WAPET GROUP	Barrow Island, 1964	1,200-6,700	412	539	15,232	236,544,108	37.7
	Dongara, 1969	5,250-5,577	1	9	31	977,414	35.5
	Thevenard Island, 1985	4,600-5,750	7	7	25,848	1,163,165	48.2
WOODSIDE	● North Rankin,[c] 1972	9,800	13	18	23,200	25,950,000	52.3
COMMAND	Dilkera, 1989	3,447	1	2	15	1,781	46.0
	Fairymount, 1985	6,730	5	6	485	446,044	50.5
	Kihee, 1986	3,200	...	2	...	813	44.7
	Koora, 1985	4,822	...	1	...	952	41.0
	Maxwell, 1987	2,930	3	3	204	187,826	47.2
	Mindagabie, 1987	5,290	...	1	...	658	10.3
	Muthero, 1989	3,448	2	2	81	8,184	44.4
	Nockatunga, 1983	3,310	3	5	120	338,187	48.0
	Thungo, 1986	3,345	4	4	218	290,935	46.8
	Winna, 1985	3,300	2	2	60	126,034	15.8
ESSO-BHP	● Barracouta, 1965	4,550	7	10	2,260	35,083,000	62.8
	● Bream, 1969	6,200	15	16	19,770	9,240,000	44.0
	● Cobia, 1972	7,700	9	13	26,420	86,058,000	44.6
	● Flounder, 1968	8,200	18	22	20,040	26,764,000	46.7
	● Fortescue, 1978	7,700	17	29	56,050	191,325,000	43.0
	● Halibut, 1967	7,700	11	19	38,930	722,308,000	43.3

PRODUCTION

WORLDWIDE PRODUCTION

		Name of field, discovery date	Depth, ft	No. of wells Producing	Total	Production 1989 average B/d	Cumulative to Dec. 31, 1989 Bbl	°API gravity
		• Kingfish, 1967	7,500	20	35	39,720	966,306,000	46.9
		• Mackerel, 1969	7,700	14	18	42,690	389,364,000	45.6
		• Marlin, 1966	5,100	19	24	10,940	62,201,000	50.0
		• Snapper, 1968	4,350	11	27	6,480	16,272,000	47.0
		• Tuna, 1968	6,500	12	18	7,060	37,754,000	40.5
		• West Kingfish, 1968	7,500	14	19	35,540	83,883,000	46.9
		• Whiting, 1983	4,800	4	5	1,630	596,000	53.0
	MARATHON AUST	• Talisman, 1984	9,600	1	1	7,334	3,473,446	41.0
	OIL COMPANY OF AUSTRALIA	Riverslea, 1981	4,950	2	2	69	358,031	37.0
		Bora Creek, 1982	4,900	1	1	6	17,318	34.0
		Kincora, 1980	4,700	1	1	...	304,131	34.0
		Sandy Creek, 1982	4,900	...	...	...	8,730	34.0
		Waratah, 1982	5,350	3	4	42	299,315	34.0
		Washpool, 1985	5,250	1	1	2	2,145	35.0
	LASMO[e]	Black Stump, 1986	5,360	1	1			
		Bodalla South, 1984	4,800- 5,230	6	6			
		Glenvale, 1985	3,800	...	1			
		Kenmore, 1985	4,600- 4,980	10	11			
	BRIDGE OIL LTD.[e]	Boggo Creek, 1980	5,320	...	1			46.0
		Louise, 1986	5,910	1	2			58.0
		Narrows, 1986	5,739	1	1	4,000	9,850,000	60.0
		Taylor, 1988	5,538	5	5			47.0
		Thomby Creek, 1979	5,170	...	4			50.0
		Waggamra, 1978	7,670	1	1			55.0
		Yellow Bank Creek, 1982	5,180	2	2			52.0
	STRATA OIL[e]	Woodada,[c] 1980	7,400	...	...			54.0
	WESTERN MINING CORP.[e]	• North Herald, 1982	3,937	1	1			44.0
		• South Pepper, 1982	4,036	1				44.0
		Total Australia		**1,024**	**1,363**	**502,883**	**3,088,069,926**	
AUSTRIA	MOBIL/SHELL	Gaiselberg, 1938	5,200	32	65	709	32,651,157	27.0
		Kematinig, 1979	6,100	8	9	705	4,248,509	34.0
		Sattledt, 1971	5,700	11	20	607	5,449,472	37.0
		Other	...	62	158	1,518	46,927,432	33.0
	OEMV[e]	Hochleiten, 1977	3,100- 5,000					21.3
		Matzen, 1949	6,100					25.7
		Pirawarth, 1957	1,600- 6,500	957	1,255	18,399	590,199,421	28.4
		St. Ulrich, 1938	1,600- 4,300					31.1
		Total Austria		**1,070**	**1,507**	**21,938**	**679,475,991**	
BAHRAIN	BANOCO	Awali, 1932	1,850- 4,600	286	358	39,034	813,600,000	33.0
		Total Bahrain		**286**	**358**	**39,034**	**813,600,000**	
BARBADOS	BARB. NOC	Woodbourne, 1966	1,732- 6,701	88	142	1,066	5,406,678	27.2
		Total Barbados		**88**	**142**	**1,066**	**5,406,678**	
BENIN[e]	WILLIAMS BROTHERS ENGINEERING	• Seme, 1968	6,232	6	6	4,000	14,814,000	...
		Total Benin		**6**	**6**	**4,000**	**14,814,000**	
BOLIVIA	YPFB	Camiri, 1927	3,936	43	54	362	48,587,297	55.0
		Carand, 1960	3,940	33	43	299	54,359,920	60.0
		Cascabel, 1985	11,319	1	1	261	229,361	45.6
		Colpa, 1961	9,184	11	15	154	18,284,329	65.0
		Espino, 1979	14,655	...	3	...	966,307	56.2
		H. Suarez R., 1982	7,820	...	3	...	486,650	35.0
		La Pena, 1965	8,856	26	31	4,966	28,619,039	43.0
		Monteagudo, 1966	4,592	20	25	1,326	32,581,324	41.0
		Naranjillos, 1964	4,806	12	14	128	92,754	65.3
		Palmar, 1964	10,758	3	5	82	1,746,131	65.0
		Rio Grande, 1961	9,512	35	43	1,905	61,459,486	62.0
		San Roque, 1981	7,170	8	8	1,739	1,350,869	71.0
		Santa Cruz, 1986	9,350	8	9	540	1,739,272	65.0
		Sirari, 1985	13,999	3	3	707	313,732	65.3
		Tita, 1976	6,450	7	7	82	4,408,595	...
		Víbora, 1988	12,949	3	3	632	328,384	57.7
		Villamontes, 1987	9,770	2	3	354	554,980	57.0
		Vuelta Grande, 1978	8,220	16	18	2,448	3,273,748	67.0
		Other	...	50	73	619	31,214,320	...
	OCCIDENTAL	Porvenir, 1978	9,600	5	9	2,011	11,967,963	60.5
	TESORO BOLIVIA PETROLEUM CO.	Escondido, 1980	11,293	3	6	5	1,862	60.0
		Los Suris, 1981	13,577	...	1	...	...	50.0
		La Vertiente, 1977	15,230	6	7	1,118	4,804,834	64.1
		Taiguati, 1981	10,174	1	2	193	70,409	63.0
		Total Bolivia		**296**	**386**	**19,931**	**307,441,476**	

PRODUCTION

WORLDWIDE PRODUCTION

		Name of field, discovery date	Depth, ft	No. of wells Producing	No. of wells Total	1989 average B/d	Production Cumulative to Dec. 31, 1989 Bbl	°API gravity
BRAZIL	ALAGOAS	Furado, 1969	4,440- 5,500	51	68	2,030	15,400,380	35.0
		• Mero, 1974	6,500- 7,400	...	...	...	151,000	35.0
		Pilar, 1981	4,100- 5,100	56	62	5,580	14,089,900	40.0
		Tabuleiro de Martins, 1957	3,900- 4,950	28	29	430	4,154,450	27.0
		Other	...	18	27	560	2,169,520	...
	AMAZONAS	Igarapé Cuia, 1985	3,600- 3,700	...	...	...	25,250	41.0
		Rio Urucú, 1986	7,500- 8,300	13	13	3,200	1,254,000	44.0
	BAHIA	Agua Grande, 1951	820- 4,265	113	124	4,830	276,968,000	40.0
		Aracas, 1965	3,280- 8,856	149	181	6,370	113,834,390	41.0
		Buracica, 1959	1,476- 1,840	171	190	5,600	139,922,730	35.0
		Candeias, 1941	3,280- 7,415	36	39	2,500	79,250,420	30.0
		• Candeias Mar, 1941	3,300- 7,400	17	17	520	7,276,920	30.0
		Cassarongongo, 1959	1,475- 4,755	153	159	3,410	16,757,050	32.0
		Cexis, 1983	1,670- 5,230	34	34	2,690	4,771,900	30.0
		• Dom Joao Mar, 1947	920- 3,000	219	289	3,450	83,792,880	38.0
		Dom Joao Terra, 1947	920- 1,350	57	94	890	21,002,610	38.0
		Faz. Alvorada, 1983	2,100- 4,900	80	80	4,850	8,403,730	31.0
		Faz. Balsamo, 1983	2,500- 4,000	114	118	6,600	11,342,630	31.0
		Faz. Boa Esperanca, 1966	7,215- 8,036	21	23	1,100	15,900,040	39.0
		Faz. Imbe, 1964	2,130- 5,575	59	78	1,800	16,044,860	40.0
		Faz. Panelas, 1962	2,400- 5,100	14	27	400	6,695,900	37.0
		• Ilheus, 1979	1,500	...	...	...	895,470	41.0
		Malombe, 1966	3,280	9	15	300	7,078,300	41.0
		Mata/Remanso, 1971	4,445- 4,560	63	76	1,830	20,897,720	35.0
		Miranga, 1965	3,280- 3,936	227	322	8,570	185,196,800	40.0
		Riacho da Barra, 1982	2,910- 3,998	51	63	2,900	12,691,340	39.0
		Rio Dos Ovos, 1979	2,410- 3,750	10	10	310	2,134,110	35.0
		Rio Pojuca, 1982	5,600- 6,200	29	42	2,200	5,340,500	35.0
		Rio do Bú, 1984	2,900- 3,000	50	50	3,950	4,219,650	33.0
		Santana, 1962	2,400- 4,000	5	6	100	5,640,870	37.0
		Sesmaria, 1963	3,100- 4,600	26	26	640	3,332,070	40.0
		Taquipe, 1958	3,115- 3,935	109	138	3,750	87,104,890	40.0
		Other	...	150	239	6,040	32,273,600	...
	CEARÁ	• Atum, 1983	5,800- 6,750	20	22	5,420	9,926,100	25.0
		• Curima, 1978	6,888	15	17	3,230	19,976,850	32.0
		• Espada, 1983	5,750- 6,900	6	6	1,390	4,473,300	26.0
		Faz. Belem, 1980	1,150- 2,500	281	567	400	6,909,300	13.0
		• Xareu, 1977	5,717- 5,766	25	27	2,700	15,501,100	39.0
	ESPIRITO SANTO	• Cacao, 1978	9,184	7	12	3,450	10,993,870	32.0
		Faz. Cedro, 1962	5,248	11	14	560	10,203,790	28.0
		Lagoa Parda, 1977	5,576	49	50	3,440	23,071,160	30.0
		Lagoa Piabanha, 1984	6,500	2	2	100	1,691,290	27.0
		Lagoa Suruaca, 1982	6,400	22	26	1,650	5,722,440	30.0
		Rio Itaunas, 1977	5,400	40	40	960	4,010,260	28.0
		Rio Preto/Oeste/Sul, 1976	...	37	37	730	2,361,870	...
		Sao Mateus, 1969	5,750	45	53	1,060	4,920,730	27.0
		Other	...	125	125	4,880	8,250,830	...
	RIO DE JANEIRO	• Albacora, 1984	8,430	7	12	16,930	12,985,230	26.0
		• Anequim, 1981	8,200- 8,500	2	2	2,620	4,807,580	27.0
		• Area RJS-150, 1983	7,300	...	...	...	296,070	26.0
		• Area RJS-236 Trilha, 1983	8,800-11,000	3	3	3,370	11,577,950	28.0
		• Area RJS-387, 1987	11,300	3	3	2,030	2,818,080	34.0
		• Badejo, 1981	8,500-10,500	5	6	3,390	18,675,900	28.0
		• Bagre, 1983	8,700- 9,000	7	9	4,370	8,175,800	25.0
		• Bicudo, 1982	8,000-11,000	11	13	17,970	42,172,120	22.0
		• Bonito, 1977	6,950- 8,000	11	17	12,590	36,420,970	27.0
		• Carapeba, 1982	10,100	29	36	16,570	6,062,050	23.0
		• Cherne, 1982	9,900-10,200	29	41	37,490	66,063,850	27.0
		• Corvina, 1983	9,000-11,000	5	8	15,040	36,197,360	25.0
		• Enchova, 1976	6,973- 7,885	4	24	5,400	79,311,850	23.0
		• Garoupa, 1974	11,382	19	20	21,290	77,847,590	30.0
		• Garoupinha, 1974	12,693	3	3	2,960	12,880,240	30.0
		• Linguado, 1981	9,000-11,000	11	18	25,760	73,211,580	20.0
		• Marimba, 1983	8,900- 9,200	5	5	19,530	19,986,100	28.0
		• Moreia, 1984	8,200- 9,500	5	6	3,400	4,616,400	22.0
		• Namorado, 1975	10,083	27	45	55,320	140,641,710	28.0
		• Pampo, 1977	6,215	24	31	36,200	122,714,750	21.0
		• Parati, 1982	8,900- 9,200	1	1	1,120	5,566,630	35.0
		• Pargo, 1975	11,840	16	21	12,350	4,509,410	20.0
		• Pirauna, 1982	8,000-10,500	7	8	15,740	32,516,320	27.0
		• Sul de Pampo, 1978	9,300- 9,600	...	...	...	...	20.0
		• Vermelho, 1983	11,200	45	62	12,550	4,579,600	24.0
		• Viola, 1981	9,200- 9,900	8	8	7,310	13,898,640	13.0
	RIO GRANDE DO NORTE	• Agulha, 1975	5,412	5	6	1,000	7,761,100	33.0
		Alto do Rodrigues, 1981	990- 1,120	290	292	3,570	8,518,410	18.0
		• Area 1-RNS-36, 1982	...	...	...	...	178,290	32.0
		Canto do Amaro, 1986	2,500- 3,600	470	486	34,200	22,143,060	25.0
		Estreito/Rio Panon, 1983	650- 820	447	447	5,400	12,190,780	16.0
		Faz. Pocinho, 1982	1,450- 2,000	174	174	4,420	11,818,520	35.0
		Guamaré, 1983	1,400- 1,600	30	33	560	1,844,960	25.0

PRODUCTION

WORLDWIDE PRODUCTION

		Name of field, discovery date	Depth, ft	No. of wells Producing	No. of wells Total	1989 average B/d	Cumulative to Dec. 31, 1989 Bbl	°API gravity
		Livramento, 1986	2,500- 3,300	30	30	2,900	4,856,670	41.0
		Lorena, 1984	1,200- 1,700	12	12	240	836,420	28.0
		● Macau-Mar, 1986	3,000	5	5	170	508,370	31.0
		Macau/Terra, 1982	3,000- 3,300	2	2	50	449,040	31.0
		Palmeiras, 1982	1,450- 1,800	...	...	...	...	25.0
		Serraria, 1982	3,200- 3,900	38	38	1,320	3,374,340	27.0
		● Ubarana, 1973	7,216- 8,365	77	92	14,650	58,134,730	40.0
		Upanema, 1985	1,110- 1,800	15	18	850	2,239,380	28.0
		Other		118	119	4,850	4,220,020	...
	PARÁ	● Area 1-PAS-11, 1983	...	...	...	...	421,300	41.0
	MARANHÃO	São João, 1983	4,950- 6,200	...	...	...	75,000	40.0
	SERGIPE	Atalaia Sul, 1961	1,450	2	3	7	2,688,650	24.0
		Brejo Grande, 1967	1,670	12	12	300	2,009,740	26.0
		● Caioba, 1969	6,232- 6,890	10	14	1,770	30,996,420	36.0
		● Camurim, 1970	6,232	61	62	6,490	19,754,830	25.0
		Carmopolis, 1963	2,460	852	1,008	25,970	179,890,490	24.0
		● Dourado, 1970	3,620- 3,635	4	4	270	4,546,540	36.0
		● Guaricema, 1968	3,335- 4,430	13	16	4,040	37,959,950	41.0
		Riachuelo, 1961	1,475	235	318	3,400	23,090,630	28.0
		● Robalo, 1973	3,300- 4,500	...	...	...	828,000	30.0
		Siririzinho, 1967	1,640	177	260	4,730	38,023,300	30.0
		Other		84	85	1,420	4,735,450	...
		Total Brazil		**6,267**	**7,574**	**591,697**	**2,656,654,940**	
BRUNEI	SHELL	● Ampa SW, 1963	7,068- 8,155	116	222	41,640	543,800,000	41.0
		● Champion, 1970	4,300	113	213	41,067	315,100,000	23.0
		● Fairley, 1969	10,740	37	54	13,907	115,300,000	40.0
		● Magpie, 1975	...	17	23	16,033	79,300,000	31.0
		Rasau, 1929	...	4	13	786	5,300,000	30.0
		Seria-Total, 1929	...	269	569	18,977	988,200,000	34.0
		Total Brunei		**556**	**1,094**	**132,410**	**2,047,000**	
CAMEROON(e)	ELF	● Betika, 1972	5,300					42.5
		● Ekoundou S., 1975	6,200					24.0
		● Kole, 1974	5,500					31.0
		● Kombo Center, 1976	5,900					37.0
		● Kombo North, 1978	7,200					35.0
		● Ekoundou Horst plus Center, 1972	7,200					29.0
		● Boa, 1976	5,900					35.0
		● Bavo South, 1976	5,600					29.3
		● Betika S. Marine, 1979	6,300	175	179	169,000	501,504,900	30.0
		● Ekuondou N. Marine, 1977	5,250					36.0
		● Mokoko NE plus Abana, 1977	4,600					29.0
		● Asoma Marine, 1973	3,600					19.0
		● Kita E. Marine, 1978	6,900					40.8
		● Mokoko S. Marine, 1977	3,300					17.0
		● Inova Marine, 1976	6,200					32.8
	TOTAL-CFP	● Moudi, 1979	5,250- 5,900					40.0
		● Moudi D., 1980	3,300					21.3
		Total Cameroon		**175**	**179**	**169,000**	**501,504,900**	
CANADA	ALBERTA	Acheson, 1950	3,390	57	118	13,180	132,037,229	38.0
		Bantry, 1940	2,574- 3,300	171	210	9,270	52,741,483	...
		Bellshill Lake, 1955	2,950	425	458	11,640	67,493,984	...
		Bonnie Glen, 1952	3,959- 6,779	134	169	12,445	507,143,641	34.0-44.0
		Caroline	...	187	282	7,800	68,884,903	...
		Carson Creek N., 1958	8,787- 8,935	36	64	7,935	143,308,313	44.0
		Clive, 1951	6,074- 6,208	117	186	4,771	52,946,209	...
		Cessford, 1950	2,800- 3,028	173	301	2,867	32,012,372	...
		Countess, 1951	2,858- 4,272	134	173	6,019	66,053,533	...
		Fenn-Big Valley, 1950	3,900- 5,400	118	295	5,323	317,705,752	32.0
		Gilby, 1962	4,201- 7,000	71	162	2,899	48,510,868	...
		Golden Spike, 1949	3,570- 5,961	69	84	2,025	182,483,299	36.0
		Grand Forks		363	453	21,745	98,317,645	...
		Harmattan East, 1954	6,358- 8,766	106	147	4,409	74,137,864	...
		Harmatton Elkton, 1955	8,796-10,757	63	82	3,919	63,493,732	...
		Hayter, 1968	2,030- 2,815	258	418	7,236	15,372,066	...
		Innisfail, 1956	6,730- 8,440	27	70	4,073	76,682,493	...
		Joarcam, 1949	3,179	173	272	6,097	104,473,973	...
		Joffre, 1953	4,983- 6,779	83	187	3,784	84,639,872	...
		Judy Creek, 1951	8,307- 8,701	183	268	19,349	402,547,694	43.0
		Kaybob, 1957	4,870- 9,577	68	94	5,869	104,674,986	...
		Kaybob S., 1958	5,600-10,042	72	100	7,736	82,373,557	...
		Leduc-Woodbend, 1947	3,062- 5,380	236	708	3,586	378,143,547	40.0
		Lloydminster, 1933	1,690- 1,945	413	964	5,066	51,224,674	...
		Medicine River, 1954	5,435- 7,600	259	364	10,175	75,086,351	...
		Mitsue, 1964	5,908	293	411	31,881	293,845,831	43.0
		Nipisi, 1965	5,648- 5,726	270	359	29,124	285,619,187	41.0

PRODUCTION

WORLDWIDE PRODUCTION

		Name of field, discovery date	Depth, ft	No. of wells Producing	No. of wells Total	Production 1989 average B/d	Production Cumulative to Dec. 31, 1989 Bbl	°API gravity
		Pembina, 1953	3,000- 6,133	3,340	4,245	98,870	1,277,689,925	32.0-37.0
		Provost, 1946	2,760- 2,898	1,561	2,104	40,139	94,882,223	...
		Rainbow, 1965	4,994- 6,160	268	367	41,715	529,321,452	38.0-42.0
		Rainbow South, 1965	6,102- 6,400	47	80	5,954	66,441,552	...
		Red Earth, 1956	4,184- 4,878	186	280	4,838	39,730,041	...
		Redwater, 1948	2,012- 3,200	588	884	12,957	794,082,745	35.0
		Simonette, 1958	10,500	28	51	2,114	38,982,776	...
		Snipe Lake, 1962	8,500	39	97	3,392	55,287,679	...
		Sturgeon Lake S., 1953	4,912- 8,471	89	149	13,976	147,941,250	...
		Swan Hills, 1957	8,100	566	837	37,683	695,276,591	40.0
		Swan Hills S., 1959	8,400	169	200	14,100	349,577,767	...
		Taber N., 1966	3,110- 3,270	78	106	5,845	16,973,413	...
		Turner Valley, 1913	3,100- 9,150	111	158	2,612	140,871,493	39.0
		Utikuma Lake, 1963	5,624	109	139	11,742	52,240,255	...
		Valhalla, 1973	1,815- 7,060	258	303	8,056	13,720,829	...
		Virginia Hills, 1957	9,210	127	188	15,927	153,776,591	34.0
		Wainwright, 1925	1,903- 2,200	565	756	8,947	72,358,599	...
		Westerose, 1952	6,818	34	42	10,009	134,043,358	...
		Westpem		11	14	13,356	30,693,959	...
		Willisden Green, 1956	5,157	563	823	8,026	120,111,497	...
		Wizard Lake, 1951	4,044- 5,973	42	47	9,142	325,853,244	...
		Zama, 1966	3,702	120	243	6,719	78,026,436	33.0-37.0
		Other	...	8,362	13,061	312,623	1,594,878,246	...
	BRITISH COLUMBIA	Boundary Lake, 1957	3,418- 4,575	279	358	9,775	175,250,142	40.0
		Eagle West, 1976	2,170- 3,940	51	96	6,224	29,198,918	...
		Other	...	375	813	18,103	240,343,836	...
	MANITOBA	All fields	...	1,353	1,581	12,462	170,671,122	...
	SASKATCHEWAN	Steelman, 1950	646	731	8,499	258,451,734	...	
		Weyburn, 1955	4,600	821	895	16,692	293,279,794	...
		Other	...	12,199	16,939	175,397	1,778,953,854	...
	ONTARIO	All fields	...	1,091	1,588	4,202	63,782,474	...
	NEW BRUNSWICK	All fields	...	...	...	...	805,261	...
	NORTHWEST TERRITORIES	All fields	...	159	164	32,344	77,377,132	...

*Does not include oil sands production nor wells.

Total Canada* — 38,794 | 55,738 | 1,222,673 | 13,772,831,606

CHILE

		Name	Depth	Producing	Total	1989 avg B/d	Cumulative Bbl	°API
	ENAP-MAINLAND	Canadon, 1962	6,090	11	62	189	14,622,376	38.4
		Daniel, 1960	5,806	10	121	195	34,553,574	25.4
		Daniel Este, 1961	5,523	9	66	227	28,500,354	25.0
		Delgada Este, 1958	7,694	9	31	195	4,935,059	35.0
		Dungeness, 1962	5,251	4	19	198	4,080,927	28.6
		Faro Este, 1959	7,381	7	32	192	3,362,757	35.0
		Posesion, 1960	5,622	31	76	675	21,685,069	63.6
		Other	...	9	118	1,964	15,605,639	...
	TIERRA DEL FUEGO	Calafate, 1955	6,045	17	61	839	26,121,984	49.3
		Catalina, 1956	5,756	9	26	351	6,615,322	38.7
		Catalina Sur, 1961	5,779	4	43	45	6,385,022	44.0
		Chañarcillo, 1951	7,404	2	17	44	7,241,717	58.2
		Cullen, 1954	5,730	17	106	341	41,233,096	41.8
		Gorrión, 1986	6,025	6	10	756	1,017,551	39.0
		San Sebastián Norte, 1986	6,870	4	4	85	168,471	37.0
		Sombrero, 1951	7,126	1	23	3	13,108,843	40.2
		Tres Lagos, 1957	5,724	22	84	507	16,148,103	40.1
		Victoria Sur, 1950	7,431	4	34	52	7,876,783	41.1
		Other	...	46	255	1,272	30,664,301	...
	STRAIT OF MAGELLAN	Daniel Este-Dungeness, 1977	6,126	14	36	1,423	4,685,424	35.9
		Jaiba, 1978	6,650	3	3	188	314,213	46.5
		Ostion, 1977	7,156	13	18	644	8,314,321	33.0
		Pejerrey, 1984	7,229	20	75	1,384	9,959,809	39.0
		Posesion, 1977	6,727	17	39	493	5,555,950	34.5
		Skúa, 1988	5,932	34	39	4,903	2,965,391	32.0
		Spiteful, 1977	6,298	34	82	2,993	43,534,087	34.5
		Spiteful Norte, 1978	7,150	9	55	455	9,661,180	39.0

Total Chile — 366 | 1,535 | 20,613 | 368,917,323

CHINA(e)

		Name						
	WESTERN REGION: JUNGGAR BASIN, TARIM BASIN, TURPEN BASIN	Karamay, 1955	...					...
		Dushanzi, 1941	...					...
		Other	...					...
	CENTRAL REGION: JIUQUAN BASIN	Yumen, 1938	...					...
	CENTRAL REGION: SICHUAN BASIN	All fields	...					
	CENTRAL REGION: SHAN-GAN-NING BASIN	Chang Qing, 1978	...					
		Yanchang, 1907	...					
	CENTRAL REGION: QAIDAM BASIN	Qinghai, 1955	...					
		Lenghu 3,4,5, 1958	...					
		Other	...					

PRODUCTION

WORLDWIDE PRODUCTION

	Name of field, discovery date	Depth, ft	No. of wells Producing	No. of wells Total	Production 1989 average B/d	Production Cumulative to Dec. 31, 1989 Bbl	°API gravity
EASTERN REGION: SONGLIAO BASIN	Daqing, 1959	...	43,700	NA	2,747,507	14,699,906,000	...
	Fuyn, 1958	...					...
	Jilin, 1978	...					...
EASTERN REGION: BOHAI BASIN	Shengli, 1962	...					...
	Renqiu, 1975	...					...
	Liaohe, 1964	...					...
	Dagang, 1964	...					...
	Zhongyuan, 1981	...					...
	Itu, 1970	...					...
EASTERN REGION: JIANGHAN BASIN	Qianjiang, 1966	...					...
	Other	...					...
EASTERN REGION: NANXIANG BASIN	Nanyang, 1971	...					...
EASTERN REGION: SUBEI BASIN	All fields	...					...
	Total China		43,700	NA	2,747,507	14,699,906,000	

CHINA, TAIWAN

CHINESE PETROLEUM							
	CBK, 1986	...	5	9	767		47.0
	Chingtsohu, 1967	...	5	17	82		50.0
	Chuhuangkeng, 1960	...	30	30	232	23,565,169	50.0
	Tiechengshan, 1959	...	23	31	931		51.0
	Yunghoshan, 1971	...	25	33	314		45.0
	Total China, Taiwan		88	120	2,326	23,565,169	

COLOMBIA

ARGOSY[e]							
	Bourdine, 1974	11,000	2	4			26.7
	Máxime, 1976	11,000	1	1			11.0
	Nancy, 1974	11,000	1	1			26.0
HOCOL SA (SHELL)[e]	Palogrande, 1971	6,500	32	36			20.0
	Pijao, 1981	7,000	6	7			21.0
	San Francisco, 1985	2,800	41	41			27.0
	Hato Nuevo, 1984	6,900	2	4			36.0
	Tenay, 1985	11,500	3	3	54,888	302,478,624	36.3
	Santa Clara, 1987	2,100	5	6			19.0
	La Jagua, 1986	5,300	2	2			22.0
	Loma Larga, 1985	3,500	1	1			22.4
	Brisas, 1973	4,200	4	6			23.5
	Dina T, 1962	3,600	51	64			20.0
	Dina K, 1969	6,500	22	26			22.0
	La Canada, 1966	700	5	7			22.0
	Tello, 1972	7,500	30	34			22.0
	Cebu, 1981	6,500	8	10			19.0
ECOPETROL	Apiay, 1981	10,600	13	15	12,370	20,350,000	23.0
	Apiay Este, 1987	10,300	1	1	190	150,000	16.0
	Boquete, 1962	8,100	8	24	330	17,500,000	43.0
	Caipal/Ermitaño, 1954, 1957	3,500	3	19	60	4,280,000	15.0
	Campo Yuca, 1958	...	1	4	10	450,000	32.0
	Cantagallo/Yarigui, 1943, 1955	6,800-8,700	68	92	7,300	135,040,000	20.0
	Carbonera, 1939	...	4	8	40	490,000	32.0
	Caribe, 1970	5,400	3	4	320	3,860,000	31.0
	Casabe, 1940	3,800	909	1,039	9,700	231,750,000	19.0-22.0
	Churuyaco/Sucombios, 1967, 1969	...	6	7	510	4,750,000	31.0
	Cicuco, 1948	8,600	5	28	300	43,610,000	41.0
	Colorado, 1925	...	45	76	80	8,190,000	34.0
	Galan/San Silvestre, 1943	3,200	53	102	820	23,310,000	18.0
	Garzas, 1957	...	1	2	110	570,000	28.0
	Guatiquia, 1985	10,700	2	2	620	360,000	20.0
	Hormiga, 1982	10,300	2	3	160	710,000	29.0
	Infantas, 1918	3,200	305	568	1,750	226,070,000	28.0
	La Cira, 1918	3,250	707	951	7,610	468,670,000	23.0
	La Cristalina, 1950	10,000	1	8	200	5,690,000	30.0
	Llanito/Gala/Yuma/Cardales, 1955, 1985, 1988, 1988	9,500	145	219	6,480	41,970,000	28.0-32.0
	Lisama/Nutria/Tesoro Petroles, 1936, 1966, 1977	9,500	145	219	6,480	41,970,000	28.0-32.0
	Loro, 19067	10,300	7	12	710	11,570,000	30.0
	Orito, 1948	600- 5,200	51	104	8,790	196,310,000	25.0-35.0
	Ortega, 1951	6,500	5	15	380	11,750,000	28.0
	Palagua, 1954	3,500	96	186	3,840	83,330,000	15.0
	Peñas Blancas, 1957	...	6	7	380	8,910,000	23.0
	Petrolea, 1933	1,200	5	34	150	36,970,000	46.0
	Puerto Barco, 1958	...	...	4	20	600,000	50.0-60.0
	Puerto Colon, 1967	9,600	7	10	4,030	13,730,000	29.0
	Quilili, 1988	...	2	3	610	250,000	21.0
	Rio de Oro, 1933	1,600-8,500	20	54	60	10,670,000	32.0-43.0
	San Antonio (S. Miguel), 1967	7,820	2	4	80	4,740,000	28.0
	Sardinata Norte/Sur, 1939, 1942	900- 4,100	14	39	260	8,480,000	30.0-52.0
	Sucio, 1971	5,500	1	2	70	1,480,000	31.0
	Suria, 1985	10,500	7	8	2,290	850,000	35.0
	Tibu, 1939	4,500- 9,300	250	508	4,980	230,310,000	32.0-45.0
	Toldado, 1987	6,200	6	7	2,680	1,050,000	20.0

289

PRODUCTION

WORLDWIDE PRODUCTION

	Name of field, discovery date	Depth, ft	No. of wells Producing	Total	1989 average B/d	Cumulative to Dec. 31, 1989 Bbl	°API gravity
	Other...	...	7	71	...	990,000	
	Aguas Blancas, 1962........................	...	4	7 ⎤		650,000	20.0
	Barrancas Lebrija, 1953...................	8,000	2	11		140,000	25.0
	Pavas, 1981.....................................	...	1	1		40,000	21.0
	San Luis, Totomal, 1927, 1950........	...	2	7	3,100	1,200,000	24.0-35.0
	Sogamoso, 1958...............................	...	2	4		850,000	29.0
	Tenerife, 1967.................................	...	1	3 ⎦		390,000	18.0
ESSO COLOMBIANA LTD.	Arauca, 1980...................................	18,000	3	4	1,775	5,913,510	40.0
	Bonanza, 1963.................................	4,000	17	17	646	13,246,795	27.0
	Provincia, 1962...............................	8,000	152	172	14,926	166,964,261	30.0
	Sabana, 1989..................................	14,000	3	3	186	103,697	25.0
ELF AQUITAINE	Barquerena, 1982............................	9,300	3	3	1,408	3,327,205	33.5
	Cano Garza, 1979............................	7,900	4	4	2,281	2,123,708	33.0
	Cano Garza Norte, 1983..................	7,900	2	2	915	2,009,235	38.5
	Cravo Este, 1987.............................	11,376	...	1	289	108,833	40.1
	Cravo Sur, 1982..............................	10,600	2	2	1,865	2,925,853	39.0
	Gloria Norte, 1983...........................	12,100	2	2	2,031	1,605,104	18.0
	Morichal, 1984................................	15,500	1	1	273	99,542	30.0
	Tierra Blanca, 1987........................	14,493	1	1	358	130,506	32.0
	Tocaria, 1980..................................	13,500	2	3	720	1,179,647	32.2
	Trinidad, 1974................................	9,000	4	4	3,788	5,090,719	33.0
OCCIDENTAL	Caño Limón, 1984............................	7,600	47	54	172,219	230,057,128	29.0
	Cano Yarumal, 1987........................	7,600	2	2	12,256	4,630,143	29.0
	La Salina, 1936...............................	2,500	38	71	3,105	14,494,004	28.0
	Payoa, 1962....................................	8,000	24	39	953	70,022,300	27.0
	Redondo, 1984................................	7,600	3	4	5,678	5,998,120	30.2
TEXACO	Cocorná, 1963.................................	2,100	46	54	1,726	14,539,000	12.0-13.0
	Tisquirama, 1963............................	8,000	5	6	422	4,977,000	16.0-28.0
	Velasquez, 1946..............................	7,500	92	100	3,123	166,961,000	19.0-24.0
TEXACO-ECOPETROL	Los Angeles, 1983...........................	6,400	4	6	669	784,000	14.0-15.0
	Nare, 1981......................................	2,100	49	49	1,792	974,000	11.0-12.0
	Teca, 1981......................................	2,100	187	200	14,241	29,976,000	12.0-13.0
CHEVRON	Castilla, 1969..................................	6,500	17	19	12,528	29,430,090	13.5
	Chichimene, 1969	6,500	1	2	466	770,488	21.3
LASMO	Santiago, 1985................................	9,600	4	4	2,343	2,093,945	29.0
	Total Colombia...........................		**3,771**	**5,442**	**404,000**	**2,971,904,457**	

CONGO

AGIP	● Loango, 1972..................................	3,000	52	62	7,823	87,479,524	6.8
	Zatchi...	1,980	8	8	3,296	1,486,665	5.9
ELF CONGO	Bindi, 1982.....................................	4,400	1	1	40	143,000	34.2
	● Emeraude, 1969..............................	820	92	99	20,300	175,400,000	22.3
	Kunji, 1980....................................	4,000	2	2	130	586,000	35.8
	● Likouala, 1972................................	4,270	27	44	15,300	62,800,000	32.3
	Mengo, 1979..................................	6,500	3	3	150	742,000	34.0
	Pointe Indienne, 1951.....................	5,000	3	5	130	7,108,000	36.9
	● Sendji, 1973...................................	3,940	50	76	29,100	65,600,000	29.5
	● Tchibouela, 1983.............................	2,110	26	29	46,000	26,500,000	27.0
	● Yanga, 1979...................................	2,950	48	63	17,000	75,200,000	29.4
	Total Congo.................................		**312**	**392**	**139,269**	**503,045,189**	

DENMARK

DUC	● Dan, 1971.......................................	6,000- 6,400	41	49	25,386	57,274,190	29.5
	● Gorm, 1971....................................	6,500- 7,000	20	26	23,263	84,058,230	33.9
	● Rolf, 1981.......................................	5,900	1	2	6,802	11,910,953	35.3
	● Skjold, 1977...................................	5,000	3	7	38,174	49,052,790	30.3
	● Tyra, 1968.....................................	6,500	30	38	18,094	24,131,390	50.4
	Total Denmark.............................		**95**	**122**	**111,719**	**226,427,553**	

DUBAI

DUPETCO[e]	● Fateh, 1966.....................................	7,900- 9,000 ⎤					31.8
	● S.W. Fateh, 1970............................	7,500- 9,000					30.3
	● Falah, 1972....................................	8,100 ⎬	141	156	400,880	2,169,802,600	25.5
	● Rashid, 1973..................................	9,400-11,500 ⎦					38.0
ARCO	Margham, 1981...............................	11,200	8	14	14,520	34,560,000	43.5
	Total Dubai..................................		**149**	**170**	**415,400**	**2,204,362,600**	

ECUADOR

TEXACO-PETROECUADOR	Atacapi, 1968..................................	9,855	4	5	1,505	12,762,667	31.0
	Auca, 1970.....................................	10,578	23	24	15,701	76,770,667	28.0
	Auca Sur, 1980...............................	9,100-10,300	2	2	614	1,634,667	25.6
	Cononaco, 1972..............................	11,233	9	11	12,785	30,736,000	33.3
	Culebra, 1973.................................	10,626	2	2	1,125	2,578,667	17.5
	Dureno/Guanta, 1968.....................	9,575-10,290	10	10	7,883	8,613,333	31.0
	Lago Agrio, 1967............................	10,175	18	24	8,738	126,482,667	28.5
	Parahuacu, 1968.............................	10,173	3	5	1,607	7,578,667	30.0
	Rumiyacu, 1982..............................	10,191	...	1	...	184,000	21.0
	Sacha, 1969....................................	10,160	96	103	63,613	406,144,000	28.0

290

PRODUCTION

WORLDWIDE PRODUCTION

	Name of field, discovery date	Depth, ft	No. of wells Producing	No. of wells Total	1989 average B/d	Cumulative to Dec. 31, 1989 Bbl	°API gravity
	Shushufindi/Aguarico, 1959	9,772	52	60	100,953	630,930,667	30.0
	Yuca, 1970	10,423	6	7	2,681	12,352,000	28.0
	Yuca Sur, 1979	9,600-10,000	1	1	351	1,242,667	14.5
	Yulebra, 1980	8,800-10,000	3	3	1,753	3,589,333	21.4
CICL	Fanny, 1972	7,700	8	8	4,685	12,900,120	22.5
	Mariann, 1971	7,700	5	5	1,048	4,302,980	20.0
	Tarapoa, 1973	8,300	1	2	333	1,150,807	20.0
CEPE-ORIENT	Bermejo Sur + Norte, 1967	5,000	21	26	5,475	7,572,353	32.5
	Charapa, 1967	10,000	1	4	291	754,353	29.5
	Cuyabeno, 1972	8,050	15	19	7,872	15,752,162	27.4
	Libertador	...	44	50	37,306	77,644,900	29.7
	Tetete-Tapi, 1980	9,300	6	11	2,738	5,961,647	29.1
CEPE-PENINSULA	Ancon, 1921	4,000	439	439	652		35.7
	Carpet	...	99	99	124	111,600,000	36.4
	Cautivo	...	79	79	112		36.4
	Total Ecuador		947	1,000	279,945	1,559,239,324	

EGYPT

	Name of field, discovery date	Depth, ft	Producing	Total	1989 avg B/d	Cumulative Bbl	°API
GUPCO/AMOCO	Abu Gharadig, 1972	9,300-12,000	17	27	11,005	58,691,385	35.0-55.0
	• Badri (315), 1981	6,400	22	38	28,935	67,516,833	27.0
	• El Morgan, 1965	5,000- 7,500	109	126	95,142	1,083,276,893	27.0-32.0
	• GH (376), 1981	8,200- 9,750	1	5	291	6,153,319	24.0-28.0
	• GS (277) 1980	8,610	1	1	1,315	3,331,520	34.0
	• GS (345/346), 1982	7,500	1	2	104	1,225,935	35.0-38.0
	• GS (365), 1980	10,140	1	1	314	679,175	42.0
	• GS (381), 1985	8,700	1	1	1,123	1,333,873	37.0
	• GS-327, 1983	5,600	1	1	1,493	2,621,126	20.0
	• Hilal (404), 1982	9,200-10,000	8	9	30,562	57,463,365	32.0
	• July, 1973	8,100-10,000	27	43	50,979	490,674,420	33.0
	NAG-9/9-1, 1985	10,000	1	2	886	1,624,363	35.0
	NEAG-9/15, 1984	10,635-12,128	...	3	608	1,051,734	35.0
	• Nessim (336), 1981	6,700	1	3	48	8,884,592	29.0
	• October + 173 + 183, 1978	11,000	28	35	128,096	389,875,769	27.0-31.0
	• Ramadan, 1974	11,400	26	30	45,857	395,890,301	30.5
	Razzak and E. Razzak, 1972	5,730- 7,710	12	17	3,145	53,292,886	38.0
	• SB (305), 1981	8,100	...	1	...	2,043,837	27.0
	• SB (339), 1981	4,600	1	1	384	839,058	29.5
	• SB (367), 1984	7,500	1	1	160	1,312,941	39.0
	• SB-294, 1986	11,600	1	2	2,675	2,014,927	30.5
	• SB-374, 1981	...	...	...	...	68,506	...
	• SG-300, 1976	5,720- 6,200	2	4	5,453	22,193,965	24.0
	• Shoab Ali, 1978	4,700- 5,500	19	47	11,115	82,719,926	28.0-33.0
	• Sidki (382), 1977	10,500	5	7	5,642	42,761,720	34.0
	• Waly (356), 1982	6,900	1	3	854	4,248,192	36.0
	WD-33, 1972	9,700	...	1	...	2,161,591	35.0
	WD-33/15, 1985	5,800	2	2	1,908	2,273,047	30.0
	• Younis (347), 1981	4,300	2	6	4,343	15,576,279	32.0
AGIP	• Belayim Land + Minori, 1954	7,000- 8,500	105	111	12,471	432,550,120	2.5
	Belayim Marine, 1961	8,500	66	74	48,685	627,689,302	9.5
	Meleiha, 1972	5,950	45	54	4,490	11,880,858	1.6
WEPCO/PHILLIPS PETROLEUM	El Alamein, 1966	8,700	18	19	1,660	67,537,121	34.0
	Umbarka, 1976	10,700	4	5	2,739	7,814,579	44.0
	Yidma, 1971	9,000	3	5	275	21,996,603	43.6
KHALDA	Hayat, 1986	9,000	5	5	887	750,513	40.0
	Khalda, 1983	7,000	6	13	2,852	2,800,466	35.0
	Safir, 1986	12,000	1	1	528	408,234	36.0
	Safir North, 1986	11,600	3	3	1,863	1,573,662	36.0
	Salam, 1985	8,000	18	29	13,420	14,680,848	35.0
	Tut, 1987	9,000	5	5	1,805	1,301,398	35.0
	Tut West, 1988	9,000	1	1	278	170,132	35.0
BADR PETROLEUM/SHELL	Badr el din, 1982	10,800	10	10	17,800	27,300,000	38.0
SUESSO (ESSO)[e]	• East Zeit, 1982	11,000	6	6			38.0
SUCO[e]	• Ras Budran, 1978	12,500	17	20			23.0-24.0
	• LL 87-2, 1975	12,000	...	1			34.0
	• Ras Fanar, 1978	3,000	5	6			30.0
	• Zeit Bay, 1981	6,000	22	32			34.5
OSOCO/TOTAL[e]	• Shukheir Bay, 1978	6,300	1	1			34.5
	• S. Ramadan, 1982	10,000	2	2			36.5
GEYSUM[e]	• Geysum, 1980	5,500	5	5			17.0
EGPC[e]	Gharib, 1938	2,200- 4,500	81	215			24.3
	Bakr/W. Bakr, 1958/1978	3,200- 1,500	48	83			18.0-19.0
	Umel Yusr, 1968	3,000- 3,300	19	37			21.5
	Amer, 1965	3,500	21	39			21.5
	El Ayoun, 1968	3,300	5	12			22.0
	Kareem, 1958	2,300	6	35			16.6
	Kheir, 1973	1,050	1	4			28.5
	Shukheir, 1966	2,500	5	11			34.0

PRODUCTION

WORLDWIDE PRODUCTION

	Name of field, discovery date	Depth, ft	No. of wells Producing	No. of wells Total	Production 1989 average B/d	Production Cumulative to Dec. 31, 1989 Bbl	°API gravity
	Ras El Behar, 1983	4,500	6	11	308,810	873,086,730	32.5
	Esh Elmalaha	3,700	2	2			44.0
	Sudr, 1946	2,600	12	20			21.6
	Asl, 1947	3,000	3	6			21.6
	Matarma, 1948	2,000	2	2			14.6
	El Khaligue, 1980	1,600	...	1			29.4
	Abu Senan T&Y, 1981	6,000	12	20			...
	Harghada, 1913	...	NA	NA			...
EPEDECO[e]	West Bakr, 1978	1,300- 1,500	20	25			18.0
EGYPTCO[e]	Meleha, 1972	6,000	12	12			40.0
SHUKAIR MARINE CO.[e]	O. Suco	...	1	1			...
EL ALAMEIN[e]	Mors	...	1	1			...
	Total Egypt		896	1,364	851,000	4,893,342,044	

FRANCE

ALSACE	All fields	...	14	21	290	3,158,398	...
	Abandoned	...	...	...	...	24,337,343	...
AQUITAINE BASIN	Castéra-Lou, 1976	9,600	8	11	382	2,614,350	21.7
	Cazaux, 1959	7,550	36	73	4,451	71,106,450	34.2-37.3
	Lacq Superieur, 1949	2,170	45	71	969	26,213,311	22.3
	Lagrave, 1984	6,500	3	3	6,010	8,017,894	42.5
	Lavergne, 1962	11,480	4	6	338	12,395,156	41.7
	Lugos, 1956	5,250	16	25	461	10,297,730	20.8
	Parentis, 1954	7,710	47	101	5,250	199,906,994	33.4
	Pecorade, 1974	10,660	7	20	1,901	10,631,201	29.3
	Vic Bilh, 1979	7,100	30	47	4,024	18,079,737	26.6
	Other	...	7	12	209	10,386,526	...
	Abandoned	...	...	...	...	8,848,690	...
PARIS BASIN	Champotran, 1985	8,400	8	13	877	881,462	31.1
	Châteaurenard, 1958	1,800	24	46	371	5,021,062	27.5
	Chaunoy, 1983	7,350	53	78	15,040	27,589,733	37.1
	Chuelles, 1961	1,800	31	68	372	7,135,011	27.5
	Coulommes, 1958	6,230	28	49	345	13,472,866	32.6
	Donmartin-Lettree, 1985	7,020	2	3	578	585,756	35.0
	Donnemarie, 1979	8,040	1	7	395	2,196,663	36.2
	Fontaine-au-Bron, 1986	6,140	10	12	2,280	1,366,704	35.4
	L'ile Du Gord, 1986	7,260	6	6	539	337,188	35.0
	Soudron, 1976	4,600	22	29	381	2,223,037	34.4
	St. Firmin, 1960	1,800	32	62	497	10,148,632	27.5
	St. Germain, 1984	7,200	7	15	955	2,285,843	36.0
	St. Martin, 1959	4,590	9	13	271	8,559,539	33.0
	Vert Le Grand, 1986	6,230	5	6	3,213	2,418,782	34.8
	Villeperdue, 1959	6,100	99	139	12,695	21,412,160	34.4
	Other	...	78	156	2,461	13,241,027	...
	Abandoned	...	...	...	...	14,293,157	...
SOUTHEAST + JURA	All fields abandoned	...	...	...	...	223,063	...
	Total France		632	1,092	65,555	539,385,465	

GABON

ELF	• Anguille, 1962	9,200					31.0
	• Anguille W, 1968	7,500					32.3
	• Anguille SE, 1962	8,200					33.0
	• Ayol Marine, 1979	7,700					19.8
	• Baliste, 1974	4,900					27.5
	• Barbier, 1972	6,000					32.6
	• Batanga, 1960	5,900					39.2
	• Baudroie Marine, 1968	7,500					29.7
	• Baudroie N., 1980	7,500					31.5
	• Breme, 1976	5,000					37.0
	• Clairette, 1956	4,400					32.5
	• Dorée, 1972	7,300					30.4
	• Girelle, 1973	9,400					31.1
	• Gonelle, 1972	5,800	218	305	101,215	983,072,000	25.7
	• Grand-Anguille, 1983	8,400					31.1
	• Grondin Marine, 1971	7,500					26.6
	• Konzi, 1962	3,600					39.0
	• M'Bya Sud, 1969	6,700					36.2
	• Mandaros, 1973	4,900					20.5
	• Mérou-Sardine, 1981	9,900					30.2
	Olendé, 1976	6,700					34.0
	• Pageau, 1972	11,500					19.8
	• Port-Gentil Sud Marine, 1975	7,500					32.1
	Tchengué, 1957	4,100					31.0
	• Tchengue-Océan, 1962	6,600					30.2
	• Torpille, 1968	8,200					34.0
	• Torpillo NE, 1968	8,200					32.3
SHELL GABON	Gamba/Ivinga, 1963	3,000- 3,100	51	84	10,685	262,600,000	31.6
	Lucina, 1971	5,900	15	22	7,123	53,800,000	38.4
	Lucina W., 1982	...	...	...	822	2,300,000	38.4

PRODUCTION

WORLDWIDE PRODUCTION

		Name of field, discovery date	Depth, ft	No. of wells Producing	Total	1989 average B/d	Cumulative to Dec. 31, 1989 Bbl	°API gravity
	AMOCO	Oguendjo B, C, and Z, 1981	5,000- 6,000	17	19	13,127	38,091,408	36.0
	BRITISH GAS	• Obando, 1984	8,650	2	2	3,016	1,912,000	32.0
		• Octopus, 1984	10,300	3	3	711	1,166,000	33.0
		• Pelican, 1985	9,300	4	4	7,432	6,004,000	31.0
		Total Gabon		310	439	144,131	1,348,945,408	
GHANA	AGRI-PETCO	• Saltpond, 1977	8,500	...	3	...	3,938,000	36.0-39.0
		Total Ghana		...	3	...	3,938,000	
GREECE	N. AEGEAN PET. CORP.	• Prinos, 1974	8,000- 9,700	13	15	18,222	70,372,217	28.5
		Total Greece		13	15	18,222	70,372,217	
GUATEMALA	MINISTERIO DE ENERGIA Y MINAS	Caribe, 1981	6,830	3	3	1,117	3,053,148	22.0-24.0
		Rubelsanto, 1974	5,300	5	7	1,072	6,127,188	26.0-30.0
		San Diego, 1981	17,150	..	1	...	135,865	35.0
		Tierra Blanca	11,230	1	1	760	209,097	19.5-23.0
		Tortugas, 1973	2,300	...	...	...	67,800	25.0
		West Chinaja, 1977	3,200- 3,800	3	3	466	7,337,899	31.0-35.0
		Xan, 1985	8,427	1	1	361	131,773	17.0
		Yalpemech, 1980	15,351	1	1	84	136,957	31.0-35.0
		Total Guatemala		14	17	3,860	17,199,727	
INDIA	GUJARAT STATE	Ankleshwar, 1960	1,969- 6,564	138	262	16,962	449,130,000	45.0
		Gandhar, 1984	9,846-12,145	30	52	15,037	9,320,000	47.0
		Kadi, N., 1967	3,610- 7,220	160	227	13,557	11,578,000	21.0
		Kadi, S., 1968	6,892- 9,846	6	39	641	3,812,000	24.0
		Kalol, 1961	4,595-10,830	152	322	11,719	42,278,000	38.0
		Nawagam, 1963	4,595-10,174	74	130	2,733	26,832,000	31.0
		Sanand, 1962	4,267-10,174	22	88	2,095	10,317,000	26.0
		Sobhasan, 1968	4,595- 9,846	59	124	12,460	25,714,000	26.0-40.0
		Viraj, 1978	4,595- 4,923	20	37	3,700	5,148,000	20.0
	ASSAM STATE	Geleki, 1968	9,518-14,769	62	161	9,144	40,579,000	30.0
		Lakwa-Lakhmani, 1964	6,564-14,113	211	344	37,419	154,469,000	33.0
		Rudrasagar, 1960	7,877-13,128	50	111	10,689	37,708,000	25.0
	WEST COAST	• B-38 (Heera), 1977	4,267- 6,564	40	76	28,836	56,379,000	37.0
		• Bombay High, 1974	3,282- 7,877	304	450	420,179	1,255,890,000	39.0
		• Panna (North Bassein), 1976	4,923- 7,549	7	23	2,978	1,691,000	38.0
		• Ratna, 1979	7,500	5	14	1,307	957,000	38.0
	NAGALAND & TAMIL NADU	Changpang, 1973	8,861-10,502	9	17	3,002	4,857,000	34.0
		Narimanam, 1985	6,236- 7,549	8	25	2,818	1,258,000	45.0
	OTHER FIELDS	Bakrol, 1966	2,626- 9,846	...	3	...	37,000	...
		Digboi, 1980	300- 7,500	234	462	570	86,156,400	...
		Jorajan, 1967	7,000-11,500	126	187	23,248	52,560,000	25.0-34.0
		Kathana, 1965	5,250- 5,906	13	33	547	5,103,000	25.0
		Kholka, 1966	5,579- 7,877	12	19	144	636,000	...
		Kosamba, 1963	1,313- 8,533	6	23	393	978,000	...
		Moran, 1958	11,000	61	89	8,286	109,287,600	...
		Nakhorkatiya, 1953	9,500	135	325	21,367	367,974,000	...
		Wavel, 1962	4,595- 7,220	...	...	...	168,000	...
		Total India		1,944	3,643	649,831	2,760,817,000	
INDONESIA	PERTAMINA UEP I[e]	Besitang, 1977	3,973					51.5
		Gebang, 1977	2,917					54.0
		K. Simpang, 1971	3,740- 5,019					45.0
		P. Tabuhan, 1937	3,350- 5,893	204	226			50.8
		Rantau, 1929	2,764					45.3
		Serang Jaya, 1926	3,701					47.5
	PERTAMINA UEP II[e]	Benuang, 1942	6,103					35.1
		Benakat East, 1973	NA					NA
		Tempino, 1931	1,647					23.0
		Gn. Kemala, 1938	5,448					38.0
		Limau, 1928	3,896- 4,626					NA
		Ogan Timur, 1943	3,793					27.42
		Talang Jimar, 1937	4,123	412	1,419			31.0
		T. Miring, 1935	640					NA
		Tanjung Tiga, 1938	3,750					28.0
		Kenali Asam, 1931	2,077					29.0
		Belimbing	5,371					
		Kuang, 1940	5,249					25.1
	PERTAMINA UEP III[e]	Cemara, 1976	3,964- 7,104					...
		Jatibarang, 1969	3,063	88	184			29.0
		Tugu Barat, 1979	3,014					...
	PERTAMINA UEP IV[e]	Bunyu, 1923	2,077					33.2
		Sangatta, 1939	1,842					33.0

PRODUCTION

WORLDWIDE PRODUCTION

	Name of field, discovery date	Depth, ft	No. of wells Producing	No. of wells Total	1989 average B/d	Production Cumulative to Dec. 31, 1989 Bbl	°API gravity
	Tanjung, 1938	2,296	161	334	7,701	2,955,200,548	39.9
	Tapian Timur, 1967	4,396					38.8
	Warukin S., 1966	2,067					26.9
PERTAMINA UEP V[e]	Klamono, 1936	574					19.0
	Linda, 1977	3,281	47	55			19.0
	Sele, 1951	2,156					35.0
PERTAMINA (FROM STANVAC)[e]	Sago, 1940	1,800					34.0
	Lirik, 1939	1,600					34.0
	Molek, 1956	2,600					34.0
	N. Pulai, 1941	1,800					34.0
	S. Pulai, 1941	1,800	354	354			34.0
	Talang Akar, 1922	2,800					35.0
	Benakat, 1932	1,600					35.0
	Jirah, 1930	2,300					35.0
	Rajah, 1940	6,000					38.0
	Abab, 1951	6,000					36.0
PERTAMINA (N.E. KALIMANTAN)[e]	Sembakung, 1976	2,600	17	19			36.0
	Bangkudulis, 1980	3,250	...	1			39.5
PERTAMINA (FROM PHILLIPS)[e]	Salawati, 1976	5,749	7	13			39.2
LEMIGAS[e]	All fields, 1893-1909	2,155- 2,297	71	419			...
TOTAL CFP[e]	• Tambora, 1980	8,000-13,000	15	34			33.0
	• Handil, 1974	2,900- 9,000	219	272			33.0
	• Bekapai, 1972	4,600- 5,600	42	45			40.0
ARCO	Ardjuna, 1969	2,380- 7,250	300	356	99,438	722,948,500	37.0
	Arimbi, 1972	2,943	27	30	9,919	29,953,200	33.0
	NWC, 1974	5,411-5,800	17	30	5,746	7,041,700	31.5
	Bima, 1983	3,000- 3,888	48	60	11,947	15,896,200	34.0
AGL PETROLEUM (SERAM) LTD.	Bula, 1897	280- 1,300	131	69	1,295	15,542,593	23.0
CHEVRON & TEXACO	Batang, 1975	690	12	13	701		19.0
	Benua, 1978	2,400	21	25	10,114		37.0
	Beruk, 1974	1,750	19	28	8,813		38.0
	Beruk NE, 1976	1,800	3	6	138		40.0
	Beruk North, 1985	1,900	5	7	1,174		38.0
	Buaya, 1978	2,035	...	2	...		28.0
	Butun, 1982	4,000	9	9	5,781		44.0
	Damar, 1974	4,650	5	6	511		34.0
	Dusun, 1979	2,550	6	7	1,008		42.0
	Gatam, 1977	2,385	...	3	...		34.0
	Idris, 1982	2,450	1	3	815		39.0
	Kasikan, 1972	660	20	27	1,156		28.0
	Jingga, 1984	6,500	3	3	1,596		45.0
	Langgak, 1976	1,380	26	27	2,340		32.0
	Lindai, 1971	1,100	18	20	1,386		32.0
	Menggala S., 1968	3,765	2	3	755		34.0
	Mengkapan, 1981	4,000	1	1	767		44.0
	Osam, 1978	900	6	6	298		32.0
	Paitan, 1978	2,315	1	1	232		32.0
	Pedada, 1973	950	30	31	11,032		33.0
	Pusaka, 1977	2,250	20	23	5,985		36.0
	Rintis, 1984	6,450	1	1	387		39.7
	Sabak, 1974	2,350	15	17	1,650		36.0
	Tanjung Medan, 1976	2,770	2	7	846		34.0
	Terantam, 1973	900	2	4	109		32.0
	Zamrud, 1975	3,600	52	54	19,295		40.0
CALTEX PACIFIC INDONESIA	Central Sumatra, Aman, 1974	4,700	10	10	1,360		39.0
	Ampuh, 1981	6,000	12	12	2,825		38.0
	Antara, 1978	1,425	5	6	939		36.0
	Balam S., 1969	1,600	59	62	9,141		33.0
	Balam SE, 1972	4,500	13	13	666		29.0
	Bangko, 1970	1,950	86	96	28,311		34.0
	Batang CPI, 1975	690	6	7	370		19.0
	Bekasap, 1955	2,950	56	63	18,708		34.0
	Bekasap South, 1968	3,900	9	14	1,199		34.0
	Benar, 1973	2,450	26	30	4,778		33.0
	Cebakan, 1974	4,700	4	5	627		30.0
	Cucut, 1981	5,800	3	3	221		33.0
	Duri, 1941	770	1,465	1,844	134,071		22.0
	Garuk, 1980	4,600	...	2	...		31.0
	Hitam, 1975	6,690	4	5	979		39.0
	Hiu, 1983	5,900	4	4	1,125		40.0
	Intan, 1977	3,350	6	7	1,866		33.0
	Jorang, 1972	5,500	19	21	11,612		38.0
	Kerang, 1977	3,900	11	14	1,032		39.0
	Kopar, 1974	5,300	20	23	3,696		37.0
	Kotabatak, 1952	5,500	68	113	18,818		29.0
	Kulin, 1970	2,050	60	68	4,996		20.0
	Libo, 1968	6,000	2	12	1,401		35.0
	Libo SE, 1973	5,900	20	27	7,458		42.0

PRODUCTION

WORLDWIDE PRODUCTION

	Name of field, discovery date	Depth, ft	No. of wells Producing	No. of wells Total	Production 1989 average B/d	Production Cumulative to Dec. 31, 1989 Bbl	°API gravity
	Lincak, 1981	2,570	4	4	323		35.0
	Menggala N., 1968	3,850	16	20	4,336		33.0
	Menggala S. CPI, 1968	3,765	6	7	1,449		34.0
	Minas, 1944	2,600	467	618	232,966	6,749,434,683	35.0
	Mindal, 1971	3,400	2	4	37		37.0
	Mutiara, 1976	3,500	2	2	...		34.0
	Nella, 1977	5,000	3	5	33		35.0
	Obor, 1978	7,370	2	2	201		40.0
	Pager, 1974	4,300	25	27	6,930		38.0
	Pelita, 1977	6,600	7	7	627		39.0
	Pematang, 1959	3,750	40	46	8,538		32.0
	Pematang Bow, 1969	5,300	7	14	1,997		33.0
	Pemburu, 1981	2,300	9	10	2,738		35.0
	Perkebunan, 1977	2,300	1	1	...		29.0
	Petani, 1964	4,750	31	34	11,419		33.0
	Petapahan, 1971	4,500	23	30	5,051		28.0
	Pinang, 1971	3,950	19	22	4,690		30.0
	Pinggir, 1972	3,300	3	7	268		35.0
	Pinggir S., 1973	3,300	2	4	59		39.0
	Pudu, 1972	5,950	8	13	1,060		35.0
	Pukat, 1983	6,200	2	2	813		40.0
	Puncak, 1979	2,400	14	16	3,788		36.0
	Pungut, 1951	3,400	27	33	4,202		38.0
	Rangau, 1968	6,100	4	12	113		42.0
	Rantaubais, 1972	1,070	14	18	541		18.0
	Rokiri, 1979	6,100	...	4	...		41.0
	Seruni, 1972	3,000	23	25	6,796		35.0
	Sikladi, 1975	4,340	19	21	2,869		40.0
	Singa, 1977	3,200	1	1	32		31.0
	Sintong, 1971	3,400	22	23	3,497		32.0
	Sintong SE, 1973	4,500	1	1	78		36.0
	Suram, 1971	1,750	4	7	238		28.0
	Tandun, 1969	3,000	12	13	2,675		33.0
	Telinga, 1975	4,205	7	10	603		33.0
	Topaz, 1977	5,200	7	8	1,558		33.0
	Topi, 1979	3,630	1	1	3		37.0
	Tunas, 1983	4,819	2	2	101		35.0
	Ubi, 1976	4,350	24	25	2,878		35.0
	Waduk, 1982	6,100	9	11	212		38.0
ASAMERA (NORTH SUMATRA)	Alur Cimon, 1972	3,098	...	24	...	1,378,939	50.8
	Geudondong, 1965	3,192	...	23	...	5,199,661	53.1
	Iee Tabeu, 1971	2,721	1	47	33	11,903,005	50.3
	Julu Rayeu, 1968	2,832	7	47	138	12,422,881	53.6
	Peudawa, 1980	3,083	...	10	...	1,567,420	48.1
	Tualang, 1973	2,631	11	58	1,021	30,856,147	49.9
ASAMERA (SOUTH SUMATRA)	Bentayan, 1932	4,446	16	18	2,037	2,366,357	22.0
	Kluang, 1913	2,591	26	70	476	34,233,038	42.6
	Mangunjaya, 1934	2,700	5	102	152	23,141,435	32.0
	Panerokan, 1976	4,915	2	5	21	206,837	39.4
	Ramba, 1982	3,150	62	74	12,288	55,475,598	37.0
	Rawa, 1985	4,038	26	34	6,138	6,487,085	38.7
	Tanjung Laban, 1982	3,590	21	29	3,994	8,162,791	38.1
	Tempino, 1931	4,493	3	6	128	1,447,940	40.9
	Keri, 1985	3,957	2	2	67	103,940	41.5
	Supat, 1984	4,626	11	11	1,679	984,286	33.6
	South Rawa, 1985	4,086	2	3	76	69,888	37.0
MAXUS ENERGY	• Cinta, 1970	3,500	41	50	11,689	180,237,816	34.0
	• Duma, 1983	2,300	2	6	135	747,577	24.0
	• Farida, 1982	8,000	15	22	3,348	12,854,871	32.0
	• Gita, 1972	5,000	5	9	623	7,732,447	34.5
	• Intan, 1989	...	9	9	17,910	6,537,400	32.6
	• Karmila, 1983	4,800	6	9	7,095	32,317,970	36.0
	• Kitty, 1973	2,700	7	7	1,120	14,382,111	38.0
	• Krisna, 1979	4,500	31	41	7,017	62,216,480	37.0
	• Nora, 1973	3,200	4	5	424	8,962,055	29.0
	• Rama, 1974	3,200	40	76	6,498	93,558,502	31.0
	• Selatan, 1971	4,000	8	19	1,030	21,173,656	19.0
	• Sundari, 1982	4,400	12	15	2,630	11,825,368	25.0
	• Titi, 1982	8,000	1	4	10	1,636,220	33.0
	Wanda, 1984	4,800	2	5	140	1,636,933	36.0
	• Yvonne, 1980	4,500	8	11	3,023	13,724,355	36.0
	• Zelda, 1971	7,500	30	45	6,899	39,165,810	33.0
PETROMER TREND	Cenderawasih, 1976	3,500	18	24	1,554	12,861,086	26.7
	Jaya, 1973	3,050	12	19	1,033	36,178,946	39.7
	Kasim, 1972	3,350	38	54	3,475	48,509,708	36.5
	Walio, 1975	2,750	210	288	13,312	156,932,595	34.3
	Other	...	12	21	998	12,685,379	...
STANVAC (CENTRAL SUMATRA)	Binio, 1972	1,600	19	19	423	11,975,355	34.0
	E. Kayuara, 1983	3,500	45	45	4,010	8,513,939	35.0
	Gemuruh, 1983	2,800	14	16	968	2,740,506	34.0
	Kayuara, 1982	3,700	7	9	337	2,150,314	35.0

PRODUCTION

WORLDWIDE PRODUCTION

	Name of field, discovery date	Depth, ft	No. of wells Producing	No. of wells Total	1989 average B/d	Cumulative to Dec. 31, 1989 Bbl	°API gravity
	Kerumutan, 1980	3,000	19	21	764	5,399,906	35.0
	Merbau, 1979	2,700	28	28	989	7,132,070	36.0
	Mutiara, 1985	2,700	2	2	48	132,620	34.0
	N. Merbau, 1980	2,600	8	8	278	2,936,676	36.0
	Parum, 1987	4,200	9	9	2,063	1,383,769	35.0
	Panduk, 1981	2,500	6	6	224	1,609,536	46.0
	Pekan, 1976	3,700	18	18	393	7,983,900	35.0
	Other	4,600	...	...	...	9,175	—
STANVAC (SOUTH SUMATRA)	Ibul, 1970	5,700	17	17	780	22,193,660	36.0
	Jene, 1985	6,500	22	25	23,765	21,443,720	35.0
	• Kerang (PSC), 1984	2,900	4	5	1,165	1,173,757	35.0
	Lagan, 1986	2,400	4	4	270	303,574	56.8
	Pian, 1986	6,000	3	3	92	97,189	36.0
	Rambutan, 1972	4,500	6	6	66	3,246,970	42.0
	Rimbabat (PSC), 1984	3,200	3	3	366	160,980	34.0
	• Tabuan (PSC)	4,000	9	9	748	1,817,026	34.0
	Other	4,500	6	6	1,060	2,383,653	—
UNION TEXAS	Badak, 1972	5,190-10,600	122	146	5,389	412,604,000	21.0
	Mutiara, 1984	900- 5,800	9	31	2,948	5,012,000	30.0
	Nilam, 1974	7,500-11,500	72	101	5,088	22,231,000	32.0
	Pamaguan, 1974	900- 5,100	13	26	2,197	7,424,000	30.0
	Semberah, 1974	2,900-10,100	...	29	...	7,000	35.0
UNOCAL	• Attaka, 1970	7,500	78	101	51,117	470,523,474	39.0
	• Kerindingan, 1972	7,000	...	8	255	5,038,870	23.8
	• Melahin, 1972	7,500	4	6	750	6,787,008	23.1
	• Rajah, 1981	5,500	6	6	4,481	6,023,749	33.8
	• Sepinggan, 1973	11,150	16	20	4,707	45,421,669	34.0
	• Yakin, 1976	5,100	21	28	7,194	23,614,688	18.8
CONOCO GROUP	• Ikan Pari, 1983	4,500	4	4	2,164	790,031	47.0
	• Kepeting, 1982	6,000	...	2	826	2,563,162	37.0
	• Udang, 1974	5,595	18	28	5,812	59,583,697	40.0
MOBIL OIL	Arun, 1971	10,000	46	62	124,267	381,192,417	55.0
MARATHON	• Kakap KH, 1980	8,500	8	8	6,900	15,559,000	45.0
	• Kapap KF, 1984	8,100	1	1	2,800	67,000	49.0
HUDBAY (MALACCA STRAIT)	• Kurau, 1986	4,500- 9,000	15	17	26,680	12,929,507	47.4
	• Lalang, 1980	3,100- 3,600	11	18	7,485	36,401,391	39.6
	• Melibur, 1984	900- 1,300	12	19	8,740	9,134,245	35.6
	• Mengkapan, 1981	3,850- 4,450	9	10	15,396	20,498,509	42.0
TESORO INDONESIA PETROLEUM	Samboja, 1909	360- 4,380	26	99	363	64,762,820	25.0
	Sanga Sanga, 1897	85- 6,000	108	244	2,540	268,598,560	31.0
	Tarakan, 1906	200- 6,540	122	1,167	1,785	211,485,507	19.0
TARAKAN PETROLEUM	Tarakan (Mamburungan field), 1984	5,020- 5,131	2	4	471	99,522	31.1
	Total Indonesia		**6,963**	**11,568**	**1,231,000**	**13,229,992,628**	

IRAN

	Name of field, discovery date	Depth, ft	Producing	Total	1989 avg B/d	Cumulative Bbl	°API
NIOC	• Aboozar, 1969	2,800	...	...			27.0
	Agha Jari, 1936	7,440	58	89			34.0
	Ahwaz Asmari, 1958	8,150	73	85			32.0
	Ahwaz Bangestan, 1958	12,000	...	7			24.3
	• Bahregansar, 1960	9,000	...	...			30.0
	Bibi Hakimeh, 1961	4,570	19	33			29.9
	Binak, 1959	10,440	...	4			29.9
	Chesmeh Khush, 1967	10,700	...	4			27.7
	Chillingar, 1974	3,350	...	1			39.3
	Dehluran, 1972	11,900	...	6			31.0
	• Dorood, 1961	11,000	...	...			34.0
	• Forozan, 1966	7,000	...	...			28.5
	Gachsaran, 1937	3,400	46	62			31.0
	Haft Kel, 1927	1,980	2	11			37.7
	• Hendijan, 1968	11,000	...	...			23.0
	Karanj, 1963	5,360	7	7			33.9
	Kharg, 1961	10,900	...	...			31.6
	Kupal, 1965	10,500	...	7			32.0
	Lab-e Safid, 1968	4,350	...	7			35.3
	Mansuri, 1963	7,100	4	8			28.4
	Marun, 1963	9,400	55	112			32.6
	Masjid-e Suleiman, 1908	1,600	7	24			39.3
	Naft Safid, 1938	5,000	5	19	2,801,370	37,992,437,094	36.8
	Naft-Shahre, 1923	2,000- 3,000	...	...			43.0
	• Nowruz, 1966	8,200	...	...			21.0
	Parsi, 1964	4,670	22	27			38.8
	Par-e Siah, 1964	4,250	...	1			37.7
	Pazanan, 1961	7,250	44	71			37.2
	Rag-e Safid, 1964	7,230	19	29			28.6
	Ramin, 1966	12,200	...	...			32.9
	Ramshir, 1962	8,750	...	3			27.6
	• Resalat, 1969	7,650	...	...			35.8
	• Reshadat, 1966	7,000	...	...			34.4

PRODUCTION

WORLDWIDE PRODUCTION

		Name of field, discovery date	Depth, ft	No. of wells Producing	No. of wells Total	Production 1989 average B/d	Production Cumulative to Dec. 31, 1989 Bbl	°API gravity
		• Soroush, 1962	7,000	...	...			18.3
		• Salman, 1965	7,500- 8,100	...	...			34.0
		Shadegan, 1989	...	...	...			...
		Sulabedar, 1971	3,400		1			30.4
		Total Iran		361	618	2,801,370	37,992,437,094	
IRAQ(e)	INOC	Abu Ghurab, 1971	9,800	12	16			24.0
		Ain Zalah, 1939	5,200- 6,500	15	28			31.0
		Bai Hassan, 1953	4,800- 5,400	56	84			34.0
		Balad, 1983	6,000- 7,000	4	8			25.0
		Butmah, 1953	3,900	1	17			31.0
		Buzurgan, 1970	12,000	20	22			24.0
		East Baghad, 1975	6,000- 7,000	32	79			23.0
		Hamrin, 1973	2,000	...	14			30.0
		Jabal Fauqui, 1974	10,000	20	23			25.5
		Jambur, 1954	5,500-12,500	27	50	2,897,000	21,671,213,000	38.0
		Kirkuk, 1927	2,800- 4,200	71	269			36.0
		Luhais, 1961	8,200	12	21			33.0
		Naft Khaneh, 1909	3,500	7	11			42.5
		Nahr Umar, 1949	8,200	5	13			42.0
		Qaiyanah, 1929	650	33	50			16.0
		Rumaila North, 1958	10,300	303	443			33.0
		Rumaila South, 1953	10,200	133	208			34.0
		Suffayah, 1978	3,900	11	39			25.0
		West Tikrit, 1985	8,200	2	6			23.0
		Zubair, 1949	10,200	56	110			35.0
		Total Iraq		820	1,511	2,897,000	21,671,213,000	
ISRAEL(e)	LAPIDOTH	Heletz-Brur, 1955	5,200	4	18			29.2
		Kochav, 1962	5,500	6	13			27.9
	OEIL	Ashdod, 1976	8,400	2	2	280	16,840,000	32.5
	EMOG	Gurim, 1984	3,280	...	...			17.0
		Total Israel		12	33	280	16,840,000	
ITALY	ELF	• Rospo Mare, 1975	4,600	14	16	20,576	24,547,151	12.2
		• Santa Maria a Mare, 1974	7,900	8	12	833	17,986,522	22.3
		• Sarago Mare, 1979	5,200	5	6	4,519	11,333,805	8.1
	AGIP	Benevento, 1974	6,599	1	1	493	325,713	39.0
		Cavone, 1979	10,663-18,068	13	15	2,993	8,713,259	24.0
		Cela Mare,	10,300	12	18	1,623	15,289,898	10.0
		Emilio S., 1971	9,215	...	1	...	337,266	13.0
		Gaggiano, 1982	13,766	2	2	260	455,537	37.0
		Gela, 1956	11,280-11,350	53	68	6,995	105,404,448	10.0
		Malossa, 1973	17,881-21,231	5	8	628	26,138,309	55.0
		Narciso, 1985	2,650	...	1	...	337,622	21.1
		Nilde, 1977	4,545	2	3	240	19,716,499	38.9
		Perla, 1976	7,850	4	4	404	1,921,622	13.3
		Pisticci-S. Cataldo, 1960	6,820- 6,880	8	12	1,248	8,356,978	11.4
		Prezloso, 1982	16,400	3	3	3,004	1,342,326	10.0
		Ragusa, 1954	5,760	37	37	4,054	129,361,392	19.5
		S. Maria Inbaro, 1985	6,520	3	4	130	360,273	15.9
		V. Fortuna, 1984	19,828	3	4	9,929	5,263,951	46.0
		Other	...	25	63	1,959	21,508,279	...
	MONTEDISON(e)	• Vega	...	...	...	23,112	21,635,880	...
		Total Italy		198	346	83,000	420,336,730	
IVORY COAST	SIOP	• Belier, 1975	6,500- 7,000	13	18	2,112	18,391,119	34.5
	PETROCI	• Espoir, 1980	6,500-11,000	...	...	...	31,102,952	32.3
		Total Ivory Coast		13	18	2,112	49,494,071	
JAPAN	JAPEX	• Aga-Oki, 1972	7,590	7	7	710	3,293,421	24.8
		Higashi-Niigata, 1965	4,500- 9,200	19	29	577	7,856,914	53.9
		Katakai, 1960	3,310-16,300	5	16	498	775,384	50.0
		Sarukawa, 1958	2,000- 3,200	62	134	1,280	18,704,187	32.3
		Yoshii, 1968	8,000	9	25	661	9,110,857	61.8
		Yurihara, 1976	1,310- 7,610	10	15	266	494,299	43.5
		Other	...	56	75	321	21,199,489	...
	IDEMITSU OIL DEVELOPMENT CO. LTD.	• Agaoki-Kita, 1982	4,950	10	10	2,364	6,500,000	37.0
	TEIKOKU	Hagashi-Kashiwazaki, 1970	6,000- 8,900	5	10	495	8,292,629	62.0
		Kubiki, 1959	1,100- 6,400	95	134	699	11,825,071	29.7
		Minami-Aga, 1964	7,300- 9,600	11	35	512	14,010,670	38.5
		Minami-Nagaoka, 1984	13,200-15,100	5	13	542	1,098,164	55.7
		Yabase, 1933	1,100- 5,800	59	80	394	33,359,850	33.0
		Other	...	15	15	121	3,340,006	...
		Total Japan		368	598	9,440	139,860,491	

PRODUCTION

WORLDWIDE PRODUCTION

		Name of field, discovery date	Depth, ft	No. of wells Producing	Total	Production 1989 average B/d	Cumulative to Dec. 31, 1989 Bbl	°API gravity
JORDAN	NRA	Hamzah, 1984	9,500	4	17	400	57,400	30.0
		Total Jordan		4	17	400	57,400	
KUWAIT(e)	KOC	Burgan, 1938	4,800	210	393			30.8
		Ahmadi, 1952	4,800	11	84			31.7
		Magwa, 1951	4,800	71	113			33.0
		Raudhatain, 1955	8,600	41	53			34.4
		Bahra, 1956	8,500	...	2	1,593,000	23,720,053,000	...
		Sabriya, 1957	8,300	9	44			36.3
		Minagish, 1959	10,000	1	21			34.4
		Umm Gudair, 1962	9,000	20	33			26.9
		Total Kuwait		363	743	1,593,000	23,720,053,000	
LIBYA	AGIP	Bu Attifel, 1968	14,300	33	40	77,464	856,523,977	41.0
		Rimal Katib, 1965	14,000	6	7	2,335	7,405,965	35.0
		● Bouri	8,700	24	24	7,348	4,359,195	26.0
	ELF(e)	El Meheiriga	10,500	1	2			46.0
	LNOC(e)	Arshad, 1966	10,100					42.0
		Jebel, 1962	8,100					38.0
		Majid, 1968	9,000- 9,800					...
		Don Mansour Unit	5,600					...
		Nasser, 1959	5,500- 7,600					38.0
		Lehib-Dor Marada Unit, 1965	9,500					50.0
		Raguba, 1961	5,400					43.0
		Beda, 1959	4,000					36.0
		Bualawn, 1972	6,000					40.0
		Dor, 1967	6,400					33.2
		Kotla, 1963	5,500					34.0
		Sahabi B & D, 1961	9,000-10,000					39.0
		Sarir, 1961	9,000					37.2
		Amal, 1959	9,900					37.5
		Ora, 1962	7,250					40.2
		Ghani, 1978	6,000					40.0
		Bahi, 1968	3,500- 3,750					43.4
		Belhedan, 1962	7,000					33.6
		Dahra, 1959	3,100- 3,700					37.0-41.0
		Defa, 1969	5,400- 5,900					35.6
		Gialo, 1961	2,200- 6,300					35.7
		Harsh, 1978	7,400	491	491	1,042,853	16,053,641,115	34.6
		Khalifa, 1978	5,400- 7,000					34.6
		Masrab, 1978	10,000					39.6
		Samah, 1961	9,000					33.4
		Waha, 1960	6,000					36.0
		Zaggut, 1962	6,000					36.4
		Balat, 1983	5,100- 5,400					32.2
		Ali, 1975	3,900	1	2			50.0
		Almas, 1975	4,000	5	8			49.5
		Intisar A, 1967	9,750	8	19			45.0
		Intisar B, 1974	9,600	2	2			50.0
		Intisar C, 1967	9,650	1	6			37.5
		Intisar D, 1967	9,400	19	27			39.0
		Intisar E, 1968	5,300- 7,100	8	11			34.7
		Nafoora-Augila, 1966	2,200-10,000	13	55			35.5
		Aswad, 1978	6,250	8	8			46.0
		Sabah, 1977	5,500	36	38			42.0
		Zella, 1977	6,750	27	31			48.1
		Other	7,900- 8,900	1	2			40.4
		Fidaa, 1979	7,000	9	9			41.7
		Hakim, 1978	6,700	3	3			48.5
		Total Libya		696	783	1,130,000	16,921,930,252	
MALAYSIA	ESSO MALAYSIA	● Bekok, 1976	5,500	22	33			47.0
		● Guntong, 1978	5,200	56	91			44.0
		● Irong Barat, 1979	3,300	9	18			35.0
		● Kepong, 1979	6,000	...	11			47.0
		● Palas, 1979	6,300	13	18			49.0
		● Pulai, 1973	4,300	22	32			42.3
		● Seligi, 1971	4,300	16	22			50.0
		● Semangkok, 1980	3,700	31	41	315,000	820,000,000	42.5
		● Tabu, 1978	5,600	15	24			44.0
		● Tapis, 1975	6,700	41	80			45.0
		● Tinggi, 1980	4,400	21	27			48.0
		● Tiong, 1978	5,800	20	41			47.0
	SARAWAK SHELL BERHAD	● D-18, 1981	4,000	5	5	3,550	2,899,000	39.9
		● Bayan, 1976	4,200	7	9	13,110	26,498,000	37.3
		● Temana, 1972	3,100	19	29	9,560	40,199,000	35.0
		● Other	5,600-11,000	29	33	10,930	268,726,000	

PRODUCTION

WORLDWIDE PRODUCTION

		Name of field, discovery date	Depth, ft	No. of wells Producing	No. of wells Total	Production 1989 average B/d	Production Cumulative to Dec. 31, 1989 Bbl	°API gravity
SABAH SHELL	•	Barton, 1971	2,400	4	4	3,010	10,402,000	37.0
	•	Erb West, 1977	6,785	15	19	6,630	21,620,000	30.0
	•	Ketam, 1977	4,800	1	4	270	999,000	30.0
	•	Samarang, 1972	5,700	56	80	59,840	257,301,000	37.0
	•	South Furious, 1974	3,300	7	15	3,010	13,499,000	30.0
	•	St. Joseph, 1975	2,200	11	14	16,360	20,271,000	32.0
PETRONAS CARIGALI[e]	•	Tembungo, 1971	6,000	8	16			40.0
SARAWAK SHELL/PETRONAS CAG.[e]	•	Bakau	...	...	...			...
	•	Baram	...	...	...			...
	•	Baronia, 1967	7,900	19	40			42.0
	•	Betty, 1967	7,900	13	17			38.0
	•	Bokor, 1972	2,000	18	19	125,730	577,182,000	35.0
	•	Fairley Baram, 1973	8,655	...	2			39.5
	•	Siwa, 1973	4,100	4	4			25.0
	•	Tukau, 1966	5,000	26	48			29.2
	•	West Lutong	...	...	...			39.5
		Total Malaysia		**508**	**796**	**567,000**	**1,990,000,000**	

MEXICO[e]	NORTH ZONE, NE	Misión, 1945	5,440					...
		Monterrey, 1950	6,510					...
		Tigrillo, 1971	3,342					...
	NORTH ZONE, N.	• Arenque, 1970	11,362					26.0
		Constituciones, 1956	6,350					17.0
		Ebano-Pánuco, 1901	1,450					12.0
		Tamaulipas	4,200			41,200		19.0
	NORTH ZONE, S.	Chicontepec, 1973	4,264					34.1
		• Isla de Lobos, 1963	6,875					40.0
		• Marsopa, 1974	10,198					39.0
		Naranjos-C. Azul, 1909	1,800					22.0
		Tres Hermanos, 1959	6,960					21.0
		Other	...					...
	CENTRAL ZONE, POZA RICA	• Atún, 1966	9,040					37.0
		• Bagre, 1973	10,919					33.0
		Escolín, 1942	7,216					26.3
		Hallazgo, 1955	10,170					25.0
		Jiliapa, 1958	7,390					24.0
		Mecatepec, 1941	7,544					29.8
		Miquetla, 1959	6,480					35.0
		Poza Rica, 1930	7,090					35.0
		Pres. Alemán, 1949	8,036					26.0
		Remolino, 1962	10,745			57,700		29.5
		San Andrés, 1956	10,410					29.0
		Other	...					...
	CENTRAL ZONE, CUENCA DEL PAPALOAPAN	Matapionche, 1974	11,129					35.6
		Other	...					...
	CENTRAL ZONE, N. FAJA DE ORO	Acuatempa, 1955	4,085					21.0
		Muro, 1965	3,966					17.0
		Ordóñez, 1952	5,220					21.0
		Ocotepec, 1953	3,737					20.0
		Santa Agueda, 1953	4,780					16.0
		Other	...					...
	SOUTHERN ZONE, NANCHITAL	Ixhuatlán Ote., 1965	1,960					22.6
		Moloacán, 1962	500					22.3
		Santa Rosa, 1952	1,312					25.4
	SOUTHERN ZONE, EL PLAN	Agata, 1956	3,830					...
		Bacal, 1976	3,500					35.0
		Concepción, 1974	1,600					31.0
		Cuichapa, 1935	2,200					30.0
		El Plan, 1931	1,700					30.0
		Lacamango, 1973	1,700					26.3
		Los Soldados, 1953	4,492					32.0
		Other	...					...
	SOUTHERN ZONE, AGUA DULCE	Blasillo, 1967	7,216			65,400		40.0
		Cinco Presidentes, 1960	6,862					35.0
		Burro, 1931	2,200					26.0
		Venta, 1954	4,730					41.0
		Ogarrio, 1957	5,790					38.0
		Otates, 1965	7,469					39.0
		Rodador, 1971	11,398					26.0
		Sánchez Magallanes, 1957	4,240					27.0
		San Ramón, 1967	9,517	4,740	4,740		16,547,594,500	30.0
		Tonalá, 1928	1,770					28.0
		Other	...					...
	SOUTHEAST ZONE, COMALCALCO	Agave, 1977	13,450					41.7
		Arroyo Zanapa, 1978	14,599					38.9
		Arteza, 1977	11,800					26.4

PRODUCTION

WORLDWIDE PRODUCTION

		Name of field, discovery date	Depth, ft	No. of wells Producing	No. of wells Total	Production 1989 average B/d	Cumulative to Dec. 31, 1989 Bbl	°API gravity
		Ayapa, 1972	8,200					7.2
		Cacho López, 1977	14,250					29.0
		Cáctus, 1972	14,100					31.5
		Carrizo, 1962	3,500					23.3
		Castarrical, 1967	10,080					29.3
		Comoapa, 1979	14,432					44.5
		Copano, 1977	11,890					43.9
		Cunduacán, 1974	13,775					28.9
		El Golpe, 1963	3,500					25.7
		Giraldas, 1977	15,225					37.0
		Iride, 1974	13,775					28.5
		Muspac, 1982	9,676					57.0
		Mecoacán, 1957	2,200					8.6
		Mora, 1981	...					38.1
		Mundo Nuevo, 1977	11,800					46.0
		Níspero, 1974	14,100					34.4
		Paredón, 1978	15,690					39.8
		Oxiacaque	11,150			607,200		29.1
		Platanal, 1978	15,900					30.2
		Río Nuevo, 1975	14,950					34.8
		Samaria (Cret), 1973	14,209					31.0
		Santuario, 1966	9,617					37.0
		Sitio Grande, 1972	13,766					35.0
		Sunuapa, 1978	12,887					32.8
		Tapijualpa, 1982	15,613					39.8
		Tintal, 1968	5,904					22.0
		Topén, 1978	11,172					25.4
		Tupilco, 1959	9,685					27.0
		Bellota, 1982	17,056					41.0
		Cardenas, 1979	17,548					38.0
		Chiapas, 1979	12,136					42.0
		Fenix, 1979	17,876					40.0
		Iris, 1979	14,432					39.2
		Jujo, 1980	17,548					32.0
		Caparroso, 1982	18,368					41.2
		Tecominoacan, 1983	19,519					37.5
		Jolote, 1983	18,119					39.4
		Artesa Terc., 1984	...					28.0
		Carmito, 1980	10,050					45.6
		Edén, 1983	17,590					38.5
		Samaria Terc., 1965	4,920					13.9
MARINE ZONE GULF OF CAMPECHE	•	Cantarell, 1976	8,528					21.3
	•	Abkatun, 1978	11,800					30.0
	•	Ku, 1979	10,000					22.0
	•	Pol, 1979	12,600			1,741,800		35.0
	•	Chuc, 1982	13,100					35.0
	•	Ixtoc, 1984	...					35.0
		Total México		4,740	4,740	2,513,300	16,547,594,500	
MOROCCO(e)		Essaouira	...	4	4	270	16,365,350	...
		Rharb, 1951	4,590	1	1			...
		Total Morocco		5	5	270	16,365,350	
MYANMAR(e)	MYANMA	Chauk-Lanywa, 1902	1,500- 4,000					36.0
		Mann, 1970	1,600- 7,000					37.5
		Myanaung, 1964	2,000- 4,000	450	450	12,000	499,555,000	40.5
		Prome, 1965	850- 5,000					36.0
		Yenangyaung, 1902	100- 6,700					36.0
		Total Myanmar		450	450	12,000	499,555,000	
NETHER-LANDS	UNOCAL	• Haven, 1980	5,300	2	2	2,746	167,484	25.5
		• Helder, 1980	4,800	11	13	5,778	21,637,264	21.6
		• Helm, 1979	4,400	9	9	1,692	8,245,485	17.9
		• Hoorn, 1981	5,000	7	8	4,564	20,943,359	25.8
	NAM	Berkel, 1954	...	17	17	2,417	6,720,000	32.1
		DeLier, 1955	5,200	22	33	472	17,120,000	33.2
		Ijeselmonde, 1956	...	59	86	4,468	92,418,000	26.8
		Pernis West, 1987	...	...	...	...	38,000	28.2
		Plinacker, 1955	...	4	5	421	6,340,000	36.3
		Rotterdam, 1984	...	2	2	3,946	3,719,000	30.5
		Schoonebeek, 1944	2,500	...	495	...	238,327,000	12.3
		Wassenaar, 1956	4,200	...	41	...	46,977,000	18.7
		Werkendam, 1956	...	...	...	...	252,000	33.0
		Zoetermeer, 1957	3,200	30	30	795	10,332,000	16.5
	CONOCO	• Kotter, 1980	6,500	8	11	15,729	37,215,459	33.0
		• Logger, 1982	6,500	6	9	9,913	16,012,118	33.5
	AMOCO	• Rijn, 1982	6,600	12	25	7,077	17,198,048	35.0
		Total Netherlands		188	786	60,018	543,662,217	

PRODUCTION

WORLDWIDE PRODUCTION

		Name of field, discovery date	Depth, ft	No. of wells Producing	Total	Production 1989 average B/d	Cumulative to Dec. 31, 1989 Bbl	°API gravity
NEUTRAL ZONE	AOC	• Hout, 1969	5,300-10,500	21	34	32,881	292,130,916	33.0
		• Khafji, 1961	4,400-10,300	136	165	232,777	2,526,521,038	28.2
	TEXACO & KOC[e]	S. Fuwaris, 1963	6,300	5	9			25.0
		S. Umm Gudair, 1966	8,900	16	17	131,342	1,783,940,000	24.3
		Wafra, 1953	1,100- 7,000	322	427			18.9-23.5
		Total Neutral Zone		**500**	**652**	**397,000**	**4,602,591,954**	
NEW ZEALAND	SHELL-BP-TODD	• Kapuni, 1959	11,700	11	14	6,100	41,520,000	54.0
		• Maui, 1969	11,000	14	14	15,000	34,200,000	51.0
	PCNZ	Ahuroa, 1987	8,688	...	1	...	9,800	46.1
		Kaimiro, 1982	11,900	2	2	74	145,000	42.0
		McKee, 1980	7,848- 8,251	19	22	11,000	17,680,000	38.5
		Tariki, 1986	9,065	...	1	...	1,500	52.7
		Waihapa, 1988	10,125-10,643	5	6	6,100	3,300,000	36.6
		Total New Zealand		**51**	**60**	**38,274**	**96,856,300**	
NIGERIA	AGIP	Agwe, 1975	10,800	1	1	110	2,042,504	47.9
		Akri, 1967	9,600-10,600	3	11	1,862	60,056,817	40.1
		Akri West, 1972	9,900-10,200	...	2		482,258	31.4
		Ashaka, 1968	9,700-12,000	1	1	1,145	2,762,447	38.4
		• Beniboye N., 1978	9,000-10,000	9	9	8,904	11,572,397	29.6
		Beniku, 1974	12,000	...	2		3,309,326	29.5
		Clough Creek, 1977	10,000-12,400	7	9	7,439	22,460,507	29.3
		Ebegoro, 1976	11,500-12,000	7	9	7,509	62,109,913	35.7
		Ebocha, 1965	8,000-10,900	8	16	6,959	110,337,245	35.5
		Idu, 1973	7,750-10,700	...	10		13,862,581	28.0
		Kwale, 1967	10,000-11,500	8	10	7,667	7,500,675	36.4
		M'Bede, 1966	7,300- 9,400	10	20	14,132	163,238,542	40.9
		Obama, 1973	11,600-14,300	6	8	14,577	97,780,167	37.2
		Obiafu, 1967	9,300-12,200	18	20	15,815	85,766,070	42.4
		Obrikom, 1973	8,000-10,000	6	13	3,371	55,040,901	45.1
		Odugri, 1972	12,500	2	2	634	9,788,065	47.2
		Ogbogene, 1972	10,000	...	2	...	3,302,763	73.0
		Ogbogene W., 1982	13,000	...	1	...	353,196	56.2
		Okpai, 1968	10,000-12,000	8	8	4,779	19,031,162	44.6
		Omoku W., 1975	11,500	...	1		4,746,492	27.4
		Oshi, 1972	9,600-11,300	10	11	9,890	66,552,033	34.1
		Taylor Creek, 1975	14,100	1	1	469	473,058	37.7
		Tebidaba, 1975	10,300-13,300	7	10	17,017	118,463,094	33.0
		Umuoru, 1975	14,100	3	4	2,477	5,353,184	40.0
	ASHLAND	• Adanga, 1980	10,611-17,586	9	9	12,953	15,386,228	35.0
		• Akam, 1980	6,700	10	10	7,561	5,802,362	35.2
		• Bogi, 1989	8,256	3	4	1,466	563,367	23.4
		• Ebughu, 1983	7,250	3	3	457	167,263	21.0
		• Izombe-Ossu, 1974	9,500	8	8	13,262	10,645,430	35.0
		• Mimbo, 1989	7,973	4	5	1,967	554,146	36.0
		• Ukpam, 1989	7,425	1	2	2,474	945,474	40.9
	ELF	Aghigo, 1972	7,303	17	29	4,446	24,405,554	24.8
		Erema, 1972	11,211	4	5	4,715	10,847,810	24.6
		Obagi, 1964	8,377	61	68	48,743	368,074,953	24.6
		Obodo-Jatumi, 1966	...	16	18	16,116	69,115,408	24.8
		Okpoko, 1967	6,745	16	17	7,304	32,326,957	24.8
		Upomani, 1965	6,256	14	14	3,600	16,898,167	24.8
	GULF NIGERIA	Abiteye, 1970	5,750- 9,400	10	19	5,294	64,007,793	39.7
		• Delta, 1965	5,600- 9,500	20	27	18,566	169,822,039	37.3
		• Delta South, 1965	7,100-10,179	18	28	31,483	269,735,952	38.4
		• Isan, 1970	5,900- 9,000	6	11	2,369	42,897,085	40.4
		Jisike, 1975	6,300- 7,600	1	4	1,589	6,338,968	41.1
		Makaraba	7,100-12,005	20	25	20,819	92,637,107	27.7
		• Malu, 1969	4,800- 6,300	12	19	16,371	102,394,051	40.4
		• Mefa, 1965	8,570-12,030	5	5	5,832	67,968,348	38.1
		• Meji, 1965	5,200-10,900	19	22	18,979	151,312,520	31.9
		• Meren, 1965	5,000- 7,500	31	55	57,685	493,040,316	31.9
		• Mina	...	3	3	6,161	2,248,843	40.3
		• Okan, 1964	5,500- 9,245	35	71	33,485	390,246,940	38.1
		• Parabe/Eko, 1968	4,500- 8,200	13	26	10,783	103,215,846	40.4
		Robertkiri, 1964	11,484-13,190	10	12	10,403	12,516,783	40.2
		• Tapa, 1978	8,150-10,842	8	10	9,233	28,142,426	39.5
		Utonana, 1971	7,400- 9,165	1	5	816	8,165,121	20.4
		• W. Isan, 1971	7,825-10,229	5	9	2,011	34,745,318	40.4
		Yorla South, 1974	11,389-12,635	...	2	...	834,040	41.0
	MOBIL	• Adua, 1967	6,970	5	5	10,768	46,537,232	30.3
		• Asabo, 1966	5,600	9	14	15,175	161,113,170	32.4
		• Asabo D	...	1	2	1,241	2,250,081	35.0
		• Ata	...	...	...	...	433	...
		Edop	8,528- 9,901	3	4	21,558	8,394,906	37.7
		• Ekpe, 1966	8,200	8	17	9,613	188,394,905	30.3
		• Ekpe-WW, 1977	6,810	5	6	7,449	190,3325,525	30.3

PRODUCTION

WORLDWIDE PRODUCTION

	Name of field, discovery date	Depth, ft	No. of wells Producing	Total	1989 average B/d	Cumulative to Dec. 31, 1989 Bbl	°API gravity
	• Eku, 1966	5,420	3	7	3,178	15,955,736	30.8
	• Enang, 1968	6,600	18	23	23,127	35,960,995	35.1
	• Etim, 1968	6,200	7	9	16,989	112,511,703	32.9
	• Idoho, 1966	9,020	3	4	1,017	28,842,611	30.8
	• Inim, 1966	5,850	7	10	14,963	145,437,397	37.8
	• Isobo, 1968	7,345	3	5	3,263	8,180,634	30.8
	• Iyak SE	7,441	6	5	16,155	40,682,165	38.7
	• Mfem, 1967	5,200	2	4	4,185	25,627,206	36.1
	• Oso	...	...	...	...	...	...
	• Ubit, 1968	5,400	30	45	25,322	185,674,618	36.1
	• Unam, 1968	5,180	7	8	9,744	43,881,190	33.5
	• Usari	...	...	...	...	5,885	...
	• Utue, 1966	5,700	5	7	8,653	70,170,645	36.8
PAN OCEAN (NIGERIA)	Ogharefe, 1973	9,900	7	7	2,764	9,890,444	47.4
SHELL	Abura	...	1	3	...	17,247,935	44.6
	Afremo	...	9	9	18,992	12,036,226	36.9
	Ahia	...	8	13	4,171	86,874,306	39.4
	Akaso		...	1	...	...	...
	Akpor	...	...	3	...	...	27.5
	Alakiri	9,625-10,712	12	21	7,248	68,233,759	44.7
	Amukpe	...	2	2	1,853	13,154,765	41.1
	Assa	...	2	2	1,051	5,562,686	20.16
	Bugama Creek	...	5	7	1,366	13,312,766	36.8
	Cesw-Adibawa, 1967	11,950	12	23	16,383	82,548,600	26.2
	Cesw-Adibawa NE, 1973	...	3	3	6,175	17,107,947	27.5
	Cesw-Diebu Creek, 1966	13,119	4	8	10,620	108,910,459	40.8
	Cesw-Etelebou, 1971	12,000	9	9	15,746	126,903,233	31.0
	Cesw-Kolo Creek, 1960	12,000	15	25	20,023	136,712,922	39.4
	Cesw-Nun River, 1960	14,200	4	6	2,098	32,772,212	33.7
	Cesw-Ubie, 1961	14,380	5	6	4,755	33,017,108	26.7
	Cosw-Nembe Creek, 1973	12,213	35	36	86,744	249,866,511	30.6
	Eade-Opobo South, 1974	...	4	6	4,109	20,268,510	43.1
	Eade-Utapate South, 1974	...	6	8	6,418	39,304,491	39.0
	Ebubu	...	2	8	157	18,808,065	19.8
	Egb-Oguta, 1965	10,300	8	11	6,346	132,487,741	43.9
	Egbema	...	6	7	4,306	55,381,361	34.8
	Egbema West	...	5	7	4,622	81,832,450	41.1
	Enwhe	...	...	2	...	...	...
	Forc-Ajuju, 1970	13,324	1	2	...	...	27.0
	Forc-Batan, 1968	11,000	7	8	6,285	57,482,932	21.3
	Forc-Benisede, 1973	...	11	14	31,089	114,511,463	22.0
	Forc-Egwa, 1967	9,350	17	19	18,337	140,603,754	33.4
	Forc-Escravos Beach, 1969	8,176	7	8	10,140	61,920,984	30.4
	Forc-Forcados/Yokri, 1968	10,859	51	52	69,810	571,116,961	24.4
	Forc-Jones Creek, 1967	7,000- 9,000	24	26	44,199	410,217,843	29.8
	Forc-Odidi, 1967	10,980	25	28	43,942	222,326,504	36.4
	Forc-Opuama, 1973	...	2	2	2,734	26,496,113	42.4
	Forc-Opukushi, 1963	7,823	9	10	26,104	124,368,680	28.4
	Forc-Opukushi N., 1977	...	1	2	556	2,791,314	24.7
	Forc-Otumara, 1969	8,176	21	25	41,532	171,485,364	24.9
	Forc-Saghara, 1970	8,590	5	6	7,858	34,812,316	31.9
	Ibigwe	...	1	2	...	...	...
	Kalaekule	...	6	7	13,090	8,932,181	40.3
	Korokoro	...	6	8	2,703	84,303,645	36.0
	Krakama	...	3	10	1,662	18,049,647	22.8
	Node-Oben, 1972	12,036	14	23	19,935	128,437,854	37.5
	Node-Sapele, 1970	12,788	10	11	13,982	118,226,990	43.1
	Obele	...	...	5	...	...	...
	Odeama Creek	...	4	5	20,847	28,171,077	35.4
	Osioka	...	1	2	786	7,502,989	26.2
	Otamini	...	2	2	1,649	14,564,408	20.8
	Phl-Afam, 1956	6,000	9	9	2,240	60,521,763	45.2
	Phl-Agbada, 1960	8,000-12,000	30	31	20,718	188,963,520	24.4
	Phl-Ajokpori, 1967	...	1	2	453	5,654,946	39.0
	Phl-Akuba, 1967	...	1	1	402	4,392,385	22.7
	Phl-Apara, 1960	9,000	4	4	4,935	27,913,330	38.3
	Phl-Bodo-West, 1959	9,700	9	9	4,919	82,443,920	26.5
	Phl-Bomu, 1958	6,500- 7,500	9	11	7,574	357,197,222	36.8
	Phl-Elelenwa, 1959	11,000	5	6	3,684	33,750,837	35.3
	Phl-Imo River, 1959	5,800-10,000	37	40	37,788	481,411,176	31.0
	Phl-Isimiri, 1964	5,900-11,000	1	3	1,617	38,228,064	26.0
	Phl-Nkali, 1963	12,000	3	3	3,848	30,539,652	39.3
	Phl-Obeakpu, 1975	...	2	2	3,098	18,230,182	26.4
	Phl-Obigbo-North, 1963	6,500-10,000	23	23	12,329	178,774,185	23.3
	Phl-Onne, 1965	10,384	3	3	1,210	9,804,577	34.3
	Phl-Umuechem, 1969	5,800-10,700	8	10	4,495	149,809,342	35.6
	Phl-Yorla, 1970	11,917	4	7	4,367	61,810,385	39.6
	Phs-Akaso, 1979	...	1	1	1,486	3,770,294	36.6
	Phs-Bonny, 1959	12,254	8	8	5,260	91,565,141	31.8
	Phs-Cawthorne Channel, 1963	11,000	19	21	44,904	297,078,967	38.3
	Phs-Ekulama, 1958	10,483	20	20	37,116	215,051,740	32.0
	Phs-Orubiri, 1971	...	1	2	2,933	18,120,944	37.98

PRODUCTION

WORLDWIDE PRODUCTION

	Name of field, discovery date	Depth, ft	No. of wells Producing	Total	1989 average B/d	Production Cumulative to Dec. 31, 1989 Bbl	°API gravity
	Phs-Soku, 1958	11,500	17	23	10,985	115,359,210	28.0
	Rapele	...	1	3	489	23,012,074	44.4
	Rumuekpe	...	2	3	2,217	1,829,155	29.3
	Tai	...	1	2	554	4,887,832	38.1
	Ugada	...	1	1	...	...	...
	Ugh-Afiesere, 1966	8,000- 9,000	23	27	10,466	121,140,595	18.9
	Ugh-Eriemu, 1961	12,500	12	14	5,480	45,696,354	20.1
	Ugh-Evwreni, 1968	10,900	8	9	3,497	37,874,976	25.2
	Ugh-Isoko, 1960	...	1	2	765	6,234,074	30.3
	Ugh-Kokori, 1960	8,000- 9,800	7	8	10,061	315,106,094	45.2
	Ugh-Ogini, 1964	5,860	4	5	2,944	23,389,308	18.7
	Ugh-Olomoro, 1963	7,000-10,000	16	22	15,234	286,297,772	21.4
	Ugh-Oroni, 1964	12,000	4	5	2,031	26,716,637	22.4
	Ugh-Oweh, 1964	12,300	9	9	5,841	96,495,512	26.2
	Ugh-Ughelli-East, 1959	11,800	7	7	2,571	90,512,168	38.6
	Ugh-Ughelli-West, 1963	7,400-10,200	7	9	5,483	35,202,936	20.4
	Ugh-Utorogu, 1964	9,000	16	16	7,246	133,198,137	28.3
	Ugh-Uzere-East, 1960	8,500	11	11	6,497	90,364,319	33.9
	Ugh-Uzere-West, 1964	8,500	10	10	5,723	101,566,708	24.2
	Ugh-Warri-River, 1961	12,264	1	2	795	15,366,811	29.4
TEXACO-CHEVRON-NNPC	• Funiwa, 1978	5,000- 7,000	20	22	22,177	43,904,868	38.2
	• Middleton, 1972	5,000- 7,000	3	7	2,363	15,813,022	36.6
	• North Apoi, 1973	4,000- 8,100	22	26	33,776	162,994,363	35.8
	• Pennington, 1965	5,000-10,400	1	10	2,000	35,109,345	38.2
	• Sengana, 1967	11,400-12,500	...	2	...	666,120	46.7
TENNECO	Abura	13,452	1	...	1,451	17,247,935	44.6
AGIP ENERGY	Agbara, 1988	10,699	16	16	9,500	2,365,648	42.5
DUBRI	Gilli-Gilli	...	1	...	981	8,074,119	46.9
	Total Nigeria		**1,432**	**1,898**	**1,673,809**	**12,752,358,229**	

NORWAY

AMOCO	• Valhall, 1972	8,200	22	22	73,092	129,725,993	35.0
BP	• Ula, 1976	10,100	6	13	88,037	98,724,867	38.8
CONOCO	• Murchison*, 1975	9,900	18	30	8,160	57,716,337	38.0
NORSK HYDRO	• Oseberg, 1979	7,500	8	12	233,003	85,046,019	34.0
PHILLIPS	• Albuskjell, 1972	10,400	13	19	2,447	42,162,601	45.0
	• Cod, 1968	9,500	6	8	1,191	15,659,706	55.0
	• Edda, 1972	10,480-10,600	7	7	2,948	22,722,952	41.5
	• Ekofisk, 1968	10,000	41	44	112,782	931,408,127	41.1
	• Eldfisk, 1970	9,350-10,000	37	39	76,978	262,167,186	41.1
	• Tor, 1970	10,000	12	13	8,703	110,437,189	38.0
	• West Ekofisk, 1970	10,300	9	14	3,717	71,550,961	43.9
STATOIL	• Gullfaks, 1978	9,500- 9,700	63	95	267,000	181,100,000	9.3
	• Statfjord†, 1974	9,500	59	93	585,790	1,455,195,549	38.2
	• Tommeliten, 1976	10,400	5	6	13,534	4,939,866	47.0
	• Veslefrikk, 1981	10,100	12	18	...	...	...
	Total Norway		**318**	**433**	**1,477,382**	**3,468,557,353**	

*Norway share (22.2%) of total production. †Norway share (84.09%) of total production.

OMAN

PDO	Al Burj, 1987	6,890	6	7	850	2,288,000	28.6
	Al Ghubar, 1974	1,916	3	6	5,300	6,842,000	28.4
	Al Huwaisah, 1969	5,320	43	47	16,500	177,091,000	37.1
	Amal/Amal South, 1973	4,638	34	39	11,900	15,974,000	24.0
	Anzauz, 1984	7,098	4	4	4,600	8,088,000	46.0
	Bahja, 1984	5,697	10	12	12,100	12,822,000	48.9
	Barik, 1974	8,856	3	3	2,300	3,190,000	47.8
	Birba, 1978	8,528	...	...	1,446	8,394,000	30.5
	Burhaan, 1984	4,011	2	4	900	1,981,000	36.9
	Dhiab, 1985	5,414	6	6	900	955,000	25.7
	Fahud, 1964	2,499	153	235	56,000	794,423,000	32.5
	Fayyadh, 1983	3,608	1	1	6,300	706,000	27.5
	Ghaba North, 1972	2,227	14	25	6,000	54,850,000	34.5
	Habur, 1975	5,800	1	2	250	4,481,000	15.6
	Hasirah, 1984	8,708	10	12	9,100	7,339,000	49.8
	Ishan, 1984	5,578	5	6	1,800	1,335,000	28.6
	Jalmud, 1979	4,236	3	3	270	467,000	22.5
	Jalmud North, 1980	4,265	2	2	270	351,000	22.5
	Jameel, 1984	5,906	4	5	1,500	1,025,000	28.6
	Jawdah, 1984	3,570	6	7	1,800	2,385,000	22.5
	Karim West, 1982	3,546	26	27	12,200	18,531,000	23.1
	Kaukab, 1983	8,282	1	1	700	1,401,000	31.9
	Lekhwair, 1968	4,352	47	63	26,200	75,262,000	38.6
	Mahjour, 1979	11,700	1	3	600	2,461,000	47.5
	Marmul, 1977	5,087	169	185	34,100	137,075,000	22.3
	Mawhoob, 1985	6,562	1	1	40	95,400	28.6
	Mukhaizna, 1975	3,478	...	...	...	12,400	15.0
	Natih, 1963	4,788	36	68	24,700	360,400,000	31.2
	Nimr, 1980	3,296	117	119	36,000	37,844,000	21.3

PRODUCTION

WORLDWIDE PRODUCTION

		Name of field, discovery date	Depth, ft	No. of wells Producing	No. of wells Total	Production 1989 average B/d	Production Cumulative to Dec. 31, 1989 Bbl	°API gravity
		Qaharir, 1978	4,536	27	32	7,400	26,432,000	30.4
		Qaharir East, 1978	4,936	2	4	150	791,000	30.4
		Qarn Alam, 1972	1,354	4	7	900	19,665,000	15.6
		Qata, 1979	4,707	11	16	2,100	5,023,000	28.6
		Rahab, 1977	3,237	21	28	2,900	7,523,000	25.7
		Ramlat Rawl, 1987	2,720	1	1	2,500	1,039,000	39.7
		Rasha, 1982	4,310	4	5	700	1,560,000	32.9
		Rima, 1979	3,247	47	48	65,500	153,293,000	32.9
		Runib, 1978	3,215	10	10	2,300	2,651,000	20.3
		Saih Nihaydah, 1972	8,700	21	28	30,000	168,580,000	39.4
		Saih Rawl, 1973	9,826	14	24	7,500	93,702,000	39.7
		Sayyala, 1981	3,690	14	14	22,700	38,483,000	47.8
		Shibkah, 1976	2,591	...	2	40	835,000	31.2
		Simsim, 1985	5,250	11	12	5,000	4,817,000	28.1
		Suwaihat, 1982	6,452	10	10	10,600	15,640,000	41.5
		Thuleilat, 1986	3,773	14	15	3,300	3,096,000	23.8
		Wafra, 1983	5,510	7	8	5,100	3,791,000	36.5
		Warad, 1985	4,922	5	5	1,600	1,256,000	29.3
		Yibal, 1962	8,465	194	206	149,100	722,333,000	37.9
		Zahra, 1986	6,562	5	5	1,100	839,000	29.3
		Zareef, 1985	3,264	10	11	12,000	10,405,000	47.3
		Zauliyah, 1981	8,177	10	14	5,400	14,879,000	46.9
	OCCIDENTAL	Safah, 1983	6,500	81	81	22,300	15,795,000	42.0
	ELF GROUP	Ramlat	...	1	1	80	148,000	46.0
		Sahmam, 1978	12,400	8	8	5,300	26,480,000	46.0
		Total Oman		**1,240**	**1,488**	**640,196**	**3,077,124,800**	
PAKISTAN	PAKISTAN OILFIELD LTD.	Balkassar, 1946	9,398	4	17	589	31,903,730	26.0
		Dhulian, 1936	8,899	12	49	24	41,332,384	48.0
		Meyal, 1968	12,513	6	6	2,913	31,923,697	41.3
		Joya Mair, 1944	6,900	2	4	411	6,069,324	15.0
		Khaur, 1915	9,306	13	397	19	4,177,167	33.0
	PAKISTAN PETROLEUM LTD.	Adhi, 1978	9,452- 9,613	5	5	398	1,467,721	51.0
	OGDC	Bobi, 1988	9,515- 7,897	3	3	611	223,085	52.0
		Chak-Naurang, 1986	9,187	3	4	...	322,508	12.1
		Fimkassar, 1982	11,465	1	1	425	190,952	29.4
		Ghotana, 1986	7,300	1	1	...	103,902	41.1
		Lashari Centre, 1988	6,916	1	1	1,763	682,185	44.2
		Pasahki, 1989	7,153	1	1	202	77,257	43.0
		Sono, 1988	7,618	3	4	1,758	1,045,318	40.5
		Tando Alam, 1984	6,890- 7,448	4	10	2,472	6,429,984	41.7
		Thora, 1987	7,546	5	5	4,091	2,612,488	42.2
		Toot, 1968	15,135	5	14	1,153	10,325,565	39.0
	UNION TEXAS	Dabhi, 1984	5,594	2	4	1,045	2,210,000	42.0
		Khaskeli, 1981	3,387	4	11	999	7,421,232	38.0
		Laghari, 1983	2,584	4	7	5,409	13,944,754	42.0
		Mazari, 1985	3,888	5	5	6,141	6,946,062	43.0
		South Mazari, 1985	3,920	2	4	2,463	899,021	43.0
		Other	...	7	8	549	200,230	31.0-46.0
	OCCIDENTAL	Bhangali, 1989	12,650	1	1	2,240	206,040	32.5
		Dhurnal, 1984	13,100	4	6	17,169	28,573,996	38.0
		Total Pakistan		**98**	**568**	**52,844**	**199,288,602**	
PERU	PETROMAR S.A.	• Litoral, 1955	1,500- 6,000	103	134	2,205	38,155,610	36.0-40.0
		• Lobitos, 1960	3,500- 7,500	140	268	6,672	54,791,522	38.0
		• Pena Negra, 1960	3,000- 8,500	183	304	9,876	112,559,041	38.0
		• Providencia, 1967	3,000- 7,000	47	92	1,716	21,882,501	38.0-40.0
		Other	...	4	12	98	291,705	...
	OCCIDENTAL	Bartra, 1979	8,000- 8,930	7	16	2,883	19,215,565	11.6
		Capahuari Sur, 1973	11,800-12,900	14	21	7,722	134,808,739	33.7
		Dorissa, 1978	9,800-10,600	4	13	6,175	39,421,977	32.0
		Forestal, 1973	9,000- 9,900	6	12	4,928	36,696,670	18.5
		Huayuri Sur, 1977	10,000-10,800	5	12	2,561	19,505,497	28.5
		Jibarito, 1981	9,000- 9,650	4	5	4,737	13,186,861	10.5
		Jibaro, 1974	10,110-10,155	4	6	2,306	8,330,884	10.6
		San Jacinto, 1978	7,200- 9,760	11	22	5,859	31,091,167	14.9
		Shiviyacu, 1973	9,100-10,100	18	26	17,909	71,596,811	20.3
		Other	...	3	12	701	2,307,545	21.3
	OCCIDENTAL/BRIDAS	Talara area, 1978	1,343- 6,080	757	1,244	5,647	48,614,022	34.8
	PETROPERU	Capirona, 1972	9,800-12,000	3	5	556	3,742,954	25.0
		Corrientes, 1971	10,000-12,600	28	36	19,375	101,536,099	24.9
		N. Esperanza, 1980	10,200-10,700	2	3	1,920	5,570,559	42.5
		Pavayacu, 1972	9,600-11,300	9	14	2,665	17,458,776	29.5
		Selva Central	...	36	46	2,633	27,362,437	36.0
		Talara-Lima area, 1869	2,000- 9,000	2,142	6,117	21,789	961,984,018	32.3
		Valencia, 1975	10,500-11,200	1	2	30	546,548	43.0
		Yanayacu, 1974	11,400-13,600	5	9	1,080	4,505,374	18.8
		Total Peru		**3,536**	**8,431**	**132,043**	**1,775,162,882**	

PRODUCTION

WORLDWIDE PRODUCTION

		Name of field, discovery date	Depth, ft	No. of wells Producing	No. of wells Total	1989 average B/d	Production Cumulative to Dec. 31, 1989 Bbl	°API gravity
PHILIPPINES	ALCORN PHILIPPINES	Galoc, 1981	7,260- 7,280	...	...	...	383,460	37.2
		Matinloc, 1978	6,656- 6,750	3	5	1,141	9,864,707	43.8
		Nido, 1977	6,576- 6,885	3	5	709	16,056,500	27.1
		North Matinloc, 1987	6,830- 7,021	3	3	1,652	922,917	43.2
	ALCORN PALAWAN	Cadlao, 1977	5,734- 5,881	2	2	1,445	10,446,633	45.6
	TRANS-ASIA OIL & MINERAL DEVELOPMENT CORP.	South Tara, 1987	4,335- 4,436	1	1	198	218,251	40.4
		Total Philippines		12	16	5,145	37,892,468	
QATAR[e]	QGPC	Bul Hanine, 1970	6,000	14	15			35.0
		Dukhan, 1940	6,500	116	161			41.1
		Idd El Shargi, 1960	4,500- 8,250	25	31	387,000	4,312,876,000	35.0
		Maydan-Mahzam, 1963	7,000	19	24			38.0
		Total Qatar		174	231	387,000	4,312,876,000	
RAS al KHAIMAH[e]	CHEVRON	Saleh, 1983	14,500	6	6	10,000	20,141,000	50.0
		Total Ras al Khaimah		6	6	10,000	20,141,000	
SAUDI ARABIA[e]	ARAMCO	Abqaiq, 1940	6,690					37.0
		Abu-Hadriya, 1940	9,340					35.0
		Abu Jiffan	...					...
		Abu-Safah, 1963	6,600					30.0
		Bakr	...					...
		Berri, 1964	8,300					32.0-34.0
		Dammam, 1938	4,800					35.0
		Dhib	...					...
		Dibdibah	...					...
		Duhul	...					...
		El Haba	...					...
		Fadhili, 1949	8,050					39.0
		Farhah	...					...
		Faridah	...					...
		Ghawar, 1948	6,920					34.0
		Habari	...					...
		Hamur	...					...
		Harmaliyah, 1971	8,430					35.0
		Harqus	...					...
		Hasbah	...					...
		Jaham	...					...
		Jaladi	...					...
		Jana	...					...
		Jauf	...	858	858	4,995,000	57,780,625,000	...
		Jawb	...					...
		Juraybi'at	...					...
		Jurayd	...					...
		Karan	...					...
		Khurais, 1957	5,200					33.0
		Khursaniyah, 1956	6,560					31.0
		Kurayn	...					...
		Lawhah	...					...
		Lughfah	...					...
		Maharah	...					...
		Manifa, 1957	7,950					...
		Marjan, 1967	6,800					33.0
		Mazalij	...					...
		Qatif, 1945	7,100					33.0-34.0
		Qirdi	...					...
		Ribyan	...					...
		Rimthan	...					...
		Safaniya, 1951	5,100					27.0
		Sahba	...					...
		Samin	...					...
		Sharar	...					...
		Suban	...					...
		Tinat	...					...
		Wari'ah	...					...
		Watban	...					...
		Zuluf, 1965	5,800					32.0
		Total Saudi Arabia		858	858	4,995,000	57,780,625,000	
SHARJAH	AMOCO	Saaja and Moveyid, 1980	10,200-12,300	16	16	28,114	114,661,700	50.0
	CRESCENT[e]	Mubarek,[c] 1972	...	4	4	6,886	99,450,000	37.0
		Total Sharjah		20	20	35,000	214,111,700	
SPAIN	SHELL ESPANA	Amposta Marino Norte, 1970	10,000	...	8	...	55,667,000	37.0
		Tarraco (Castellon B5/8), 1973	9,000	...	3	...	14,547,000	35.0
	CHEVRON	Ayoluengo, 1964	4,600- 5,200	34	42	750	15,379,025	37.0

PRODUCTION

WORLDWIDE PRODUCTION

		Name of field, discovery date	Depth, ft	No. of wells Producing	Total	1989 average B/d	Production Cumulative to Dec. 31, 1989 Bbl	°API gravity
	REPSOL	• Casablanca, 1975	8,900	7	9	17,531	95,422,194	33.0
		• Gaviota,(c) 1980	8,300	3	3	2,840	2,267,315	52.7
		• Angula, 1984	12,600	...	1	...		44.0
		• Dorada, 1975	6,700	...	4	...	17,843,906	21.5
		• Salmonete, 1984	12,600	...	1	...		44.0
		Total Spain		**44**	**71**	**21,121**	**201,126,440**	
SURINAME	STAATSOLIE	Tambaredjo, 1981	1,040	124	124	3,800	4,861,700	15.9
		Total Suriname		**124**	**124**	**3,800**	**4,861,700**	
SYRIA	AL FURAT OIL CO. (SHELL)	Al Ishara East, 1986	...	...	...	7,945	5,400,000	38.1
		Ash Shola, 1985	...	...	...	...	220,000	37.0
		El Ahwar, 1986	...	...	...	5,479	4,500,000	34.0
		El Wafd North, 1986	...	...	...	21,006	14,500,000	38.1
		Jido, 1986	...	...	...	2,192	2,600,000	33.0
		Thayyem, 1983	...	...	...	71,233	83,600,000	36.5
	S.P.C.(e)	Alian, 1974	5,900	25	37			19.0
		Gbebeh, 1976	2,300	54	75			18.0
		Jebisseh, 1968	2,000- 3,300	132	219			18.0
		Karachok, 1956	6,500	103	119	232,055	1,141,495,075	20.0
		Rumilan, 1962	6,500	59	66			24.0
		Soudie, 1959	6,500	532	579			24.0
		Tichrine, 1976	2,600	58	73			18.0
		Total Syria		**963**	**1,168**	**340,000**	**1,252,315,075**	
THAILAND	UNOCAL	• Baanpot,(c) 1980	7,000	22	30	1,863	2,698,500	54.0
		• Erawan,(c) 1973	7,000	93	155	7,325	19,410,200	54.6
		• Kaphong,(c) 1974	8,000	23	24	2,406	368,100	55.0
		• Platong,(c) 1976	8,500	26	49	2,885	6,576,200	59.1
		• Satun,(c) 1980	8,500	64	67	5,409	7,984,800	55.1
	THAI SHELL	Pru Krathiam, 1984	4,200	1	1	20	25,700	19.0
		Sirikit, 1981	5,258- 5,976	54	85	19,475	40,744,400	41.0
		Sirikit West, 1983	5,500	2	2	116	630,300	41.0
		Thap Raet, 1989	11,100	1	1	537	66,000	41.0
		Wat Taen, 1988	7,790	1	1	108	81,500	37.0-40.0
		Total Thailand		**287**	**415**	**40,144**	**78,585,700**	
TRINIDAD	TRINIDAD & TOBAGO OIL CO. LTD. (TRINTOC)	Area IV and Guapo, 1963	...	85	192	1,265	39,598,000	...
		Barrackpore, 1911	11,100	83	379	3,762	33,301,000	...
		Brighton, 1903	7,500	59	620	548	70,674,000	...
		Catshill, 1950	9,693	31	134	300	23,252,000	...
		Forest Reserve, 1913	11,000	289	2,042	3,738	261,724,000	...
		Grand Ravine,* 1929	11,000	51	168	594	27,077,000	...
		Guayaguayare, 1902	10,750	115	699	1,944	87,963,000	...
		Oropouche, 1944	9,100	29	128	146	6,764,000	...
		Palo Seco, 1929	12,700	77	939	1,652	94,484,000	...
		Parrylands 1-5, 1913	10,626	118	508	1,526	41,268,000	...
		Penal, 1936	11,067	78	289	982	62,763,000	...
		Point Fortin C & W, 1907	10,626	124	551	1,699	42,047,000	...
		Trinity, 1956	9,700	20	95	240	15,306,000	...
		Other	...	63	813	933	71,817,000	...
	TRINIDAD NORTHERN AREAS (TNA)	• Fortin Offshore (FOS), 1954	...	12	35	247	7,185,000	...
		• Soldado, 1955	11,000	371	673	36,679	490,015,000	...
	AMOCO TRINIDAD OIL CO. LTD.	• Cassia, 1973	12,700	9	10	4,439	14,184,000	...
		Mora, 1982	8,500	4	6	427	728,000	...
		• Poui, 1974	11,650	36	58	20,274	164,596,000	...
		• Samaan, 1971	11,780	40	57	15,530	184,737,000	...
		• Teak, 1969	15,191	53	104	31,874	251,133,000	...
	TRINIDAD & TOBAGO PETROLEUM CO. LTD. (TRINTOPEC)	Centarl Los Bajos, 1973	...	162	221	2,509	10,304,000	...
		Coora/Quarry, 1936	14,000	151	736	1,720	92,482,000	...
		Fyzabad/Apex/Quarry, 1920-38	11,000	300	1,042	3,062	173,046,000	...
		Galeota, 1963	6,304	47	105	2,177	16,893,000	...
		Palo Seco/Erin/McKenzie, 1926	12,718	493	1,581	7,761	122,996,000	...
		Other	...	196	923	2,492	72,239,000	...
	PREMIER CONSOLIDATED OILFIELDS LTD. (P.C.O.L)	All fields	...	103	503	820	21,150,000	...
		Total Trinidad		**3,199**	**13,611**	**149,340**	**2,499,726,000**	...
TUNISIA	ELF-ETAP	• Ashtart	10,500	16	21	28,675	193,045,843	29.8
	ELF-SEREPT	Douleb/Semmama, 1966	3,937	12	15	585	18,155,068	40.0
		Tamesmida, 1967	3,937	3	3	176	2,151,019	36.9
	SHELL	• Tazerka, 1979	...	...	...	...	14,621,000	30.0
	ETAP-AGIP-FINA (SODEPS)	Debbech, 1980	7,500	1	1	27	259,422	42.6
		Larich, 1979	7,550	2	2	117	313,851	40.7
		Makhrouga, 1980-89	6,230	4	4	5,318	7,568,176	45.0

PRODUCTION

WORLDWIDE PRODUCTION

		Name of field, discovery date	Depth, ft	No. of wells Producing	No. of wells Total	1989 average B/d	Production Cumulative to Dec. 31, 1989 Bbl	°API gravity
	AGIP-TUNISIA (SITEP)	Chouech es Saida, 1971	12,600	1	1	455	2,759,065	42.9
		El Borma, 1964	8,250- 8,900	87	113	28,885	420,131,958	42.5
	CFPT[e]	Sidi El Itayem, 1971	7,500- 7,800	21	23	38,762	148,143,200	44.0
		Total Tunisia		147	183	103,000	807,148,602	
TURKEY	ALADDIN MIDDLE EAST LTD.[e]	Kahta, 1958	3,500	12	25	1,993	4,778,472	11.0
		Yasince, 1974	6,400	2	4			31.0
	MOBIL	Bulgurdağ, 1961	4,500	2	2	39	2,495,807	37.2
		Selmo, 1964	5,800	21	21	3,566	68,088,457	34.2
	SHELL	Barbes Deep, 1971	...	18	22	...	50,000	37.2
		Baysu, 1985	...	6	7	1,800	1,160,000	34.5
		Bektas, 1985	...	1	1	380	440,000	34.7
		Beykan, 1964	...	38	48	2,360	60,660,000	33.2
		Cobantepe, 1975	...	...	...	...	218,000	29.5
		Katin, 1971	...	3	8	...	3,730,000	29.7
		Kayako W (OF), 1964	...	19	29	3,100	31,440,000	34.1
		Kayakoy, 1960	...	14	27	1,340	20,200,000	38.2
		Kervan, 1982	...	...	...	...	4,000	30.3
		Kurkan, 1963	...	29	35	1,420	49,620,000	31.4
		Kurkan S, 1965	...	5	10	930	7,240,000	34.5
		Malatepe, 1970	...	14	15	790	7,190,000	32.7
		Piyanko, 1968	...	1	3	...	1,520,000	35.3
		Sahaban, 1965	...	7	13	540	13,440,000	34.5
		Sahaban SE, 1983	...	2	2	190	470,000	34.5
		Sincan, 1979	...	8	9	600	3,320,000	31.4
		Yatir E, 1973	...	9	11	410	5,450,000	30.9
		Yenikoy E, 1974	...	2	2	760	4,570,000	31.3
		Yesildere, 1985	...	3	3	550	597,000	34.5
	TPAO	Adiyaman, 1971	5,800	8	8	508	8,420,930	27.6
		Akpinar, 1984	10,500	4	4	67	157,937	31.6
		Alcik, 1983	6,500	1	1	36	95,486	32.0
		B. Firat, 1986	8,150	11	11	1,137	1,754,761	34.8
		B. Kozluca, 1985	5,000	23	23	933	1,563,510	12.0
		B. Selmo, 1981	5,500	4	4	288	739,096	34.0
		Bati Raman, 1961	4,200	112	112	6,344	37,884,010	13.3
		Beycayir, 1976	7,700	1	1	29	207,804	26.4
		Bölükyayla, 1977	10,300	...	...	...	167,663	35.0
		Camurlu, 1976	5,600	14	14	420	1,186,859	12.2
		Celikli, 1964	10,500	11	11	888	3,821,585	35.2
		Cemberlitas, 1983	6,500	20	20	2,859	8,135,017	30.5
		Cukurtas, 1985	10,000	1	1	23	219,021	36.0
		Devecatak, 1973	4,800	2	2	66	272,735	37.0
		G. Adiyaman, 1977	5,000	...	...	...	115,836	20.4
		G. Dincer, 1981	5,350	15	15	902	4,861,381	16.7
		G. Karakus, 1989	7,900	1	1	265	39,691	26.5
		G. Kayaköy, 1976	8,600	4	4	416	2,347,288	30.4
		G. Sahaban, 1978	5,450	9	9	807	3,234,655	33.2
		G. Saricak, 1973	5,250	12	12	160	4,384,785	31.5
		Garzan, 1956	4,700	24	24	1,147	33,360,976	26.2
		Germik, 1968	6,500	10	10	294	3,508,066	18.3
		Ikiztepe, 1976	4,900	...	...	...	88,051	11.3
		K. Osmancik, 1971	3,800	3	3	230	2,172,980	37.6
		Karakus, 1988	8,850	10	10	8,673	3,236,560	30.0
		Karteltepe, 1982	6,550	3	3	366	1,017,971	32.0
		Kurtalan, 1961	5,750	1	1	26	252,180	31.9
		Kuzey Adiyaman, 1977	9,500	...	...	...	59,375	32.0
		Mağrip, 1961	5,700	11	11	658	15,11,439	18.4
		Mehmetdere, 1982	6,700	...	...	...	157,439	31.0
		Oyuktas, 1972	7,600	...	...	...	592,005	31.0
		Raman, 1940	4,500	80	80	4,583	52,063,105	18.7
		S. Sinan, 1987	3,900	3	3	161	119,003	13.0
		Saricak, 1973	5,250	5	5	221	2,737,116	31.5
		Sezgin, 1970	5,600	...	...	...	87,559	17.0
		Silivanka, 1962	8,200	11	11	808	8,027,847	23.0
		Sivritepe, 1977	8,200	...	...	...	591,789	33.4
		Yeniköy, 1973	6,400	23	23	1,937	9,768,385	31.4
		Total Turkey		653	734	56,000	499,243,632	
UNITED KINGDOM	AMOCO	• Montrose, 1969	8,100	13	15	3,640	79,613,099	40.0
		• NW Hutton, 1975	11,200	32	35	17,571	94,920,109	36.0
	ARCO	• Thames,[c] 1973	7,892	6	7	148	222,000	48.8
	BP[e]	• Beatrice (A & B), 1976	6,750	27	39			38.4
		• Buchan, 1974	8,700	9	9			33.0
		• Clyde, 1978	12,400	10	30			38.1
		• Deveron, 1972	9,000	3	3			38.2
		East Midlands, 1939	3,500	65	132			36.0
		• Forties and Echo, 1970	7,000	79	103			37.0
		Kimmeridge, 1959	1,705	1	1	478,888	2,918,163,942	40.0

307

PRODUCTION

WORLDWIDE PRODUCTION

	Name of field, discovery date	Depth, ft	No. of wells Producing	No. of wells Total	1989 average B/d	Production Cumulative to Dec. 31, 1989 Bbl	°API gravity
	• Magnus, 1974	9,500	11	11			39.0
	Nettleham, 1983	8,850	1	2			35.0
	Stainton, 1984	5,200	1	3			...
	Scampton, 1985	5,360	...	3			...
	• Thistle, 1973	9,000	33	52			38.0
	Welton, 1981	5,000	...	...			36.0
	Wytch Farm Bridport Sand, 1974	3,031	10	16			37.0
	Wytch Farm Sherwood Sandstone, 1978	5,198	5	5			37.9
CHEVRON	• Ninian, 1974	10,000	69	105	130,426	858,229,932	35.6
CONOCO	• Dunlin, 1973	9,100	26	41	39,408	273,335,000	36.0
	• Hutton, 1973	10,000	16	29	41,253	117,430,506	33.0
	• Murchison*, 1975	9,900	18	30	28,595	202,267,164	38.0
	• Statfjord†, 1974	9,500	59	93	110,833	275,375,974	39.0
HAMILTON BROS. OIL & GAS LTD.	• Argyll, 1971	9,000- 9,400	7	7	5,790	68,230,000	38.0
	• Crawford, 1975	6,900- 8,500	4	4	5,830	2,130,000	30.0
	• Duncan, 1980	9,600-10,000	2	2	1,440	17,980,000	36.0
	• Innes, 1983	12,300-12,600	2	2	1,190	5,440,000	45.0
MARATHON	• Central Brae crude oil, 1976	14,100-15,000	3	3	3,100	1,105,000	33.0
	• North Brae condensate, 1975	11,900-12,475	10	13	69,600	36,603,000	45.0
	• South Brae crude oil, 1975	12,700-13,437	17	27	48,800	213,114,000	35.0
MOBIL	• Beryl, 1972	9,900-11,500	38	48	98,352	427,369,427	37.0
	• Linnhe, 1988	11,500-12,065	1	1	469	171,286	37.8
	• Ness, 1986	9,900-10,112	3	4	13,524	13,164,501	37.0
OCCIDENTAL	• Claymore, 1974	8,100	5	29	4,705	324,709,332	30.1
	• Piper, 1973	8,100	...	...	...	834,416,000	37.0
	• Scapa, 1981	8,100	4	5	11,868	21,287,150	32.0
SHELL	• Auk, 1971	...	9	11	6,301	2,700,000	38.0
	• Brent, 1971	...	53	130	198,630	1,277,400,000	40.0
	• Cormorant, 1972	...	17	37	70,137	324,200,000	36.0
	• Dunlin, 1973	...	26	40	25,565	317,254,000	36.0
	• Eider, 1976	...	...	...	30,904	15,200,000	34.0
	• Fulmar, 1975	...	13	28	117,494	392,145,000	40.0
PHILLIPS	• Maureen, 1973	8,000- 8,700	7	12	48,777	154,878,640	35.9
TEXACO NORTH SEA	• Highlander, 1976	8,759- 9,372	5	7	22,100	40,623,000	34.0
	• Petronella, 1975	8,110	1	1	11,000	10,654,000	40.2
	• Tartan, 1974	10,217-12,614	7	5	16,700	72,157,000	38.0
TOTAL OIL MARINE	• Alwyn North, 1975	...	9	29	90,000	57,301,000	40.0-41.0
UNOCAL	• Heather, 1973	10,800	25	40	11,962	84,180,446	34.3
	Total United Kingdom		**762**	**1,259**	**1,773,000**	**9,613,970,508**	

*U.K. share (77.8%) of total production. †U.K. share (15.91%) of total production.

VENEZUELA

	Name of field, discovery date	Depth, ft	Producing	Total	1989 avg B/d	Cumulative Bbl	°API
PDVSA/ ZULIA	Alturitas	...	7	14	1,745	3,346,087	32.2
	Bachaquero, 1930	10,720	1,850	3,709	215,868	6,541,430,236	22.9
	Boscan, 1946	9,708	76	478	19,381	747,254,243	10.2
	Cabimas, 1917	2,200	522	1,234	48,620	1,638,449,492	22.4
	Centro, 1959	9,600-11,000	329	606	135,853	936,775,718	33.6
	Ceuta	16,380	110	179	65,565	477,291,615	30.2
	Cruces-Manueles, 1916	3,000- 8,000	41	133	1,302	177,289,790	30.2
	La Concepcion, 1953	3,148- 8,000	52	162	3,297	138,735,445	30.1
	La Paz, 1925	4,268- 8,000	57	154	10,136	862,352,054	30.1
	Lago, 1959	11,450	30	66	17,501	241,447,842	30.5
	Lagunillas, 1926	3,000	2,609	5,187	346,050	10,954,851,444	22.0
	Lama, 1957	8,320	1,941	1,941	138,246	2,629,778,032	32.2
	Lamar, 1958	13,000	108	186	70,491	1,162,688,577	33.5
	Mara, 1945	5,240	37	105	4,899	415,588,131	26.6
	Mene Grande, 1914	4,130	170	583	953	673,630,970	19.9
	Motatan, 1952	...	27	30	42,977	100,625,197	22.0
	Sibucana, 1948	13,450	...	1	...	43,418,125	...
	Tia Juana, 1928	3,000	1,801	3,079	130,114	4,058,506,604	17.9
	Urdaneta, 1970	10,000	152	283	35,989	152,274,347	22.0
	West-Tarra, 1942	4,250- 5,500	20	44	2,465	74,003,614	30.2
	Other	...	135	408	31,341	2,042,929,176	
PDVSA/ ANZOA- TEGUI	Adjuntas, 1957	5,000	...	54	...	8,714,849	13.9
	Arecuna	1,300	53	118	3,742	10,997,661	11.0
	Bare	1,110	158	275	13,800	25,850,128	10.2
	Caico Seco, 1946	6,500- 7,300	2	8	381	8,981,806	31.8
	Chimire, 1948	7,000- 7,200	50	190	2,655	122,907,767	31.1
	Dacion, 1957	6,700	70	140	7,914	229,438,085	23.0
	El Roble, 1939	3,500-11,500	24	58	2,070	43,107,831	49.3
	El Toco	2,300	27	43	866	15,795,449	23.7
	Elias, 1954	5,000- 6,470	44	220	...	...	...
	Guara, 1946	5,000-10,000	102	221	3,986	249,620,996	27.0
	Guario, 1940	3,400	18	59	743	27,232,886	43.6
	Isla, 1954	...	21	67	4,731	57,938,711	40.5
	La Ceiba, 1946	9,450	22	37	3,381	49,682,318	41.9
	La Ceibita, 1963	9,878	19	47	1,978	83,320,150	42.0
	Leona, 1938	2,200-12,800	16	63	1,222	51,881,445	26.7

PRODUCTION

WORLDWIDE PRODUCTION

	Name of field, discovery date	Depth, ft	No. of wells Producing	Total	1989 average B/d	Cumulative to Dec. 31, 1989 Bbl	°API gravity
	Mata, 1954	8,970-10,516	76	385	4,016	54,787,078	35.9
	Merey, 1937	5,400- 5,700	17	41	...	34,579,770	12.5
	Nipa, 1945	6,000- 8,500	35	187	1,800	120,273,230	29.6
	Oficina, 1937	5,900	34	104	1,806	361,270,333	26.1
	Oscurote-Norte, 1952	9,513	10	77	1,171	113,024,585	24.2
	Pato Este, 1988	...	2	...	193	78,502,618	41.9
	San Joaquin, 1939	6,550	66	108	2,160	48,780,081	51.3
	Soto, 1950	9,500	14	82	...	...	...
	Sta. Ana, 1936	8,500	48	97	2,473	61,975,184	45.6
	Sta. Rosa, 1941	8,500	166	210	7,901	291,608,541	45.7
	Yopales, 1937	4,600	94	203	4,990	66,410,170	22.6
	Zanjas, 1958	13,270	6	17	1,359	40,741,295	38.1
	Zapatos, 1955	11,500	28	58	3,336	181,250,848	32.4
	Zorro, 1953	11,100	8	29	1,520	74,645,449	29.3
	Zulus, 1957	12,710	7	23	1,587	24,953,350	41.2
	Zumo, 1954	9,200	4	35	910	70,616,083	25.3
	Other	...	541	2,021	98,603	2,929,648,421	...
PDVSA/MONAGAS	Acema, 1960	...	26	56	11,743	57,397,773	26.4
	Acema-Casma	4,023	36	85	12,662	85,318,167	24.3
	Aguasay, 1955	8,100-13,400	21	72	1,354	81,535,764	41.3
	Aguasay Norte/Sur, 1955	4,849	6	17	1,675	4,946,943	42.4
	Carisito, 1984	4,881	7	8	6,767	6,168,079	41.7
	Carito, 1988	...	4	6	25,778	11,338,247	31.4
	Cerro Negro	1,200	43	296	6,444	24,737,881	9.0
	Furrial, 1986	...	17	18	136,959	80,392,219	26.3
	Jobo, 1956	3,600- 4,000	147	432	25,392	289,183,963	10.3
	Mata Acema, 1951	...	33	56	5,600	93,649,415	27.0
	Morichal, 1958	3,312	19	189	1,930	172,329,996	9.2
	Onado, 1971	4,791	5	25	1,152	19,594,004	28.0
	Oritupano, 1950	7,657	29	64	3,262	38,659,692	18.4
	Oritupano Norte/Sur, 1954	...	24	253	7,261	30,357,856	24.1
	Orocual, 1953	2,954	8	57	1,708	30,826,639	25.1
	Pilon, 1937	...	45	139	7,491	156,598,659	11.6
	Pirital, 1958	450- 1,100	...	84	...	27,535,135	...
	Quiriquire, 1928	7,000- 7,200	3	474	1,298	761,301,593	27.7
	Sta. Barbara, 1941	5,020- 6,500	1	331	6,716	169,782,969	31.4
	Tacat, 1953	1,820- 3,670	...	133	...	45,209,945	...
	Temblador, 1936	3,500- 4,500	3	49	837	109,621,745	18.7
	Other	...	63	305	28,295	615,933,764	...
PDVSA/BARINAS	Guafita, 1984	8,500	21	22	57,226	49,921,084	29.1
	Hato, 1965	9,550	18	16	2,349	44,525,088	22.7
	La Victoria, 1987	...	6	6	11,727	6,033,547	34.5
	Maporal, 1957	10,950	12	16	3,107	17,141,820	25.7
	Paez, 1963	...	45	67	7,164	66,895,836	19.9
	Silvan, 1949	10,860	18	22	3,142	50,844,701	26.8
	Silvestre, 1948	8,860	19	48	2,849	138,382,915	22.1
	Sinco, 1953	8,500- 9,100	49	84	5,901	262,316,155	21.5
	Other	...	3	7	342	5,285,014	...
PDVSA/GUARICO	Barzo, 1959	...	3	6	92	2,388,097	24.0
	Bella Vista, 1952	...	8	86	...	...	...
	Budare, 1958	4,523	23	44	10,977	72,274,969	31.1
	Jobal, 1960	...	8	12	74	2,200,027	33.6
	Las Mercedes, 1942	4,500	...	184	...	91,650,680	...
	Ruiz, 1949	4,500	...	42	...	33,189,615	...
	Saban, 1947	...	8	70	251	22,240,027	43.9
	Other	...	115	207	5,493	363,236,638	...
PDVSA/AMACURO	Tucupita, 1945	...	...	48	...	67,074,189	...
PDVSA/FALCON	All fields, 1920	...	...	...	...	107,226,375	27.4
	Total Venezuela		**12,752**	**27,995**	**1,909,105**	**43,926,451,082**	

WEST GERMANY

	Name of field, discovery date	Depth, ft	Producing	Total	1989 avg B/d	Cumulative Bbl	°API
BEB	Annaveen, 1963	4,428	...	4	...	500,475	33.8
	Barenburg, 1953	2,300- 3,300	43	72	1,974	40,731,540	29.0
	Barver, 1963	3,640	2	5	30	502,901	32.8
	Bodenteich, 1960	4,641	1	2	18	361,509	31.1
	Eich, 1983	5,995	7	8	1,650	2,102,345	38.9
	Eilte-West, 1956	6,051	12	16	85	3,633,888	32.8
	Eldingen, 1949	5,028	36	43	723	21,262,463	35.5
	Elsfleth, 1956	5,175	3	8	35	2,860,587	25.7
	Eystrup 2, 1960	3,526	...	1	...	444,798	17.4
	Eystrup 3-6, 1959	1,075	4	4	83	1,177,255	17.4
	Eystrup-Verden, 1977	3,526	...	4	1	180,280	17.1
	Georgsdorf, 1944	1,800- 2,600	196	369	5,300	104,005,149	24.0-29.0
	Gross Lessen, 1969	3,300	6	11	557	21,231,936	30.0
	Hankenbüttel, Mitte, 1954	5,182	5	8	287	4,496,989	33.0
	Hankenbüttel-Nord, 1955	5,182	4	4	316	6,429,186	33.0
	Hankenbüttel, Ost, 1954	5,182	1	1	46	1,290,049	33.0
	Hebelermeer, 1955	3,899	5	15	73	4,550,102	27.3
	Hillerse-Nord, 1958	2,683	1	1	9	772,698	31.5
	Hillerse-West, 1957	6,077	1	1	2	29,815	43.1
	Hohenassel, 1943	1,705	6	14	22	3,558,034	30.4

309

PRODUCTION

WORLDWIDE PRODUCTION

	Name of field, discovery date	Depth, ft	No. of wells Producing	No. of wells Total	1989 average B/d	Production Cumulative to Dec. 31, 1989 Bbl	°API gravity
	Hohnebostel, 1969	8,341	2	6	57	1,226,842	41.0
	Königsgarten, 1984	6,235	3	3	337	247,639	38.1
	Lingen-Dal. (incl. Wd), 1942	3,129	46	128	132	16,995,157	31.1
	Lüben, 1955	3,950	7	11	585	12,445,058	35.0-38.0
	Lüben-West, 1958	4,690	5	6	133	1,586,475	36.9
	Meerdorf, 1954	6,077	14	27	62	1,632,811	43.1
	Meppen-Schwefingen, 1959	4,257	22	24	843	12,818,086	29.8
	Mölme-Feldbergen, 1934	1,410	2	2	12	597,238	31.1
	Mölme-Wachtel, 1934	3,886	2	10	3	904,044	36.9
	Nienhagen, 1909	3,079	10	26	167	26,895,965	29.6
	Nienhagen-Dachtmissen, 1959	3,427	5	8	89	1,046,892	27.8
	Pfullendorf-Ostrach, 1962	3,558	15	23	131	2,575,180	37.7
	Pötrau, 1959	5,248	...	2	...	31,121	14.3
	Rühlermoor-Malm, 1962	4,519	6	10	210	6,541,938	31.7
	Rühlermoor-Valendis, 1949	2,381	248	443	10,885	139,735,749	24.6
	Rühme, 1954	1,950	38	48	682	10,719,373	33.4
	Steimbke-Alt+Nord, 1935	1,246	18	30	76	7,756,810	20.6
	Steimbke-Lichtenmoor, 1943	2,722	14	29	109	4,020,591	17.4
	Steimbke-Ost, 1959	4,428	3	9	51	3,836,847	30.5
	Stockstadt, 1952	5,402	11	21	120	7,749,995	39.6
	Suderbruch, 1949	5,175	41	64	494	23,740,686	31.1
	Sulingen, 1973	3,345	9	9	442	5,353,332	30.0
	Varel, 1957	5,084	6	15	143	5,793,000	25.5
	Varloh, 1984	4,129	6	6	138	349,590	31.5
	Volkensen, 1960	7,540	4	4	77	807,340	22.9
	Wehrbleck, 1957	3,342	50	61	623	16,867,171	32.0
	Wietingsmoor, 1954	2,820	22	30	483	13,388,438	30.5
MOBIL	Bavaria	5,640- 6,190	...	20	...	7,526,952	20.0
	Hannover	980- 6,630	49	68	810	31,296,759	25.0-35.0
	Hofolding, 1980	11,100	3	3	300	1,105,204	28.0-30.0
	Süd-Oldenburg, 1950	2,630- 6,800	84	137	1,101	40,332,556	31.0-39.0
	Voigtei, 1953	980- 1,900	74	106	633	25,385,230	32.0-34.0
	Weser Ems	2,000- 2,450	18	25	292	6,212,439	24.0
RWE-DEA A.G.	Boostedt, 1952	5,900	26	45	502	18,568,187	32.0
	Hankenbuettel-Sued, 1954	4,900	37	81	1,881	83,272,415	27.0-35.0
	Hardesse, 1957	8,200	28	67	442	9,016,774	40.0
	Hohne, 1951	5,250	27	55	737	41,208,241	33.0
	Leiferde, 1956	2,600	21	42	364	20,524,815	33.0
	• Mittelplate, 1987	7,500- 9,350	5	6	4,517	3,500,160	19.3
	Oerrel-Sued, 1954	4,600	22	47	636	11,955,623	26.0
	Ploen-Ost, 1958	8,850	29	83	870	51,398,026	33.0
	Preetz, 1962	8,200	15	33	286	11,746,832	26.0
	Schwedeneck-See, 1978	4,900	15	18	7,053	13,124,057	28.0
	Wesendorf, 1943	4,650	13	36	347	17,648,816	25.0-40.0
	Other, 1935	2,300- 8,200	17	73	420	31,289,513	19.0-37.0
WINTERSHALL A.G.	Aitingen, 1977	6,232	4	4	933	4,668,321	31.1
	Aldorf, 1952	4,100	29	103	421	17,247,936	33.0
	Arlesried, 1964	4,920- 8,550	13	18	466	14,190,315	38.5
	Bockstedt, 1954	4,264	28	48	1,026	21,197,314	27.4
	Dickel, 1953	3,936	32	59	294	6,826,501	28.3
	Dueste, 1954	3,936	28	71	271	12,250,670	33.0
	Eicklingen, 1937	2,460	9	13	51	3,680,337	30.0
	Emlichheim, 1944	3,116	87	160	2,936	45,498,458	24.5
	Hauerz, 1985	8,639	4	4	108	290,815	40.4
	Kirchdorf, 1982	5,737	1	1	12	60,955	38.5
	Landau, 1955	3,936	77	82	1,143	27,513,710	36.9
	Maximiliansau, 1959	2,100- 5,363	...	...	...	14,387	41.0
	Meckelfeld, 1954	6,396	17	28	246	14,116,718	27.5
	Moenchsrot, 1958	5,248	16	22	281	10,675,677	38.1
	Nienhagen, 1925	3,729	12	12	64	10,945,356	30.0
	Oberschwarzach, 1978	4,920	4	4	113	710,515	37.7
	Pattensen, 1954	2,165	6	7	52	1,077,932	29.3
	Rheinzabern, 1959	3,750- 5,576	...	1	...	27,041	37.9
	Ruehlertwist, 1949	2,350- 2,900	70	134	1,496	32,135,700	25.0
	Ruelzheim, 1984	7,544	...	2	...	40,818	37.7
	Schwabmuenchen, 1976	4,428	...	1	...	39,491	25.7
	Schwedeneck (only onshore), 1956	5,182	13	29	177	5,275,233	27.5
	Wathlingen, 1939	2,624	7	10	24	1,422,000	30.0
DST[e]	Bramberge, 1958	2,560- 3,300	57	82			29.0
	Fronhofen, 1964	3,800	10	15			37.0
	Illmensee, 1967	3,800	2	5			37.0
	Markdorf, 1960	7,700	...	1			37.0
	Memmingen-M, 1958	5,000	...	7			37.0
OTHER OPERATORS[e]	Scheerhorn, 1949	3,550- 3,750			14,433	391,523,950	30.0
	Reitbrook, 1937	1,450- 2,100					29.0
	Oelheim-Sued, 1968	7,650	932	1,630			33.0
	Sinstorf, 1960	6,750- 7,150					...
	Vorhop, 1952	2,950- 6,900					36.0
	Total West Germany		2,883	5,144	74,000	1,592,328,114	

310

PRODUCTION

WORLDWIDE PRODUCTION

		Name of field, discovery date	Depth, ft	No. of wells Producing	Total	Production 1989 average B/d	Cumulative to Dec. 31, 1989 Bbl	°API gravity
YEMEN	HUNT OIL	Alif, 1984	5,750	55	56	152,500	118,170,200	40.0
		Al-Shura, 1989	7,600	1	1	100	41,300	40.0
		Azal, 1987	6,850	12	12	22,900	13,507,400	40.0
		Jabal, Nuqum, 1988	3,780	1	1	1,300	466,500	44.0
		Saif Ben The-Yazen, 1988	7,600	1	1	1,300	470,500	40.0
		Total North Yemen		70	71	178,100	132,655,900	
YUGOSLAVIA	INA-NAFTAPLIN	Beničanci, 1966	6,752- 5,938	34	70	5,583	106,071,189	30.2
		Bilogora, 1966	3,346- 5,052	43	60	822	8,359,466	40.8
		Bizovac, 1989	5,483- 5,499	1	5	264	32,235	33.8
		Bokšić-Klokočevci, 1975	7,218-10,171	12	19	119	234,820	31.0
		Bunjani, 1951	1,800- 2,550	19	36	69	2,514,916	30.2
		Crnac, 1977	6,814- 8,061	2	4	286	208,827	34.6
		Deletovci, 1982	3,143- 3,422	44	51	4,036	6,180,054	28.9
		Dugo Selo, 1953	3,445- 4,183	15	26	469	9,785,176	31.1
		Ferdinandovac, 1959	5,625- 5,915	5	6	102	1,286,776	36.0
		Ilača, 1982	3,445- 3,855	9	14	362	407,404	31.1
		Ivanič Grad, 1959	5,282- 5,709	46	85	3,479	44,149,246	33.4
		Jagnjedovac, 1961	2,864- 3,570	26	32	475	*6,399,537	33.0
		Jamarice, 1965	3,281- 5,512	90	124	2,672	14,229,838	37.6
		Ježevo, 1963	6,234- 7,382	7	9	91	859,548	36.7
		Kloštar, 1952	2,297- 4,593	74	137	1,010	40,424,057	36.3
		Kozarice, 1972	3,281- 3,609	16	23	407	1,632,613	23.6
		Križ, 1948	1,600- 2,250	32	110	90	6,494,756	17.4
		Kučanci-Kapelna, 1973	6,562- 7,218	12	14	1,359	1,378,110	25.2
		Lacići, 1979	7,605- 8,012	8	20	2,248	4,342,624	35.6
		Lepavina, 1969	1,780- 2,200	8	11	297	838,671	22.8
		Letičani, 1989	4,584- 4,817	5	9	108	39,285	35.0
		Lipovljani, 1960	3,937- 5,709	81	112	2,553	22,147,253	36.7
		Lupoglav, 1971	3,860- 5,600	5	6	93	140,549	33.4
		Mihovljani, 1973	5,150- 5,760	4	5	92	59,452	32.5
		Molve,[c] 1974	10,263-11,345	16	20	1,175	2,178,050	57.2
		Mramor Brdo, 1949	2,625- 4,101	4	30	49	4,539,835	31.1
		Obod, 1967	6,378- 7,550	3	8	156	2,224,427	32.5
		Okoli, 1962	7,382- 8,366	4	9	60	538,895	35.5
		Okoli,[c] 1962	7,218- 7,874	12	16	740	830,814	80.6
		Privlaka, 1983	2,933- 3,041	13	17	942	1,383,584	25.9
		Sandrovac, 1962	2,250- 3,700	105	156	4,237	37,966,186	37.8
		Stevkovica, 1978	7,710- 9,843	22	33	883	3,842,985	31.9
		Struzec, 1956	2,362- 3,773	66	107	5,520	101,300,919	38.8
		Vezišće, 1965	6,135- 6,562	5	5	19	47,989	31.6
		Žutica, 1963	5,742- 6,923	154	194	7,696	86,049,116	33.8
		Other,[c]	5,250-11,800	10	19	1,375	2,599,630	87.9
	NAFTA-GAS	Boka, 1956	3,969- 4,100	25	43	544	7,815,100	32.7
		Čoka, 1975	4,888- 5,046	9	23	205	3,665,890	44.3
		Elemir, 1959	5,002- 5,376	29	63	520	17,377,400	32.7
		Itebej, 1978	6,973- 7,327	5	16	43	204,290	26.4
		Janošik, 1958	2,148- 3,293	1	4	16	144,766	24.2
		Jermenovci, 1952	2,919- 3,148	56	100	731	9,415,246	24.2
		Karadjordjevo, 1973	8,308- 8,846	6	22	146	1,069,360	26.4
		Kelebija, 1970	2,061- 2,788	12	28	546	6,461,960	21.2
		Kikinda, 1960	3,543- 5,735	166	229	4,632	38,380,480	34.9
		Kikinda-varoš, 1963	3,740- 6,490	50	89	2,028	28,532,000	27.9
		Lokve, 1953	2,909- 3,099	9	26	42	731,325	25.1
		Morkrin, 1961	6,232- 7,216	105	186	2,525	19,858,100	35.1
		Mokrin-jug, 1982	6,658- 6,740	16	23	556	1,011,460	41.1
		Majdan, 1972	7,708- 7,790	14	29	270	565,300	37.4
		Palić, 1968	3,345- 3,427	6	34	55	1,668,950	45.9
		V. Greda-jug, 1981	3,237- 3,427	10	22	145	292,250	23.8
		Velebit, 1965	2,434- 2,575	122	159	7,830	45,461,540	22.8
		Zrenjanin, 1974	6,829- 6,986	5	11	321	142,260	25.0
		Total Yugoslavia		1,671	2,728	77,178	709,768,649	
ZAIRE	CHEVRON	• GCO, 1970	7,000- 9,000	5	7	169	6,394,909	38.7
		• Libwa, 1981	5,500	...	2	...	...	33.0
		• Lukami, 1983	12,400	4	5	1,388	7,396,373	33.2
		• Mibale, 1973	5,000- 6,000	8	9	10,460	89,011,881	30.0
		• Motoba, 1976	11,098	7	7	5,439	2,152,540	31.5
		• Mwambe, 1979	6,150	1	1	143	736,610	29.7
	ZAIREP (FINA)	East Mibale, 1978	6,000	...	1	...	397,600	34.0
		Kifuku, 1983	3,400	1	1	16	51,000	31.0
		Kinkasi, 1972	3,700	37	42	2,411	8,182,400	31.0
		Liawenda, 1972	3,900	29	36	1,659	4,020,500	31.0
		Makelekese, 1983	3,600	8	8	388	10,357,000	31.0
		Muanda, 1972	3,600	6	7	371	821,500	31.0
		Nsiamfumu, 1977	8,300	...	1	...	5,000	34.0
		Tshiende, 1978	6,000	4	8	4,341	7,612,900	34.0
		Total Zaire		110	135	29,385	127,818,913	

PRODUCTION

U.S. fields with reserves exceeding 100 million bbl

State Field	Disc. date	1990 prod. (1,000 bbl)	Cum. prod. 1-1-91 (1,000 bbl)	Est. rem. reserves (1,000 bbl)	Est. No. wells
ALABAMA*					
Citronelle,	1955	1,596	151,242	3,619	443
ALASKA					
Endicott,	1978	37,740	120,083	255,260	55
Granite Point,	1965	1,533	113,270	15,467	26
Kuparuk River,	1969	109,029	725,268	775,971	345
Lisburne,	1967	14,827	64,126	146,173	70
McArthur River,	1965	4,540	540,098	35,460	82
Middle Ground Shoal,	1962	3,179	161,655	4,821	42
Prudhoe Bay,	1967	*473,607	†7,078,307	4,421,693	769
Swanson River,	1957	1,852	213,192	6,148	30

*Includes 33.18 million bbl of condensate †Includes 201.17 million bbl of condensate

ARKANSAS*					
Smackover,	1922	2,292	561,971	5,708	1,975
CALIFORNIA					
San Joaquin Valley					
Belridge South,	1911	54,181	791,489	320,880	4,301
Buena Vista,	1909	1,602	650,446	35,099	764
Coalinga,	1890	11,509	775,574	139,459	2,042
Coalinga Nose,	1938	883	500,663	7,451	75
Coles Levee North,	1938	384	161,290	3,383	69
Cuyama South,	1949	380	219,041	6,088	92
Cymric,	1909	8,919	218,038	76,302	959
Edison,	1928	861	136,529	17,503	594
Elk Hills,	1911	31,298	958,982	528,894	1,006
Fruitvale,	1928	447	116,511	12,751	217
Greeley,	1936	120	112,964	1,446	25
Kern Front,	1912	2,191	176,717	51,579	799
Kern River,	1899	44,080	1,297,104	650,375	6,779
Kettleman North Dome,	1928	172	456,998	948	45
Lost Hills,	1910	6,403	190,280	60,280	1,650
McKittrick,	1896	1,648	270,873	86,868	760
Midway-Sunset,	1894	58,906	1,997,711	255,499	9,385
Mount Poso,	1926	5,617	275,072	67,458	468
Rio Bravo,	1937	57	116,220	507	11
Yowlumne,	1974	4,722	92,112	17,199	63
Coastal Area					
Carpinteria,	1966	1,878	92,184	15,632	125
Cat Canyon E. & W.,	1908	1,332	291,721	29,203	325
Dos Cuadras,	1969	3,468	220,392	46,552	138
Elwood,	1928	209	106,206	1,782	7
Hondo,	1969	8,240	107,986	94,128	24
Orcutt,	1901	858	167,526	8,586	137
Point Pedernales,	1982†	6,392	24,312	*48,171	10
Rincon,	1927	980	150,834	14,095	180
San Ardo,	1947	4,110	416,038	115,216	430
Santa Maria Valley,	1934	1,037	200,635	18,209	125
South Mountain,	1916	719	147,396	10,623	361
Ventura,	1919	6,534	908,069	83,699	468
Los Angeles Basin					
Beta,	1976	4,613	50,601	104,326	71
Beverly Hills,	1900	1,888	126,329	38,643	91
Brea Olinda,	1880	1,948	388,032	50,649	727
Coyote East,	1909	556	109,766	12,109	91
Coyote West,	1909	613	251,284	6,166	65
Dominguez,	1923	569	270,419	6,295	106
Huntington Beach,	1920	3,842	1,075,320	63,092	562
Inglewood,	1924	2,380	350,534	49,068	262
Long Beach,	1921	2,189	914,172	12,801	338
Montebello,	1917	565	193,520	9,104	127
Richfield,	1919	1,152	191,737	24,840	192
Santa Fe Springs,	1919	1,012	615,060	18,379	140
Seal Beach,	1924	746	204,641	12,595	157
Torrance,	1922	1,259	214,857	33,073	278
Wilmington,	1932	26,118	2,344,984	442,878	1,695

*Revised to reflect only developed portion of field. Original estimate of total field reserves is 350 million bbl.

COLORADO*					
Rangely,	1933	12,125	763,829	37,000	521

State Field	Disc. date	1990 prod.	Cum. prod. 1-1-91	Est. rem. reserves	Est. No. wells
FLORIDA*					
Jay,	1970	4,335	369,712	46,844	55
ILLINOIS*					
Clay City,	1938	2,092	398,160	3,908	2,860
Lawrence,	1906	2,546	399,887	7,000	831
Louden,	1938	876	390,337	1,924	739
Main,	1906	1,710	237,023	4,500	1,487
New Harmony,	1938	800	155,293	3,600	569
Salem,	1938	865	228,312	3,300	129
KANSAS*					
Bemis-Shutts,	1928	1,120	246,359	3,200	807
Chase-Silica,	1931	850	303,808	2,730	914
El Dorado,	1917	840	300,226	1,500	776
Hall-Gurney,	1931	905	148,038	2,500	1,021
Trapp,	1937	1,050	228,497	2,526	844
LOUISIANA*					
Offshore					
•Bay Marchand Blk. 2,	1949	8,825	615,097	41,175	223
•Eugene Island Blk. 330,	1930	6,665	284,395	40,603	131
•Grande Isle Blk. 16,	1948	3,373	270,714	78,660	53
•Grande Isle Blk. 43,	1956	3,900	281,328	76,616	91
•Mississippi Canyon Blk. 194,	1980	—	—	—	35
•Main Pass Blk. 41,	1957	2,025	242,831	18,407	77
•Main Pass Blk. 306,	1969	1,560	205,145	75,059	92
•South Marsh Island Blk. 128,	1974	—	93,063	15,464	79
•South Pass Blk. 27,	1954	3,650	135,410	65,250	315
•South Pass Blk. 61,	1968	4,600	165,007	35,400	151
•South Pass Blk. 62,	1965	3,225	111,804	76,154	87
•South Pass Blk. 65,	1965	2,550	107,361	82,764	60
•Ship Shoal Blk. 204,	1968	990	68,068	36,932	31
•Ship Shoal Blk. 207,	1967	750	89,384	36,434	25
•Ship Shoal Blk. 208,	1962	2,811	166,532	58,911	49
•South Timbalier Blk. 21,	1939	1,836	220,103	43,269	46
•South Timbalier Blk. 135,	1956	1,107	141,956	23,044	35
•West Delta Blk. 30,	1949	5,339	457,119	36,336	144
•West Delta Blk. 73,	1962	2,950	195,766	79,225	72
Onshore South					
Bay de Chene,	1941	305	97,131	17,203	27
Bay St. Elaine,	1928	396	165,720	24,380	25
Bayou Sole,	1941	502	162,639	2,147	27
Black Bay West,	1953	1,780	148,739	6,492	68
Caillou Island,	1930	1,765	606,492	69,759	146
Cote Blanche Bay West,	1940	550	183,512	44,456	79
Delta Farms,	1944	590	116,241	6,129	33
Garden Island Bay,	1934	1,085	223,486	29,729	143
Golden Meadow,	1938	568	136,679	3,050	156
Grand Bay,	1938	450	171,372	2,902	37
Hackberry, East,	1927	586	110,129	6,561	62
Hackberry, West,	1928	1,075	144,446	3,941	113
Iowa,	1931	97	99,544	456	26
Jennings,	1901	210	116,859	600	183
Lafitte,	1935	1,221	257,845	7,009	111
Lake Barre,	1929	885	205,940	18,144	42
Lake Pelto,	1929	402	117,990	16,425	27
Lake Washington,	1931	2,409	247,439	11,327	93
Leeville,	1931	556	142,766	6,101	44
Paradis,	1939	607	127,905	7,345	36
Quarantine Bay,	1937	883	174,311	14,117	42
Timbalier Bay, 1938 onshore		567	297,734	22,607	64
Venice,	1937	669	183,423	6,045	60

312

PRODUCTION

State Field Disc. date	1990 prod.	Cum. prod. 1-1-91 1,000 bbl	Est. rem. reserves	Est. No. wells
Vinton, 1910	297	161,610	775	106
Weeks Island, 1945	930	227,925	19,370	36
West Bay, 1940	845	230,339	14,262	69
North Caddo-Pine Island 1905	2,820	367,007	8,900	11,608
Delhi, 1944	530	212,789	33,569	65
Haynesville, 1921	608	169,697	2,000	142
Homer, 1919	355	99,138	2,375	196
Rodessa, 1935	350	106,785	2,461	89

MICHIGAN*

Albion-Scipio, 1957	43	>132,173	9,957	382

MISSISSIPPI*

Baxterville, 1944	1,986	243,366	6,189	192
Heidelberg, 1944	2,390	179,630	7,350	234
Tinsley, 1939	649	220,985	1,889	99

MONTANA*

Bell Creek, 1967	601	130,770	21,670	101
Cut Bank, 1926	765	163,912	35,655	615
Pine, 1951	1,170	107,571	3,238	97

NEW MEXICO

Denton, 1949	810	140,363	5,000	175
Empire-Abo, 1957	1,231	222,783	47,217	398
Eunice-Monument, 1929	2,260	130,594	5,714	857
Hobbs, 1928	6,590	311,019	17,500	583
Maljamar, 1926	1,653	148,408	4,800	848
Vacuum, 1929	9,327	452,107	30,673	1,245

NORTH DAKOTA*

Beaver Lodge, 1951	1,520	114,490	13,666	65
Billings Nose, 1978	1,941	65,097	47,767	204
Little Knife, 1977	2,158	55,751	53,254	135
Mondak, 1976	295	13,643	82,326	113

OKLAHOMA*

Burbank, 1920	890	538,530	6,593	1,031
Eola-Robberson, 1921	825	134,805	6,529	507
Fitts, 1933	2,292	204,692	7,208	568
Glenn Pool, 1905	815	329,581	3,285	618
Golden Trend, 1945	3,489	481,275	23,750	1,238
Healdton, 1913	1,527	337,949	6,259	968
Hewitt, 1919	1,970	272,323	10,030	740
Oklahoma City, 1928	984	818,103	4,033	193
Postle, 1960	1,050	108,569	12,745	218
Sho-Vel-Tum, 1919	16,595	1,200,895	34,405	6,982
Sooner Trend, 1965	3,466	302,872	12,672	3,404

TEXAS

Railroad Commission District 2-South

Greta, 1933	474	148,558	11,542	97
Lake Pasture, 1953	1,698	91,222	9,206	197
Tom O'Connor, 1934	5,950	760,230	42,618	648
West Ranch, 1938	1,901	384,115	6,599	304

District 3-Gulf Coast

Anahuac, 1935	610	286,230	13,770	94
Conroe, 1931	2,484	732,474	28,469	263
Giddings, 1971	9,593	280,418	175,750	2,380
Hastings, 1934	1,819	697,023	68,912	188
Magnet Withers, 1936	1,411	113,824	2,797	141
Oyster Bayou, 1941	670	161,619	16,681	35
Thompson, 1931	2,319	478,390	21,510	199
Tomball, 1933	279	121,629	9,295	66
Webster, 1937	3,708	580,906	12,292	192

District 4-South

Agua Dulce-Stratton, 1928	340	147,323	24,051	132
Borregos, 1945	138	114,340	19,867	39
Kelsey, 1938	178	115,253	35,853	59
Plymouth, 1925	205	123,268	2,855	65
Seeligson, 1925	179	271,869	55,158	64
TCB, 1944	48	113,088	52,150	16

State Field Disc. date	1990 prod.	Cum. prod. 1-1-91 1,000 bbl	Est. rem. reserves	Est. No. wells
White Point, E, 1938	90	104,217	6,157	58

District 5-East

Alabama Ferry, 1983	2,111	19,263	81,737	272
Van, 1928	2,547	527,258	11,320	432

District 6-East

East Texas, 1930	34,704	5,078,708	918,797	9,501
Fairway, 1960	1,830	196,602	13,866	100
Hawkins, 1940	6,350	834,798	29,218	436
Neches, 1953	960	105,911	4,105	172
Quitman, 1948	1,211	123,817	5,969	173

District 7C-West Central

Big Lake, 1923	519	130,935	2,505	70

District 8-West

Andector, 1946	591	183,124	5,264	42
Block 31, 1945	2,747	225,865	9,253	271
Cowden, N., 1930	14,897	516,488	45,000	1,352
Cowden, S. Foster, Johnson, 1932	8,425	517,944	27,000	1,678
Dollarhide, 1945	2,585	198,896	9,000	221
Dune, 1938	2,368	187,696	14,008	750
Fullerton, 1942	7,575	362,975	7,225	912
Goldsmith, 1934	5,690	767,662	21,606	1,562
Howard Glasscock, 1925	6,009	415,125	15,383	2,157
Iatan, E., 1926	3,016	148,768	7,215	1,314
Jordan, 1937	824	130,676	2,050	240
Keystone, 1930	1,511	316,163	6,291	756
McElroy, 1926	8,888	481,085	41,264	1,954
Means, 1934	6,265	240,671	30,000	744
Midland Farms, 1944	2,655	248,211	12,385	292
Sand Hills, 1931	2,030	252,553	17,254	1,305
TXL, 1944	1,415	266,431	3,585	600
Waddell, 1927	555	101,855	2,649	193
Ward Estes, N., 1929	4,213	371,258	69,355	1,621
Westbrook, 1923	2,110	92,114	11,890	667
Yates, 1926	22,001	1,221,306	733,199	1,023

District 8A-West

Anton-Irish, 1944	2,978	182,423	18,092	245
Cogdell Area, 1949	1,199	294,722	38,700	208
Diamond M., 1948	1,845	242,595	12,886	317
Kelly-Snyder, 1948	8,608	1,262,074	87,926	787
Levelland, 1938	17,105	517,105	55,200	3,250
Prentice, 1951	5,002	171,294	22,300	413
Salt Creek, 1950	10,335	268,091	27,665	196
Seminole, 1936	16,845	557,909	110,383	449
Slaughter, 1936	17,898	1,065,551	135,550	3,018
Spraberry Trend, 1951	18,490	690,622	37,809	7,546
Wasson, 1936	25,937	1,764,036	140,961	2,046
Welch, 1942	3,106	150,334	25,911	651

District 10-Panhandle

Panhandle, 1921	6,859	1,436,157	30,000	12,263

UTAH*

Altamont, 1955	2,815	94,614	235,000	232
Aneth, 1956	5,513	365,083	23,225	463
East Anschutz Ranch, 1979	7,320	97,066	710,417	40
Red Wash, 1951	914	79,526	11,413	154

WYOMING*

Brady, 1973	1,514	57,990	45,705	22
Byron, 1918	830	122,457	5,828	61
Elk Basin, 1915	2,695	437,536	20,080	179
Frannie, 1928	730	117,601	3,338	62
Garland, 1906	2,495	167,933	7,790	229
Grass Creek, 1914	2,114	189,626	5,850	283
Hamilton Dome, 1918	2,390	235,210	5,800	245
Hartzog Draw, 1976	4,875	77,141	272,983	160
Hilite, 1969	467	76,392	54,906	100
Lance Creek, 1918	146	108,075	250	27
Little Buffalo Basin, 1914	2,121	123,308	6,879	174
Lost Soldier, 1916	3,000	213,724	7,500	92
Oregon Basin, 1912	7,511	403,895	17,489	506
Painter Reservoir, 1979	1,320	47,559	77,780	35
Salt Creek, 1889	4,162	638,489	16,041	1,218
Wertz, 1921	2,500	106,267	15,000	65
Whitney Canyon-Carter Creek, 1978	1,286	12,296	102,262	32

*Petroleum Information data

•Offshore

Crude oil prices

	$/bbl											
	U.S. refiner acquisition cost			U.S. landed cost of imports – Selected countries							Average export prices	
Year	Domestic	Imported	Composite	Total	Canada	Mexico	United Kingdom	Nigeria	Saudi Arabia	Venezuela	Worldwide	OPEC	Non-OPEC
1976	8.84	13.48	10.89	13.34	13.57	NA	NA	13.80	13.04	11.80	NA	NA	NA
1977	9.55	14.53	11.96	14.31	14.21	13.75	NA	15.25	13.61	13.13	NA	NA	NA
1978	10.61	14.57	12.46	14.38	14.50	13.54	NA	14.86	13.92	12.83	NA	NA	NA
1979	14.27	21.67	17.72	21.65	20.43	20.86	22.16	22.96	19.15	18.18	NA	NA	NA
1980	24.29	33.89	28.07	33.95	30.47	31.80	35.88	37.05	30.02	25.86	NA	NA	NA
1981	34.93	37.05	35.24	36.52	32.16	33.78	37.24	39.70	34.19	29.87	34.81	34.42	36.16
1982	31.22	33.55	31.87	33.18	26.92	26.64	34.28	36.17	35.00	24.82	33.33	33.71	32.3
1983	28.87	29.30	28.99	28.93	25.63	25.78	30.87	30.84	29.76	22.94	29.66	29.89	29.13
1984	28.53	28.88	28.63	28.46	26.59	26.87	29.60	30.50	29.50	25.15	28.60	28.61	28.6
1985	26.66	26.99	25.75	26.66	25.71	25.63	28.35	28.96	24.72	24.43	27.60	28.00	26.98
1986	14.82	14.00	14.55	13.49	13.43	12.17	14.63	15.29	12.84	11.52	14.98	15.45	14.21
1987	17.76	18.13	17.90	17.65	17.04	16.69	18.78	19.32	16.81	15.76	17.41	17.31	17.57
1988	14.74	14.56	14.67	14.08	15.15	12.58	15.82	15.88	13.37	13.66	13.79	13.67	14.04
1989	17.88	18.08	17.97	17.68	18.35	16.35	18.74	19.19	17.33	16.78	16.65	16.40	17.11

From: Oil & Gas Journal Energy Database
Source: U.S. Department of Energy

Petroleum product prices

	Motor gasoline-pump price (¢/gallon)				Heating Oil Residential	Refiner resale (wholesale) product prices							
Year	Leaded Regular	Unleaded Regular	Unleaded Premium	Average All-Types		Motor Gasoline	Aviation Gasoline	Kerosene Jet Fuel	Kerosene	No. 2 Fuel Oil	No. 2 Diesel Oil	Propane	Resid
1978	62.60	67.00	NA	65.20	49.00	43.40	53.70	38.60	40.40	36.90	36.50	23.70	26.30
1979	85.70	90.30	NA	88.20	70.40	63.70	72.10	66.00	62.40	56.90	57.40	29.10	39.90
1980	119.10	124.50	NA	122.10	97.40	94.10	112.80	86.80	86.40	80.30	80.10	41.50	52.80
1981	131.10	137.80	147.00	135.30	119.40	106.40	125.00	101.20	106.60	97.60	97.20	46.60	66.30
1982	122.20	129.60	141.50	128.10	116.00	97.30	122.80	95.30	101.80	91.40	91.40	42.70	61.20
1983	115.70	124.10	138.30	122.50	107.80	88.20	117.80	85.40	89.20	81.50	80.80	48.40	60.90
1984	112.90	121.20	136.60	119.80	109.10	83.20	116.50	83.00	91.60	82.10	80.30	45.00	65.40
1985	111.50	120.20	134.00	119.60	105.30	83.50	113.00	79.40	87.40	77.60	77.20	39.80	57.70
1986	85.70	92.70	108.50	93.10	83.60	53.10	91.20	49.50	60.60	48.60	45.20	29.00	30.50
1987	89.70	94.80	109.30	95.70	80.30	58.90	85.90	53.80	59.20	52.70	53.40	25.20	38.50
1988	89.90	94.60	110.70	96.30	81.30	57.70	85.00	49.50	54.90	47.30	47.30	24.00	30.00
1989	99.80	102.10	119.70	106.00	90.00	65.50	95.00	58.40	66.90	56.50	56.80	24.60	35.80

From: Oil & Gas Journal Energy Database
Source: U.S. Department of Energy

Comparative energy prices

	Natural gas delivered to consumers ($/MCF)					Cost of fossil fuels delivered to utilities (¢/Million BTU)				Electricity retail prices (¢/Kilowatt Hour)			
Year	Residential	Commercial	Industrial	Electric Utilities	Average	Coal	Heavy oil	Natural gas	All fuels	Residential	Commercial	Industrial	Total
1973	1.29	0.94	0.50	0.38	0.68	40.5	78.5	33.8	47.6	2.54	2.41	1.25	1.96
1974	1.46	1.07	0.67	0.51	0.84	70.9	189.0	48.2	91.4	3.10	3.04	1.69	2.49
1975	1.71	1.35	0.96	0.77	1.12	81.4	200.5	75.2	104.4	3.51	3.45	2.07	2.92
1976	1.98	1.64	1.24	0.98	1.37	84.8	195.2	103.4	111.9	3.73	3.69	2.21	3.09
1977	2.35	2.04	1.50	1.32	1.66	94.7	219.8	129.1	129.7	4.05	4.09	2.50	3.42
1978	2.56	2.23	1.70	1.48	1.85	111.6	212.5	142.2	141.1	4.31	4.36	2.79	3.69
1979	2.98	2.73	1.99	1.81	2.21	122.4	298.3	174.9	163.9	4.64	4.68	3.05	3.99
1980	3.68	3.39	2.56	2.27	2.80	135.1	426.7	219.9	192.8	5.36	5.48	3.69	4.73
1981	4.29	4.00	3.14	2.89	3.39	153.2	533.4	280.5	225.6	6.20	6.29	4.29	5.46
1982	5.17	4.82	3.87	3.48	4.15	164.70	483.0	337.6	224.9	6.86	6.86	4.95	6.13
1983	6.06	5.59	4.18	3.58	4.82	165.60	457.0	347.4	220.6	7.18	7.01	4.97	6.29
1984	6.12	5.55	4.22	3.70	4.85	166.40	481.0	360.3	219.1	7.54	7.33	5.04	6.52
1985	6.12	5.50	3.95	3.55	4.72	164.8	424.4	344.4	209.4	7.79	7.47	5.16	6.71
1986	5.83	5.08	3.23	2.43	4.13	157.9	240.1	235.1	175.0	7.41	7.13	4.90	6.42
1987	5.54	4.77	2.94	2.32	4.05	150.2	297.6	224.0	170.6	7.41	7.01	4.72	6.32
1988	5.47	4.63	2.95	2.33	4.09	146.6	240.5	226.3	164.3	7.49	7.07	4.62	6.31
1989	5.64	4.74	2.97	2.43	4.22	144.5	284.6	235.5	167.5	7.64	7.23	4.69	6.43

From Oil & Gas Journal Energy Database
Source: U.S. Department of Energy

Price history, crude oil, natural gas and motor gasoline

	Actual current prices						Inflation adjusted real prices				
	Crude oil		Motor	Natural gas			Crude oil		Motor	Natural gas	
	U.S. average wellhead	West Texas Intermediate posted	Gasoline leaded pump price	U.S. average wellhead	Consumer average	Producer Price index 1982-100	U.S. average wellhead	West Texas Intermediate posted	Gasoline Leaded pump price	U.S. average wellhead	Consumer average
Year	$/bbl	$/bbl	$/gal	$/Mcf	$/Mcf		$/bbl	$/bbl	$/gal	$/Mcf	$/Mcf
1925	1.68	1.79	0.222	0.094	0.223	17.8	9.438	10.056	1.247	0.528	1.253
1926	1.88	1.90	0.234	0.095	0.229	17.2	10.930	11.047	1.359	0.552	1.311
1927	1.30	1.28	0.211	0.088	0.220	16.5	7.879	7.758	1.278	0.533	1.333
1928	1.17	1.36	0.209	0.089	0.232	16.7	7.006	8.144	1.254	0.533	1.389
1929	1.27	1.45	0.214	0.082	0.216	16.4	7.744	8.841	1.306	0.500	1.317
1930	1.19	0.95	0.200	0.076	0.214	14.9	7.987	6.376	1.339	0.510	1.436
1931	0.65	0.77	0.170	0.070	0.233	12.6	5.159	6.111	1.348	0.556	1.849
1932	0.87	0.69	0.179	0.064	0.247	11.2	7.768	6.161	1.601	0.571	2.205
1933	0.67	1.00	0.178	0.062	0.237	11.4	5.877	8.772	1.563	0.544	2.079
1934	1.00	1.00	0.189	0.060	0.223	12.9	7.752	7.752	1.461	0.465	1.729
1935	0.97	1.00	0.188	0.058	0.224	13.8	7.029	7.246	1.365	0.420	1.623
1936	1.09	1.10	0.195	0.055	0.220	13.9	7.842	7.914	1.399	0.396	1.583
1937	1.18	1.22	0.200	0.051	0.220	14.9	7.919	8.188	1.342	0.342	1.477
1938	1.13	1.02	0.195	0.049	0.218	13.5	8.370	7.556	1.445	0.363	1.615
1939	1.02	1.02	0.188	0.049	0.216	13.3	7.669	7.699	1.410	0.368	1.624
1940	1.02	1.02	0.184	0.045	0.217	13.5	7.556	7.556	1.364	0.333	1.607
1941	1.14	1.17	0.192	0.049	0.221	15.1	7.550	7.748	1.274	0.325	1.464
1942	1.19	1.17	0.204	0.051	0.227	17.0	7.000	6.882	1.202	0.300	1.335
1943	1.20	1.17	0.205	0.052	0.223	17.8	6.742	6.573	1.153	0.292	1.253
1944	1.21	1.17	0.206	0.051	0.215	17.9	6.760	6.536	1.150	0.285	1.201
1945	1.22	1.17	0.205	0.049	0.214	18.2	6.703	6.429	1.126	0.269	1.176
1946	1.41	1.36	0.208	0.053	0.220	20.8	6.779	6.524	0.999	0.255	1.058
1947	1.93	1.84	0.231	0.060	0.232	25.6	7.539	7.188	0.903	0.234	0.906
1948	2.60	2.57	0.259	0.065	0.241	27.7	9.386	9.278	0.934	0.235	0.870
1949	2.54	2.57	0.268	0.063	0.254	26.3	9.658	9.772	1.019	0.240	0.966
1950	2.51	2.57	0.268	0.065	0.266	27.3	9.194	9.414	0.980	0.238	0.974
1951	2.53	2.57	0.272	0.073	0.298	30.4	8.322	8.454	0.893	0.240	0.980
1952	2.53	2.57	0.276	0.078	0.332	29.6	8.547	8.682	0.931	0.264	1.122
1953	2.68	2.72	0.287	0.092	0.355	29.2	9.178	9.298	0.983	0.315	1.216
1954	2.78	2.82	0.290	0.101	0.381	29.3	9.488	9.625	0.991	0.345	1.300
1955	2.77	2.82	0.291	0.104	0.400	29.3	9.454	9.625	0.992	0.355	1.365
1956	2.79	2.82	0.299	0.108	0.415	30.3	9.208	9.307	0.988	0.356	1.370
1957	3.09	3.04	0.310	0.113	0.431	31.2	9.904	9.753	0.992	0.362	1.381
1958	3.01	3.06	0.304	0.119	0.462	31.6	9.525	9.677	0.961	0.377	1.462
1959	2.90	2.98	0.305	0.129	0.477	31.7	9.148	9.385	0.962	0.407	1.505
1960	2.88	2.97	0.311	0.140	0.500	31.7	9.085	9.369	0.982	0.442	1.577
1961	2.89	2.97	0.308	0.151	0.510	31.6	9.146	9.399	0.973	0.478	1.614
1962	2.90	2.97	0.306	0.155	0.514	31.7	9.148	9.369	0.967	0.489	1.621
1963	2.89	2.97	0.304	0.158	0.512	31.6	9.146	9.399	0.963	0.500	1.620
1964	2.88	2.95	0.304	0.154	0.519	31.6	9.114	9.320	0.962	0.487	1.642
1965	2.86	2.92	0.312	0.156	0.522	32.3	8.854	9.040	0.966	0.483	1.616
1966	2.88	2.94	0.321	0.157	0.523	33.3	8.649	8.817	0.964	0.471	1.571
1967	2.92	3.03	0.332	0.160	0.520	33.4	8.743	9.060	0.994	0.479	1.557
1968	2.94	3.07	0.337	0.164	0.504	34.2	8.596	8.977	0.985	0.480	1.474
1969	3.09	3.30	0.348	0.167	0.515	35.6	8.680	9.256	0.978	0.469	1.447
1970	3.18	3.35	0.357	0.171	0.550	36.9	8.618	9.079	0.967	0.463	1.491
1971	3.39	3.56	0.364	0.182	0.590	38.0	8.921	9.368	0.958	0.479	1.553
1972	3.39	3.56	0.361	0.186	0.630	39.8	8.518	8.945	0.907	0.467	1.583
1973	3.89	3.87	0.388	0.216	0.680	45.0	8.644	8.604	0.862	0.480	1.511
1974	6.74	10.37	0.532	0.304	0.840	53.5	12.598	19.383	0.994	0.568	1.570
1975	7.56	11.16	0.567	0.445	1.120	58.4	12.945	19.110	0.971	0.762	1.918
1976	8.14	12.65	0.590	0.580	1.370	61.1	13.322	20.704	0.966	0.949	2.242
1977	8.57	14.30	0.622	0.790	1.660	64.9	13.205	22.034	0.958	1.217	2.558
1978	8.96	14.85	0.626	0.905	1.850	69.9	12.818	21.245	0.896	1.295	2.647
1979	12.51	22.40	0.857	1.178	2.210	78.7	15.896	28.463	1.089	1.497	2.808
1980	21.59	37.37	1.191	1.590	2.800	89.7	24.069	41.661	1.328	1.773	3.122
1981	31.77	36.67	1.311	1.980	3.390	98.0	32.418	37.418	1.338	2.020	3.459
1982	28.52	32.75	1.222	2.460	4.150	100.0	28.520	32.750	1.222	2.460	4.150
1983	26.19	30.25	1.157	2.590	4.820	101.2	25.879	29.891	1.143	2.559	4.763
1984	25.88	29.83	1.129	2.660	4.850	103.6	24.981	28.793	1.090	2.568	4.681
1985	24.09	28.08	1.115	2.510	4.720	103.1	23.366	27.236	1.081	2.435	4.578
1986	12.51	16.44	0.857	1.940	4.130	100.1	12.498	16.421	0.856	1.938	4.046
1987	15.40	18.21	0.897	1.670	4.050	102.8	14.981	17.714	0.873	1.625	3.940
1988	12.58	15.52	0.899	1.690	4.090	106.9	11.768	14.518	0.841	1.581	3.826
1989	15.85	18.29	0.998	1.690	4.220	112.2	14.127	16.301	0.889	1.506	3.761

From: Oil & Gas Journal Energy Database
Source: U.S. Department of Energy

Worldwide production statistics

NORTH AMERICA

Upper figures = 1,000 b/d
Lower figures = million t/y

	1940	1950	1960	1970	1980	1982	1983	1984	1985	1986	1987	1988	1989	1990
Canada	23.5	79.6	525.6	1,263.6	1,412.0	1,233.0	1,396.0	1,430.0	1,453.0	1,475.3	1,508.3	1,604.6	1,588.0	1,508.0
	1.2	4.0	26.2	62.9	70.3	61.4	69.5	71.2	72.4	73.5	75.1	79.9	79.1	75.1
United States	3,707.4	5,407.1	7,054.6	9,630.0	8,569.0	8,655.0	8,669.0	8,750.0	8,919.0	8,790.0	8,276.7	8,165.9	7,675.6	7,220.0
	184.6	269.3	351.3	479.6	426.7	431.0	431.7	435.8	444.2	437.7	412.2	406.7	382.2	359.6
Total	3,730.9	5,486.7	7,580.2	10,893.6	9,981.0	9,888.0	10,065.0	10,180.0	10,372.0	10,265.3	9,785.0	9,770.5	9,263.6	8,728.0
	185.8	273.3	377.5	542.5	497.0	492.4	501.2	507.0	516.6	511.2	487.3	486.6	461.3	434.7

MIDDLE EAST

Upper figures = 1,000 b/d
Lower figures = million t/y

	1940	1950	1960	1970	1980	1982	1983	1984	1985	1986	1987	1988	1989	1990
Abu Dhabi	...	...	...	693.8	1,350.0	883.0	757.0	750.0	705.1	949.0	969.4	1,012.6	1,359.0	1,587.0
	...	...	...	34.6	67.2	44.0	37.7	37.4	35.1	47.3	48.3	50.4	67.7	79.0
Bahrain	19.3	30.2	45.1	76.6	48.0	45.0	41.0	41.0	41.9	44.0	43.2	42.3	42.2	42.0
	1.0	1.5	2.2	3.8	2.4	2.2	2.0	2.0	2.1	2.2	2.2	2.1	2.1	2.1
Dubai	...	...	...	...	349.0	357.0	327.0	324.0	348.0	350.0	381.2	355.3	410.6	469.0
	...	...	...	...	17.4	17.8	16.3	16.1	17.3	17.4	19.0	17.7	20.4	23.4
Iran	181.2	664.3	1,067.6	3,328.8	1,467.0	1,896.0	2,606.0	2,166.0	2,279.0	1,806.3	2,341.7	2,207.5	2,934.3	3,120.0
	9.0	33.1	53.2	165.8	73.1	94.4	129.8	107.9	113.5	90.0	116.6	109.9	146.1	155.4
Iraq	60.7	136.2	1,004.2	1,548.6	2,638.0	914.0	905.0	1,218.0	1,396.8	1,787.7	2,095.8	2,679.2	2,830.0	2,083.0
	3.0	6.8	50.0	77.1	131.4	45.5	45.1	60.7	69.6	89.0	104.4	133.4	140.9	103.7
Israel	...	...	2.5	70.2	1.0	0.3	0.2	0.1	0.1	0.1	1.0	0.3	0.3	0.3
	...	...	0.1	3.5	...	...	...	...	...	...	...	...	...	...
Jordan	...	...	...	...	...	...	...	...	...	...	4.0	0.5	0.3	0.4
	...	...	...	...	...	...	...	...	...	...	0.2	...	...	...
Kuwait	...	344.4	1,628.2	2,734.5	1,382.0	675.0	912.0	925.0	823.3	1,202.0	1,095.8	1,254.2	1,542.6	1,080.0
	...	17.2	81.1	136.2	68.8	33.6	45.4	46.1	41.0	59.9	54.6	62.5	76.8	53.8
Neutral Zone	...	...	136.2	500.6	540.0	315.0	398.0	420.0	357.8	325.7	399.2	316.3	396.6	315.0
	...	...	6.8	24.9	26.9	15.7	19.8	20.9	17.8	16.2	19.9	15.8	19.8	15.7
Oman	...	...	...	332.0	283.0	328.0	378.0	404.0	484.3	540.8	565.0	596.7	622.7	658.0
	...	...	...	16.6	14.1	16.3	18.8	20.1	24.1	26.9	28.1	29.7	31.0	32.8
Qatar	...	33.6	175.1	362.4	472.0	340.0	270.0	395.0	297.0	332.3	284.4	349.2	394.6	387.0
	...	1.7	8.7	18.0	23.5	16.9	13.4	19.7	14.8	16.5	14.2	17.4	19.7	19.3
Ras Al Khaimah	...	...	...	...	...	...	...	5.8	9.1	11.0	10.0	10.0	10.0	10.0
	...	...	...	...	...	...	...	0.3	0.5	0.5	0.5	0.5	0.5	0.5
Saudi Arabia	13.9	546.7	1,247.1	3,548.9	9,630.0	6,484.0	4,872.0	4,545.0	3,295.3	4,719.7	4,054.2	4,708.3	4,935.6	6,215.0
	0.7	27.2	62.1	176.7	479.6	322.9	242.6	226.3	164.1	235.0	201.9	234.5	245.8	309.5
Sharjah	...	...	...	...	10.0	7.0	35.0	62.0	66.1	65.0	66.9	65.0	65.0	35.0
	...	...	...	...	0.5	0.3	1.7	3.1	3.3	3.2	3.3	3.2	3.2	1.7
Syria	...	...	...	83.1	165.0	175.0	165.0	161.0	161.7	185.0	231.7	273.3	300.6	385.0
	...	...	...	4.1	8.2	8.7	8.2	8.0	8.1	9.2	11.5	13.6	15.0	19.2
Turkey	...	0.3	7.0	67.9	44.0	45.0	45.0	41.0	41.5	46.8	52.0	50.0	55.5	70.0
	...	...	0.3	3.4	2.2	2.2	2.2	2.0	2.1	2.3	2.6	2.5	2.8	3.5
Yemen	...	...	...	...	...	...	...	...	...	...	10.0	10.0	172.4	186.6	179.0
	...	...	...	...	...	...	...	...	...	...	0.5	0.5	8.6	9.3	8.9
Total	275.1	1,755.7	5,313.0	13,347.8	18,379.0	12,464.3	11,711.2	11,457.9	10,307.0	12,375.4	12,605.5	14,093.1	16,086.5	16,635.7
	13.7	87.5	264.5	664.7	915.3	620.5	583.0	570.6	513.4	616.1	627.8	701.8	801.1	828.5

WORLDWIDE STATISTICS

EUROPE

Upper figures = 1,000 b/d
Lower figures = million t/y

	1940	1950	1960	1970	1980	1982	1983	1984	1985	1986	1987	1988	1989	1990
Austria	7.7 0.4	27.9 1.4	45.4 2.3	53.3 2.7	29.0 1.4	25.0 1.2	24.0 1.2	23.0 1.1	21.7 1.1	22.2 1.1	21.6 1.1	23.2 1.2	23.3 1.2	24.0 1.2
Denmark	... ...	... ...	... ...	... ...	6.0 0.3	38.0 1.9	46.0 2.3	45.0 2.2	58.0 2.9	74.7 3.7	94.0 4.7	95.5 4.8	111.0 5.5	118.0 5.9
France	1.4 0.1	2.5 0.1	38.6 1.9	45.8 2.3	26.0 1.3	33.0 1.6	32.0 1.6	35.0 1.7	49.0 2.4	59.0 2.9	63.5 3.2	68.7 3.4	65.3 3.3	62.0 3.1
Germany	20.1 1.0	23.2 1.2	108.3 5.4	148.4 7.4	100.0 5.0	94.0 4.7	91.0 4.5	89.0 4.4	90.5 4.5	90.0 4.5	83.8 4.2	85.9 4.3	81.7 4.1	78.0 3.9
Greece	... ...	... ...	... ...	... ...	... ...	21.0 1.0	25.0 1.2	27.0 1.3	26.4 1.3	26.4 1.3	24.4 1.2	21.8 1.1	17.0 0.8	15.0 0.7
Italy	0.2 ...	0.2 ...	37.5 1.9	26.3 1.3	39.0 1.9	28.0 1.4	37.0 1.8	45.0 2.2	41.7 2.1	49.0 2.4	50.0 2.5	92.9 4.6	94.7 4.7	97.0 4.8
Netherlands	... ...	13.4 0.7	36.0 1.8	35.8 1.8	25.0 1.2	28.0 1.4	42.0 2.1	61.0 3.0	69.0 3.4	79.0 3.9	91.3 4.5	83.7 4.2	87.0 4.3	71.0 3.5
Norway	... ...	... ...	... ...	... ...	528.0 26.3	488.0 24.3	600.0 29.9	688.0 34.3	771.0 38.4	823.3 41.0	973.3 48.5	1,069.1 53.2	1,469.8 73.2	1,574.0 78.4
Spain	... ...	... ...	... ...	3.1 0.2	32.0 1.6	30.0 1.5	59.0 2.9	45.0 2.2	45.3 2.3	37.2 1.9	32.8 1.6	30.4 1.5	18.0 0.9	14.3 0.7
United Kingdom	0.3 ...	0.9 0.1	1.7 0.1	1.7 0.1	1,619.0 80.6	2,050.0 102.1	2,260.0 112.5	2,452.0 122.1	2,519.8 125.5	2,602.0 129.6	2,445.9 121.8	2,376.3 118.3	1,743.0 86.8	1,860.0 92.6
Total	29.7 1.5	68.1 3.4	267.5 13.4	314.4 15.8	2,404.0 119.6	2,835.0 141.1	3,216.0 160.0	3,510.0 174.5	3,692.4 183.9	3,862.8 192.3	3,880.6 193.3	3,947.5 196.6	3,710.8 184.8	3,913.3 194.8

AFRICA

Upper figures = 1,000 b/d
Lower figures = million t/y

	1940	1950	1960	1970	1980	1982	1983	1984	1985	1986	1987	1988	1989	1990
Algeria	... ...	... ...	182.7 9.1	1,029.1 51.2	1,016.0 50.6	750.0 37.4	686.6 34.2	608.0 30.3	643.3 32.0	600.7 29.9	647.9 32.3	666.8 33.2	698.0 34.8	797.0 39.7
Angola (inc. Cabinda)	... ...	... ...	1.3 0.1	13.7 0.7	150.0 7.5	122.0 6.1	174.0 8.7	207.0 10.3	228.5 11.4	280.7 14.0	341.3 17.0	449.3 22.4	454.7 22.6	480.0 23.9
Cameroon	... ...	... ...	... ...	... ...	58.0 2.9	109.0 5.4	114.0 5.7	125.0 6.2	133.5 6.6	180.0 9.0	170.0 8.5	170.0 8.5	172.7 8.6	164.0 8.2
Congo	... ...	... ...	... ...	... ...	56.0 2.8	87.0 4.3	95.0 4.7	118.0 5.9	115.1 5.7	115.0 5.7	118.0 5.9	134.8 6.7	150.0 7.5	161.0 8.0
Egypt	18.1 0.9	44.9 2.2	61.6 3.1	327.3 16.3	596.0 29.7	667.0 33.2	690.0 34.4	790.0 39.3	885.8 44.1	773.7 38.5	899.3 44.8	851.3 42.4	854.3 42.5	873.0 43.5
Gabon	... ...	... ...	16.1 0.8	108.8 5.4	180.0 9.0	130.0 6.5	150.0 7.5	150.0 7.5	150.0 7.5	146.2 7.3	155.8 7.8	175.0 8.7	197.3 9.8	269.0 13.4
Ivory Coast	... ...	... ...	... ...	... ...	... ...	9.0 0.4	24.0 1.2	22.0 1.1	28.0 1.4	20.0 1.0	16.0 0.8	12.9 0.6	2.0 0.1	2.1 0.1
Libya	... ...	... ...	... ...	3,318.0 165.2	1,785.0 88.9	1,127.0 56.1	1,020.0 50.8	1,090.0 54.3	1,045.3 52.1	1,030.7 51.3	1,019.7 50.8	1,012.5 50.4	1,100.6 54.8	1,369.0 68.2
Nigeria	... ...	... ...	17.3 0.9	1,083.3 53.9	2,057.0 102.4	1,324.0 65.9	1,232.0 61.4	1,414.0 70.4	1,445.5 72.0	1,464.0 72.9	1,238.6 61.7	1,358.3 67.6	1,605.3 79.9	1,808.0 90.0
Tunisia	... ...	... ...	3.0 0.1	87.7 4.4	110.0 5.5	106.0 5.3	115.0 5.7	114.0 5.7	125.0 6.2	106.0 5.3	104.3 5.2	102.8 5.1	102.0 5.1	93.0 4.6
Zaire	... ...	... ...	... ...	... ...	20.0 1.0	22.0 1.1	26.0 1.3	27.0 1.3	30.5 1.5	32.1 1.6	31.8 1.6	29.5 1.5	27.5 1.4	28.0 1.4
Others	... ...	0.8 ...	3.0 0.1	1.5 0.1	4.0 0.2	1.7 0.1	6.6 0.3	7.7 0.4	8.9 0.4	7.6 0.4	12.3 0.6	5.0 0.2	4.8 0.2	4.3 0.2
Total	18.1 0.9	45.7 2.2	285.0 14.2	5,969.4 297.2	6,032.0 300.5	4,454.7 221.8	4,333.2 215.9	4,672.7 232.7	4,839.4 240.9	4,756.7 236.9	4,755.0 237.0	4,968.2 247.3	5,369.2 267.3	6,048.4 301.2

DRILLING at Egypt's Ras Budran field. (Photo couresty OPEC Bulletin).

ASIA-PACIFIC

Upper figures = 1,000 b/d
Lower figures = million t/y

	1940	1950	1960	1970	1980	1982	1983	1984	1985	1986	1987	1988	1989	1990
Australia	...	...	...	175.6	379.0	353.0	405.0	481.0	556.0	472.0	566.1	552.3	488.8	582.0
	...	...	...	8.7	18.9	17.6	20.2	24.0	27.7	23.5	28.2	27.5	24.3	29.0
Bangladesh	...	...	...	...	...	...	...	...	...	...	0.5	0.9	0.9	0.9
	...	...	...	...	...	...	...	...	...	...	...	...	...	...
Brunei	19.3	84.8	92.9	148.0	230.0	155.0	155.0	160.0	149.5	170.0	133.7	138.3	143.3	143.0
	1.0	4.2	4.6	7.4	11.5	7.7	7.7	8.0	7.4	8.5	6.7	6.9	7.1	7.1
China	...	4.0	100.0	500.0	2,119.0	2,042.0	2,120.0	2,250.0	2,496.0	2,620.0	2,700.0	2,733.0	2,770.0	2,755.0
	...	0.2	5.0	24.9	105.5	101.7	105.6	112.1	124.3	130.5	134.5	136.1	137.9	137.2
China, Taiwan	...	...	0.4	1.8	5.0	3.0	2.3	2.6	2.7	1.9	2.6	2.6	2.6	2.5
	...	...	...	0.1	0.2	0.1	0.1	0.1	0.1	0.1	0.1	0.1	0.1	0.1
India	3.9	5.1	9.0	138.6	182.0	384.0	390.0	543.0	614.0	622.7	608.7	631.8	673.3	679.0
	0.2	0.3	0.4	6.9	9.1	19.1	19.4	27.0	30.6	31.0	30.3	31.5	33.5	33.8
Indonesia	169.4	132.6	411.2	853.6	1,576.0	1,341.0	1,292.0	1,332.0	1,219.5	1,243.8	1,185.8	1,137.5	1,208.6	1,274.0
	8.4	6.6	2.05	42.5	78.5	66.8	64.3	66.3	60.7	61.9	59.1	56.6	60.2	63.4
Japan	7.2	5.6	10.2	16.3	10.0	6.0	6.9	6.4	10.0	12.8	12.1	12.3	11.3	10.5
	0.4	0.3	0.5	0.8	0.5	0.3	0.3	0.3	0.5	0.6	0.6	0.6	0.6	0.5
Malaysia	...	...	...	...	288.0	306.0	370.0	462.0	432.8	503.3	467.4	540.0	557.3	605.0
	...	...	...	...	14.3	15.2	18.4	23.0	21.6	25.1	23.3	26.9	27.8	30.1
Myanmar (Burma)	21.1	1.5	11.1	16.4	30.0	30.0	30.0	30.0	30.0	30.0	24.0	15.0	12.0	13.0
	1.1	0.1	0.6	0.8	1.5	1.5	1.5	1.5	1.5	1.5	1.2	0.7	0.6	0.6
New Zealand	...	...	...	...	7.0	15.0	15.0	18.0	18.0	29.3	28.3	28.0	28.0	39.0
	...	...	...	...	0.3	0.7	0.7	0.9	0.9	1.5	1.4	1.4	1.4	1.9
Pakistan	2.4	3.5	7.2	9.9	10.0	12.0	13.0	18.0	34.3	41.0	41.6	47.0	45.0	60.0
	0.1	0.2	0.4	0.5	0.5	0.6	0.6	0.9	1.7	2.0	2.1	2.3	2.2	3.0
Papua New Guinea	...	4.8	4.2	...	...	...	...	...	...	...	...	...	...	...
	...	0.2	0.2	...	...	...	...	...	...	...	...	...	...	...
Philippines	...	...	...	...	15.0	7.0	15.0	12.0	9.3	6.3	5.1	9.2	4.0	5.0
	...	...	...	...	0.7	0.3	0.7	0.6	0.5	0.3	0.3	0.5	0.2	0.2
Thailand	...	...	...	0.4	...	6.0	11.0	19.0	34.3	35.6	31.3	38.4	39.0	41.0
	...	...	...	...	...	0.3	0.5	0.9	1.7	1.8	1.6	1.9	1.9	2.0
Viet Nam*	...	...	...	...	...	...	...	...	...	...	...	...	...	40.0
	...	...	...	...	...	...	...	...	...	...	...	...	...	2.0
Total	223.3	241.9	646.2	1,860.6	4,851.0	4,660.0	4,825.2	5,334.0	5,606.4	5,788.7	5,807.2	5,886.3	5,984.1	6,249.9
	11.2	12.1	32.2	92.6	241.5	231.9	240.0	265.6	279.2	288.3	289.4	293.0	297.8	310.9

*Prior to 1990, included in other Eastern Europe - U.S.S.R.

LATIN AMERICA

Upper figures = 1,000 b/d
Lower figures = million t/y

	1940	1950	1960	1970	1980	1982	1983	1984	1985	1986	1987	1988	1989	1990
Argentina	56.3	64.0	171.9	382.9	487.0	483.0	481.0	467.0	447.5	430.2	419.1	449.6	459.0	473.0
	2.8	3.2	8.6	19.1	24.3	24.1	24.0	23.3	22.3	21.4	20.9	22.4	22.9	23.6
Bolivia	0.8	1.7	9.8	16.3	30.0	24.0	22.0	20.0	20.0	18.7	18.5	18.8	20.3	19.2
	...	0.1	0.5	0.8	1.5	1.2	1.1	1.0	1.0	0.9	0.9	0.9	1.0	1.0
Brazil	...	0.9	80.9	160.5	182.0	252.0	315.0	437.0	540.5	578.0	562.8	555.6	597.1	633.0
	...	...	4.0	8.0	9.1	12.5	15.7	21.8	26.9	28.8	28.0	27.7	29.7	31.5
Chile	...	1.7	19.8	34.6	29.0	41.0	39.0	38.0	32.5	34.4	28.3	24.8	23.7	20.3
	...	0.1	1.0	1.7	1.4	2.0	1.9	1.9	1.6	1.7	1.4	1.2	1.2	1.0
Colombia	69.9	93.3	152.7	214.0	125.0	140.0	155.0	165.0	178.8	324.2	387.8	346.7	405.0	445.0
	3.5	4.6	7.6	10.7	6.2	7.0	7.7	8.2	8.9	16.1	19.3	17.3	20.2	22.2
Cuba*	...	...	...	...	...	...	...	...	...	...	...	...	...	15.0
	...	...	...	...	...	...	...	...	...	...	...	...	...	0.7
Ecuador	6.4	7.2	7.7	4.1	222.0	215.0	236.0	254.0	273.3	270.1	157.2	310.1	286.6	287.0
	0.3	0.4	0.4	0.2	11.1	10.7	11.8	12.6	13.6	13.5	7.8	15.4	14.3	14.3
Guatemala	...	...	...	...	5.0	6.4	7.5	5.2	3.1	4.9	3.9	3.7	3.7	4.0
	...	...	...	...	0.2	0.3	0.4	0.3	0.2	0.2	0.2	0.2	0.2	0.2
Mexico	120.3	198.5	270.6	430.2	1,936.0	2,734.0	2,702.0	2,743.0	2,797.0	2,468.0	2,537.8	2,527.3	2,618.0	2,633.0
	6.0	9.9	13.5	21.4	96.4	136.2	134.6	136.6	139.3	122.9	126.4	125.9	130.4	131.1
Peru	33.2	41.1	52.6	72.2	191.0	198.0	171.0	201.0	189.0	179.0	166.7	141.7	129.0	132.0
	1.7	2.0	2.6	3.6	9.5	9.9	8.5	10.0	9.4	8.9	8.3	7.1	6.4	6.6
Trinidad & Tobago	60.9	56.5	115.7	139.8	211.0	182.0	158.0	169.0	179.0	167.9	163.1	149.3	153.7	151.0
	3.0	2.8	5.8	7.0	10.5	9.1	7.9	8.4	8.9	8.4	8.1	7.4	7.7	7.5
Venezuela	507.0	1,498.0	2,846.1	3,708.0	2,167.0	1,826.0	1,791.0	1,724.0	1,669.0	1,664.9	1,591.9	1,658.0	1,731.6	2,118.0
	25.2	74.6	141.7	184.7	107.9	90.9	89.2	85.9	83.1	82.9	79.3	82.6	86.2	105.5
Others	...	...	...	...	...	0.6	0.5	2.2	2.9	3.6	3.6	3.7	4.9	5.2
	...	...	...	...	...	...	...	0.1	0.1	0.2	0.2	0.2	0.2	0.3
Total	854.8	1,962.9	3,727.8	5,162.6	5,585.0	6,102.0	6,078.0	6,225.4	6,332.6	6,143.9	6,040.7	6,189.3	6,432.6	6,935.7
	42.5	97.7	185.7	257.2	278.1	303.9	302.8	310.1	315.3	305.9	300.8	308.3	320.4	345.5
World Total, excl. E. Europe - U.S.S.R.	5,131.9	9,561.0	17,819.7	37,048.4	47,232.0	40,404.0	40,228.6	41,380.0	41,149.8	43,192.8	42,874.0	44,854.9	46,846.8	48,511.0
	255.6	476.2	887.5	1,870.0	2,352.0	2,011.6	2,002.9	2,060.5	2,049.3	2,150.7	2,135.6	2,233.6	2,332.7	2,415.6

*Prior to 1990, included in other Eastern Europe - U.S.S.R.

EASTERN EUROPE — U.S.S.R.

Upper figures = 1,000 b/d
Lower figures = million t/y

	1940	1950	1960	1970	1980	1982	1983	1984	1985	1986	1987	1988	1989	1990
Bulgaria	...	...	4.0	6.7	6.0	3.0	3.0	3.0	3.0	3.0	6.0	5.0	5.0	5.0
	...	...	0.2	0.3	0.3	0.1	0.1	0.1	0.1	0.1	0.3	0.2	0.2	0.2
Czechoslovakia	...	0.8	2.7	4.1	1.9	2.0	2.0	2.0	2.0	2.0	2.0	3.0	3.0	3.0
	...	...	0.1	0.2	0.1	0.1	0.1	0.1	0.1	0.1	0.1	0.1	0.1	0.1
Hungary	...	10.1	24.3	38.7	40.6	41.0	40.0	40.0	40.0	40.0	40.0	40.0	39.0	40.0
	...	0.5	1.2	1.9	2.0	2.0	2.0	2.0	2.0	2.0	2.0	2.0	1.9	2.0
Poland	...	3.3	3.9	8.5	6.6	4.0	5.0	5.0	5.0	4.0	4.0	4.0	3.0	3.0
	...	0.2	0.2	0.4	0.3	0.2	0.2	0.2	0.2	0.2	0.2	0.2	0.1	0.1
Romania	...	87.7	231.0	268.6	230.9	234.0	240.0	229.0	226.0	220.0	215.0	188.0	180.0	160.0
	...	4.4	11.5	13.4	11.5	11.7	12.0	11.4	11.3	11.0	10.7	9.4	9.0	8.0
U.S.S.R.*	624.9	760.6	2,969.1	7,089.2	12,031.0	12,260.0	12,320.0	12,260.0	11,900.0	12,300.0	12,480.0	12,452.0	12,140.0	11,500.0
	31.1	37.9	147.9	353.0	599.1	610.5	613.5	610.5	592.6	612.5	621.5	620.1	604.6	572.7
Yugoslavia	0.3	2.1	18.8	57.8	84.5	87.0	82.0	80.0	81.0	83.0	84.0	73.0	70.0	65.0
	...	0.1	0.9	2.9	4.2	4.3	4.1	4.0	4.0	4.1	4.2	3.6	3.5	3.2
Others	...	2.7	14.9	37.1	50.0	81.0	86.0	96.0	86.0	77.0	80.0	85.0	86.0	30.0
	...	0.1	0.7	1.8	2.5	4.0	4.3	4.8	4.3	3.8	4.0	4.2	4.3	1.5
Total	625.2	867.3	3,268.7	7,510.7	12,451.5	12,712.0	12,778.0	12,715.0	12,343.0	12,729.0	12,911.0	12,850.0	12,526.0	11,806.0
	31.1	43.2	162.7	373.9	620.0	632.9	636.3	633.1	614.6	633.8	643.0	639.8	623.7	587.8

*Includes gas liquids

World Total	5,757.1	10,428.3	21,088.4	45,059.1	59,683.5	53,116.0	53,006.6	54,095.0	53,492.8	55,921.8	55,785.0	57,704.9	59,372.8	60,317.0
	286.7	519.4	1,050.2	2,243.9	2,972.0	2,644.5	2,639.2	2,693.6	2,663.9	2,784.5	2,778.6	2,873.4	2,956.4	3,003.4

Worldwide oil consumption

Upper figures = 1,000 b/d
Lower figures = million t/y

	1987	1988	1989		1987	1988	1989
U.S.A.	16,035 767.3	16,525 793.4	16,585 792.6	United Kingdom	1,610 75.2	1,690 79.2	1,730 81.2
Canada	1,515 70.2	1,590 73.7	1,660 76.8	West Germany	2,445 115.2	2,445 115.2	2,295 107.4
Total North America	17,550 837.5	18,115 867.1	18,245 869.4	Other	55 2.8	60 3.1	65 3.1
Latin America	4,985 236.4	5,070 240.3	6,265 248.5	Total Western Europe	12,410 587.7	12,575 596.4	12,560 594.1
Total Western Hemisphere	**22,535 1,073.9**	**23,185 1,107.4**	**23,510 1,117.9**	U.S.S.R.	8,985 444.2	8,885 439.1	8,795 434.7
Austria	220 10.7	215 10.6	215 10.4	Eastern Europe	2,165 105.5	2,170 106.2	2,160 105.2
Belgium and Luxembourg	490 23.7	500 24.4	480 23.1	Middle East	2,705 130.1	2,805 135.1	2,920 140.3
Denmark	200 9.6	195 9.5	185 9.1	Africa	1,770 84.8	1,820 87.3	1,830 87.5
Finland	230 11.2	230 11.0	230 11.0	South Asia	1,245 60.4	1,360 65.8	1,450 69.7
France	1,845 86.6	1,830 86.0	1,875 88.4	Other Asia	2,080 101.2	2,285 111.6	2,530 123.4
Greece	270 13.2	280 13.7	275 13.6	China	2,110 105.3	2,210 110.2	2,360 117.1
Ireland	85 4.1	80 3.9	85 4.0	Japan	4,500 209.2	4,805 224.7	5,005 232.8
Italy	1,850 90.1	1,880 91.7	1,940 94.3	South Korea	605 29.0	720 34.7	825 39.5
Netherlands	705 32.4	740 34.5	740 34.3	Australasia	730 33.0	745 34.3	790 35.6
Norway	215 9.7	205 9.2	195 8.9	**Total Eastern Hemisphere**	**39,305 1,890.4**	**40,380 1,945.4**	**41,225 1,979.9**
Portugal	190 8.9	185 8.7	195 9.3	**World Total**	**61,840 2,964.3**	**63,565 3,052.8**	**64,735 3,097.8**
Spain	945 43.8	970 44.7	1,010 46.6	Of which OECD and LDCs*.	48,205 2,290.9	49,920 2,378.3	51,020 2,420.7
Sweden	355 17.1	340 16.3	325 15.4		Organisation for Cooperation and Development and lesser developed countries. Source: British Petroleum Co. plc South Asia—Afghanistan, Bangladesh, India, Myanmar, Nepal, Pakistan, Sri Lanka, Other Asia—Brunei, Cambodia, Hong Kong, Indonesia, Laos, Malaysia, Mongolia, North Korea, Papua New Guinea, Philippines, Singapore, SW Pacific Islands, Taiwan, Thailand, Australasia—Australia, New Zealand.		
Switzerland	265 12.4	265 12.4	255 11.9				
Turkey	435 21.0	465 22.3	465 22.1				

WORLDWIDE REFINING STATISTICS

Worldwide refining capacity

EUROPE

Upper figures = 1,000 b/d
Lower figures = million t/y

	1940	1950	1960	1970	1980	1983	1984	1985	1986	1987	1988	1989	1990	1991
Austria	... / ...	26 / 1.3	47 / 2.3	94 / 4.7	280 / 13.9	269 / 13.4	268 / 13.3	273 / 13.6	204 / 10.2	204 / 10.2	204 / 10.2	204 / 10.2	204 / 10.2	204 / 10.2
Belgium	12 / 0.6	16 / 0.8	171 / 8.5	669 / 33.3	1,064 / 53.0	693 / 34.5	694 / 34.6	693 / 34.5	652 / 32.5	648 / 32.3	631 / 31.4	631 / 31.4	614 / 30.6	602 / 30.0
Cyprus	... / ...	... / ...	... / ...	... / ...	16 / 0.8	16 / 0.8	15 / 0.7	16 / 0.8	16 / 0.8	16 / 0.8	16 / 0.8	17 / 0.8	18 / 0.9	19 / 0.9
Denmark	1 / ...	1 / ...	1 / ...	186 / 9.3	214 / 10.7	215 / 10.7	174 / 8.7	176 / 8.8	166 / 8.3	166 / 8.3	177 / 8.8	177 / 8.8	187 / 9.3	186 / 9.3
Finland	... / ...	... / ...	25 / 1.2	174 / 8.6	336 / 16.7	299 / 14.9	299 / 14.9	299 / 14.9	241 / 12.0	241 / 12.0	241 / 12.0	241 / 12.0	241 / 12.0	241 / 12.0
France	152 / 7.6	306 / 15.2	769 / 38.3	2,317 / 115.4	3,385 / 168.6	2,871 / 143.0	2,670 / 133.0	2,386 / 118.8	1,947 / 97.0	1,834 / 91.3	1,941 / 96.7	1,876 / 93.4	1,820 / 90.6	1,816 / 90.4
Germany	48 / 2.4	81 / 4.0	573 / 28.5	2,359 / 117.5	2,986 / 148.7	2,471 / 123.1	2,386 / 118.8	2,172 / 108.2	1,933 / 96.3	1,720 / 85.7	1,648 / 82.1	1,518 / 75.6	1,507 / 75.0	2,065 / 102.8
Greece	... / ...	... / ...	30 / 1.5	102 / 5.1	431 / 21.5	422 / 21.0	369 / 18.4	390 / 19.4	390 / 19.4	385 / 19.2	385 / 19.2	385 / 19.2	385 / 19.2	395 / 19.7
Ireland	11 / 0.5	... / ...	40 / 2.0	55 / 2.7	56 / 2.8	56 / 2.8	56 / 2.8	56 / 2.8	56 / 2.8	56 / 2.8	56 / 2.8	56 / 2.8	56 / 2.8	56 / 2.8
Italy	41 / 2.0	107 / 5.3	773 / 38.5	2,963 / 147.5	4,131 / 205.7	3,283 / 163.5	3,050 / 151.9	3,095 / 154.1	2,738 / 136.4	2,679 / 133.4	2,563 / 127.6	2,450 / 122.0	2,804 / 139.6	2,385 / 118.8
Netherlands	15 / 0.7	77 / 3.8	359 / 17.9	1,362 / 67.8	1,828 / 91.0	1,552 / 77.3	1,552 / 77.3	1,499 / 74.7	1,468 / 73.1	1,401 / 69.8	1,381 / 68.8	1,381 / 68.8	1,381 / 68.8	1,197 / 59.6
Norway	1 / ...	1 / ...	2 / 0.1	123 / 6.1	264 / 13.1	243 / 12.1	243 / 12.1	244 / 12.2	240 / 12.0	240 / 12.0	240 / 12.0	239 / 11.9	295 / 14.7	288 / 14.3
Portugal	4 / 0.2	8 / 0.4	23 / 1.1	37 / 1.8	378 / 18.8	365 / 18.2	282 / 14.0	290 / 14.4	296 / 14.7	294 / 14.6	294 / 14.6	313 / 15.6	313 / 15.6	294 / 14.6
Spain	11 / 0.5	17 / 0.8	146 / 7.2	685 / 34.1	1,455 / 72.5	1,522 / 75.8	1,493 / 74.4	1,493 / 74.4	1,367 / 68.1	1,305 / 65.0	1,305 / 65.0	1,285 / 64.0	1,293 / 64.4	1,321 / 65.8
Sweden	4 / 0.2	26 / 1.3	48 / 2.4	230 / 11.5	458 / 22.8	453 / 22.6	453 / 22.6	439 / 21.9	429 / 21.4	437 / 21.8	437 / 21.8	427 / 21.3	428 / 21.3	428 / 21.3
Switzerland	... / ...	... / ...	... / ...	106 / 5.3	137 / 6.8	137 / 6.8	137 / 6.8	127 / 6.3	137 / 6.8	65 / 3.2	132 / 6.6	132 / 6.6	132 / 6.6	132 / 6.6
United Kingdom	133 / 6.6	197 / 9.8	963 / 47.9	2,301 / 114.6	2,527 / 125.8	2,260 / 112.5	2,092 / 104.2	2,007 / 99.9	1,792 / 89.2	1,780 / 88.6	1,803 / 89.8	1,803 / 89.8	1,831 / 91.2	1,867 / 93.0
Yugoslavia	5 / 0.2	8 / 0.4	10 / 0.5	179 / 8.9	296 / 14.7	297 / 14.8	297 / 14.8	297 / 14.8	302 / 15.0	483 / 24.1	608 / 30.3	609 / 30.3	609 / 30.3	609 / 30.3
Total Europe	438 / 21.5	871 / 43.1	3,980 / 198.1	13,942 / 694.4	20,242 / 1,007.9	17,424 / 867.8	16,530 / 823.3	15,952 / 794.5	14,374 / 716.0	13,954 / 695.1	14,062 / 700.5	13,744 / 684.5	14,118 / 703.1	14,105 / 702.4

NORTH AMERICA

Upper figures = 1,000 b/d
Lower figures = million t/y

	1940	1950	1960	1970	1980	1983	1984	1985	1986	1987	1988	1989	1990	1991
Canada	203 / 10.1	329 / 16.4	920 / 45.8	1,400 / 69.7	2,222 / 110.7	2,020 / 100.6	1,807 / 90.0	1,869 / 93.1	1,856 / 92.4	1,760 / 87.6	1,869 / 93.1	1,856 / 92.4	1,852 / 92.2	1,882 / 93.7
United States	4,460 / 222.1	6,696 / 333.5	10,250 / 510.5	12,600 / 627.5	17,720 / 882.5	16,800 / 836.6	15,930 / 793.3	15,400 / 766.9	15,182 / 756.1	15,258 / 759.8	15,288 / 761.3	15,557 / 774.7	16,244 / 809.0	15,559 / 774.8
Total	4,663 / 232.2	7,025 / 349.9	11,170 / 556.3	14,000 / 697.2	19,942 / 993.2	18,820 / 937.2	17,737 / 883.3	17,269 / 860.0	17,038 / 848.5	17,018 / 847.4	17,157 / 854.4	17,413 / 867.1	18,096 / 901.2	17,441 / 868.5

WORLDWIDE REFINING STATISTICS

LATIN AMERICA

Upper figures = 1,000 b/d
Lower figures = million t/y

	1940	1950	1960	1970	1980	1983	1984	1985	1986	1987	1988	1989	1990	1991
Argentina	103 5.1	152 7.6	238 11.9	457 22.8	676 33.7	676 33.7	678 33.8	678 33.8	667 33.2	670 33.4	690 34.4	690 34.4	689 34.3	696 34.7
Bahamas	... ...	... ...	... ...	... ...	500 24.9	500 24.9	500 24.9	350 17.4	... ...	... ...	... ...	... ...	... ...	... ...
Barbados	... ...	... ...	... ...	3 0.1	3 0.1	3 0.1	3 0.1	3 0.1	3 0.1	3 0.1	3 0.1	3 0.1	3 0.1	3 0.1
Bolivia	1 ...	7 0.3	12 0.6	12 0.6	74 3.7	61 3.0	47 2.3	47 2.3	47 2.3	47 2.3	47 2.3	58 2.9	58 2.9	45 2.2
Brazil	4 0.2	12 0.6	156 7.8	502 25.0	1,205 60.0	1,219 60.7	1,301 64.8	1,305 65.0	1,305 65.0	1,321 65.8	1,407 70.1	1,407 70.1	1,397 69.6	1,412 70.3
Chile	1 ...	4 0.2	24 1.2	91 4.5	139 6.9	141 7.0	141 7.0	141 7.0	141 7.0	150 7.5	141 7.0	147 7.3	147 7.3	146 7.3
Colombia	15 0.7	24 1.2	51 2.5	137 6.8	193 9.6	214 10.7	211 10.5	211 10.5	211 10.5	226 11.3	226 11.3	227 11.3	227 11.3	247 12.3
Costa Rica	... ...	... ...	... ...	8 0.4	11 0.5	16 0.8	17 0.8	16 0.8	15 0.7	16 0.8	16 0.8	16 0.8	16 0.8	15 0.7
Cuba	4 0.2	7 0.3	87 4.3	93 4.6	160 8.0	160 8.0	160 8.0	160 8.0	160 8.0	160 8.0	160 8.0	160 8.0	160 8.0	280 13.9
Ecuador	2 0.1	5 0.2	7 0.4	33 1.6	86 4.3	79 3.9	84 4.2	82 4.1	88 4.4	88 4.4	122 6.1	123 6.1	145 7.2	142 7.1
El Salvador	... ...	... ...	... ...	14 0.7	16 0.8	16 0.8	16 0.8	16 0.8	16 0.8	16 0.8	17 0.8	17 0.8	17 0.8	16 0.8
Guatemala	... ...	... ...	... ...	21 1.0	16 0.8	16 0.8	16 0.8	16 0.8	16 0.8	16 0.8	16 0.8	16 0.8	16 0.8	16 0.8
Honduras	... ...	... ...	... ...	10 0.5	14 0.7	14 0.7	14 0.7	14 0.7	14 0.7	14 0.7	14 0.7	14 0.7	14 0.7	14 0.7
Jamaica	... ...	... ...	... ...	32 1.6	34 1.7	36 1.8	36 1.8	36 1.8	36 1.8	36 1.8	36 1.8	36 1.8	34 1.7	34 1.7
Mexico	98 4.9	160 8.0	357 17.8	495 24.6	1,394 69.4	1,289 64.2	1,269 63.2	1,269 63.2	1,269 63.2	1,349 67.2	1,354 67.4	1,354 67.4	1,514 75.4	1,679 83.6
Netherlands Antilles	495 24.7	617 30.7	680 33.9	840 41.8	792 39.4	782 38.9	740 36.9	740 36.9	320 15.9	320 15.9	320 15.9	320 15.9	320 15.9	320 15.9
Nicaragua	... ...	... ...	... ...	22 1.1	14 0.7	14 0.7	15 0.7	15 0.7	15 0.7	15 0.7	15 0.7	15 0.7	15 0.7	16 0.8
Panama	... ...	... ...	... ...	78 3.9	100 5.0	100 5.0	100 5.0	100 5.0	100 5.0	100 5.0	100 5.0	100 5.0	100 5.0	100 5.0
Paraguay	... ...	... ...	... ...	5 0.2	7 0.3	8 0.4	8 0.4	8 0.4	8 0.4	8 0.4	8 0.4	8 0.4	8 0.4	8 0.4
Peru	22 1.1	35 1.7	49 2.5	92 4.6	170 8.5	168 8.4	169 8.4	176 8.8	176 8.8	176 8.8	182 9.1	172 8.6	172 8.6	189 9.4
Puerto Rico	... ...	... ...	84 4.2	155 7.7	284 14.1	123 6.1	121 6.0	121 6.0	121 6.0	121 6.0	123 6.1	123 6.1	123 6.1	125 6.2
Trinidad & Tobago	49 2.4	104 5.2	182 9.1	430 21.4	456 22.7	375 18.7	375 18.7	320 15.9	260 12.9	300 14.9	300 14.9	300 14.9	300 14.9	246 12.3
Uruguay	5 0.2	16 0.8	28 1.4	40 2.0	40 2.0	45 2.2	45 2.2	45 2.2	45 2.2	45 2.2	45 2.2	33 1.6	33 1.6	33 1.6
Venezuela	55 2.7	254 12.6	886 44.1	1,313 65.4	1,446 72.0	1,284 63.9	1,224 61.0	1,224 61.0	1,230 61.3	1,225 61.0	1,201 59.8	1,201 59.8	1,201 59.8	1,167 58.1
Virgin Islands	... ...	... ...	... ...	220 11.0	728 36.3	600 29.9	600 29.9	545 27.1	600 29.9	600 29.9	545 27.1	545 27.1	545 27.1	545 27.1
Others	... ...	... ...	... ...	11 0.5	62 3.1	77 3.8	55 2.7	57 2.8	57 2.8	57 2.8	61 3.0	60 3.0	61 3.0	60 3.0
Total	854 42.3	1,397 69.4	2,841 141.5	5,114 254.5	8,620 429.2	8,016 399.1	7,945 395.6	7,695 383.1	6,920 344.4	7,080 352.5	7,149 355.8	7,143 355.5	7,315 364.0	7,554 376.0

HERE IS a section of the Luanda refinery in Angola. (Photo courtesy OPEC Bulletin).

ASIA-PACIFIC

Upper figures = 1,000 b/d
Lower figures = million t/y

	1940	1950	1960	1970	1980	1983	1984	1985	1986	1987	1988	1989	1990	1991
Australia	4 0.2	14 0.7	237 11.8	594 29.6	725 36.1	716 35.7	722 36.0	697 34.7	623 31.0	626 31.2	637 31.7	644 32.1	675 33.6	706 35.2
Bangladesh	...	...	...	...	31 1.5	31 1.5	22 1.1	31 1.5	31 1.5	31 1.5	31 1.5	31 1.5	31 1.5	31 1.5
Brunei-Malaysia-Singapore	10 0.5	10 0.5	46 2.3	320 15.9	1,093 54.4	1,271 63.3	1,313 65.4	1,286 64.0	1,240 61.8	1,183 58.9	1,080 53.8	1,071 53.3	1,049 52.2	1,098 54.7
China	...	8 0.4	98 4.9	300 14.9	1,600 79.7	2,000 99.6	2,050 102.1	2,150 107.1	2,150 107.1	2,200 109.6	2,200 109.6	2,200 109.6	2,200 109.6	2,200 109.6
China, Taiwan	...	...	28 1.4	119 5.9	425 21.2	515 25.6	515 25.6	543 27.0	543 27.0	543 27.0	600 29.9	570 28.4	570 28.4	543 27.0
Guam*	...	...	...	...	44 2.2	44 2.2	44 2.2	44 2.2	...	...	...	...	...	...
India	10 0.5	6 0.3	112 5.6	446 22.2	557 27.7	753 37.5	779 38.8	705 35.1	867 43.2	991 49.4	1,059 52.7	1,051 52.3	1,080 53.8	1,122 55.9
Indonesia	170 8.5	203 10.1	274 13.6	268 13.4	528 26.3	341 17.0	387 19.3	631 31.4	636 31.7	636 31.7	714 35.6	714 35.6	714 35.6	814 40.5
Japan	53 2.6	34 1.7	640 31.9	3,140 156.4	5,509 274.3	5,548 276.3	5,020 250.0	4,813 239.7	4,613 229.7	4,790 238.5	4,567 227.4	4,363 217.3	4,198 209.1	4,383 218.3
Korea, North†	...	...	...	...	...	...	...	...	...	...	...	...	...	42 2.1
Korea, South	...	...	...	180 9.0	601 29.9	755 37.6	776 38.6	776 38.6	782 38.9	862 42.9	820 40.8	880 43.8	867 43.2	867 43.2
Myanmar (Burma)	...	...	9 0.5	26 1.3	26 1.3	26 1.3	26 1.3	26 1.3	26 1.3	26 1.3	26 1.3	26 1.3	32 1.6	32 1.6
New Zealand	...	...	...	66 3.3	74 3.7	74 3.7	74 3.7	53 2.6	53 2.6	54 2.7	82 4.1	88 4.4	95 4.7	95 4.7
Okinawa*	...	...	...	...	198 9.9	183 9.1	153 7.6	153 7.6	110 5.5	...	...	...	...	...
Pakistan	...	5 0.2	6 0.3	107 5.3	98 4.9	133 6.6	126 6.3	129 6.4	130 6.5	130 6.5	130 6.5	130 6.5	121 6.0	121 6.0
Philippines	...	...	22 1.1	182 9.1	253 12.6	286 14.2	286 14.2	286 14.2	216 10.8	286 14.2	286 14.2	254 12.6	284 14.1	279 13.9
Sri Lanka	...	...	...	44 2.2	38 1.9	50 2.5	50 2.5	50 2.5	50 2.5	50 2.5	50 2.5	50 2.5	50 2.5	50 2.5
Thailand	...	...	1	62 3.1	186 9.3	176 8.8	176 8.8	172 8.6	192 9.6	192 9.6	192 9.6	191 9.5	215 10.7	221 11.0
Others	27 1.3	4 0.2	...	12 0.6	...	...	...	...	...	...	...	...	...	...
Total	274 13.6	284 14.1	1,472 73.3	5,866 292.1	11,986 596.9	12,902 642.5	12,519 623.5	12,545 624.5	12,262 610.7	12,600 627.5	12,474 621.2	12,263 610.7	12,181 606.6	12,604 627.7

*Prior to 1980, included in others
†Prior to 1991, included in other U.S.S.R.—Eastern Europe

WORLDWIDE REFINING STATISTICS

AFRICA

Upper figures = 1,000 b/d
Lower figures = million t/y

	1940	1950	1960	1970	1980	1983	1984	1985	1986	1987	1988	1989	1990	1991
Algeria	... / ...	... / ...	... / ...	48 / 2.4	122 / 6.1	137 / 6.8	137 / 6.8	464 / 23.1	465 / 23.2	465 / 23.2	465 / 23.2	465 / 23.2	465 / 23.2	465 / 23.2
Angola-Cabinda	... / ...	... / ...	2 / 0.1	14 / 0.7	36 / 1.8	32 / 1.6	32 / 1.6	32 / 1.6	32 / 1.6	32 / 1.6	32 / 1.6	32 / 1.6	32 / 1.6	38 / 1.9
Cameroon	... / ...	... / ...	... / ...	... / ...	... / ...	43 / 2.1	41 / 2.0	43 / 2.1	43 / 2.1	43 / 2.1	43 / 2.1	43 / 2.1	42 / 2.1	43 / 2.1	42 / 2.1
Congo	... / ...	... / ...	... / ...	14 / 0.7	... / ...	... / ...	21 / 1.0	21 / 1.0	21 / 1.0	21 / 1.0	21 / 1.0	21 / 1.0	21 / 1.0	21 / 1.0
Egypt	17 / 0.8	39 / 1.9	87 / 4.3	175 / 8.7	234 / 11.7	341 / 17.0	369 / 18.4	369 / 18.4	434 / 21.6	452 / 22.5	452 / 22.5	489 / 24.4	489 / 24.4	523 / 26.0
Ethiopia	... / ...	... / ...	... / ...	11 / 0.5	14 / 0.7	14 / 0.7	15 / 0.7	14 / 0.7	18 / 0.9	18 / 0.9	18 / 0.9	18 / 0.9	18 / 0.9	18 / 0.9
Gabon	... / ...	... / ...	... / ...	13 / 0.6	20 / 1.0	20 / 1.0	20 / 1.0	20 / 1.0	23 / 1.1	16 / 0.8	23 / 1.1	24 / 1.2	24 / 1.2	24 / 1.2
Ghana	... / ...	... / ...	... / ...	29 / 1.4	26 / 1.3	27 / 1.3	28 / 1.4	28 / 1.4	28 / 1.4	28 / 1.4	28 / 1.4	27 / 1.3	27 / 1.3	27 / 1.3
Ivory Coast	... / ...	... / ...	... / ...	19 / 0.9	50 / 2.5	90 / 4.5	90 / 4.5	90 / 4.5	90 / 4.5	73 / 3.6	60 / 3.0	60 / 3.0	69 / 3.4	69 / 3.4
Kenya	... / ...	... / ...	... / ...	44 / 2.2	95 / 4.7	79 / 3.9	79 / 3.9	95 / 4.7	95 / 4.7	95 / 4.7	90 / 4.5	90 / 4.5	95 / 4.7	90 / 4.5
Liberia	... / ...	... / ...	... / ...	10 / 0.5	15 / 0.7	15 / 0.7	15 / 0.7	15 / 0.7	15 / 0.7	15 / 0.7	15 / 0.7	15 / 0.7	15 / 0.7	15 / 0.7
Libya	... / ...	... / ...	... / ...	10 / 0.5	138 / 6.9	130 / 6.5	125 / 6.2	330 / 16.4	330 / 16.4	329 / 16.4	329 / 16.4	329 / 16.4	329 / 16.4	348 / 17.3
Madagascar	... / ...	... / ...	... / ...	13 / 0.6	11 / 0.5	16 / 0.8	16 / 0.8	16 / 0.8	16 / 0.8	16 / 0.8	16 / 0.8	16 / 0.8	16 / 0.8	16 / 0.8
Morocco	... / ...	... / ...	2 / 0.1	35 / 1.7	72 / 3.6	74 / 3.7	74 / 3.7	80 / 4.0	81 / 4.0	81 / 4.0	155 / 7.7	155 / 7.7	155 / 7.7	155 / 7.7
Mozambique	... / ...	... / ...	... / ...	20 / 1.0	16 / 0.8	16 / 0.8	17 / 0.8	17 / 0.8	17 / 0.8	17 / 0.8	... / ...	... / ...	... / ...	... / ...
Nigeria	... / ...	... / ...	... / ...	46 / 2.3	160 / 8.0	260 / 12.9	247 / 12.3	250 / 12.5	250 / 12.5	250 / 12.5	270 / 13.4	415 / 20.7	433 / 21.6	433 / 21.6
Senegal	... / ...	... / ...	... / ...	13 / 0.6	20 / 1.0	18 / 0.9	18 / 0.9	30 / 1.5	30 / 1.5	30 / 1.5	30 / 1.5	30 / 1.5	24 / 1.2	24 / 1.2
Sierra Leone	... / ...	... / ...	... / ...	10 / 0.5	10 / 0.5	10 / 0.5	10 / 0.5	10 / 0.5	10 / 0.5	10 / 0.5	10 / 0.5	10 / 0.5	10 / 0.5	10 / 0.5
South Africa	1 / ...	... / ...	25 / 1.2	184 / 9.2	478 / 23.8	424 / 21.1	389 / 19.4	389 / 19.4	389 / 19.4	389 / 19.4	434 / 21.6	434 / 21.6	434 / 21.6	431 / 21.5
Sudan	... / ...	... / ...	... / ...	21 / 1.0	26 / 1.3	24 / 1.2	24 / 1.2	24 / 1.2	24 / 1.2	24 / 1.2	21 / 1.0	21 / 1.0	25 / 1.2	22 / 1.1
Tanzania	... / ...	... / ...	... / ...	14 / 0.7	17 / 0.8	17 / 0.8	17 / 0.8	14 / 0.7	14 / 0.7	14 / 0.7	14 / 0.7	14 / 0.7	14 / 0.7	17 / 0.8
Togo	... / ...	... / ...	... / ...	... / ...	20 / 1.0	20 / 1.0	20 / 1.0	20 / 1.0	20 / 1.0	20 / 1.0	20 / 1.0	... / ...	... / ...	... / ...
Tunisia	... / ...	... / ...	... / ...	23 / 1.1	34 / 1.7	34 / 1.7	34 / 1.7	34 / 1.7	34 / 1.7	34 / 1.7	34 / 1.7	34 / 1.7	34 / 1.7	34 / 1.7
Zaire	... / ...	... / ...	... / ...	21 / 1.0	16 / 0.8	17 / 0.8	17 / 0.8	17 / 0.8	17 / 0.8	17 / 0.8	17 / 0.8	17 / 0.8	17 / 0.8	17 / 0.8
Others	... / ...	... / ...	... / ...	... / ...	35 / 1.7	35 / 1.7	35 / 1.7	35 / 1.7	35 / 1.7	35 / 1.7	34 / 1.7	31 / 1.5	35 / 1.7	35 / 1.7
Total	18 / 0.8	39 / 1.9	116 / 5.7	787 / 38.8	1,665 / 82.9	1,893 / 94.0	1,890 / 93.8	2,457 / 122.2	2,531 / 125.8	2,524 / 125.5	2,631 / 130.8	2,789 / 138.8	2,824 / 140.4	2,874 / 142.9

MIDDLE EAST

Upper figures = 1,000 b/d
Lower figures = million t/y

	1940	1950	1960	1970	1980	1983	1984	1985	1986	1987	1988	1989	1990	1991
Abu Dhabi	...	...	...	...	14	135	185	185	185	185	180	180	180	193
	...	...	...	...	0.7	6.7	9.2	9.2	9.2	9.2	9.0	9.0	9.0	9.6
Bahrain	33	150	187	205	250	250	250	250	250	250	250	243	243	243
	1.6	7.5	9.3	10.2	12.5	12.5	12.5	12.5	12.5	12.5	12.5	12.1	12.1	12.1
Iran	333	502	495	645	921	530	530	530	530	530	530	530	530	720
	16.6	25.0	24.7	32.1	45.9	26.4	26.4	26.4	26.4	26.4	26.4	26.4	26.4	35.9
Iraq	4	9	56	104	168	169	319	319	319	319	319	319	319	319
	0.2	0.4	2.8	5.2	8.4	8.4	8.4	15.9	15.9	15.9	15.9	15.9	15.9	15.9
Israel	...	83	87	107	195	190	190	170	170	180	180	180	180	180
	...	4.1	4.3	5.3	9.7	9.5	9.5	8.5	8.5	9.0	9.0	9.0	9.0	9.0
Jordan	...	...	...	7	21	100	100	100	100	100	100	100	100	100
	...	...	...	0.4	1.0	5.0	5.0	5.0	5.0	5.0	5.0	5.0	5.0	5.0
Kuwait*	...	25	270	569	645	623	623	669	634	618	628	817	819	819
	...	1.2	13.5	28.4	32.1	31.0	31.0	33.3	31.6	30.8	31.3	40.7	40.8	40.8
Lebanon	...	6	24	38	53	52	52	17	17	37	37	37	37	38
	...	0.3	1.2	1.9	2.6	2.6	2.6	0.8	0.8	1.8	1.8	1.8	1.8	1.9
Qatar	...	...	1	1	11	12	63	56	56	62	62	62	62	62
	...	...	...	...	0.5	0.6	3.1	2.8	2.8	3.1	3.1	3.1	3.1	3.1
Saudi Arabia*	...	140	189	377	487	705	860	840	1,115	1,125	1,375	1,375	1,484	1,863
	...	7.0	9.4	18.8	24.3	35.1	42.8	41.8	55.5	56.0	68.5	68.5	73.9	92.8
Syria	...	...	...	59	223	229	229	228	229	229	229	244	244	237
	...	...	...	2.9	11.1	11.4	11.4	11.4	11.4	11.4	11.4	12.2	12.2	11.8
Turkey	...	1	7	149	356	467	472	460	460	460	676	725	725	729
	...	...	0.3	7.4	17.7	23.3	23.5	22.9	22.9	22.9	33.7	36.1	36.1	36.3
Yemen	...	...	120	178	175	178	130	178	162	170	170	162	162	172
	...	...	6.0	8.9	8.7	8.9	6.5	8.9	8.1	8.5	8.5	8.1	8.1	8.6
Others	...	...	...	...	...	46	48	50	48	58	87	87	87	76
	...	...	...	...	...	2.3	2.4	2.5	2.4	2.9	4.3	4.3	4.3	3.8
Total	370	916	1,436	2,439	3,519	3,686	3,901	4,052	4,275	4,323	4,823	5,061	5,172	5,751
	18.4	45.5	71.5	121.4	175.2	183.7	194.3	201.9	213.0	215.4	240.4	252.2	257.7	286.6
Total, exclu. U.S.S.R., E. Europe	6,617	10,532	21,015	42,148	65,974	62,741	60,522	59,970	57,400	57,499	58,296	58,413	59,706	60,329
	328.8	523.9	1,046.4	2,098.4	3,285.3	3,124.3	3,013.8	2,986.2	2,858.4	2,863.4	2,903.1	2,908.8	2,973.0	3,004.1

*Includes Neutral Zone.

U.S.S.R.-EASTERN EUROPE

Upper figures = 1,000 b/d
Lower figures = million t/y

	1940	1950	1960	1970	1980	1983	1984	1985	1986	1987	1988	1989	1990	1991
Albania*	...	...	...	...	...	...	...	...	...	...	...	...	...	40
	...	...	...	...	...	...	...	...	...	...	...	...	...	2.0
Bulgaria*	...	...	...	...	...	...	...	...	...	...	...	...	...	300
	...	...	...	...	...	...	...	...	...	...	...	...	...	14.9
Czechoslovakia	...	16	46	220	455	455	455	455	455	455	455	455	455	455
	...	0.8	2.3	11.0	22.7	22.7	22.7	22.7	22.7	22.7	22.7	22.7	22.7	22.7
Hungary	...	20	58	130	290	311	312	242	242	242	220	220	220	220
	...	1.0	2.9	6.5	14.4	15.5	15.5	12.1	12.1	12.1	11.0	11.0	11.0	11.0
Poland	...	4	20	156	385	385	385	385	385	385	385	385	385	385
	...	0.2	1.0	7.8	19.2	19.2	19.2	19.2	19.2	19.2	19.2	19.2	19.2	19.2
Romania	...	182	260	356	608	617	617	617	617	617	617	617	617	617
	...	9.1	12.9	17.7	30.3	30.7	30.7	30.7	30.7	30.7	30.7	30.7	30.7	30.7
U.S.S.R.	...	900	2,800	5,640	10,950	11,750	12,000	12,200	12,260	12,260	12,260	12,300	12,300	12,300
	...	44.8	139.4	280.9	545.3	585.2	597.6	607.6	610.5	610.5	610.5	612.5	612.5	612.5
Others	...	4	40	332	852	898	894	894	894	894	852	910	852	...
	...	0.2	2.0	16.5	42.4	44.7	44.5	44.5	44.5	44.5	42.4	45.3	42.4	...
Total	...	1,126	3,224	6,834	13,540	14,416	14,663	14,793	14,853	14,853	14,789	14,887	14,829	14,317
	...	56.1	160.5	340.4	674.3	718.0	730.2	736.8	739.7	739.7	736.5	741.4	738.5	713.0
World Total	6,617	11,658	24,239	48,982	79,514	77,157	75,185	74,763	72,253	72,352	73,085	73,300	74,535	74,646
	328.8	580.0	1,206.9	2,438.8	3,959.6	3,842.3	3,744.0	3,723.0	3,598.1	3,603.1	3,639.6	3,650.2	3,711.5	3,717.1

*Prior to 1991, included in others.

World 1990: Oil production, refining capacity, oil movements, reserves

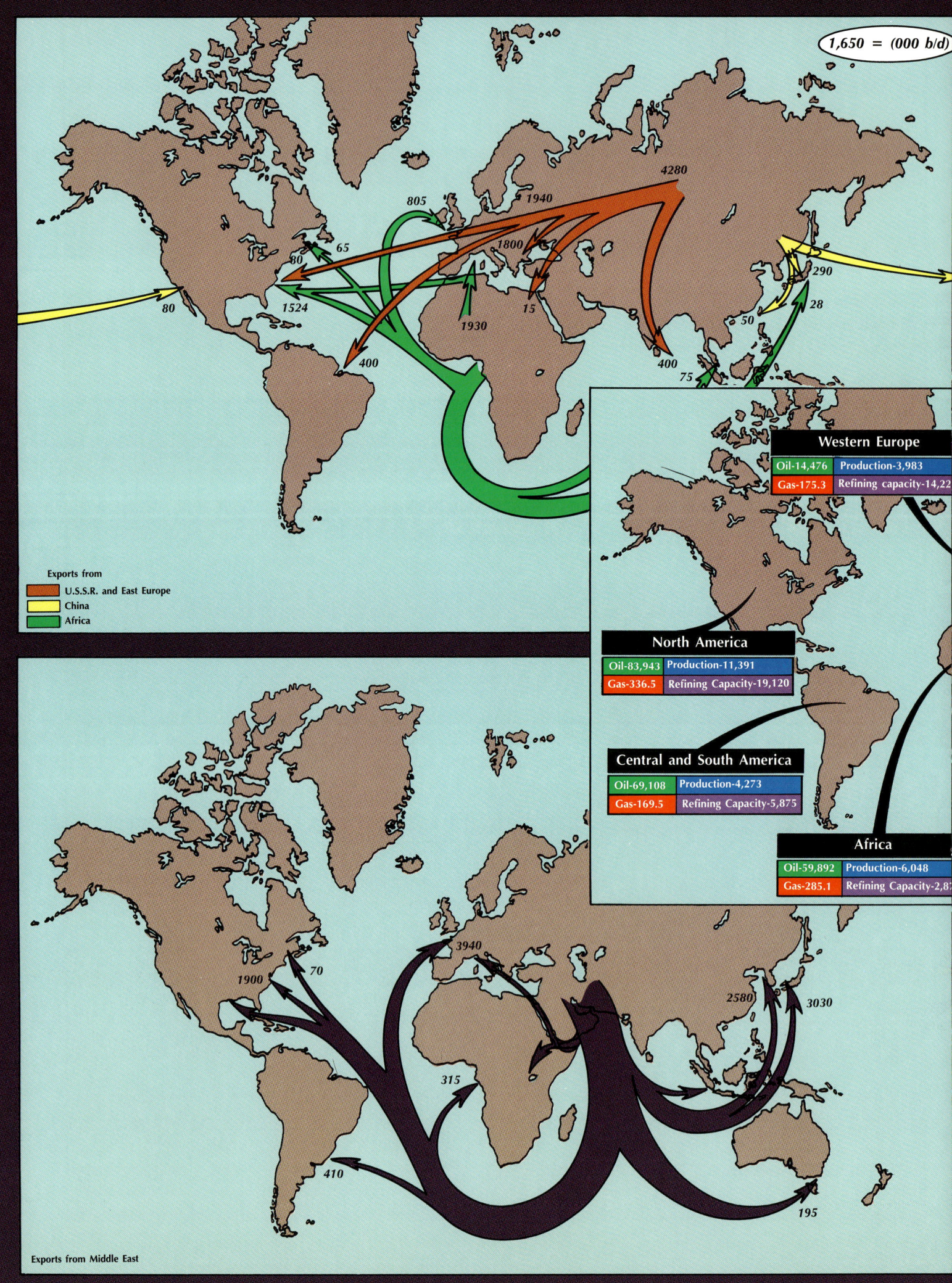

1,650 = (000 b/d)

Exports from
- U.S.S.R. and East Europe
- China
- Africa

Western Europe
| Oil-14,476 | Production-3,983 |
| Gas-175.3 | Refining capacity-14,22... |

North America
| Oil-83,943 | Production-11,391 |
| Gas-336.5 | Refining Capacity-19,120 |

Central and South America
| Oil-69,108 | Production-4,273 |
| Gas-169.5 | Refining Capacity-5,875 |

Africa
| Oil-59,892 | Production-6,048 |
| Gas-285.1 | Refining Capacity-2,8... |

Exports from Middle East

*CPE's-centrally planned economics

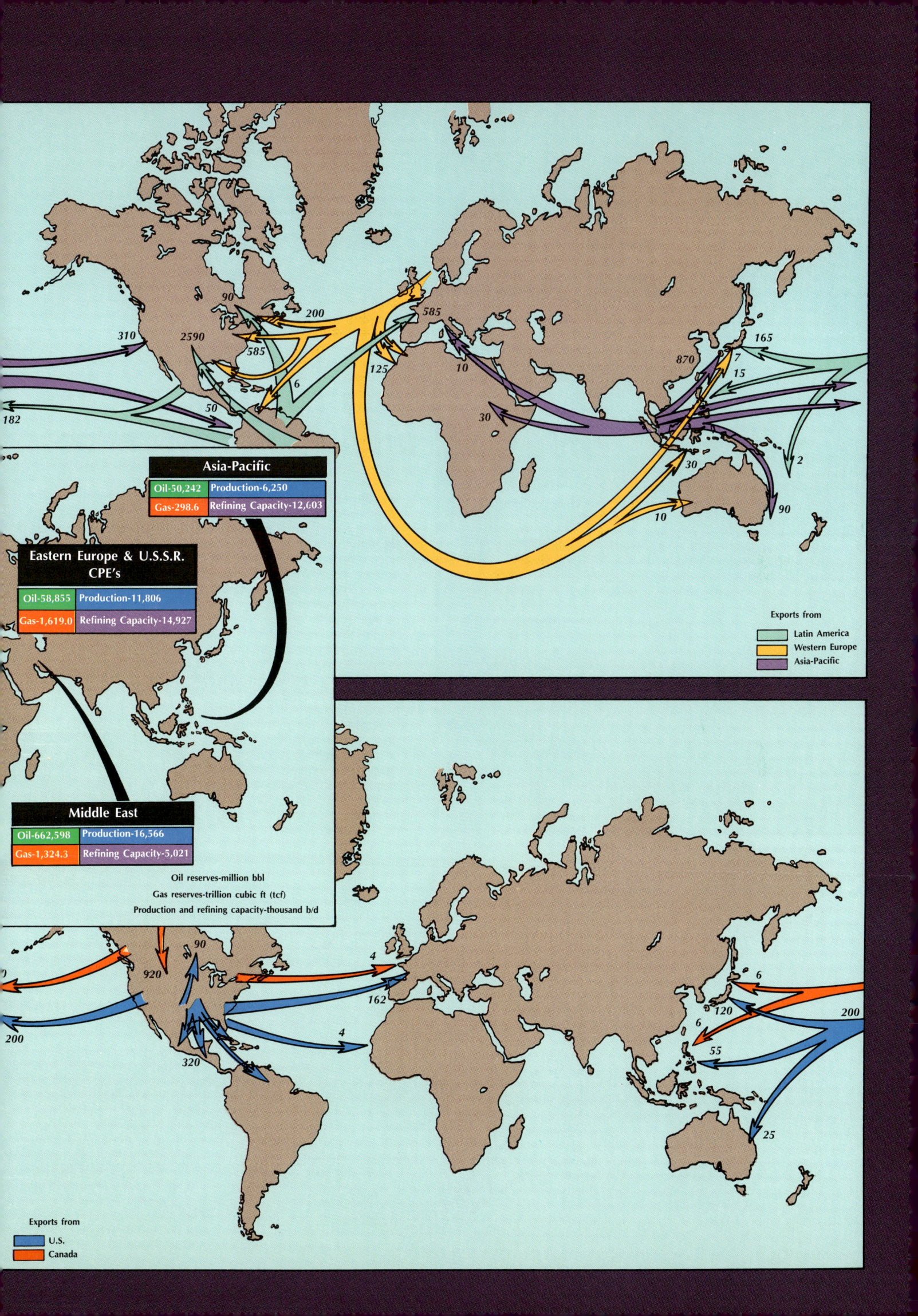

Asia-Pacific
| Oil-50,242 | Production-6,250 |
| Gas-298.6 | Refining Capacity-12,603 |

Eastern Europe & U.S.S.R. CPE's
| Oil-58,855 | Production-11,806 |
| Gas-1,619.0 | Refining Capacity-14,927 |

Middle East
| Oil-662,598 | Production-16,566 |
| Gas-1,324.3 | Refining Capacity-5,021 |

Oil reserves-million bbl
Gas reserves-trillion cubic ft (tcf)
Production and refining capacity-thousand b/d

Exports from
- Latin America
- Western Europe
- Asia-Pacific

Exports from
- U.S.
- Canada

SCHWEDT REFINERY in eastern Germany is among former eastern block plants joining Europe's refining network. The massive changes accompanying the introduction of a market economy in eastern Europe will include a dramatic overhaul of the formerly Communist downstream industry to accommodate western level products needs and environmental concerns. (Photo courtesy PCK AG Schwedt.)

High margins, healthy profits seen for refiners

REFINERS UNDERWENT A WATERSHED YEAR IN 1990.

Mounting environmental concerns forced refiners worldwide to earmark billions of dollars to reduce the environmental effects of their operations and to revamp plants to accommodate a changing product slate to meet air quality concerns.

U.S. refiners mounted a widespread campaign to introduce reduced emission gasolines and other products touted as being more environmentally benign. The effort probably helped them avoid a more draconian Clean Air Act, although there remain pitfalls in the final legislation President Bush signed in November 1990.

In Europe, the quickly expanding demand for unleaded gasoline is forcing refiners to accelerate spending for upgrading refineries to produce more of the product.

European refiners are leading the frenzied rush into downstream joint ventures in eastern Europe, where a promised renaissance of the formerly Communist bloc's dilapidated refining/marketing sector will provide major business opportunities in the 1990s. Even the Soviet Union's bedraggled downstream sector offers the promise of a huge modernization program, assuming chaos doesn't overtake that troubled nation.

Table 1
U.S. crude capacity changes by type

Process	Capacity 1,000 b/d As of 1-1-91	As of 1-1-90	% change from 1-1-90	Vol. chg. from 1-1-90	Construction, 1,000 b/d 1991 Completion	Est. % change by 1-1-92	1992 Completions*
Crude, b/cd	15,478.9	15,426.7	0.3	52.2	94.5	0.61	97.0
Feed, b/sd							
Thermal operations	2,016.9	1,975.6	2.0	41.3	1.6	0.1	10.0
Cat. reforming	3,901.5	3,862.0	1.0	39.5	89.5	2.3	94.0
Cat. cracking (FF)	5,294.4	5,372.1	−1.5	−77.7	51.2	1.0	0.0
Cat. Hydrocracking	1,260.8	1,209.5	4.1	51.3	25.5	2.0	37.0
Hydrorefining	2,446.9	2,410.0	1.5	36.9	12.0	0.5	6.5
Hydrotreating	7,270.1	7,137.8	1.8	132.3	59.9	0.8	78.0
Products, b/sd							
Alkylation	1,072.4	1,031.3	3.8	41.1	11.7	1.1	12.0
Aromatics/Isom.	836.3	784.9	6.2	51.4	19.7	2.4	12.5
Lubes	224.7	236.9	−5.4	−12.2	3.8	1.7	5.9
Coke, 1,000 tons/day	77.5	74.4	3.9	3.1	0.0	0.0	0.8
Asphalt	748.7	754.5	−0.8	−5.8	13.7	1.8	0.0

*Based on reported projects known to date.

REFINING

How U.S. crude and gasoline prices have varied

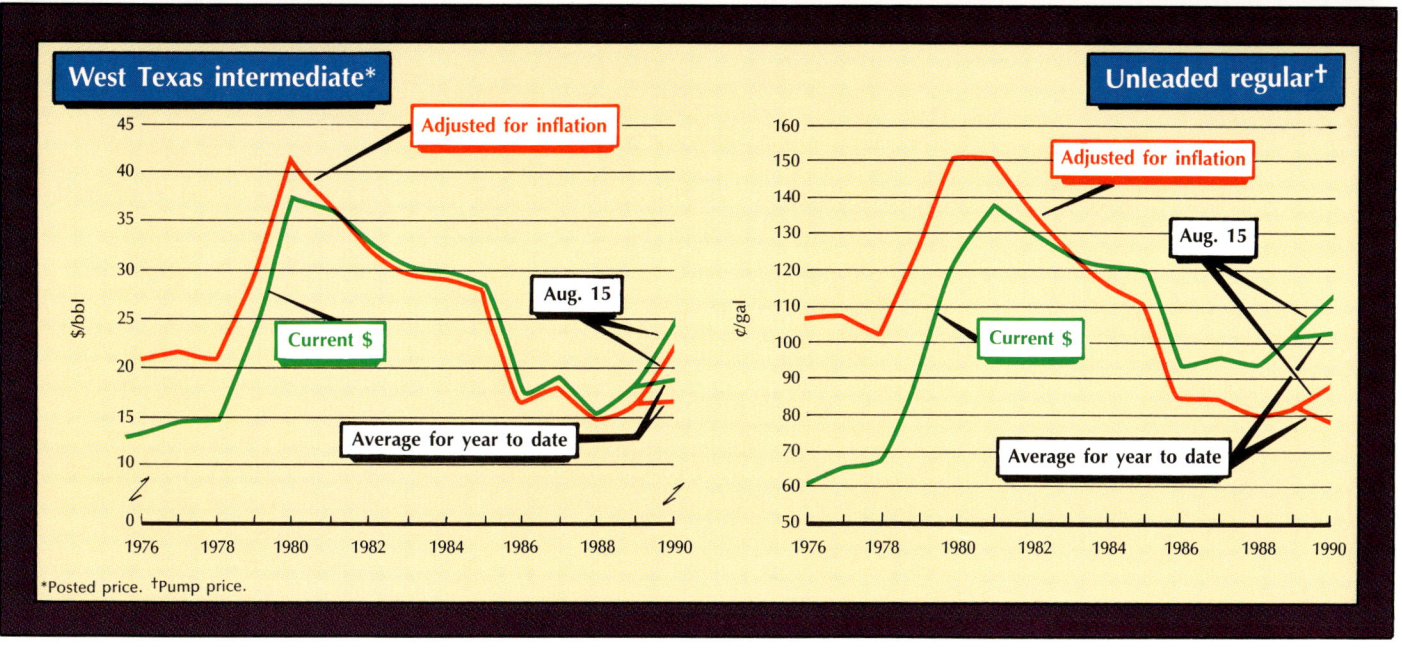

Sweeping change in Latin America, paced by privatization and economic reform among South America's state oil companies, offers a no less exciting prospect for investment in refining operations there. Economic reform has included moves to end subsidized fuels prices in the region, which has hamstrung the state companies in the past.

Because the fastest growing demand for petroleum products during the 1990s is expected to occur in the Asia-Pacific region, there is a strong effort to boost downstream capacity to better meet domestic demand and increase products exports as well.

Members of the Organization of Petroleum Exporting Countries continued their efforts to acquire downstream interests in other nations as well as increase domestic capacity. But that push was dealt a severe blow—as were refiners' margins and profits—by Iraq's blitzkrieg invasion and takeover of Kuwait and the subsequent breakout of war in the Persian Gulf stemming from that action.

Refiners in 1991 will be especially hard pressed to sustain healthy margins if the Persian Gulf conflict manages to result in higher crude oil prices. Oil companies were blistered by political and consumer/environmentalist lobbies after a rapid runup in crude and products prices during the Middle East crisis led to charges of price-gouging. In fact, refiner/marketers efforts to restrain product price increases while crude prices remained high caused refining profits to plummet in second half 1990.

Refineries worldwide were expected to continue operating all out in 1991, mainly because of the loss of sophisticated Kuwait refining capacity dismantled by the invading Iraqis. Refiners also took heart when the prospect suddenly loomed early in 1991 that crude oil prices might again

U.S. capacity changes by region*

Table 2

Process	PADD I*	PADD II	PADD III	PADD IV	PADD V	Total
No. refineries	20	41	66	18	49	194
Crude, b/cd	1,523,630	3,393,475	6,874,811	548,900	3,138,133	15,478,949
% of U.S. capacity (b/cd basis)	9.8	21.9	44.4	3.5	20.3	
Crude, b/sd	1,618,947	3,554,867	7,243,686	578,400	3,308,054	16,303,954
Cat. cracking (FF), b/sd	602,500	1,290,500	2,405,600	201,300	794,500	5,294,400
% U.S. capacity	11.4	24.4	45.4	3.8	15.0	
Cat. reforming, b/sd	344,670	933,300	1,795,900	128,650	699,000	3,901,520
% U.S. capacity	8.8	23.9	46.0	3.3	17.9	
Hydrocracking, b/sd	74,500	161,890	522,500	9,900	492,000	1,260,790
% U.S. capacity	5.9	12.8	41.4	0.8	39.0	
Hydrorefining & hydrotreating, b/sd†	521,400	944,500	2,898,250	150,750	1,099,900	5,614,800
% U.S. capacity	9.3	16.8	51.6	2.7	19.6	
Alkylation, b/sd	85,500	272,000	526,100	34,100	154,700	1,072,400
% U.S. capacity	8.0	25.4	49.1	3.2	14.4	
Coking, b/sd	30,040	114,382	241,726	7,342	183,831	577,320
% U.S. capacity	5.2	19.8	41.9	1.3	31.8	

*PADD I includes East Coast and Southeast, PADD II includes North Central states, PADD III includes Gulf Coast and southern Midcontinent, PADD IV includes Rocky Mountain states, and PADD V includes West Coast, Hawaii, and Alaska. †Excludes pretreating catalytic reformer feed, naphtha desulfurization, and naphtha, olefin, and aromatics saturation.

REFINING

Table 3
U.S. upgrading capacity changes

	Capacity 1-1-91	Capacity 1-1-90	% Change
	—1,000 b/d—		
Hydrorefining			
Resid & heavy gas oil	779.8	852.6	−9.3
FCC & cycle stock pretreatment	1,170.5	1,061.5	9.3
Mid distillate	437.3	421.4	3.6
Other	59.3	74.5	−25.6
Hydrotreating			
Cat. reformer feed & naphtha desulfurization	3,951.8	3,959.0	−0.2
Straight run distillate	1,546.4	1,369.4	11.4
Other distillate & material	1,400.3	1,429.8	−2.1
Lube oil polishing	221.2	223.7	−1.1
Naphtha olefin & aromatics saturation	150.4	177.5	−18.0

collapse during the year. The harbinger of that development came with the market response to the outbreak of hostilities in the Persian Gulf. After a rapid buildup in Saudi and other OPEC productive capacity, coupled with reduced demand owing to higher oil prices, the market essentially was in balance ahead of war in the Middle East. Only the fear of a further loss of production capability kept oil prices higher than normal. When it quickly became apparent that Iraq's threat to the rest of the region's oil productive capacity was greatly diminished following a massive U.S. led allied offensive against Iraq's war machine, and the International Energy Agency and U.S. government took steps to tap strategic oil reserves, oil prices dropped like a rock. There emerged a surprising new consensus in oil markets ahead of the seasonally weak second quarter: even with a war under way in the Middle East, a crude oil price collapse was a strong possibility because of a surfeit of crude on the market amid softer demand.

That paradox only underscores the range of pitfalls and opportunities facing refiners during the 1990s.

U.S. refining changes

U.S. crude distillation capacity rose slightly in 1990, although conversion and treating capacity posted sizable gains. That trend tracked buoyant demand for gasoline, diesel, and other high quality products in most of 1990.

Absent further market dislocations such as occurred in the wake of the Middle East crisis, industry capacity should remain closely balanced with demand. That will mean high operating utilization rates and healthy margins during the early 1990s.

U.S. refinery utilization rose during the latter half of the 1980s after a huge shakeout had stripped much of the U.S. capacity during the early 1980s. With only slight increases in capacity and steadily rising demand during the last half of the 1980s, U.S. refinery utilization climbed to 80-90% by 1990.

Capacities of the primary downstream refining processes that yield gasoline and diesel logged respectable gains in 1990, while processes that treat feeds to the primary processes jumped sharply because of increased feed demand for conversion and light fuels production processes.

The gains in downstream capacity also reflect industry's efforts to meet increasingly stringent air quality rules that require higher oxygen content gasolines in winter in some areas of the U.S. and that mandate reduced volatility of summertime gasoline across the U.S.

Still tougher rules governing air emissions of motor fuels by specifying their formulas are on the horizon. Even with

Table 4
U.S. capacity by refiner size

	Companies with more than 200,000 b/cd crude capacity		Companies with less than 200,000 b/cd crude capacity	
	1-1-91	1-1-90	1-1-91	1-1-90
No. companies	20	19	92	87
No. refineries	88	89	106	102
Crude capacity, b/cd	11,692,275	11,947,325	3,786,674	3,611,598
Crude capacity, b/sd	12,295,520	12,596,372	4,008,434	3,815,506
% U.S. capacity (b/cd basis)	75.5	76.8	24.5	23.2
Cat. cracking (FF), b/sd	4,299,900	4,333,700	1,144,500	1,070,400
% on crude	35.0	34.4	28.6	28.0
Cat. reforming, b/sd	3,083,000	3,106,000	818,520	815,970
% on crude	25.1	24.7	20.4	21.4
Hydrocracking, b/sd	1,134,600	1,089,600	126,190	153,800
% on crude	9.2	8.7	3.1	4.0
Hydrorefining & hydrotreating, b/sd*	4,693,450	4,693,450	817,650	787,400
% on crude	39.0	37.3	20.4	20.6
Alkylation, b/sd	835,700	817,600	240,000	213,700
% on crude	7.0	6.5	6.0	5.6
Coking, b/sd†	1,376,200	1,321,900	239,100	216,400
% on crude	11.2	10.5	6.0	5.7

*Excludes pretreating cat. reformer feed, naphtha, desulfurization and naphtha, olefin, or aromatics saturation. †Includes fluid and delayed coking.

REFINING

Table 5
Breakout of large U.S. refiners' capacities

Company	No. Refineries	Crude Capacity b/cd	Crude Capacity b/sd	Cat Cracking Fresh Feed b/sd	Cat Reforming b/sd	Hydro-cracking b/sd	Hydrorefining Hydrotreating b/sd*	Alkylation b/sd	Coking b/sd†
Chevron U.S.A.	12	1,575,700	1,650,000	421,000	334,600	223,500	700,100	80,400	178,600
Exxon Co.	5	1,147,000	1,197,000	574,000	283,000	80,900	679,100	93,100	153,200
Shell Oil Co.	7	1,096,200	1,142,000	319,500	298,000	161,500	507,300	77,000	95,000
Amoco Oil Co.	7	1,002,000	1,048,500	430,000	282,000	120,000	370,600	101,800	84,800
Mobil Oil Corp.	5	838,000	902,000	357,000	255,500	71,700	421,550	80,500	169,500
BP Oil Co.	5	744,800	784,000	259,500	199,500	80,000	147,000	64,300	55,000
ARCO	5	684,500	726,500	172,000	214,000	74,000	298,000	28,000	14,600
Marathon Petroleum Co.	5	636,000	661,000	217,500	162,000	24,000	151,000	59,000	22,000
Star Enterprise	3	615,000	663,000	264,000	142,000	84,000	256,900	42,200	46,000
Sun Refining & Marketing Co.	4	505,000	538,000	206,000	159,200	58,200	117,900	26,800	0
Conoco Inc.	5	406,500	424,000	133,500	88,700	0	207,300	25,500	80,500
Texaco Refining & Marketing Inc.	4	351,000	365,000	109,500	103,500	40,300	148,000	28,800	123,200
Ashland Petroleum Co.	3	346,500	357,220	148,000	95,500	0	184,000	24,500	0
Koch Refining Co.	2	343,500	360,000	95,000	80,500	0	117,500	16,900	70,000
Citgo Petroleum Corp.	1	320,000	330,000	0	106,000	45,000	68,000	23,000	88,000
Phillips 66 Co.	3	305,000	331,000	155,400	68,000	0	216,600	31,100	0
Hill Petroleum Co.	4	275,500	291,700	139,000	48,500	0	59,000	10,000	0
Coastal Refining and Marketing Inc.	6	274,075	288,500	102,000	76,500	10,000	84,800	12,800	17,500
Unocal Corp.	2	226,000	236,100	47,000	86,000	61,500	62,500	10,000	46,900
Total	**88**	**11,692,275**	**12,295,520**	**4,299,900**	**3,083,000**	**1,134,690**	**4,797,150**	**835,700**	**1,376,200**

*Excludes pretreating cat. reformer feed, naphtha desulfurization, and naphtha, olefin, or aromatics saturation. †Includes fluid and delayed coking.

increases in crude and downstream processing capacity, the more stringent rules might squeeze U.S. refiners' ability to meet demand.

Those concerns notwithstanding, the tighter rules' crunch on capacity will tend to keep refiners' margins healthy through much of the 1990s—assuming crude costs remain near an equilibrium level.

Capacity changes

U.S. crude charging capacity inched up only 52,200 b/cd to 15,478,900 b/cd as of Jan. 1, 1991, compared with its prior year level.

Fluid catalytic cracking capacity slipped 1.5% in 1990 to about 5.294 million b/sd. That trend is likely to reverse in 1991 with the expected addition of 51,200 b/sd in FCC capacity.

Catalytic hydrocracking capacity increased 4.1% to about 1.26 million b/sd as of Jan. 1, 1991, while alkylation capacity rose 3.8% to 1.072 million b/d, and aromatics/isomerization capacity shot up 6.2% to about 836,000 b/d.

Feed treating process capacities jumped accordingly. Overall U.S. hydrotreating capacity climbed 1.8% to about 7.27 million b/sd in 1990. As of Jan. 1, 1991, FCC and cycle stock capacity posted a 9.3% rise to about 1.17 million b/sd from the previous year's level.

During the same period, catalytic naphtha reformer feed treatment and naphtha desulfurization capacity slipped 0.2% to about 3.95 million b/sd.

The biggest increase among U.S. refining capacity types in 1990 occurred in straight run distillate hydrotreating, up 11.4% to about 1.55 million b/d. That is partly due to new construction and partly reflects a shift or a redefinition of the types of materials being hydrotreated. That can be seen in the decline in other distillate and material hydrotreating capacity, which fell 2.1% to about 1.41 million b/sd.

The trend toward converting heavier feedstocks into light products was slowed with the reversal in resid/heavy gas oil hydrorefining capacity, which fell 9.3% to 779,800 b/sd as of Jan. 1, 1991, following a 2% gain in 1989.

There is not likely to be much in the way of grassroots refinery construction in the U.S. during the 1990s. Although the economics are better than in the early 1980s with operating utilization rates expected to remain high and crude costs—absent more turmoil in the Middle East—expected to remain relatively low, any incremental distillation capacity probably will be added only through debottlenecking and efficiency improvements.

Margins in the 1990s, while expected to remain healthy, still are not likely to justify the economics of a new grassroots refinery of any significant scale or sophistication in the U.S.

The same environmental impetus that will keep refinery margins strong in the 1990s through new air quality rules also contributes to a regulatory thicket that makes it all but impossible to obtain construction permits for a new refinery

Table 6
Alkylation growth

	1985	1986	1987	1988	1989	1990	1991
Sulfuric acid	481,875	477,400	490,600	485,700	498,200	499,500	516,500
Hydrofluoric acid	430,950	456,150	481,350	498,400	515,500	528,800	555,900
Total	**912,825**	**933,550**	**971,950**	**984,100**	**1,013,700**	**1,028,300**	**1,072,400**

B/sd

REFINING

U.S. MTBE capacity — Table 7

Company	Location	(B/d)
Amoco Oil Co.	Whiting, Indiana	4,000
Amoco Oil Co.	Yorktown, Virginia	800
Ashland Petroleum Co.	Catlettsburg, Kentucky	3,200
Atlantic Richfield Co.	Carson, California	2,400
Champlin Refining & Chemicals Inc.	Corpus Christi, Texas	1,700
Chevron U.S.A. Inc.	El Segundo, California	2,300
Citgo Petroleum Corp.	Lake Charles, Louisiana	2,100
Coastal Refining & Marketing Inc.	Corpus Christi, Texas	12,000
Crown Central Petroleum Corp.	Houston, Texas	5,000
Hill Petroleum Co.	Houston, Texas	1,300
Lyondell Petrochemical Co.	Houston, Texas	2,100
Marathon Petroleum Co.	Detroit, Michigan	1,100
Mobil Oil Corp.	Beaumont, Texas	2,200
Phillips 66 Co.	Sweeny, Texas	3,000
Sun Refining & Marketing Co.	Marcus Hooks, Pennsylvania	2,600
Valero Refining Co.	Corpus Christi, Texas	2,200
TOTAL		**48,000**

in the U.S.

Although products demand and refinery margins plunged after the onset of the Middle East crisis in second half 1990, an end to the crisis is likely to keep capacity in close balance with demand, which should return to its normal growth patterns after the war and its market aftermath have run their course.

World refining trends

Worldwide refining capacity as of Jan. 1, 1991, was 588,000 b/d higher than the previous year's total.

Two Persian Gulf refineries accounted for most of that increase. Saudi Arabia started up the 332,500 b/d Petromin Petrola Rabigh Refining Co. plant at Rabigh in 1990. Iran brought its 190,000 b/d refinery at Abadan—shut down for repairs of damage from attacks during the Iran-Iraq war—back on stream.

In Europe, the idled 78,000 b/d refinery Holborn Europa Raffinerie BmbH operates at Harburg, Germany, also restarted.

Environmental concerns will spur big outlays for process changes and tighter safeguards outside the U.S. in the 1990s.

Royal Dutch/Shell Group plans big investments in refining capacity in France and Switzerland to accommodate product and environmental needs in the 1990s.

Shell France tentatively plans to spend $200 million on its 128,000 b/d Berre refinery in southern France. About a fourth of that expenditure will go for environmental protection.

Shell Raffinerie Cressier, after completing a $4.4 million program to meet local air quality rules in Switzerland, will spend another $40 million to upgrade environmental protection systems by 2000.

Another key trend outside the U.S. is the push toward privatization and market economies. Hungary was among the first of the former Soviet Satellites to do so, ending its state monopoly on oil imports and allowing market forces to set gasoline prices effective Jan. 1, 1991.

Europe

The big refining story in Europe at the close of the last decade was the rapid market penetration of unleaded gasoline.

Unleaded's market share in Europe climbed to more than 30% by midyear 1990 from 27% at yearend 1989.

At the start of 1990, Europremium 95 RON/85 RON was available at more than 81,400 service stations, superpremium 98 RON at 25,580 sations, and regular 91/92 RON at more than 21,100 stations.

Unleaded's growth has proven to have a sharp regional bias in Europe. In Germany, where the environmental lobby is the strongest in Europe, unleaded commands a market share of about 65%. In Spain and Portugal, however, the unleaded market share remains less than 1%. There has been even less progress in the newly emerging democracies of eastern Europe, although that will change soon.

The dramatic growth in unleaded gasoline demand in Europe stems from the introduction of tax incentives in countries where there is ample supply of the fuel.

In Denmark, where there is the equivalent of 10¢/1. tax differential between leaded and unleaded, more than 51% of gasoline sales are unleaded.

Spain, without incentives, has the lowest level of unleaded sales in Europe.

However, Greece, with a strong differential of 7¢/1., has an unleaded market share of less than 1%. That's because only 6.4% of that country's service stations offers unleaded gasoline.

Even with unleaded's market share expected to continue expanding in the 1990s, spending to accommodate that growth won't account for a major role in European refiners' investments in the 1990s.

An Arthur D. Little study forecasts European refiners will have to spend $6 billion to upgrade capacity by 2000.

By far the biggest component of that sum will entail reducing the sulfur content of diesel fuel and positioning refiners to meet a lighter barrel in the 1990s.

It is not likely that Europe's refiners will have to go as far as their U.S. counterparts to reformulate fuels to meet air quality needs. The gasoline pool in Europe has a much lower level of olefins than does that in the U.S. In addition, Europe does not have the severe smog problems seen in the U.S., and the trend toward managing evaporative emissions is through big carbon canisters on vehicles.

Another key refining trend in Europe during the 1990s will be the total rehabilitation of the refining/marketing sector of the former Soviet satellites.

Companies in former West Germany are grappling with expanding operations in the former East Germany while at the same time trying to provide a western style level of service.

The post-reunification eastern Germany will have about 3,000 retail outlets, compared with 1,300 previously, companies in the western part of the country estimate.

Hungary will be another prime target for western company expansion. Total/CFP, Mobil Europe Inc., and Royal Dutch/Shell Group in 1990 took steps toward establishing downstream joint ventures in that country.

Asia

Several developing nations in Asia are in the midst of a refining capacity expansion boom. Foreign and private domestic companies are expected to play a major role in that effort.

Thailand in late 1990 gave a green light for more active participation by foreign companies in its downstream petroleum industry. The government approved removing conditions hindering major refinery projects planned by Royal Dutch/Shell and Caltex Petroleum Corp.

Thailand plans to boost refining capacity to more than 600,000 b/d by the mid-1990s from 230,000 b/d in 1990. Thai products consumption is expected to climb to 900,000 b/d by 2000 from about 400,000 b/d in 1990.

The government approvals allow Caltex to proceed with plans for a $520 million refinery in the eastern seaboard province of Rayong and Shell to proceed with its plans for a $760 million refinery.

India still is grappling with how best to meet its expected growth in products demand. The government has been unable to decide whether to proceed with grassroots refineries at Karnal in Haryana, Mangalore in Karnataka, and Gulaghat in Assam.

Government estimates put India's total demand for petroleum products at 2 million b/d by the end of the 1990s. The country's refining capacity stood at less than 1 million b/d in 1990.

Plans called for installation of four 120,000 b/d grassroots refineries to be on stream by 2000. But those plans ran into a series of political obstacles.

Allowing capacity expansions totaling 240,000 b/d at India's 12 refineries—in addition to those already under way—would hike its refining capacity to almost 1.3 million b/d by 1995 at a cost of about $335 million. In addition to representing a major cost savings, the incremental expansions would allow the controversial Karnal and Mangalore expansions to be put on hold.

Indonesia is another Asian hotbed of refining capacity growth. A Japanese joint venture in mid-1990 agreed to provide state owned Pertamina a loan of $2.8 billion to finance construction of the planned $1.8 billion, 125,000 b/d grassroots refinery at Balongan in Central Java. Construction on the Java refinery began in early 1990. The plant is due to start up in 1994.

Latin America

Latin American's downstream won't be spared the upheaval sweeping that region in the 1990s.

A broad push toward privatization is shaking that region's oil sector to its foundations. Many state oil companies are faced, with varying degrees of urgency, with selling off assets, participating in joint ventures with private and foreign companies, and with allowing market forces to set product prices. The last element is the most controversial.

Brazil is attempting to relax monopolies on products distribution and improve operating efficiencies and eliminate bureaucratic red tape. At the same time, state owned Petroleos Brasileiro SA is considering participating in foreign downstream operations while it also seeks to step up products exports.

The prime target for Petrobras foreign downstream investment is the U.S. In fall 1990, Petrobras was reported negotiating a possible U.S. joint venture with an undisclosed U.S. company. Also under consideration was a direct investment in as many as 100 Petrobras owned service stations in the U.S.

LYONDELL PETROCHEMICAL CO.'S 265,000 b/d Houston refinery is among those in the U.S. that operated at utilization rates approaching 90% in 1990, a situation likely to continue to the 1990s as U.S. refiners cope with toughening air quality rules. (Photo courtesy Lyondell.)

The push downstream in Latin America is strongest in Venezuela, where state owned Petroleos de Venezuela SA plans 1 million b/d in added refining capacity by 1996, including some foreign capacity interests. Pdvsa plans break out as: a new 200,000 b/d high conversion unit, 400,000 b/d added to existing domestic capacity, and another 400,000 b/d of new domestic capacity plus stakes in foreign refineries.

Reformulated fuels

The critical trend of the 1990s for U.S. refiners will be how they cope with the push for reformulating gasoline and other fuels to meet tough new standards enacted under the reauthorized Clean Air Act of 1990.

Many industry observers said at the outset that U.S. refiners would not be able to meet the standards set under the act, signed in fall 1990 by President Bush, within the given time frame.

The critical elements of reformulation will focus on new gasoline specifications to meet air quality concerns and development of low sulfur diesel fuel.

U.S. refiners could see product supply shortfalls and price spikes in some parts of the country as they seek to meet clean air goals. The concern is that resulting consumer resistance will turn into reduced demand growth as spending on process changes grows.

Correspondingly, U.S. refiners will continue to press government for more flexibility in timing and fuel formulas to meet air quality goals.

John Doscher, vice-president of Pace Consultants Inc., Houston, cited these characteristics likely to be incorporated in reformulated gasolines in the 1990s: Reid vapor pressure of 8 psi, benzene content not to exceed 0.8 vol %, aromatics content of 25 vol % with xylene capped at 5 vol %, olefins content of not more than 5 vol %, oxygen content of 2.7 wt %, no lead, sulfur content of 250 ppm, and additives to

REFINING

Table 8
Unleaded gasoline tax incentives*

	Grade (RON)	Local currency (LC)	Tax incentive LC/liter	Exhange rate	U.S.¢/liter
Austria	92	schilling	0.64	11.807	5.78
Austria	95	schilling	0.53	11.807	4.79
Belgium	98	franc	1.65	35.02	4.71
Belgium	95	franc	1.81	35.02	5.17
Belgium	92	franc	1.88	35.02	5.37
Denmark	All	krone	0.69	6.47	10.66
Finland	92	mark	0.325	4.04	8.04
France	All	franc	0.41	5.69	7.21
Greece	95	drachma	11	164.3	6.70
Iceland	92	known	3.88	61.41	6.32
Ireland	All	pence	2.2	0.632	3.48
Italy	95	lire	71	1,244.12	5.71
Netherlands	All	cents	9	1.89	4.76
Norway	All	oere	51	6.47	7.88
Portugal	95	escudos	6.2	197.63	4.20
Spain	95	pesetas	0	108.28	0
Sweden	95	krone	0.24	6.14	3.91
Switzerland	95	cents	8.68	1.49	5.83
West Germany	All	pfennig	9.1	1.68	5.42
U.K.	98	pence	3.07	0.589	5.21
U.K.	95	pence	2.79	0.589	4.74

*Superpremium unleaded (98 RON) is available in many countries. Tax incentives are the same as for Europremium grade.

Source: Concawe

Table 9
Market shares of unleaded gasoline

	Regular 92 RON	Euro-premium 95 RON	Super-premium 98 RON
Austria	30	15	—
Belgium	0.5	19	3
Denmark	8	30	13
Finland	—	50	—
France	—	15	9
Greece	—	0.5	—
Ireland	—	12	0.6
Italy	—	3	—
Netherlands	—	34	8.5
Norway	—	28	1.5
Portugal	—	0.3	—
Spain	—	<1	—
Sweden	—	46	0.2
Switzerland	—	45	—
West Germany	34	23	8
U.K.	—	27	2

Source: Concawe

prevent and remove deposits.

Doscher expects reformulated gasolines will be required for all cars in nonattainment areas but relatively minor changes will be needed for gasoline sold in standards that meet federal air quality standards. About one fourth of U.S. gasoline is used in nonattainment areas.

Doscher estimates reformulated gasoline will cost 7-12¢/gal more to produce than conventional gasoline.

Sulfur probably will be limited to 0.05 wt % in diesel fuel by 1993.

Arthur D. Little sees reformulation as more pervasive. The analyst expects most or all U.S. gasoline will have to be reformulated by 2000, calling for a $22 billion investment by U.S. refiners.

Crisis effects

The Persian Gulf crisis turned what had been for refiners an excellent second half 1990 into a nightmare.

Oil markets responded to Iraq's seizure of Kuwait on Aug. 2, 1990, by sending spot and futures prices to their highest levels in almost 5 years.

West Texas intermediate for next month delivery on the New York Mercantile Exchange soared to $28.05/bbl on closing Aug. 6, up more than $8/bbl on the week. Brent blend spot 15 day deliveries closed Aug. 7 almost $7 higher on the week at $25.90/bbl.

Products prices on spot and futures markets followed suit. Nymex unleaded gasoline closed Aug. 7 almost 17¢ higher at 82.39¢/gal, and heating oil futures rose about 10¢ to 75.14¢/gal in a comparison of the same period.

Prices skyrocketed at the pump as well. The average price of unleaded self-serve regular gasoline jumped more than 16¢ in the week following the invasion, reaching $1.237/gal, the highest in 5 years.

Refiner/marketers, especially in the U.S., were hit by claims of price gouging and profiteering related to the price hikes. Subcommittees in Congress launched investigations into the charges, as did the U.S. Justice Department.

President Bush and the Department of Energy called on U.S. consumers to begin conservation measures. Bush also called on U.S. oil companies to show restraint in gasoline pricing.

Price freezes

Oil companies responded to the President's request, with ARCO the first.

It froze gasoline, diesel, and jet fuel prices Aug. 8. ARCO warned at the time that if its prices were significantly out of line with world market prices during that time it would have raised them again, especially if the result were runouts at its service stations.

Quickly following ARCO's lead were Amoco Oil Co., Phillips Petroleum Co., Chevron Corp., BP America, Unocal Corp., and Getty Petroleum Corp.

However, ARCO's warning came to pass. The company's price freeze lasted 2 weeks before a flurry of service station runouts forced ARCO to abandon the freeze. ARCO then hiked gasoline prices by an average 2.5¢/gal in response to a 19% jump in sales volumes and gasoline runouts at more than 160 stations every day. During the price freeze, ARCO's regular unleaded sold for as much as 13¢/gal below the general market.

With that kind of panic buying pressuring the market, other refiner/marketers followed suit, and the average retail price of gasoline shot up about 8-10¢/gal as the freezes came off after only 2 weeks. That sparked a fresh outcry by U.S. oil industry critics.

Industry's image took a hammering in Europe as well. U.K. companies, after keeping a lid on retail prices for several weeks, boosted prices for premium unleaded 4 pence/imp. gal to 217 pence/imp. gal.

That elicited calls from Conservative as well as Labour party politicians for an investigation by Britain's Office of Fair Trading and from Prime Minister Thatcher for price restraint.

In addition to concerns over crude supplies arising from the international embargo against Iraq and Kuwait, markets remained skittish about the flexibility of the world refining system in the crisis. Supply problems had been expected for the loss of about 600,000 b/d of sophisticated refining capacity in Kuwait.

REFINING

Table 10
Venezuela's refining interests

DOMESTIC REFINERIES

	Atmospheric distillation	Capacities Effective conversion (1,000 b/d)	Crude processed
Amuay	571	340	428
Cardon	286	260	235
Puerto La Cruz	195	100	126
El Palito	105	105	102
El Toreno/San Roque	10	10	10
Total	**1,167**	**815**	**901**

REFINERIES OUTSIDE VENEZUELA

Location/ company name	Pdvsa share (Percent)	Installed capacity	Venezuelan crude processed (1,000 b/d)
Belgium			
Antwerp/Nynas	50	15	10
Germany			
Schloven/Ruhr Oel	50	110	97
Horstadt/Ruhr Oel	50	100	97
Neustadt/Ruhr Oel	12.5	144	38
Karlsruhe/Ruhr Oel	16	142	38
Netherlands Antilles			
Curacao/Refineria	(Leased)	310	179
Sweden			
Nynashamn/AB Nynas Petroleum	50	25	12
Gothenberg/AB Nynas Petroleum	50	12	5
U.S.			
Lake Charles/Citgo	100	320	136
Corpus Christi/Champlin	100	165	131
Chicago/Uno-Ven	50	153	56
Total		**1,496**	**664**

Table 11
India's refining capacity, location

Company, location	Distillation	Delayed coker	Visbreaker	FCC
		1,000 b/d		
Indian Oil corp Ltd.				
Digboi	10	0.8	—	—
Gauhati	17	6.0	—	—
Barauni	66	22.0	—	—
Haldia	50	—	9.2	—
Koyali	146	—	20.0	20.0
Mathura	120	—	20.0	20.0
Hindustan Petroleum Corp. Ltd.				
Bombay	110	—	—	8.0
Visakhapatnam	90	—	—	20.0
Bharat Petroleum Corp. Ltd.				
Bombay	120	—	—	24.0
Madras Refineries Ltd.				
Madras	112	7.4	1.2	—
Cochin Refinereis Ltd.				
Ambalamugal	90	—	24.6	20.0
Bongaigaon Refineries & Petrochemicals Ltd.				
Bongaigaon	27	10.0	—	—
Total	**958**	**46.2**	**75.0**	**112.0**

Source: East-West Center

The political uproar over alleged price gouging notwithstanding, the price restraint that refiner/marketers showed during the crisis slashed margins, and thus downstream profits plunged in fourth quarter 1990.

In fact, gasoline remained a bargain during a period when crude oil prices soared to record levels in the weeks and months following the onset of the crisis. Even after gasoline prices jumped about 18¢/gal in the 2 weeks following the crisis, the inflation adjusted price of gasoline was only about 8¢ higher than its post-price collapse low in 1988. Adjusted for inflation, the price of leaded regular gasoline at the time was lower than anytime since 1918 except for 1972-73 and 1986-89. By contrast, the posted price of about $25/bbl for WTI adjusted for inflation would be $21.66/bbl, or about half the inflation adjusted peak in 1980 and more than double the inflation adjusted average price during 1950-1973.

That would suggest that the competitive nature of the U.S. gasoline market—when free of regulatory intervention—has kept U.S. gasoline consumers mostly shielded from spikes in the price of crude oil.

Still, refiner/marketers resorted to the same efforts at self-imposed price restraint when the Persian Gulf crisis turned into war Jan. 16, 1991. Most U.S. refiner/marketers announced retail and wholesale price freezes shortly after news came of an air attack upon military targets in Iraq and Kuwait in response to Iraq's disregard of a U.N. deadline of Jan. 15 for withdrawing from Kuwait. Then, the market responded to the astonishing speed with which uncertainty over the security of Persian Gulf oil supplies was resolved. After jumping about $8 during the first few hours of the onset of hostilities, oil prices plunged. The awareness that Iraq's ability to further disrupt the region's oil supplies had been all but eliminated by the allied air attacks within the first few days of the conflict, the surprising effectiveness of the Saudis and others in OPEC to boost production to compensate for the loss of Iraqi/Kuwaiti supplies, and the decisions by the International Energy Agency and U.S. DOE to draw down strategic reserves calmed a nervous market. Within a few weeks, oil prices again were at precrisis levels.

Unfortunately, refiner/marketers' efforts to show price restraint by implementing products price freezes blew up in their faces. Freezing products prices when crude prices were falling was seen as greed instead of just inept public relations.

With a surplus of crude supplies and refining capacity tight worldwide, refiner/marketers approached second quarter 1991 with the realization that no one had predicted: That oil prices could collapse even with the destruction of crude productive capacity in two major OPEC exporting nations as a result of a Persian Gulf war.

That points to a likely healthy outlook for world refining in the 1990s. Crude prices are likely to remain generally low—albeit with brief flurries of volatility.

Refining capacity is likely to remain tight in response to rebounding demand—notably in the Asia Pacific region—and reconfigured process schemes caused by accommodating new environmental rules. At the same time, adding grassroots refining capacity in the U.S. and Europe will prove extremely difficulty in the 1990s, because of overwhelming permitting obstacles. Capacity additions in these two key refining areas are expected to be incremental—"capacity creep."

Although refining capacity is expected to grow outside the U.S. and Europe, especially in the developing nations of the Asia-Pacific region, projections of demand in many of those countries fall short of planned capacity additions.

Together, these trends point to a generally sustained period of high margins and healthy profits for the refining industry in the 1990s.

GAUGER measures contents of storage tanks at Conoco's Lake Charles, La., refinery. (Photo courtesy Conoco).

Refining survey

YOUR USE OF THE INTERNATIONAL REFINing survey may be aided by the following information:

All country-by-country volumes are presented in calendar-day figures except those for the U.S. In that tabulation, only crude is reported in calendar-day figures. Cat cracking and cat reforming are given in stream-day figures.

In the U.S. tabulation, when an asterisk (*) appears beside a number in the crude column, this indicates that the figure was reported as a stream-day figure and has been converted to a calendar-day figure using the crude conversion factor of 0.95.

Calendar-day figures are the average volume a refinery unit processes each day including downtime used for turnarounds. This is actual total volume for the year divided by 365.

Steam-day figures represent the amount a unit can process when running full capacity for short periods.

LEGEND

CAT CRACKING
  1. Fluid
  2. Other

CAT REFORMING
  Semiregenerative:
    1. Conventional catalyst
    2. Bimetallic catalyst
  Cyclic:
    3. Conventional catalyst
    4. Bimetallic catalyst
  Other:
    5. Conventional catalyst
    6. Bimetallic catalyst

Cat reforming definitions are:

Semiregenerative. Characterized by shutdown to the reforming unit at specified intervals or at the operator's convenience for regeneration of catalyst in situ.

Cyclic. Characterized by continuous or continual regeneration of catalyst in situ in any one of several reactors that can be isolated from and returned to the reforming operation. This is accomplished without changing feed rate or octane.

Other includes:

Nonregenerative—catalyst replaced by fresh catalyst.

Continuous—continuous regeneration of a part of the catalyst in a special regenerator followed by continuous addition to the reactor.

Moving-bed catalyst systems.

REFINING SURVEY

Survey of operating refineries worldwide
(capacities of January 1, 1991)

Country	No. plants	Crude (b/cd)	Catalytic cracking (b/cd)	Catalytic reforming (b/cd)
Abu Dhabi	2	192,500	....	30,100
Algeria	4	464,700	....	55,600
Angola	1	37,630	....	1,714
Argentina	11	696,285	171,300	41,300
Australia	10	705,500	205,200	160,900
Austria	1	204,000	24,000	32,000
Bahrain	1	243,000	39,000	18,000
Bangladesh	1	31,200	....	1,650
Barbados	1	3,000	....	....
Belgium	4	602,000	104,700	81,900
Bolivia	3	45,250	....	11,300
Brazil	13	1,411,520	316,585	23,610
Brunei	1	10,000	....	....
Cameroon	1	42,000	....	7,000
Canada	28	1,882,060	387,870	367,060
Chile	3	145,800	37,870	10,300
Colombia	4	247,400	91,000	6,000
Congo	1	21,000	....	2,000
Costa Rica	1	15,000	....	1,200
Cyprus	1	18,600	....	4,300
Denmark	3	185,500	....	31,100
Dominican Republic	2	48,000	....	9,100
Ecuador	5	141,800	16,000	2,780
Egypt	8	523,153	....	33,540
El Salvador	1	16,300	....	3,000
Ethiopia	1	18,000	....	2,400
Finland	2	241,000	41,800	42,900
France	14	1,815,630	349,480	246,760
Gabon	1	24,000	....	1,400
Germany	20	2,065,400	240,250	376,500
Ghana	1	26,600	....	6,175
Greece	4	395,300	59,900	47,000
Guatemala	1	16,000	....	3,000
Honduras	1	14,000	....	1,800
Hungary	3	220,000	20,000	23,000
India	12	1,122,360	136,970	28,233
Indonesia	6	813,600	12,600	61,500
Iran	5	720,000	....	105,397
Iraq	8	318,500	....	43,500
Ireland	1	56,000	....	11,000
Israel	2	180,000	22,000	25,000
Italy	21	2,385,358	306,300	315,050
Ivory Coast	2	69,000	....	13,600
Jamaica	1	34,200	....	3,240
Japan	41	4,383,400	629,975	607,970
Jordan	1	100,000	4,250	8,640
Kenya	1	90,000	....	9,000
Korea	6	867,000	....	62,450
Kuwait	4	819,000	42,000	33,000
Lebanon	2	37,500	7,250	7,442
Liberia	1	15,000	....	2,000
Libya	3	347,600	....	13,882
Madagascar	1	16,350	....	2,600
Malaysia	4	209,500	....	30,300
Martinique	1	12,000	....	3,000
Mexico	9	1,679,000	267,000	157,800
Morocco	2	154,600	5,600	26,800
Myanmar	2	32,000	....	....
Netherlands	7	1,196,700	128,000	182,600
Netherlands Antilles	2	470,000	42,000	15,000
New Zealand	1	95,100	....	26,600
Nicaragua	1	16,000	....	3,000
Nigeria	4	433,250	82,700	70,070
Norway	3	288,000	41,000	47,000
Oman	1	76,932	....	15,386
Pakistan	3	120,975	....	4,800
Panama	1	100,000	....	7,500
Paraguay	1	7,500	....	....
Peru	6	188,820	24,600	1,760
Philippines	4	279,300	26,900	37,500
Poland	9	385,454	48,705	42,186
Portugal	3	294,300	10,100	51,000
Puerto Rico	2	125,000	12,000	25,500
Qatar	1	62,000	....	12,030
Saudi Arabia	8	1,862,500	87,541	183,922
Senegal	1	24,000	....	2,600
Sierra Leone	1	10,000	....	....
Singapore	5	878,000	....	63,600
Somalia	1	10,000	....	....
South Africa	4	430,500	84,400	62,600
Spain	10	1,321,000	157,300	191,000
Sri Lanka	1	50,000	....	3,750
Sudan	1	21,700	....	1,900
Sweden	5	427,500	25,000	72,000
Switzerland	2	132,000	....	26,000
Syria	2	237,394	....	31,352
Taiwan	2	542,500	22,600	51,960
Tanzania	1	17,000	....	6,700
Thailand	3	220,550	26,100	27,550
Trinidad	2	246,000	26,000	14,400
Tunisia	1	34,000	....	3,300
Turkey	5	728,644	40,575	66,557
United Kingdom	15	1,866,940	444,500	366,300
Uruguay	1	33,000	4,100	2,500
Venezuela	6	1,167,000	139,000	6,000
Virgin Islands	1	545,000	....	125,000
Yemen	2	171,500	....	13,300
Yugoslavia	7	609,135	52,050	74,612
Zaire	1	17,000	....	3,500
Zambia	1	24,500	....	5,600
Total	**433**	**42,997,790**	**5,064,171**	**5,205,728**

REFINING SURVEY

WORLDWIDE REFINING	Crude capacity	b/cd Catalytic cracking	Cat reforming
ABU DHABI			
Abu Dhabi National Oil Co.–Ruwais	120,000	....	[2]19,200
Umm Al-Nar 2	72,500	....	[4]10,900
Total	**192,500**	**....**	**30,100**
ALBANIA			
Government-owned refineries at Ballsh, Cerrik, Stalin.			
Total	**40,000**		
ALGERIA			
Sonatrach: Arzew	60,000	....	8,600
Hassi Messaoud	23,700	....	2,000
Maison Carree	58,000	....	15,000
Skikda	323,000	....	30,000
Total	**464,700**	**....**	**55,600**
ANGOLA			
Fina Petroleos De Angola–Luanda	37,630	....	[1]1,714
Total	**37,630**	**....**	**1,714**
ARGENTINA			
Destileria Argentina de Petróleo SA–Lomas de Zamora	3,585	....	....
Esso SAPA–Campana[1]	93,000	[1]29,600	[2]8,300
Galvan	18,400	....	....
Isaura SA–Bahia Blanca	12,000	....	....
Shell Cia. Argentina de Petróleo SA–Buenos Aires	121,700	[1]29,200	[1]12,000
Sol Petroleo SA—San Francisco Solana, Quilmes	6,000	....	....
Yacimientos Petroliferos Fiscales–Campo Duran	32,000	....	....
Dock Sud	4,000	....	....
La Plata	216,000	[1]71,900	[1]9,000
Lujan de Cuyo	129,000	[1]40,600	[1]9,000
Plaza Huincul	23,000	....	[1]3,000
San Lorenzo	37,600	....	....
Total	**696,285**	**171,300**	**41,300**

Solvent Extract: [1]3,600 b/cd

WORLDWIDE REFINING	Crude capacity	b/cd Catalytic cracking	Cat reforming
AUSTRALIA			
Ampol Refineries Ltd.–Lytton	72,000	[1]28,000	[1]15,000
Australian Lubricating Oil Refinery Ltd.–Kurnell[1]	....	....	....
BP Australia–Brisbane	57,000	[1]17,300	[2]12,200
Kwinana	118,000	[1]28,000	[2]14,000
Caltex Refining CPL–Kurnell	100,000	[1]44,000	[2]24,000
Mobil Oil Australia Ltd.–Adelaide	....	....	....
Petroleum Refineries Australia PL– Adelaide	65,500	....	[2]20,000
Altona	108,000	[2]28,000	[1]31,000
Shell Refining (Australia) PL.–Clyde	75,000	[1]35,000	[2]13,700
Geelong	110,000	[1]25,000	[2]11,000
			[6]20,000
Total	**705,500**	**205,300**	**160,900**

Solvent Extract: [1]6,400 b/cd

WORLDWIDE REFINING	Crude capacity	b/cd Catalytic cracking	Cat reforming
AUSTRIA			
OeMV–Schwechat	204,000	[1]24,000	[2]14,000
	[6]6,426		
Total	**204,000**	**24,000**	**32,000**
BAHRAIN			
Bahrain Petroleum Co. BSC (Closed)–Sitra	243,000	[1]39,000	[2]18,000
Total	**243,000**	**39,000**	**18,000**
BANGLADESH			
Eastern Refinery Ltd.–Chittagong	31,200	....	[1]1,650
BARBADOS			
Mobil Oil Barbados Ltd.–Bridgetown	3,000	....	....
BELGIUM			
Belgian Refining Corp. NV–Antwerp	80,000	....	[3]12,500
Esso Belgium–Antwerp	239,000	[1]28,200	[4]39,200
Nynas Petroleum NV—Antwerp	15,000	....	....
Fina Raffinaderij Antwerpen—Antwerp	268,000	[1]76,500	[2]30,200
Total	**602,000**	**104,700**	**81,900**
BOLIVIA			
Yacimientos Petroliferos Fiscales Bolivianos– Cochabamba[1]	27,250	....	[2]4,900
Santa Cruz	15,000	....	[2]6,400
Sucre	3,000	....	....
Total	**45,250**	**....**	**11,300**

[1]Solvent extraction: 166 b/cd.

WORLDWIDE REFINING	Crude capacity	b/cd Catalytic cracking	Cat reforming
BRAZIL			
Petróleo Brasileiro SA: Araucária, Paraná[1]	144,800	[1]42,300	...
Betim, Minas Gerais	144,700	[1]32,100	....
Canoas, Rio Grande do Sul	72,400	[1]13,700	....
Capuava, São Paulo	30,200	[2]17,300	....
Cubatão, São Paulo	155,325	[1]50,600	[1]10,500
Duque de Caxias, Rio de Janeiro[2]	209,190	[1]35,815	[1]13,110
Fortaleza, Ceará	....	...	....
Manaus, Amazonas	8,500	[1]1,900	
Mataripe, Bahia[3]	113,525	[1]25,000	....
Paulínia, São Paulo	289,500	[1]41,600	....
São José dos Campos, São Paulo	215,080	[1]53,770	
Refinaria de Petróleo Ipiranga SA–Rio Grande do Sul	9,300	[1]2,500	
Refinaría de Petroleos de Manguinhos SA–Rio de Janeiro	10,000	...	
	[3]4,400		
Total	**1,411,520**	**316,585**	**23,610**

[1]Solvent extraction: 27,800 b/cd. [2]30,200 b/cd. [3]3,400 b/cd.

338

REFINING SURVEY

WORLDWIDE REFINING	Crude capacity	b/cd Catalytic cracking	Cat reforming
BRUNEI			
Brunei Shell Petroleum CL–Seria	10,000	....	....
BULGARIA			
Government-owned refineries at Burgas, Pleven, Ruse.			
Total	300,000		
CAMEROON			
Sonara–National Refining CL–Cape Limboh, Limbe	42,000	....	[2]7,000
CANADA			
Alberta			
Husky Oil Operations Ltd.–Lloydminster	23,500	....	....
Imperial Oil Ltd.–Edmonton	164,900	[1]47,000	[4]20,900
Parkland Refining Ltd.–Bowden	6,700	....	[1]3,000
Petro-Canada Products Inc.–Edmonton	115,500	[1]33,300	[2]9,000
Shell Canada Ltd.–Scotford	56,810	....	[6]20,425
Turbo Resources Ltd.–Balzac	27,500	[1]11,570	[2]7,550
Total	394,910	91,870	60,875
British Columbia			
Chevron Canada Ltd.–North Burnaby	45,000	[1]10,500	[2]10,000
Husky Oil Operations Ltd.–Prince George	9,500	[1]3,300	[2]1,400
Imperial Oil Ltd.–Ioco	42,800	[1]12,300	[4]7,000
Petro-Canada Products Inc.–			
Port Moody	25,500	....	[2]8,600
Taylor	15,500	[1]6,300	[2]3,000
Shell Canada Ltd.–Shellburn, Burnaby	22,800	[1]5,320	[2]3,420
Total	161,100	37,720	33,420
New Brunswick			
Irving Oil Ltd.–St. John	237,500	[1]17,100	[2]34,650
Total	237,500	17,100	34,650
Newfoundland			
Newfoundland Processing Ltd.–Come By Chance	105,000	....	[2]26,000
Northwest Territories			
Imperial Oil Ltd.–Norman Wells	3,500	....	....
Nova Scotia			
Imperial Oil Ltd.–Dartmouth	82,300	[1]23,000	[4]9,500
McColl Frontenac Inc.–Halifax	20,000	[1]7,200	[2]3,600
Total	102,300	30,200	13,100
Ontario			
Imperial Oil Ltd.–Nanticoke	106,400	[1]37,600	[6]24,700
Sarnia	122,600	[1]25,000	[2]13,300
			[4]14,300
Petro-Canada Products Inc.–			
Clarkson	41,500	....	[6]9,800
Oakville	80,500	[1]25,400	[2]13,200
Shell Canada Ltd.–Sarnia	67,450	[1]13,680	[2]19,950
Suncor Inc.–Sarnia	83,000	[2]16,300	[2]26,700
Total	501,450	117,980	121,950
Quebec			
Petro-Canada Products Inc.–Montreal	87,400	[1]17,200	[2]31,300
Shell Canada Ltd.–Montreal	114,000	[1]20,900	[2]19,665
Ultramar Canada Inc.–St. Romuald	116,400	[1]37,300	[2]17,100
Total	317,800	75,400	68,065
Saskatchewan			
Consumers' Cooperative Refineries Ltd.–Regina	45,200	[1]17,600	[2]9,000
Producers Pipeline–Moose Jaw	13,300	....	....
Total	58,500	17,600	9,000
Total Canada	1,882,060	387,870	367,060
CHILE			
ENAP–Gregorio-Magallanes	8,800	....	....
Petrox SA–Talcahuano	72,000	[1]18,870	[2]4,500
Refinería de Concón–Concón	65,000	[1]19,000	[2]5,800
Total	145,800	37,870	10,300
CHINA			
Government-owned refineries at Anshan, Beijing, Dalian, Daqing, Fushun, Hangzhou, Jinxi, Karamai-Dushanzi, Lanchow, Lenghu, Maoming, Nanchong, Nanjing, Shanghai, Shengli, Tianjin, Yumen.			
Total	2,200,000		
COLOMBIA			
Empresa Colombiana de Petróleos –Barrancabermeja-Santander[1]	170,000	[1]62,000	[2]6,000
Cartagena, Bolivar	70,000	[1]29,000	....
Orito, Putumayo	2,400	....	....
Tibu, N. de Santander	5,000	....	....
Total	247,400	91,000	6,000

Solvent extraction: [1]35,000 b/cd.

CONGO			
Coraf–Pointe-Noire	21,000	....	[2]2,000
COSTA RICA			
Refinadora Costarricense de Petroleo SA–Limón	15,000	....	[2]1,200
CUBA			
Government-owned refineries at Cabaiguan, Cienfuegos, Havana, Santiago de Cuba.			
Total	280,000		
CYPRUS			
Cyprus Petroleum Refinery Ltd.–Larnaca	18,600	....	[2]4,300
CZECHOSLOVAKIA			
Government owned refineries at Kolin, Pardubice, Strazke, Zaluzi, Zyolen.			
Crude capacity	240,200		
Kaucuk s.p.–Kralupy	64,800	....	[2]7,500
Slovnaft s.p.–Bratislava[1]	150,000	....	[2]22,700
Total	455,000	....	30,200

Solvent extraction: [1]700 b/cd.

DENMARK			
AS Dansk Shell–Fredericia	62,000	....	[2]12,000
Statoil AS–Kalundborg	65,000	....	[2]8,600
Kuwait Petroleum Refining (Danmark) A/S–Skaelskoer	58,500	....	[2]10,500
Total	185,500	....	31,100
DOMINICAN REPUBLIC			
Falconbridge Dominicana C por A–La Bonao	16,000	....	....
Refineria Dominicana de Petróleo SA–Haina	32,000	....	[1]9,100
Total	48,000	....	9,100
ECUADOR			
CEPE–Esmeraldas	90,000	[1]16,000	[2]2,780
Sta. Elena Peninsula	32,300	....	....
Refieria Amazonas	10,000	....	....
Repetrol–Santa Elena	8,500	....	....

REFINING SURVEY

WORLDWIDE REFINING	Crude capacity	b/cd Catalytic cracking	Cat reforming
Texaco–Lago-Agrio	1,000	....	....
Total	**141,800**	**16,000**	**2,780**

EGYPT
Alexandria Petroleum Co.– Alexandria–(El-Mex)	97,850	....	....
Cairo Oil Refining Co.–Mostorod	156,750	....	[2]9,000
Tanta	21,800	....	....
Amerya Oil Refining Co.– Alexandria	64,000	....	[2]12,000
El-Nasr Petroleum Co.–El-Suez	58,400	....	....
Wadi-Feran	9,680	....	....
Suez Oil Processing Co.–El-Suez	62,530	....	[2]12,540
Assiout	52,143	....	....
Total	**523,153**	....	**33,540**

EL SALVADOR
Refinería Petrólera Acajutla SA– Acajutla	16,300	....	[1]3,000
Total	**16,300**	....	**3,000**

ETHIOPIA
Ethiopian Petroleum Corp.–Assab	18,000	....	[3]2,400

FINLAND
Neste Oy–Naantali	44,000	[2]13,300	[5]6,800
Porvoo	197,000	[1]28,500	[5]36,100
Total	**241,000**	**41,800**	**42,900**

FRANCE
CRD–Total France– Gonfreville L'Orcher[1]	309,000	[1]31,500	[2]45,400
La Mède	136,000	[1]29,700	[4]23,400
Mardyck	122,000	[1]29,800	[4]20,500
Cie. Rhénane de Raffinage– Reichstett-Vendenheim	80,000	[1]13,760	[1]12,900
Elf France–Donges	200,000	[1]66,000	[2]25,600
Feyzin	120,000	[1]22,700	[2]8,600
Grandpuits	92,000	[1]28,800	[2]12,900
Esso SAF–Fos sur Mer	100,000	[1]22,000	[1]16,000
Port Jerome	136,000	[1]24,000	[1]16,000
Mobil Oil Francaise– Notre Dame de Gravenchon	57,000	....	[2]13,300
Shell Francaise–Berre l'Etang	128,000	[1]38,000	[2]19,000
Petit Couronne	158,000	[1]20,000	[2]28,000
Ste. Francaise des Petroles BP– Dunkirk	....	....	....
Lavera	177,630	[1]23,220	[1]5,160
Total	**1,815,630**	**349,480**	**246,760**

Solvent extraction: [1]9,800 b/cd.

GABON
Ste. Gabonaise de Raffinage– Port Gentil	24,000	....	[3]1,400

GERMANY
Addinol Mineralol GmbH Lutzkendorf—Krumpa, Kreis Merseburg[1]	16,500	....	....
BP Oiltech GmbH–Hamburg	....	....	....
DEA Mineraloel AG–Heide	80,000	[1]9,000	[4]20,000
UK-Wesseling	100,000	....	[4]19,000
DMP Mineralol Petrochemie GmbH–Burghausen	72,000	....	....
Deutsche Shell AG–Godorf	170,000	....	[2]17,000 [6]21,000
Harburg-Grasbrook	86,000	[1]15,000	[6]16,000
Erdoel Raffinerie Neustadt GmbH–Neustadt-Donau	144,000	[1]26,000	[1]17,700
Esso AG–Ingolstadt	95,000	[1]24,500	[2]15,600
Karlsruhe	150,000	....	[2]28,000
Holborn Europa Raffinerie GmbH— Harburg	78,000	[1]18,750	[2]14,700
Hydrierwerk Zeitz GmbH–Zeitz	73,700	....	....
Leune-Werke AG–Leuna	100,000	....	[2]15,000
Mobil Oil AG–Woerth	93,000	[1]19,000	[2]15,900

WORLDWIDE REFINING	Crude capacity	b/cd Catalytic cracking	Cat reforming
Oberrheinische Mineralolwerke GmbH–Karlsruhe	174,000	[1]66,000	[2]11,400 [6]21,300
PCK Schwedt AG–Schwedt	230,000	[1]37,000	[2]23,000 [6]18,000
Raffineriegesellschaft– Vohburg/Ingolstadt	120,000	[1]19,000	[2]18,000 [6]18,800
Ruhr Oel GmbH–Gelsenkirchen	215,400	[1]6,000	[1]11,900 [5]27,000
Wintershall AG–Lingen	65,000	....	[2]12,200 [4]15,000
Salzbergen	2,800	....	....
Total	**2,065,400**	**240,250**	**376,500**

Solvent extraction: [1]5,000 b/cd.
Note: Some of the smaller east german refineries are not accounted for.

GHANA
Ghanaian Italian Petroleum CL–Tema	26,600	....	[2]6,175

GREECE
Hellenic Aspropyrgos Refinery SA–Aspropyrgos	120,800	[1]32,600	[2]9,800 [6]20,600
Motor Oil (Hellas) Corinth Refineries SA–Aghii Theodori	100,000	[1]27,300	[2]8,500
Petrola Hellas SA–Elefsis	108,000	....	....
Thessaloniki Refining Co. AE– Thessaloniki	66,500	....	[2]8,100
Total	**395,300**	**59,900**	**47,000**

GUATEMALA
Texas Petroleum Co.–Escuintla	16,000	....	[2]3,000
Total	**16,000**	....	**3,000**

HONDURAS
Refinería Texas de Honduras SA– Puerto Cortes	14,000	....	[1]1,800
Total	**14,000**	....	**1,800**

HUNGARY
Dunai KV–Százhalombatta	150,000	[1]20,000	[2]23,000
Tiszai KV–Leninváros	60,000	....	....
Zalai KV–Zalaegerszeg	10,000	....	....
Total	**220,000**	**20,000**	**23,000**

INDIA
Bharat Petroleum CL– Mahul, Bombay	156,600	[1]33,800	[1]5,400
Bongaigaon Refinery & Petrochemicals Ltd.–Bongaigaon, Assam	27,610	....	[2]1,833
Cochin Refineries Ltd.– Ambalamugal	90,000	[1]20,000	[1]5,000
Hindustan Petroleum CL– Mahul, Bombay[1]	111,650	[1]12,000	....
Visakhapatnam	90,000	[1]19,200	....
Indian Oil CL–Barauni, Bihar[2]	70,200	....	....
Digboi, Assam	11,700	....	....
Guwahati, Assam	18,100	....	....
Koyali, Gujarat	205,700	[1]20,100	[2]8,300
Haldia, West Bengal[3]	61,100	....	[1]5,300
Mathura, Uttar Pradesh	165,000	[1]20,100	....
Madras Refineries Ltd.–Madras	114,700 [3]3,000	[1]11,770	[1]2,400
Total	**1,122,360**	**136,970**	**28,233**

Solvent extraction: [1]10,000 b/cd, [2]4,000 b/cd, [3]16,000 b/cd.

INDONESIA
Pertamina–Balikpapan, Kalimantan	231,500	....	[6]18,000
Cilacap, Central Java	300,400	....	[2]12,000 [6]18,000
Dumai, Central Sumatra	114,400	....	[1]5,600 [6]7,900

REFINING SURVEY

WORLDWIDE REFINING	Crude capacity	b/cd Catalytic cracking	Cat reforming
Musi, South Sumatra	114,300	[1]12,600	....
Pangakalan Brandan, North Sumatra	5,000	....	....
Sungai Pakning, Central Sumatra	48,000	....	....
Total	**813,600**	**12,600**	**61,500**

IRAN
National Iranian Oil Co.–

Abadan	190,000	....	[6]41,555
Esfahan	200,000	....	[6]29,600
Shiraz	40,000	....	[5]6,245
Tabriz	80,000	....	....
Tehran	210,000	....	[6]28,000
Total	**720,000**	....	**105,400**

IRAQ
Oil Refineries Administration–Baiji	150,000	....	22,000
			16,500
Basra	70,000	....	....
Daura	71,000	....	5,000
K3-Haditha	7,000	....	....
Khanaqin	12,000	....	....
Mufthia	4,500	....	....
Qaiyarah, Mosul	2,000	....	....
Iraqi Company for Oil Operations– Kirkuk	2,000	....	....
Total	**318,500**	....	**43,500**

Present status uncertain due to Middle East crisis.

IRELAND
Irish National Refining plc– Whitegate	56,000	....	[4]11,000

ISRAEL
Oil Refineries Ltd.–Ashdod	70,000	....	[2]10,000
Haifa	110,000	[1]22,000	[2]15,000
Total	**180,000**	**22,000**	**25,000**

ITALY
Agip Plas SpA–Livorno	84,000	....	[2]9,500
			[6]12,000
Agip Raffinazione SpA–			
Porto Marghera	70,000	....	[2]15,000
Rho, Milan	80,000	[1]14,300	[2]17,000
Sannazzaro, Pavia	200,000	[1]30,000	[2]12,000
Taranto	84,000	....	[2]17,000
			[3]37,250
Anonima Petroli Italiana–Falconara, Marittima	63,000	....	[2]11,000
Arcola Petrolifera SpA–La Spezia	17,358	....	....
Esso Italiana SpA– Augusta, Siracusa	180,000	[1]45,300	[2]22,300
Industrie Chimiche Italiane de Petrolio SpA-Frassino, Mantova	55,000	....	[2]7,000
Iplom SpA–Busalla	46,500	....	....
Isab–Priolo Gargallo	220,000	....	[4]34,500
Kuwait Raffinazione E Chimica– Naples	100,000	[2]20,000	[2]26,000
Raffineria di Roma SpA–Rome	81,500	....	[2]14,200
Raffineria Mediterranea Spa– Milazzo	160,000	[1]40,000	[2]12,000
Raffineria Siciliana SrL–Gela	105,000	[1]35,000	[2]13,850
Saras SpA–Sarroch	285,000	[1]72,000	[2]27,000
Sarpom–S. Martino Di Trecate	219,000	[1]17,700	[2]18,500
			[4]8,100
ENI–Priolo	240,000	[1]32,000	[2]9,000
Tamoil Italia SpA–Cremona	95,000	....	[2]12,400
Total	**2,385,358**	**306,300**	**315,050**

IVORY COAST
Ste. Ivoirienne de Raffinage– Abidjan	59,000	....	[2]13,600
Ste. Multinationale de Bitumes– Abidjan	10,000	....	....
Total	**69,000**	....	**13,600**

JAMAICA
Petrojam Ltd.–Kingston	34,200	....	[1]3,240
Total	**34,200**	....	**3,240**

JAPAN
Cosmo Oil CL–Chiba	209,000	[1]30,400	[2]34,675
Sakaide	133,000	[1]16,150	[2]9,500
Sakai	104,500	[1]18,525	[2]6,650
Yokkaichi City	166,250	[1]23,750	[2]5,700
			[4]10,450
Fuji Oil CL–Sodegaura, Chiba	126,600	[1]14,000	[2]16,300
General Sekiyu Seisei KK–Sakai	131,400	[1]32,500	[2]26,100
Idemitsu Kosan CL–Chita, Aichi	123,500	[1]27,000	[5]16,200
Himeji, Hyogo	104,500	....	[2]12,600
Ichihara, Chiba	199,500	[1]37,800	[2]15,300
Tokuyama, Yamaguchi	95,000	[1]20,700	....
Tomakomai, Hokkaido	85,500	....	[2]16,200
Kainan Petroleum Refining CL– Kaiwan City, Wakayama	64,000	....	....
Kashima Oil CL–Kashima, Ibaragi	150,000	[1]17,500	[2]7,500
			[5]14,000
Koa Oil CL–Marifu	110,000	[1]20,000	[2]11,000
Osaka	80,000	[1]23,500	[2]11,000
Kyenus Sekiyu Seisei KK– Kawasaki	66,500	....	....
Kyokuto Petroleum Ltd.–Chiba	125,000	....	[2]20,000
Kyushu Oil CL–Oita	130,000	[1]16,500	[2]8,000
Mitsubishi Oil CL–Kawasaki	55,000	....	[2]10,000
Mizushima	220,000	[1]34,000	[2]25,000
			[6]18,500
Nansei Sekiyu KK–Nishihara, Okinawa	54,000	....	[2]8,600
Nihonkai Oil CL–Toyama	43,700	....	[2]4,950
Nippon Mining CL–Chita	85,000	[1]13,000	[2]22,000
Funakawa	6,000	....	....
Mizushima[1]	190,200	[1]38,000	[2]20,000
Nippon Oil CL–Niigata	26,000	[2]4,000	....
Nippon Petroleum Refining CL– Muroran	150,000	[1]21,000	[2]21,000
Nakagusuku, Okinawa	....	....	[2]2,500
Negishi	305,000	[1]39,000	[2]20,000
			[6]30,000
Yokohama	....	....	[2]1,300
Okinawa Sekiyu Seisei– Yonashiro, Okinawa	57,000	....	....
Seibu Oil CL–Yamaguchi	110,100	[1]19,900	[2]20,000
Showa Shell Sekiyu KK–Kawasaki	96,400	....	[2]13,800
Niigata	27,500	....	[2]4,300
Showa Yokkaichi Sekiyu CL– Yokkaichi	165,200	[1]22,600	[2]40,600
Taiyo Oil CL–Ehime	61,750	....	[1]5,700
Tonen Corp.–Kawasaki	190,000	[1]74,100	[2]18,600
			[4]22,320
Wakayama	157,700	[1]34,200	[2]34,875
Toa Oil CL–Kawasaki	59,700	[1]27,100	[2]7,100
Toho Oil CL–Owase	35,000	....	....
Tohoku Oil CL–Sendai	83,900	....	[2]12,000
Total	**4,383,400**	**629,975**	**607,970**

Solvent extraction: [1]9,000 b/cd.

JORDAN
Jordan Petroleum Refinery–Zerka	100,000	[1]4,250	[2]8,640
Total	**100,000**	**4,250**	**8,640**

KENYA
Kenya Petroleum Refineries Ltd.– Mombasa	90,000	....	[1]3,800
			[2]5,200
Total	**90,000**	....	**9,000**

KOREA, N.
Government-owned, Unggi	42,000		

KOREA, S.
Honam Oil Refinery CL–Yocheon	342,000	....	[2]15,300

REFINING SURVEY

WORLDWIDE REFINING	Crude capacity	b/cd Catalytic cracking	Cat reforming
Kukdong Oil CL–Busan[1]	10,000	....	....
Daesan	60,000	....	[2]3,000
Kyung In Energy CL–Inchon	60,000	....	[2]3,600
Ssangyong Oil Refining CL–Onsan[2]	90,000	....	[1]4,000
Yukong Ltd.–Ulsan	305,000	....	[2]17,650 [6]18,900
Total	**867,000**	**....**	**62,450**

Solvent extraction: [1]2,100 b/d. [2]8,000 b/d.

KUWAIT
Getty Oil Co. (subsidiary of Texaco Inc.)–Mina Al-Zour	72,000	....	....
Kuwait National Petroleum Co. –Mina Abdulla	190,000	....	....
Mina Al-Ahmadi	370,000	[1]28,000	[2]33,000
Shuaiba	187,000	[2]14,000	....
Total	**819,000**	**42,000**	**33,000**

Present status unclear due to Middle East Crisis.

LEBANON
Tripoli Oil Installations–Tripoli	20,000	[1]7,250	[2]4,392
Zahrani Oil Installations–Sidon	17,500	....	[1]3,050
Total	**37,500**	**7,250**	**7,442**

LIBERIA
Liberia Petroleum Refining–Monrovia	15,000	....	[1]2,000
Total	**15,000**	**....**	**2,000**

LIBYA
Azzawiya Oil Refining Co.–Azzawiya	120,000	....	[2]12,982
Ras Lanuf Oil & Gas Processing Co.–Ras Lanuf	220,000	....	....
Sirte Oil Co.–Brega	7,600	....	[1]900
Total	**347,600**	**....**	**13,882**

MADAGASCAR
Solima–Managareza, Tamatave	16,350	....	[1]2,600

MALAYSIA
Esso Malaysia Berhad–Port Dickson	47,500	....	[2]13,300
Petronas Penapisan Sdn. Berhad –Kerteh, Kemaman, Terengganu	27,000	....	....
Sarawak Shell Berhad–Luton	45,000	....	....
Shell Refining Co. Berhad–Port Dickson	90,000	....	[1]4,000 [4]13,000
Total	**209,500**	**....**	**30,300**

MARTINIQUE
Ste. Anonyme de la Raffinerie des Antilles–Fort-de-France	12,000	....	[2]3,000
Total	**12,000**	**....**	**3,000**

MEXICO
Petroleos Mexicanos–Azcapotzalco	105,000	[1]24,000	....
Cadereyta	235,000	[1]40,000	[2]20,000
Ciudad Madero[1]	195,000	[1]43,000	[2]15,000
Minatitlan	200,000	[1]24,000 [2]16,000	[2]48,000
Poza Rica	50,000	....	....
Reynosa	9,000	....	....
Salamanca	235,000	[1]40,000	[2]24,800
Salina Cruz	330,000	[1]40,000	[6]20,000
Tula, Hidalgo	320,000	[1]40,000	[2]30,000
Total	**1,679,000**	**267,000**	**157,800**

Solvent extraction: [1]38,000 b/cd.

MOROCCO
Samir–Mohammedia	129,000	....	[2]24,000
Ste. Cherifienne des Petroles –Sidi Kacem	25,600	[2]5,600	[1]2,800
Total	**154,600**	**5,600**	**26,800**

MYANMAR
Myanma Petrochemical Enterprise– Chauk	6,000	....	....
Thanlyin	26,000	....	....
Total	**32,000**	**....**	**....**

Formerly Burma

NETHERLANDS
Esso Nederland BV–Rotterdam	153,200	....	[4]26,600
Kuwait Petroleum Europoort BV–Rotterdam[1]	75,500	....	[4]19,000
Netherlands Refining Co.–Europoort & Pernis	460,000	[1]50,000	[2]58,000
Shell Nederland Raffinaderij BV–Pernis	348,000	[1]78,000	[2]17,000 [6]42,000
Smid & Hollander Raffinaderij BV–Amsterdam	10,000	....	....
Total Raffinaderij Nederland NV–Vlissingen	150,000	....	[3]20,000
Total	**1,196,700**	**128,000**	**182,600**

Solvent extraction: [1]3,200 b/cd.

NETHERLANDS ANTILLES
Refineria Isla Curazao SA–Emmastad	320,000	[1]42,000	[2]15,000
Coastal Aruba Refining Co. N.V.–San Nicolas	150,000	....	....
Total	**470,000**	**42,000**	**15,000**

NEW ZEALAND
New Zealand Refining–Whangarei[1]	95,100	....	[2]26,600
Total	**95,100**	**....**	**26,600**

Solvent extraction: [1]10,000 b/cd.

NICARAGUA
Esso Standard Oil SA Ltd.–Managua	16,000	....	[1]3,000
Total	**16,000**	**....**	**3,000**

NIGERIA
Kaduna Refinery & Petrochemical Co. (NNPC)–Kaduna[1]	104,500	[1]18,000	[2]15,300
Port Harcourt Refining Co. (NNPC)–Alesa Eleme	60,000	....	[1]6,000
Rivers State	150,000	[1]40,000	[6]33,000
Warri Refinery & Petrochemical Co. (NNPC)–Warri	118,750	[1]24,700	[2]15,770
Total	**433,250**	**82,700**	**70,070**

Solvent extraction: [1]9,492 b/cd.
†Dimersol

NORWAY
Esso Norge AS–Slagen-Toensberg	90,000	....	[2]10,000
Norske Shell AS–Sola	53,000	....	[2]11,000
Statoil Division Mongstad–Mongstad	145,000	[2]41,000	[2]10,000 [5]16,000
Total	**288,000**	**41,000**	**47,000**

OMAN
Oman Refinery Co.–Mina Al Fahal	76,932	....	[2]15,386

REFINING SURVEY

WORLDWIDE REFINING	Crude capacity	b/cd Catalytic cracking	Cat reforming
PAKISTAN			
Attock Refinery Ltd.–Rawalpindi	30,500	....	....
National Refinery Ltd.–			
Korangi, Karachi[1]	44,175	....	[2]2,100
Pakistan Refinery Ltd.–Karachi	46,300	....	[4]2,700
Total	120,975	....	4,800

[1]Solvent Extraction: 5,605

PANAMA			
Refinería Panama SA–Las Minas	100,000	....	[2]7,500
Total	100,000	....	7,500
PARAGUAY			
Petroleos Paraguayos–Villa Elisa	7,500	....	....
PERU			
Petróleos del Peru:			
Conchan	6,300	....	....
Iquitos, Loreto	10,500	....	....
La Pampilla, Lima	105,000	[1]8,000	[1]1,760
Marsella, Loreto	2,170	....	....
Pucallpa	2,850	....	....
Talara	62,000	[1]16,600	....
Total	188,820	24,600	1,760
PHILIPPINES			
Caltex (Philippines) Inc.–Batangas	58,500	[1]11,900	[2]5,700
Petron Corp.–Limay, Bataan	155,000	[2]15,000	[1]6,800
			[2]17,000
Philippine Petroleum Corp.–Pililla	....	....	....
Pilipinas Shell Petroleum–			
Tabangao	65,800	....	[1]8,000
Total	279,300	26,900	37,500

POLAND
Government-owned refineries at Czechowice, Glinik Mariampolski, Jasto, Jealicze, Kralaty, L. Warynski, Trzebinia.

Crude capacity	73,000		
Mazovian Refinery & Petrochemical Works–Plock	252,454	[1]48,705	[2]31,186
Petroleum Refinery Gdansk–			
Gdansk[1]	60,000	....	[4]11,000
Total	385,454	48,705	42,186

Solvent extraction: [1]6,800 b/cd.

PORTUGAL			
Petrogal–Lisboa	....	[2]10,100	....
Leça da Palmeira, Porto	88,300	....	[2]15,600
			[4]10,500
Sines	206,000	....	[6]24,900
Total	294,300	10,100	51,000
PUERTO RICO			
Caribbean Petroleum Corp.–			
Bayamón	40,000	[1]12,000	[2]6,500
Puerto Rico Sun Oil Co.–Yabucoa	85,000	....	[2]19,000
Total	125,000	12,000	25,500
QATAR			
National Oil Distribution Co.–			
Umm Said	62,000	....	[2]2,200
			[6]9,830
Total	62,000	....	12,030

ROMANIA
Government-owned refineries at Bacau, Borzesti, Brazi, Brazov, Cimpina, Darmanesti, G. Gheorghiu Dej, Navodari, Onesti, Pitesti, Ploesti, Sulpacu, and Telaejen.

WORLDWIDE REFINING	Crude capacity	b/cd Catalytic cracking	Cat reforming
Total	617,000		
SAUDI ARABIA			
Saudi Arabian Oil Co. (Saudi Aramco)–Ras Tanura	530,000	....	[2]50,000
Arabian Oil CL–Ras Al Khafji	30,000	....	....
Jeddah Oil Refinery–Jeddah	82,000	[1]9,041	[2]2,622
Petromin-Mobil–Yanbu	300,000	[1]78,500	[5]43,300
Petromin Petrola Rabigh Refinery Co.–Rabigh	332,500	....	....
Petromin-Shell-Al-Jubail[1]	284,000	....	[6]19,000
Riyadh Oil Refinery–Riyadh[2]	134,000	....	....
Yanbu Petromin Refinery–Yanbu	170,000	....	[2]35,000
Total	1,862,500	87,541	183,922

Solvent extraction: [1]14,000 b/cd. [2]8,800 b/cd.

SENEGAL			
Ste. Africaine de Raffinage–M'Bao (Dakar)	24,000	....	[4]2,600
SIERRA LEONE			
Sierra Leone Petroleum Refining Co.–Freetown	10,000	....	....
SINGAPORE			
BP Refinery Singapore PL–Pasir Panjang	28,000	....	....
Esso Singapore PL–Pulau Ayer Chawan	190,000	....	[2]11,400
Mobil Oil Singapore PL–Jurong	223,000	....	[2]18,000
Shell Eastern Petroleum Ltd.–Pulau Bukom	258,000	....	[2]6,800
			[6]13,900
Singapore Petroleum CPL†–Pulau Merlimau	179,000	....	[4]13,500
Total	878,000	....	63,600

†Singapore Petroleum CPL shares percentages of refinery capacities.

SOMALIA			
Iraqsoma Ref. Co.–Mogadishu	10,000	....	....
SOUTH AFRICA			
Caltex Oil SA PL–Cape Town	90,000	[1]25,000	[2]10,500
Gencor–Durban	65,000	[1]15,000	[2]15,000
National Petroleum Refiners of South Africa PL–Sasolburg OFS	75,500	[1]17,400	[2]11,100
Shell and BP South Africa Petroleum Refineries PL–Durban	200,000	[1]27,000	[2]26,000
Total	430,500	84,400	62,600
SPAIN			
Asfaltos Españoles SA–Tarragona	21,000	....	....
Cia. Española de Petróles–			
San Roque (Cádiz)	160,000	[1]36,000	[2]35,000
Tenerife	130,000	....	[2]14,000
Ertoil, SA–La Rábida, Huelva	80,000	....	[2]14,100
Petromed–Castellón de la Plana	120,000	[1]23,300	[2]13,400
Petronor SA–Somorrostro, Vizcaya	240,000	[1]40,000	[2]33,500
Repsol Petroleo SA–			
Cartagena, Murcia	120,000	....	[2]25,000
La Coruña	135,000	[1]28,000	[2]22,000
Puertollano, Ciudad Real	135,000	[1]30,000	[2]18,000
Tarragona	180,000	....	[2]16,000
Total	1,321,000	157,300	191,000
SRI LANKA			
Ceylon Petroleum Corp.–Sapugaskanda	50,000	....	[2]3,750
Total	50,000	....	3,750

343

REFINING SURVEY

WORLDWIDE REFINING	Crude capacity	b/cd Catalytic cracking	Cat reforming
SUDAN			
Port Sudan Refinery Ltd.–			
Port Sudan	21,700	....	[1]1,900
SWEDEN			
AB Nynas Petroleum–Gothenburg	12,500	....	....
Nynäshamn	28,000	....	....
BP Raffinaderi–Gothenburg	106,000	....	[2]20,500
Shell Raffinaderi BV–Gothenburg	81,000	....	[2]17,000
Skandinaviska Raffinaderi AB–			
Brofjorden-Lysekil	200,000	[1]25,000	[4]34,500
Total	427,500	25,000	72,000
SWITZERLAND			
Raffinerie de Cressier SA–Cressier	60,000	....	[2]16,000
Raffinerie du Sud-Ouest SA–			
Collombey	72,000	....	[2]10,000
Total	132,000	....	26,000
SYRIA			
Banias Refining Co.–Banias	120,000	....	[6]20,000
Homs Refinery Co.–Homs	117,394	....	[2]2,266
			[4]9,086
Total	237,394	....	31,352
TAIWAN,			
Chinese Petroleum Corp.–			
Kaohsiung	425,000	[1]22,600	[2]18,080
			[6]18,080
Tao-Yuan	117,500	....	[2]15,800
Total	542,500	22,600	51,960
TANZANIA			
Tanzanian & Italian Petroleum Refining CL–Kigamboni, Dar es Salaam	17,000	....	[2]6,700
THAILAND			
Esso Standard Thailand Ltd.–			
Sriracha	75,000	....	[5]8,800
Petroleum Authority of Thailand–			
Bangchak, Bangkok	61,750	....	[2]9,450
Thai Oil CL–Sriracha	83,800	[1]10,200	[1]2,700
		[2]15,900	[6]6,600
Total	220,550	26,100	27,550
TRINIDAD			
Trinidad and Tobago Oil CL–			
Pointe-a-Pierre	116,000	[1]26,000	[4]8,700
Point Fortin	80,000	....	[4]5,700
Total	246,000	26,000	14,400
TUNISIA			
Ste. Tunisienne Industries des Raffinage–Bizerte	34,000	....	3,300
TURKEY			
Anadolu Tasfiyehanesi AS–Mersin	90,000	....	[2]12,000
Turkish Petroleum Refineries			
Corp.–Aliaga, Izmir[1]	234,120	[1]15,100	[2]12,830
Batman, Siirt	24,350	....	[2]1,300
Izmit	267,325	[1]25,475	[2]20,085
Kirikkale	112,849	....	[2]20,342
Total	728,644	40,575	66,557

Solvent extraction: [1]11,700 b/cd (furfural), 6,700 b/cd (propane deasphalting).

UNITED KINGDOM			
England			
Conoco Ltd.–South Killingholme	130,000	[1]45,000	[2]53,000
Eastham Refinery Ltd.–Eastham,			

WORLDWIDE REFINING	Crude capacity	b/cd Catalytic cracking	Cat reforming
Cheshire	12,000	....	....
Ellesmere Port	22,000	....	....
Esso Petroleum CL–Fawley	300,200	[1]76,500	[2]27,900
			[4]22,500
Lindsey Oil Refinery Ltd.–			
Killingholm, South Humberside	190,000	[1]40,000	[2]33,600
Mobil Oil CL–Coryton, Essex	152,000	[1]52,000	[4]32,000
Phillips Imperial Petroleum Ltd.–			
Port Clarence	100,000	....	....
Shell U.K. Ltd.–Shell Haven	92,000	....	[6]32,000
Stanlow	262,000	[1]59,000	[2]29,000
			[6]27,000
Total	1,260,200	272,500	257,000
Scotland			
BP Refinery Grangemouth Ltd.–			
Grangemouth	198,500	[1]21,000	[2]32,000
Briggs Oil Ltd.–Dundee	10,240	....	....
Total	208,740	21,000	32,000
Wales			
BP Refinery Llandarcy Ltd.–			
Llandarcy, Neath	....	....	....
Elf Oil (G.B.) Ltd.–Milford Haven	108,000	[1]32,500	[2]17,800
Gulf Oil–GB–Milford Haven	110,000	[1]28,500	[2]20,500
Pembroke Cracking Co. (65% Texaco, 35% Gulf Oil-GB)—Milford Haven	....	[1]90,000	....
Texaco Ltd.–Pembroke, Dyfed	180,000	....	[6]39,000
Total	398,000	151,000	77,300
Total United Kingdom	1,866,940	444,500	366,300
URUGUAY			
Ancap–La Teja, Montevideo	33,000	[1]4,100	[1]2,500
U.S.S.R.			

Government-owned refineries at Achinsk, Angarsk, Baku, Batumi, Chimkent, Drogobych, Fergana, Gorki, Grozny, Guryev, Ishimbai, Khabarovsk, Kherson, Kirishi, Komsomolsk, Krasnovodsk, Kremenchug, Kuibyshev, Lisichansk, Mazheikiai, Moscow, Mozyr, Nadvornaya, Nizhnekamsk, Odessa, Omsk, Orsk, Pavlodar, Perm, Polotsk, Ryazan, Saratov, Syzran, Tuapse, Ufa, Ukhta, Vannovskiy, Volgograd, and Yaroslavl.

Total	12,300,000		
VENEZUELA			
Corpoven–El Palito, Carabobo	105,000	[1]41,000	[2]6,000
El Toreño, Barinas	4,800	....	....
Puerto La Cruz, Anzoategui	195,000	[1]12,000	....
San Roque, Anzoategui	5,200	....	....
Lagoven–Judibana, Falcon	571,000	[1]26,500	....
Maraven–Punta Cardon, Falcon	286,000	[1]59,500	....
Total	1,167,000	139,000	6,000
VIRGIN ISLANDS			
Hess Oil Virgin Islands Corp.–			
St. Croix	545,000	....	[2]125,000
Total	545,000	....	125,000
YEMEN			
Aden Refinery Co.–Little Aden[1]	161,500	....	[1]10,800
Yemen Hunt Oil Co.–Marib	10,000	....	[1]2,500
Total	171,500	....	13,300

Solvent extraction: [1]7,920 b/cd.

YUGOSLAVIA			
Bosanski Brod[1]	112,835	....	[2]7,612
			[6]13,500
Lendava	14,600	....	....
Novi Sad	63,000	....	[2]8,200
Pancevo	108,000	[1]20,750	[2]8,800
Rijeka	160,000	[1]21,300	[2]17,500

REFINING SURVEY

WORLDWIDE REFINING	Crude capacity	b/cd Catalytic cracking	Cat reforming
Sisak	150,000	[2]10,000	[2]19,000
Zagreb	700	...	...
Total	**609,135**	**52,050**	**74,612**

Solvent extraction: [1]3,300 b/cd (deasphalting).

ZAIRE

Sozir–Muanda	17,000	....	[2]3,500

ZAMBIA

Indeni Petroleum Refinery CL–Bwana Nkubwa Area, Ndola	24,500	....	[2]5,600

U.S. refineries

Company and refinery location	Crude b/cd	Catalytic cracking b/sd	Catalytic reforming b/sd
ALABAMA			
Coastal Mobile Refining Co.—Mobile Bay	*14,250	...	...
Gamxx Energy Inc.—Theodore	26,500	....	[2]5,000
Hunt Refining Co.—Tuscaloosa	45,000	......	[2]6,000
Louisiana Land & Exploration Co.—Saraland	80,000	......	[2]20,000
Total	**165,750**	**......**	**31,000**
ALASKA			
ARCO Alaska Inc.—Kuparuk	12,000	......	......
Prudhoe Bay	17,500	......	......
Chevron U.S.A. Inc.—Kenai	22,000	......	......
Mapco Alaska Petroleum—North Pole	112,000	......	......
Petro Star Inc.—North Pole	7,200	......	......
Tesoro Petroleum Corp.—Kenai	72,000	......	[4]12,000
Total	**242,700**	**......**	**12,000**
ARIZONA			
Intermountain Refining Cl—Fredonia	5,710	......	......
Sunbelt Refining Co.—Randolph	8,500	...	...
Total	**14,210**	**......**	**.......**
ARKANSAS			
Berry Petroleum Co.—Stevens	*5,700	......	......
Cross Oil & Refining Co. Inc.—Smackover	6,770	......	......
Lion Oil Co.—El Dorado[1]	48,000	[1]18,500	[2]9,000
Total	**60,470**	**18,500**	**9,000**

Solvent extraction: [1]5,500 b/d.

CALIFORNIA			
Anchor Refining Co.Inc.—McKittrick	10,000	......	......
Atlantic Richfield Co.—Carson	223,000	[1]82,000	[2]48,000
Chemoil Refining Corp.—Signal Hill	14,200	......	......
Chevron U.S.A. Inc.—El Segundo	254,000	[1]62,000	[2]51,000
Richmond[1]	270,000	[1]63,000	[2]60,000
Conoco Inc.—Santa Maria	9,500	......	......
Edgington Oil Co. Inc.—Long Beach	41,600	......	......
Exxon Co.—Benicia	128,000	[1]65,000	[3]32,000
Fletcher Oil & Refining Co.—Carson	28,750	[1]12,000	[4]5,000
Golden West Refining Co.—Santa Fe Springs	44,000	[1]13,500	[2]19,000

Company and refinery location	Crude b/cd	Catalytic cracking b/sd	Catalytic reforming b/sd
Huntway Refining Co.—Benicia	8,400	......	......
Wilmington	5,500	......	......
Kern Oil & Refining Co.—Bakersfield	20,000	......	[2]3,000
Lunday-Thagard Co.—South Gate	7,000	......	......
Mobil Oil Corp.—Torrance	123,000	[1]63,000	[2]36,000
Pacific Refining Co.—Hercules	*52,250	......	[2]15,000
Paramount Petroleum Corp.—Paramount	*37,050	......	[2]8,500
Powerine Oil Co.—Santa Fe Springs	46,550	[1]12,500	[2]10,000
San Joaquin Refining Cl—Bakersfield	18,000	......	......
Shell Oil Co.—Martinez	140,100	[1]66,000	[3]28,000
Wilmington	133,300	[1]42,000	[2]24,000
Sunland Refining Corp.—Bakersfield	15,000	......	[2]1,500
Ten By, Inc.—Oxnard	4,000	......	......
Texaco Refining & Marketing Inc.—Bakersfield	48,000	......	[2]22,000
Wilmington	*95,000	[1]30,000	[2]39,000
Tosco Corp.—Martinez	131,900	[1]60,000	[2]20,000 [5]23,000
Ultramar, Inc.—Wilmington	68,000	[1]38,000	[6]14,500
Unocal Corp.—Los Angeles	108,000	[2]47,000	[2]52,000
San Francisco (includes Santa Maria)	118,000	......	[2]34,000
Witco Chemical Corp., Golden Bear Division—Oildale	10,348	......	......
Total	**2,212,448**	**656,000**	**545,500**

Solvent extraction: [1]50,000 b/sd.

COLORADO			
Colorado Refining Co.—Commerce City	28,000	[1]8,500	[2]9,500
Conoco Inc.—Denver	48,000	[1]19,000	[2]10,000
Landmark Petroleum Inc.—Fruita	15,200	......	[2]3,400
Total	**91,200**	**27,500**	**22,900**

DELAWARE			
Star Enterprise—Delaware City	140,000	[1]69,000	[2]18,000 [5]38,000
Total	**140,000**	**69,000**	**56,000**

GEORGIA			
Amoco Oil Co.—Savannah	28,000	......	......
Young Refining Corp.—Douglasville	7,500	......	......
Total	**35,500**	**......**	**......**

HAWAII			
Chevron U.S.A. Inc.—Barber's Point	52,800	[1]20,000	......
Hawaiian Independent Refinery Inc.—Ewa Beach	*90,250	......	[2]13,000
Total	**143,050**	**20,000**	**13,000**

ILLINOIS			
Clark Oil & Refining Corp.—Blue Island	66,500	[1]25,000	[2]30,500
Hartford	57,000	[1]26,000	[2]12,000
Indian Refining Co.—Lawrenceville	54,000	[1]34,000	[2]15,000
Marathon Petroleum Co.—Robinson	195,000	[1]43,000	[4]36,500 [5]41,000
Mobil Oil Corp.—Joliet	180,000	[1]98,000	[2]46,000
Shell Oil Co.—Wood River	274,000	[1]94,000	[2]18,000 [3]75,000
Uno-Ven Co.—Lemont	147,000	[1]58,000	[2]29,800
Total	**973,500**	**378,000**	**303,800**

INDIANA			
Amoco Oil Co.—Whiting	350,000	[1]145,000	[4]85,000
Indiana Farm Bureau Cooperative Association Inc.—Mt. Vernon	20,600	[1]7,000	[2]4,000
Laketon Refining Corp.—Laketon	8,300	......	......
Marathon Petroleum Co.—			

345

REFINING SURVEY

Company and refinery location	Crude b/cd	Catalytic cracking b/sd	Catalytic reforming b/sd
Indianapolis[1]	48,000	[1]19,500	[1]10,500
Total	**426,900**	**171,500**	**99,500**

Solvent extraction: [1]4,500.

KANSAS

Company and refinery location	Crude b/cd	Catalytic cracking b/sd	Catalytic reforming b/sd
Coastal Refining and Marketing Inc.—Augusta	......	......	[4]10,000
El Dorado[1]	*30,400	[1]14,500	[2]4,500
Wichita	*29,925	[1]19,000	[2]6,500
Farmland Industries Inc.—Coffeyville	59,600	[1]25,000	[2]16,000
Phillipsburg	26,400	......	[2]5,300
National Cooperative Refinery Association—McPherson	70,900	[1]20,000	[4]15,000
Texaco Refining & Marketing Inc.—El Dorado	78,000	[1]31,500	[2]18,500
Total Petroleum Inc.—Arkansas City	56,000	[1]19,500	[2]18,000
Total	**351,225**	**129,500**	**93,800**

Solvent extraction: [1]5,500 b/sd.

KENTUCKY

Company and refinery location	Crude b/cd	Catalytic cracking b/sd	Catalytic reforming b/sd
Ashland Petroleum Co.—Catlettsburg	213,400	[1]60,000 §[2]40,000	[2]25,000
Somerset Refinery Inc.—Somerset	5,500	......	[2]1,000
Total	**218,900**	**100,000**	**53,000**

§Reduced crude converter.

LOUISIANA

Company and refinery location	Crude b/cd	Catalytic cracking b/sd	Catalytic reforming b/sd
American International Refinery, Inc.—Lake Charles	*28,500	.....	.....
Atlas Processing Co., Division of Pennzoil—Shreveport	46,200	......	[2]5,000
BP Oil Co.—Belle Chasse	*220,400	[1]92,000	[2]38,000
Calcasieu Refining Co.—Lake Charles	13,500	......	......
Calumet Lubricants Co.—Princeton	4,376	......	......
Canal Refining Co.—Church Point	9,865	......	[1]1,900
CAS Refining Inc.—Mermentau	15,000	......	......
Citgo Petroleum Corp.—Lake Charles	320,000	[1]150,000	[2]52,000 [5]54,000
Conoco Inc.—Lake Charles	159,500	[1]42,500	[2]16,000 [5]12,000
Exxon Co.—Baton Rouge[1]	421,000	[1]188,000	[4]90,000
Hill Petroleum Co.—Krotz Springs	56,700	[1]28,000	[2]12,000
St. Rose	28,300	......	......
Kerr-McGee Refining Corp.—Cotton Valley	7,800	......	......
Marathon Petroleum Co.—Garyville[4]	255,000	[1]90,000	[5]45,000
Mobil Oil Corp.—Chalmette	160,000	[1]58,000	[2]28,000 [4]19,000
Murphy Oil USA Inc.—Meraux[2]	97,000	[1]35,000	[6]23,000
Placid Refining Co.—Port Allen[3]	47,000	[1]19,000	[2]10,000
Shell Oil Co.—Norco	215,000	......	[2]16,000 [3-4]38,000
Star Enterprise—Convent	225,000	[1]85,000	[2]40,000
Total	**2,330,141**	**787,500**	**499,900**

Solvent extraction: [1]6,900 b/sd. [2]12,000 b/sd. [3]6,000 b/sd. [4]30,000 b/sd.

MICHIGAN

Company and refinery location	Crude b/cd	Catalytic cracking b/sd	Catalytic reforming b/sd
Crystal Refining Co.—Carson City	4,000	......	......
Lakeside Refining Co.—Kalamazoo	5,600	......	[1]1,000
Marathon Petroleum Co.—Detroit	68,500	[1]27,000	[1]18,500
Total Petroleum Inc.—Alma	45,600	[1]19,500	[2]14,000
Total	**123,700**	**46,500**	**33,500**

MINNESOTA

Company and refinery location	Crude b/cd	Catalytic cracking b/sd	Catalytic reforming b/sd
Ashland Petroleum Co.—St. Paul Park	67,100	[1]23,000	[2]23,500
Koch Refining Co.—Rosemount	*218,500	[1]55,000	[6]26,000 [2]6,000
Total	**285,600**	**78,000**	**55,500**

MISSISSIPPI

Company and refinery location	Crude b/cd	Catalytic cracking b/sd	Catalytic reforming b/sd
Amerada-Hess Corp.—Purvis	30,000	[2]16,000	[2]5,800
Chevron U.S.A. Inc.—Pascagoula	295,000	[1]64,000	[2]90,000
Ergon Refining Inc.—Vicksburg	16,800	......	......
Southland Oil Co.—Lumberton	5,800	......	......
Sandersville	11,000	......	......
Total	**358,600**	**80,000**	**95,800**

MONTANA

Company and refinery location	Crude b/cd	Catalytic cracking b/sd	Catalytic reforming b/sd
Cenex—Laurel[1]	40,400	[1]12,000	[2]12,000
Conoco Inc.—Billings[2]	49,500	[1]19,000	[2]14,700
Exxon Co.—Billings	42,000	[1]21,000	[1]10,000
Montana Refining Co.—Great Falls	6,700	[1]2,400	[2]1,000
Total	**138,600**	**54,400**	**37,700**

Solvent extraction: [1]4,000 b/sd. [2]7,500 b/sd (PDA).

NEVADA

Company and refinery location	Crude b/cd	Catalytic cracking b/sd	Catalytic reforming b/sd
Petro Source Refining Partners—Tonopah	4,500	......	......

NEW JERSEY

Company and refinery location	Crude b/cd	Catalytic cracking b/sd	Catalytic reforming b/sd
Amerada-Hess Corp.—Port Reading	......	[1]50,000	......
Chevron U.S.A. Inc.—Perth Amboy	80,000	......	......
Coastal Eagle Point Oil Co.—Westville	*109,250	[1]50,000	[2]27,000
Exxon Co.—Linden	130,000	[1]120,000	[2]28,000
Mobil Oil Corp.—Paulsboro[1]	100,000	[1]36,000	[2]23,500
Seaview Petroleum Co. L.P.—Thorofare	75,000	......	......
Total	**494,250**	**256,000**	**78,500**

Solvent extraction: [1]9,500 b/sd.
* Distillation

NEW MEXICO

Company and refinery location	Crude b/cd	Catalytic cracking b/sd	Catalytic reforming b/sd
Bloomfield Refining Co.—Bloomfield	16,800	[1]6,000	[2]4,000
Giant Industries Inc.—Gallup	20,000	[1]7,800	[2]6,800
Navajo Refining Co.—Artesia	*38,000	[1]20,000	[2]8,000
Thriftway Marketing Corp.—Farmington	2,500	......	[2]500
Total	**77,300**	**33,800**	**19,800**

NEW YORK

Company and refinery location	Crude b/cd	Catalytic cracking b/sd	Catalytic reforming b/sd
Cibro Petroleum Products Co.—Albany	*39,900	......	......

NORTH DAKOTA

Company and refinery location	Crude b/cd	Catalytic cracking b/sd	Catalytic reforming b/sd
Amoco Oil Co.—Mandan	58,000	[1]26,000	[2]12,100
Total	**58,000**	**26,000**	**12,100**

OHIO

Company and refinery location	Crude b/cd	Catalytic cracking b/sd	Catalytic reforming b/sd
Ashland Petroleum Co.—Canton	66,000	[1]25,000	[5]20,000
BP Oil Co.—Lima	*142,500	[1]34,000	[4]55,000
Toledo	*120,650	[1]55,000	[3]23,000 [4]19,000
Sun Refining & Maketing Co.—Toledo[1]	125,000	[1]60,000	[2]45,600
Total	**454,150**	**174,000**	**162,600**

REFINING SURVEY

Company and refinery location	Crude b/cd	Catalytic cracking b/sd	Catalytic reforming b/sd
Solvent extraction: [1]9,000 b/sd.			

OKLAHOMA
Barrett Refining Corp.—Thomas	11,000	......	......
Conoco Inc.—Ponca City	140,000	[1]53,000	[2]36,000
Cyril Petrochemical Corp.—Cyril	12,500	......	......
Kerr-McGee Refining Corp.—Wynnewood[1]	43,000	[1]20,000	[5]12,500
Sinclair Oil Corp.—Tulsa	50,000	[1]18,000	[1]12,000
Sun Refining & Marketing Co.—Tulsa[2]	85,000	[1]30,000	[2]24,000
Total Petroleum Inc.—Ardmore	68,000	[1]25,000	[6]17,000
Total	**409,500**	**146,000**	**101,500**

Solvent extraction: [1]5,000 b/sd. [2]5,800 b/sd.

OREGON
Chevron U.S.A. Inc.—Portland	.....	.....	.....

PENNSYLVANIA
BP Oil Co.—Marcus Hook	*171,000	[1]50,000	[6]48,000
Chevron U.S.A. Inc.—Philadelphia	175,000	[1]62,000	[2]34,000
Pennzoil Products Co.—Rouseville	15,700	......	[2]5,820
Sun Refining & Marketing Co.—			
Marcus Hook	170,000	[1]87,000	[2]39,600
Philadelphia	125,000	[1]29,000	[2]50,000
United Refining Co.—Warren	*64,600	[1]22,000	[2]16,000
Witco Chemical Co.—Bradford	10,000	......	[2]1,650
Total	**731,300**	**250,000**	**195,070**

Solvent extraction: [1]730 b/sd

TENNESSEE
Mapco Petroleum Inc.—Memphis	60,000	[1]30,000	[2]10,000
Total	**60,000**	**30,000**	**10,000**

TEXAS
Amoco Oil Co.—Texas City	433,000	[1]200,000	[2]160,000
Champlin Refining & Chemicals, Inc.—Corpus Christi	130,000	[1]70,000	[5]52,000
Chevron U.S.A. Inc.—El Paso	66,000	[1]22,000	[2]25,000
Port Arthur	315,900	[1]110,000	[2]23,000 [4]44,100
Coastal Refining & Marketing Inc.—Corpus Christi	*90,250	[1]18,500	[2]11,000 [6]17,500
Crown Central Petroleum Corp.—Houston	100,000	[1]56,000	[2]14,000 [5]22,000
Diamond Shamrock Corp.—Sunray[1]	111,000	[1]45,000	[2]15,000 [5]25,000
Three Rivers[2]	53,000	[1]20,000	[2]11,000
El Paso Refining Co. Ltd.—El Paso	26,000	[1]10,800	[2]7,400
Exxon Co. U.S.A.—Baytown[3]	426,000	[1]180,000	[3]60,000 [4]63,000
Fina Oil & Chemical Co.—Big Spring[4]	55,000	[1]22,000	[2]20,000
Port Arthur[5]	110,000	[1]36,000	[5]34,000
Hill Petroleum Co.—Houston[6]	67,000	[1]61,000	[2]13,500
Texas City	123,500	[1]50,000	[2]11,000 [6]12,000
Howell Hydrocarbons Inc.—San Antonio	2,900	......	[1]1,200
Koch Refining Co.—Corpus Christi	125,000	[1]40,000	[3]15,000 [6]33,500
LaGloria Oil & Gas Co.—Tyler	49,500	[1]17,000	[2]4,000 [5]11,700
Leal Petroleum Corp.—Nixon	12,000	......	......
Liquid Energy Corp.—Bridgeport	10,000	......	......
Lyondell Petrochemical Co.—Houston[7]	265,000	[1]90,000	[2]110,000
Marathon Petroleum Co.—Texas City	69,500	[1]38,000	[1]10,500
Mobil Oil Corp.—Beaumont	275,000	[1]102,000	[2]57,000 [5]46,000
Phillips 66 Co.—Borger	105,000	[1]60,000	[2]26,000
Sweeny	175,000	[1]87,000	[2]36,000
Pride Refining Inc.—Abilene	45,500	......	......
Shell Oil Co.—Deer Park	215,900	[1]65,000	[2]20,000 [4]43,000
Odessa	28,600	[1]10,500	[2]10,000
Southwestern Refining Co., Inc.—Corpus Christi	104,000	[1]50,000	[2]30,000
Star Enterprise—Port Arthur	250,000	[1]110,000	[5]46,000
Trifinery—Corpus Christi	18,000	......	......
Valero Refining Co.—Corpus Christi	25,000	†[2]65,000	......
Total	**3,882,550**	**1,635,800**	**1,140,400**

Solvent extraction: [1]15,000 b/sd. [2]6,000 b/sd. [3]53,000 b/sd. [4]10,000 b/sd. [5]18,000 b/sd. [6]17,000 b/sd. [7]6,000 b/sd.
†Heavy oil cracker.

UTAH
Amoco Oil Co.—Salt Lake City	40,000	[1]18,000	[4]7,600
Big West Oil Co.—Salt Lake City	24,000	[2]5,000	[2]5,000
Chevron U.S.A.—Salt Lake City	45,000	[1]11,000	[2]7,500 [7]7,000
Crysen Refining Inc.—Woods Cross	12,500	......	[2]3,000
Pennzoil Products Co.—Roosevelt	8,000	[1]6,000	[2]2,000
Phillips 66 Co.—Woods Cross[1]	25,000	[2]8,400	[2]6,000
Total	**154,500**	**55,400**	**31,100**

Solvent extraction: [1]5,000 b/sd.

VIRGINIA
Amoco Oil Co.—Yorktown	53,000	[1]27,500	[4]10,200
Total	**53,000**	**27,500**	**10,200**

WASHINGTON
Atlantic Richfield Co.—Ferndale	167,000	......	[2]56,000
BP Oil Co.—Ferndale	*90,250	[2]28,500	[4]16,500
Chevron U.S.A. Inc.—Seattle	......	......	......
Shell Oil Co.—Anacortes	89,300	[1]42,000	[3]26,000
Sound Refining Inc.—Tacoma	11,900	......	......
Texaco Refining & Marketing Inc.—Anacortes[1]	130,000	[1]48,000	[1]8,000 [2]16,000
U.S. Oil & Refining Co.—Tacoma	*32,775	......	[2]6,000
Total	**521,225**	**118,500**	**128,500**

Solvent extraction: [1]4,500 b/sd.

WEST VIRGINIA
Phoenix Refining Co.—St. Mary's[1]	19,180	......	[1]1,500
Quaker State Oil Refining Corp.—Newell	10,500	......	[2]3,400
Total	**29,680**	**......**	**4,900**

Solvent extraction: [1]1,000 b/sd.

WISCONSIN
Murphy Oil USA Inc.—Superior	32,000	[1]11,000	[2]8,000

WYOMING
Amoco Oil Co.—Casper	40,000	[1]13,500	[4]7,100
Frontier Oil & Refining Co.—Cheyenne	*36,100	[1]11,500	[2]6,600
Little America Refining Co.—Casper	22,000	[1]14,000	[3]6,000
Sinclair Oil Corp.—Sinclair	54,000	[1]21,000	[2]14,500
Wyoming Refining Co.—Newcastle	12,500	[2]4,000	[1]2,750
Total	**164,600**	**64,000**	**36,950**

UNITED STATES
Survey of operating refineries in the U.S. (state capacities as of January 1, 1991)

State	No. plants	Crude capacity b/cd	Cat cracking Fresh feed b/sd	Cat reforming b/sd
Alabama	4	165,750	......	31,000
Alaska	6	242,700	......	12,000
Arizona	2	14,210	......	......
Arkansas	3	60,470	18,500	9,000
California	30	2,212,448	656,000	545,500
Colorado	3	91,200	27,500	22,900
Delaware	1	140,000	69,000	56,000
Georgia	2	35,500	......	......
Hawaii	2	143,050	20,000	13,000
Illinois	7	973,500	378,000	303,800
Indiana	4	426,900	171,500	99,500
Kansas	8	351,225	129,500	93,800
Kentucky	2	218,900	100,000	53,000
Louisiana	19	2,330,141	787,500	499,900
Michigan	4	123,700	46,500	33,500
Minnesota	2	285,600	78,000	55,500
Mississippi	5	358,600	80,000	95,800
Montana	4	138,600	54,400	37,700
Nevada	1	4,500	......	......
New Jersey	6	494,250	256,000	78,500
New Mexico	4	77,300	33,800	19,800
New York	1	39,900	......	......
North Dakota	1	58,000	26,000	12,100
Ohio	4	454,150	174,000	162,600
Oklahoma	7	409,500	146,000	101,500
Oregon	1	......	......	......
Pennsylvania	7	731,300	250,000	195,070
Tennessee	1	60,000	30,000	10,000
Texas	31	3,882,550	1,635,800	1,140,400
Utah	6	154,500	55,400	31,100
Virginia	1	53,000	27,500	10,200
Washington	7	521,225	118,500	128,500
West Virginia	2	29,680	......	4,900
Wisconsin	1	32,000	11,000	8,000
Wyoming	5	164,600	64,000	36,950
Total	**194**	**15,478,949**	**5,444,400**	**3,901,520**

Refinery throughput

	1980	1981	1982	1983	1984	1985	1986	1987	1988	1989
					(Thousand barrels per calendar day)					
U.S.	13,520	12,470	11,775	11,685	12,045	12,005	12,715	12,855	13,445	13,555
Canada	1,915	1,780	1,520	1,455	1,460	1,430	1,395	1,460	1,535	1,550
Latin America	6,585	6,700	6,205	5,955	5,845	5,585	5,615	5,665	5,590	5,825
Western Europe	12,330	10,970	10,075	9,710	9,790	9,490	10,060	9,960	10,310	10,340
U.S.S.R & Eastern Europe	11,735	12,620	12,720	12,710	12,470	12,185	12,440	12,455	12,485	12,370
Middle East	2,445	2,205	2,245	2,515	2,665	2,795	2,950	2,970	3,140	3,355
Africa	1,460	1,525	1,735	1,920	1,975	2,015	2,095	2,170	2,200	2,175
China	1,430	1,395	1,520	1,600	1,765	1,785	1,865	1,950	2,035	2,150
Japan	4,015	3,630	3,360	3,255	3,355	3,120	2,990	2,910	2,990	3,175
South Asia	705	795	840	880	885	1,020	1,120	1,150	1,160	1,215
Other Asia	2,400	2,510	2,365	2,410	2,485	2,685	2,800	2,785	2,915	3,090
Australasia	640	655	625	605	615	580	580	645	680	695
Total world	**59,180**	**57,255**	**54,985**	**54,700**	**55,355**	**54,695**	**56,625**	**56,975**	**58,485**	**59,495**
Of which OECD & LDCs*	45,860	43,085	40,590	40,230	40,960	40,555	42,150	42,395	43,790	44,705

*Organisation for Economic Cooperation and Development and lesser developed countries
Source: British Petroleum plc

Historical reserve data (billion bbl oil)

	1960	1970	1975	1980	1985	1988	1989	1990	1991
Asia-Pacific									
Australia	—	2.0	1.7	2.4	1.4	1.7	1.7	1.7	1.6
Brunei	—	1.0	2.0	1.7	1.4	1.4	1.4	1.4	1.4
India	—	—	—	2.5	3.5	4.3	6.4	7.5	7.9
Indonesia	9.5	10.0	14.0	9.5	8.6	8.4	8.2	8.2	11.0
Malaysia	—	—	2.5	3.0	3.0	2.9	2.9	3.0	2.9
Europe									
Norway	—	1.0	7.0	5.5	8.3	14.8	10.4	11.5	7.6
United Kingdom	—	1.0	16.0	14.8	13.6	5.2	5.2	4.2	3.8
Middle East									
Abu Dhabi	—	11.8	29.5	29.0	30.5	92.2	92.2	92.2	92.2
Divided Zone	6.0	25.7	6.4	6.0	5.4	5.2	5.2	5.2	5.0
Dubai	—	—	1.0	1.4	1.4	4.0	4.0	4.0	4.0
Iran	35.0	70.0	64.5	57.5	48.5	92.8	92.8	92.8	92.9
Iraq	27.0	32.0	34.3	30.0	44.5	100.0	100.0	100.0	100.0
Kuwait	62.0	67.1	68.0	64.9	90.0	91.9	91.9	94.5	94.5
Oman	—	—	5.0	2.3	3.5	4.0	4.0	4.2	4.3
Qatar	2.5	3.5	5.8	3.6	3.4	3.2	3.1	4.5	4.5
Saudi Arabia	50.0	128.5	148.6	165.0	169.0	166.9	169.9	254.9	257.5
Syria	1.0	1.0	1.0	1.3	1.4	1.7	1.7	1.7	1.7
Yemen	—	—	—	—	—	—	—	3.0	4.0
Africa									
Algeria	5.2	7.0	7.4	8.2	9.0	8.5	8.4	9.2	9.2
Angola-Cabinda	—	—	1.3	1.2	1.8	1.1	2.0	2.0	2.1
Egypt	—	4.5	3.9	2.9	3.2	4.3	4.3	4.5	4.5
Gabon	—	—	—	—	—	0.6	0.7	0.7	0.7
Libya	2.0	29.2	26.0	23.0	21.1	21.0	22.0	22.8	22.8
Nigeria	—	9.3	20.2	16.7	16.7	16.0	16.0	16.0	17.1
Tunisia	—	—	1.0	1.7	1.5	1.8	1.8	1.8	1.7
Western Hemisphere									
Argentina	2.2	4.5	2.5	2.5	2.3	2.3	2.3	2.3	2.3
Brazil	—	—	—	1.3	1.9	2.3	2.5	2.8	2.8
Canada	—	10.7	7.1	6.4	7.1	6.8	6.8	6.1	5.8
Colombia	—	1.7	1.7	1.0	0.624	1.6	2.0	2.0	2.0
Ecuador	—	—	2.5	1.1	1.4	1.6	1.4	1.5	1.4
Mexico	2.3	3.2	9.5	44.0	48.6	48.6	54.1	56.4	51.9
Venezuela	18.5	14.0	17.7	17.9	25.8	56.3	58.0	58.5	59.0
U.S.A.	33.5	37.0	33.0	26.4	27.3	25.3	26.5	25.9	26.2
Subtotal	**268.5**	**511.4**	**555.7**	**562.2**	**614.6**	**808.1**	**823.6**	**917.5**	**916.3**
U.S.S.R.	33.5	77.0	80.4	63.0	63.0	59.0	58.5	24.0	57.0
China	—	—	—	—	19.1	18.4	23.6	24.0	24.0
Other	—	—	—	—	—	1.8	1.7	1.7	1.8
Total world	**302.0**	**611.4**	**658.7**	**648.5**	**698.7**	**887.3**	**907.4**	**1,001.6**	**999.1**

Gas pipeline construction leads world activity

WORLDWIDE PIPELINE CONSTRUCTION ACTIVITY IN 1990 was steady and promised to remain that way the next few years.

Growth of gas pipeline systems was especially strong.

This trend toward more gas pipeline construction continued a pattern evident for several years.

As a clean burning fuel, gas stands to be the big winner as the world searches for fuels that are more environmentally acceptable.

Construction plans, estimated costs

In the U.S., gas pipeline projects in several major areas started up or were due to get under way by yearend 1991. The effect was to clarify the gas transportation and supply pictures for California and areas along the western slopes of the Rocky Mountains and for Oklahoma's Arkoma basin, Florida, and the Northeast.

In Canada, a major gas system to serve Vancouver Island is completing a 2 year installation program. The project had been on the books for more than 30 years. TransCanada PipeLines Ltd. embarked on its major expansion program to move more western Canadian gas east and into U.S. markets.

In Europe, gas holds the promise of reversing much of the environmental harm caused by the burning of high sulfur coal and fuel oil in countries of the former Eastern Bloc. Germany is making by far the more rapid progress as it began plans this year to extend gas transportation and distribution systems into the states of the former East Germany.

In addition, plans for another gas pipeline across the Mediterranean Sea moved ahead in 1991.

Reports to Oil & Gas Journal earlier this year showed pipeline operators planned to lay more than 43,000 miles of crude oil, natural gas, and petroleum products line in 1991 and later.

Those reports indicated that 13,860 miles of line either were being laid as the year began or were part of projects that will begin this year, all to be completed by yearend 1991.

Investment in U.S. liquids pipelines

	Company and investment, $					Total, $	%
	A	B	C	D	E		
CRUDE PIPELINES INVESTMENT BY FIVE COMPANIES							
Land	2,153,078	814,497	292,759	1,029,353	1,236,081	5,525,768	0.32
Right-of-way	12,288,388	1,006,210	316,392	16,578,568	15,997,228	56,186,786	2.64
Line pipe	138,203,476	25,703,218	7,797,471	98,166,788	160,965,781	430,836,734	24.65
Line pipe fittings	10,387,524	955,700	1,386,293	18,173,126	29,552,566	60,455,209	3.46
Pipeline construction	204,045,722	27,837,621	10,931,879	223,969,191	236,543,625	703,328,038	40.23
Buildings	24,171,592	5,013,673	2,517,063	14,446,340	11,470,520	57,619,188	3.30
Boilers	—	—	—	—	—	—	
Pumping equipment	24,646,028	5,524,747	2,414,882	17,621,540	48,174,143	98,381,340	5.63
Machine tools and machinery	—	—	—	33,938	−9,742	24,196	0.00
Other station equipment	96,402,176	22,635,571	4,602,358	48,121,742	36,323,807	208,085,654	11.90
Oil tanks	14,381,418	6,079,983	5,678,530	17,132,528	28,805,137	72,077,596	4.12
Delivery facilities	—	3,973	5,382,864	1,572,342	6,849,372	13,808,551	0.79
Communications systems	1,015,203	3,278,416	71,545	11,853,339	3,757,003	19,975,506	1.14
Office furniture and equipment	3,145,205	295,563	255,012	2,830,149	767,911	7,293,840	0.42
Vehicles and other work equip.	9,934,692	1,078,032	518,113	6,676,245	416,984	18,624,066	1.07
Other property	—	1,399,696	—	—	4,432,008	5,831,704	0.33
Total	540,774,502	101,626,900	42,165,161	478,205,189	585,282,424	1,748,054,176	100.00
PRODUCT PIPELINES INVESTMENT BY FIVE COMPANIES							
Land	4,535,405	1,763,961	572,137	1,073,091	3,247,187	11,191,780	0.38
Right-of-way	29,674,813	10,829,964	16,907,192	8,777,681	6,839,606	73,029,256	2.48
Line pipe	345,579,941	67,590,791	99,942,839	87,385,038	98,402,711	698,901,320	23.76
Line pipe fittings	72,723,899	12,899,770	14,370,980	1,743,152	22,323,267	124,061,268	4.22
Pipeline construction	584,068,773	109,254,525	184,761,177	102,011,453	153,971,129	1,134,067,057	38.55
Buildings	24,001,307	4,101,502	4,749,308	10,994,655	17,091,775	60,938,547	2.07
Boilers	—	—	—	—	—	—	
Pumping equipment	45,177,720	6,182,370	28,552,138	28,849,924	20,176,729	128,938,881	4.38
Machine tools and machinery	—	—	—	—	441,510	441,510	0.02
Other station equipment	122,325,732	21,791,408	34,910,983	25,585,067	126,711,226	331,324,416	11.26
Oil tanks	48,938,264	17,525,411	9,065,619	20,147,096	66,656,498	162,332,888	5.52
Delivery facilities	—	—	8,729,825	10,871,129	50,835,414	70,436,368	2.39
Communications systems	1,700,822	644,086	2,366,238	9,175,498	—	13,886,644	0.47
Office furniture and equipment	11,634,736	2,538,234	3,095,206	652,793	1,104,306	19,025,275	0.65
Vehicles and other work equip.	6,148,086	1,930,413	8,668,353	2,594,054	2,655,948	21,996,854	0.75
Other property	73,352,479	—	18,092,213	—	—	91,444,692	3.11
Total	1,369,861,977	257,052,635	434,784,208	309,860,631	570,457,305	2,942,016,756	*100.00

* Source: U.S. FERC Form 6, Annual Report of Oil Pipeline Companies, Dec. 31, 1989.

MISSING SOMETHING?

Complete your collection of International Petroleum Encyclopedias by ordering past editions.

The **International Petroleum Encyclopedia** is the industry's most widely read annual report. Year after year, it keeps you up-to-date on developments in the global petroleum industry with statistical information, hundreds of colorful atlas maps, and feature articles. Widely regarded as the most comprehensive worldwide petroleum resource, each edition of the **International Petroleum Encyclopedia** is a vital part of any library. A complete collection will enable you to fill in the blanks easily when it comes to research, forecasts, and follow-up. As a buyer of the current edition you are eligible to purchase past issues of the **International Petroleum Encyclopedia** for 50% off the regular price. This allows you to complete your library of worldwide petroleum information at a substantially reduced cost. Check and see which of these important editions you are missing and contact us for information on ordering replacements. Quantities are extremely limited, so call today.

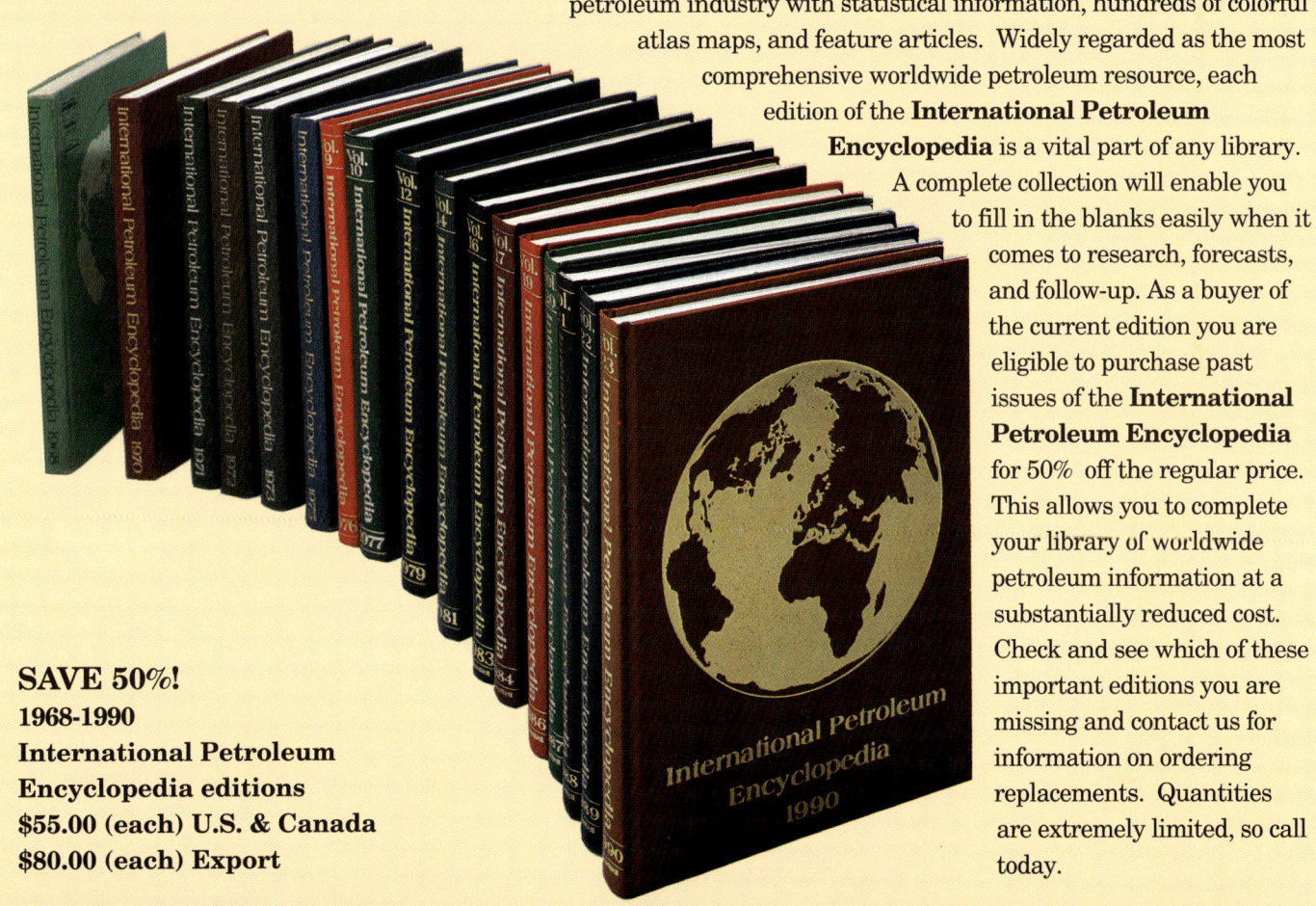

SAVE 50%!
1968-1990
International Petroleum Encyclopedia editions
$55.00 (each) U.S. & Canada
$80.00 (each) Export

International Petroleum Encyclopedia
P. O. Box 21288, Tulsa, OK 74121

CALL 1-800-752-9764 TOLL-FREE IN U.S.A. ◆ PHONE 918-831-9421 ◆ FAX 918-831-9555 ◆ TELEX 211012

PIPELINING

Projects excluded from these projections were those only under study.

In addition, projections were adjusted to eliminate mileage duplications caused by competing projects in which only one or two were likely to be built.

Cost of 1991's work alone could reach $11.5 billion. For 1991 and beyond, total land and offshore construction costs could exceed $36 billion.

Cost estimates were based on recent U.S. average cost per mile data for onshore and offshore gas pipeline construction as filed with the Federal Energy Regulatory Commission.

Cost projections assumed that 90% of all construction will be onshore and 10% offshore. The exceptions were pipelines 32 in. in diameter or larger, which were assumed to be solely onshore projects.

Here is a breakout of costs:
- Land construction for 1991 will reach almost $10.7 billion—$3.5 billion for 4-10 in. pipelines, $2.3 billion for 12-20 in., $1.9 billion for 22-30 in., and $3 billion for 32 in. and larger.
- Offshore construction for 1991 will reach $854.1 million—$388.7 million for 4-10 in., $257.7 million for 12-20 in., and $207.7 million for 22-30 in.
- Onshore construction for 1991 and beyond will reach $33.9 billion—$6.6 billion for 4-10 in., $7.4 billion for 12-20 in., $6 billion for 22-30 in., and $13.9 billion for 32 in. and larger.
- Offshore construction for 1991 and beyond will reach $2.2 billion—$738.9 million for 4-10 in., $826.6 million for 12-20 in., and $673.1 million for 22-30 in.

California projects sift out

By far the most significant news in U.S. pipeline constuction in 1991 is start of construction of a gas pipeline from the Wyoming Overthrust region to southern California.

Construction began in January in Southwest Utah and southern Nevada on Spreads 5 and 6 of Kern River Gas Transmission Co.'s 904 mile pipeline from Wyoming to near Bakersfield, Calif. In all there are eight spreads at work.

From Opal, Wyo., to near Barstow, Calif., the line will consist of 676 miles of 36 in. At that point, 121 miles of 42 in. pipe will extend into Kern County, southeast of Bakersfield, where the system will spilt into two 30 in. laterals, 55 and 48 miles long.

The 42 in. segment will be jointly owned and operated with Mojave Pipeline Co.

When complete, Kern River will be California's only direct link to Rocky Mountain gas. The $934 million project is the largest gas pipeline built in the U.S. since the groundbreaking of Northern Border in 1981.

Capacity will be 700 MMcfd, expandable to 1.2 bcfd with the addition of two compressor stations.

Initially, 500 MMcfd of gas will originate from Wyoming, with remaining volumes coming from Utah and from Canada via Altamont Gas Transmission Co.'s proposed line.

FERC commissioners also gave preliminary approval to Altamont's plan to transport 719 MMcfd of Canadian gas to the Kern River system in Southwest Wyoming. Altamont's 30 in., 620 mile pipeline would receive gas at Port of Wild Horse, Mont., on the Canadian border.

Altamont officials expected FERC's environmental reviews of its project to be completed in 1991. Federal regulators said the plan would enhance competition for markets in southern California.

Apparently a loser in the race to build a gas pipeline from Wyoming to California, Coastal Corp., Houston, shelved plans late in 1990 for its competing $576 million Wyoming-California Pipeline Co. (WyCal) pipeline.

That line would have shipped 600 MMcfd of mostly Rocky Mountain gas 670 miles from Hams Fork, Wyo. It was to have connected with intrastate systems operated by Southern California Gas Co. and Pacific Gas & Electric Co. (PG&E) at Piute Junction, near Needles, Calif.

Kern River and Mojave are to merge near Needles.

Another project to move gas to California markets received final approval from FERC in 1991. Pacific Gas Transmission Co. (PGT) and PG&E plan an 845 mile, 42 and 36 in. pipeline expansion to transport as much as 755 MMcfd of Canadian gas to California and 148 MMcfd to the U.S. Pacific Northwest.

Construction of the $1.2 billion project is to begin early in 1992 and be complete in fourth quarter 1993.

U.S. Northeast, Arkoma, Mobile Bay

At the other end of the country, wrangling among proponents of Iroquois Gas Transmission Co.'s 370 mile pipeline, Canadian producers, U.S. gas producer opponents, and regional and environmental factions came to an end when FERC and the U.S. Army Corps of Engineers formally approved the project.

The line will transport as much as 575.9 MMcfd of gas from near Waddington, N.Y., through New York State and Connecticut to Long Island.

TransCanada will spend $12.2 million (U.S.) to lay a 2.8 mile, 30 in. pipeline to connect with the Iroquois system. Supplying all the gas for Iroquois could generate $800 million in revenues for Canadian producers.

Back in the middle of the country, Arkla Energy Resources, a division of Arkla Inc., Shreveport, La., began interstate deliveries late in 1990 out of the Arkoma basin in Oklahoma on its 42 and 36 in. Line AC.

Without compression, Line AC was shipping 800 MMcfd of gas 221 miles from Arkla's Chandler compression station, west of Wilburton, Okla., to Mississippi River Transmission Co.'s Glendale compression station south of Pine Bluff, Ark.

By late summer 1991, Arkla is to have expanded capacity to 1 bcfd when it brings on line 14,750 hp of compression near Malvern, Ark.

Shares of capacity on Line AC after compression is added will be ANR Pipeline Co. 250 MMcfd, Texas Gas Transmission 300 MMcfd, Arkla 280 MMcfd, and Tennessee Gas Pipeline and Columbia Gulf 85 MMcfd each.

In other Arkoma activity, Natural Gas Pipeline of America

Pipeline construction, 1991 and beyond*

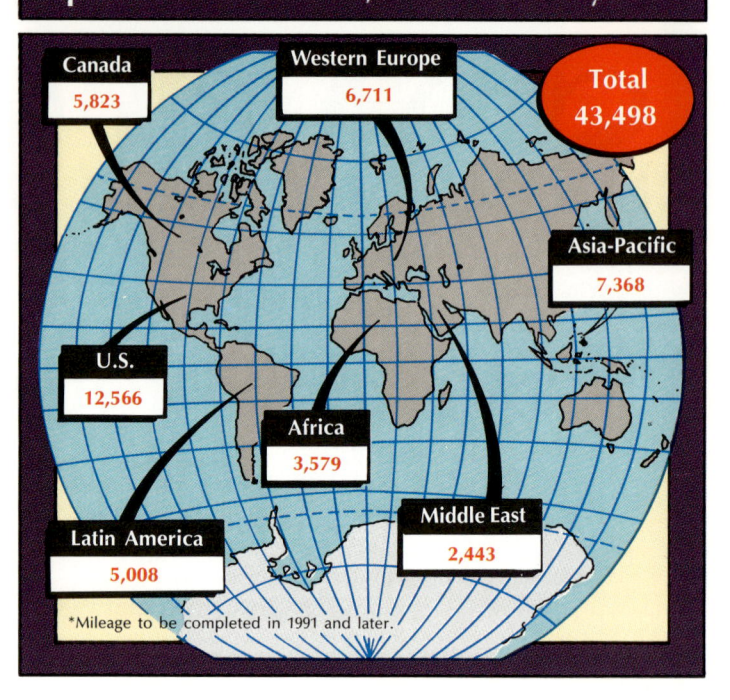

Region	Mileage
Canada	5,823
Western Europe	6,711
Asia-Pacific	7,368
U.S.	12,566
Africa	3,579
Latin America	5,008
Middle East	2,443
Total	**43,498**

*Mileage to be completed in 1991 and later.

PIPELINING

U.S. natural gas pipeline transmission expenses*

	Expenses, $ 1987-majors†	Expenses, $ 1988-majors§	Cost/mile, $ 1987-majors†	Cost/mile, $ 1988-majors§	Cost/MMcf sold, $ 1987-majors†	Cost/MMcf sold, $ 1988-majors§
Operation expenses						
Supervision and engineering	151,679,238	166,584,029	610.48	661.37	23.45	26.16
System control and load dispatching	26,858,420	29,284,519	108.10	116.26	4.15	4.60
Communication system expenses	29,310,758	30,224,686	117.97	120.00	4.53	4.75
Compressor station labor and expenses	251,651,867	269,976,844	1,012.85	1,071.86	38.90	42.39
Gas for compressor station fuel	367,406,730	375,892,286	1,478.75	1,492.36	56.80	59.02
Other fuel and power for compressor stations	63,842,107	65,481,393	256.95	259.97	9.87	10.28
Mains	168,358,688	182,166,091	677.61	723.23	26.03	28.60
Measuring and regulating station expenses	61,623,174	61,872,085	248.02	245.64	9.53	9.72
Transmission and compression of gas by others	1,049,122,569	936,299,992	4,222.53	3,717.28	162.18	147.02
Other transmission expenses	38,278,436	40,409,766	154.06	160.43	5.92	6.35
Rents	18,421,763	19,796,436	74.14	78.60	2.85	3.11
Total operation expenses	**2,226,553,750**	**2,177,988,127**	**8,961.49**	**8,657.00**	**344.19**	**341.98**
Maintenance expenses						
Supervision and engineering	34,596,228	31,519,887	139.24	125.14	5.35	4.95
Structures and improvements	21,181,592	22,517,885	85.25	89.40	3.27	3.54
Mains	91,003,249	109,462,229	366.27	434.58	14.07	17.19
Compressor station equipment	197,326,728	218,175,493	749.21	866.20	30.50	34.26
Measuring and regulating station equipment	12,117,702	13,202,645	48.77	52.42	1.87	2.07
Communication equipment	12,077,950	11,906,439	48.61	47.27	1.87	1.87
Other equipment	3,259,046	4,036,815	13.12	16.03	0.50	0.63
Total maintenance expenses	**371,562,495**	**410,821,393**	**1,495.47**	**1,631.03**	**57.44**	**64.51**
Total transmission expenses	**$2,598,116,245**	**2,588,809,520**	**10,456.96**	**10,278.03**	**401.63**	**406.49**
Total miles of transmission pipeline	248,458	251,878				
Total natural gas sold, MMcf	6,468,964	6,368,682				

* U.S. interstate pipelines for calendar years 1987 & 1988. † 43 of 126 companies filing Form 2's or 2A's with the U.S. FERC for 1982; majors are companies whose combined gas sold for resale and gas transported or stored for a fee exceed 50 bcf (at 14.7 psi; 60° F.) in each of the 3 previous calendar years. §46 of 135 companies filing Form 2 and 2A with the U.S. FERC for 1988; majors defined in previous note and in U.S. FERC Accounting and Reporting Requirements for Natural Gas Companies, para. 20-011, effective Feb. 2, 1985, beginning with the 1984 reporting year (OGJ, Nov. 25, 1985, p. 80).
Source: Statistics of Interstate Natural Gas Pipeline Companies—1987 & 1988, U.S. Department of Energy.

(NGPL) began service on its $51 million Arkoma basin pipeline in southern Oklahoma.

Farther south, construction was expected to begin by midyear 1991 on a $230 million onshore and offshore pipeline system to transport as much as 1.2 bcfd of gas from Mobile Bay off Alabama to interstate markets.

The Alabama system is a consolidation of competing plans of six interstate pipeline companies. The companies sought FERC approval in 1991 to proceed with a compromise plan.

Parties to the FERC filings were ANR Pipeline Co., Florida Gas Transmission Co., Southern Natural Gas Co., Tennessee Gas Pipeline Co., Texas Eastern Transmission Corp., and Transcontinental Gas Pipeline Corp.

ANR, Southern, Tennessee, Texas Eastern, and TGPL are jointly to own the onshore Mobile Bay pipeline. Florida Gas will lease capacity.

Onshore facilities include TGPL's existing 30 in., 123 mile line from the gas processing plant in Mobile County, Ala., owned by Mobil Exploration & Producing U.S. Inc. to an interconnect with TPGL's mainline near Butler, Ala.

The system is to connect near Citronelle, Ala., with Florida Gas' mainline, where ANR, Southern, and Tennessee agreed to lease capacity to move gas on Florida Gas to their mainline systems.

Texas Eastern is to lease capacity from TPGL to move gas from the interconnect at Butler to its mainline.

Primary project constructor of the $150 million onshore segment of the project, SNG will lay loops and add compression to increase capacity to 1.2 bcfd from 300 MMcfd. It also will lay lines connecting existing onshore facilities with new facilities offshore.

Florida Gas, Southern, Tennessee, Texas Eastern, and TGPL are jointly to construct and own a proposed 77 mile, 600 MMcfd offshore pipeline in Mobile Bay, extending from Main Pass Block 252 to Shell Offshore Inc.'s gas plant being built onshore near Mobil's plant.

The offshore system is to consist of a 30 in. line extending 27 miles from shore to a platform on Mobile Block 955 and a 26 in. line running 50 miles from Mobile 955 to Main Pass 252.

TGPL agreed to build the $80 million offshore system, Florida Gas to operate it.

Vancouver Island, Arctic projects

After more than 30 years of planning, a gas pipeline now connects the Canadian mainland to Vancouver Island, B.C.

Westcoast Energy Inc., Vancouver, B.C., and Alberta Energy Co., Edmonton, Alta., expect to start gas deliveries

PIPELINING

by early fall 1991 through the 366 mile pipeline. The $330 million system will receive gas from a Westcoast line at Coquitlam, B.C.

Contractors completed span adjustments and testing in winter 1990 on the Strait of Georgia crossing to Vancouver Island. Much of the onshore portion runs over mountainous terrain and includes about 250 stream and river crossings.

First deliveries to seven pulp mills and, through local distribution companies (LDCs), to 14 communities along the pipeline's route will amount to about 37 MMcfd. Projections to the end of the first decade of service place deliveries at more than 65 MMcfd.

In Alaska, Yukon Pacific Corp. (YPC) in November 1990 began the first phase of a plan to build an 800 mile pipeline for Alaskan North Slope (ANS) gas and an LNG export facility. Plans call for export of as much as 14 million metric tons/year of LNG.

The Trans-Alaska Gas System (TAGS) would transport gas to Anderson Bay, Alas., for shipment overseas.

Under a contract awarded late in 1990, Bechtel Corp. is designing a gas liquefaction plant and marine terminal to be located 3 miles west of Prudhoe Bay oil storage and loading facilities at Valdez.

YPC budgeted $300 million to design the export facility and pipeline.

When all phases of the $11 billion project to liquify and export ANS gas are complete, YPC's LNG plant will be the largest of its kind.

Scheduled to break ground in 1993 for the $3 billion plant, YPC was trying to sign up customers to take 7 million tons/year of LNG. To ensure the project will be profitable, it must increase export rates within 5 years to 14 million tons/year, nearly twice the export volume of the world's largest existing LNG plant.

Markets in Japan, Korea, and Taiwan, which expect to require an additional 24 million tons/year of LNG by 2000, were prime candidates to take North Slope exports. As planned, TAGS can provide access by 1998 to 37 tcf of proved gas reserves on Alaska's North Slope.

Also in the region, sponsors of the Polar Gas Project planned to apply to Canada's National Energy Board (NEB) for permission to build and operate a large diameter pipeline to transport gas from the Mackenzie Delta on Canada's arctic frontier to markets in southern Canada and the U.S.

Polar Gas sponsors TransCanada, Tenneco Gas, and Panarctic Oils Ltd. propose to lay a 36 in., 1,421 mile, buried line from gas fields on the delta, through the Mackenzie River Valley, to Caroline, Alta.

The Mackenzie Valley Gas Pipeline would move as much as 1.2 bcfd of gas for 20 years after completion.

Mackenzie Delta producers Esso Resources Canada Ltd., Shell Canada Ltd., and Gulf Canada Resources Ltd. hold proved, combined Mackenzie Delta gas reserves of about 10.5 tcf. The Geological Survey of Canada estimates total reserves in the area at 68 tcf.

Europe, Far East gas projects

In late 1990, Den norske stats oljeselskap AS (Statoil) began urging installation by 1995 of Europipe, the third Norwegian gas trunkline linking Troll and Sleipner gas fields in the North Sea with continental Europe.

Initial capacity of the 450 km, 36 in. and 40 in. pipeline would be 12 billion cu m/year. The $1.4 billion system would begin at Statpipe's riser platform in Block 16/11, terminating either in The Netherlands or Germany.

Elsewhere in Europe, Portugal is expected to begin construction in 1991 of a gas pipeline system the government hopes to have on stream in 1995.

The proposed system would link with the European network through the Spanish system. The first stage called for laying of pipe from Setubal, south of Lisbon, to Braga, in northern Portugal near the Spanish border.

Included in the project would be construction of an LNG receiving terminal at Setubal, a trunkline, and distribution lines. When operational, the system would supply gas to about 2 million residential, 80,000-100,000 commercial, and 4,000-5,000 industrial customers.

Also in late 1990, the U.S.S.R. signed a contract with Greece for Soviet crews in 1993 to begin laying a 500 km line that would deliver gas from Izmail on the Soviet-Romanian border, through Romania and Bulgaria, to Thessalonika in northern Greece.

Greece in 1988 signed a 25 year agreement to accept initial Soviet gas deliveries amounting to 35 bcf/year.

In Germany, competition to supply the newly unified country with gas has been steadily heating up.

Ruhrgas AG purchased a 35% interest in Verbundnetz Gas AG, the transmission company for the former East Germany. And Ruhrgas announced plans to construct several pipelines to connect its supply system with the eastern states.

Wintershall AG teamed up with Kombinate Gasanlagen to build the Stegal line. This will be an east-west pipeline from the Czechoslovak border to Heringen near the former boundary between East and West Germany. Initial capacity is set at 71 bcf/year.

This line is to connect with another Wintershall project, the 580 mile Midal north-south line from Tysum on the North Sea to Ludwigshafen, south of Frankfurt.

In another part of Europe, steps toward a second trans-Mediterranean gas pipeline advanced early in 1991.

In February, six companies formed a joint venture to study the feasibility of a 1,250 mile line from Algeria to Spain. If built by 1995-96, the line could supply 105-175 bcf/year of gas to Spain, 35-70 bcf/year to Morocco, and 35 bcf/year to Portugal.

In the Far East, Malaysia's Petroliam Nasional Bhd. (Petronas) was nearing completion of its Peninsula Gas Utilization (PGU) pipeline system.

Initially, the 452 mile, mostly 36 in. PGU line is to transport 750 MMcfd of gas from Malaysia's east coast to a power plant on the west side of the Malaysian peninsula near the capital city of Kuala Lumpur, down the western coast to Johor Bahru in southern Malaysia, and to Singapore by way of a water crossing.

PGU throughput is expandable to 1 bcfd.

PGU project manager Novacorp (Malaysia) Sdn. Bhd. expected to begin testing by midyear 1991 of the PGU segment linking Kerteh on Malaysia's east coast to the west coast.

U.S. interstate pipeline mileage

Year	Gas	Liquid	Total
1980	274,248	172,673	446,921
1981	274,634	172,815	447,449
1982	286,999	172,549	459,548
1983	285,204	167,819	453,023
1984	*258,379	173,922	*432,301
1985	*279,395	171,401	*450,796
1986	*281,881	170,014	*451,895
1987	*280,085	167,865	*447,950
1988	*281,381	170,457	*451,838
1989	*273,401	168,637	*442,038

*Reflects mileage operated as reported under the FERC's classification system for natural gas pipeline companies effective beginning with the 1984 reporting year. Only major gas pipeline companies are required to file mileage.
Source: U.S. FERC Form 6, annual reports for oil-pipeline companies; Forms 2 & 2A, annual reports for natural-gas pipeline companies.

PIPELINING

Pipeline construction in 1991 and beyond*

Area	†4-10 in.	12-20 in.	22-30 in. Miles	Larger than 30 in.	Total
GAS PIPELINES					
U.S.	106	569	1,881	3,515	6,071
Canada	—	22	491	3,487	4,000
Latin America	789	519	519	—	1,827
Asia-Pacific	112	860	441	227	1,640
Western Europe	201	287	258	2,350	3,096
Middle East	—	—	—	880	880
Africa	3	9	17	741	770
Total gas	1,211	2,266	3,607	11,200	18,284
CRUDE PIPELINES					
U.S.	—	139	122	130	391
Canada	—	—	130	150	280
Latin America	—	271	—	199	470
Asia-Pacific	338	361	—	—	699
Western Europe	22	137	161	—	320
Middle East	—	—	—	1,340	1,340
Africa	1,096	632	43	23	1,794
Total crude	1,456	1,540	456	1,842	5,294
PRODUCT PIPELINES					
U.S.	246	45	—	—	291
Canada	—	—	—	—	—
Latin America	688	1,098	—	—	1,786
Asia-Pacific	70	1,337	1,517	—	2,924
Western Europe	525	500	—	—	1,025
Middle East	—	—	—	—	—
Africa	—	34	—	—	34
Total product	1,529	3,014	1,517	—	6,060
WORLD TOTALS					
Gas	1,211	2,266	3,607	11,200	18,284
Crude	1,456	1,540	456	1,842	5,294
Product	1,529	3,014	1,517	—	6,060
Total	4,196	6,820	5,580	13,042	29,638

*Projects planned to commence in 1991 and be completed in 1992 or later. Includes some probable major gas projects whose installation will begin later than 1991. † Includes some projects of less than 4 in. OD.

By October 1991, it will test the entire system from Kerteh to Johor Bahru.

Novacorp Malaysia, which took over as project manager in December 1986, will operate the system for 2 years after completion. MMC Gas consortium is contractor on the PGU project.

Costs continue to moderate

Construction cost estimates for planned U.S. interstate gas pipeline construction provide one measure of the cost of building a pipeline for any service anywhere in the world.

Regulated U.S. interstate gas pipeline companies report cost estimates to FERC as part of filings to gain approval to add or replace segments in the U.S. The latest available figures for such estimates are current through mid-1990.

For proposed U.S. gas pipeline projects in the most recent 12 month period surveyed, the average onshore construction cost was $830,882/mile, a slight decrease for the 12 month period immediately preceding.

For offshore projects, the more current figure was $834,057/mile, an increase of 13.1% from the period immediately preceding.

Combined land and offshore projects show an average cost of $830,908/mile.

Analyses of the four major categories of pipeline construction costs—material, labor, right-of-way, and miscellaneous—can also reveal trends within each group.

Right-of-way figures include the cost of obtaining right-of-way and of surveying, along with allowance for damages.

Miscellaneous generally includes engineering, supervision, contingencies, allowances for funds used during construction, administration and overhead, and FERC filing fees.

For the 171 projects surveyed for the most recent period, cost per mile data for the four categories are as follows:
• Material—Land, $342,912/mile; offshore, $345,566/mile; total, $342,934/mile.
• Labor—Land, $298,043/mile; offshore, $283,654/mile; total, $297,925/mile.
• Right-of-way and damages—Land, $159,081/mile; offshore, $18,295/mile; total, $159,305/mile.
• Miscellaneous—Land, $30,846/mile; offshore, $18,543/mile; total, $30,744/mile.

Although generalities are difficult to make concerning trends in construction costs, the cost per mile figures within a given diameter category shows that the longer the pipeline, the lower the incremental cost for construction.

Nonetheless, road, highway, and river crossings and marshy or rocky terrain each strongly affects pipeline construction costs.

Material and labor make up almost 78% of the cost of constructing land pipelines and almost 76% of the cost for offshore pipelines.

Compressor station costs make up another major cost element of gas pipelines.

PIPELINING

Pipeline construction in 1991*

Area	†4-10 in.	12-20 in.	22-30 in.	Larger than 30 in.	Total
		Miles			
GAS PIPELINES					
U.S.	1,011	1,135	930	1,782	4,858
Canada	764	102	89	564	1,519
Latin America	—	—	—	236	236
Asia-Pacific	122	329	629	485	1,565
Western Europe	236	174	290	436	1,136
Middle East	—	67	93	—	160
Africa	—	22	—	—	22
Total gas	**2,133**	**1,829**	**2,031**	**3,503**	**9,496**
CRUDE PIPELINES					
U.S.	209	159	13	60	441
Canada	—	16	—	—	16
Latin America	37	49	290	—	376
Asia-Pacific	293	210	—	—	503
Western Europe	9	39	47	—	95
Middle East	—	—	—	63	63
Africa	19	84	57	—	160
Total crude	**567**	**557**	**407**	**123**	**1,654**
PRODUCT PIPELINES					
U.S.	388	126	—	—	514
Canada	—	8	—	—	8
Latin America	194	82	37	—	313
Asia-Pacific	22	—	15	—	37
Western Europe	864	175	—	—	1,039
Middle East	—	—	—	—	—
Africa	490	309	—	—	799
Total product	**1,958**	**700**	**52**	—	**2,710**
WORLD TOTALS					
Gas	2,133	1,829	2,031	3,503	9,496
Crude	567	557	407	123	1,654
Product	1,958	700	52	—	2,710
Total	**4,658**	**3,086**	**2,490**	**3,626**	**13,860**

* Projects under way at the start of 1991 or set to begin and be completed that year. † Includes some projects of less than 4 in. OD.

Costs for new land compressor stations for the most recent period ranged from a low of $667/hp for a 12,600 hp station in Colorado to a high of $4,551/hp for a 1,200 hp station in Massachusetts. Cost per horsepower figures show no particular correlation with compressor station size or location.

After a gas pipeline is laid and compressor stations constructed, operating costs become the next consideration.

As an aid in estimating this element, transmission expenses for interstate gas pipelines for 1987 and 1988, the most recent years for which the FERC has data available, have been included in an accompanying table.

Its data are based on 248,458 miles of pipeline operated and 6.5 tcf of gas sold for 1987 majors and on 251,878 miles of pipeline operated and 6.4 tcf of gas sold for 1988 majors.

• The highest component of transmission operating expenses is transmission and compression of gas by others. For majors in 1987 this component was $4,222/mile or $162/MMcf. For majors in 1988 this component was $/3,717mile or $147/MMcf.

• The second highest cost component of transmission operating expense is gas for compressor station fuel. For majors in 1987 this component was $1,479/mile or $57/MMcf, for majors in 1988 $1,492/mile or $59/MMcf.

• The highest cost component of maintenance expense covers compressor station equipment: for majors in 1987 $794/mile or $30/MMcf, for majors in 1988 $866/mile or $34/MMcf.

• The second highest cost component for maintenance is maintenance of pipelines: for majors in 1987 $366/mile or $14/MMcf, for 1988 majors, $435/mile or $17/MMcf.

• The table indicates that the highest component of transmission expenses for natural gas pipelines is the operation expense: for 1987 majors $8,961/mile or $344/MMcf, for 1988 majors $8,647/mile or $342/MMcf;

• Unit maintenance expenses for 1987 majors were $1,495/mile or $402/MMcf, for 1988 majors $1,631/mile or $406/MMcf.

• During 1987 major natural gas pipeline companies spent slightly more than $2.2 billion for operating expenses and $371.6 million on maintenance. Total spending by 1987 majors for operation and maintenance expenses reached almost $2.6 billion.

Total operation expenses for majors in 1988 fell slightly less than 2.2 billion, a drop of almost $48.6 million (−2.2%). This follows a sharp drop of 14.3% from 1986 to 1987 noted a year earlier.

Total maintenance expenses in 1988 rose to slightly more than $410 million from a 1987 cost of $371 million.

Total transmission expenses showed a slight decline in 1988 to $2.588 billion.

• On a unit basis, the total 1987 transmission expense for majors was $10,457/mile or $402/MMcf, for 1988 $10,278/mile or $407/MMcf of gas sold.

WORLDWIDE GAS PROCESSING

Capacities as of January 1, 1990, and average production

Country	No. plants	Gas capacity (MMcfd)	Gas through-put (MMcfd)	Ethane	Prop.	Isobut.	Process or unsplit butane	LP-gas mix	Raw NGL mix	Debut. nat. gaso.	Other	Total products
United States												
Alabama	7	255.0	200.7	....	164.3	....	129.1	6.4	39.9	93.2	434.7	867.6
Alaska	4	4,073.5	3,946.0	....	2.6	....	....	....	2,356.6	....	....	2,359.2
Arkansas	4	933.0	578.4	0.3	7.5	1.6	6.5	....	48.8	6.0	3.0	73.7
California	34	1,032.7	655.3	....	416.0	36.8	245.6	77.0	331.2	83.2	11.4	1,201.2
Colorado	42	1,101.5	687.8	63.1	248.0	....	98.9	44.4	655.8	57.6	18.3	1,186.1
Florida	2	890.0	755.0	71.1	115.8	22.0	79.0	....	....	70.1	....	358.0
Kansas	25	5,448.1	3,786.6	360.2	937.0	95.5	408.5	47.2	378.2	302.7	1.9	2,531.2
Kentucky	2	71.0	45.6	....	....	....	....	....	117.5	....	....	117.5
Louisiana	80	18,720.5	11,059.8	1,152.6	1,094.5	344.5	361.0	35.5	5,698.3	668.3	320.1	9,674.8
Michigan	31	4,824.1	1,703.9	....	24.6	....	....	61.6	665.1	15.0	38.4	804.7
Mississippi	7	902.2	373.7	....	9.3	....	8.0	....	15.8	9.1	2.3	44.5
Montana	7	27.0	10.6	....	29.6	6.8	5.0	17.7	18.2	11.1	1.0	89.4
Nebraska	1	1.5	0.9	....	....	....	....	....	0.4	....	....	0.4
New Mexico	28	3,176.0	2,210.4	160.4	130.6	11.5	41.9	34.3	4,633.9	44.0	35.3	5,091.9
North Dakota	8	165.0	108.9	....	150.4	....	84.1	....	113.4	23.6	13.2	384.7
Oklahoma	103	4,433.5	2,644.7	782.2	679.3	100.4	259.8	188.4	4,230.5	303.9	220.9	6,765.4
Pennsylvania	2	14.0	8.3	....	4.7	....	3.0	....	....	4.1	....	11.8
Texas	315	16,621.4	10,691.3	3,477.7	3,325.5	563.5	1,544.1	3967.7	15,270.6	2,249.6	1,420.1	28,247.8
Utah	12	537.5	266.9	....	95.6	....	22.8	....	135.1	43.8	1.7	299.0
West Virginia	7	397.0	324.9	235.6	143.7	24.1	44.8	88.2	256.2	....	....	792.6
Wyoming	40	3,489.5	2,687.0	....	236.8	10.8	155.5	73.5	2,518.4	135.1	257.4	3,387.5
Total U.S.	761	67,114.0	42,756.2	6,303.2	7,815.8	1,217.5	3,497.6	1,070.9	37,483.9	4,120.4	2,779.7	64,289.0
Abu Dhabi	5	1,602.0	1,269.0	....	....	....	....	....	....	546.0	....	546.0
Algeria	4	5,335.2	4,268.2	....	....	....	....	632.8	....	....	5,449.2	6,082.0
Argentina	14	1,488.4	1,498.9	687.7	497.2	....	334.8	....	....	319.8	0.8	1,840.3
Australia	5	2,573.0	1,376.0	279.0	441.0	....	249.0	610.3	....	1,058.5	850.0	3,487.8
Austria	1	17.5	13.7	....	....	....	....	26.7	....	28.4	....	55.1
Bahrain	1	168.0	155.0	....	....	....	....	25.6	....	....	....	25.6
Bangladesh	2	240.0	178.0	....	....	....	....	....	....	....	8.2	8.2
Bolivia	2	474.0	177.3	....	1.4	2.5	....	161.7	....	61.5	136.9	364.0
Brazil	11	449.7	433.6	0.9	....	....	....	519.1	209.6	135.3	9.6	874.5
Brunei	2	1,032.0	832.4	....	9.4	....	....	....	50.1	252.1	....	311.6
Canada												
Alberta	473	22,630.2	14,033.4	6,419.9	3,248.3	870.4	814.2	1,214.8	4,160.0	4,123.1	671.4	21,522.1
British Columbia	14	1,780.3	856.5	....	192.5	....	65.9	3.5	....	31.0	....	292.9
Northwest Territories	2	219.9	28.3	....	....	....	....	....	....	....	....	....
Saskatchewan	23	295.4	194.3	....	....	....	....	12.9	104.7	0.3	18.0	135.9
Chile	2	470.0	423.0	....	190.0	....	120.0	....	....	106.0	....	416.0
China, Taiwan	4	270.3	60.3	13.2	26.4	0.8	16.8	....	8.7	....	12.3	78.2
Colombia	6	348.0	218.0	132.8	52.0	....	58.5	43.5	....	70.7	1.1	358.6
Denmark	1	450.0	240.0	....	....	....	....	....	....	....	100.0	100.0
Dominican Republic	1	....	....	....	....	....	....	....	....	....	277.2	277.2
Dubai	1	330.0	330.0	....	....	....	....	....	....	....	650.5	650.5
Ecuador	2	41.0	20.5	....	....	....	....	97.1	....	37.3	....	134.4
Egypt	8	1,320.0	1,188.0	....	....	....	....	691.7	....	....	1,158.5	1,850.2
France	1	570.0	563.0	....	139.0	....	132.0	....	....	270.0	....	541.0
Germany	3	366.0	274.0	....	....	....	....	....	....	....	....	....
Greece	2	39.0	31.2	....	....	....	....	....	72.0	....	....	72.0
Hungary	12	1,097.2	475.3	....	65.7	6.3	70.0	....	367.4	819.2	103.6	1,432.2
Indonesia	10	3,900.5	3,496.2	17.3	345.6	113.4	0.9	51.6	1,462.4	24.0	5.2	2,020.4
Italy	17	7,614.0	1,587.0	....	....	....	....	2.7	8.0	7.5	....	18.2
Kuwait	1	1,815.0	1,452.0	....	1,722.0	....	980.0	....	....	811.8	....	3,513.8
Libya	7	1,851.0	1,227.0	....	....	....	....	37.8	5,577.0	....	551.0	6,165.8
Malaysia	2	282.0	105.0	161.4	23.2	136.3	28.2	....	....	154.1	....	503.2
Mexico	11	4,649.0	3,370.4	12,704.9	....	....	....	....	....	....	9.5	12,714.4
Netherlands	1	350.0	35.0	....	....	....	....	....	....	....	4.2	4.2
New Zealand	6	1,152.0	533.0	....	39.1	....	1.8	134.7	....	18.6	733.7	927.9
Norway	2	950.0	507.0	....	682.0	108.0	256.0	....	200.0	193.0	....	1,439.0
Oman	8	626.0	309.4	....	....	....	....	....	202.0	43.0	....	245.0
Pakistan	7	430.0	187.7	....	....	....	....	48.8	6.1	....	9.1	64.0
Peru	2	70.0	37.0	....	3.4	....	1.0	5.6	28.9	....	4.0	42.9
Qatar	2	579.0	232.0	....	434.6	....	255.5	....	....	173.0	....	863.1
Saudi Arabia	5	4,300.0	....	....	....	....	....	....	....	....	....	....
Sharjah	2	990.0	925.0	....	290.0	....	214.0	....	....	....	1,239.0	1,743.0
Spain	3	428.0	177.0	....	....	....	....	....	....	....	73.0	73.0
Thailand	2	850.0	550.0	....	....	....	....	400.4	8.4	84.0	....	492.8
United Kingdom	9	3,732.0	1,537.0	....	332.0	....	291.0	....	150.0	1,748.5	51.8	2,573.3
Venezuela	14	2,913.0	2,557.0	213.3	197.4	....	113.4	231.3	3,274.6	171.9	....	4,201.9
Yemen	1	400.0	324.0	....	....	....	....	....	120.0	189.0	....	309.0
Total, outside U.S.	712	81,488.6	48,286.6	20,630.4	8,932.2	1,237.7	4,003.0	4,952.6	16,009.9	11,477.6	12,127.8	79,371.2
World Total	1,473	148,602.6	91,042.8	26,933.6	16,748.0	2,455.2	7,500.6	6,023.5	53,493.8	15,598.0	14,907.5	143,660.2

INDEX

A

28th of April oil field, 148-149
Abderrahman formation, 138
Abu Dhabi, 105, 195, 282, 338
Abu Qir gas field, 126
Abu Safah oil field, 94
Accidents (refinery/petrochemical plant, U.S.), 54-55
Afia oil field, 138
Africa, 124-139, 317, 324: production statistics, 317; refining capacity, 324. SEE ALSO individual countries.
Agogo oil/gas field, 89
Aguaragua gas field, 106
Aguaytia gas/condensate field, 120
Ahuroa/Tariki gas condensate field, 18
Air pollution, 53-54
Alabama, 10, 36, 45, 53, 345
Alaska, 18, 20, 32, 34, 36-37, 39, 42, 53, 197-199, 345
Alba gas/condensate field, 128
Albacora oil field, 111
Albania, 156, 338: Durres Basin, 156
Alberta, 16, 60-61
Algeria, 23, 124-125, 195-196, 282, 338: LNG/LPG, 124; seismic activity, 124-125
Alkylation growth, 331
Allies attack (Iraq), 205
Almy formation, 18, 52
Alternate motor fuels, 261-262
Amauligak oil field, 58
Amethyst gas field, 171, 182
Anglia gas field, 182
Angola, 5, 125, 282-283, 338
Aoshan Island, 64
Appalachian Basin, 19
Aptian formation, 138
Arab limestone, 94
Arabian Gulf, 274, 277
Arctic National Wildlife Refuge Coastal Plain, 36
Argentina, 106-107, 283, 338: exploration/production activities, 106; gas reserve development, 106; production statistics, 107; foreign investment, 107
Arizona, 345
Arkansas, 10, 12, 345
Arkoma Basin, 53
Arkoma Basin pipeline, 53
Arkticheskoye gas field, 145
Aromatics contents (fuel), 54
Arun gas field, 91

Asab oil field, 105
Ash Coulee oil field, 14
Ash Shaer gas field, 104
Ashtart oil field, 138
Asia-Pacific, 74-93, 318, 323, 333: production statistics, 318; refining capacity, 323; refining industry, 333. SEE ALSO individual countries.
Atascosa Co, TX, 5
Atoka Co, OK, 53
Auca oil field, 114
Austin Chalk, 5-8, 10, 19, 45, 48, 53
Austral Basin, 106
Australia, 5, 74-75, 194-195, 283-285, 338: Timor Sea, 74; Western Australia, 74; Skua oil field, 74; Cowie oil field, 74; Yammaderry oil field, 74; Great Barrier Reef, 75
Austria, 158, 285, 338
Awali oil field, 94
Ayoluengo off field, 171
Azerbaijan, 148
Azua Basin, 114

B

Bach Ho (White Tiger) oil field, 93
Bahrain, 94, 285, 338: Khuff formation, 94; Jarim reef, 94
Bakken shale, 12
Baltic Sea, 150
Bangladesh, 75-76, 338
Barbados, 285, 338
Barents Sea, 143-145, 168, 184, 188
Barque gas field, 182
Barrow Island oil field, 74
Bassein gas field, 77
Bay of Campeche, 118
Bayou Choctaw, 32
Bearpaw arch, 14
Beaufort Sea, 36, 58
Beaver Co, OK, 15
Beghraji oil field, 77
Beibu Gulf, 67
Belco oil field, 120
Belgium, 158, 335, 338: refining industry, 335
Belize, 107: Peten Basin, 107
Belyi Island, 143, 145
Ben Nevis-Avalon sandstone, 57-58
Benin, 285
Benzene, 54
Bermejo oil field, 114
Berthoud oil field, 14
Big Injun formation, 19

Billings Co, ND, 12, 14
Bioco Island, 128
Birsa oil field, 138-139
Bitumen recovery, 61
Black Sea, 143, 157
Blaine Co, MT, 14
Bohai Sea, 66-67, 69
Bolivia, 107, 285, 338
Bombay High gas field, 77
Bon Secour Bay gas field, 220
Boulder Co, CO, 14
Boulder oil field, 14
Bouri oil field, 131-132, 165-166
Bovanenkovskoye gas/condensate field, 145
Bowes-Sawtooth formation, 14
Bozhong 28-1 oil field, 67
Bozhong 34-2 oil field, 67BP America (public relations response), 254-255
Brazil, 5, 107-108, 111, 286-287, 338: financial problems, 108; Campos Basin, 108; oil import reduction, 108, 111; exploration/production activities, 111; Santos Basin, 111
Brazos Co, TX, 7, 10, 48
British Gas plc, 56-57
Bromide sand, 15
Brookeland oil field, 10
Brown dolomite, 48
Brunei, 287, 339
Bryan Co, OK, 53
Bryan Mound, 34
Bu Attifel gas field, 130
Bu Hasa oil field, 105
Buckrange (Ozan) formation, 12
Buda formation, 7
Bufaloa oil field, 125
Bukha gas/condensate field, 101
Bulgaria, 156, 339
Buoyant turret mooring, 216
Burleson Co, TX, 5, 7, 45, 48
Byeloostrovskaya formation, 145

C

Calhoun Co, WV, 19
California, 18, 34, 36, 42-45, 53, 345
California-West Texas pipeline, 53
Camar oil field, 79
Cameroon, 126, 287, 339
Camisea gas/condensate field, 119-120
Campos Basin, 108, 111
Canada, 5, 16-18, 23, 55-61, 287-288: gas exports, 55-56; offshore development, 57-59; petroleum industry per-

formance, 59; reserves replacement, 59-60; carbon dioxide flooding 60-61
Candon Alfa gas field, 106
Cano Limon oil field, 112
Cape Taran, 148
Cape York Peninsula, 75
Capline pipeline, 34
Carbon dioxide flooding, 60-61
Carbon dioxide injection, 227-228, 242-245
Carson Co, TX, 48
Carter Co, OK, 15
Casabe oil field, 112
Casablanca oil field, 171
Caspian Sea, 143, 147-149, 153
Cauvery Basin, 77
Center oil field, 10
Central (Salin) Basin, 87
Chambira oil field, 119-120
Chemical EOR projects (U.S.), 230-231, 244-247
Chengbei oil field, 67
Cherrife gas field, 104
Chester formation, 15
Chicontepec oil field, 118
Chile, 106, 111, 288, 339: Magallanes Basin, 111
China (Taiwan). SEE Taiwan.
China, 5, 62-72, 288-289, 339: production decline, 62-63; exports, 63-64; refining, 64, 72; Tarim Basin, 64-65; onshore potential, 65-66; Shengli producing area, 66; offshore potential, 67; Huizhou producing area, 67; Bohai Sea, 67, 69; Dongting Basin, 69; Hainan Island, 69-70; natural gas, 70-72; Sichuan producing area, 70-72; ethylene projects, 72
Chindwin Basin, 87
Chronology of events, 190
Clean Air Act, 53-54
Clean coal, 260-261, 264-268
Clean Coal Technology program, 264
Clear Springs oil field, 12
Clipper gas field, 182
Coal, 264-268: Clean Coal Technology program, 264; demonstration projects, 264-267; emissions control, 265; cofiring, 265-268; combustion methods of coal cleaning, 266-267; flue gas scrubber, 267-268; post-combustion innovations, 268; coal conversion, 268
Coal cleaning (combustion methods), 266-267
Coalbed methane, 30, 53
Cochin High oil field, 77
Cofiring process, 260-261, 265-268
Cohasset oil field, 57-58
Colombia, 112, 289-290, 339: Llanos Basin, 112
Colorado, 14, 45, 52-53, 345
Columbia Co, AR, 12
Colville Delta oil field, 42
Communications recommendations, 252-253
Completed interval length, 8
Completions/footage (Gulf of Mexico), 223
Computerized fracturing, 30
Con Son Island, 93
Congo, 126, 290, 339
Connecticut, 55
Consumers Gas Co. Ltd., 56-57
Consumption statistics, 320
Cooper Basin, 74
Coral oil field, 111
Corrosion resistant alloys, 222
Costa Rica, 112-113, 339: Limon Sur Basin, 113; Limon Norte Basin, 113
Cowie oil field, 74
Cozzette formation, 19
Creek Co, OK, 15
Crisis management, 253-254
Crude capacity changes (U.S. refining), 328-332, 335
Crude distillation unit, 54
Crude oil prices, 314
Cruse formation, 10
Cuba, 113-114, 339
Cullen report, 182
Cummings sand, 16
Cuntala oil field, 125
Cuyabeno oil field, 114
Cuyana Basin, 106
Cyprus, 339
Czechoslovakia, 156, 339: Gbely oil field, 156; Vienna Basin, 156

D

Dagang oil field, 66
Daqing area, 63, 65-66, 70
Das Island, 105
Dauletabad-Donmez oil/gas field, 143
Davis oil field, 15
Deep water pipelines, 225
Deep water port study, 214-215
Deep water projects, 222-223
Dekalb gas field, 19
Delaware, 54, 345
Denmark, 5, 158-159, 181, 290, 339: Greenland exploration, 160
Denver Basin, 15, 52
Devonian shale, 19
Diesel fuel, 53-54
Dimmit Co, TX, 5, 7
Divide Creek gas field, 19
Dominican Republic, 114, 339: Azua Basin, 114; San Pedro Basin, 114
Dongting Basin, 69
Downhole motors, 19
Drainage radius, 5
Drawdown test (Strategic Petroleum Reserve), 20, 30, 32, 34
Drilling activity (U.S.), 30
Drilling program, 5
Dubai, 105, 290
Dunn CO, ND, 14
Duri oil field, 228

Durres Basin, 156

E

EA oil field, 136
East Brae gas/condensate field, 181
East Cameron Block 265, 219
East China rift system, 69
East Govi Basin, 85
East Lombo oil field, 125
East Shetlands Basin, 17
Eastern Europe, 154-157, 206: Albania, 156; Bulgaria, 156; Czechoslovakia, 156; Hungary, 156-157; Poland, 157; Romania, 157; Yugoslavia, 157; war effects on, 206
Eastern Europe-U.S.S.R., 319, 325: production statistics, 319; refining capacity, 325
Economic growth (U.S.), 23, 26, 30
Ecuador, 114-115, 290-291, 339-340: exploration/production activity, 114-115
Edikan oil field, 138
Edop oil field, 136
Egypt, 126, 291-292, 340
Ekofisk gas field, 180-181
Ekofisk oil field, 188
El Borma formation, 138
El Borma oil field, 138
El Gueria formation, 138
El Salvador, 340
Elaine oil field, 7
Electric power supply/demand (U.S.), 256-258
Emissions control, 53-54, 265: coal combustion
Endicott oil field, 39
Energy consumption (U.S.), 20, 23, 26, 30, 43
Energy prices comparison, 314
Enhanced oil recovery, 227-249: carbon dioxide injection, 227-228, 242-245; hydrocarbon gas injection, 227-228, 242-245; Duri oil field, 228; U.S. steam projects, 230, 238-241; U.S. chemical projects, 230-231, 244-247; U.S. EOR projects, 230-233, 238-249; Canadian
EOR projects, 232-235; other EOR projects, 234-239; U.S. thermal EOR, 238-241; U.S. gas injection EOR, 242-245; U.S. chemical EOR, 244-247; U.S. completed/terminated projects, 246-249; Canadian heavy oil EOR projects, 248-249
Enriquillo Basin, 114
Entrada sandstone, 15
Environmental issues, 36, 44, 140, 220, 222, 225, 250-255: petroleum industry image, 250-252; media relations, 252; communications recommendations, 252-253; crisis management, 253-254; Mobil's public relations

359

relations response, 254-255
Environmental Protection Agency, 53-54
EOR projects (Canada), 232-235
EOR projects (U.S.), 230-233, 238-249
Equatorial Guinea, 126, 128: Alba gas-/condensate field, 128
Erawan gas field, 91-92
Eren oil field, 66
Eromanga Basin, 74
Escambia Co, AL, 10, 48
Espirito Santo Basin, 111
Essungo oil field, 125
Estrela do Mar oil field, 111
Ethiopia, 126, 340
Ethylene, 72, 96
Europe/North Sea, 158-188, 317, 321, 332-333: production statistics, 317; refining capacity, 321; refining industry.
SEE ALSO individual countries.
Events chronology (1990), 190-191
Exploration economics, 272-273
Exploration/production replacement efficiency (Canada), 59-60
Exploration/production spending, 272-273
Exxon Valdez oil spill, 42
Ezzaouia oil field, 139

F

F3 oil/gas field, 188
Fahud oil field, 100
Fairway gas field, 223
Fayette Co, TX, 7, 48
Fertilizer, 72
Field experience (horizontal drilling), 5-18
Field history, 5-6
Fines/penalties, 55
Finland, 160, 340
Floating production storage and offloading systems, 215-216
Flue gas scrubber, 267-268
Fort Jessup oil field, 10
France, 5, 160, 162, 292, 340: refining-/marketing, 162
Franklin Co, MS, 12
Freight rates, 212
Frio Co, TX, 5, 7-8, 48
Frontera oil field, 114
Fuel oil demand, 20
Fuel reformulation, 53-54
Fushan Basin, 69

G

Gabon, 5, 128, 130, 292-293, 340
Gallup siltstone, 15
Garfield Co, OK, 15
Gas applications (horizontal drilling), 19
Gas exports (Canada), 55-56
Gas handling expansion project (Prudhoe Bay), 37, 39
Gas imports (U.S.), 55-56

Gas injection, 227-228, 242-245: U.S. projects, 242-245
Gas pipeline projects (U.S.), 53
Gas processing capacities, 357
Gas producing statistics, 269
Gas reserve, 6
Gas surplus, 262-263
Gasoline demand, 20
Gasoline price, 54
Gasoline tax, 53
Gavilan-Mancos oil field, 15
Gbely oil field, 156
Georgetown formation, 7
Georgia, 53, 345
Geothermal power, 30
Germany, 5, 177, 309-310, 340: refining, 177
Ghadames Basin, 124
Ghana, 293, 340
Ghawar oil field, 102
Giddings oil field, 5, 8, 45, 48
Gingin (Bullsbrook) anticline, 74
Golden Valley Co, ND, 12
Gonzales Co, TX, 7-8
Grand Banks area, 57
Grant Co, OK, 15
Grantsville gas field, 19
Great Barrier Reef, 75
Greece, 163, 293, 340: Prinos oil field, 163
Green Canyon Block 6, 219-220
Green Canyon Block 18, 220
Green Canyon Block 29 project, 219, 223-224
Greenland, 160
Grondin oil field, 130
Groningen gas field, 166
Guatemala, 293, 340
Guizhou Province, 72
Gulf Coast, 30
Gulf of Gabes, 138
Gulf of Hammamet, 139
Gulf of Martaban, 91
Gulf of Mexico, 36, 53, 218-225: well completion, 218; rig activity, 218; upstream work increasing, 218-219; deep water technology, 219-220; Mobile Bay, 220, 222; corrosion resistant alloys, 222; ultradeepwater projects, 222-223; Green Canyon Block 29 project failure, 223-224; deep water pipelines, 225; offshore acreage access, 225
Gulf of Suez, 126
Gulf of Thailand, 91-92
Gulf of Valencia Basin, 171
Guyana, 115: Takutu graben area, 115
Gwydyr Bay oil field, 42
Gyda oil/gas field, 188

H

Hainan Island, 69-70
Hainan Province, 69
Haltenbanken area, 184

Hanna Basin, 15, 52
Harken formation, 94
Hawaii, 345
Hay Draw oil field, 12
Heavy oil EOR projects (Canada), 248-249
Heavy oil steamflooding, 18
Heidrun oil/gas field, 168
Heimdal gas field, 181
Hemphill oil field, 10
Hereford oil field, 14
Hibernia oil field, 57-59: development schedule, 59
Hibernia sandstone, 57-58
Hides gas field, 89
High Island A-573 oil field, 220
High Island Block A-382, 220
Historical reserve data, 349
Hith formation, 94
Honduras, 340
Horizontal drilling, 5-19, 30, 45, 48, 52: field experience, 5-18; technology evolution/concerns, 18-19; steamflooding, 18; gas applications, 19; forecast, 19
Horizontal wells, 5-20, 30, 45, 48, 52: completion, 19; performance, 45, 48, 52
Horndean oil field, 177
Hostile environment, 222
Huallaga Basin, 119
Huizhou area, 67
Huizhou oil field, 67, 217
Hunan Province, 69
Hungary, 156-157, 340: Pannonian Basin, 156
Hunton formation, 15
Huxford oil field, 10, 48
Hydraulic fracturing, 37
Hydroelectric power, 26, 30

I

Iagifu-Hedinia oil/gas field, 89
Illinois, 53-54, 345
Illizi Basin, 124
Ime oil field, 138
Imported oil/gas, 20, 23
India, 76-79, 293, 335, 340: energy plan, 76; petroleum industry performance, 76-77; gas utilization, 77; exploration/development programs, 77-78; refining, 78-79; refining industry, 335
Indian Ocean, 101, 138
Indiana, 345-346
Indonesia, 5, 18, 79, 82, 193-194, 293-296, 340-341: Camar oil field, 79; Intan oil field, 79; Widuri oil field, 79; refining, 82; liquefied natural gas, 193-194
Indus Basin, 88
Infantas oil field, 112
Injection well, 18
Inner Mongolia area, 66, 70

Intan oil field, 79
Internal turret mooring, 216-217
International Energy Agency (war), 205
Iowa, 53
Iran, 94, 96, 296-297, 341: reconstruction, 94, 96; refinery, 96
Iraq, 43, 97, 208, 341: Yamama formation, 97; Suba oil field, 97
Iraq-Kuwait war, 201-209: Iraqi invasion, 201-202; Iraqi facilities damage, 208
Ireland, 163, 165, 341: Kinsale Head gas field, 163; Celtic Sea, 165
Iroquois Gas Transmission System, 55
Isongo sandstone, 128
Israel, 341
Italy, 5, 165, 341: production, 165; refining, 165
Ivory Coast, 297, 341
Iyak oil field, 137

J

Jabiru Venture platform, 74
Jamaica, 341
Japan, 63, 82-83, 194, 206, 297, 341: war effects on, 206
Jarim reef, 94
Jasper Co, MS, 12
Java Sea, 79
Jawf oil field, 105
Jefferson Co, OK, 16
Jerneh gas field, 84
Jinfeng oil field, 69
Jingzhou 21-1 oil field, 69
Jinzhou 20-2 oil field, 69
Joffre Viking oil field, 60-61
Jones Co, MS, 12
Jordan, 97, 99, 298, 341: Risheh gas field, 99
Junggar Basin, 66

K

Kandkhot gas field, 88
Kansas, 346
Kaptev Sea, 145
Kara Sea, 143-145
Karnes Co, TX, 8
Kentucky, 346
Kenya, 130, 340
Kern River oil field, 44
Kern River pipeline, 53
Khabbaz oil field, 97
Kharasavei gas/condensate field, 145
Kharg Island, 96
Khuff formation, 94, 105
Kingfisher Co, OK, 15
Kittiwake oil field, 182
Kolguyev Island, 145
Kotter oil field, 17-18
Krishna-Godavari Basin, 77-78
Kruzenshtern gas/condensate field, 145
Kuldiga oil field, 148
Kuparuk River oil field, 39
Kuwait, 43, 99, 201-209, 298, 342: Iraqi invasion, 201-202; United Nations, 202; oil supply, 202-203; OPEC response, 203; production increases, 203; market allocation, 203; Kuwaiti Petroleum Co., 203-204; U.S. reaction, 204-205; Allies attack, 205; International Energy Agency, 205; Strategic Petroleum Reserve drawdown (U.S.), 205; Saudi Arabia production cuts, 205; effect on Japan, 206; effect on Eastern Europe, 206; U.S. refining response, 207; Kuwaiti refinery damage, 207-208; Iraqi facilities damage, 208; shipping threats, 208-209; oil spill, 209; oil price, 209; OPEC's future, 209
Kuwaiti Petroleum Co. (war), 203-204

L

La Barge formation, 52
La Barge platform, 18
La Cira oil field, 112
Lafayette Co, AR, 12
Lake Maracaibo, 123
Laos, 83-84
Laramie Co, WY, 14
Larimer Co, CO, 14
LaSalle Co, TX, 5, 7
LaSalle Parish, LA, 10
Latimer Co, OK, 53
Latin America, 106-123, 319, 322, 333: production statistics, 319; refining capacity, 322; refining industry, 333. SEE ALSO individual countries.
Latvia, 148
Lebanon, 342
Lee Co, TX, 7, 48
Lekhwair oil/gas field, 99-101
Leningradskaya formation, 143
Liaodong Bay area, 67
Liaohe area, 65-66
Liaoning province, 66
Liberia, 342
Libertador oil field, 114
Libya, 130-132, 298, 342: gas projects, 130; Tahaddi gas field, 130-131; Mediterranean pipeline, 131-132
Lifting cost, 6
Limon Norte Basin, 113
Limon Sur Basin, 113
Liquefied natural gas, 23, 192-200:
LNG/LPG trade outlook, 192-193; Indonesia, 193-194; Japan, 194; Australia, 194-195; Abu Dhabi, 195; Algeria, 195-196; Nigeria, 196-197; Norway, 197; Venezuela, 197; Alaska, 197-199; U.S. imports, 199-200; Taiwan imports, 200
Lisburne oil field, 39
Lithuania, 147-148
Liuhua 11-1 oil field, 67
Llanos Basin, 112
LNG/LPG imports (Taiwan), 200
LNG/LPG imports (U.S.), 199-200
LNG/LPG trade outlook, 192-193
Logger oil field, 18
Loma de la Lata gas field, 106
Lorena oil field, 15
Los Angeles Basin, 45
Louisiana, 18, 30, 45, 54, 346
Love Co, OK, 15
Lower Amazon Basin, 111
Luna gas field, 165

M

Mackenzie Delta, 58
Macrofracture, 8
Madagascar, 132, 342
Madison formation, 14
Madre de Dios Basin, 119
Magallanes Basin, 106, 111
Malaysia, 5, 84, 298-299, 342: Jerneh gas field, 84; South China Sea, 84
Malta, 165-166
Mancos shale, 15
Manzanares pipeline, 53
Maranon Basin, 119
Marib oil field, 105
Market allocation (war), 203
Market shares (unleaded gasoline), 334
Marlim oil field, 108, 111
Martaban Basin, 91
Martaban Bay, 91
Martinique, 342
Mary Ann gas field, 220
Massachusetts, 55
Maverick Co, TX, 5
McKee formation, 18
McKee oil field, 18
McKenzie Co, ND, 12
McKinley Co, NM, 15
Media relations, 252
Mediterranean Sea, 126
Medium radius well, 18
Meghalaya area, 76
Merluza gas/condensate field, 111
Mesa Co, CO, 19
Methanol, 96, 101, 168
Methyl tertiary butyl ether (MTBE), 54
Mexico, 115, 117-118, 299-300: exploration/production activities, 115, 117; capital needs, 117; domestic demand, 117-118; hydrocarbon potential, 118; U.S.-Mexico relations, 118
Michigan, 45, 56, 346
Midale oil field, 60
Middle East, 23, 94-105, 316, 325: production statistics, 316; refining capacity, 325. SEE ALSO individual countries.
Middle East crisis, 30, 44-45, 201-209
Middle Magdalena Basin, 112
Midway-Sunset oil field, 274, 276-277
Miller gas field, 176
Mime oil field, 184
Minnesota, 56, 346
Mishrif oil field, 97
Miskar gas field, 138

Mission Canyon formation, 12, 14
Mississippi, 10, 45, 346
Mississippi Canyon Block 118, 224
Mississippi limestone, 15
Mizoram area, 76
Mobil Oil Corp. (public relations campaign), 254
Mobile Bay, 53, 220, 222
Molve gas field, 157
Mond oil field, 96
Mongolia, 84-85, 87: sedimentary basins, 85; hydrocarbon potential, 85; exploration history, 85, 87
Montana, 12, 45, 346
Moran oil field, 77
Morecambe Bay, 176
Morocco, 132, 300
Mountrail Co, ND, 14
MTBE capacity (U.S.), 332
Muharraq Island, 94
Murchison oil field, 17
Murray Co, OK, 15
Myanmar, 87, 91, 274, 276, 300: hydrocarbon potential, 87; field data, 87

N

Nagaland area, 76
Nahorkatiya oil field, 77
Narimanam oil field, 77
Nassan Island, 94
Natih gas field, 100
National Energy Board, 55-56
Natural gas, 256-263: electric power supply/demand (U.S.), 256-258; regulations, 258-260; clean coal industry, 260; cofiring process, 260-261; alternate motor fuels, 261-262; gas surplus, 262-263
Natural gas liquids, 20
Neitinskoye gas field, 145
Netherlands, 5, 17-18, 166, 181, 188, 300: Groningen gas field, 166
Netherlands Antilles, 335, 342: refining industry, 335
Neuquen Basin, 106
Neutral Zone, 5, 301
Nevada, 36, 53, 346
New Hampshire, 55
New Jersey, 54-55, 346
New Mexico, 15, 45, 346
New York, 54-56, 346
New Zealand, 5, 18, 87-88, 301, 342: Waihapa gas/condensate field, 87
Newfoundland, 57-58
Niakuk oil field, 42
Nicaragua, 342
Nido-Galoc formation, 89
Niger Delta, 135
Nigeria, 132-138, 196-197, 301-303, 342: exploration/production activities, 132-134; spending plan, 134-135; production plan, 135-136; gas development, 136; refining, 138
Nile Delta, 126
Ninian oil field, 182
Niobrara formation, 14, 52
Nitrogen jetting, 12
Noroeste Basin, 106
Norphlet formation, 219, 222
North America, 20-61, 316, 321: production statistics, 316; refining capacity, 321. SEE ALSO Canada AND United States.
North Central Gulf gas field, 220
North Carolina, 53
North Dakota, 10, 12, 45, 346
North gas field, 102
North Korea, 341
North Louisiana Basin, 10
North Ravenspurn gas field, 182
North Sea, 17-19, 168, 177-188: United Kingdom cogeneration, 177, 180; Norway, 180, 184, 188; Troll gas field, 180; Ekofisk gas field, 180-181; Denmark, 181; Netherlands, 181, 188; field development, 181-183; Cullen report, 182
North Slope, 36-37, 39, 42
North Valiant gas field, 17, 19
Northeast Clay gas field, 19
Northwest Gulf gas field, 220
Northwest Shelf, 74
Norway, 166, 168, 180, 184, 188, 197, 303, 342: onshore facilities, 168
Norwegian Sea, 168
Nova Scotia, 57
Novaya Zemlya Island, 144
Novaya Zemlya trough, 145
Nowata Co, OK, 15
Nuclear power, 30
Nyalga Basin, 87

O

Oakdale oil field, 15
Observation well, 18Occupational Safety and Health Administration (OSHA), 54-55
Odudu oil field, 138
Offshore acreage access, 225
Offshore loading systems, 215-217: turret systems, 215-217
Ohio, 54, 346-347
Oil consumption statistics, 320
Oil inventories, 206
Oil price, 36, 209: war effects, 209
Oil production statistics. SEE Production statistics.
Oil reserves, 5, 34, 36-37, 59-60, 280-281, 312-313: United States, 34, 36-37; Canada, 59-60; statistics, 280-281, 312-313
Oil sands recovery project, 61
Oil spill liability law (U.S.), 213-214
Oil spill (war), 209
Oil supply (war), 202-203, 206
Oil/gas supply and demand (U.S.), 20, 23, 26, 30, 32, 43
Oklahoma, 15-16, 45, 53-54, 347
Oklahoma-Arkansas pipeline, 53
Olla oil field, 10
Oman, 5, 99-101, 303-304, 342: Lekhwair oil/gas field, 99-101
Omar oil field, 104
Ontario, 55-56
OPEC response (war), 203
OPEC's future, 209
Oregon, 347
Orimulsion, 123
Orinoco Belt, 123
Osage Co, OK, 15
OSLO oil sands plant, 61
Oso gas condensate field, 136
Overbalanced drilling, 19

P

Pakistan, 88, 304, 343: Indus Basin, 88; Sui gas field, 88
Palanca oil field, 125
Palawan Island, 89
Palawan shelf, 274, 279
Panama, 343
Pannonian Basin, 156-157
Panuke oil field, 57-58
Papers Wash oil field, 15
Papua New Guinea, 88-89: Hides gas field, 89
Papuan Basin, 89
Paraguay, 118, 343
Parana Basin, 111
Pars gas field, 96
Pavayacu formation, 119
Pearsall oil field, 5, 7-8, 48
Pechora Sea, 145
Pelican Lake oil field, 16
Pendleton-Many oil field, 10
Pennsylvania, 54, 347
Persian Gulf crisis, 26, 63, 334: refining industry effects, 334
Perth Basin, 74
Peru, 118-121, 304, 343: investment bariers lifted, 119; production statistics, 119; exploration/production activities, 119-121; refining, 121
Peten Basin, 107
Petroleum industry image, 250-252
Petroleum product prices, 314
Philippines, 63, 89-90, 305, 343: Nido-Galoc formation, 89
Piceance Basin, 19
Pipeline capacity expansion, 55-56
Pipeline construction, 53, 350-356: United States, 53, 350-353; U.S. liquids pipelines investment, 350; U.S. gas transmission expenses, 353; Canada, 353-354; Europe, 354; Far East, 354-355; construction costs, 355-356; construction projects, 355-356
Piper Alpha accident study, 182
Piper formation, 14
Point Arguello oil field, 43
Point McIntyre oil field, 42
Point Thomson-Flaxman Island oil field, 42

Poland, 157, 343
Polyethylene, 54, 82-83
Polystyrene, 82
Pontotoc Co, OK, 15
Portugal, 168-170, 343
Post-combustion processes, 268
Potiguar Basin, 111
Pranhita-Godavari Basin, 78
Price freezes (refining industry), 334-335
Price history (oil/gas/gasoline), 315
Prinos oil field, 163
Production allowable, 5, 36
Production decline (U.S.), 20, 34, 36-37, 39, 42-45
Production incentive, 36
Production increases (war), 203
Production logging, 16
Production rate, 34, 36-37, 326: United States, 34, 36-37
Production statistics, 20, 34, 36-37, 280-311, 316-319: United States, 34, 36-37; worldwide, 280-281, 316-319; Abu Dhabi, 282; Algeria, 282; Angola, 282-283; Argentina, 283; Australia, 283-285; Austria, 285; Bahrain, 285; Barbados, 285; Benin, 285; Bolivia, 285; Brazil, 286-287; Brunei, 287; Cameroon, 287; Canada, 287-288; Chile, 288; China, 288-289; China (Taiwan), 289; Colombia, 289-290; Congo, 290; Denmark, 290; Dubai, 290; Ecuador, 290-291; Egypt, 291-292; France, 292; Gabon, 292-293; Ghana, 293; Greece, 293; Guatemala, 293; India, 293; Indonesia, 293-296; Iran, 296-297; Ivory Coast, 297; Japan, 297; Jordan, 298; Kuwait, 298; Libya, 298; Malaysia, 298-299; Mexico, 299-300; Morocco, 300; Myanmar, 300; Netherlands, 300; Neutral Zone, 301; New Zealand, 301; Nigeria, 301-303; Norway, 303; Oman, 303-304; Pakistan, 304; Peru, 304; Philippines, 305; Qatar, 305; Ras al Khaimah, 305; Saudi Arabia, 305; Sharjah, 305; Spain, 305-306; Suriname, 306; Syria, 306; Thailand, 306; Trinidad, 306; Tunisia, 306-307; Turkey, 307; United Kingdom, 307-308; Venezuela, 308-309; West Germany, 309-310; Yemen, 311; Yugoslavia, 311; Zaire, 311; North America, 316; Middle East, 316; Europe, 317; Africa, 317; Asia-Pacific, 318; Latin America, 319; Eastern Europe-U.S.S.R., 319
Production survey, 282
Productivity, 5-19
Prudhoe Bay oil field, 18, 37, 39, 42
Puerto Rico, 343

Q

Qatar, 101-102, 305, 343: North gas field, 102; petrochemicals, 102
Quebec, 56

R

Rabi-Kounga oil field, 128, 130
Raia oil/gas field, 125
Ramos gas field, 106
Ras al Khaimah, 101, 305
Ratawi oil field, 97
Ratcliff formation, 14
Ratka oil field, 104
Rawson Basin, 106
Reed Tool Co. rig census, 52-53
Refining capacity, 321-325, 337, 348: Europe, 321; North America, 321; Latin America, 322; Asia-Pacific, 323; Africa, 324; Middle East, 325; U.S.S.R.-Eastern Europe, 325; United States, 348 Refining cost, 53-54
Refining industry, 53-54, 328-335: United States, 53-54, 328-332, 335; profitability, 328-330; U.S. crude capacity changes, 328-332, 335; alkylation growth, 331; U.S. MTBE capacity, 332; worldwide refining trends, 332; Europe, 332-333; Asia, 333; Latin America, 333; reformulated fuels, 333-334; unleaded gasoline tax incentives, 334; market shares (unleaded gasoline), 334; Persian Gulf crisis effects on, 334; price freezes, 334-335; Venezuela, 335; India, 335; Belgium, 335; Germany, 335; Netherlands Antilles, 335; Sweden, 335
Refining statistics, 321
Refining survey, 336-348: refining capacity, 337; Abu Dhabi, 338; Albania, 338; Algeria, 338; Angola, 338; Argentina, 338; Australia, 338; Austria, 338; Bahrain, 338; Bangladesh, 338; Barbados, 338; Belgium, 338; Bolivia, 338; Brazil, 338; Brunei, 339; Bulgaria, 339; Cameroon, 339; Canada, 339; Chile, 339; China, 339; Colombia, 339; Congo, 339; Costa Rica, 339; Cuba, 339; Cyprus, 339; Czechoslovakia, 339; Denmark, 339; Dominican Republic, 339; Ecuador, 339-340; Egypt, 340; El Salvador, 340; Ethiopia, 340; Finland, 340; France, 340; Gabon, 340; Germany, 340; Ghana, 340; Greece, 340; Guatemala, 340; Honduras, 340; Hungary, 340; India, 340; Indonesia, 340-341; Iran, 341; Iraq, 341; Ireland, 341; Israel, 341; Italy, 341; Ivory Coast, 341; Jamaica, 341; Japan, 341; Jordan, 341; Kenya, 340; North Korea, 341; South Korea, 341-342; Liberia, 342; Libya, 342; Madagascar, 342; Malaysia, 342; Martinique, 342; Mexico, 342; Morocco, 342; Myanmar, 342; Netherlands, 342; Netherlands Antilles, 342; New Zealand, 342; Nicaragua, 342; Nigeria, 342; Norway, 342; Oman, 342; Pakistan, 343; Panama, 343; Paraguay, 343; Peru, 343; Philippines, 343; Poland, 343; Portugal, 343; Puerto Rico, 343; Qatar, 343; Romania, 343; Saudi Arabia, 343; Senegal, 343; Sierra Leone, 343; Singapore, 343; Somalia, 343; South Africa, 343; Spain, 343; Sri Lanka, 343; Sudan, 344; Sweden, 344; Switzerland, 344; Syria, 344; Taiwan, 344; Tanzania, 344; Thailand, 344; Trinidad, 344; Tunisia, 344; Turkey, 344; United Kingdom, 344; Uruguay, 344; U.S.S.R., 344; Venezuela, 344; Virgin Islands, 344; Yemen, 344; Yugoslavia, 344-345; Zaire, 345; Zambia, 345; United States, 345-348
Refining throughput, 349
Refining trends, 332
Refining update, 328
Reformulated fuels, 53-54, 333-334
Reserves statistics, 349
Reservoir depletion, 8
Reservoir modeling, 30
Rhode Island, 55
Richland Co, MT, 14
Ridgelawn oil field, 14
Rig activity, 270
Rig count, 270-271
Rig utilization (U.S.), 52-53
Rio Arriba Co, NM, 15
Rio Guadalquivir Basin, 171
Rio Puerco oil field, 15
Riser turret mooring, 216
Risheh gas field, 99
Robertson Co, TX, 7
Rogers Co, OK, 15
Romania, 152, 157, 343
Romashkino oil field, 146
Roslaval oil field, 17
Rospo Mare oil field, 217
Rotliegendes sandstone, 17
Rusanovskoye gas field, 143-144
Russett oil/gas field, 19
Russia, 140-153: gas resource development, 140-142; competitive bidding, 143; exploration activities, 143-145; Tengiz oil field, 145-147; Lithuanian production, 147-148; Caspian Sea, 148-149; natural gas pipeline, 149-150; refining, 150; development strategy, 150-151; equipment needs, 151-152; petrochemicals, 152-153

S

Saban oil field, 104
Sabine Co, TX, 10
Sabine Parish, LA, 10
Sable Island, 57
Saddam oil field, 97
Safah oil field, 101
Safaniya oil field, 102

Safety recommendations, 182
Safety violations, 54-55
Sahil oil field, 105
Sahl gas field, 130
Saih Nihayda gas field, 101
Sajaa oil field, 105
Saladin oil field, 74
Saleh gas field, 101
Salman oil field, 96
Samotlor oil field, 146
San Francisco oil field, 112
San Joaquin Valley, 18, 45
San Jorge Gulf Basin, 106
San Juan Basin, 15, 53
San Juan Co, NM, 15
San Martin/Cashiriari gas/condensate field, 120
San Pedro Basin, 114
San Sebastian gas field, 106
Sand Wash Basin, 15, 52
Sandoval Co, NM, 15
Sansahuari oil field, 114
Santa Barbara Basin, 44
Santa Maria Basin, 44
Santa Ynez oil field, 45
Santos Basin, 111
Sapale oil field, 136
Saratoga Chalk, 10
Saskatchewan, 16, 60
Satpura Basin, 78
Saudi Arabia, 102-104, 205, 305, 343: producing capacity, 102-104; production rate, 205
Saudi Aramco budget, 96
Sawtooth formation, 14
Scotland, 175
Scott oil field, 181
Sea of Okhotsk, 143
Seal Island/Northstar oil field, 42
Seasonal demand, 23
Seasonal fuel formulation, 54
Seaway pipeline, 34
Secura Basin, 119
Sedano Basin, 171
Seismic crew count, 271
Seismic/borehole geophysics, 30
Senegal, 343
Sergipe-Alagoas Basin, 111
Severance tax, 42
Severo-Bovanenkovskoye gas field, 145
Shaanxi Province, 70
Shaanxi-Ninqxia-Inner Mongolia Basin, 70
Shabwa oil field, 105
Shah oil field, 105
Shams gas/condensate field, 101
Sharjah, 105, 305
Shaybah oil field, 102
Shdeha oil field, 104
Shelby Co, TX, 10
Shengli area, 65-66
Shipping threats (war), 208-209
Shtokmanovskoye gas field, 144-145
Shuizhong 36-1 oil field, 67
Shushuqui oil field, 114

Shutin wells (California), 43-44
Siberia, 5, 17, 141
Siberian craton, 85
Sichuan Province, 66, 70-72
Sidi El Itayem oil field, 138
Sierra Leone, 343
Sijan oil field, 104
Silo oil field, 14
Singapore, 63, 343
Sirri oil field, 96
Sirte Basin, 130
Skua oil field, 74
Slant hole optimization, 19
Sleipner gas field, 184
Slotted liner completion, 19
Smackover limestone, 10, 12, 48
Smith Co, MS, 12
Snorre oil field, 184Solimoes Basin, 111, 274, 277
Somalia, 138, 343
Sooner Trend oil field, 15
Soso oil field, 12
South Africa, 343
South Bassein gas field, 77
South Borie-Rawhide oil field, 14
South Carolina, 53
South China Sea, 63, 67, 84, 93
South Korea, 63, 83, 341-342
South Marsh Island Block 239-241, 219
South Rewa Basin, 78
South Sulele oil field, 125
Southeast Cornish oil field, 16
Spain, 170-171, 174, 305-306, 343: exploration/production activities, 170-171; pipeline projects, 170-171; refining, 171; petrochemicals, 171, 174
Spratly Islands, 93
Sri Lanka, 343
Statfjord gas field, 181
Steam projects (U.S.), 230, 238-241
Steerable drilling systems, 30
Stephens oil field, 12
Strait of Hormuz, 96, 101
Strait of Malacca, 91
Strategic Petroleum Reserve, 20, 30, 32, 34, 205: drawdown, 205
Stratigraphic charts, 274-279: Myanmar, 274, 276; Midway-Sunset oil field, 274, 276-277; Solimoes Basin, 274, 277; Arabian Gulf, 274, 277; Venezuela, 274, 277, 279; Palawan shelf, 274, 279
Stripper well, 36
Styrene monomer, 54
Suba oil field, 97
Sublette Co, WY, 18, 52
Success ratio, 6
Sudan, 344
Sui gas field, 88
Suizhong 36-1 oil field, 69
Sulfur content, 54
Sumatra, 91
Suriname, 121-122, 306: Tambaredjo oil field, 121
Sweden, 174, 335, 344: refining industry, 335
Switzerland, 174, 344
Syria, 104-105, 306, 344

T

Tagrin oil field, 17
Tahaddi gas field, 130-131
Taimyr Peninsula, 145
Taiwan, 90-90; gasoline/ diesel fuel, 90; fuel oil, 90; refining, 90-91
Takutu graben area, 115
Talara Basin, 119
Tanker construction cost, 211-212
Tanker fleet aging, 211, 213
Tanker orders, 210-211
Tanker supply/demand, 212-213
Tankers, 210-217: Middle East crisis effect, 210; tanker orders, 210-211; fleet aging, 211, 213; construction costs, 211-212; freight rates, 212; tanker supply/demand, 212-213; U.S. oil spill liability law, 213-214; deep water port study, 214-215; offshore loading systems, 215-217 Tanzania, 138, 344
Taranaki Basin, 87
Tarim Basin, 62, 64-65
Tatipaka-Pasarlapudi structure, 77
Tazerka oil field, 138-139
Tengiz oil field, 64, 140, 145-147, 150
Tennessee, 347
Tetete oil field, 114
Texas, 30, 34, 45, 54-55, 347
Texoma pipeline, 34
Thailand, 63, 91-93, 306, 344: price deregulation, 91; oil import reduction, 91; gas grid expansion, 91; development strategy, 91; pipeline project, 91-92; gas importation, 92; refining, 92-93
Thermal EOR projects (U.S.), 238-241
Tierra del Fuego, 106
Tikorangi formation, 18
Timan-Pechora Basin, 153
Timor Sea, 74
Toni oil field, 181
Tordis oil field, 184
Trans-Alaska Pipeline System, 42
Trinidad, 5, 122, 306, 344
Trinidad and Tobago, 122
Troll gas field, 180
Troll oil field, 180
Tsagaan-Els oil field, 85
Tubarao oil field, 111, 125
Tubular goods market (U.S.), 30
Tubular goods/drilling activity/footage statistics (U.S.), 52
Tuktoyaktuk Peninsula, 58
Tulsa Co, OK, 15
Tunisia, 138-139, 306-307, 344: exploration history, 138-139
Turkey, 307, 344
Turkmen Soviet Socialist Republic, 143

MODERN TURBOEXPANDER is part of Western Gas Processors Ltd's reconditioning at Midkiff, Tex., gas processing plant.

Turpan-Hami depression, 66
Turret mooring systems, 215-217
Tyra gas field, 181

U

U.S.S.R., 17, 140-153, 344
Ucayali Basin, 119
Ultra-short radius well, 18
Underbalanced drilling, 19
Underground gas storage, 53
United Arab Emirates, 105
United Kingdom, 5, 17-18, 174-177, 180, 307-308, 344: pipeline projects, 175-176; refining, 176; Wytch Farm oil field, 176; exploration/production activities, 176-177
United Nations (war), 202
United States, 5, 20-55, 63, 204-205, 207, 213-214, 345-348: production statistics, 20; oil/gas supply and demand, 20, 23, 26, 30, 32, 43; energy consumption, 20, 23, 26, 30, 43; drilling activity, 30; SPR drawdown, 30, 32, 34; production decline, 34, 36-37, 39, 42-45; horizontal drilling, 45, 48-52; rig utilization, 52-53; pipeline construction, 53; refining outlook, 53-54; accidents, 54-55; reaction (war), 204-205; refining response (war), 207; oil spill liability law, 213-214; refining survey, 345-348
Unleaded gasoline tax incentives, 334
Upper Zakum oil field, 105
Urengoi gas field, 141, 146
Uruguay, 122, 344
Usano oil/gas field, 89
Ushakov oil field, 148
Utah, 53, 347
Uthmaniyah oil field, 102-103

V

Vega oil field, 165
Velasques oil field, 112
Venezuela, 5, 18, 123, 197, 274, 277, 279, 308-309, 335, 344: development strategy, 123; exploration/production activities, 123; natural gas, 123; refining, 123; petrochemicals, 123; refining industry, 335
Verde oil field, 15
Veslefrikk oil field, 180
Vienna Basin, 156
Viet Nam, 93: sovereignty issue, 93; petrochemicals, 93
Vilkichiai oil field, 148
Vindhyan Basin, 78
Viola limestone, 16
Viosca Knoll Block 957, 219
Virgin Islands, 344
Virginia, 53, 56, 347
Vlieland sandstone, 18
Volcanic feature, 7-8
Vulcan gas field, 17

W

Wabiskaw formation, 16
Wadi Rafash gas/condensate field, 101
Waihapa oil/gas field, 18, 87
Wasatch formation, 18, 52
Washington, 347
Washington Co, TX, 7, 48
Washita-Fredericksburg formation, 12
Wayne Co, WV, 19
Weeks Island, 32
Weiyuan gas field, 71
Weld Co, CO, 14
Well completion, 269: statistics, 269
Well performance, 5-19
Well spacing, 5
Welland gas field, 182
Wellbore placement, 8
Wellington oil field, 14
West Germany. SEE Germany.
West Hackberry, 32
West Panhandle-Carson oil field, 48
West Qurna oil field, 97
West Sak oil field, 42
West Sulele oil field, 125
West Varyegan oil field, 17
West Virginia, 54, 347
Western Australia, 74
Whisby oil field, 177
White Night joint venture, 17
Widuri oil field, 79
Wilcox formation, 10, 12
Williston Basin, 12
Wilson Co, TX, 7-8
Winter oil field, 16
Wisconsin, 53, 56, 347
Woods Co, OK, 15
Worker safety, 54-55
Wyoming, 14, 18-19, 45, 52-53, 347
Wyoming-California pipeline, 53
Wytch Farm oil field, 163, 176

X

Xinjiang Uygur area, 64, 66

Y

Yacheng gas field, 69
Yamal Peninsula, 143-144
Yamama formation, 97
Yamburg gas field, 141
Yammaderry oil field, 74
Yangtze River Valley area, 66
Yasmin oil field, 138-139
Yellow Creek gas field, 19
Yellow River delta area, 66
Yemen, 105, 311, 344
Yibal gas field, 100
Yuca oil field, 114
Yugoslavia, 157, 311, 344-345: Molve gas field, 157
Yunnan province, 66, 72

Z

Zaire, 311, 345
Zambia, 345
Zapadno-Sharapovskaya formation, 143
Zavala Co, TX, 5
Zeit Bay, 126
Zeit Bay oil field, 126
Zelten gas field, 130
Zhejiang province, 64
Zimbabwe, 139
Zubair oil field, 97
Zueitina 103-D gas field, 130
Zuunbayan oil field, 85

52 reasons why Oil & Gas Journal is must reading for anyone with oil and gas interests... ...anywhere in the world.

Every one of OGJ's 52 weekly issues is must reading because each issue contains a careful balance of international news and technology to give you everything about the petroleum industry you need to know to make important personal and business decisions.

EXPERIENCE EDITORIAL FIREPOWER!

Another OGJ advantage is editorial firepower! More editors with more experience reporting more news and technology than any other oil industry publication... in fact, almost four times the number of technical pages than the next best publication.

Journal editors make this pledge: If you read all 52 issues in the year, you won't miss any news or technical development of significance to your job in the petroleum industry.

These are just a few of the reasons why operating oil industry people around the world make OGJ required reading every week. And, you'll get one more reason with every one of the 52 issues each year.

ORDER YOUR 30-WEEK INTRODUCTORY SUBSCRIPTION TODAY!

But, don't take our word for it. Order your 30-week introductory subscription today (USA $30, Overseas $52). Then, you be the judge. This introductory offer comes with this unconditional guarantee: If you're not satisfied, tell us. We'll stop service and return your money. It's that simple.

Start receiving Oil & Gas Journal with the next weekly issue. Mail your order to: Circulation Services, P.O. Box 2002, Tulsa, OK 74101, U.S.A. or call (918) 832-9255. Fax (918) 831-9497. Telex 203604 PPUB UR.

OIL & GAS JOURNAL

NOTES

ADVERTISERS' INDEX INTERNATIONAL PETROLEUM ENCYCLOPEDIA 1991

C

CHEVRON CHEMICAL COMPANY 229
Enhanced Oil Recovery Chemicals
Head office
Post Office Box 5047
San Ramon, CA 94583-0947
Tel: 800-553-1196
FAX: 415-842-1520

Sales Office
5080 California Avenue
Bakersfield, CA 93309
Tel: 805-955-4434
FAX: 805-334-4559

Sales Office
1301 McKinney
Houston, TX 77010
Tel: 713-754-4523
FAX: 713-754-2016

E

EUROPIPE GmbH 161
Postfach 40 55
D-4030 Rotingen
Tel: (21 02) 857-0
Fax: (21 02) 857-285
Telex via Teletex
211 4550-MRW

H

HALLIBURTON SERVICES 40, 41
P. O. Drawer 1431
Duncan, Oklahoma 73536 USA
Tel: 405-251-3760
Telex: 796023-796348

L

LURGI GAS-UND MINERALÖLTECHNIK
GmbH 226
Lurgi-Allee 5
Postfach 11 12 31
D-6000 Frankfurt am Main
Tel. (0 69) 58 08-0
Tx: 41236-0 lg d
Fax: (0 69) 58 08-38 88

M

MANNESMANN ANLAGENBAU AG 186
TheodorstraBe 90
D-4000 Dusseldorf 30
Federal Republic of Germany
Tel: (211) 659-1, Telex: 8 586 677
Fax: (211) 659-2372

MANNESMANN DEMAG 185
Compressors and Pneumatic Equipment
P.O. Box 10 15 07, Wolfgang-Reuter-Platz
D-4100 Duisburg 1,
Federal Republic of Germany
Tel: Germany (203) 605-1
Fax: (0203) 6 1061-63

MANNESMANNROHREN-WERKE AG .187
Postfach 11 04, Mannesmannufer 3
D-4000 Dusseldorf 1
Federal Republic of Germany
Tel: (211) 875-0
Fax: (211) 875-3245
Telex: 8 581 421

O

OMV Aktiengesellschaft 183
Otto-Wagner-Platz 5
A-1090 Wien
Austria
Tel: (02 22) 404 40/0
Fax: (02 22) 404 40/91
Telex: 135 704

R

ROSEMOUNT CONTROL SYSTEMS
DIVISION 13
12000 Portland Avenue South
Burnsville, MN 55337
Tel: 612-828-3568
Fax: 612-828-3236

S

SARAS SPA 159
Galleria de Cristoforis, 8
20122 Milan, Italy
Tel: (02) 77371
Telex: 311273
Fax: (02) 76020640

SERVO DELDEN BV
P.O. Box 1
7490 AA Delden
The Netherlands
Tel: 05407-63535
Telex: 44347
Fax: 05407-64125

SOJUZNEFTEEXPORT 189
32/34 Smolenskaya Sq.
121200 Moscow, U.S.S.R.
Tel: 253-94-88, 253-94-89
Telefax: 244-22-91
Telex: 411148A

T

TOTAL CORPORATE
COMMUNICATIONS 73
Tour Total, CEDEX 47
92069 Paris LaDefense, France
Tel: 1 42 91 40 00
Fax: 1 42 91 42 91
Telex: 615 700

V

VALLOUREC INDUSTRIES 11
130 Rue de Silly
92100 Boulogne-Billancourt
France
Tel: (33-1) 49.09.35.00
FAX: (33-1) 49.09.37.13

W

WESTERN RESEARCH 33
1313-44 Ave. N.E.
Calgary, Alberta
Canada TZE 6L5
Tel: 1 (403) 291-1313
Fax: 1 (403) 250-2610

WILLBROS GROUP, INC 9
2431 East 61st Street, Suite 700
Tulsa, Oklahoma 74136 USA
Phone: 918-748-7000
Telex: 79-6660

22-23

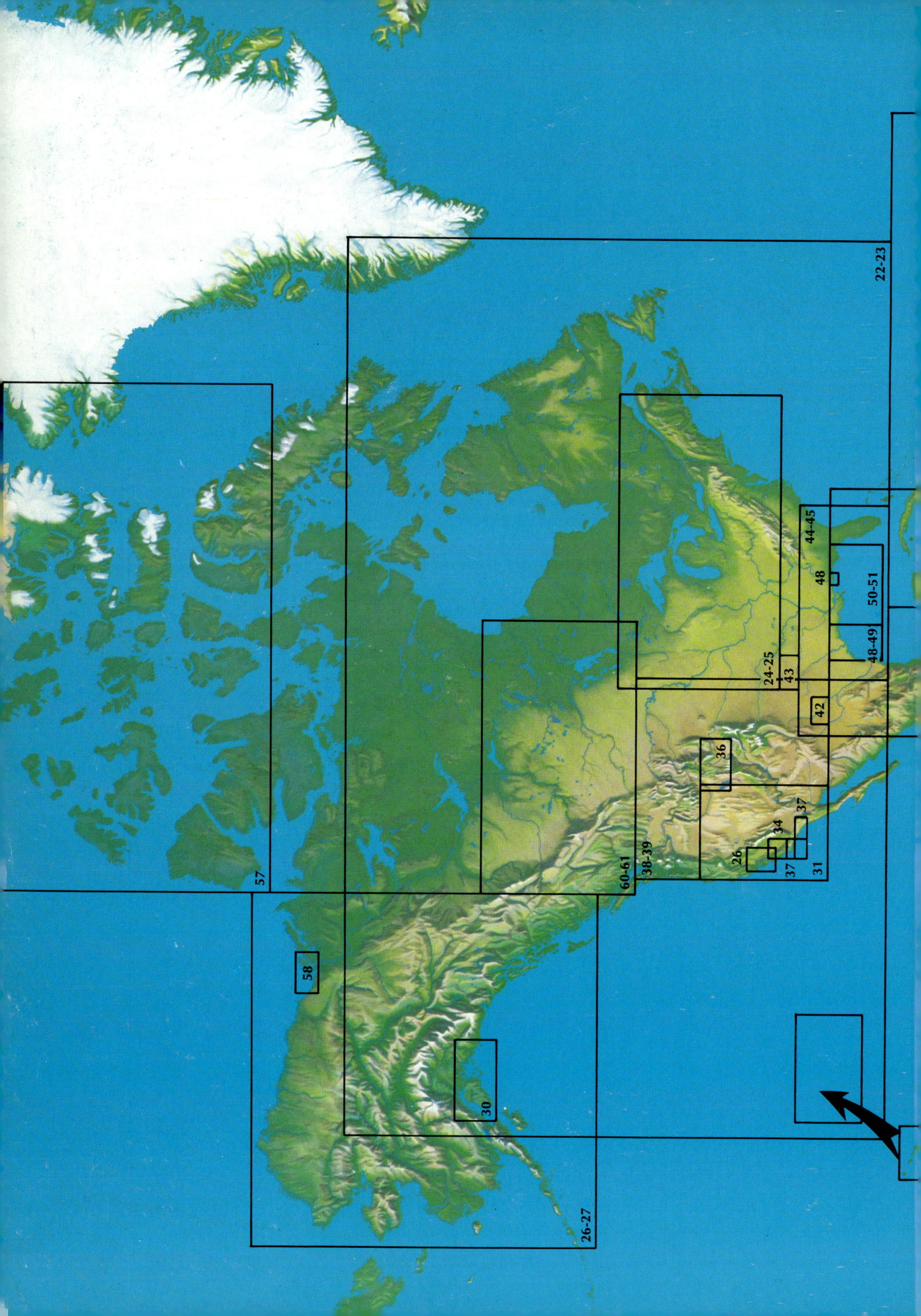

44-45

48

50-51

48-49

24-25
43

42

57

36

37

60-61
38-39
34
37
26
37
31

58

30

26-27